International Conference "Advanced Element Base of Micro- and Nano-Electronics with Using of ToDate Achievements of Theoretical Physics" (MRSU 2021)

Journal of Physics: Conference Series Volume 2056

Moscow, Russia
20 – 23 April 2021

ISBN: 978-1-7138-5027-4
ISSN: 1742-6588

Printed from e-media with permission by:

Curran Associates, Inc.
57 Morehouse Lane
Red Hook, NY 12571

Some format issues inherent in the e-media version may also appear in this print version.

Institute of Physics
Dirac House, Temple Back
Bristol BS1 6BE UK

Phone: 44 1 17 929 7481
Fax: 44 1 17 920 0979

techtracking@iop.org

Additional copies of this publication are available from:

Curran Associates, Inc.
57 Morehouse Lane
Red Hook, NY 12571 USA
Phone: 845-758-0400
Fax: 845-758-2633
Email: curran@proceedings.com
Web: www.proceedings.com

TABLE OF CONTENTS

International Conference "Advanced Element Base of Micro- and Nano-Electronics with Using of To-Date Achievements of Theoretical Physics" 20-23 April 2021, Moscow, Russia

V V Belyaev[1,2], D N Chausov[1,3], M M Kuznetsov[1]

[1]Moscow Region State University «MRSU», Very Voloshinoy str., 24, Mytishi, 141014, Russia}

[2]RUDN University (Peoples' Friendship University of Russia), 6 Miklukho-Maklay St., Moscow, 117198, Russia}

[3]Prokhorov General Physics Institute of the Russian Academy of Sciences, 119991, Russia, Moscow, 38 Vavilova St.}

vic_belyaev@mail.ru, d.chausov@yandex.ru

Abstract

A brief description of scientific program and papers of International Conference "Advanced Element Base of Micro- and Nano-Electronics with Using of To-Date Achievements of Theoretical Physics" is presented. It is an annual conference of Moscow Region State University (MRSU).

Introduction

International Conference "Advanced Element Base of Micro- and Nano-Electronics with Using of To-Date Achievements of Theoretical Physics" is an annual conference of Faculty for Physics and Mathematics of Moscow Region State University (MRSU). MRSU is a leading Russian regional university. This conference is traditionally organized by Department of Theoretical Physics. The Department and its Head Prof. Victor Belyaev were awarded by a Grant of President of Russian Federation for Russian Leading Scientific Schools.

The conference attracted 100+ participants from many countries (Germany, Canada, Byelorussia, Pakistan, Sudan, Lebanon et al.) and 10+ Russian regions. Special session was devoted to R&D of graduate and post-graduate students.

The Conference program included sessions as follows:

1. Theoretical physics
2. Physics and application of condensed medium including such sub-sessions as
   - Liquid crystals
   - Organic electronics
   - Display devices and components
   - Quantum effects
3. Physics of plasma
4. Physical gasodynamics
5. Physics and application of dispersed media

The first talk of the plenary session was made by Prof. V. Belyaev who celebrated his 70th anniversary. He said about his 50 years long way in the science, about raising of scientific ideas and projects. Prof. V. Rudyak, Novosibirsk State Architecture Construction University, presented a new direct stochastic molecular model of transport processes in gases. Corresponding Member of RAS Prof. N. Bogolyubov Jr and Prof. A. Soldatov, Department of Mechanics, V.A. Steklov Mathematics Institute of RAS presented a talk “Low-frequency spectrum of a laser driven polar quantum emitter damped by squeezed vacuum with nite bandwidth”. Prof. V. Chigrinov, Foshan University, China, told on-line about new photonic crystal materials and devices on the base of liquid crystals and photoanisotropic substances. The report by V. Chertopolokhov, Lomonosov Moscow State University, was “Providing of immersion effect in to-date virtual reality systems by high quality tuning of different parameters”.

The plenary session was concluded by PhD Maria Burdanova, MRSU, former post-graduate of Warwick University, UK, who told about ultrafast spectroscopy of one dimension Van-der-Waals nanomaterials.

The MRSU physicists presented a set of talks on optical properties of the liquid crystals (LC) and their nanocomposite systems. Dr. E. Vasilchikova et al. presented their investigation of polarized light propagation in wedge mesogen layers.

A few talks on the LC and other organic materials have been presented by Prof. Pozhidaev’s group of physicists from Levedev Physics Institute of RAS (LPI): “The concept of biaxial surface potential and polymer aligning layers in electro-optical cells based on helical nanostructures of ferroelectric liquid crystal” by T. Tkachenko et al., “Ferroelectric Smectic C* LC induced in mixtures of nematic LC and non-mesogenic chiral substances” by V. Barbashov et al.,

“Quadratic orientational electro-optic effect in ferrielectric LC with sub-wave pitch of the chiral structure” by A. Kuznetsov et al.

A few talks have been presented on materials of organic electronics, electro-optic materials and their physical properties: “Intermolecular interactions of layers octa-phenyl-2,3-naphthalocyaninato zinc” by D. Chausov et al.; “Influence of substituted naphthalocyanines structure on their optical properties in thin films” by A. Kazak et al.; “Diamagnetic anisotropy and rotational viscosity of 6CHBT nematic liquid crystal” by A. Kurilov et al.

A team from Institute of Chemistry of New Materials of National Academy of Sciences of Belarus, Byelorussian State University, MRSU, Kazan National Reserarch Technology University presented a talk about control of QD radiation by light beam and tuning of selective reflection of a chiral LC.

Similar objects have been investigated in G.A. Krestov Institute of Solution Chemistry of RAS. L A Antina report was “The aggregation behavior and spectral properties of β, β′-benzil-BODIPY in Langmuir-Schaefer and poly(methyl methacrylate) thin films”.

T. Khantuleva from Saint-Petersburg State University presented specific features of fast processes in condensed media.

Many interesting talks were presented by employees of other institutions.

N. Klassen and big group of young researchers from Institute of Solid State Physics of RAS - described their experiments in different types of the condensed materials: deformation of organic-inorganic compositions (I. Tsebruk, A. Azhgalieva et al.), super-long nano-chains formation during evaporation of metals with a help of intensive impacts (P. Provotorov et al.), formation of strip-like vortex motion by surface gravity waves (A. Poplevin et al.), structural transformations in the system lanthanum bromide – water – electrical field (S. Vinokurov et al.), liquid nitrogen luminescence under IR excitation (P. Provotorov et al.), second sound standing waves in a medium with NPs (A. Orlova) and many other interesting results.

Some investigations have been devoted to thermal properties of different media under extremal ambient conditions including open space – diamond-like carbon coatings under cyclic temperature changes (M. Makeev, Bauman Moscow State Technical University and RUDN University), thermoelectric generator (Pakistan, post-graduate of RUDN University).

M. Frah et al. (Sudan, post-graduate of RUDN University) evaluated electromagnetic pollution index (EMPI) emitted from multiple sources and investigated a few metal coated shields to protect from the EMP.

A group by M. Sychov (Saint-Petersburg State Institute of Technology (Technical University), Russia) in collaboration with Khlopin Radium Institute (Saint-Petersburg) and Shizuoka University (Japan) synthesized and investigated new efficient and stable phosphors for electroluminescent displays (ZnS:Cu,Br radioluminescent phosphors, Ce3+-doped fluoride nanophosphors et al.).

A group by O. Maximova (Ulyanovsk Civil Aviation Institute) proposed a few technical solutions of thin film electroluminescent indicating devices to enhance their efficiency and manufacturing cost.

A. Abduev from Peoples' Friendship University of Russia (RUDN University) presented a conceptual talk "Finding routes for low-temperature formation of functional metal oxide based thin films" about promising technology that can replace ITO electrodes in electro-optic and opto-electronic devices.

V. Chigrinov (Foshan University, China) and A. Kudreyko (Bashkir State Medical University, Department of Medical Physics and Informatics, Ufa) proposed to improve characteristics of optically rewritable electronic paper.

In the Sessions "Theoretical physics" and "Plasma physics" new RnD results were presented on electromagnetic wave propagation in layers of both conductors and dielectrics.

Dr. N. Zverev and a graduate student A. Zotov (both MRSU) gave a talk "Longitudinal Plasmons in a Thin Flat Conductive Film". They described a mode allowing measure thickness of nanometer metal and semiconductor layers.

Prof. I. Kuznetsova and her disciples from P.G. Demidov Yaroslavl State University got a talk "The influence of Fermi surface anisotropy and the charge carrier surface scattering kinetics on the electrical conductivity of a thin metal film in the view of the quantum size effect".

Associated Professor of MRSU T. Kamalov and his post-graduate students presented reports on quantum computing to create efficient algorithms.

A group by V. Popov from Northern (Arctic) Federal University in Arkhangelsk presented a few reports on flow the binary gas mixture in channels of different shape.

E. Kozhanova (Yuri Gagarin State Technical University of Saratov), I. Tkachenko (École de Technologie Superieure, Montreal, Canada) proposed a few methods of analysis of the energy spectrum of an electron in a one-dimensional quantum well with infinitely high walls in the Al-Ga-As system using Wavelet analysis and MATLAB.

In the Session "Plasma physics" Prof. A. Bishaev, MIPT and Dr. M. Abgaryan, Moscow Aviation Institute (Technology University) investigated influence of magnetic field on jet from stationary plasma engine.

In the Session "Physics and application of dispersed media" a few talks were presented by MRSU post-graduate students. Ms. V.Askerova and her supervisor E. Kalashnikov described moving an atom by swapping places across a crystal lattice without vacancies. They proposed a model of these materials heat and mechanical properties.

Physical properties of biological objects have been investigated in a group from Semenov Federal Research Center of Chemical Physics, RAS. E. Tereshkin et al. presented a talk "Molecular dynamics of DNA-binding protein and its 2D-crystals".

The International Conference "Advanced Element Base of Micro- and Nano-Electronics with Using of To-Date Achievements of Theoretical Physics" became very important scientific event with significant results of big importance for basic science and practical application.

Organizing Committee

V.V. Belyaev, MRSU – Chairman

D.N. Chausov, MRSU – Vice-Chairman

A.P. Chernyaev, MIPT, Russia

A.G. Smirnov, Byelorussian State University of IT and Radio-Electronics, Director of Byelorussia SID Chapter

S.P. Palto, A.V. Shubnikov Institute of Crystallography, Russia

S.I. Torgova, LPI, Russia

M.M. Sychov, Saint-Petersburg Technology Institute (Technological University), Russia

A.D. Kurilov, MRSU, Russia

Program Committee

M.M. Kuznetsov, MRSU – Chairman

A.V. Kazak, Ivanovo State University – Vice-Chairman

N.S. Erokhin, RUDN University, Russia

V.G. Chigrinov, Foshan University, China

M. Osipov, Strathclyde University, UK

S. Kumar, Nitte Meenakshi Institute, India

A.G. Smirnov, Byelorussian State University of IT and Radio-Electronics, Director of Byelorussia SID Chapter

H.L. Margaryan, Erevan State University, Armenia

A. Chanishvili, V. Chavchanidze Institute of Cybernetics, Georgia Technology University, Georgia

Peer review declaration

All papers published in this volume of Journal of Physics: Conference Series have been peer reviewed through processes administered by the Editors. Reviews were conducted by expert referees to the professional and scientific standards expected of a proceedings journal published by IOP Publishing.

- **Type of peer review: Single-blind / Double-blind / Triple-blind / Open / Other (please describe)**

Single-blind

- **Conference submission management system:**

Morressier virtual conference and publishing platform

- **Number of submissions received:**

76

- **Number of submissions sent for review:**

76

- **Number of submissions accepted:**

71

- **Acceptance Rate (Number of Submissions Accepted / Number of Submissions Received X 100):**

93.4

- **Average number of reviews per paper:**

1

- **Total number of reviewers involved:**

8

- **Any additional info on review process:**

Typical review questionnaire like in leading scientific journals and detailed review about value and novelty of the publications reviewed.
The Referees are from universities and scientific organizations from Russia, Byelorussia, China, Canada, India.

- **Contact person for queries:**
  Name : Professor Victor Belyaev
  Affiliation: Moscow Region State University (MRSU)
  Email : vic_belyaev@mail.ru

Low-frequency fluorescence spectrum of a laser driven polar quantum emitter damped by squeezed vacuum with finite bandwidth

N N Bogolyubov Jr, and A V Soldatov

Department of Mechanics,V.A. Steklov Mathematical Institute of the Russian Academy of Sciences, 8, Gubkina str., Moscow, 119991, Russia

E-mail: `bogolubv@mi-ras.ru` `soldatov@mi-ras.ru`

Abstract. A two-level quantum emitter with broken inversion symmetry driven by external semiclassical monochromatic high-frequency electromagnetic (e.g., laser) field and damped by squeezed vacuum reservoir with finite bandwidth is presented. The squeezed vacuum source is assumed to be either degenerate parametric oscillator (DPO) or a non-degenerate parametric oscillator (NDPO). It is shown that the shape of low-frequency fluorescence spectrum of the emitter can be effectively alternated by controlling the degree of the squeezed vacuum source degeneration and phase of the squeezing.

Keywords: polar emitter; fluorescence spectrum; squeezed vacuum; two-level atom; broken inversion symmetry; asymmetric quantum dot; polar molecule

1. Introduction

As a rule, theoretical and experimental studies have been carried out under the assumption of symmetric quantum emitters where inversion symmetry is assumed. Nevertheless, violation of this symmetry is common in such natural systems as polar molecules as well as in artificially manufactured systems, like quantum dots. The cause for this violation is different for different systems. For example, in polar molecules its origin is due to the parity mixing of the molecular states [1], while in quantum dots it is induced by the asymmetry of the confining potential of the dot. However, in all cases this violation results in nonzero permanent dipole matrix elements of the ground and excited states. To our knowledge, the idea that a simple two-level quantum system driven by high-frequency classical electromagnetic (EM) field can emit EM field of much lower frequency if its dipole operator possesses permanent non-equal diagonal matrix elements, was introduced in [2]. This phenomenon was further studied in [3, 4, 5], where the properties of this low-frequency radiation were thoroughly investigated for the case of a two-level system driven by external EM field and damped by a dissipative reservoir. It was also found that the same system can also be used for the amplification of the weak low-frequency EM field [4]. This property of a relatively simple quantum system may help develop useful techniques for manipulation of the low-frequency EM radiation, especially in the terahertz range of frequencies. Present research aims to expand the investigation of the low-frequency EM fluorescence radiation phenomenon to the case of an externally driven two-level system with broken inversion symmetry interacting with a finite-band squeezed vacuum dissipative reservoir, which properties can be tuned appropriately in order to control the shape of the fluorescence radiation spectrum. The case of interaction with a broadband squeezed vacuum dissipative reservoir was studied earlier for weak driving EM field in [6, 7].

2. Model Hamiltonian

In this study we consider a two-level atom with ground state $|g\rangle$, excited state $|e\rangle$, transition frequency ω_0 and the electric dipole moment $\hat{\mathbf{d}}$, driven by external classical monochromatic field $\mathbf{E}(t) = \mathbf{E}\cos(\omega_f t)$ with an amplitude $\mathbf{E}$ and frequency ω_f. The atom is also coupled to a reservoir B made of a plurality of modes of quantized electromagnetic field being in the squeezed vacuum state. It is assumed that the frequency Lamb shift due to interaction with the reservoir is already incorporated into the atomic transition frequency ω_0. Thus, the model Hamiltonian reads

$$H = H_S(t) + \hbar\sum_k \omega_k b^+(\omega_k)b(\omega_k) + \sum_k \left(g(\omega_k)S^+b(\omega_k) + g^*(\omega_k)b^+(\omega_k)S^-\right). \tag{1}$$

Here $S^+ = |e\rangle\langle g|$ and $S^- = |g\rangle\langle e|$ are the usual raising and lowering atomic operators and $S^z = \frac{1}{2}(|e\rangle\langle e| - |g\rangle\langle g|)$ is the atomic population inversion operator.

The operators $b(\omega_k)$ and $b^+(\omega_k)$ are the annihilation and creation operators for the vacuum modes satisfying the commutation relations

$$[b(\omega), b^+(\omega')] = \delta(\omega - \omega'),\ [b(\omega), b(\omega')] = 0,\ [b^+(\omega), b^+(\omega')] = 0, \tag{2}$$

and the term

$$H_S(t) = \hbar\omega_0 S^z + \frac{\hbar}{2}\Omega_R(S^- e^{i\omega_f t} + S^+ e^{-i\omega_f t}) + \frac{\hbar}{2}(e^{i\omega_f t} + e^{-i\omega_f t})\left[\delta_a S^z - \frac{\delta_s}{2}(|e\rangle\langle e| + |g\rangle\langle g|)\right] \tag{3}$$

contains an interaction between the driving field and the atom in the rotating wave approximation (RWA). Here $\Omega_R = -\mathbf{E}\mathbf{d_{eg}}/\hbar$ is the Rabi frequency being made real and positive by the appropriate choice of the phase factors of the states $|e\rangle$ and $|g\rangle$, and $\mathbf{d}_{eg} = e\langle e|\hat{\mathbf{r}}|g\rangle$, $\mathbf{d}_{ge} = e\langle g|\hat{\mathbf{r}}|e\rangle$, $\mathbf{d}_{ee} = e\langle e|\hat{\mathbf{r}}|e\rangle$, $\mathbf{d}_{gg} = e\langle g|\hat{\mathbf{r}}|g\rangle$ are the atomic dipole moment operator matrix elements. As a rule, it is assumed that $\mathbf{d}_{ee} = \mathbf{d}_{gg} = 0$, because typical physical systems, like atoms and molecules, possess the inversion symmetry, and each of the states $|g\rangle$ and $|e\rangle$ is either symmetric or antisymmetric. Contrary to this view, we assume that the inversion symmetry of the system in question is violated, $\mathbf{d}_{ee} \neq \mathbf{d}_{gg}$, so that $\delta_a = \mathbf{E}(\mathbf{d}_{gg} - \mathbf{d}_{ee})/\hbar$ and $\delta_s = \mathbf{E}(\mathbf{d}_{gg} + \mathbf{d}_{ee})/\hbar$. The term proportional to δ_s does not influence the dynamics of the system and can be omitted, while the term proportional to the symmetry violation parameter δ_a is retained. The squeezed vacuum reservoir source is assumed to be either a degenerate parametric oscillator (DPO) or a non-degenerate parametric oscillator (NDPO). The output fields from this oscillators are characterized by the following correlation functions [8]

$$\langle b^+(\omega_k)b(\omega_{k'})\rangle_{svac} = N(\omega_k)\delta(\omega_k - \omega_{k'}),\ \langle b(\omega_k)\, b(\omega_{k'})\rangle_{svac} = -M(\omega_k,\theta)\delta(\omega_k + \omega_{k'} - 2\omega_s), \tag{4}$$

$$\langle b(\omega_k)b^+(\omega_{k'})\rangle_{svac} = (N(\omega_k)+1)\delta(\omega_k - \omega_{k'}),\ \langle b^+(\omega_k')b^+(\omega_k)\rangle_{svac} = -M^*(\omega_k,\theta)\delta(\omega_{k'} + \omega_k - 2\omega_s), \tag{5}$$

where ω_s is the carrier frequency of the squeezed field, θ is the phase of squeezing, $N(\omega)$ is related to the mean number of photons at frequency ω and $M(\omega,\theta)$ is characteristic of the squeezed vacuum field and describes the correlation between the two photons created in the down-conversion process. They satisfy the inequality $|M(\omega,\theta)| \leq \sqrt{N(\omega)(N(\omega)+1)}$, which in the case of ideal squeezed state produced by an optical parametric oscillator transforms into the equality. The frequency dependencies of $N(\omega)$ and $M(\omega,\theta)$ for an optical parametric oscillator below threshold for the ideal DPO are given by [9]

$$N(\omega) = \frac{\lambda^2 - \mu^2}{4}\left[\frac{1}{(\omega - \omega_s)^2 + \mu^2} - \frac{1}{(\omega - \omega_s)^2 + \lambda^2}\right], \tag{6}$$

$$M(\omega,\theta) = e^{i\theta}\frac{\lambda^2-\mu^2}{4}\left[\frac{1}{(\omega-\omega_s)^2+\mu^2}+\frac{1}{(\omega-\omega_s)^2+\lambda^2}\right], \tag{7}$$

while for the ideal NDPO they assume more general form [10]

$$N(\omega) = \frac{\lambda^2-\mu^2}{8}\left[\frac{1}{(\omega-\omega_s-\alpha)^2+\mu^2}+\frac{1}{(\omega-\omega_s+\alpha)^2+\mu^2}-\right.$$
$$\left.-\frac{1}{(\omega-\omega_s-\alpha)^2+\lambda^2}-\frac{1}{(\omega-\omega_s+\alpha)^2+\lambda^2}\right], \tag{8}$$

$$M(\omega,\theta) = e^{i\theta}\frac{\lambda^2-\mu^2}{8}\left[\frac{1}{(\omega-\omega_s-\alpha)^2+\mu^2}+\frac{1}{(\omega-\omega_s+\alpha)^2+\mu^2}+\right.$$
$$\left.+\frac{1}{(\omega-\omega_s-\alpha)^2+\lambda^2}+\frac{1}{(\omega-\omega_s+\alpha)^2+\lambda^2}\right]. \tag{9}$$

Here the parameters λ and μ are expressed in terms of the parametric oscillator cavity damping rate γ_c and the effective pump amplitude ϵ of the coherent field driving the parametric oscillator

$$\lambda = \gamma_c + \epsilon,\ \ \mu = \gamma_c - \epsilon,\ \ \epsilon = E_s/E_c, \tag{10}$$

where E_s is the amplitude of the pump coherent field and E_c is its threshold value for parametric oscillator. In optical parametric oscillator (OPO) the amplitude E_s is related to the power of pumping P [11, 12], so the effective pump amplitude is related to the ratio of input pump power to the critical power, $r = P/P_c$, and we have $\epsilon = \sqrt{r}\gamma_c/2$. The noise spectrum and the squeezing level of the output light from OPO is related to $\epsilon/(\gamma_c/2)$. When this ratio goes to 1 and therefore $r \to 1$, the threshold happens in OPO. The parameter $\alpha = (\omega_1 - \omega_2)/2$ is the displacement from the central frequency of the squeezing at which the two-mode squeezed vacuum is maximally squeezed. It is worth noticing that Eqs.(6,7,8,9) are only valid sufficiently below of the threshold, i.e. when $0 < \epsilon < \gamma_c/2$, both parameters λ and μ are positive and $\lambda > \mu$, and the squeezing values are not too large. When the parameters λ and μ are much greater than all other relaxation rates in the problem, the frequency dependence of $N(\omega)$ and $M(\omega,\theta)$ can be neglected. This case is referred to as broadband squeezed vacuum in which there is no difference between the output fields from DPO and NDPO.

3. Equations of Motion for Atomic Variables

In what follows, it is assumed that $\delta_a \ll \Omega_R$, so that the interaction of the driving field with the permanent dipole moment is much weaker than its interaction with the transitional dipole moment while, at the same time, the driving field is strong enough to be viewed as a dressing field for the two-level system. To derive the master equation, two successive unitary transformations are performed. The first one

$$U_1(t) = \exp\left[-i\left(\omega_f S^z + \sum_k \omega_k b^+(\omega_k)b(\omega_k)\right)t\right] \tag{11}$$

transforms the Hamiltonian (1) to the frame rotating with the driving field frequency ω_f and to the interaction picture with respect to the vacuum field modes, thus yielding the resulting Hamiltonian

$$H_1(t) = U_1^+(t)HU_1(t) - i\hbar U_1^+(t)\frac{\partial U_1(t)}{\partial t} = -\frac{\hbar}{2}\Delta S_z + \frac{\hbar}{2}\Omega_R(S^+ + S^-) + \frac{\hbar}{2}\delta_a(e^{i\omega_f t} + e^{-i\omega_f t})S^z +$$
$$+\sum_k\left(g(\omega_k)S^+b(\omega_k)e^{i(\omega_f-\omega)t} + g^*(\omega_k)b^+(\omega_k)S^-e^{-i(\omega_f-\omega)t}\right), \tag{12}$$

where $\Delta = \omega_f - \omega_0$ is the detuning of the driving field frequency ω_f from the atomic frequency ω_0. The second transformation is the unitary transformation

$$U_2(t) = \overleftarrow{T}\exp\left[-\frac{i}{\hbar}\int_0^t d\tau\left(-\frac{\hbar}{2}\Delta S_z + \frac{\hbar}{2}\Omega_R(S^+ + S^-) + \frac{\hbar}{2}\delta_a(e^{i\omega_f\tau} + e^{-i\omega_f\tau})S^z\right)\right] =$$

$$= \exp\left[-\frac{i}{2}\left(\Omega_R(S^+ + S^-) - \Delta S_z\right)t\right]\overleftarrow{T}\exp\left[-\frac{i}{2}\int_0^t d\tau\delta_a(e^{i\omega_f\tau} + e^{-i\omega_f\tau})\hat{\tilde{S}}^z(t)\right], \quad (13)$$

where

$$\hat{\tilde{S}}^z(t) = \exp\left[\frac{i}{2}\left(\Omega_R(S^+ + S^-) - \Delta S_z\right)t\right]S^z\exp\left[-\frac{i}{2}\left(\Omega_R(S^+ + S^-) - \Delta S_z\right)t\right]. \quad (14)$$

As a result of this transformation, the model Hamiltonian takes the form

$$H_2(t) = U_2^+(t)H_1(t)U_2(t) - i\hbar U_2^+(t)\frac{\partial U_2(t)}{\partial t} =$$

$$= \sum_k\left(g(\omega_k)\hat{S}(t)^+ b(\omega_k)e^{i(\omega_f-\omega)t} + g^*(\omega_k)b^+(\omega_k)\hat{S}^-(t)e^{-i(\omega_f-\omega)t}\right), \quad (15)$$

where

$$\hat{S}^\pm(t) = U_2^+(t)S^\pm U_2(t) \approx e^{\frac{i}{2}(\Omega_R(S^+ + S^-) - \Delta S_z)t}S^\pm e^{-\frac{i}{2}(\Omega_R(S^+ + S^-) - \Delta S_z)t}, \quad (16)$$

if the condition of weak inversion symmetry violation $\delta_a \ll \Omega_R$ is accounted for. Therefore,

$$\hat{S}^\pm(t) \approx \frac{1}{2}\left[S_a \pm (1 \mp \tilde{\Delta})S_b e^{-i\Omega' t} \pm (1 \pm \tilde{\Delta})S_c e^{i\Omega' t}\right], \quad (17)$$

where

$$S_a = \tilde{\Omega}\left[\tilde{\Omega}(S^+ + S^-) - 2\tilde{\Delta}S^z\right], \quad S_b = \frac{1}{2}\left[(1 - \tilde{\Delta})S^+ - (1 + \tilde{\Delta})S^- - 2\tilde{\Omega}S^z)\right], \quad (18)$$

$$S_c = \frac{1}{2}\left[(1 + \tilde{\Delta})S^+ - (1 - \tilde{\Delta})S^- + 2\tilde{\Omega}S^z)\right], \quad (19)$$

and $\tilde{\Omega} = \Omega_R/\Omega'$, $\tilde{\Delta} = \Delta/\Omega'$, $\Omega' = \sqrt{\Omega_R^2 + \tilde{\Delta}^2}$. As a consequence of the performed transformations, the approximate master equation for the transformed atomic reduced density operator $\rho_S^{trans}(t)$ can be derived using standard methods [13], and in the Born approximation is given by

$$\frac{\partial\rho_S^{trans}(t)}{\partial t} = -\frac{1}{\hbar^2}\int_0^t d\tau Sp_R\left([H_2(t), [H_2(t-\tau), \rho_R(0)\rho_S^{trans}(t-\tau)]]\right), \quad (20)$$

where $\rho_S^{trans}(t) = Sp_R\left(U_2^+(t)U_1^+(t)\rho(t)U_1(t)U_2(t)\right)$, $\rho(t)$ is the density operator for the total system (1), $\rho_R(0)$ is the density operator for the vacuum field reservoir and $Sp_R(...)$ is the trace over the reservoir states. As usual, in the Markoff approximation the upper limit of the integration over τ can be extended to infinity, and the master equation in the frame rotating with the driving field frequency ω_f can be written, under the assumption that the carrier frequency ω_s of the squeezed field coincides with the frequency ω_f, as

$$\begin{aligned}
\tfrac{\partial\rho_S^{rf}(t)}{\partial t} &= i\Gamma\delta[S^z, \rho_S^{rf}(t)] - \tfrac{i}{2}\delta_a(e^{i\omega_f t} + e^{-i\omega_f t})[S^z, \rho_S^{rf}(t)] + \\
&+ \tfrac{1}{2}\Gamma\tilde{N}(2S^+\rho_S^{rf}(t)S^- - S^-S^+\rho_S^{rf}(t) - \rho_S^{rf}(t)S^-S^+) + \\
&+ \tfrac{1}{2}\Gamma(\tilde{N}+1)(2S^-\rho_S^{rf}(t)S^+ - S^+S^-\rho_S^{rf}(t) - \rho_S^{rf}(t)S^+S^-) - \\
&- \Gamma\tilde{M}S^+\rho_S^{rf}(t)S^+ - \Gamma\tilde{M}^*S^-\rho_S^{rf}(t)S^- - \\
&- \tfrac{1}{2}i\Omega_R[S^+ + S^-, \rho_S^{rf}(t)] + \tfrac{1}{2}i(\beta[S^+, [S^z, \rho_S^{rf}(t)]] - \beta^*[S^-, [S^z, \rho_S^{rf}(t)]]),
\end{aligned} \quad (21)$$

where

$$\rho_S^{rf}(t) = U_2(t)\rho_S^{trans}(t)U_2^+(t) = e^{i\omega_f S^z t}\rho_S(t)e^{-i\omega_f S^z t}, \quad \tilde{N} = N(\omega_f + \Omega') + \tfrac{1}{2}(1-\tilde{\Delta}^2)\gamma_n, \tag{22}$$

$$\tilde{M} = (|M(\omega_f + \Omega', \theta)| + i\tilde{\Delta}\delta_M)\exp(i\theta) - \tfrac{1}{2}(1-\tilde{\Delta}^2)(\gamma_n - i\delta_n), \tag{23}$$

$$\delta = \Delta/\Gamma + \tilde{\Delta}\delta_N + \tfrac{1}{2}(1-\tilde{\Delta}^2)\delta_n, \quad \beta = \Gamma\tilde{\Omega}[\delta_N + \delta_M\exp(i\theta) - i\tilde{\Delta}(\gamma_n - i\delta_n)], \tag{24}$$

$$\gamma_n = N(\omega_f) - N(\omega_f + \Omega') - (|M(\omega_f, \theta)| - |M(\omega_f + \Omega', \theta)|)\cos(\theta), \tag{25}$$

$$\delta_n = (|M(\omega_f, \theta)| - |M(\omega_f + \Omega', \theta)|)\sin(\theta), \tag{26}$$

$$\delta_N = \tfrac{1}{\pi}\mathcal{P}\int_{-\infty}^{\infty} d\omega \frac{N(\omega)}{\omega - \omega_s + \Omega'}\Bigg|_{\omega_s=\omega_f}, \quad \delta_M = \frac{1}{\pi}\mathcal{P}\int_{-\infty}^{\infty} d\omega \frac{|M(\omega,\theta)|}{\omega - \omega_s + \Omega'}\Bigg|_{\omega_s=\omega_f}. \tag{27}$$

Here Γ is the radiative damping constant. The principal values of the integrals (27) can be evaluated by means of the contour integration [14] as

$$\delta_N = \delta_\mu - \delta_\lambda, \quad \delta_M = \delta_\mu + \delta_\lambda, \tag{28}$$

where for the degenerate DPO case

$$\delta_\mu^{dpo} = \Omega' \frac{\lambda^2 - \mu^2}{4} \frac{1}{\mu(\Omega'^2 + \mu^2)}, \quad \delta_\lambda^{dpo} = \Omega' \frac{\lambda^2 - \mu^2}{4} \frac{1}{\lambda(\Omega'^2 + \lambda^2)}, \tag{29}$$

and for the non-degenerate NDPO case

$$\delta_\mu^{ndpo} = \frac{\lambda^2 - \mu^2}{8\mu}\left[\frac{(\Omega' + \alpha)}{(\Omega' + \alpha)^2 + \mu^2} + \frac{(\Omega' - \alpha)}{(\Omega' - \alpha)^2 + \mu^2}\right], \tag{30}$$

$$\delta_\lambda^{ndpo} = \frac{\lambda^2 - \mu^2}{8\lambda}\left[\frac{(\Omega' + \alpha)}{(\Omega' + \alpha)^2 + \lambda^2} + \frac{(\Omega' - \alpha)}{(\Omega' - \alpha)^2 + \lambda^2}\right]. \tag{31}$$

This equation is similar to the master equation obtained earlier in [15] for a two-level non-polar emitter without broken inversion symmetry and is derived assuming that the system-reservoir and the system-field interactions are weak and the reservoir correlation time is small compared with the time t of observation. So, a closed set of equations follows from Eq.(21):

$$\frac{d}{dt}\langle\tilde{S}^-(t)\rangle = -\Gamma\left(\frac{1}{2} + \tilde{N} - i\delta + i\frac{\delta_a}{2\Gamma}(e^{i\omega_f t} + e^{-i\omega_f t})\right)\langle\tilde{S}^-(t)\rangle + \Gamma\tilde{M}\langle\tilde{S}^+(t)\rangle + \Omega_R\langle S^z(t)\rangle, \tag{32}$$

$$\frac{d}{dt}\langle\tilde{S}^+(t)\rangle = -\Gamma\left(\frac{1}{2} + \tilde{N} + i\delta - i\frac{\delta_a}{2\Gamma}(e^{i\omega_f t} + e^{-i\omega_f t})\right)\langle\tilde{S}^+(t)\rangle + \Gamma\tilde{M}^*\langle\tilde{S}^-(t)\rangle + \Omega_R\langle S^z(t)\rangle, \tag{33}$$

$$\frac{d}{dt}\langle S^z(t)\rangle = -\frac{1}{2}(\Omega_R + \beta^*)\langle S^-(t)\rangle - \frac{1}{2}(\Omega_R + \beta)\langle S^+(t)\rangle - \Gamma(2\tilde{N} + 1)\langle S^z(t)\rangle - \Gamma/2, \tag{34}$$

where $\langle\tilde{S}^\pm(t)\rangle = \pm i\langle S^\pm(t)e^{\mp i\omega_f t}\rangle$ are slowly varying parts of the atomic operators. The system of equations (32-34) can be solved numerically by means of the technique employed earlier in [19], where the components of the vector $\vec{X}(t) = (\langle\tilde{S}^-(t)\rangle, \langle\tilde{S}^+(t)\rangle, \langle S^z(t)\rangle)$ are decomposed as $X_i(t) = \sum\limits_{l=-\infty}^{+\infty} X_i^{(l)}(t)e^{il\omega_f t}$, $i = 1, 2, 3$, and the slowly varying amplitudes $X_i^{(l)}(t)$ obey the system of equations

$$\frac{d}{dt}X_1^{(l)}(t) = -\Gamma\left(\frac{1}{2} + \tilde{N} - i\delta + il\frac{\omega_f}{\Gamma}\right)X_1^{(l)}(t) -$$
$$-i\frac{\delta_a}{2}(X_1^{(l-1)}(t) + X_1^{(l+1)}(t)) + \Gamma\tilde{M}X_2^{(l)}(t) + \Omega_R X_3^{(l)}(t), \tag{35}$$

$$\frac{d}{dt}X_2^{(l)}(t) = -\Gamma\left(\frac{1}{2} + \tilde{N} - i\delta + il\frac{\omega_f}{\Gamma}\right)X_2^{(l)}(t) +$$
$$+i\frac{\delta_a}{2}(X_2^{(l-1)}(t) + X_2^{(l+1)}(t)) + \Gamma\tilde{M}^*X_1^{(l)}(t) + \Omega_R X_3^{(l)}(t), \tag{36}$$

$$\frac{d}{dt}X_3^{(l)}(t) = -\frac{\Gamma}{2}\delta_{l,0} - (\Gamma(2\tilde{N} + 1) + il\omega_f)X_3^{(l)}(t) - \frac{\Omega_R + \beta^*}{2}X_1^{(l)}(t) - \frac{\Omega_R + \beta}{2}X_2^{(l)}(t). \tag{37}$$

Fluorescence spectrum

The incoherent part of the steady-state fluorescence spectrum is given by[16, 17]

$$F_{inc}(\omega) = \frac{\Gamma}{\pi}\mathrm{Re}\int_0^{\infty} d\tau \lim_{t\to\infty}[\langle\tilde{S}^+(t)\tilde{S}^-(t+\tau)\rangle - \langle\tilde{S}^+(t)\rangle\langle\tilde{S}^-(t+\tau)\rangle]e^{i(\omega-\omega_f)\tau}, \quad (38)$$

where the coherent contribution from the incident driving field scattered by the atom is subtracted, as usual. In accordance with the so-called quantum regression hypothesis [13, 18], the fluctuation correlation functions $Y_1(t,t+\tau) = \langle\tilde{S}^+(t)\tilde{S}^-(t+\tau)\rangle - \langle\tilde{S}^+(t)\rangle\langle\tilde{S}^-(t+\tau)\rangle$, $Y_2(t,t+\tau) = \langle\tilde{S}^+(t)\tilde{S}^+(t+\tau)\rangle - \langle\tilde{S}^+(t)\rangle\langle\tilde{S}^+(t+\tau)\rangle$, $Y_3(t,t+\tau) = \langle\tilde{S}^+(t)\tilde{S}^z(t+\tau)\rangle - \langle\tilde{S}^+(t)\rangle\langle\tilde{S}^z(t+\tau)\rangle$, satisfy virtually the same set of equations of motion (32-34) for the correspondent averages $\langle\tilde{S}^-(\tau)\rangle$, $\langle\tilde{S}^+(\tau)\rangle$ and $\langle\tilde{S}^z(\tau)\rangle$ with the only difference that the inhomogeneity $-\Gamma/2$ disappears due to the subtraction of the mean. These correlation functions can be decomposed as $Y_i(t,t+\tau) = \sum_{l=-\infty}^{+\infty} Y_i^{(l)}(t,\tau)e^{il\omega_f(t+\tau)}$, $i = 1,2,3$, so that

$$\frac{d}{d\tau}Y_1^{(l)}(t,\tau) = -\Gamma\left(\frac{1}{2}+\tilde{N}-i\delta+il\frac{\omega_f}{\Gamma}\right)Y_1^{(l)}(t,\tau) -$$
$$-i\frac{\delta_a}{2}(Y_1^{(l-1)}(t,\tau)+Y_1^{(l+1)}(t,\tau)) + \Gamma\tilde{M}Y_2^{(l)}(t,\tau) + \Omega_R Y_3^{(l)}(t,\tau), \quad (39)$$

$$\frac{d}{d\tau}Y_2^{(l)}(t,\tau) = -\Gamma\left(\frac{1}{2}+\tilde{N}-i\delta+il\frac{\omega_f}{\Gamma}\right)Y_2^{(l)}(t,\tau) +$$
$$+i\frac{\delta_a}{2}(Y_2^{(l-1)}(t,\tau)+Y_2^{(l+1)}(t,\tau)) + \Gamma\tilde{M}^*Y_1^{(l)}(t,\tau) + \Omega_R Y_3^{(l)}(t,\tau), \quad (40)$$

$$\frac{d}{d\tau}Y_3^{(l)}(t,\tau) = -(\Gamma(2\tilde{N}+1)+il\omega_f)Y_3^{(l)}(t,\tau) - \frac{\Omega_R+\beta^*}{2}Y_1^{(l)}(t,\tau) - \frac{\Omega_R+\beta}{2}Y_2^{(l)}(t,\tau), \quad (41)$$

and the Laplace transforms $\bar{Y}_i^{(l)}(t,z) = \int_0^{\infty} e^{-z\tau}Y_i^{(l)}(t,\tau)d\tau$ of the components $Y_i^{(l)}(t,\tau)$ will satisfy the following set of equations:

$$z\bar{Y}_1^{(l)}(t,z) + \Gamma\left(\frac{1}{2}+\tilde{N}-i\delta+il\frac{\omega_f}{\Gamma}\right)\bar{Y}_1^{(l)}(t,z) +$$
$$+i\frac{\delta_a}{2}\left(\bar{Y}_1^{(l-1)}(t,z)+\bar{Y}_1^{(l+1)}(t,z)\right) - \Gamma\tilde{M}\bar{Y}_2^{(l)}(t,z) - \Omega_R\bar{Y}_3^{(l)}(t,z) =$$
$$= \frac{1}{2}\delta_{l,0} + X_3^{(l)}(t) - \sum_{r=-\infty}^{\infty} X_1^{(l-r)}(t)X_2^{(r)}(t), \quad (42)$$

$$z\bar{Y}_2^{(l)}(t,z) + \Gamma\left(\frac{1}{2}+\tilde{N}-i\delta+il\frac{\omega_f}{\Gamma}\right)\bar{Y}_2^{(l)}(t,z) -$$
$$-i\frac{\delta_a}{2}\left(\bar{Y}_2^{(l-1)}(t,z)+\bar{Y}_2^{(l+1)}(t,z)\right) - \Gamma\tilde{M}^*Y_1^{(l)}(t,z) - \Omega_R\bar{Y}_3^{(l)}(t,z) = -\sum_{r=-\infty}^{\infty} X_2^{(l-r)}(t)X_2^{(r)}(t), \quad (43)$$

$$z\bar{Y}_3^{(l)}(t,z) + (\Gamma(2\tilde{N}+1)+il\omega_f)\bar{Y}_3^{(l)}(t,z) + \frac{\Omega_R+\beta^*}{2}\bar{Y}_1^{(l)}(t,z) + \frac{\Omega_R+\beta}{2}\bar{Y}_2^{(l)}(t,z) =$$
$$= -\sum_{r=-\infty}^{\infty}\left(\frac{1}{2}\delta_{r,0}+X_3^{(r)}(t)\right)X_2^{(l-r)}(t). \quad (44)$$

In the steady state limit ($t\to\infty$) only the zero-order component $\bar{Y}_1^{(0)}(t,z)$ contributes to $F_{inc}(\omega)$, and the incoherent part of the spectrum (38) reads as

$$F_{inc}(\omega) = \frac{\Gamma}{\pi}\mathrm{Re}\lim_{t\to\infty}\bar{Y}_1^{(0)}(t,z)\Big|_{z=-i(\omega-\omega_f)}. \quad (45)$$

4. Numerical Results

Equations (35)-(37) and (42)-(44) were solved numerically, as usual [19], in the steady state limit ($t \to \infty$) by truncation of the number of the harmonic amplitudes $X_i^{(l)}(t)$ and $\bar{Y}_i^{(l)}(t,z)$ involved, and the case of the driving laser field frequency ω_f and the carrier frequency of the squeezed field ω_s being simultaneously in resonance with the atomic transition frequency ω_0 was studied. It was already found [3] in the case of a two-level system with broken symmetry interacting with non-squeezed vacuum reservoir that for $\delta_a \neq 0$ a low-frequency radiation peak centered nearly exactly at the frequency $\omega = \Omega_R$ appears in the fluorescence spectrum. The amplitude of the peak increases steadily with the increase of the symmetry violation parameter δ_a and decreases with the increase of the driving field frequency ω_f. The central frequency of the peak is defined for the most part by the Rabi frequency Ω_R and depends weakly on the symmetry violation parameter δ_a, so that it drifts very slowly toward $\omega = 0$ along with the increase of this parameter. Here it was found that the interaction with the finite bandwidth squeezed vacuum reservoir does not change these aspects of the spectral peak behavior. At the same time, the amplitude of this peak strongly depends on the NDPO non-degeneracy parameter α. It is maximal in the DPO case $\alpha = 0$ and decreases with the increase of the non-degeneracy while the width of the peak increases, see Fig.1. At the same time, the position of the peak is not significantly affected and, consequently, the NDPO degeneracy can be employed to control the fluorescence intensity output of the polar emitter at the fixed frequency Ω_R. The same goal of the fluorescence intensity output control can be achieved for fixed non-degeneracy parameter α by varying the phase of squeezing θ instead, see Fig.2. It was also found that the fluorescence peak is most pronounced for the case of the OPO at the threshold when $2\epsilon/\gamma_c = 1$. Below the threshold the amplitude of the peak decreases while the position of the peak is not shifted, see Fig.3 and Fig.4. This property provides additional means for the control of the fluorescence intensity output.

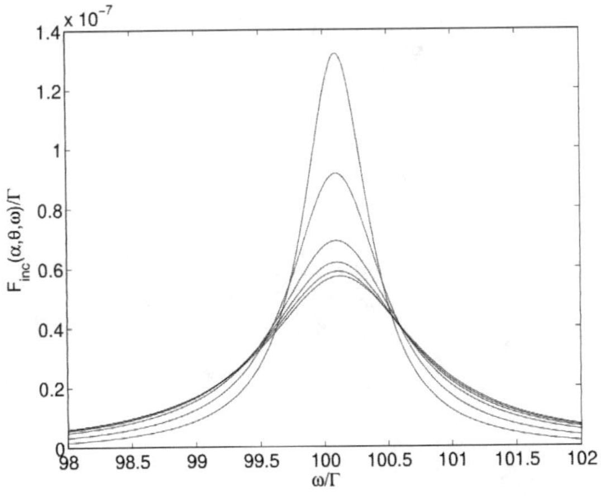

Figure 1. Fluorescence spectrum at $\omega = \Omega_R$ for various values of α at the OPO threshold. $\Gamma = 1$, $\alpha = 0 : 10 : 50$, $\gamma_c = 10, \epsilon = 5, \theta = 0$, $\omega_f = \omega_s = \omega_0 = 5000, \Omega_R = 100, \delta_a = 10$.

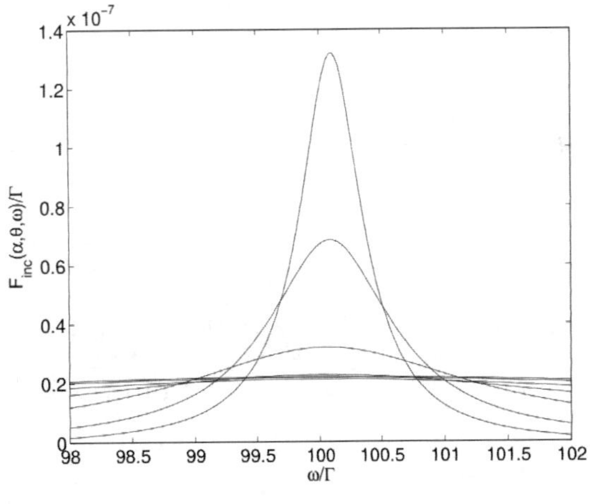

Figure 2. Fluorescence spectrum at $\omega = \Omega_R$ for various values of θ at the OPO threshold. $\Gamma = 1, \alpha = 0, \gamma_c = 10, \epsilon = 5$, $\theta = 0 : \pi/6 : \pi$, $\omega_f = \omega_s = \omega_0 = 5000, \Omega_R = 100, \delta_a = 10$.

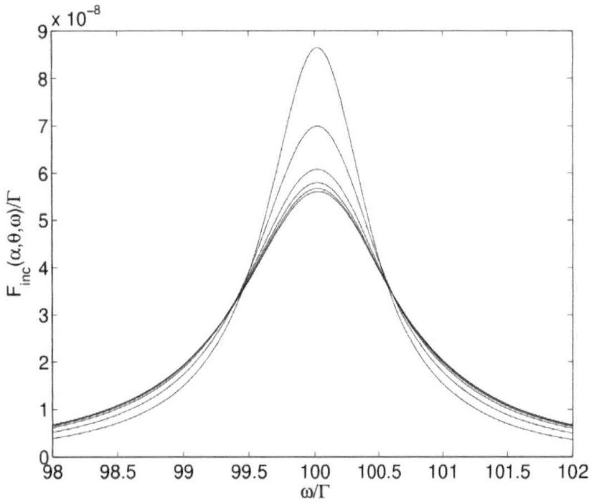

Figure 3. Fluorescence spectrum at $\omega = \Omega_R$ for various values of α below the OPO threshold. $\Gamma = 1$, $\alpha = 0 : 10 : 50$, $\gamma_c = 10, \epsilon = 2, \theta = 0$, $\omega_f = \omega_s = \omega_0 = 5000, \Omega_R = 100, \delta_a = 10$.

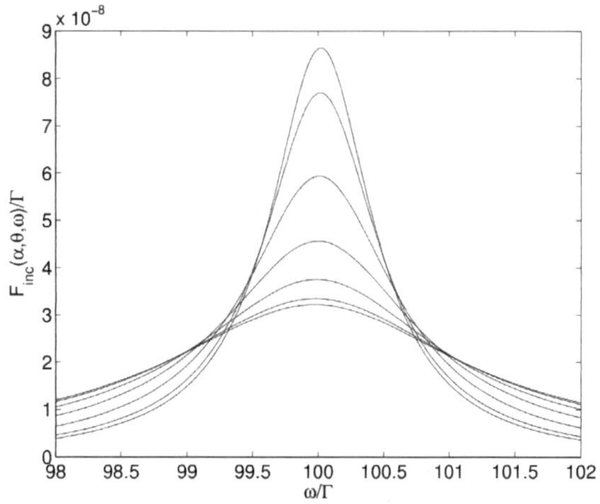

Figure 4. Fluorescence spectrum at $\omega = \Omega_R$ for various values of θ below the OPO threshold. $\Gamma = 1, \alpha = 0, \gamma_c = 10, \epsilon = 2, \theta = 0 : \pi/6 : \pi$, $\omega_f = \omega_s = \omega_0 = 5000, \Omega_R = 100, \delta_a = 10$.

5. Conclusion

In conclusion, we investigated the effect of the vacuum dissipative reservoir finite bandwidth squeezing on the phenomenon of the low-frequency fluorescence by a damped quantum two-level polar system with broken inversion symmetry driven by external high-frequency classical EM (laser) field. The source of the squeezed vacuum was represented by the OPO being in either the NDPO or the DPO mode of operation. It was shown that the parameters of the OPO, such as the non-degeneracy degree and the phase of squeezing, provide efficient means for the control of the intensity and spectral width of the fluorescence radiation output.

References

[1] Kovarskii V A 1999 *Phys. Usp.* **42** 797
[2] Kibis O V, Slepyan G Ya, Maksimenko S A and Hoffmann A 2009 *Phys. Rev. Lett.* **102** 023601
[3] Soldatov A V 2016 *Mod. Phys. Lett.* B **30** no.27 1650331
[4] Soldatov A V 2017 *Mod. Phys. Lett.* B **34** no.4 1750027
[5] Bogolyubov N N Jr and Soldatov A V 2018 *Mosc. Univ. Phys. Bull.* **73** no.2 154-61
[6] Bogolyubov (Jr.) N.N. and Soldatov A.V. 2020 *Phys. Part. Nucl.* **51:4** 762
[7] Bogolyubov (Jr.) N.N. and Soldatov A.V. 2020 *Journ. of Phys.: Conf. Ser.* **1560** 12001
[8] Gardiner C W 1986 *Phys. Rev. Lett.* **56** 1917
[9] Collett M J and Gardiner C W 1984 *Phys. Rev.* A **30** 1386
[10] Drummond P D and Reid M D 1984 *Phys. Rev.* A **41** 3930
[11] Drummond P D McNeil K J and Walls D F 1980 *Opt. Acta.* **27** 321
[12] Drummond P D McNeil K J and Walls D F 1981 *Opt. Acta.* **28** 211
[13] Puri R R 2001 *Mathematical Methods of Quantum Optics* (Springer Series in Optical Sciences vol. 79) (Berlin Heidelberg: Springer-Verlag)
[14] Legua M and Luis M Sánchez-Ruiz 2017 *Entropy* **19** 215
[15] Tanaś R Ficek Z Messikh A and El-Shahat T 1998 *J. Mod. Opt.* **45** No. **9** 1859-1883
[16] Louisell W H 1973 *Quantum Statistical Properties of Radiation* Wiley, New York
[17] Agarwal G S 1974 *Quantum statistical theories of spontaneous emission and their relation to other approaches* (Quantum Optics, Springer Tracts in Modern Physics vol. 70) Ed. by G. Höhler, Springer-Verlag, Berlin, Heidelberg
[18] Carmichael H 1993 *An Open Systems Approach to Quantum Optics* (Lecture Notes in Physics vol. 18) (Berlin: Springer-Verlag)
[19] Ficek Z, Seke J, Soldatov A V and Adam G 2001 *Phys. Rev.* A **64** no.1 013813

Superconductivity of twisted double bi-layer graphene in chiral model

Yu P Rybakov [1], **M Umar** [1]

[1] Peoples' Friendship University of Russia (RUDN University), Moscow, 117198, Russia

e-mail: soliton4@mail.ru, rybakov-yup@rudn.ru

Abstract. Taking into account the sp – hybridization effect for valence electrons in carbon atoms, a very simple chiral model of graphene was suggested some years ago [1]. This model used as a special order parameter the unitary $SU(2)$ – matrix U with the kink-like (or domain wall) structure for the description of electrons in a mono-atomic graphene sheet. However, later the new graphene physics began since studying twisted multilayer configurations revealing unconventional superconducting properties (twist-tronics) [2]. Within the scope of the chiral model these twisted configurations can be described by the so-called "product ansatz" $U = U_1 U_2 \cdots U_n$. As an example, the special case of twisted double bi-layer graphene (TDBG) configuration is studied and the corresponding twist angles, for which the superconductivity takes place, are found.

1. Introduction. Structure of the chiral model

Mono-atomic carbon lattices called graphenes attracted a deep interest of researchers due to their extraordinary electronic and mechanical properties [3], which can be effectively described within the scope of the chiral model based on the sp – hybridization effect. As is well known, the strong coupling between carbon atoms forming hexagon lattices is created by three covalent bonds with the neighbors, the fourth valence electron being "free" for each atom. Therefore, it seems natural to combine scalar a_0 and vector $\vec{a}$ fields corresponding to the s – orbital and the p – orbital states of the "free" electron, respectively, into the unitary matrix $U \in SU(2)$ serving as the order parameter in our model:

$$U = a_0 \tau_0 + \mathrm{i}\, \vec{a} \cdot \vec{\tau}\,. \tag{1}$$

Here τ_0 is the unit 2×2 – matrix and $\vec{\tau}$ stands for the three Pauli matrices with the subsidiary $SU(2)$ – condition being imposed: $a_0^2 + \vec{a}^2 = 1$.

The Lagrangian density of the model can be constructed through the left chiral currents

$$l_\mu = U^+ \partial_\mu U, \quad \mu = 0, 1, 2, 3, \tag{2}$$

involving the coordinate x^1, x^2, x^3 and the time $x^0 = ct$ derivatives:

$$L = -\frac{I}{4}\,\mathrm{tr}\,(l_\mu l^\mu) - \frac{\lambda^2}{2}\vec{a}^{\,2}. \tag{3}$$

The parameter I in (3) can be interpreted as the exchange energy between carbon atoms (per spacing). The equations of motion corresponding to (3) admit the kink-like solution [1]:

$$U = \exp\,(\mathrm{i}\,\hat{n}\,\Theta), \quad \hat{n} = \vec{n}\cdot\vec{\tau}, \ \Theta = \Theta(z) = 2\arctan\exp\,(-z/l_0)\,, \tag{4}$$

describing the electrons distribution in an ideal graphene plane oriented along the unit vector $\vec{n}$ and orthogonal to the $z-$axis. The configuration (4) contains the characteristic length $l_0 = I^{1/2}/\lambda$, which can be identified with the diameter of the carbon atom.

It is worth-while to underline that the interaction with an external electromagnetic field can be included via extending the derivatives in accordance with the gauge invariance principle:

$$\partial_\mu U \Rightarrow D_\mu U = \partial_\mu U - \mathrm{i}\,e_0 A_\mu [\tau_3, U]\,, \tag{5}$$

where e_0, τ_3, A_μ denote the electromagnetic coupling constant, the charge operator and the 4-potential, respectively. In particular case of the interaction with the uniform magnetic field oriented along the $y-$axis the Lagrangian density reads:

$$L = -\frac{I}{4}\,\mathrm{tr}\,(L_\mu L^\mu) - \frac{\lambda^2}{2}\vec{a}^{\,2} - \frac{\vec{B}^2}{8\pi}, \tag{6}$$

where

$$L_\mu = U^+ D_\mu U, \quad \vec{B} = (0, B, 0), \quad B = A'(z), \quad A(z) = A_1, \quad B(\pm\infty) = B_0 = const.$$

2. Strongly correlated states in multi-layer graphene and the Meissner effect

The main concept behind this research concerns the ample justification of the connection between the formation of twisted graphene configurations and the emergence of the orbital superconductivity confirmed by the Meissner effect, that is vanishing magnetic intensity within the sample. Our approach builds on estimating the internal magnetic field for the TDBG configuration interacting with the uniform magnetic field parallel to the graphene sheets.

To this end, let us consider the product ansatz $U = U_{-2}U_{-1}U_1U_2$ with the fixed positions of the sheets along the $z-$axis:

$$z_{-2} = -3l, \quad z_{-1} = -l, \quad z_1 = l, \quad z_2 = 3l,$$

where $2l$ stands for the distance between the sheets and the matrices $U_k = \exp\,(\mathrm{i}\,\hat{n}_k \Theta_k)$ contain the following unit vectors:

$$\vec{n}_{-2} = (\cos\beta/2, \sin\beta/2, 0), \quad \vec{n}_{-1} = (\cos\alpha/2, \sin\alpha/2, 0), \quad \vec{n}_1 = \vec{n}_{-1}(-\alpha), \quad \vec{n}_2 = \vec{n}_{-2}(-\beta)$$

and the unknown phases $\Theta_k(z-z_k)$. In this case the Lagrangian density (6) includes only two components of the chiral current:

$$L_1 = -\mathrm{i}\, e_0 A(U^+\tau_3 U - \tau_3), \quad L_3 = U^+\partial_3 U$$

and the vector $\vec{a} = -(\mathrm{i}/2)\ \mathrm{tr}(\vec{\tau}\, U)$. The corresponding calculations appear to be drastically simplified through using the formula:

$$\exp(\mathrm{i}\hat{m}\Theta)\,\hat{n}\exp(-\mathrm{i}\hat{m}\Theta) = \hat{n}\cos 2\Theta + (1-\cos 2\Theta)\,\hat{m}\,(\vec{m}\cdot\vec{n}) - \sin 2\Theta\,(\vec{m}\times\vec{n})\cdot\vec{\tau}\,.$$

The resulting Lagrangian density takes the form:

$$L = \frac{I}{4}(\,\mathrm{tr}\, L_3^2 + \mathrm{tr}\, L_1^2) - \frac{\lambda^2}{2}\vec{a}^{\,2} - \frac{A'^2}{8\pi}. \tag{7}$$

Using the denotations $4X_k = -\,\mathrm{tr}\, L_k^2$, the corresponding terms read:

$$X_3 = \frac{1}{2}\sum_i \Theta'^2_i + \cos\frac{\beta-\alpha}{2}(\Theta'_{-2}\Theta'_{-1} + \Theta'_2\Theta'_1) + \cos\alpha\,\Theta'_{-1}\Theta'_1 +$$
$$\Theta'_2\Theta'_{-1}\left(\cos\frac{\beta+\alpha}{2} + 2\sin^2\Theta_1\sin\alpha\sin\frac{\beta-\alpha}{2}\right) + \Theta'_{-2}\Theta'_1\left(\cos\frac{\beta+\alpha}{2} + 2\sin^2\Theta_{-1}\sin\alpha\sin\frac{\beta-\alpha}{2}\right) +$$
$$\Theta'_{-2}\Theta'_2\cos^2\frac{\beta-\alpha}{2}\left[\begin{matrix}\cos\alpha - \sin\alpha\tan\frac{\beta-\alpha}{2}(\cos 2\Theta_{-1} + \cos 2\Theta_1) + \\ \tan^2\frac{\beta-\alpha}{2}(\sin 2\Theta_{-1}\sin 2\Theta_1 - \cos\alpha\cos 2\Theta_{-1}\cos 2\Theta_1)\end{matrix}\right];$$

$$X_1 = e_0^2 A^2\left\{1 - \prod_i \cos 2\Theta_i\left[1 - \cos^2\frac{\beta-\alpha}{2}\prod_k \tan 2\Theta_k - \cos\alpha\tan 2\Theta_{-1}\tan 2\Theta_1 - \cos\beta\tan 2\Theta_{-2}\tan 2\Theta_2\right.\right.$$
$$\left. - \cos\frac{\beta+\alpha}{2}(\tan 2\Theta_{-2}\tan 2\Theta_1 + \tan 2\Theta_{-1}\tan 2\Theta_2) - \cos\frac{\beta-\alpha}{2}(\tan 2\Theta_1\tan 2\Theta_2 + \tan 2\Theta_{-1}\tan 2\Theta_{-2})\right] +$$
$$\prod_k \sin 2\Theta_k \sin\frac{\beta-\alpha}{2}\left[\sin\frac{\beta+\alpha}{2}(\tan^{-1}2\Theta_{-1}\tan\Theta_1 + \tan^{-1}2\Theta_1\tan\Theta_{-1}) + \cos\alpha\sin\frac{\beta-\alpha}{2}\tan\Theta_{-1}\tan\Theta_1 +\right.$$
$$\left.\left.\sin\alpha(\tan^{-1}2\Theta_2\tan\Theta_{-1} + \tan^{-1}2\Theta_{-2}\tan\Theta_1)\right]\right\} \equiv e_0^2 A^2 V_0;$$

$$\vec{a}^{\,2} = M_\alpha^2\left[\cos^2\Theta_{-2}\cos^2\Theta_2 - M_\beta^2 + \frac{1}{2}(\sin^2\Theta_{-2} + \sin^2\Theta_2)\right] + 2\sin\alpha\sin\beta\prod_k \sin\Theta_k(M_{\alpha\beta} - M_\alpha M_\beta) +$$
$$2M_{\alpha\beta}M_\alpha M_\beta - \frac{1}{2}\sin\alpha\sin\beta\,(\sin^2\Theta_{-2} - \sin^2\Theta_2)(\sin^2\Theta_{-1} - \sin^2\Theta_1) +$$
$$\frac{1}{16}(1-P_1)[4(1+P_2) - 2S_2(\cos\alpha + \cos\beta) + (1-P_2)(1-\cos\alpha\cos\beta)].$$

Here the following denotations were introduced:

$$M_\alpha = \cos\Theta_{-1}\cos\Theta_1 - \cos\alpha \sin\Theta_{-1}\sin\Theta_1; \quad M_\beta = \cos\Theta_{-2}\cos\Theta_2 - \cos\beta\sin\Theta_{-2}\sin\Theta_2;$$

$$M_{\alpha\beta} = \prod_k \cos\Theta_k \tan\Theta_{-1}\left(\tan\Theta_{-2}\cos\frac{\beta-\alpha}{2} + \tan\Theta_2\cos\frac{\beta+\alpha}{2}\right);$$

$$P_1 = \cos 2\Theta_{-1}\cos 2\Theta_1; \quad P_2 = \cos 2\Theta_{-2}\cos 2\Theta_2; \quad S_2 = \sin 2\Theta_{-2}\sin 2\Theta_2.$$

3. Structure of solutions to the equations of motion

Let us now fix the boundary conditions for our problem:

$$\Theta_i(-\infty) = \pi; \quad \Theta_i(\infty) = 0; \quad \Theta_i(0) = \pi/2 \tag{8}$$

and make the substitution $\tan\Theta_i = u_i = \sinh^{-1} w_i$. First, to study the behaviour of solutions at large $z \to \infty$, it is convenient to consider the asymptotic limit of the Lagrangian as $u_i \to 0$:

$$L = -I\left[\sum_i u'^2_i + \cos\alpha\, u'_{-1}u'_1 + \cos\beta\, u'_{-2}u'_{21} + \cos\frac{\beta-\alpha}{2}(u'_{-2}u'_{-1} + u'_2 u'_1) + \cos\frac{\beta+\alpha}{2}(u'_2 u'_{-1} + u'_{-2}u'_1)\right] -$$
$$4Ie_0^2A^2\left[\frac{1}{2}\sum_i u_i^2 + \cos\alpha\, u_{-1}u_1 + \cos\beta\, u_{-2}u_2 + \cos\frac{\beta-\alpha}{2}(u_{-2}u_{-1} + u_2u_1) + \cos\frac{\beta+\alpha}{2}(u_2u_{-1} + u_{-2}u_1)\right] - \frac{A'^2}{8\pi},$$

where it was taken into account that $A = B_0 z + a(z) \approx B_0 z$. Therefore, the equations for u_i admit the symmetric solution of the form:

$$u_i = 2\exp(-e_0 B_0 z^2). \tag{9}$$

As follows from (9), the equation for the magnetic field reads:

$$B' = a'' = 256\,\pi\gamma I e_0^2 B_0 z \exp(-2e_0 B_0 z^2), \quad \gamma \equiv (\cos\beta/2 + \cos\alpha/2)^2, \tag{10}$$

and admits the evident solution:

$$B = B_0 - 64\,\pi\gamma I e_0 \exp(-2e_0B_0z^2); \quad A = B_0 z - 64\,\pi\gamma I e_0 \int_0^z \exp(-2e_0B_0t^2)\,dt.$$

Let us now consider the central part of the sample, where the reflection symmetry holds:

$$z \Rightarrow -z: \quad w_{-1}(-z) = k(l-z) = -w_1(z); \quad w_{-2}(-z) = k(3l-z) = -w_2(z); \quad A(-z) = -A(z),$$

and the vector potential satisfies in the first approximation the equation:

$$A'' = 8\pi I e_0^2 V_0 A, \quad V_0 = const,$$

with the following approximate solution:

$$B = C(1 + z^2 4\pi I e_0^2 V_0); \quad A = Cz[1 + z^2(4\pi/3)I e_0^2 V_0]; \quad C = const.$$

A global behavior of the magnetic field can be obtained by matching the asymptotic configurations at some intermediate point $z = \bar{l}$. Introducing the denotations:

$$x^2 = 2e_0 B_0 \bar{l}^2, \quad y = C / B_0, \quad \Gamma = 8\pi I e_0 / B_0, \quad \gamma = (\cos\beta/2 + \cos\alpha/2)^2$$

and using the special representation for the error function [4]:

$$\mathrm{erf}(x) = \frac{2}{\sqrt{\pi}} \int_0^x \mathrm{d}s \exp(-s^2) = \frac{2x}{\sqrt{\pi}} [1 + g(x^2)] \exp(-x^2); \quad g = \sum_{n=1}^{\infty} \frac{2^n x^{2n}}{(2n+1)!!} = \frac{2}{3} h x^2; \quad h(0) = 1,$$

one finds the following matching conditions:

$$y(1 + x^2 \Gamma V_0 / 12) = 1 - 8\Gamma\gamma(1 + g)\exp(-x^2); \quad y(1 + x^2 \Gamma V_0 / 4) = 1 - 8\Gamma\gamma \exp(-x^2). \tag{11}$$

It should be underlined that the system (11) determines the conditions under which our sample reveals superconducting properties. It means that the magnetic intensity in the central part ($z \to 0$) of the TDBG configuration should be vanishing (the Meissner effect), i. e. $y << 1$. Taking into account that for realistic magnetic fields $\Gamma >> 1, \quad x^2 << 1, \quad h \approx 1$, one gets from (11) the value

$$y \approx (1 + \Gamma V_0 / 4)^{-1} << 1. \tag{12}$$

Eliminating the variable y between the equations (11), one derives the constraint on α, β:

$$\left(\cos\frac{\beta}{2} + \cos\frac{\alpha}{2} \right)^2 \approx \frac{V_0}{8(\Gamma V_0 + 4)}. \tag{13}$$

In view of the restriction (12) caused by the Meissner effect, this amounts to searching the maximal value of $V_0(\alpha, \beta)$ under the condition (13). Due to the Lagrangian structure (7), one finds

$$V_0 = 4 \frac{\sinh^2 w_1 \sinh^2 w_2}{\cosh^4 w_1 \cosh^4 w_2} \left\{ 4 \frac{\sinh^2 w_2 - 1}{\sinh w_1 \sinh w_2} \left[\cos\frac{\alpha}{2} \cos\frac{\beta}{2} (\sinh^2 w_1 - 1) - \sin\alpha \sin\frac{\beta - \alpha}{2} \right] + \right.$$
$$2[1 + \cos(\beta - \alpha) - 2\cos\alpha] + \sinh^2 w_1 (1 + \cos\beta) + \sinh^{-2} w_1 [1 + \cos(2\alpha - \beta)] + \tag{14}$$
$$\left. (\sinh^2 w_2 + \sinh^{-2} w_2)(1 + \cos\alpha) \right\},$$

where in accordance with the kink structure [1] $w_1 = kl \approx 1, \quad w_2 = 3kl \approx 3$. As follows from (14), the maximal value of V_0 corresponds to the choice $\beta = \alpha = \pi - \mu, \quad \mu << 1$, that implies the structure:

$$V_0(\mu) = M(1 + N\mu^2); \quad M \approx 0.0761; \quad N \approx 6.511. \tag{15}$$

Solving equations (13) and (15) with respect to μ^2, one derives the dependence of the "magic" angle on the external magnetic field:

$$\mu^2 = [8(\Gamma + \rho)]^{-1}; \quad \rho = 4 / M - N / 8 \approx 51.715. \tag{16}$$

For numerical illustration of the twist effect one can use the following parameters of the chiral model [5]: the spacing $a = 0.287$ nm, the exchange energy between atoms $E_0 = 2.9$ eV with the value

$I = E_0 / a = 1.619$ nN, the coupling constant $e_0 = e/(\hbar c)$ with the value $Ie_0 = 0.246$ T, the distance between the sheets $2l = 0.34$ nm. The dependence of the relative magnetic intensity (12) and the "magic" angle (16) on the intensity of the external magnetic field is represented in the Table 1.

Table 1. The relative magnetic intensity y inside the TDBG and the twisting angle μ as functions of B_0.

B_0 (mT)	$\Gamma \times 10^{-3}$	$y \times 10^2$	μ (deg)
5	1.2364	4.08	± 0.56
10	0.6182	7.85	± 0.78
15	0.4121	11.3	± 0.94

4. Conclusion

To sum up, notice that within the scope of the chiral graphene model the special TDBG sample was studied and proved to reveal superconducting properties for some critical "magic" twisting angles [6, 7]. This fact confirms the idea of forming strongly correlated electronic states as a result of twisting graphene sheets. Another consequence of this twist effect appears to be a creation of giant hexagon super-lattices, so-called moiré patterns [8], that facilitates the transitions of electrons between the sheets due to flat bands in dispersion curves of the model (3).

Acknowledgements

This publication has been supported by the RUDN University Strategic Academic Leadership Program. The paper is dedicated to the 70th anniversary of outstanding physicist Professor Victor Belyaev.

References

[1] Rybakov Yu P 2012 On chiral model of graphene *Solid State Phenomena* **190** 59-62
[2] Cao Y et al 2018 Unconventional superconductivity in magic-angle graphene superlattices *Nature* **556** 43-50
[3] Castro Neto A N et al 2000 The electronic properties of graphenes *Reviews of Modern Physics* **81** 109-162
[4] Abramowitz M and Stegun I A 1964 *Handbook of Mathematical Functions* (Washington: National Bureau of Standards)
[5] Sheka E 2018 *Spin Chemical Physics of Graphene* (Singapore: Pan Stanford Publishing)
[6] Magera U and Kulkarui G U 2020 A new twist in graphene research: twisted graphene *Carbon* **156** 470-487
[7] Podaro J F et al 2018 Phases of a phenomenological model of twisted bi-layer graphene *Physical Review* **B98** 075154
[8] Chen G et al 2019 Signatures of tunable superconductivity in a tri-layer graphene moiré super-lattice *Nature* **572** 215-219

Direct stochastic molecular modelling of transport processes in gases

V Ya Rudyak[1,2]* **and E V Lezhnev**[2]

[1]Department of Theoretical Mechanics, Novosibirsk State University of Architecture and Civil Engineering, Novosibirsk, Russia
[2]Department of Nanocomposite Material, Novosibirsk State University, Russia

*Corresponding author email: rudyak@sibstrin.ru

Abstract. The stochastic molecular modeling method (SMM) of transport processes in rarefied gases developed by the authors is systematically discussed in this paper. It is shown that, it is possible to simulate the transport coefficients of rarefied gas with high accuracy, using a relatively small number of molecules. The data of modeling the thermal conductivity coefficient are presented for the first time. The second part of the paper is devoted to the generalization of the SMM method for modeling transport processes in confined conditions. To describe the dynamics of molecules in this case, the splitting of their evolution by processes is used: first, the movement of molecules in the configuration space is simulated, and then their dynamics in the velocity space is imitated. Anisotropy of viscosity and thermal conductivity in nanochannels has been established. The interaction of gas molecules with walls is described by specular or specular-diffuse reflection laws. Gas viscosity can be either greater than in the bulk or less, depending on the law of gas interaction with the channel walls.

Keywords: molecular modeling; rarefied gas; transport coefficients; nanochannel.

1. Introduction

Most of the actually occurring flows of gases and liquids can be described within the framework of continuum mechanics. However, the application of the corresponding equations requires knowledge of the transport coefficients. The transport coefficients have the molecular nature. Therefore these coefficients can be derived from first principles only. A consistent theory of transport processes has been developed only for a rarefied gas on the basis of the Boltzmann kinetic theory. One of its main achievements is the derivation of explicit formulas for the transport coefficients [1, 2]. However, the calculating these coefficients requires solving a system of integral equations, which is not an easy task.

A constructive kinetic theory cannot be created for a dense gas and liquid in principle. Despite of this fact, a consistent theory of transport processes by methods of nonequilibrium statistical mechanics has been developed. However, the calculation of the transport coefficients comes up against a well-known problem, a regular method for solving the dynamic many-body problem does not exist. The situation is even more complicated when it is necessary to find the transport coefficients of nanofluids or other complex fluids. In these cases, the only alternative to the experimental determination of the transport coefficients is their direct numerical molecular modeling. This idea is implemented in the molecular dynamics method [3, 4]. However, its limitations are well known. This method manages to simulate systems of a small number of molecules and at relatively short times. It is practically inapplicable for modeling the properties of a rarefied gas up to now.

In recent years, a new important field of application for molecular modeling has opened up: the description of the transport processes and fluid flows in nanochannels and nanoporous media. Here, the measurement of transport coefficients is generally impossible, as a rule. Therefore, it is extremely important to create a method that would retain the main advantage of the molecular dynamics method – the transport properties calculation, based on processing the phase variables of the system under consideration (coordinates and velocities of all its particles), on the one hand, and would be much more economical and effective, on the other. This work was devoted to the development of such a method, which was called the method of stochastic molecular modeling (SMM). A rarefied gas in bulk and in nanochannels is considered here.

2. SMM method for rarefied gases in the bulk

The molecular dynamics method has extremely successfully used to solve a variety of problems in physics, mechanics, chemistry, and biology. However, it is known that it does not give true phase trajectories [4-6]. Adequate data on observables are obtained only by averaging over a large number of independent phase trajectories of the modeled system. That stimulated the idea of the SMM method constructing, where the phase trajectories of the system were simulated stochastically. The easiest way to implement this idea was for a rarefied gas. That was due to two circumstances. First, the molecules interaction in such gas does not contribute to the transport coefficients and the equation of state [7]. In particular, it means that all fluctuation-dissipation theorems (FDT) [4, 8], that determine these coefficients, contain correlation functions that depend only on the velocities of molecules. For this reason, the transport coefficients depend only on the velocities of the molecules, and it is necessary to have only information about the evolution of molecules in the velocity space for their calculation. An important circumstance is that the corresponding correlation functions are calculated by meanы of the equilibrium Maxwellian distribution function. This means that it is necessary to simulate the equilibrium state of gas molecules in order to calculate the transport coefficients. The equivalence of FDTs to the formulas of the gases kinetic theory for calculating the transport coefficients [1] was proved in [9, 10] (see also [4]).

The second circumstance is determined by the fact that only pair collisions of molecules take place in a rarefied gas. Since the momenta of molecules change only in the process of their collisions, it is not difficult to determine the probability of each molecule collision in a given time interval using the kinetic theory [1]. Then, if the collision should occur, randomly find a molecule, which this collision is realized with.

This idea was implemented in our papers [11-13]. The SMM method allows using any interparticle interaction potentials. In this paper the Lennard-Jones potential was used. The corresponding algorithm is quite simple. First, the initial state of the simulated gas is created. In order to do this, N molecules with a given density n are evenly distributed in a given volume of modeling. Molecular velocities are plotted according to the Maxwellian distribution function. Thus, at initial moment of time t, the molecules have velocities $(\boldsymbol{v}_1, \boldsymbol{v}_2, \dots \boldsymbol{v}_N)$. Then the time interval t_s is determined, where the dynamics of the system will be studied. This time interval is divided into steps of duration $\tau_i = \sigma/v_{i,max}$, where σ is the effective size of molecules, and $v_{i,max}$ is the maximum value of the molecules velocity at a given time step.

Next step is the imitation of the molecules collisions. In rarefied gas the probability of molecule i collision with any other in time τ_i is calculated on the basis of the kinetic theory, $P_{ci} = 4\tau_i n\sigma^2\sqrt{\pi kT/m}$ [1], here T is the gas temperature, k is Boltzmann constant and m is the molecule mass. For each molecule, the probability of its collision is sequentially played out. If collision should occur, the molecule j is randomly selected from the remaining $(N-1)$ molecules. Then the velocities of the molecules i and j are changed in accordance with the conservation laws. This procedure is performed for all N at each time step.

It is shown in the papers [11-13] that the described algorithm permit to simulate with high accuracy the diffusion and viscosity of rarefied gases. However, the possibility of modeling the most complex transport coefficient, the thermal conductivity coefficient, has not yet been considered. It is determined by the following FDT [4, 14]

$$\lambda = \frac{k}{3VT^2}\int_0^{\tau_p}\langle \boldsymbol{j}(0)\cdot \boldsymbol{j}(t)\rangle dt = \frac{k}{3VT^2}\int_0^{\tau_p}\chi(0,t)dt, j(t) = 0.5\sum_i m_i \boldsymbol{v}_i v_i^2(t). \quad (1)$$

Here V is volume, $\boldsymbol{v}_{ij}$ is the vector of the relative velocity of molecules i and j.

The correlation functions must exponentially decay in a rarefied gas [4]. This is also confirmed by the data of calculations using SMM, presented in Fig. 1a. Here is a calculation of the correlation function normalized to the initial values for xenon at a temperature of 300 K and atmospheric pressure, the time is normalized to the mean free path time of xenon molecules. The decay of the correlation function occurs over several free path times.

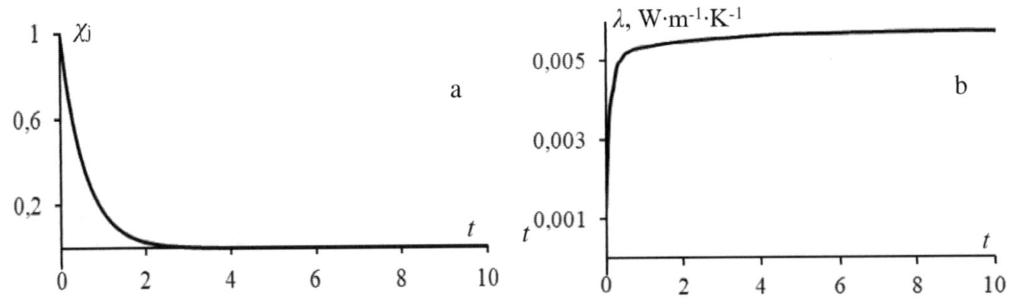

Figure 1. Evolution of normalized correlation function of xenon heat flux (a) and corresponding thermal conductivity coefficient (b).

The exponential decay of the correlation function means a fairly fast approach to the plateau value of the thermal conductivity coefficient. Fig. 1b illustrates this fact, reaching the plateau value occurs in 8–10 times of free path of molecules. This plateau value actually determines the experimentally measurable value of the thermal conductivity coefficient.

The accuracy of this method was investigated by calculating various transport coefficients for both monatomic and polyatomic gases. It was shown that in all cases it increased with an increase in both the number of particles N used and the number of phase trajectories l, which were taken for the averaged results. The relative error turned out to be of the order of $\Delta \sim 1/\sqrt{Nl}$. In particular, therefore, the number of molecules can be "exchanged" for the number of trajectories to achieve a given accuracy.

Calculation of the thermal conductivity coefficient requires more resources in comparison with other transport coefficients. However, the error in determining the thermal conductivity coefficient turns out to be less than one percent (averaging was carried out over 1000 phase trajectories) even for polyatomic gases, using 6400 molecules in the calculations. Thus, the relative calculation error for oxygen was 0.93%, and it was 0.67% for methane (the error was determined by comparison with experimental data [15]).

3. SMM algorithm for nanochannels

The SMM method, described in the previous part, simulates only the dynamics of molecules in the velocity space. In confined conditions, let us speak about channels for the sake of definiteness, it is necessary to take into account the interaction of gas molecules with the channel walls. For this reason, despite the fact that rarefied gas is simulated, the transport coefficients of which depend only on the molecules velocities, it is now necessary to simulate the evolution of the considered system not only in the velocity space, but also in the configuration space. The most important element of modeling such a system is the interaction of its molecules with the boundary. Below the interaction of molecules with the channel walls is described using specular or specular-diffuse reflection laws. This approach is usually used to solve various problems of the rarefied gas dynamics [16, 17]. In the case of specular reflection, the molecule velocity component does not change along the surface, but the normal velocity component changes a sign. In the case of diffuse reflection, the reflected molecule velocity is played out according to the Maxwell distribution function with a given temperature. This temperature must correspond to the channel wall's temperature, which may differ from the gas temperature in the general case. In specular-diffuse reflection, the fraction of molecules θ interacts with the wall diffusely, and $(1-\theta)$ is specular, where θ is so-called accommodation coefficient.

In confined conditions (particular in nanochannels) at initial moment of time, the coordinates $\boldsymbol{r}_1, \boldsymbol{r}_2, \dots, \boldsymbol{r}_N$ and velocities $\boldsymbol{v}_1, \boldsymbol{v}_2, \dots, \boldsymbol{v}_N$ of all molecules considered are determined. The simulation of each molecule dynamics is split into processes. First, the position of molecule i is determined, it shifts in the configuration space in accordance with its velocity: $\boldsymbol{r}'_i = \boldsymbol{r}_i + \boldsymbol{v}_i \tau_n$ during the time step n. If a fluid molecule reaches the wall, then its further interaction with the wall is realized, while the velocity of this molecule changes. The velocity of the reflected molecule is determined in accordance with the law of interaction with the wall adopted in this particular calculation: specular, diffuse, or specular-diffuse.

If molecule i does not reach the wall, its collision with other gas molecules is played out. This procedure is realized by the same way as in the bulk (see previous section). The described procedure is repeated until the specified computation time t_s, which is equal to: $t_s = \tau_1 + \tau_2 + \dots + \tau_k$. The result of the calculation is a complete set of coordinates and velocities of all modeled system molecules at successive times.

The SMM method [11–13] has proven itself well in modeling the transport coefficients of rarefied gases and their mixtures both, monoatomic, and also polyatomic. In all cases, the simulation accuracy was assessed by comparison with known experimental data. This accuracy was quite high even using a relatively small number of molecules, and it could be enhanced by increasing the number of molecules or independent phase trajectories used. In the case of flow in nanochannels, there are no reliable experimental data on measuring the transport coefficients. In this paper, we use the modeling of the self-diffusion coefficient D of rarefied argon at atmospheric pressure and a temperature of 273 K in order to test the SMM algorithm. Testing is possible, since the value of the self-diffusion coefficient along the channel must correspond to its value in the bulk.

The self-diffusion coefficient was calculated on the basis of FDT [4, 8, 14], which connected the self-diffusion coefficient D with the velocity autocorrelation function (VACF) χ of the simulated gas molecules by the following relation

$$D = \frac{1}{3}\int_0^{\tau} \chi(0,t)dt, \quad \chi_{vv}(t) = \frac{1}{Nl}\sum_{i=1}^{N}\sum_{\alpha=1}^{l}[\boldsymbol{v}_i(t)\cdot\boldsymbol{v}_i(t+\alpha\Delta t)], \tag{2}$$

here l is the number of phase trajectories, which were used to calculate the VACF and Δt is time integration step.

Diffusion along the channel should occur in the same way as in the bulk, since in this direction the channel is unlimited. Indeed, the VACF of gas molecules decays exponentially along the channel. The relaxation time is of the order of the molecules free path time. On the other hand, diffusion in the channel is anisotropic. The interaction of molecules with the channel walls leads to the appearance of a negative branch of the VACF. The depth of this negative tail depends on the channel height. It increases with decreasing the channel height.

Like the VACF, the self-diffusion coefficient (2) is a function of time. Under the normal conditions (in the bulk), the actual experimentally measured value of the diffusion coefficient is obtained only when function (2) reaches the plateau value. The time for reaching plateau value corresponds to the decay time of the VACF. The evolution of the argon self-diffusion coefficient along the channel, calculated using formula (2), is shown in Fig. 2a. Coefficient (2) along the channel reaches a plateau value during 15–20 free path times of the molecules. On the contrary, the evolution of the self-diffusion coefficient across the channel (see Fig. 2b) is much more complicated: first, some growth is observed, and then it decreases monotonically and tends to zero. Moreover, this happens quite quickly, the corresponding time will be determined by the channel height.

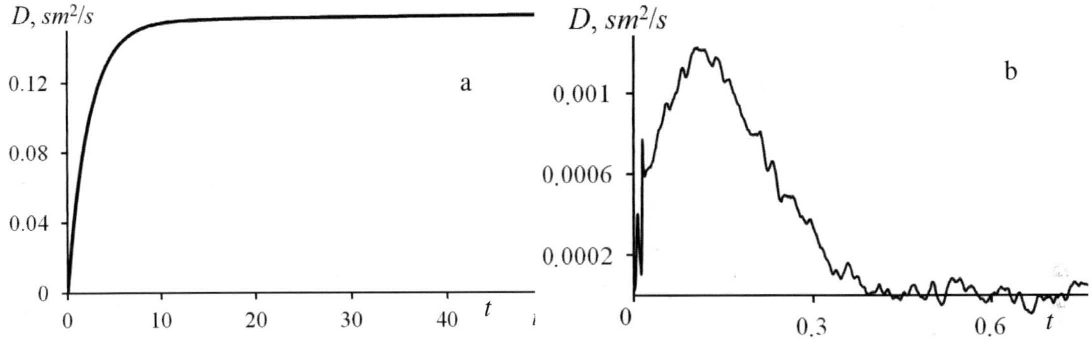

Figure 2. Evolution of the self-diffusion coefficient along a) and across b) the channel.

The isotropy absence of the molecules self-diffusion coefficient in nanochannel, described above, is quite expected. So, the SMM method gives a qualitatively physically quite reasonable result. However, its use requires a clear understanding, on the one hand, the accuracy of the calculation of the transport coefficients. On the other, it is necessary to understand how effective this algorithm is, in particular, how many molecules in a cell should be used to achieve acceptable accuracy, what should be its minimum possible size, etc. Some of the posed questions can be answered fairly accurately. It is clear that the simulation accuracy will depend on the channel length. As an example of studying this dependence in Table 1, the data of calculating argon self-diffusion coefficient in a channel with a square cross section of 10 nm^2 under normal conditions were compared. In this Table, the second row shows the calculated value of the self-diffusion coefficient, and the last one shows the relative error in comparison with the experimental value [15]. The channel length L varies from 311 to 9952 nm (i.e., approximately from 0.3 to 100 μm). In the latter case, one could already expect that the self-diffusion coefficient would be close to the corresponding value in the bulk. It really turned out to be like that. The experimental value of the self-diffusion coefficient in the bulk is 0.156 sm^2/s [15], and the simulation accuracy at the maximum channel length turned out to be about one percent.

Table 1. Accuracy of the self-diffusion coefficient calculation for different channel lengths L.

***L*, nm**	311	622	933	9952
***D*, sm²/s**	0.192	0.175	0.168	0.158
Δ, %	22.95	12.36	7.46	1.23

Table 2. Dependence of simulation accuracy on l

l	100	200	500	1000	10000
***D*, sm²/s**	0.1674	0.1633	0.1627	0.1623	0.1601
Δ, %	7.28	4.69	4.3	4.04	2.57

It is clear that, in the general case, the accuracy should essentially depend on the number of independent phase trajectories l used for averaging in stochastic modeling of phase trajectories. To study the effect of the phase trajectories number used let us consider again the self-diffusion in a square channel with a cross section of 777.5 nm^2 and a length of 15 nm only. The number of phase trajectories varied from 100 to 10000. The corresponding data are given in Table 2. The result is quite obvious. The simulation data significantly improve with an increase in the number of ensemble members of the phase trajectories used for averaging.

4. Anisotropy of the gas viscosity in nanochannels

The anisotropy of molecules diffusion is a natural consequence of the anisotropy of nanochannel geometry (its length is much greater than the height and width). In an unrestricted environment, the root-mean-square movement of molecules grows in proportion to time. On the other hand, this distance cannot be greater than the characteristic size of the system in a limited environment. On the contrary, viscosity is usually understood as some universal characteristic of a given fluid. However, viscosity is a redistribution of momentum in a system from a molecular kinetic point of view. In confined conditions, for example, in a nanochannel, the most important factor in the redistribution of the momentum of gas molecules is their interaction with the boundaries of the system. Anisotropy also naturally arises here: the redistribution of the momentum along and across the channel should be different. This means that the viscosity must also be anisotropic.

In order, to verify this, let us consider modeling the viscosity of argon in nanochannels with a square cross section. The modeling cell was selected in the form of a rectangular parallelepiped as it had been before. The viscosity coefficient is calculated using the Green–Kubo formula

$$\eta = \frac{1}{3VkT}\int_0^\tau \langle \boldsymbol{\sigma}(0) : \boldsymbol{\sigma}(t)\rangle dt, \quad \boldsymbol{\sigma}(t) = \frac{1}{Nl}\sum_{j=1}^{l}\sum_{i=1}^{N} m[\mathbf{v}_i(j\Delta t)\,\mathbf{v}_i(j\Delta t)]. \tag{3}$$

The viscosity coefficient η of rarefied argon was simulated at atmospheric pressure and a temperature of 273 K. The channel had a square cross section, its height varied from 6.22 to 500 nm. The viscosity coefficient was averaged over a thousand independent phase trajectories, each of which was calculated for about 1 nanosecond.

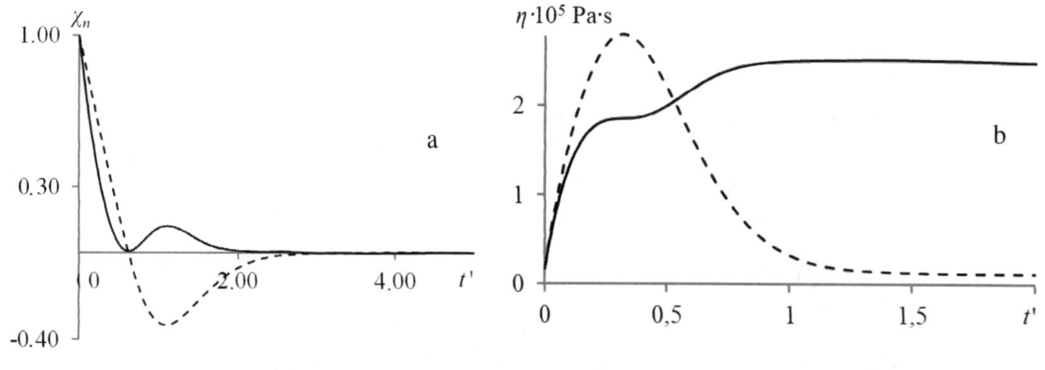

Figure 3. Evolution of the normalized correlation function a) and the viscosity coefficient b) of argon along (solid line) and across (dashed line) of nanochannel.

The viscosity coefficient (3) is determined by the integral of the corresponding correlation function. The evolution of this correlation function χ_n normalized to the initial values along and across the channel is shown in Fig. 3a. The correlation function across the channel has a characteristic negative tail caused by the molecules interaction with the wall. In this example (in Fig.

3), the interaction with the wall was described by a specular law, and the channel height was 15.55 nm. The presence of a negative tail led to the fact that the contribution to the viscosity coefficient of molecules across the channel was almost twenty times smaller than along. The corresponding values of the viscosity coefficients are shown in Fig. 3b. Thus, the total gas viscosity coefficient in the channel was almost three times lower than in the bulk.

The law of gas molecules interaction with the wall will essentially change the momentum transfer in the system. Changing the accommodation coefficient of the channel walls (their material) one may transform the viscosity of the gas in channel. The value of the viscosity coefficient across the channel for purely diffuse reflection of molecules from the wall is almost 50% higher than along, and the total coefficient of viscosity is almost three times higher than in the bulk. The gas viscosity coefficient changes more than six times, when the accommodation coefficient changes from zero to unity. The excess of the total viscosity naturally depends on the high of the channel.

5. Conclusion

This work is devoted to the development of the SMM method for modeling transport processes in rarefied gases. This method is quite simple and reproducible by almost any user. Its performance was previously demonstrated by the example of calculating the coefficients of self-diffusion, diffusion and viscosity of both, monoatomic and polyatomic gases [11-13]. This method is shown to be successfully applied to simulate the thermal conductivity coefficient in this work.

The study of the gas viscosity in nanochannel shows that the processes of momentum transfer, and hence the viscosity in such channels, are anisotropic. The average gas viscosity in a channel can be controlled by varying the wall material. Real channels usually have roughness. Their presence will significantly affect the gas flow character. In particular, the presence of roughness leads to an increase in the channel surface area. This means that anisotropy of momentum transfer in gas in such channels will be more pronounced than in a channel with smooth walls. Therefore, the presence of roughness can lead to both, an increase in viscosity or a decrease in it. The similar effect may have place when the nano-structuring of the channel walls.

It is shown that the anisotropy of momentum transfer increases with decreasing depth of the Lennard-Jones potential well and with an increase in the characteristic size of molecules. The latter one is generally a rather important factor in confined conditions, since the numerical density of the gas will significantly change at given pressure and temperature. Indeed, the characteristic volume of molecules, such as neon and xenon, differs by more than three times.

Anisotropy is characteristic of all transport processes in nanochannels. The presence of such anisotropy will significantly change the thermal conductivity of the gas. It is important to note that, with a specular law of reflection from the walls, the thermal conductivity of the gas in the channel will be lower than the corresponding value in the volume.

The main conclusion that follows from the obtained modeling data is that both, the viscosity and the thermal conductivity of gases in nanochannels are a property of the entire system "gas + channel walls", and not just only gas. This was consistent with the conclusions, obtained by the methods of nonequilibrium statistical mechanics, where generalized fluctuation-dissipation theorems were derived that described transport processes in confined conditions [18,19].

Finally, it is important to have an answer to the question when the processes of momentum transfer become practically isotropic. Our calculations show that viscosity anisotropy can be observed not only in nano, but also in microchannels with character size of the order of certain micrometers.

6. Acknowledgments

This paper was carried out with financial support from the Russian Foundation for Basic Research (grants No. 19-01-00399 and No. 20-01-00041) and Russian megagrant No. 2020-220-08-1436.

7. References

[1] Chapman S and Cowling T G 1990 *The Mathematical Theory of Non-Uniform Gases* (Cambridge: Cambridge University Press)
[2] Burnett D 1935 Proc. London Math. Soc. **39**(6) 385
[3] Rapaport D C 1995 *The Art of Molecular Dynamics Simulation* (Cambridge: Cambridge University Press)
[4] Rudyak V Ya 2005 *Statistical Aerohydromechanics of Homogeneous and Heterogeneous Media Pt 2 Hydromechanics* (Novosibirsk: NSUACE)
[5] Norman G E and Stegailov V V 2002 Comp. Physics Comm. **147**(4) 678
[6] Norman G E and Stegailov V V 2013 Math. Models and Computer Simulations **5**(4) 305
[7] Klimontovich Yu L 1982 *Kinetic Theory of Non-Ideal Gases and Non-Ideal Plasmas* (Oxford: Pergamon Press)
[8] Rudyak V Ya 2020 J. of Physics: Conference Series **1560** 012002
[9] Ernst M H 1966 Physica **32** 209
[10] Khon'kin D 1970 Theoret. and Math. Phys. **5** 1029
[11] Rudyak V Ya and Lezhnev E V 2016 J. of Physics: Conference Series **738** 012086
[12] Rudyak V Ya and Lezhnev E V 2018 J. of Computational Physics **355** 95
[13] Rudyak V Ya and Lezhnev E V 2020 Nanosystems: Phys. Chem. Math. **11**(3) 285
[14] Zubarev D 1974 *Nonequilibrium Statistical Thermodynamics* (New York: Consultants Bureau)
[15] Grigor'ev S and Meilikhov E Z (Eds) 1996 *Physical Quantities: Handbook* (Moscow: Energoatomizdat)
[16] Goodman F O and Wachman H Y 1976 *Dynamics of Gas-Surface Scattering* (New York San Francisco London: Academic Press)
[17] Cercignani C 1975 *Theory and Application of the Boltzmann Equation* (Edinburgh London: Scottlish Academic Press)
[18] Rudyak V Ya and Belkin A A 2014 Doklady Physics **59**(12) 604
[19] Rudyak V Ya and Belkin A A 2015 Nanosystems: Phys. Chem. Math. **6** 366

Analytical solution to the problem of diffusion of the light component of the binary gas mixture in a plane channel

V N Popov [1,*] **and I V Popov** [2]

[1] Northern (Arctic) Federal University named after M.V. Lomonosov, 17, Severnaya Dvina Emb., Arkhangelsk, 163002, Russia
[2] Peter the Great St. Petersburg Polytechnic University, 29, Polytechnicheskaya St., St. Petersburg, 195251, Russia

E-mail: *v.popov@narfu.ru

Abstract. Within the framework of the kinetic approach, an analytical solution to the problem of diffusion of the light component of a binary mixture in a flat channel with infinite parallel walls is constructed. It is assumed that the mass of light component molecules and their concentration is much less than the mass of molecules and the concentration of heavy components. The flow rate of the heavy component is assumed to be zero. The change in the state of a light gas component is described on the basis of the BGK (Bhatnagar, Gross, Kruk) model of the Boltzmann kinetic equation. The diffuse reflection model is used as a boundary condition on the channel walls. The mass velocity profile of the light gas component is constructed. The flow rate of the light gas component per unit channel width is calculated. A comparison with similar results presented in open sources was done.

1. Introduction

In boundary layer theory [1] often there is a case of binary gas, the mass of a molecule of one of the components which is much less than the mass of the molecule of the other component. If the partial density of the light component is significantly less than the heavy one, the motion of the heavy component can be neglected, considering it as stationary. For the half-space case (i.e. the case when a binary mixture fills a half-space bounded by a solid flat surface lying in the plane of a rectangular Cartesian coordinate system), an analytical solution to this problem is obtained in [2]. No numerical analysis of the final expressions describing the diffusion process was performed in [2]. Using the method of discrete ordinates solution of the problem of flow of a binary mixture in the channel, the thickness of which is comparable to the average free path length of gas molecules, constructed in [3]. In [4] the flow of binary mixtures of neon-argon, helium-xenon in the presence of pressure, temperature and concentration gradients was considered. In the first case, the mass of the molecules of both components are approximately equal, in the second – the mass of the molecule of one component is much greater than the mass of the molecule of the other.In the article [3] the problem of the Couette for the mixture of helium-argon in the channel, the walls of which are made of different materials. A study of the gas mixture flow in ultrathin capillaries, the size of which is comparable to the radius of action of surface forces, was carried out in [5]. The aim of the present work was to construct an analytical solution to the problem of diffusion of a light component of a binary mixture relatively heavier in a channel with infinite parallel walls.

Currently, there are two main methods to build exact analytical solutions to boundary value problems of the kinetic theory of gas and plasma using linearized model kinetic equations. This is the Case method (the method of decomposition of the solution by eigenvular generalized functions of the characteristic equation corresponding to the given kinetic equation) and the Wiener-Hopf method (the method of reducing the kinetic equation to unilateral equation of the convolution type with an integral taken along the real positive half-axis). The main advantage of the Case method is that it allows to obtain in an explicit form the distribution function of gas molecules by coordinates and velocities, and, as a consequence, to calculate any macroparameters characterizing the state of the gas. In this paper, a generalization of the Case method to the case of problems with bounded geometry presented in [6] is used as the main method. The general solution of the inhomogeneous integro differential equation, to which the model kinetic equation is reduced, is constructed in the space of generalized functions. The substitution of boundary conditions in the General solution leads to a system of two coupled singular integral equations with a Cauchy-type kernel, which after the transformation are reduced to the Riemann boundary value problem on a real positive half-axis. The use of Sohotsky's formulas for finding coefficients in the decomposition of the solution of the problem by eigenvectors of the continuous spectrum leads to the Fredholm integral equation of the second kind. The solution of the latter is sought in the form of Neumann series.

2. Materials and methods

Let us consider the diffusion of the light component of a binary gas mixture with respect to heavy in the channel with its walls located in planes $x' = \pm d$ in the rectangular Cartesian coordinate system. Let us suppose that the mixture consists of gas a and b (respectively light and heavy component) with numerical densities (concentrations) n_a and n_b and mass densities $\rho_a = m_a n_a$ and $\rho_b = m_b n_b$. Here m_a and m_b are masses of gas molecules a and b. We assume that a temperature is constant equal to T. We assume that the Oy' axis is directed along the concentration gradient of light component. If the partial density of the light component is substantially less than heavy, then we can neglected by movement of the heavy component and considering it as motionless. Under these assumptions the BGK-equation for the light component of gas mixture is written as:

$$v_x \frac{\partial f}{\partial x'} + v_y \frac{\partial f}{\partial y'} = \nu_a (f_{eq} - f) + \nu_{ab}(f_0 - f), \tag{1}$$

Here $f_{eq} = n(y')(\beta/\pi)^{3/2} \exp[-\beta(\mathbf{v} - \mathbf{u}(x'))^2]$ – locally equilibrium distribution function, $\mathbf{u}(\mathbf{x}')$ – the mass average velocity of the lighter component, $f_0 = n(y')(\beta/\pi)^{3/2} \exp(-\beta v^2)$, ν_a – collisions frequency of light molecules among themselves, ν_{ab} – collisions frequency of light molecules with heavy

$$\nu_{ab} = \nu_a \left(\frac{d_{ab}}{d_a}\right)^2 \sqrt{\frac{m_a + m_b}{2m_b} \frac{n_b}{n_a}}, \qquad \nu_a = \frac{p_a}{\mu_a} \Omega(\omega_a, \alpha_a),$$

$$d_{ab} = \frac{d_a + d_b}{2} \qquad \Omega(\omega_a, \alpha_a) = \frac{5(\alpha_a + 1)(\alpha_a + 2)}{\alpha_a (7 - \omega_a)(5 - \omega_a)}.$$

Here according to [7] d_a d_b – diameters of molecules of gas components a and b, μ_a and ω_a – coefficient of dynamic viscosity and indicator of temperature dependence of the viscosity of the gas a. We assume later that $\alpha_a = 1$. This assumption corresponds to the solid sphere model

(VHS). In this case

$$\Omega(\omega_a, 1) = \Omega(\omega_a) = \frac{30}{(7-\omega_a)(5-\omega_a)}.$$

If $\beta u^2(x') << 1$, we can linearize the distribution function and write it as:

$$f = f_0(y', v)(1 + h(x', \mathbf{v})). \tag{2}$$

Then

$$\frac{\partial f}{\partial x'} = f_0(y, v)\frac{\partial h}{\partial x'},$$

$$\frac{\partial f}{\partial y'} = \frac{dn}{dy'}(\beta/\pi)^{3/2}\exp(-\beta v^2) = \frac{1}{n}\frac{dn}{dy'}f_0(y, v) = f_0(y', v)\frac{d\ln n}{dy'},$$

$$f_{eq} = n(y')(\beta/\pi)^{3/2}\exp[-\beta \mathbf{v}](1 + 2\beta v_y u_y(x')) = f_0(y', v)(1 + 2\beta v_y u_y(x')).$$

Substituting the obtained results in (1), we arrive at the equation

$$v_x f_0(y', v)\frac{\partial h}{\partial x'} + v_y f_0(y', v)\frac{d\ln n}{dy'} =$$

$$= \nu_a f_0(y', v)(1 + 2\beta v_y u_y(x') - 1 - h(x', \mathbf{v})) + \nu_{ab} f_0(y', v)(1 - 1 - h(x', \mathbf{v})),$$

or

$$v_x\frac{\partial h}{\partial x'} + v_y\frac{d\ln n}{dy'} = 2\nu_a\beta v_y u_y(x') - \nu h(x', \mathbf{v}), \tag{3}$$

$\nu = \nu_a + \nu_{ab}$.

Let us introduce the notation: $C_x = \beta^{-1/2}v_x$, $C_y = \beta^{-1/2}v_y$, $U(x) = \beta^{-1/2}u_y(x)$, $x = x'\beta^{1/2}\nu$, $y = y'\beta^{1/2}\nu$, $G_n = d\ln n/dy$, $c = \nu_a/\nu$. In the accepted notations, we rewrite the equation (3) as

$$C_x\frac{\partial h}{\partial x} + C_y G_n = 2cC_y U(x) - h(x, \mathbf{v}). \tag{4}$$

The solution of (4) is sought in the form

$$h(x, \mathbf{C}) = \frac{C_y G_n}{1-c}\left[Z(x, C_x) - 1\right]. \tag{5}$$

Substituting (5) in (4) and considering that from the statistical the meaning of the distribution function

$$U(x) = \pi^{-3/2}\int \exp(-C^2)C_y h(x, \mathbf{C})\, d^3C = \frac{G_n}{2(1-c)}\left[\frac{1}{\sqrt{\pi}}\int\limits_{-\infty}^{+\infty}\exp(-\tau^2)Z(x, \tau)\, d\tau - 1\right], \tag{6}$$

we arrive to the equation ($\mu = C_x$)

$$\frac{\mu}{1-c}\frac{\partial Z}{\partial x} + 1 = \frac{c}{1-c}\left[\frac{1}{\sqrt{\pi}}\int\limits_{-\infty}^{+\infty}\exp(-\tau^2)Z(x, \tau)\, d\tau - 1\right] - \frac{1}{1-c}\left[Z(x, \mu) - 1\right],$$

or, after reduction of similar terms to the homogeneous equation $c \in (0, 1)$

$$\mu\frac{\partial Z}{\partial x} + Z(x, \mu) = \frac{c}{\sqrt{\pi}}\int\limits_{-\infty}^{+\infty}\exp(-\tau^2)Z(x, \tau)\, d\tau. \tag{7}$$

As a boundary condition on the walls of the channel, we use diffuse reflection model. In this case, the boundary conditions for function $Z(x,\mu)$ on the channel walls taking into account (2) and (5), we written in the form

$$Z(\pm d, \mp\mu) = 1, \qquad \mu > 0. \tag{8}$$

Thus, the solution of the considered problem is reduced to the solution equations (7) with boundary conditions (8).

We seek a solution of (7) as a decomposition of the spectral parameter η

$$Z_\eta(x,\mu) = \exp(-\frac{x}{\eta})\, F(\eta,\mu). \tag{9}$$

Substituting (9) into (7) and introducing the normalization integral

$$\int\limits_{-\infty}^{+\infty} \exp(-\tau^2)\, F(\eta,\tau)\, d\tau = 1, \tag{10}$$

we arrive at the characteristic equation

$$(\eta - \mu)\, F(\eta,\tau) = \frac{1}{\sqrt{\pi}} c\eta. \tag{11}$$

Solution of (11) in the space of generalized functions has the form

$$F(\eta,\mu) = \frac{1}{\sqrt{\pi}} c\eta\, \mathrm{P}\frac{1}{\eta-\mu} + g(\eta)\delta(\eta-\mu). \tag{12}$$

Here $\mathrm{P}(1/z)$ and $\delta(z)$ are the distribution in the sense of a principal value of the integral of $1/z$ and the Dirac delta function, respectively.

Substituting (12) for the normalization integral (10), we find an explicit form of the function $g(\eta)$.

$$\frac{1}{\sqrt{\pi}} c\eta \int\limits_{-\infty}^{+\infty} \frac{\exp(-\tau^2)}{\tau-\mu}\, d\tau + g(\eta)\exp(-\eta^2) = 1.$$

Thus,

$$g(\eta) = \exp(\eta^2)\lambda(\eta),$$

where

$$\lambda(z) = 1 + \frac{1}{\sqrt{\pi}} cz \int\limits_{-\infty}^{\infty} \frac{\exp(-\mu^2)}{\mu - z}\, d\mu \tag{13}$$

is the dispersion function.

On the real positive half-axis

$$\lambda(\mu) = 1 - 2c\mu\exp(-\mu^2) \int\limits_{0}^{\mu} \exp(u^2)\, du. \tag{14}$$

In the complex plane the dispersion function is calculated by the formula

$$\lambda(z) = 1 - 2cz\exp(-z^2) \int\limits_{0}^{z} \exp(u^2)\, du \pm ic\sqrt{\pi}z\exp(-z^2), \quad \pm Im(z) > 0.$$

Substituting (13) into (12), we find the distinct form of the function $F(\eta,\mu)$

$$F(\eta,\mu)=\frac{1}{\sqrt{\pi}}\,c\eta\,\mathrm{P}\frac{1}{\eta-\mu}+\exp(\eta^2)\lambda(\eta)\,\delta(\eta-\mu). \tag{15}$$

Let us estimate the behavior of $\lambda(z)$ in the vicinity of the infinitely remote points. Decomposing (13) into asymptotic series, we obtain

$$\lambda(z)=1-\frac{1}{\sqrt{\pi}}c\int\limits_{-\infty}^{\infty}\frac{\exp(-\mu^2)}{1-\mu/z}\,d\mu=1-c\left[1+\frac{1}{2z^2}+\frac{3}{4z^4}+\ldots\right],\quad |z|\to\infty. \tag{16}$$

From the decomposition (16) it is seen that $\lambda(\infty)=1-c>0$. Imaginary part of the function $\lambda^+(\mu)$ for all $\mu>0$ is positive, and for all $\mu<0$ – is negative. The real part of this function generally has four roots $\pm\tau_0(c)$ and $\pm\tau_1(c)$. Denote by $\theta(\mu)$ the main value of the function argument $\lambda^+(\mu)$, fixed at zero by the condition $\theta(0)=0$. The value of $\theta(\mu)$ is defined by the equality

$$\theta(\mu)=\frac{\pi}{2}-\mathrm{arctg}\frac{\lambda(\mu)}{\sqrt{\pi}c\mu\exp(-\mu^2)}. \tag{17}$$

The increment of the main value $\theta(\mu)=\arg\lambda^+(\mu)$ is zero, both on the positive half-axis and the entire axis. Hence, as follows from the argument principle, the function $\lambda(z)$ has no zeros in the complex plane with a cut along the real numerical line. Thus, the discrete spectrum in the considered problem is an empty set.

Thus, we seek a solution of (7) in the form

$$Z(x,\mu)=\int\limits_{-\infty}^{+\infty}\exp(-\frac{x}{\eta})\,F(\eta,\mu)a(\eta)\,d\eta. \tag{18}$$

Here $a(\eta)$ is an unknown function to be further defined. Substituting the decomposition (18) into boundary conditions (8) and taking into account (15) and (13), we arrive at a system of coupled integral equations

$$\frac{1}{\sqrt{\pi}}c\int\limits_{-\infty}^{+\infty}\frac{\eta\,b(\eta,d)}{\eta-\mu}\,d\eta+\exp(\mu^2)b(\mu,d)\lambda(\mu)=1,\qquad \mu<0, \tag{19}$$

$$\frac{1}{\sqrt{\pi}}c\int\limits_{-\infty}^{+\infty}\frac{\eta\,b(\eta,-d)}{\eta-\mu}\,d\eta+\exp(\mu^2)b(\mu,-d)\lambda(\mu)=1,\qquad \mu>0. \tag{20}$$

Here we have introduced the notation

$$b(\eta,x)=\exp\left(-\frac{x}{\eta}\right)\,a(\eta). \tag{21}$$

Replace in (19) μ with $-\mu$ and present the obtained integral as the sum of two

$$\frac{1}{\sqrt{\pi}}c\int\limits_{-\infty}^{0}\frac{\eta\,b(\eta,d)}{\eta+\mu}\,d\eta+\frac{1}{\sqrt{\pi}}c\int\limits_{0}^{+\infty}\frac{\eta\,b(\eta,d)}{\eta+\mu}\,d\eta+\exp(\mu^2)b(-\mu,d)\lambda(\mu)=1,\qquad \mu>0. \tag{22}$$

By writing (22) we took into account that the function $\lambda(\mu)$ is even on a real numeric line. The first integral in the obtained equation is singular, and the second is regular. Replace the integration variable η with $-\eta$ in the first integral and rewrite it as

$$\int_{-\infty}^{0} \frac{\eta\, b(\eta,d)}{\eta+\mu}\, d\eta = \int_{+\infty}^{0} \frac{-\eta\, b(-\eta,d)}{-\eta+\mu}\, d(-\eta) = \int_{0}^{+\infty} \frac{\eta\, b(-\eta,d)}{\eta-\mu}\, d\eta.$$

Taking into account the performed transformations, the equation (19) takes the form

$$\frac{1}{\sqrt{\pi}} c \int_{0}^{+\infty} \frac{\eta\, b(-\eta,d)}{\eta-\mu}\, d\eta + \frac{1}{\sqrt{\pi}} c \int_{0}^{+\infty} \frac{\eta\, b(\eta,d)}{\eta+\mu}\, d\eta + \exp(\mu^2) b(-\mu,d)\lambda(\mu) = 1, \qquad \mu > 0. \qquad (23)$$

Similarly, we transform the integral in (19). In the end, the equation (19) can be written in the form

$$\frac{1}{\sqrt{\pi}} c \int_{0}^{+\infty} \frac{\eta\, b(\eta,-d)}{\eta-\mu}\, d\eta + \frac{1}{\sqrt{\pi}} c \int_{0}^{+\infty} \frac{\eta\, b(-\eta,-d)}{\eta+\mu}\, d\eta + \exp(\mu^2) b(\mu,-d)\lambda(\mu) = 1, \qquad \mu > 0. \qquad (24)$$

Let us transform the resulting equations. Folding (23) and (24), taking into account (21) we come to the equation

$$\frac{1}{\sqrt{\pi}} c \int_{0}^{+\infty} \frac{\eta\,[a(\eta)+a(-\eta)]\, \exp(d/\eta)\, d\eta}{\eta-\mu} + \frac{1}{\sqrt{\pi}} c \int_{0}^{+\infty} \frac{\eta\,[a(\eta)+a(-\eta)]\, \exp(-d/\eta)\, d\eta}{\eta+\mu} +$$

$$+ \exp(\mu^2)[a(\eta)+a(-\eta)]\, \exp(d/\eta)\lambda(\mu) = 2, \qquad \mu > 0. \qquad (25)$$

Similarly, subtracting (23) from (24), we obtain

$$\frac{1}{\sqrt{\pi}} c \int_{0}^{+\infty} \frac{\eta\,[a(\eta)-a(-\eta)]\, \exp(d/\eta)\, d\eta}{\eta-\mu} + \frac{1}{\sqrt{\pi}} c \int_{0}^{+\infty} \frac{\eta\,[a(\eta)-a(-\eta)]\, \exp(-d/\eta)\, d\eta}{\eta+\mu} +$$

$$+ \exp(\mu^2)[a(\eta)-a(-\eta)]\, \exp(d/\eta)\lambda(\mu) = 0, \qquad \mu > 0. \qquad (26)$$

It is easy to see that (26) for $a(\eta) = a(-\eta)$ refers to the identity. In this case (25) can be rewritten as

$$\frac{1}{\sqrt{\pi}} c \int_{0}^{+\infty} \frac{\eta\, b(\eta,-d)}{\eta-\mu}\, d\eta + \exp(\mu^2) b(\mu,-d)\lambda(\mu) = f(\mu), \qquad \mu > 0, \qquad (27)$$

where

$$f(\mu) = 1 - \frac{1}{\sqrt{\pi}} c \int_{0}^{+\infty} \frac{\eta\, b(\eta,d)}{\eta+\mu}\, d\eta. \qquad (28)$$

We seek a solution of (27) using the methods of boundary value problems of the theory of functions of complex variable. For this purpose, we introduce an auxiliary function given by the Cauchy type integral

$$N(z) = \frac{1}{\sqrt{\pi}} c \int_{0}^{+\infty} \frac{\eta\, b(\eta,-d)}{\eta-z}\, d\eta. \qquad (29)$$

On the real positive numerical axis, the function $N(z)$ has a cut. On the upper and lower banks of the cut are fair equalities

$$N^+(\mu) - N^-(\mu) = 2\sqrt{\pi} i c \mu b(\mu, -d), \qquad 0 < \mu < +\infty, \tag{30}$$

$$N^+(\mu) + N^-(\mu) = \frac{2}{\sqrt{\pi}} c \int\limits_0^{+\infty} \frac{\eta\, b(\eta, -d)}{\eta - \mu}\, d\eta, \tag{31}$$

Similar relations for $\lambda(z)$, defined by equality (13), have the form

$$\lambda^+(\mu) - \lambda^-(\mu) = 2\sqrt{\pi} i c \mu \exp(-\mu^2), \qquad -\infty < \mu < +\infty, \tag{32}$$

$$\lambda^+(\mu) + \lambda^-(\mu) = 2\lambda(\mu). \tag{33}$$

Here, the cut coincides with the entire real number line. Taking into account (30)-(33), we reduce the integral equation (27) to the Riemann boundary value problem on the real positive half-axis

$$N^+(\mu)\,\lambda^+(\mu) - N^-(\mu)\,\lambda^-(\mu) = 2\sqrt{\pi} i c \mu f(\mu) \exp(-\mu^2), \qquad \mu > 0, \tag{34}$$

The functions $N(z)$ and $\lambda(z)$ have different cuts. To eliminate this feature it is necessary to solve the problem of factorization, that is, to construct a function(let's call it $X(z)$), which has the following properties:

– the function does not vanish at any end point of the complex plane;

– on the real positive axis for it is true equality

$$\frac{X^+(\mu)}{X^-(\mu)} = \frac{\lambda^+(\mu)}{\lambda^-(\mu)}, \qquad \mu > 0, \tag{35}$$

– function $X(z)$ is analytic at all other points of the complex plane.

Taking into account the symmetry of the graph of the function $\lambda(z)$ with respect to the real axis, we can write

$$\frac{X^+(\mu)}{X^-(\mu)} = \frac{|\lambda^+(\mu)| \exp(i\theta(\mu))}{|\lambda^+(\mu)| \exp(-i\theta(\mu))} = \exp(2i\theta(\mu)). \tag{36}$$

Let us logarithm (36)

$$\ln X^+(\mu) - \ln X^-(\mu) = 2i(\theta(\mu) + \pi k), \qquad \mu > 0, \quad 0, \pm 1, \pm 2, \ldots \tag{37}$$

Thus, we obtain a problem of defining a piecewise analytic function from a given jump. We write its solution using the Sokhotski formula.

$$\ln X(z) = \frac{1}{\pi} \int\limits_0^{+\infty} \frac{[\theta(\tau) + \pi k]\, d\tau}{\tau - z}. \tag{38}$$

The changing of the angle $\theta(\mu)$ on the real half-axis is zero. Hence, the integral in the right part (38) converges only at $k = 0$. Thus, the factorizing function is constructed and has the form

$$X(z) = \exp\left[V(z)\right], \qquad V(z) = \frac{1}{\pi} \int\limits_0^{+\infty} \frac{\theta(\tau)\, d\tau}{\tau - z}, \tag{39}$$

Note that

$$V(z) = -\frac{\theta(0)}{\pi} \ln z + O_1(z) = O_1(z), \qquad z \to 0.$$

Thus, the function $X(z)$ defined by the equality (39) does not vanish in any end point of the complex plane, including the origin. In order to estimate the behavior of $X(z)$ in the vicinity of an infinitely remote point, we use the decomposition of the function $V(z)$

$$V(z) = -\frac{1}{z\pi}\int\limits_0^{+\infty}\frac{\theta(\tau)\,d\tau}{1-\tau/z} = -\frac{1}{z\pi}\int\limits_0^{+\infty}\theta(\tau)\left[1+\frac{\tau}{z}+\frac{\tau^2}{z^2}+\ldots\right]d\tau = O\left(\frac{1}{z}\right), \qquad |z|\to\infty.$$

Thus,

$$X(\infty) = 1. \tag{40}$$

Taking into account the solved factorization problem, the boundary value problem (34) is written as

$$N^+(\mu)\,X^+(\mu) - N^-(\mu)\,X^-(\mu) = \frac{X^-(\mu)}{\lambda^-(\mu)}\,2\sqrt{\pi}ic\mu f(\mu)\,\exp(-\mu^2), \qquad \mu > 0. \tag{41}$$

The jump lines of $N(z)$ and $X(z)$ coincide with the contour of the boundary condition. Therefore, we have reduced singular integral equation (27) to the Riemann boundary value problem. We write its solution using the Sokhotski formula

$$N(z) = \frac{1}{X(z)}\,c\frac{1}{\sqrt{\pi}}\int\limits_0^{+\infty}\frac{X^-(\eta)}{\lambda^-(\eta)}\eta f(\eta)\,\exp(-\eta^2)\,\frac{d\eta}{\eta - z},$$

or, taking into account (28)

$$N(z) = \frac{1}{X(z)}\,c\frac{1}{\sqrt{\pi}}\int\limits_0^{+\infty}\frac{X^-(\eta)}{\lambda^-(\eta)}\eta\left[1-\frac{1}{\sqrt{\pi}}\,c\int\limits_0^{+\infty}\frac{\mu\, b(\mu,d)}{\mu+\eta}\,d\mu\right]\exp(-\eta^2)\,\frac{d\eta}{\eta - z}. \tag{42}$$

Convert the resulting expression. To do this, we use the integral representation of the function $X(z)$

$$X(z) - X(\infty) = \frac{1}{2\pi i}\int\limits_0^{+\infty}\frac{X^+(u)-X^-(u)}{u-z}\,du.$$

Take into account that

$$X^+(u) - X^-(u) = X^-(u)\left[\frac{X^+(u)}{X^-(u)}-1\right] = X^-(u)\left[\frac{\lambda^+(u)}{\lambda^-(u)}-1\right] = \frac{X^-(u)}{\lambda^-(u)}\,2\sqrt{\pi}icu\exp(-u^2).$$

Thus, the integral representation of $X(z)$ has the form

$$X(z) = 1 + \frac{1}{\sqrt{\pi}}\,c\int\limits_0^{+\infty}\frac{X^-(u)}{\lambda^-(u)}\,\frac{u\exp(-u^2)}{u-z}\,du. \tag{43}$$

Taking into account the constructed integral representation of the function $X(z)$, we can write

$$\frac{1}{X(z)}\,c\frac{1}{\sqrt{\pi}}\int\limits_0^{+\infty}\frac{X^-(\eta)}{\lambda^-(\eta)}\,\frac{\exp(-\eta^2)\,\eta}{\eta - z} = \frac{1}{X(z)}\,[X(z)-1] = 1 - \frac{1}{X(z)}. \tag{44}$$

To convert the second term to (42), we will take advantage of the obvious equality

$$\frac{1}{\eta - z}\,\frac{1}{\mu+\eta} = \frac{1}{\mu+z}\left[\frac{1}{\eta-z} - \frac{1}{\eta+\mu}\right].$$

Then

$$\frac{1}{\sqrt{\pi}}c\int\limits_0^{+\infty}\frac{X^-(\eta)}{\lambda^-(\eta)}\eta\exp(-\eta^2)\,\frac{d\eta}{\eta-z}\,\frac{1}{\sqrt{\pi}}c\int\limits_0^{+\infty}\frac{\mu\, b(\mu,d)\,d\mu}{\mu+\eta} =$$

$$= X(z)\,\frac{1}{\sqrt{\pi}}c\int\limits_0^{+\infty}\frac{\mu\, b(\mu,d)\,d\mu}{\mu+z} - \frac{1}{\sqrt{\pi}}c\int\limits_0^{+\infty}\frac{\mu\, X(-\mu) b(\mu,d)\,d\mu}{\mu+z}.$$

Thus, taking into account the obtained results, we rewrite the expression for $N(z)$ as

$$N(z) = 1 - \frac{1}{\sqrt{\pi}}c\int\limits_0^{+\infty}\frac{\mu\, b(\mu,d)\,d\mu}{\mu+z} - \frac{1}{X(z)}\left[1 - \frac{1}{\sqrt{\pi}}c\int\limits_0^{+\infty}\frac{\mu\, X(-\mu)b(\mu,d)\,d\mu}{\mu+z}\right]. \tag{45}$$

For the constructed function $N(z)$ according to Sokhotsky's formulas we can write

$$N^+(\eta) - N^-(\eta) = -\left[\frac{1}{X^+(\eta)} - \frac{1}{X^-(\eta)}\right]\left[1 - \frac{1}{\sqrt{\pi}}c\int\limits_0^{+\infty}\frac{\mu\, X(-\mu)b(\mu,d)\,d\mu}{\mu+\eta}\right]. \tag{46}$$

Note that

$$\frac{1}{X^+(\eta)} - \frac{1}{X^-(\eta)} = \frac{1}{X^-(\eta)}\left[\frac{X^-(\eta)}{X^+(\eta)} - 1\right] = \frac{1}{X^-(\eta)}\left[\frac{\lambda^-(\eta)}{\lambda^+(\eta)} - 1\right] = \frac{\lambda^-(\eta) - \lambda^+(\eta)}{X^-(\eta)\lambda^+(\eta)} =$$

$$= -\frac{2\sqrt{\pi}ic\eta\exp(-\eta^2)}{X^-(\eta)\lambda^+(\eta)} = -\frac{2(1-c)\sqrt{\pi}ic\eta X(-\eta)\exp(-\eta^2)}{|\lambda^+(\eta)|^2}.$$

When writing the last equality, we used the formula of factorization of the dispersion function

$$X^-(\eta)X(-\eta) = \frac{\lambda^-(\eta)}{\lambda(\infty)} = \frac{\lambda^-(\eta)}{1-c}.$$

Comparing (46) with (30) and taking into account (21), we find the coefficients in the decomposition of the solution of the boundary value problem (7), (8) on eigenvectors of the continuous spectrum

$$a(\eta) = \frac{(1-c)X(-\eta)\exp(-\eta^2 - d/\eta)}{|\lambda^+(\eta)|^2} \times \left[1 - \frac{1}{\sqrt{\pi}}c\int\limits_0^{+\infty}\frac{\mu\, X(-\mu)a(\mu)\exp(-d/\mu)\,d\mu}{\mu+\eta}\right].$$

Let us denote

$$h(\eta) = \frac{(1-c)X(-\eta)\exp(-\eta^2 - d/\eta)}{|\lambda^+(\eta)|^2}, \tag{47}$$

and consider the equation

$$a(\eta) = h(\eta)\left[1 + \lambda\int\limits_0^{+\infty}\frac{\mu\, X(-\mu)a(\mu)\exp(-d/\mu)\,d\mu}{\mu+\eta}\right], \tag{48}$$

which at $\lambda = -c/\sqrt{\pi}$ passes into the equation (47)

We seek a solution of (48) in the form

$$a(\eta) = \sum_{k=0}^{+\infty} \lambda^k a_k(\eta). \tag{49}$$

Substituting (49) into (48) and equating the coefficients at of equal degrees λ, we obtain a system of recursive relations for finding $a_k(\eta)$ $(\eta > 0)$

$$a_0(\eta) = h(\eta), \tag{50}$$

$$a_1(\eta) = h(\eta) \int\limits_0^{+\infty} \frac{g(\mu_1)\,d\mu_1}{\mu_1 + \eta}, \tag{51}$$

$$a_2(\eta) = h(\eta) \int\limits_0^{+\infty} \frac{g(\mu_1)\,d\mu_1}{\mu_1 + \eta} \int\limits_0^{+\infty} \frac{g(\mu_2)\,d\mu_2}{\mu_2 + \mu_1}, \tag{52}$$

$$\cdots \tag{53}$$

$$a_k(\eta) = h(\eta) \int\limits_0^{+\infty} \frac{g(\mu_1)\,d\mu_1}{\mu_1 + \eta} \int\limits_0^{+\infty} \frac{g(\mu_2)\,d\mu_2}{\mu_2 + \mu_1} \cdots \int\limits_0^{+\infty} \frac{g(\mu_k)\,d\mu_k}{\mu_k + \mu_{k-1}}, \tag{54}$$

$$g(\mu) = \mu\, X(-\mu) \exp(-d/\mu) h(\mu) = \frac{(1-c)\mu X^2(-\mu)\exp(-\mu^2 - 2d/\mu)}{|\lambda^+(\mu)|^2}. \tag{55}$$

Finding the explicit form of the $a_k(\eta)$ completes the construction of the distribution function of the molecules of the light component of the binary gas. Substituting (18) into an expression for mass velocity (6) and taking into account the normalization integral (10) we find

$$U(x) = \frac{G_n}{2(1-c)} \left[\frac{1}{\sqrt{\pi}} \int\limits_{-\infty}^{+\infty} \exp(-\mu^2) Z(x,\mu)\, d\mu - 1 \right] =$$

$$= \frac{G_n}{2(1-c)} \left[\frac{1}{\sqrt{\pi}} \int\limits_{-\infty}^{+\infty} \exp\left(-\frac{x}{\eta}\right) a(\eta)\, d\eta - 1 \right]. \tag{56}$$

Transform the integral in the expression (56)

$$\int\limits_{-\infty}^{+\infty} \exp(-\frac{x}{\eta})\, a(\eta)\, d\eta = \int\limits_{-\infty}^{0} \exp\left(-\frac{x}{\eta}\right) a(\eta)\, d\eta + \int\limits_{0}^{+\infty} \exp\left(-\frac{x}{\eta}\right) a(\eta)\, d\eta =$$

$$= -\int\limits_{+\infty}^{0} \exp\left(\frac{x}{\eta}\right) a(-\eta)\, d\eta + \int\limits_{0}^{+\infty} \exp\left(-\frac{x}{\eta}\right) a(\eta)\, d\eta = \int\limits_{0}^{+\infty} \left[\exp\left(\frac{x}{\eta}\right) + \exp\left(-\frac{x}{\eta}\right)\right] a(\eta)\, d\eta.$$

When writing the last equality, we took into account that $a(-\eta) = a(\eta)$. Thus,

$$U(x) = \frac{G_n}{2(1-c)} \left[\frac{1}{\sqrt{\pi}} \int\limits_{0}^{+\infty} \left[\exp\left(\frac{x}{\eta}\right) + \exp\left(-\frac{x}{\eta}\right)\right] a(\eta)\, d\eta - 1 \right]. \tag{57}$$

Substituting in (57) the decomposition (49) and changing the order of summation and integration, we obtain

$$U(x) = \frac{G_n}{2(1-c)} \left[(1-c) \sum_{k=0}^{+\infty} \lambda^k J_k(x) - 1 \right]. \tag{58}$$

Here introduced the notation

$$J_k(x) = \frac{1}{\sqrt{\pi}} \int\limits_0^{+\infty} \left[\exp\left(\frac{x}{\eta}\right) + \exp\left(-\frac{x}{\eta}\right) \right] a_k(\eta)\, d\eta. \tag{59}$$

Taking into account the form of functions $a(\eta)$ given by equations (50)–(54) we can write

$$J_0(x) = \frac{1}{\sqrt{\pi}} \int\limits_0^{+\infty} \gamma(x,\eta)\, d\eta, \tag{60}$$

$$J_1(x) = \frac{1}{\sqrt{\pi}} \int\limits_0^{+\infty} \gamma(x,\eta)\, d\eta \int\limits_0^{+\infty} \frac{g(\mu_1)\, d\mu_1}{\mu_1 + \eta}, \tag{61}$$

$$J_2(x) = \frac{1}{\sqrt{\pi}} \int\limits_0^{+\infty} \gamma(x,\eta)\, d\eta \int\limits_0^{+\infty} \frac{g(\mu_1)\, d\mu_1}{\mu_1 + \eta} \int\limits_0^{+\infty} \frac{g(\mu_2)\, d\mu_2}{\mu_2 + \mu_1}, \tag{62}$$

$$\cdots$$

$$J_k(x) = \frac{1}{\sqrt{\pi}} \int\limits_0^{+\infty} \gamma(x,\eta)\, d\eta \int\limits_0^{+\infty} \frac{g(\mu_1)\, d\mu_1}{\mu_1 + \eta} \int\limits_0^{+\infty} \frac{g(\mu_2)\, d\mu_2}{\mu_2 + \mu_1} \cdots \int\limits_0^{+\infty} \frac{g(\mu_k)\, d\mu_k}{\mu_k + \mu_{k-1}}, \tag{63}$$

$$\gamma(x,\eta) = \frac{X(-\eta)\exp(-\eta^2)}{|\lambda^+(\eta)|^2} \left[\exp\left(\frac{x-d}{\eta}\right) + \exp\left(-\frac{x+d}{\eta}\right) \right]. \tag{64}$$

The value of the specific mass flow J_M of the light component in the direction of the axis Oy' we define as follows

$$J_M = \frac{1}{2d^2} \int\limits_{-d}^{d} U(x)\, dx. \tag{65}$$

Substituting (58) in the decomposition (65) and changing the order of summation and integration, we obtain

$$J_M = \frac{G_n}{2(1-c)d^2} \left[(1-c) \sum_{k=0}^{+\infty} \lambda^k K_k - d \right]. \tag{66}$$

Here

$$K_0 = \frac{1}{\sqrt{\pi}} \int\limits_0^{+\infty} \zeta(\eta)\, d\eta, \qquad K_1 = \frac{1}{\sqrt{\pi}} \int\limits_0^{+\infty} \zeta(\eta)\, d\eta \int\limits_0^{+\infty} \frac{g(\mu_1)\, d\mu_1}{\mu_1 + \eta},$$

$$K_2 = \frac{1}{\sqrt{\pi}} \int\limits_0^{+\infty} \zeta(\eta)\, d\eta \int\limits_0^{+\infty} \frac{g(\mu_1)\, d\mu_1}{\mu_1 + \eta} \int\limits_0^{+\infty} \frac{g(\mu_2)\, d\mu_2}{\mu_2 + \mu_1},$$

$$\cdots$$

$$K_k = \frac{1}{\sqrt{\pi}} \int\limits_0^{+\infty} \zeta(\eta)\, d\eta \int\limits_0^{+\infty} \frac{g(\mu_1)\, d\mu_1}{\mu_1 + \eta} \int\limits_0^{+\infty} \frac{g(\mu_2)\, d\mu_2}{\mu_2 + \mu_1} \cdots \int\limits_0^{+\infty} \frac{g(\mu_k)\, d\mu_k}{\mu_k + \mu_{k-1}},$$

$$\zeta(\eta) = \frac{\eta X(-\eta)}{|\lambda^+(\eta)|^2} \exp(-\eta^2) \left[1 - \exp\left(-\frac{2d}{\eta}\right) \right].$$

3. Results and discussion

The values of J_M/G_n for the mixture of He-Xe and the relative concentration of the light component $C = n_a/(n_a + n_b)$ are given in the table

$2d$	0.01	0.1	0.5	1.0	2.0	5.0	10.0	100.0
$C = 0.01$								
$-J_M$ (66)	3.1866	1.9082	1.0757	0.7637	0.5013	0.2524	0.1380	0.0150
$C = 0.05$								
$-J_M$ (66)	3.1852	1.9133	1.0843	0.7724	0.5088	0.2572	0.1409	0.0153
$C = 0.1$								
$-J_M$ (66)	3.1827	1.9200	1.0957	0.7840	0.5192	0.2638	0.1449	0.01576
$-J_M$ [4]	2.8760	1.7652	1.0507	0.7717	0.5236	0.2707	0.1491	0.0162

When calculating J_M we assumed that $m_b = 131.30$, $m_a = 4.0026$, $d_b/d_a = 2.226$, $d = (2d'/3)\beta^{1/2}\nu$. As follows from the results given in the table, the approach used in the work leads to results that are in good agreement with the results from [4].

4. Conclusion

In the present work, an analytical solution of the problem of diffusion of the light component of a binary gas mixture in a long flat channel with infinite parallel walls is obtained in the linearized approximation. Expressions describing the mass velocity profile and specific flow of the light component of the mixture, depending on its concentration in the mixture and the channel thickness, are obtained.

References

[1] Sharipov F 2016 *Rarefied gas dynamics: Fundamentals for research and practice* (Wiley-VCH: Weinheim)
[2] Latyshev A 1996 *Vvedeniye v Keysologiyu. Monografiya* Otdel teoreticheskikh problem RAN
[3] Garcia R and Siewert C 2007 *Journal of Applied Mathematics* **67** 1041
[4] Siewert C and Valougeorgis D 2004 *European Journal of Mechanics - B/Fluids* **23** 709
[5] Roldugin V and Zhdanov V 2006 it Technical Physics **51** 436
[6] Latyshev A and Yushkanov A 2005 *Theoretical and Mathematical Physics* **143** 854
[7] Popov V, Testova I and Yushkanov A 2011 *Technical Physics* **56(1)** 49
[8] Podryga V and Polyakov S 2016 *Vychislitel'nyye metody i programmirovaniye* **17** 147

Journal of Physics: Conference Series **2056** (2021) 012005 doi:10.1088/1742-6596/2056/1/012005

A collocation method for the Williams equation with Chebyshev polynomials

O V Germider [1,*] and V N Popov [2]

[1,2] Northern (Arctic) Federal University named after M.V. Lomonosov, 17, Severnaya Dvina Emb., Arkhangelsk, 163002, Russia

E-mail: *o.germider@narfu.ru

Abstract. The linearized problem of gas flow in plane channel with infinite walls has been solved in the kinetic approximation. The flow in the channel is caused by a constant pressure gradient parallel to the walls of the channel. The Williams equation has been used as a basic equation, and the boundary condition has been set in terms of the diffuse reflection model. The collocation method for Chebyshev polynomials has been applied to construct the solution of the equation of Williams with the given boundary conditions. The mass flux of the gas in the channel has been calculated.

1. Introduction

The gas flow through a long channel is one of the most important application of the MEMS and vacuum technology [1]. In the intermediate range of Knudsen numbers, where the concepts of continuous medium and free-molecular regime are inapplicable, the function determining the macroscopic (mass) velocity can be obtained by integrating the Boltzmann kinetic equation or model kinetic equations. The BGK (Bhatnagar, Gross, Krook) and the S (Shahov) models of the Boltzmann kinetic equation, and the Williams equation are often used as model equations [1-7]. The basis for the construction of BGK and S models is the constancy of the collision frequency of gas molecules. However, the assumption on the independence of the speed of the frequency of gas molecule collisions is quite a strong simplification [8]. It would be more realistic to assume the constancy of the length of the free path of gas molecules, at least for molecules, whose mutual interaction can be approximated by the model of solid spheres. This assumption is equivalent to the assumption that the frequency of molecule collisions is proportional to the absolute value (module) of their heat velocity. The latter assumption leads to the following the Williams equation [8]. In this paper, the Williams equation is used to describe the kinetics of the processes under consideration.

At the present time a large number of methods and their variations are proposed to construct solutions of model kinetic equations. Thus, the method of discrete ordinates and velocities was used in [1]. The finite-volume method based on a TVD scheme is applied in [7]. In [4, 5] the results are obtained by the use the Neumann series. Here we propose an efficient approximate method for solving the Williams model kinetic equation by expansion of the solution in a series of Chebyshev polynomials [9, 10]. Using the Chebyshev collocation points, we transform the Williams integro-differential equation to a matrix equation which corresponds to a system of linear algebraic equations with unknown Chebyshev coefficients. Therefore this allows us much to decrease the number of grid nodes in the physical space. Methods related to Chebyshev polynomial interpolation also have received much interest recently in the areas of signal processing and applied mathematics [11]. In [12] by applying the shifted Chebyshev polynomials and rational Chebyshev functions, the authors solved the fractional diffusion equation with initial-boundary conditions over a long time domain. Their approach was based on collocation method [10].

In this paper the collocation method for Chebyshov polynomials is used to construct the solution of the Williams equation in the problem of the slow isothermal flow of a rarefied gas due to a small pressure gradient in a plane channel with infinite parallel walls (Poiseuille flow). It is assumed that the walls of the channel are in the planes $y' = \pm b'$ of a rectangular Cartesian coordinate system, the axis Oz' of which is parallel to the channel walls. The diffuse reflection model of the boundary condition on the walls of channel is used.

By following [4, 8] we write the Williams equation and the boundary conditions in the following form

$$C_y \frac{\partial Z}{\partial y}\gamma Kn + CZ(y, C_y) + 1 = \frac{3C}{4\pi}\int C' \exp(-C'^2) C_z'^2 Z(y, C_y') d^3\mathbf{C}', \quad (1)$$

$$Z(\pm 1, C_y) = 0, \quad \pm C_y < 0. \quad (2)$$

Here $\mathbf{C} = \beta^{1/2}\mathbf{v}$ is the dimensionless velocity of the gas molecules, $\beta = m/(2k_B T_0)$, m is the molecular mass of the gas, k_B is the Boltzmann constant, T_0 is gas temperature at the origin of coordinate system, $y = y'/b'$, and $z = z'/b'$ are the coordinates of the dimensionless radius-vector $\mathbf{r}$, $Kn = l_g/b'$ is Knudsen number, l_g is the mean free path of gas molecules, $\gamma = 15\sqrt{\pi}/16$. The walls of the channel situate in the planes $y' = \pm b'/2$ of the Cartesian coordinate system with the origin located in the mid-section $z' = 0$. The axis z' of is parallel to the channel walls. The mass velocity profile we will calculate according to [13]

$$U_z(y) = \frac{\gamma Kn G_p}{\pi^{3/2}} \int \exp\left(-C^2\right) C_z^2 Z(y, C_y) d^3\mathbf{C}, \quad (3)$$

where $G_p = p_0^{-1} dp/dz$ is the dimensionless pressure gradient ($G_p \ll 1$), p_0 is the gas pressure at the origin of coordinate system.

The mass flux through a channel of finite width $2a'$ is calculated as

$$J'_{M,2a} = 2\beta^{1/2} p_0 b'^2 \int_{-a}^{a}\int_{-1}^{1} U_z(x, y) dx dy,$$

where $a = a'/b'$, $x = x'/b'$. In theoretical calculations, for channel with infinite width ($a' \to \infty$) the mass flux is determined per unit height, i.e., as limits [13]

$$J'_M = \lim_{a' \to \infty} \frac{1}{2a'} J'_{M,2a} = 2\beta^{1/2} p_0 b' \int_{-1}^{1} U_z(y) dy. \quad (4)$$

Taking into account (4) the reduced mass flux we can calculate as [13]:

$$J_M = \frac{J'_M}{2b'\beta^{1/2} p_0} = \int_{-1}^{1} U_z(y) dy. \quad (5)$$

To find the solution of the integro-differential equation (1) the unknown function $Z(y, C_y)$ is expanded in the orthogonal functions and solution of the resulting equation is represented by Chebyshev polynomials in the matrix form.

2. Materials and methods

Let us introduce in the velocity space the scalar product of functions depending on the molecular velocity modulus as follows

$$(e_1, e_2) = \int\limits_0^{+\infty} g(C)e_1(C)e_2(C)dC.$$

Let us choose two orthogonal functions $e_1 = 1$ and $e_2 = 1/C - 3\sqrt{\pi}/8$ (here, orthogonality is meant as the vanishing of the above scalar product) and expand function $Z(y, C_y)$ by the orthogonal functions:

$$Z(y, C_y) = Z_1(y, \cos\theta) + \left(\frac{1}{C} - \frac{3\sqrt{\pi}}{8}\right) Z_2(y, \cos\theta). \tag{6}$$

In (6), we passed to the velocity space in the spherical coordinate system ($C_z = C\cos\varphi\sin\theta$, $C_x = C\sin\varphi\sin\theta$, $C_y = C\cos\theta$). Substituting expansion (6) into (1) and taking into account the orthogonality of functions $e_1(C)$ and $e_2(C)$, we come to a set of equations

$$\tau\frac{\partial Z_1}{\partial y}\gamma Kn + Z_1(y,\tau) + \frac{3\sqrt{\pi}}{8} = \frac{3}{4}\int\limits_{-1}^{1}\left(1-\tau'^2\right) Z_1(y,\tau')d\tau', \tag{7}$$

$$\tau\frac{\partial Z_2}{\partial y}\gamma Kn + Z_2(y,\tau) + 1 = 0, \quad \tau = \cos\theta. \tag{8}$$

From (2) and (6) it follows that boundary conditions for $Z_1(y,\tau)$ and $Z_2(y,\tau)$ have the form

$$Z_i(\pm 1, \tau) = 0, \quad -1 \le \pm\tau < 0, \quad i = 1, 2. \tag{9}$$

The solution of equation (8) is sought by applying the method of characteristics. Taking into account (9), we obtain

$$Z_2(y,\tau) = \exp\left(-\frac{y-1}{\gamma Kn\tau}\right) H_+(-\tau) + \exp\left(-\frac{y+1}{\gamma Kn\tau}\right) H_+(\tau) - 1. \tag{10}$$

Here H_+ is the stepwise Heaviside function.

The solution of equation (7) responsible for boundary condition (9) is written as

$$Z_1(y,\tau) = W_1(y,\tau)H_+(-\tau) + W_2(y,\tau)H_+(\tau). \tag{11}$$

Substituting expression (11) into equation (7) and boundary condition (9), for $-1 \le \tau \le 0$ we obtain

$$\tau\frac{\partial W_1}{\partial y}\gamma Kn + W_1(y,\tau) + \frac{3\sqrt{\pi}}{8} =$$

$$= \frac{3}{4}\int\limits_{-1}^{0}\left(1-\tau'^2\right) W_1(y,\tau')d\tau' + \frac{3}{4}\int\limits_{0}^{1}\left(1-\tau'^2\right) W_2(y,\tau')d\tau', \tag{12}$$

$$W_1(1,\tau) = 0, \quad -1 \le \tau < 0. \tag{13}$$

For $0 \le \tau \le 1$ we obtain

$$\tau\frac{\partial W_2}{\partial y}\gamma Kn + W_2(y,\tau) + \frac{3\sqrt{\pi}}{8} =$$

$$= \frac{3}{4}\int_{-1}^{0}\left(1-\tau'^2\right)W_1(y,\tau')d\tau' + \frac{3}{4}\int_{0}^{1}\left(1-\tau'^2\right)W_2(y,\tau')d\tau', \tag{14}$$

$$W_2(-1,\tau) = 0, \quad 0 < \tau \le 1. \tag{15}$$

The functions $W_1(y,\tau)$ and $W_2(y,\tau)$ may be expanded in Chebyshev polynomials as

$$W_i(y,\tau) = \sum_{j_1,j_2=0}^{\infty} a_{j_1j_2}^{(i)} T_{j_1}(y)T_{j_2}(z_i(\tau)),\ z_i = 2\tau-(-1)^i,\ y, z_i \in [-1,\,1], \tag{16}$$

where $a_{jk}^{(i)}$ are unknowns parameters, to be determined. The following recurrence relation, generate these polynomials as [12]

$$T_0(y) = 1, \quad T_1(y) = y, \quad T_j(y) = 2yT_{j-1}(y) - T_{j-2}(y), \quad j \ge 2. \tag{17}$$

The Chebyshev polynomials satisfy at the following properties [10]

$$\frac{dT_j(y)}{dt} = 2j\sum_{k=1}^{j/2} T_{2k-1}(y), \quad j \textit{ even}, \tag{18}$$

$$\frac{dT_j(y)}{dt} = j + 2j\sum_{k=1}^{(j-1)/2} T_{2k}(y), \quad j \textit{ odd}; \tag{19}$$

$$2T_j(y)T_k(y) = T_{j+k}(y) + T_{|j-k|}(y); \tag{20}$$

$$T_j(-y) = (-1)^j T_j(y); \tag{21}$$

$$\int_{-1}^{1} T_j(y)dy = \begin{cases} \dfrac{2}{1-j^2}, & j \textit{ even}, \\ 0, & j \textit{ odd}. \end{cases} \tag{22}$$

For $0 \le \tau \le 1$ from equality (21) we deduce that

$$W_1(1,-\tau) = \sum_{j_1,j_2=0}^{\infty} (-1)^{j_1+j_2} a_{j_1j_2}^{(1)} T_{j_1}(-1)T_{j_2}(z_2(\tau)). \tag{23}$$

In this case from boundary conditions (13) and (15) we have the equality $W_2(-1,\tau) = W_1(1,-\tau)$. Then, taking into account the uniqueness of the expansion $W_2(-1,\tau)$, we have

$$a_{j_1j_2}^{(2)} = (-1)^{j_1+j_2} a_{j_1j_2}^{(1)}, \quad j_{1,2} = \overline{0, n_{1,2}}. \tag{24}$$

Let the approximation of $W_i(y,\tau)$ be obtained by truncating the series (16) as

$$W_i(y,\tau) = \sum_{j_1=0}^{n_1}\sum_{j_2=0}^{n_2} a_{j_1j_2}^{(i)} T_{j_1}(y)T_{j_2}(z_i(\tau)), \quad i = 1, 2. \tag{25}$$

The matrix form of expression (25) can be written as

$$W_i(y,\tau) = \mathbf{T}(y)\mathbf{Q}(z_i(\tau))\mathbf{A}_i, \quad i = 1, 2. \tag{26}$$

Here $\mathbf{T}(y)$ is the $1 \times (n_1+1)$ matrix as

$$\mathbf{T}(y) = (T_0(y)\, T_1(y) \dots T_{n_1}(y)), \tag{27}$$

and $\mathbf{A}_i$ is the $(n_1+1)(n_2+1) \times 1$ matrix as

$$\mathbf{A}_i = \left(a_{00}^{(i)}\, a_{01}^{(i)} \dots a_{0n_2}^{(i)}\;\; a_{10}^{(i)}\, a_{11}^{(i)} \dots a_{1n_2}^{(i)} \dots\;\; a_{n_1 0}^{(i)} \dots a_{n_1 n_2}^{(i)}\right)^T, \quad i = 1, 2; \tag{28}$$

$\mathbf{Q}(z_i(\tau))$ is $(n_1+1) \times (n_1+1)(n_2+1)$ block matrix

$$\mathbf{Q}(z_i(\tau)) = \begin{pmatrix} \hat{\mathbf{T}}(z_i(\tau)) & \mathbf{0} & \mathbf{0} & \dots & \mathbf{0} \\ \mathbf{0} & \hat{\mathbf{T}}(z_i(\tau)) & \mathbf{0} & \dots & \mathbf{0} \\ \vdots & & \ddots & & \vdots \\ \mathbf{0} & \mathbf{0} & \dots & \mathbf{0} & \hat{\mathbf{T}}(z_i(\tau)) \end{pmatrix}, \tag{29}$$

$$\hat{\mathbf{T}}(z_i(\tau)) = (T_0(z_i(\tau))\, T_1(z_i(\tau)) \dots T_{n_2}(z_i(\tau))), \tag{30}$$

and $\mathbf{0}$ is the $1 \times (n_2+1)$ zero matrix.

From equalities (18) and (19), we deduce that

$$\frac{d\mathbf{T}(y)}{dy} = \mathbf{T}(y)\mathbf{J}^T, \tag{31}$$

in which $\mathbf{J}$ is the $(n_1+1) \times (n_1+1)$ matrix as [14]

$$\mathbf{J} = \begin{pmatrix} 0 & 0 & 0 & 0 & 0 & \dots & 0 & 0 \\ 1 & 0 & 0 & 0 & 0 & \dots & 0 & 0 \\ 0 & 2\cdot 2 & 0 & 0 & 0 & \dots & 0 & 0 \\ 3 & 0 & 2\cdot 3 & 0 & 0 & \dots & 0 & 0 \\ 0 & 2\cdot 4 & 0 & 2\cdot 4 & 0 & \dots & 0 & 0 \\ 5 & 0 & 2\cdot 5 & 0 & 2\cdot 5 & \dots & 0 & 0 \\ \vdots & & & \ddots & & & & \vdots \\ & & & \textit{if } n_1 \textit{ even} & & & & \\ 0 & 2\cdot n_1 & 0 & 2\cdot n_1 & 0 & \dots & 2\cdot n_1 & 0 \\ & & & \textit{if } n_1 \textit{ odd} & & & & \\ n_1 & 0 & 2\cdot n_1 & 0 & 2\cdot n_1 & \dots & 2\cdot n_1 & 0 \end{pmatrix}. \tag{32}$$

In view of the designations (26)-(32), equation (12) can be rewritten in the form

$$\tau\gamma Kn\mathbf{T}(y)\mathbf{J}^T\mathbf{Q}(z_1(\tau))\mathbf{A}_1 + \mathbf{T}(y)\mathbf{Q}(z_1(\tau))\mathbf{A}_1 + \frac{3\sqrt{\pi}}{8} = \frac{3}{4}\mathbf{T}(y)\sum_{i=1}^{2}\mathbf{G}_i\mathbf{A}_i, \tag{33}$$

where $\mathbf{G}_i$ is $(n_1+1) \times (n_1+1)(n_2+1)$ block matrix

$$\mathbf{G}_i = \begin{pmatrix} \mathbf{K}_i & \mathbf{0} & \mathbf{0} & \dots & \mathbf{0} \\ \mathbf{0} & \mathbf{K}_i & \mathbf{0} & \dots & \mathbf{0} \\ \vdots & & \ddots & & \vdots \\ \mathbf{0} & \mathbf{0} & \dots & \mathbf{0} & \mathbf{K}_i \end{pmatrix}, \tag{34}$$

$$\mathbf{K}_i = (K_{i,0}\, K_{i,1} \dots K_{i,n_2}), \tag{35}$$

$$K_{1,k} = \int\limits_{-1}^{0} \left(1-\tau'^2\right) T_k(2\tau'+1)d\tau', \quad K_{2,k} = \int\limits_{0}^{1} \left(1-\tau'^2\right) T_k(2\tau'-1)d\tau'. \tag{36}$$

In (36), we passe variable z_i as (16)

$$2K_{i,k} = \int\limits_{-1}^{1} \left(1 - \frac{(z_i+(-1)^i)^2}{4}\right) T_k(z_i)dz_i. \tag{37}$$

The functions $1-(z_i+(-1)^i)^2/4$ may be expanded in Chebyshev polynomials as

$$1 - \frac{(z_i+(-1)^i)^2}{4} = \frac{5}{8}T_0(z_i) - \frac{1}{2}T_1(z_i)(-1)^i - \frac{1}{8}T_2(z_i), \quad i=1,2. \tag{38}$$

Substituting (38) into (37), from equalities (20)-(22) we deduce that

$$K_{2,k} = (-1)^k K_{1,k}, \; 2K_{1,k} = \begin{cases} \dfrac{12-k^2}{k^4-10k^2+9}, & k\,even, \\ \dfrac{1}{4-k^2}, & k\,odd, \end{cases} \quad k=\overline{0,n_2}. \tag{39}$$

In view of the first equality (39) and taking into account (24), we can rewrite the equation (33) in the following way

$$\tau\gamma Kn\mathbf{T}(y)\mathbf{J}^T\mathbf{Q}(z_1(\tau))\mathbf{A}_1 + \mathbf{T}(y)\mathbf{Q}(z_1(\tau))\mathbf{A}_1 + \frac{3\sqrt{\pi}}{8} = \frac{3}{2}\mathbf{T}(y)\mathbf{G}\mathbf{A}_1. \tag{40}$$

Here $\mathbf{G}$ is $(n_1+1)\times(n_1+1)(n_2+1)$ block matrix of the form

$$\mathbf{G} = \begin{pmatrix} \mathbf{K}_1 & \mathbf{0} & \mathbf{0} & \mathbf{0} & \dots & \mathbf{0} \\ \mathbf{0} & \mathbf{0} & \mathbf{0} & \mathbf{0} & \dots & \mathbf{0} \\ \mathbf{0} & \mathbf{0} & \mathbf{K}_1 & \mathbf{0} & \dots & \mathbf{0} \\ \vdots & & & \ddots & & \vdots \\ & & & if\ n_1\ even & & \\ \mathbf{0} & \mathbf{0} & \dots & \mathbf{0} & \mathbf{0} & \mathbf{K}_1 \\ & & & if\ n_1\ odd & & \\ \mathbf{0} & \mathbf{0} & \dots & \mathbf{0} & \mathbf{0} & \mathbf{0} \end{pmatrix}. \tag{41}$$

We choose the collocation points as the zeros of the Chebyshev polynomials $T_{n_1+1}(y)$, $T_{n_2+1}(2\tau+1)$ from (25) for $i=1$ [10]

$$y_{j_1}^{(1)} = \cos\left(\frac{\pi\left(n_1-j_1+\frac{1}{2}\right)}{n_1+1}\right), \quad \tau_{j_2}^{(1)} = \frac{1}{2}\left(\cos\left(\frac{\pi\left(j_2+\frac{1}{2}\right)}{n_2+1}\right)-1\right), \tag{42}$$

$j_{1,2}=\overline{0,n_{1,2}}$ and $T_{n_1+1}(y)$, $T_{n_2+1}(2\tau-1)$ for $i=2$

$$y_{j_1}^{(2)} = \cos\left(\frac{\pi\left(j_1+\frac{1}{2}\right)}{n_1+1}\right), \quad \tau_{j_2}^{(2)} = \frac{1}{2}\left(\cos\left(\frac{\pi\left(n_2-j_2+\frac{1}{2}\right)}{n_2+1}\right)+1\right). \tag{43}$$

By substituting the collocation points (42) and (43) into (40), we obtain the system of matrix equations ($j_{1,2} = \overline{0, n_{1,2}}$) as follows ($i = 1, 2$)

$$\mathbf{\Lambda}_i(y_{k_1}^{(i)}, \tau_{k_2}^{(i)})\mathbf{A}_i = -\frac{3\sqrt{\pi}}{8}, \tag{44}$$

$$\mathbf{\Lambda}_i(y_{k_1}^{(i)}, \tau_{k_2}^{(i)}) = \left(\tau_{k_2}^{(i)}\gamma Kn\mathbf{T}(y_{k_1}^{(i)})\mathbf{J}^T + \mathbf{T}(y_{k_1}^{(i)})\right)\mathbf{Q}(z_i(\tau_k^{(i)})) - \frac{3}{2}\mathbf{T}(y_{k_1}^{(i)})\mathbf{G}. \tag{45}$$

The boundary condition (9) is discretized in the following form

$$\mathbf{\Omega}_i(y_{n_1}^{(i)}, \tau_{k_2}^{(i)})\mathbf{A}_i = 0, \tag{46}$$

$$\mathbf{\Omega}_i(y_{n_1}^{(i)}, \tau_{k_2}^{(i)}) = \mathbf{T}(y_{n_1}^{(i)})\mathbf{Q}(z_i(\tau_k^{(i)})), \quad i = 1, 2. \tag{47}$$

Hence, equations (44) and (46) can be rewritten in the form

$$\mathbf{B}_i\mathbf{A}_i = \mathbf{F}, \tag{48}$$

where $\mathbf{B}_i$ is the $(n_1+1)(n_2+1) \times (n_1+1)(n_2+1)$ matrix ($i = 1, 2$), F is the $(n_1+1)(n_2+1) \times 1$ matrix as

$$\mathbf{B}_i = \begin{pmatrix} \mathbf{\Lambda}_i(y_0^{(i)}, \tau_0^{(i)}) \\ \mathbf{\Lambda}_i(y_0^{(i)}, \tau_1^{(i)}) \\ \dots \\ \mathbf{\Lambda}_i(y_{n_1-1}^{(i)}, \tau_{n_2-1}^{(i)}) \\ \mathbf{\Lambda}_i(y_{n_1-1}^{(i)}, \tau_{n_2}^{(i)}) \\ \mathbf{\Omega}_i(y_{n_1}^{(i)}, \tau_0^{(i)}) \\ \dots \\ \mathbf{\Omega}_i(y_{n_1}^{(i)}, \tau_{n_2-1}^{(i)}) \\ \mathbf{\Omega}_i(y_{n_1}^{(i)}, \tau_{n_2}^{(i)}) \end{pmatrix}, \quad \mathbf{F} = \begin{pmatrix} -\frac{3\sqrt{\pi}}{8} \\ -\frac{3\sqrt{\pi}}{8} \\ \dots \\ -\frac{3\sqrt{\pi}}{8} \\ -\frac{3\sqrt{\pi}}{8} \\ 0 \\ \dots \\ 0 \\ 0 \end{pmatrix}. \tag{49}$$

Substituting (6) into (3) and taking into account (10) and (11), we have

$$U_z(y) = \frac{\gamma KnG_p}{2\sqrt{\pi}}\left(1 - \frac{9\pi}{32}\right)\int\limits_{-1}^{1} Z_2(y, \tau)\left(1 - \tau^2\right)d\tau + \frac{3\gamma KnG_p}{4}\mathbf{T}(y)\mathbf{G}\mathbf{A}_1. \tag{50}$$

Hence, the mass flux (50) is obtained as

$$J_M = -\frac{\gamma KnG_p}{\sqrt{\pi}}\left(1 - \frac{9\pi}{32}\right)\left(\frac{4}{3} - \int\limits_{-1}^{1} dy \int\limits_{-1}^{0}\left(1 - \tau^2\right)\exp\left(-\frac{y-1}{\gamma Kn\tau}\right)d\tau\right) + \\ + \frac{3\gamma KnG_p}{4}\mathbf{L}\mathbf{G}\mathbf{A}_1. \tag{51}$$

Here $\mathbf{L}$ is the $1 \times (n_1 + 1)$ matrix as $\mathbf{L} = \begin{pmatrix} 1 & 0 & -\frac{1}{3} & \dots & 0 & \frac{1}{1-n_1^2} \end{pmatrix}$ if n_1 is even and $\mathbf{L} = \begin{pmatrix} 1 & 0 & -\frac{1}{3} & \dots & \frac{1}{1-(n_1-1)^2} & 0 \end{pmatrix}$ if n_1 is odd.

Following [15, 16], we write the error of calculating the second term in (51) as

$$E_{\mathbf{n}} = I_{\mathbf{n}} - I_{\mathbf{n}/2}, \tag{52}$$

where $I_{\mathbf{n}}$ and $I_{\mathbf{n}/2}$ are the values of this term for $\mathbf{n} = (n_1, n_2)^T$ and $\mathbf{n}/2$.

3. Results and discussion

The solution of the boundary value problem (1), (2) can also be solved by applying the method used in [4], which the expansion of the solution in the eigenfunctions can be reduced to solving the Riemannian boundary value problem. In [4] mass flux values were obtained depending on the parameter introduced as $D = 2b'/l_g$. This definition of D is determined by the choice of the size scale $2b'$. In this case, the reduce dimensionless pressure gradient is calculated according to the formula $\hat{G}_p = 2p_0^{-1}dp/dz = 2G_p$. The mass flow values $J_M/\hat{G}_p$ given by (51) with $Kn = 2/D$ are presented in the table 1. The comparison between values $J_M/\hat{G}_p$ calculated by the proposed method and method [4] are listed in the table 2. The results show that the values $J_M/\hat{G}_p$ obtained according to (51) for $D = 1$ and 10 with $n_1 = 2n_2 = 36$ and for $D = 0.1$ with $2n_1 = n_2 = 36$ are in good agreement with the corresponding values of mass flux given in [4]. The using the Chebyshev polynomials and collocation points (42) and (43) in the matrix equation (49) allows to reduce the calculations thus accelerating the convergence. In [4] for $D = 1$ a uniform grid consisting of 13500 cells was used. Numerical investigation according to (49) makes it possible to grid refinement near the ends of the interval of the definition of variables y and τ thus minimizing the problem of Runge's phenomenon [11]. It should also be noted that the proposed method can be used to solve the Williams equation in the Poiseuille flow problem in a long channel with a much complex cross section configuration. Analogous results of mathematical simulation of the mass transfer process in the channel between two parallel walls obtained in [2, 17, 18] by the discrete coordinate method using the BGK model of the Boltzmann kinetic equation (BGK), the S-model (S), the model with combined kernel (CES), and the Boltzmann linearized equation for hardsphere molecules (LBE). As can be seen from the table 2, the results obtained in this work are in good agreement with the analogous results from [2, 17, 18]. The differences can be explained by a weak dependence of the results on the collision integral model.

Table 1. The mass flux $-J_M/\hat{G}_p$ for various values of D

D	$n_1 = 2n_2 = 18$	$n_1 = 2n_2 = 36$	D	$2n_1 = n_2 = 18$	$2n_1 = n_2 = 36$
0.5	1.6373	1.6372	0.1	2.1289	2.1295
1.0	1.5446	1.5446	0.2	1.8801	1.8807
2.0	1.5738	1.5738			
5.0	1.9448	1.9445			
10.0	2.7159	2.7151			

Table 2. Comparison of the mass flux $-J_M/\hat{G}_p$ for $D = 0.1$, 1, 10

D	W (51)	W [4]	BGK [2]	S [17]	CES [18]	LBE [18]
0.1	2.1295	2.130	2.0323	2.0395	1.9259	1.9499
1.0	1.5446	1.545	1.5387	1.5536	1.4863	1.5067
10.0	2.7151	2.715	–	2.7799	2.7220	2.7296

4. Conclusion

In the presented article we solved the Williams equation with diffuse boundary conditions by applying the Chebyshev polynomials. Thus, and after using a set of collocation points, the required numerical solution is found to be equivalent to the solution of a linear system of algebraic equations. The value of the mass flux per unit height of the channel has been calculated. The resultant expressions are analyzed numerically. From the conducted comparison, it turns out that the proposed method converges quickly and can be used to solve the Williams equation in the Poiseuille flow problem in a long channel with a more complex cross section configuration.

References

[1] Graur I and Ho M 2014 *Vacuum* **101** 328
[2] Barihcello L, Camargo M Rodrigues P and Siewert C 2001 *Zeitschrift fur angewandte Mathematik und Physik ZAMP* **52** 517
[3] Germider O and Popov V 2020 *Fluid Dynamics* **55** 407
[4] Gulakova S and Popov V 2015 *Technical Physics* **60** 477
[5] Latyshev A and Yushkanov A 2005 *Theoretical and Mathematical Physics* **143** 854
[6] Sharipov F 2016 *Rarefied gas dynamics: Fundamentals for research and practice* (Wiley-VCH: Weinheim)
[7] Valougeorgis D, Vasileiadis N and Titarev V 2017 *European Journal of Mechanics - B/Fluids* **64** 2
[8] Cercignani C 1969 *Mathematical methods in kinetic theory* (New York: Plenum)
[9] Boyd J 2000 *Chebyshev and Fourier spectral methods. Second Edition* (Mineola: DOVER Publications)
[10] Mason J and Handscomb D 2003 *Chebyshev polynomials* (Florida: CRC Press)
[11] Li B, Zhang Y and Wang X 2016 *Circuits, Systems, and Signal Processing* **35** 719
[12] Baseri A, Abbasbandy S and Babolian E 2018 *Applied Mathematics and Computation* **322** 55
[13] Sharipov F and Seleznev V 1998 *Journal of Physical and Chemical Reference Data* **27** 657
[14] Tohidi E 2015 *Ain Shams Engineering Journal* **6** 373
[15] Berntsen J, Espelid T and Genz A 1991 ACM TOMS **17** 437
[16] Berntsen J, Espelid T and Genz A 1991 ACM TOMS **17** 452
[17] Siewert C 2002 *European Journal of Mechanics - B/Fluids* **21** 579
[18] Siewert C 2003 *Journal of Applied Mathematics and Physics* **54** 273

Study of the pressure distribution along a concentric annular channel

O V Germider [1,*] and V N Popov [2]

[1,2] Northern (Arctic) Federal University named after M.V. Lomonosov, 17, Severnaya Dvina Emb., Arkhangelsk, 163002, Russia

E-mail: *o.germider@narfu.ru

Abstract. A study of an isothermal rarefied gas flow through a long concentric annular channel is carried out. The solution is based on the linearized BGK model with diffuse boundary conditions and it is valid in the whole range of the rarefaction parameter. The pressure distribution along the channel is obtained and investigated depending on the values of the pressure maintained at the channel ends.

1. Introduction

During the last years it is possible by development of the kinetic approach made to solve in a computationally multidimensional problems in complex geometries [1-3]. The main advantage of the kinetic approach is that the solution is valid in the whole range of the the rarefaction parameter from the free molecular to the hydrodynamic regimes. Using the solution is critical when carefully and systematically studying of the pressure distribution. However, finding solution of the model kinetic equation and analysis are challenging computational tasks in a rarefied gas dynamics [1]. The present work is devoted to investigating the gas pressure along a concentric annular channel applying kinetic approach. Here, we consider the Maxwell diffuse reflection model with complete accommodation at the bounding walls of the channel and assume, that the rarefied gas flow is due to a pressure gradient imposed in the longitudinal direction. The methodology of obtaining the pressure distribution along the channel is proposed and results are provided due to a ratio of the pressure maintained at the channel ends in the wide range of gas rarefaction. The proposed approach is accomplished through Chebyshev approximation [4] for the solution of the linearized BGK (Bhatnagar, Gross and Krook [5]) model kinetic equation and applying the discrete orthogonality relations and Kronecker and Hadamard products [6] to minimize computational error. The realization of the polynomial approach during the study can be applied to any cross section.

2. Materials and methods

Consider a long concentric annular channel of length L' and radius R_1' and R_2' ($R_2' > R_1' > 0$), connecting two reservoirs maintained at pressures p_1' and p_2' ($p_2' > p_1' > 0$). Under the influence of the pressure gradient there is an isothermal (Poiseuille) flow from the high towards the low pressure reservoir. Assume that the radius $R_2' \ll L'$ and the non-dimensionless pressure gradient $|G_p| = R_2'|dp'/(p_0'dz')| \ll 1$, where p_0' the gas pressure at a certain point taken as the origin. The pressure p' vary only along the flow direction z'. Since a kinetic approach is followed, the main unknown is the distribution function $f'(\mathbf{r}', \mathbf{v})$, where $\mathbf{r}'$ is the radius-vector and $\mathbf{v}$ is the molecular velocity. In the configuration and velocity spaces are use the cylindrical coordinates $\mathbf{r}' = (\rho', r'_\varphi, r'_z)$ and $\mathbf{v} = (v_\perp, v_\psi, v_z)$. Here φ and ψ are polar angles.

The dimensionless radius-vector $\mathbf{r}'$, radii R'_1 and R'_2, length L', distribution function f', molecular velocity $\mathbf{v}$, mass velocity $\mathbf{u}'$, gas number density n', temperature T' and mass flux J'_M are written as

$$\mathbf{r} = \frac{\mathbf{r}'}{R'_2},\ R_1 = \frac{R'_1}{R'_2},\ R_2 = 1,\ L = \frac{L'}{R'_2},\ f = \frac{f'}{n'_0 \beta^{3/2}},$$

$$\mathbf{C} = \beta^{1/2}\mathbf{v},\ \mathbf{u} = \beta^{1/2}\mathbf{u}',\ n = \frac{n'}{n'_0},\ T = \frac{T'}{T'_0},\ J_M = \frac{J'_M}{\pi(R'^2_2 - R'^2_1)p'_0\beta^{1/2}}. \tag{1}$$

Here n'_0 and T'_0 are the gas number density and temperature at a certain point taken as the origin, $\beta = m'/(2k_B T'_0)$, k_B is the Boltzmann constant, m' is mass of a molecule, $T = 1$. Below, the non-dimensional variables are denoted by the same letters as the dimensional ones.

We will consider gas flow in the middle part of channel ($z^*_0 = 0$, where $z^* = z'/L'$). In the case, when the pressure and temperature gradients are small then, we may assume linear distribution along the channel and write

$$p'_0 = p'_{av},\quad p'(z) = p'_{av}(1 + G_p z),\quad G_p = \frac{p'_2 - p'_1}{L p'_{av}}, \tag{2}$$

where $p'_{av} = (p'_2 + p'_1)/2$ and G_p is the dimensionless pressure gradient.

However, when the pressure ratio $p^*_2 = p'_2/p'_1$ is large, the pressure distribution along the channel is not linear. Also, the rarefaction parameter $\delta = \mathrm{Kn}^{-1}$ varies significantly along the channel. Here $\mathrm{Kn} = l'_g/R'_2$ is the Knudsen number, l'_g is the mean free path of gas molecules. For the hard-sphere model it depends on the local pressure and temperature according to [7]

$$\delta = \frac{R'_2 p'}{\eta'_g(T')}\sqrt{\frac{m'}{2k_B T'}}. \tag{3}$$

Here η'_g is the dynamic gas viscosity. Using the equation of continuity (the total mass flux is a constant) we introduce the new reduced mass flux whose value does not vary along the channel as [7]

$$J^*_M = \frac{L}{\pi(R'^2_2 - R'^2_1)p'_1} J'_M \sqrt{\frac{2k_B T'_1}{m'}}. \tag{4}$$

Linearize the distribution function as [8]

$$f(\mathbf{r}, \mathbf{C}) = f_0(C)\left(1 + G_p(z + h(\rho, \mathbf{C}))\right),\ f_0(C) = \pi^{-3/2}\exp\left(-C^2\right),$$

where $\zeta = \cos\psi$. The component u_z can be expressed in terms of the function $f(\mathbf{r}, \mathbf{C})$ as follows:

$$u_z = G_p U_z,\ U_z(\rho) = \frac{2}{\pi}\int_0^{+\infty} \exp(-C^2_\perp)C_\perp \int_{-1}^{1} \frac{1}{\sqrt{1-\zeta^2}} Z(\rho, C_\perp, \zeta)\, d\zeta\, dC_\perp, \tag{5}$$

$$Z(\rho, C_\perp, \zeta) = \frac{1}{\sqrt{\pi}}\int_{-\infty}^{+\infty} \exp(-C^2_z)C_z h(\rho, \mathbf{C})\, dC_z.$$

Then the dimensionless mass flux may be computed according to

$$J_M = G_p J_{M,p},\ J_{M,p} = \frac{4}{(1-R_1^2)} \int\limits_{R_1}^{1} U_z(\rho)\rho d\rho. \tag{6}$$

Substituting (1) and (6) in (4) and taking into account that $G_p = p_1 dp^*/(Ldz^*)$, we have

$$J_M^* dz^* = J_{M,p} dp^*, \tag{7}$$

where J_M^* is the parameter of the differential equation; $p^* = p'/p_1'$. At the inlet ($z^* = -1/2$) and outlet ($z^* = 1/2$) boundaries the pressure is equal to the inlet $p_1^* = 1$ and to the outlet p_2^* reservoirs pressure accordingly.

Expressing $p^* = p'/p_1'$ from (3) and substituting in (7), we will integrate the equation (7). As a result, we get

$$J_M^* = \int\limits_{1}^{p_2^*} J_{M,p} dp^* = \frac{1}{\delta_1} \int\limits_{\delta_1}^{\delta_2} J_{M,p} d\delta. \tag{8}$$

Using the expression of the local rarefaction parameter (3), the values $J_{M,p}(\delta)$ and the value of the parameter J_M^* from (8) we can numerically obtain a solution off the differential equation (7) with the boundary condition $p^*(-1/2) = 1$.

For restoration of $Z(\rho, C_\perp, \zeta)$ the equation takes the form [8]:

$$\left(\frac{\partial Z}{\partial \rho}\zeta + \frac{\partial Z}{\partial \zeta}\frac{(1-\zeta^2)}{\rho} \right) C_\perp + \delta Z(\rho, C_\perp, \zeta) + \frac{1}{2} = \delta U_z(\rho). \tag{9}$$

As the boundary conditions on the channel walls, we will use the diffuse reflection model. Then, for the function $Z(\rho, C_\perp, \zeta)$ boundary conditions can be written as

$$Z(1, C_\perp, \zeta) = 0,\ \zeta < 0, \quad Z(R_1, C_\perp, \zeta) = 0, \quad \zeta > 0. \tag{10}$$

We will obtain the solution to the equation (9) with the boundary conditions (10). We expand the unknown function $Z(\rho, C_\perp, \zeta)$ in the truncated series in the Chebyshev polynomials T_{k_i} ($k_i \le n_i$; $i = \overline{1,3}$) of the first kind defined on $\rho \in [0,\ 1]$, $C_\perp \in [0, +\infty)$ and $\zeta \in [-1,\ 1]$ as

$$Z(\rho, C_\perp, \zeta) = \mathbf{T_1}(x_1) \otimes \mathbf{T_2}(x_2) \otimes \mathbf{T_3}(x_3)\mathbf{A}, \tag{11}$$

where $x_1 = (2\rho - 1 - R_1)/(1 - R_1)$, $x_2 = (C_\perp - 1)/(C_\perp + 1)$, $x_3 = \zeta$, $\mathbf{T_i}$ is the $1 \times n_i'$ matrix ($n_i' = n_i + 1$, $i = \overline{1,3}$) as

$$\mathbf{T}_i(x_i) = \left(T_0(x_i)\, T_1(x_i) \ldots T_{n_i-1}(x_i)\, T_{n_i}(x_i)\right), \tag{12}$$

and $\otimes$ denotes the Kronecker product defined for two matrices $\mathbf{T_1}(x_1)$ and $\mathbf{T_2}(x_2)$ as [6]

$$\mathbf{T_1}(x_1) \otimes \mathbf{T_2}(x_2) = \left(T_0(x_1)T_0(x_2)\, T_0(x_1)T_1(x_2) \ldots T_{n_1}(x_1)T_{n_2-1}(x_2)\, T_{n_1}(x_1)T_{n_2}(x_2)\right), \tag{13}$$

and $\mathbf{A}$ is the unknown $n_1'n_2'n_3' \times 1$ matrix as

$$\mathbf{A} = \left(a_{000}\, a_{001} \ldots a_{n_1n_2n_3-1}\, a_{n_1n_2n_3}\right)^T. \tag{14}$$

Substituting (5) and (11) in (9), we have

$$\mathbf{B}(x_1, x_2, x_3)\mathbf{A} = -\frac{1}{2}, \tag{15}$$

$$\mathbf{T_1}(-1)\otimes\mathbf{T_2}(x_2)\otimes\mathbf{T_3}(x_3)\mathbf{A}=0,\quad x_2>0, \tag{16}$$

$$\mathbf{T_1}(1)\otimes\mathbf{T_2}(x_2)\otimes\mathbf{T_3}(x_3)\mathbf{A}=0,\quad x_2<0, \tag{17}$$

where

$$\mathbf{B}(x_1,x_2,x_3)=\frac{2}{1-R_1}\left(\frac{d\mathbf{T_1}(x_1)}{dx_1}\otimes\mathbf{T_2}(x_2)x_2+\right.$$
$$\left.+\mathbf{T_1}(x_1)\otimes\frac{d\mathbf{T_2}(x_2)}{dx_2}\frac{(1-x_2^2)}{x_1+(1+R_1)/(1-R_1)}\right)\otimes$$
$$\otimes\mathbf{T_3}(x_3)\frac{1+x_3}{1-x_3}+\delta\mathbf{T_1}(x_1)\otimes(\mathbf{T_2}(x_2)\otimes\mathbf{T_3}(x_3)-\mathbf{P_2}\otimes\mathbf{P_3}), \tag{18}$$

$$\mathbf{P_2}=\frac{2}{\pi}\int_{-1}^{1}\frac{\mathbf{T_2}(x_2)}{\sqrt{1-x_2^2}}dx_2=(2\,0\,\ldots 0\,0), \tag{19}$$

$$\mathbf{P_3}=2\int_{-1}^{1}\frac{1+x_3}{(1-x_3)^3}\mathbf{T_3}(x_3)\exp\left(-\frac{(1+x_3)^2}{(1-x_3)^2}\right)dx_3. \tag{20}$$

As the collocation points in the equation (15) we will use the extreme points of the polynomial $T_{n_1}(x_1)$ on the segment $[-1;1]$

$$x_{1,k_1}=\cos\left(\frac{\pi(n_1-k_1)}{n_1}\right),\quad k_1=\overline{0,n_1}, \tag{21}$$

and the roots of the polynomials $T_{n_2'}(x_2)$ and $T_{n_3'}(x_3)$ on the segment $[-1;1]$

$$x_{i,k_i}=\cos\left(\frac{\pi(2n_i-2k_i+1)}{2(n_i+1)}\right),\quad k_i=\overline{0,n_i},\ i=2,3. \tag{22}$$

Substituting (21) and (22) in (15), we arrive at the system of the $n_1'n_2'n_3'$ linear equations, in which we replace the equations with $x_{1,0}$ and x_{2,k_2} $(k_2=\overline{n_2'/2,n_2})$ by the equations following from the boundary condition (10) as

$$\mathbf{T_1}(-1)\otimes\mathbf{T_2}(x_{2,k_2})\otimes\mathbf{T_3}(x_{3,k_3})\mathbf{A}=0,\quad k_3=\overline{0,n_3}, \tag{23}$$

with x_{1,n_1} and x_{2,k_2} $(k_2=\overline{0,n_2'/2-1})$ by the equations following from the boundary condition (10) as

$$\mathbf{T_1}(1)\otimes\mathbf{T_2}(x_{2,k_2})\otimes\mathbf{T_3}(x_{3,k_3})\mathbf{A}=0,\quad k_3=\overline{0,n_3}. \tag{24}$$

Here and below, we assume that, n_2 is an odd number. To find the values of the Chebyshev polynomials at points (23) and (24), we use the geometric form: $T_{j_i}(x_i)=\cos(j_i\arccos x_i)$, where $x_i\in[-1,1]$ [5]. Then

$$T_{j_1}(x_{1,k_1})=\cos\left(\frac{\pi j_1(n_1-k_1)}{n_1}\right),\quad j_1,k_1=\overline{0,n_1},$$

$$T_{j_i}(x_{i,k_i})=\cos\left(\frac{\pi j_i(2n_i-2k_i+1)}{2(n_i+1)}\right),\quad j_i,k_i=\overline{0,n_i},\ i=2,3.$$

Now, if we consider the extreme points (23) of the polynomial $T_{n_1}(x_1)$, then we have the discrete orthogonality relation [4]

$$\frac{2}{n_1}\sum_{k_1=0}^{n_1}{}''\,T_{j_1}(x_{1,k_1})T_{l_1}(x_{1,k_1})=\gamma_{j_1}\delta_{j_1,l_1}, \tag{25}$$

where $\sum\limits_{k_1=0}^{n_1}{}''$ is meant the sum, in which the first and last terms are multiplied by 1/2, the points x_{1,k_1} $(k_1 = \overline{0, n_1})$ are given by (23),

$$\gamma_{j_i} = \begin{cases} 2, & j_i = 0 \vee (i = 1 \wedge j_i = n_i) \\ 1 & j_i > 0 \wedge (i = 1 \wedge j_i < n_i), \end{cases} \qquad \delta_{j_i,l_i} = \begin{cases} 1, & j_i = l_i, \\ 0, & j_i \neq l_i, \end{cases} \quad i = \overline{1,3}. \tag{26}$$

If we now consider the roots (24) of the polynomials $T_{n_i}(x_i)$ $(i = 2, 3)$, then we obtain

$$\frac{2}{n_i + 1} \sum_{k_i=0}^{n_i} T_{j_l}(x_{i,k_i}) T_{l_l}(x_{i,k_i}) = \gamma_{j_i} \delta_{j_i,l_i}, \ i = 2, 3, \tag{27}$$

$$\frac{2}{n_i + 1} \sum_{k_i=0}^{n_i}{}' T_{k_i}(x_{i,j_1}) T_{k_i}(x_{i,l_i}) = \delta_{j_1,l_1}, \ i = 2, 3, \tag{28}$$

where $\sum\limits_{k_i=0}^{n_i}{}'$ is meant the sum in which the first term is multiplied by 1/2.

The relations (25)-(28) allow as the $n_1' n_2' n_3' \times 1$ unknown matrix we will use

$$\mathbf{Z} = (Z_{000}, Z_{001} \ldots Z_{n_1 n_2 n_3 - 1} Z_{n_1 n_2 n_3})^T, \ Z_{k_1 k_2 k_3} = Z(\rho_{k_1}, C_{\perp,k_2}, \zeta_{k_3}), \tag{29}$$

In this case the matrix **A** depend on the unknown matrix **Z** as

$$\mathbf{A} = \varrho \mathbf{J_1} \otimes \mathbf{H_1} \otimes \mathbf{G_1 Z}, \quad \varrho = \frac{8}{n_1 n_2' n_3'}, \tag{30}$$

where $\mathbf{J_1}$ is the $n_1' \times n_1'$ matrix, in which

$$J_{1,i_1,i_2} = T_{i_1}(x_{1,i_2})/4, \ (i_1, i_2 = 0, n_1),$$

$$J_{1,i_1,i_2} = T_{i_1}(x_{1,i_2})/2, \ (i_1 = 0, n_1 \wedge i_2 \neq 0, n_1) \vee (i_2 = 0, n_1 \wedge i_1 \neq 0, n_1),$$

and $J_{1,i_1,i_2} = T_{i_1}(x_{1,i_2})$ for other values of the indices i_1 and i_2; $\mathbf{H_1}$ and $\mathbf{G_1}$ are the $n_k' \times n_k'$ matrices, respectively $(k = 2, 3)$, in which if $i_{1,k} = 0, n_k$ or $i_{2,k} = 0, n_k$ then

$$H_{1,i_{1,2},i_{2,2}} = T_{i_{1,2}}(x_{2,i_{2,2}})/2, \quad G_{1,i_{1,3},i_{2,3}} = T_{i_{1,3}}(x_{3,i_{2,3}})/2,$$

else $H_{1,i_{1,2},i_{2,2}} = T_{i_{1,2}}(x_{2,i_{2,2}})$, $G_{1,i_{1,3},i_{2,3}} = T_{i_{1,3}}(x_{3,i_{2,3}})$.

Taking into consideration (23)-(30), the system of the $n_1' n_2' n_3'$ linear equations can be written in the form:

$$\mathbf{LZ} = \mathbf{F}, \tag{31}$$

$$\mathbf{L} = \frac{2}{1 - R_1} \mathbf{B_1} + \delta \mathbf{I} - \delta \mathbf{B_2}, \ \mathbf{F} = -\frac{1}{2} \mathbf{E_{JH,1}} \otimes \mathbf{G_2},$$

$$\mathbf{B_1} = \mathbf{E_{JH}} \circ \left(\frac{2}{n_1} \mathbf{J_2 J_1} \otimes \mathbf{I_{d_2}} + \frac{2}{n_2'} \mathbf{I_{d_1}} \otimes (\mathbf{H_2 H_1}) \right) \otimes \mathbf{I_{d_3}},$$

$$\mathbf{B_2} = \frac{4}{n_2' n_3'} \mathbf{E_{JH}} \circ (\mathbf{I_1} \otimes \mathbf{E_H}) \otimes \mathbf{E_G}.$$

Here, $\circ$ denotes the Hadamard product of two matrices [6], $\mathbf{I_1}$ and $\mathbf{I}$ are $n_1' \times n_1'$ and $n_1' n_2' n_3' \times n_1' n_2' n_3'$ identity matrices, respectively; $\mathbf{I_{d_1}}$, $\mathbf{I_{d_2}}$ and $\mathbf{I_{d_3}}$ are $n_k' \times n_k'$ diagonal matrices $(k = \overline{1,3})$ as

$$I_{d_1,i_1,i_1} = \frac{1}{x_{1,i_1} + (1 + R_1)/(1 - R_1)}, \ I_{d_2,i_2,i_2} = x_{2,i_2}, \ I_{d_3,i_3,i_3} = \frac{1 + x_{3,i_3}}{1 - x_{3,i_3}};$$

and $\mathbf{E_{JH}}$ is $n_1'n_2' \times n_1'n_2'$ matrix such that the null rows have numbers l ($l = \overline{n_2'/2, n_2}$, $l = \overline{n_1 n_2 - n_2'/2 - 2, n_1 n_2}$) and all other rows consist of ones, $\mathbf{E_{JH,1}}$ is the first column of the matrix $\mathbf{E_{JH}}$, $\mathbf{J_2}$ and $\mathbf{H_2}$ are $n_k' \times n_k'$ matrices ($k = 1, 2$) respectively as

$$J_{2,i_1,j_1} = \frac{dT_{1,j_1}(x_{1,i_1})}{dx_1}, \ (i_1, j_1 = \overline{0, n_1}),$$

$$H_{2,i_2,j_2} = \frac{dT_{2,j_2}(x_{2,i_2})}{dx_2}(1 - x_{2,i_2}^2) \ (i_2, j_2 = \overline{0, n_2}),$$

and $\mathbf{E_H}$ and $\mathbf{G_2}$ are $n_2' \times n_2'$ and $n_3' \times 1$ matrices with ones; $\mathbf{E_G}$ is $n_3' \times n_3'$ matrix as

$$\mathbf{E_G} = \mathbf{G_2} \otimes (\mathbf{P_3 G_1}).$$

The system of the $n_1'n_2'n_3'$ linear equations (31) with unknown matrix $\mathbf{Z}$ has been solved by the LU-method. Based on the obtained elements of the matrix, we restore the dimensionless component of the mass velocity $U_z(\rho)$ and mass flux $J_{M,p}$:

$$U_z(\rho) = \varrho \mathbf{T_1}\left(\frac{2\rho - 1 - R_1}{1 - R_1}\right)\mathbf{J_1} \otimes \mathbf{E_{H,r}} \otimes \mathbf{E_{G,r}}\mathbf{Z}, \tag{32}$$

$$J_{M,p} = \frac{\varrho}{1 + R_1}\mathbf{P_1 J_1} \otimes \mathbf{E_{H,1}} \otimes \mathbf{E_{G,1}}\mathbf{Z}, \tag{33}$$

$$\mathbf{P_1} = \int_{-1}^{1} \mathbf{T_1}(x_1)\ (x_1(1 - R_1) + 1 + R_1)\, dx_1.$$

Here, $\mathbf{E_{H,r}}$ and $\mathbf{E_{G,r}}$ are first rows of the matrices $\mathbf{E_H}$ and $\mathbf{E_G}$ respectively.

3. Results and discussion

Numerical results for the dimensionless mass flux $J_{M,p}$ obtained by the LU-method using the formula (33) are presented figures 1 and 2 for $R_1 = 0.1$ and $R_1 = 0.5$ consideration, with $0 \le \delta \le 10$. For the sake of comparison, in the figures 1 and 2 we have given the values of the $J_{M,p}$ borrowed from [9] by the method of discrete velocities with the use of cubic splines. These are curves 2. It is seen that in figure 2 the agreement between the results for $R_1 = 0.5$ is good. The differences between the results for $R_1 = 0.5$ are due to the hydraulic diameter $2(R_2' - R_1')$ of the concentric annular channel has been taken as the dimensional length scale in [9]. In figures 3 and 4 we have given the distributions $p^*(z^*)$ obtained by the Runge-Kutta method for $\delta_1 = 0.1$ and $\delta_1 = 1$ with $R_1 = 0.1$ and $p_2^* = 10$. These are curves 1. From the figure 3 it can be seen that for $\delta_1 = 0.1$ the distribution $p^*(z^*)$ approaches the linear distribution. The results for $R_1 = 0.5$ are almost identical (figures 5 and 6).

Next, a comparison between the kinetic and free molecular solutions is performed. In the free molecular regime the equation (9) can be solved analytically. In this case we obtain

$$Z(\rho, C_\perp, \zeta) = -\frac{\rho\zeta + \sqrt{1 - \rho^2(1 - \zeta^2)}}{2C_\perp}, \quad -1 \le \zeta \le \frac{\sqrt{R_1^2 - \rho^2}}{\rho}, \tag{34}$$

$$Z(\rho, C_\perp, \zeta) = -\frac{\rho\zeta - \sqrt{R_1^2 - \rho^2(1 - \zeta^2)}}{2C_\perp}, \quad \frac{\sqrt{R_1^2 - \rho^2}}{\rho} < \zeta \le 1. \tag{35}$$

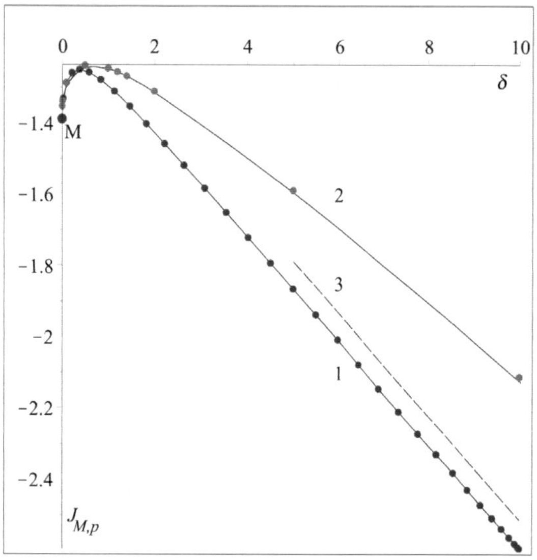

Figure 1. Dimensionless mass flux $J_{M,p}$ for $R_1 = 0.1$.

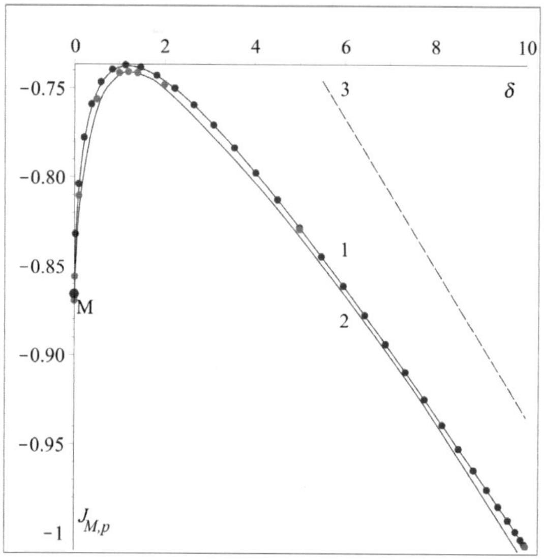

Figure 2. Dimensionless mass flux $J_{M,p}$ for $R_1 = 0.5$.

Substituting (34) and (35) in (5), we have

$$U_z(\rho) = \frac{1}{2\sqrt{\pi}} \left(\int\limits_{\frac{\sqrt{R_1^2-\rho^2}}{\rho}}^{1} \frac{\sqrt{R_1^2 - \rho^2(1-\zeta^2)}}{\sqrt{1-\zeta^2}} d\zeta - \int\limits_{-1}^{\frac{\sqrt{R_1^2-\rho^2}}{\rho}} \frac{\sqrt{1 - \rho^2(1-\zeta^2)}}{\sqrt{1-\zeta^2}} d\zeta \right). \tag{36}$$

We will integrate (8). In the figures 1 and 2, for $R_1 = 0.1$ and $R_1 = 0.5$, values of the free molecular mass flux $J_{M,p}$ shown by points M. Taking (3) and (36) into account, we obtain $J_M^* = (p_2^* - 1)J_{M,p}$ and the solution of the differential equation (7) with the boundary conditions $p^*(-1/2) = 1$ and $p^*(1/2) = p_2^*$ takes the form

$$p^*(z) = \frac{1 + p_2^*}{2} + (p_2^* - 1)z^*, \tag{37}$$

which corresponds to the expression (2). For comparison, let us consider the values of J_M^* found by means of the equation (31) using the formulas (8) and (33) at $\delta_1 = 0.1$ and the linearized Boltzmann equation for collisionless gas in the channels with $R_1 = 0.1$ and $R_1 = 0.5$ at $p_2^* = 10$, taking $J_M^* = (p_2^* - 1)J_{M,p}$ into account in this case. For $R_1 = 0.1$, these values are -11.3970 and -12.5010, while the corresponding values for $R_1 = 0.5$ are -6.7940 and -7.7922, respectively. It can be seen that for $\delta_1 = 0.1$ the obtained values is close to corresponding free molecular limits. From figures 3 and 5 it follows that for $\delta_1 = 0.1$ and $p_2^* = 10$ the kinetic solutions (curves 1) almost coincide free molecular ones (lines 2).

In the hydrodynamic flow regime ($\delta^{-1} \ll 1$) the gas flow is described by the Stokes equation. The analytical solution of this equation for U_z with the nonslip boundary conditions may be found in [10]

$$U_z(\rho) = \frac{\delta\,(\rho^2 - 1)}{4} \left(1 - \frac{\ln \rho}{\ln R_1}\right), \tag{38}$$

Therefore the function $Z(\rho, C_\perp, \zeta)$ take the form

$$Z(\rho, C_\perp, \zeta) = \frac{\delta\,(\rho^2 - 1)}{4}\left(1 - \frac{\ln\rho}{\ln R_1}\right). \tag{39}$$

Substituting (38) in (6), we obtain

$$J_{M,p} = \frac{\delta\,(R_1^2(1 - \ln R_1) - 1 - \ln R_1)}{4\ln R_1}. \tag{40}$$

The Stokes equation for U_z with the slip boundary conditions [12]

$$U_z(R_i) = \frac{\sigma_p}{\delta}\frac{\partial U_z}{\partial \mathbf{n}_{R_i}}(R_i),\ i = 1, 2, \tag{41}$$

can also be solved analytically

$$U_z(\rho) = U_{H,z}(\rho) + \sigma_p U_{s,z}(\rho), \tag{42}$$

$$U_{s,z}(\rho) = \frac{(1 + 2R_1^2\ln R_1 - R_1^3 + 2R_1\ln R_1 - R_1^2 + R_1)\ln\rho}{4R_1\ln^2 R_1} + \frac{1}{4}\frac{R_1^2 - 2\ln R_1 - 1}{\ln R_1}. \tag{43}$$

Here, $\mathbf{n}_{R_i}$ is the unit vector normal to cylinder with the radius R_i directed towards the gas $(i = 1, 2)$, $U_{H,z}(\rho)$ is the hydrodynamic solution (38) and $U_{s,z}(\rho)$ is the perturbation of the mass velocity and σ_p is the viscous slip coefficient. For the diffuse interaction of gas molecules the coefficient σ_p is given by 1.016 [12]. Substituting (42) in (6) and using the notations (40) and (43), we have

$$J_M = \frac{\delta\,(R_1^2(1 - \ln R_1) - 1 - \ln R_1)}{4\ln R_1} +$$

$$+\sigma_p\frac{4R_1^3\ln^2 R_1 - 4R_1^2\ln^2 R_1 - 4R_1^3\ln R_1 + R_1^4 + 4R_1\ln^2 R_1 + 4R_1\ln R_1 - 2R_1^2 + 1}{4R_1(R_1 - 1)\ln^2 R_1}. \tag{44}$$

In the figures 1 and 2, for $R_1 = 0.1$ and $R_1 = 0.5$, values of the slip mass flux $J_{M,p}$ shown by broken curves 3. By comparing the kinetic and the slip results for $R_1 = 0.5$, it is found that the relative error in the slip solution at $\delta = 5$ and 10 is 12.5% and 7.2% respectively. Also, as R_1 is decreased the discrepancy of the slip solution is decreased. Substituting (44) in the differential equation (7) for p^* with the boundary conditions $p^*(-1/2) = 1$ and $p^*(1/2) = p_2^*$, we can be solved analytically the boundary value problem. From figure 4 it can be seen that even for $\delta_1 = 1$ at $p_2^* = 10$ the agreement between the kinetic and slip solutions is good. This is due to the fact that for $R_1 = 0.1$ the Knudsen minimum of $J_{M,p}$ appears at about $\delta = 0.5$ (figure 1), while for $R_1 = 0.5$ the Knudsen minimum of $J_{M,p}$ appears at about $\delta = 1$ (figure 2). The value of J_M^* calculated from the formula (8) for $R_1 = 0.1$ at $\delta_1 = 1$ and $p_2^* = 10$ is -17.4534 and corresponds value -16.7507 obtained in the slip regime, while the values for $R_1 = 0.5$ are -7.6692 and -6.7123, respectively.

The pressure distributions agree qualitatively with the corresponding ones obtained in [12] for microchannels of triangular and trapezoidal cross sections and in [2] for a long cylindrical channel. They approach linear distributions for small values of δ_1 and then, as δ_1 is increased, they become nonlinear.

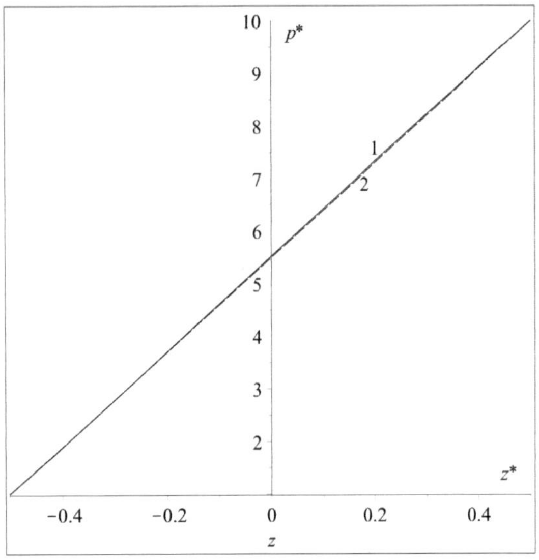

Figure 3. Pressure distribution along a concentric annular channel, with $R_1 = 0.1$, for $\delta_1 = 0.1$ and $p_2^* = 10$.

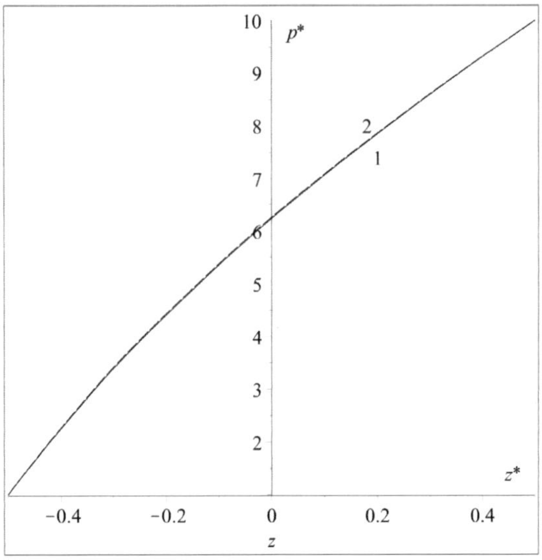

Figure 4. Pressure distribution along a concentric annular channel, with $R_1 = 0.1$, for $\delta_1 = 1$ and $p_2^* = 10$.

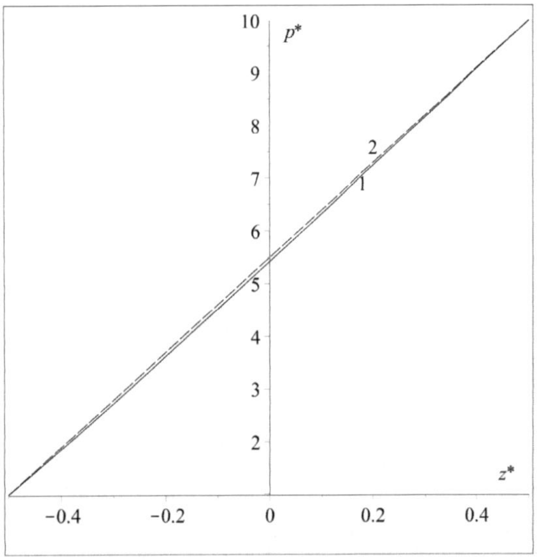

Figure 5. Pressure distribution along a concentric annular channel, with $R_1 = 0.5$, for $\delta_1 = 0.1$ and $p_2^* = 10$.

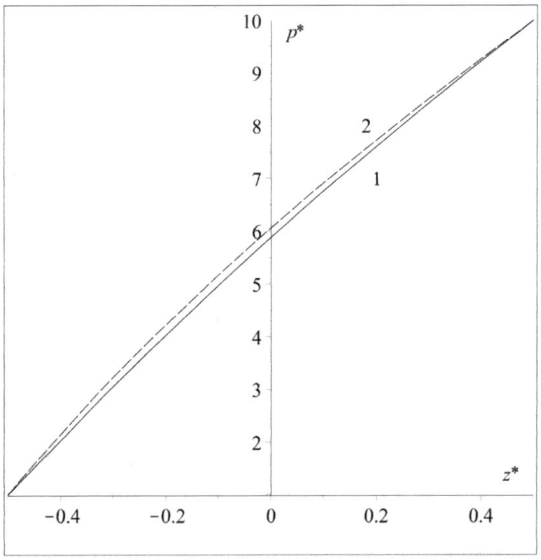

Figure 6. Pressure distribution along a concentric annular channel, with $R_1 = 0.5$, for $\delta_1 = 1$ and $p_2^* = 10$.

4. Conclusion

The flow of rarefied gas in a long concentric annular channel due to an imposed pressure gradient has been investigated implementing a kinetic approach. The linearized BGK model kinetic equation with the diffuse boundary conditions has been solved by the method Chebyshev approximation using the discrete orthogonality relations and Kronecker and Hadamard products. Results are provided for the in the pressure distributions through a long concentric annular channel. It has been found that in the transition regime ($\delta \geq 1$) the pressure distributions obtained on the basis of the BGK model of the kinetic equation at the large pressure drops at the channel ends approach the corresponding distributions in the slip regime. The validity and the accuracy of the kinetic results have been verified in several ways including the recovery of the well known solutions at the hydrodynamic and free molecular limits. The proposed methodology may be applied to other concentric channels, which are of some interest in several technological fields including nano- and micro-fluidics and vacuum technology.

References

[1] Ambrus V and Sharipov F and Sofonea V 2020 *Computers and Fluids* **211** 104637
[2] Germider O and Popov V 2020 *Fluid Dynamics* **55** 407
[3] Valougeorgis D and Vasileiadis N and Titarev V 2017 *European Journal of Mechanics - B/Fluids* **64** 2
[4] Mason J and Handscomb D 2003 *Chebyshev polynomials* (Florida: CRC Press)
[5] Bhatnagar P and Gross E and Krook M 1954 *Phys Rev* **94** 511
[6] Liu S and Trenkler G 2008 *International Journal of Information and Systems Sciences* **4** 160
[7] Sharipov F and Seleznev V 2008 *Motion of rarefied gases in channels and microchannels* (Yekaterinburg: UrO RAN)
[8] Germider O and Popov V 2019 *Siberian electronic mathematical reports* **16** 1947
[9] Breyiannis G and Varoutis S and Valougeorgis D 2008 *European Journal of Mechanics B/Fluids* **27** 609
[10] Landau L and Lifshitz E 1989 *Fluid Mechanics* (New York: Pergamon)
[11] Savaro C and Bonnet J and Johansson M, Perrier P and Graur I 2020 *European Journal of Mechanics - B/Fluids* **79** 44
[12] Ritos K and Lihnaropoulos Yia and Naris S and Valougeorgis D 2009 *2nd Micro and Nano Flows Conference. Brunel University. West London. UK.*

Simulation of a two-dimensional binary gas mixture outflow into vacuum through a thin slit on the basis of direct solution of the Boltzmann kinetic equation

S S Sitnikov[1,2], F G Tcheremissine[1,3] and T A Sazykina[2]

[1]Moscow Institute of Physics and Technology, Russia
[2]National Research Center "Kurchatov Institute", Russia
[3]Federal Research Center "Computer Science and Control", Dorodnicyn Computing Centre of RAS, Russia

E-mail: mailtosergeysitnikov@yandex.ru

Abstract. Two-dimensional binary gas mixture outflow from a vessel into vacuum through a thin slit is studied on the basis of direct solution of the Boltzmann kinetic equation. For evaluation of collision integrals in the Boltzmann equation a conservative projection method is used. Numerical simulation of a two-dimensional argon-neon gas mixture outflow from a vessel into vacuum was performed. Graphs of mixture components flow rate dependence on time during the flow formation, as well as fields of molecular density and temperature for steady-state regime, were obtained.

1. Introduction

Rarefied gas flow through a thin orifice is of importance in science and technology [1-4]. For one component gas, the problem has been studied rather well [5, 6]. The case of a gas mixture is examined much less [7]. At the same time gas mixture flows are important for engineering applications [8]. In this paper a two-dimensional binary gas mixture outflow into vacuum through a thin slit is considered.

Three regimes of flow are commonly distinguished with respect to gas rarefaction: free molecular, transitional and hydrodynamic. For free-molecular [9] and hydrodynamic [10] gas flow regimes, the solution to the problem is relatively easy to obtain analytically or by numerical simulation. In the case of transitional regime solving the problem is much more complicated. Gas flow in transitional regime can be studied using model kinetic equations, i.e. simplified Boltzmann equation, or direct simulation Monte Carlo (DSMC) technique [11]. Both of the above approaches have known limitations by applicability and accuracy of simulation. For instance, the uncertainty of model kinetic equations is estimated in [7] of about 5%. The DSMC technique is a powerful tool for solving many practical problems of rarefied gas dynamics. However, it is not so efficient for unsteady flows and flows with small Knudsen numbers. In our work an unsteady process of a gas mixture outflow into vacuum through a thin slit in a transitional regime is considered, which requires the direct solution of the Boltzmann kinetic equation to be obtained. The collision integrals are calculated with the conservative projection method [12, 13]. This method is conservative and ensures the exact equality of the collision integral of the Maxwellian distribution function to zero, and is also not affected by statistical noise that makes it possible to analyze the flow not only in a steady state regime, but during the flow formation as well. In addition, the implementation of the conservative projection method is relatively simple, and its computational complexity is quite acceptable for numerical simulation even in case of a binary mixture.

Thus, the aim of this work is to perform numerical simulation of a two dimensional problem of the outflow of a binary gas mixture through a slit into vacuum with good accuracy. Also, a study is made of the unsteady process of the flow formation.

2. Statement of the problem

A gas mixture is contained in a large vessel, which is surrounded by vacuum. The mixture consists of argon and neon gases in equal proportions, $n_1 = n_2, n_0 = n_1 + n_2$. The gas is in thermodynamic equilibrium with the temperature $T_0 = 273K$ and a given molecular density $n_0 = 10^{22} m^{-3}$. Molecular distribution functions of both components are Maxwellian by momentum variable. For the convenience of calculations, the vacuum space is modeled also by Maxwellian distributions of components with negligibly small densities. At some initial time, gas mixture outflow through the thin slit begins. The problem is two-dimensional in physical space with symmetry along the straight line traced from the middle of the slit width.

The problem setting scheme is shown in Figure 1. Here LPP_1L_1 is the vessel with binary gas mixture and PRR_1P_1 is the area of vacuum. Due to symmetry with the axis of symmetry MM', counting is performed only in the upper half-plane of MLRM'. Mirror conditions are set on the axis of symmetry MM'. On the walls PD, LP, LM diffuse boundary conditions are posed, and on PR and RM' – the condition for the absence of gas flowing into the region.

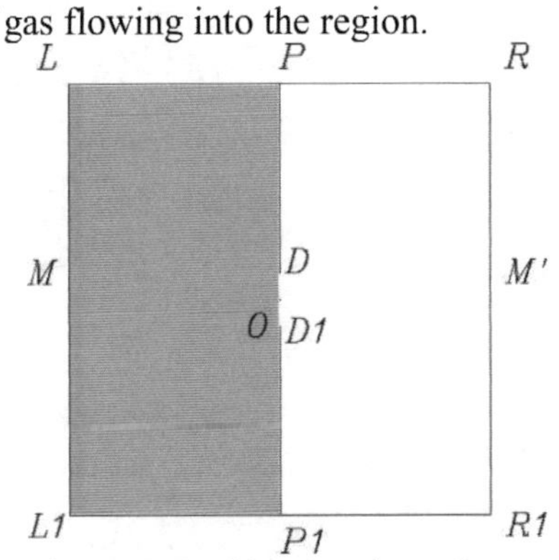

Figure 1. Problem setting scheme.

3. Mathematical model

To describe a behavior of a binary gas mixture, we introduce distribution functions for each component: $f_i(t,\mathbf{r},\mathbf{p}),\ i = 1,2$. Macroscopic quantities describing the state of each component of a mixture are expressed in terms of distribution functions as follows: molecular density

$$n_i(t,\mathbf{r}) = \int_{\mathbb{R}^3} f_i \, d^3 p, \quad (1)$$

average gas velocity

$$\mathbf{u}_i(t,\mathbf{r}) = \frac{1}{n_i m_i} \int_{\mathbb{R}^3} \mathbf{p}_i f_i \, d^3 p, \quad (2)$$

temperature

$$T_i(t,\mathbf{r}) = \frac{1}{3k_B n_i m_i} \int_{\mathbb{R}^3} (\mathbf{p}_i - m_i \mathbf{u}_i)^2 f_i \, d^3 p. \quad (3)$$

The evolution of the distribution functions in time is described by the Boltzmann kinetic equation [14]

$$\frac{\partial f_i}{\partial t}+\frac{\mathbf{p}_i}{m_i}\frac{\partial f_i}{\partial \mathbf{r}}=\sum_{j=1,2} I_{ij}, \quad i=1,2 \quad (4)$$

where $I_{ij}\left(t,\mathbf{r},\mathbf{p}_i\right)=\int_0^{2\pi} d\varphi \int_0^{b_{\max}} bdb \int_{\mathbb{R}^3} d^3p_j\left(f_i(t,\mathbf{r},\mathbf{p}_i')f_j(t,\mathbf{r},\mathbf{p}_j')-f_i(t,\mathbf{r},\mathbf{p}_i)f_j(t,\mathbf{r},\mathbf{p}_j)\right)g$ are the collision integrals written for interaction of the i-th and j-th gas components.

In this paper, the Lennard-Jones potential is used as an interaction potential of molecules

$$U(\mathbf{r})=4\varepsilon\left(\left(\frac{\sigma}{r}\right)^{12}-\left(\frac{\sigma}{r}\right)^{6}\right) \quad (5)$$

Parameters ε, σ for each component of the gas mixture are given in [15]. Parameters of the potential for interaction of different molecules are found using the combination relations [16]

$$\sigma_{ij}=\frac{\sigma_i+\sigma_j}{2}, \quad \varepsilon_{ij}\sigma_{ij}^{\ 6}=\sqrt{\varepsilon_i\sigma_i^{\ 6}\varepsilon_j\sigma_j^{\ 6}} \quad (6)$$

For convenience, dimensionless variables are used. Let the characteristic parameters $m_0,\sigma_0,\varepsilon_0$ be equal to those of argon, $n_0=n_1+n_2$, and $T_0=T_1=T_2$ be the initial temperature of the mixture. We define the characteristic velocity, time and mean free path of molecules as follows:

$$v_0=\sqrt{\frac{kT_0}{m_0}}, \tau=\frac{\lambda}{v_0}, \lambda=\frac{1}{\sqrt{2}\pi n_0\sigma_0^{\ 2}\Omega_1^{(2,2)}}, \quad (7)$$

where $\Omega_1^{(2,2)}=\Omega_1^{(2,2)}\left(\frac{kT_0}{\varepsilon_0}\right)$ (see [15]). Let us make the transition to dimensionless variables using the following scheme: $f\rightarrow\frac{f}{n_0(m_0v_0)^{-3}}$, $t\rightarrow\frac{t}{\tau}$, $\mathbf{p}\rightarrow\frac{\mathbf{p}}{m_0v_0}$, $m\rightarrow\frac{m}{m_0}$, $\mathbf{r}\rightarrow\frac{\mathbf{r}}{\lambda}$, $b\rightarrow\frac{b}{\sigma_0}$, $g\rightarrow\frac{g}{v_0}$. In dimensionless variables, system (4) takes the following form:

$$\frac{\partial f_i}{\partial t}+\frac{\mathbf{p}_i}{m_i}\frac{\partial f_i}{\partial \mathbf{r}}=\frac{1}{\sqrt{2}\pi\,\Omega_1^{(2,2)}}\sum_{j=1,2} I_{ij}, \quad i=1,2. \quad (8)$$

Here $I_{ij}=\int_0^{2\pi} d\varphi \int_0^{b_{\max}/\sigma_0} bdb \int_{\mathbb{R}^3} d^3p_j\left(f_i(\mathbf{p}_i')f_j(\mathbf{p}_j')-f_i(\mathbf{p}_i)f_j(\mathbf{p}_j)\right)g$.

4. Transition from mathematical model to numerical simulation

For the numerical solution of system (8), we use the method of splitting by physical processes [17]. The essence of this method lies in the fact that at each iteration in time, the advection equations are first solved

$$\frac{\partial f_i}{\partial t}+\frac{\mathbf{p}_i}{m_i}\frac{\partial f_i}{\partial \mathbf{r}}=0, \quad i=1,2, \quad (9)$$

and then we substitute the obtained intermediate values as initial conditions for the relaxation equations

$$\frac{\partial f_i}{\partial t}=\frac{1}{\sqrt{2}\pi\,\Omega_1^{(2,2)}}\sum_{j=1,2} I_{ij}, \quad i=1,2 \quad (10)$$

For the convenience of solving the advection equations, we apply the method of splitting along spatial directions [18], which allows to reduce the solution of the multidimensional advection equation to the sequential solution of one-dimensional advection equations along the coordinate axes.

The main difficulty for carrying out numerical simulation presents the solution of the relaxation equations (10). The conservative projection method [12, 13] is used to calculate the collision integrals on the right-hand side of (10), the calculation of which is resource-intensive.

To perform numerical simulation, we have developed a simulation environment software that uses the Message Passing Interface (MPI) [19] and OpenMP [20] code parallelization technologies. The use of code parallelization made it possible to speed up the computations up to 100 times.

5. Results

Numerical simulation of argon-neon gas mixture outflow for Knudsen numbers equal to $1.27\cdot10^{-1}, 5.99\cdot10^{-1}, 1.63, 16.6$ has been carried out. The grid step in momentum space was the same in all experiments and equaled 0.2. The radius of the sphere in momentum space, within which the distribution functions are assumed to be nonzero, was 5.7. A non-uniform grid was used in coordinate space to improve the accuracy of numerical simulations. The spatial grid step in the area of the slit was constant for each specific width of the slit and varied depending on geometry of the problem.

The flow establishing process was studied by analyzing the dependence of the output flow on time. Figure 2 represents time dependence of the output flow rate of gas mixture components for various Knudsen numbers. The flow rate is measured in dimensionless units $J_0 = n_0 \cdot v_0 \cdot \lambda^2$, and time is given in τ.

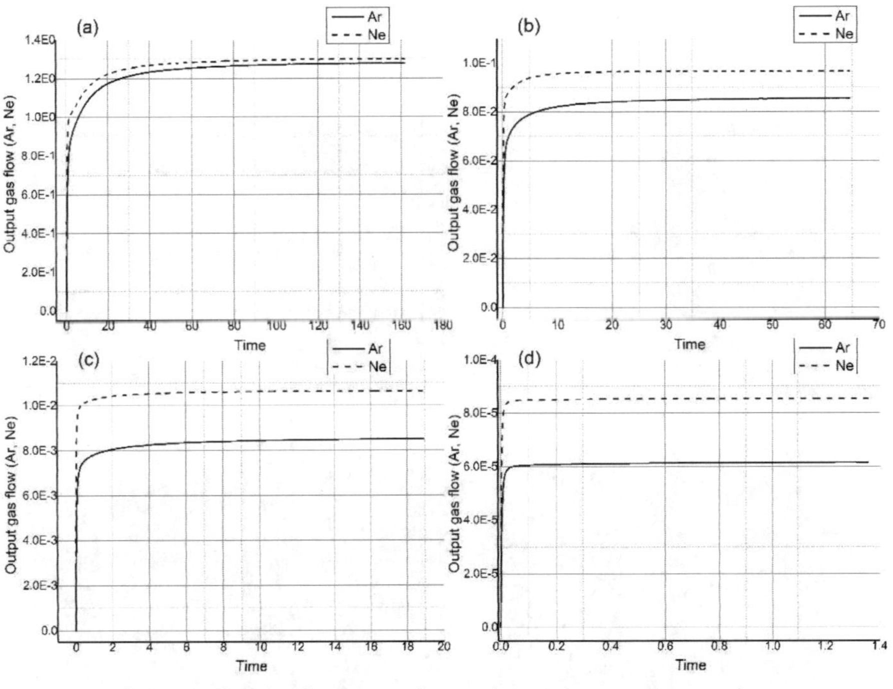

Figure 2. Output flow rate of gas mixture components on time,
$(a)\ Kn = 1.27\cdot10^{-1}$, $(b)\ Kn = 5.99\cdot10^{-1}$, $(c)\ Kn = 1.63$, $(d)\ Kn = 16.6$.

Figures 3-5 represent distributions of molecular density and temperature of gas mixture components near the slit in a steady-state flow for various flow regimes. It can be seen that all the fields substantially depend on the flow regime.

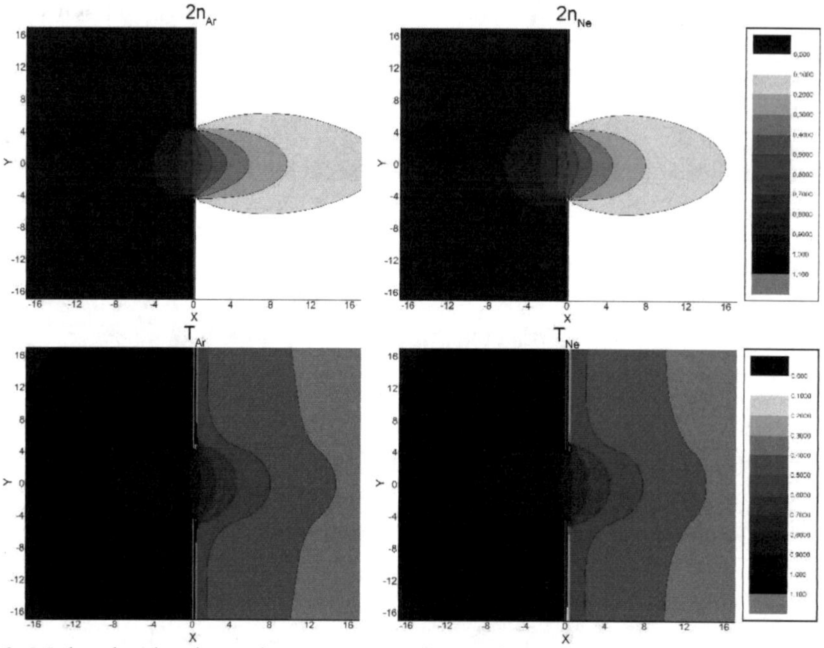

Figure 3. Molecular density and temperature of gas mixture components near the slit in the steady-state flow, $Kn = 1.27 \cdot 10^{-1}$.

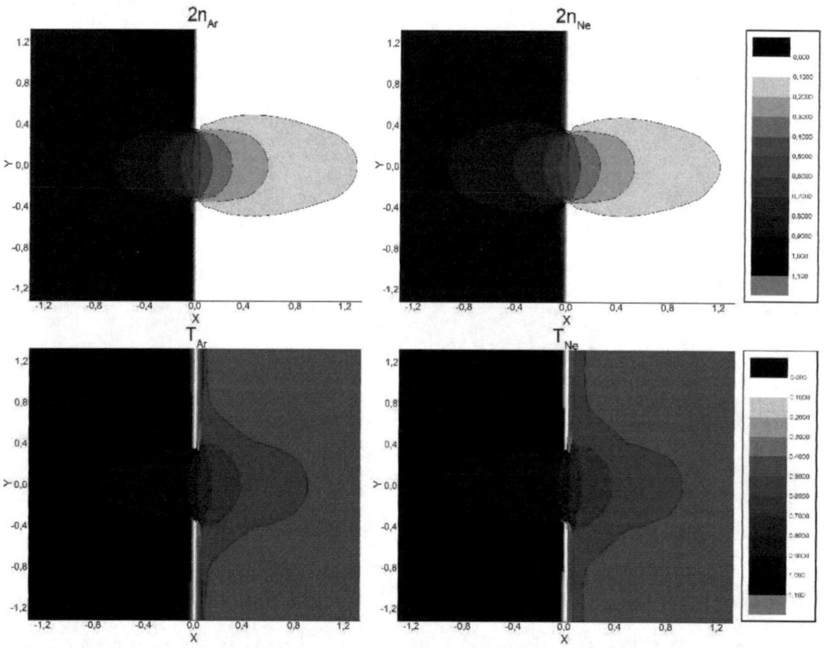

Figure 4. Molecular density and temperature of gas mixture components near the slit in the steady-state flow, $Kn = 1.63$.

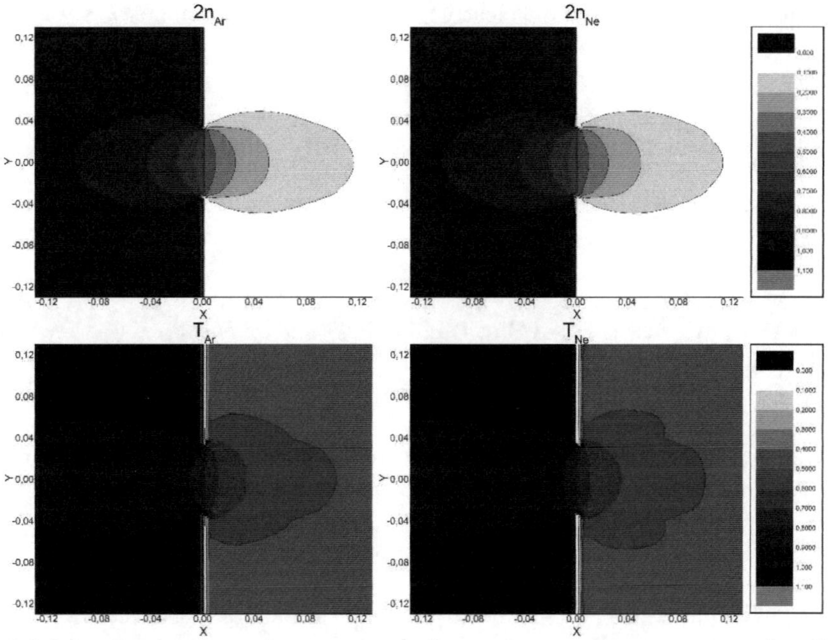

Figure 5. Molecular density and temperature of gas mixture components near the slit in the steady-state flow, $Kn = 16.6$.

6. Conclusion

Two-dimensional binary gas mixture outflow from a vessel into vacuum through a thin slit is studied on the basis of direct solution of the Boltzmann kinetic equation. Numerical simulation of a two-dimensional argon-neon gas mixture outflow from a vessel into vacuum in the transitional regime and also near hydrodynamic and near free-molecular ones was performed. Graphs of the gas mixture components flow rate dependence on time during the flow formation, as well as the fields of molecular density and temperature for the steady-state flow, were obtained. The obtained results may serve as a benchmark problem in rarefied gas dynamics [5].

This work was supported by the Russian Foundation for Basic Research, the Government of Krasnoyarsk krai, and Krasnoyarsk Regional Fund for Supporting Scientific and Technologies Activities project nos. 20-48-242918 “Computer modeling and analysis of gas-dynamic processes in multi-element thermoemission energy conversion systems.”

References

[1] Aktas O, Aluru N R and Ravaioli U 2001 Application of a parallel DSMC techique to predict flow characteristics in microfluidic filters *J. Microelectromech. Systems* **10** 538–49
[2] Jitschin W, Ronzheimer M and Khodabakhshi S 1999 Gas flow measurement by means of orifices and venturi tubes *Vacuum* **53** 181–5
[3] Wang M R and Li Z X 2004 *Microfluid Nanofluid* **1** 62
[4] Jamison A J, Ketsdever A D and Muntz E P 2002 Gas dynamic calibration of a nano-Newton thrust stand *Rev. Sci. Instrum.* **73** 3629–37
[5] Sharipov F Benchmark problems in rarefied gas dynamics 2012 *Vacuum* **86** 1697–700
[6] Sharipov F and Seleznev V 1998 Data on internal rarefied gas flows *J. Phys. Chem. Ref. Data* **27** 657-706
[7] Sharipov F 2017 Ab initio simulation of gaseous mixture ow through an orifice *Vacuum* **143** 106–18
[8] Sharipov F 2013 Gaseous mixtures in vacuum systems and microfluidics *J. Vac. Sci. Technol.* A **31**

[9] Rotenberg A and Weitzner H 1969 Nearly free flow through an orifice *Phys. Fluids* **12** 1573–81
[10] Alder G M 1979 The numerical solution of choked and supercritical ideal gas flow through orifices and convergent conical nozzles *J. Mech. Engng Sci.* **21** 197–203
[11] Bird G A 1976 *Molecular Gas Dynamics* (Oxford: Clarendon Press)
[12] Tcheremissine F G 1997 Conservative method for calculating the Boltzmann collision integral *Reports of the USSR Academy of Sciences* **357** 53–6 [in Russian]
[13] Tcheremissine F G 2015 *Introduction to the conservative projection method for calculating the collision integral and the method for solving the Boltzmann equation* (Moscow: MIPT) [in Russian]
[14] Kogan N M 1969 *Rarefied Gas Dynamics* (New York: Plenum)
[15] Hirschfelder J O, Curtiss C F and Bird R B 1954 *Molecular Theory of Gases and Liquids* (New York: John Wiley & Sons, Inc.)
[16] Ferziger J H and Kaper H G 1972 *Mathematical Theory of transport processes in gases* (Amsterdam: North-Holland)
[17] Aristov V V and Tcheremissine F G 1980 The conservative splitting method for solving of the Boltzmann equation *Computational Mathematics and Mathematical Physics* **20** 208–25 [in Russian]
[18] Samarskii A A and Vabishchevich P N 2001 *Additive schemes for problems of mathematical physics* (Moscow: Nauka) [in Russian]
[19] Message Passing Interface Forum. https://www.mpi-forum.org/ (last accessed 02/15/2020).
[20] The OpenMP API specification for parallel programming. https://www.openmp.org/ (last accessed 02/15/2020).

Analytical theory of high-speed nonequilibrium in a binary mixture of gases with a predominant light component

I V Demidov, M M Kuznetsov, Y D Kuleshova, A V Tikhonovets

Moscow Region State University, ul. Vera Voloshina, 24, Mytishchi, Moscow Region, 141014, Russia

Moscow Aviation Institute (National Research University), Volokolamskoye sh., 4, Moscow, 125993, Russia

E-mail: kuznets-omn@yandex.ru, juliaybogdanova@mail.ru, tikhonovets.a.v@gmail.com

Abstract. Analytical estimates are obtained that allow, first of all, to determine the fundamental conditions for the existence of a high-speed effect. In addition, an analytical assessment of the greatest value of this effect is given.

1. Introduction

The effect of high-speed nonequilibrium, discovered earlier in numerical studies, was directly related to the values of the distribution function of pairs of colliding molecules. The effect is that the value of this function inside the wave significantly exceeds its value behind the wave when the gas is in thermodynamic equilibrium [1].

In the works of the authors [2-5] this effect was studied purely analytically.

The main efforts of the authors in the implementation of this work were aimed at identifying the main dimensionless parameters of the problem, which determine both the condition for the existence of the effect and its largest value.

The predominance of the light component in the binary mixture is a necessary condition for such a study due to two factors:

1) as numerical studies have shown, the effect of high-speed disequilibrium becomes significant only if the concentration of the light component n_l prevails over the concentration of the heavy n_h. In most calculations, the concentration of the heavy component n_h was several percent of the concentration of the light component and less;

2) the constancy of temperatures and macrospeeds included in the bimodal Tamm – Mott-Smith distribution, namely, in the second "hot" Maxwell function associated with the values of these parameters in thermodynamic equilibrium behind the shock wave, is always valid only in a one-component gas. In mixtures of gases, however, it will not always be true, but only in the case of $n_l \gg n_h$ [6, 7].

The possibility of using a bimodal distribution at $n_l \gg n_h$ for each of the components of a binary mixture in its classical form allows us to fully use its analytical capabilities.

As it is shown in this paper this leads to the allocation of a bilinear symmetric form in the desired distribution function of pairs of molecules analytical representation. By applying the subsequent transformations associated with the bilinear form, it is possible to obtain the final result in a simple analytical form.

2. Distribution function of pairs of light and heavy components

As noted above, the distribution function of pairs of molecules $G^{(l,h)}$ is constructed based on molecules single-particle bimodal distributions for each component of the mixture

$$F^{(\gamma)} = \chi^{(\gamma)} F_0^{(\gamma)} + \left(1 - \chi^{(\gamma)}\right) F_1^{(\gamma)}, \quad (1)$$

$$\chi^{(\gamma)} = \frac{n_0^{(\gamma)}}{n_0^{(\gamma)} + n_1^{(\gamma)}} \quad (2)$$

$$F_i^{(\gamma)}(c) = \left(\frac{m_\gamma}{2\pi k T_i}\right)^{\frac{3}{2}} exp\left[-\frac{m_\gamma (c-u_i)^2}{2kT_i}\right] \quad (3)$$

Here, the superscript γ corresponds to one of the components of the mixture: light $\gamma = l$, or heavy $\gamma = h$, with $m_l < m_h$; the total volume concentrations of the mixture of components n_l and n_h are composed of their partial $n_0^{(\gamma)}$ and $n_1^{(\gamma)}$, where: $n_\gamma = n_0^{(\gamma)} + n_1^{(\gamma)}$; $\gamma = l, h$; single-particle functions $F_i^{(\gamma)}$ $(i = 0,1)$ are Maxwellian distributions; m_γ is the mass of the γ - component molecule; u_i and T_i are the velocity and temperature of the flow in front of $(i = 0)$ and behind $(i = 1)$ the wave; k is the Boltzmann constant; $(c - u_i)$ is the actual velocity of the molecule.

As in a simple one-component gas, the partial concentrations included in the numerator and denominator of formula (2) are functions of the x coordinate that varies in the wavelength. The partial parameters of the flow: the temperature T_i and the macrospeed u_i remain unchanged and coincide, respectively, with the thermodynamic temperature and the macrospeed of the flow at the entrance to the wave $(i = 0)$ and at the exit from it $(i = 1)$.

The pairs of molecules distribution function obtained on the basis of relations (1-3), as shown in [8], has the following form:

$$G^{(l,h)} = \chi^{(l)}\chi^{(h)}G_0^{(l,h)} + \chi^{(l)}\left(1-\chi^{(h)}\right)G_{01}^{(l,h)} + \left(1-\chi^{(l)}\right)\chi^{(h)}G_{10}^{(l,h)} + \left(1-\chi^{(l)}\right)\left(1-\chi^{(h)}\right)G_1^{(l,h)} \quad (4)$$

For the purposes of further discussion, we will focus a little more on the structure of the relation (4).

This relation is obtained by multiplying the single-particle distributions (1) with subsequent integration over the entire space of the velocities of both colliding molecules, except for the modulus of their relative velocity g (where $g = |\vec{c}_l - \vec{c}_h|$).

We show that formula (4) can be algebraically transformed to the form containing as one of the terms a bilinear symmetric form by the product of dimensionless coordinates $\chi^{(l)}, \chi^{(h)}$, and as the other – a linear form by these coordinates

$$G^{(l,h)} - G_1^{(l,h)} = -\left(G_{01}^{(l,h)} + G_{10}^{(l,h)} - G_1^{(l,h)} - G_0^{(l,h)}\right)\chi^{(l)}\chi^{(h)} + \left(G_{01}^{(l,h)} - G_1^{(l,h)}\right)\chi^{(l)} + \left(G_{10}^{(l,h)} - G_1^{(l,h)}\right)\chi^{(h)} \quad (5)$$

It is easy to see that in the right part of the formula (5), the negative term containing the product of dimensionless coordinates $\chi^{(l)}\chi^{(h)}$ is a symmetric bilinear form by permutation of these coordinates. The parameters: $G_0^{(l,h)}$, $G_{01}^{(l,h)}$, $G_{10}^{(l,h)}$, $G_1^{(l,h)}$ do not depend on the coordinates and are functions of the molecules relative velocity modulus g. The representation of single-particle functions in accordance with expression (1) can be understood as the decomposition of these functions into basis vectors consisting of a "cold" Maxwell distribution at the entrance to the shock wave $F_0^{(\gamma)}$ and a "hot" - $F_1^{(\gamma)}$ at the exit from it. In the bilinear term in the formula (5), each of the dimensionless parameters $G_0^{(l,h)}$, $G_{01}^{(l,h)}$, $G_{10}^{(l,h)}$, $G_1^{(l,h)}$ is the result of the action of the bilinear functional $A\left(e_i^{(\gamma)}, e_j^{(\gamma)}\right)$ on the corresponding pair of basis vectors: $e_i^{(\gamma)} = F_i^{(\gamma)}$, $e_j^{(\gamma)} = F_j^{(\gamma)}$. The bilinear functional in this problem, as already mentioned, is the integration of the mixture components single-particle functions over the velocity space of the different mixture components colliding molecules (with the exception of integration over the molecules relative velocity).

Thus, the pairs of molecules distribution function (5) depends parametrically on the molecules relative velocity modulus g and the coordinate in the shock wave "x". This coordinate is included in the partial dimensionless densities $\chi^{(l)}$ and $\chi^{(h)}$. The right part of the formula (5) depends on the variable arguments of the problem: the relative velocity g and the coordinates in the wave "x". In this case, the variable g is included only in the dimensionless parameters $G_0^{(l,h)}$, $G_{01}^{(l,h)}$, $G_{10}^{(l,h)}$, $G_1^{(l,h)}$; the coordinate "x" is only in the dimensionless densities $\chi^{(l)}$ and $\chi^{(h)}$. This is one of the advantages of the original analytical a priori representation of the single – particle distribution function (1) by Tamm-Mott-Smith.

We see that the transition to the distribution function of pairs of molecules also preserves this fundamental simplicity.

3. The centrality of the curve geometrically corresponding to the distribution function of pairs of molecules

To fulfill the goals set in the work, it is necessary to perform a number of transformations that are traditional for bilinear forms. First of all, we note that geometrically, relations (4) and (5) can be considered as equations of second-order curves in the coordinates $\chi^{(l)}$ and $\chi^{(h)}$. We show that by a linear transformation that translates the old coordinates $\chi^{(l)}$ and $\chi^{(h)}$ into the new χ_l and χ_h:

$$\chi^{(l)} = \chi_l + d_l,\ \chi^{(h)} = \chi_h + d_h, \qquad (6)$$

the curves of dependencies (4) and (5) will become symmetric in the new coordinates. This will be expressed in the fact that the left parts of the equations (4) and (5) will remain unchanged when the coordinates χ_l are replaced by $-\chi_l$, χ_h by $-\chi_h$. In this case, the center of symmetry position will be shifted by the distances d_l and d_h (in the old coordinates) relative to the origin of the old coordinates $\chi^{(l)} = \chi^{(h)} = 0$. The coordinates of the center (d_l, d_h) are determined from the condition of the absence of linear terms in equations (4) and (5) after switching to new coordinates in them.

As a result, we get

$$\overline{\Delta G}^{(l,h)} = -\chi_l\chi_h + \bar{g}_{01}^{(l,h)}\bar{g}_{10}^{(l,h)}, \quad (7)$$

where

$$\overline{\Delta G}^{(l,h)} = \frac{G^{(l,h)} - G_1^{(l,h)}}{G_{01}^{(l,h)} + G_{10}^{(l,h)} - G_1^{(l,h)} - G_0^{(l,h)}}, \qquad (8)$$

$$\bar{g}_{01}^{(l,h)} = \frac{G_{01}^{(l,h)} - G_1^{(l,h)}}{G_{01}^{(l,h)} + G_{10}^{(l,h)} - G_1^{(l,h)} - G_0^{(l,h)}}, \qquad (9)$$

$$\bar{g}_{10}^{(l,h)} = \frac{G_{10}^{(l,h)} - G_1^{(l,h)}}{G_{01}^{(l,h)} + G_{10}^{(l,h)} - G_1^{(l,h)} - G_0^{(l,h)}} \qquad (10)$$

$$d_h = \bar{g}_{01}^{(l,h)};\ d_l = \bar{g}_{10}^{(l,h)} \qquad (11)$$

Note that, as shown in [8], the equations (8-10) can be represented entirely in terms of the values $g_{01}^{(l,h)}, g_{10}^{(l,h)}, g_1^{(l,h)}$, where

$$g_{01}^{(l,h)} = G_{01}^{(l,h)} - G_1^{(l,h)};\ g_{10}^{(l,h)} = G_{10}^{(l,h)} - G_1^{(l,h)};\ g_1^{(l,h)} = G_1^{(l,h)} - G_0^{(l,h)} \qquad (12)$$

4. The region of existence of the high-speed overshoot effect

In the previous works of the authors [5, 8] necessary and sufficient conditions for the effect of high-speed overshoot were formulated.

The necessary conditions were reduced to the non-negativity of the expression $\Delta G_{ij}^{(l,h)}$, where $i,j = 0,1$, and

$$\Delta G_{ij}^{(l,h)} = (G_{01}^{(l,h)} + G_{10}^{(l,h)} - G_1^{(l,h)} - G_0^{(l,h)}) \geq 0, \qquad (13)$$

or

$$\Delta G_{ij}^{(l,h)} = g_{01}^{(l,h)} + g_{10}^{(l,h)} + g_1^{(l,h)} \qquad (14)$$

Sufficient conditions consisted in the non-negativity of each term in equation (14):

$$g_{01}^{(l,h)} \geq 0;\ g_{10}^{(l,h)} \geq 0;\ g_1^{(l,h)} \geq 0 \qquad (15)$$

Inequalities (15) can also be considered as the domain of an effect existence in a certain range of the molecules relative velocity modulus g values, since all the values standing in the left parts of the non-strict inequalities (15) are functions of g.

However, in the coordinate space $\chi^{(l)}$ and $\chi^{(h)}$, where $0 \leq \chi^{(l)} \leq 1$, $0 \leq \chi^{(h)} \leq 1$, such a region was not specified, unlike the one-component gas considered in [3].

To establish the boundaries of the desired area, we will return to the old coordinates $\chi^{(l)}$ and $\chi^{(h)}$ in the expression (7):

$$\overline{\Delta G}^{(l,h)} = -(\chi^{(l)} - \bar{g}_{10}^{(l,h)})(\chi^{(h)} - \bar{g}_{01}^{(l,h)}) + \bar{g}_{01}^{(l,h)}\bar{g}_{10}^{(l,h)}, \qquad (16)$$

It is easy to see that when the product of the "new coordinates" χ_l and χ_h is a negative value, the effect of high-speed overshoot $\overline{\Delta G}^{(l,h)} \geq 0$ always exists. In addition, it is greater than the value $\bar{g}_{01}^{(l,h)}\bar{g}_{10}^{(l,h)}$ in the original "old coordinates" $\chi^{(l)}$ and $\chi^{(h)}$.

Figure 1. The area of existence of the high-speed overshoot effect

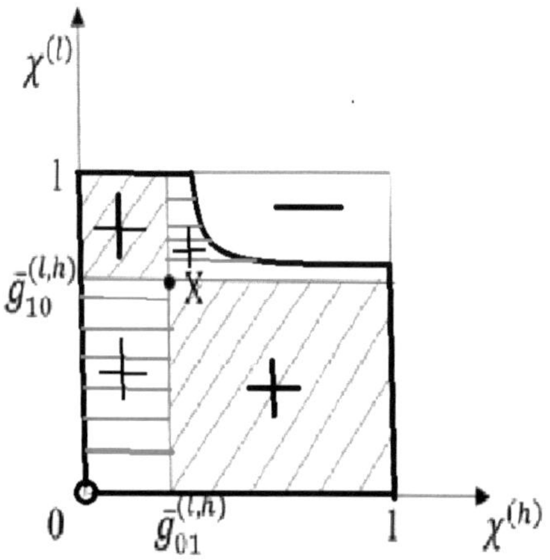

We transform the right-hand sides of equations (8-10) using the functions (12-14):

$$\overline{\Delta G}^{(l,h)} = \frac{\Delta G^{(l,h)}}{g_{01}^{(l,h)} + g_{10}^{(l,h)} + g_{1}^{(l,h)}}, \qquad (17)$$

$$\bar{g}_{01}^{(l,h)} = \frac{g_{01}^{(l,h)}}{g_{01}^{(l,h)} + g_{10}^{(l,h)} + g_{1}^{(l,h)}} < 1, \qquad (18)$$

$$\bar{g}_{10}^{(l,h)} = \frac{g_{10}^{(l,h)}}{g_{01}^{(l,h)} + g_{10}^{(l,h)} + g_{1}^{(l,h)}} < 1, \qquad (19)$$

where $\Delta \boldsymbol{G}^{(l,h)} = \boldsymbol{G}^{(l,h)} - \boldsymbol{G}_1^{(l,h)}$.

At the point "X" with coordinates $(\bar{\boldsymbol{g}}_{01}^{(l,h)}; \bar{\boldsymbol{g}}_{10}^{(l,h)})$ the overshoot effect $\overline{\boldsymbol{\Delta G}}^{(l,h)}$ is exactly equal to the product of $\bar{\boldsymbol{g}}_{01}^{(l,h)}\bar{\boldsymbol{g}}_{10}^{(l,h)}$. It has the same value on each of the lines: $\boldsymbol{\chi}^{(h)} = \bar{\boldsymbol{g}}_{01}^{(l,h)}, \boldsymbol{\chi}^{(l)} = \bar{\boldsymbol{g}}_{10}^{(l,h)}$.

In the fields: $(\boldsymbol{\chi}^{(h)} > \bar{\boldsymbol{g}}_{01}^{(l,h)}, \boldsymbol{\chi}^{(l)} < \bar{\boldsymbol{g}}_{10}^{(l,h)})$ and $(\boldsymbol{\chi}^{(h)} < \bar{\boldsymbol{g}}_{01}^{(l,h)}, \boldsymbol{\chi}^{(l)} > \bar{\boldsymbol{g}}_{10}^{(l,h)})$ the value $\overline{\boldsymbol{\Delta G}}^{(l,h)}$ will be greater than $\bar{\boldsymbol{g}}_{01}^{(l,h)}\bar{\boldsymbol{g}}_{10}^{(l,h)}$.

In the fields: $(\boldsymbol{\chi}^{(h)} > \bar{\boldsymbol{g}}_{01}^{(l,h)}, \boldsymbol{\chi}^{(l)} > \bar{\boldsymbol{g}}_{10}^{(l,h)})$ and $(\boldsymbol{\chi}^{(h)} < \bar{\boldsymbol{g}}_{01}^{(l,h)}, \boldsymbol{\chi}^{(l)} < \bar{\boldsymbol{g}}_{10}^{(l,h)})$ the value $\overline{\boldsymbol{\Delta G}}^{(l,h)}$ will be less than $\bar{\boldsymbol{g}}_{01}^{(l,h)}\bar{\boldsymbol{g}}_{10}^{(l,h)}$.

At the point $(\boldsymbol{\chi}^{(h)} = \boldsymbol{0}, \boldsymbol{\chi}^{(l)} = \boldsymbol{0})$, i.e. at the origin of the coordinates, the effect is completely absent, $\overline{\boldsymbol{\Delta G}}^{(l,h)} = \boldsymbol{0}$. The effect is also absent on the line with the equation: $-\boldsymbol{\chi}^{(h)}\boldsymbol{\chi}^{(l)} + \bar{\boldsymbol{g}}_{01}^{(l,h)}\boldsymbol{\chi}^{(l)} + \bar{\boldsymbol{g}}_{10}^{(l,h)}\boldsymbol{\chi}^{(h)} = \boldsymbol{0}$. This curve is drawn in Figure 1 and is located in the upper-right rectangle. Above and to the right of this curve is the function $\overline{\boldsymbol{\Delta G}}^{(l,h)} < \boldsymbol{0}$.

5. Estimation of the highest value of the high-speed overshoot effect

The equation (16) obtained from the expression (5) for the pairs of molecules distribution function $\boldsymbol{G}^{(l,h)}$ also allows us to estimate the largest value of the overshoot effect.

Indeed, as follows from the equation (16), a positive addition to the main value of the effect: $\overline{\boldsymbol{\Delta G}}^{(l,h)} = \bar{\boldsymbol{g}}_{01}^{(l,h)}\bar{\boldsymbol{g}}_{10}^{(l,h)}$, can only be written in the following variants:

a) $(\bar{g}_{01}^{(l,h)} - \chi^{(l)})(\chi^{(h)} - \bar{g}_{10}^{(l,h)})$;

b) $(\chi^{(l)} - \bar{g}_{01}^{(l,h)})(\bar{g}_{10}^{(l,h)} - \chi^{(h)})$.

In each of these variants, we will find the largest values of each of the multipliers and multiply them:

a) $\left(\bar{g}_{01}^{(l,h)} - \chi^{(l)}\right)\Big|_{sup} \times \left(\chi^{(h)} - \bar{g}_{10}^{(l,h)}\right)\Big|_{sup} = \left(\bar{g}_{01}^{(l,h)} - 0\right)\left(1 - \bar{g}_{10}^{(l,h)}\right) = \bar{g}_{01}^{(l,h)} - \bar{g}_{01}^{(l,h)}\bar{g}_{10}^{(l,h)}$; (20)

b) $\left(\chi^{(l)} - \bar{\boldsymbol{g}}_{01}^{(l,h)}\right)\Big|_{sup} \times \left(\bar{\boldsymbol{g}}_{10}^{(l,h)} - \chi^{(h)}\right)\Big|_{sup} = \left(1 - \bar{\boldsymbol{g}}_{01}^{(l,h)}\right)\left(\bar{\boldsymbol{g}}_{10}^{(l,h)} - 0\right) = \bar{\boldsymbol{g}}_{10}^{(l,h)} - \bar{\boldsymbol{g}}_{01}^{(l,h)}\bar{\boldsymbol{g}}_{10}^{(l,h)}$. (21)

In each of these variants, we will find the largest values of each of the multipliers and multiply them:

a) $\overline{\Delta G}_{sup}^{(l,h)} = \bar{g}_{01}^{(l,h)} - \bar{g}_{01}^{(l,h)}\bar{g}_{10}^{(l,h)} + \bar{g}_{01}^{(l,h)}\bar{g}_{10}^{(l,h)} = \bar{g}_{01}^{(l,h)}$; (22)

b) $\overline{\Delta G}_{sup}^{(l,h)} = \bar{g}_{10}^{(l,h)} - \bar{g}_{01}^{(l,h)}\bar{g}_{10}^{(l,h)} + \bar{g}_{01}^{(l,h)}\bar{g}_{10}^{(l,h)} = \bar{g}_{10}^{(l,h)}$. (23)

Thus, the expressions (22) and (23) give an estimate of the highest value of the high-speed overshoot effect $\overline{\Delta \boldsymbol{G}}_{sup}^{(l,h)}$:

1) if $\bar{\boldsymbol{g}}_{01}^{(l,h)} \geq \bar{\boldsymbol{g}}_{10}^{(l,h)}$, then $\bar{\boldsymbol{g}}_{10}^{(l,h)} \leq \overline{\Delta \boldsymbol{G}}_{sup}^{(l,h)} \leq \bar{\boldsymbol{g}}_{01}^{(l,h)}$;

2) if $\bar{\boldsymbol{g}}_{01}^{(l,h)} \leq \bar{\boldsymbol{g}}_{10}^{(l,h)}$, then $\bar{\boldsymbol{g}}_{01}^{(l,h)} \leq \overline{\Delta \boldsymbol{G}}_{sup}^{(l,h)} \leq \bar{\boldsymbol{g}}_{10}^{(l,h)}$.

In Figure 1, each of the values: $\bar{\boldsymbol{g}}_{01}^{(l,h)}$, $\bar{\boldsymbol{g}}_{10}^{(l,h)}$ corresponds to its own pair of coordinates at which they are reached. So, the values of $\bar{\boldsymbol{g}}_{01}^{(l,h)}$ correspond to $\boldsymbol{\chi}^{(h)} = \mathbf{1}$ and $\boldsymbol{\chi}^{(l)} = \mathbf{0}$, i.e. the point (1;0). The value of $\bar{\boldsymbol{g}}_{10}^{(l,h)}$ is the point (0;1).

Conclusion

The paper shows that in the case of a strong predominance of the concentration of a light component over the concentration of a heavy one, the pairs of molecules distribution function can be represented as the sum of a bilinear symmetric shape and a constant value, which is the "average measure" of the overshoot effect.

Thanks to the selection of a symmetric bilinear form in the analytical representation of the distribution function of pairs of molecules, it was possible to determine both the area of the overshoot effect existence and to estimate the greatest value of this effect. It is important to note that these significant conclusions are obtained analytically. They do not require obtaining a solution to the Boltzmann equation for a mixture of gases in a shock wave.

Acknowledgments

This study is partially supported by Russian Foundation for Basic Researches, grant 20-07-00740_A.
This study is partially supported by a Grant of President of The Russian Federation to support young scientists - candidates of sciences MK-1330.2020.9.
The part of the study was carried out within the state assignment of the Ministry of Science and Higher Education of the Russian Federation (theme FSFF-2020-0013).

References

[1] Genich A P, Kulikov S V, Manelis G B, Chereshnev S L, Izvestiya AN 1990 *SSSR Mekhanika Zhidkosti i Gaza* (Bulletin of Academy of Sciences of USSR, Mechanics of Fluid and Gas). (2) 144

[2] Kuznetsov M M, Kuleshova Ju D and Smotrova L V 2012 *Bulletin of the Moscow RegionState University. Series: Physics and Mathematics* (2) 108-115

[3] Kuznetsov M M, Kuleshova Yu D 2012 *Heat Transfer Research* (3) 228–236.

[4] Kuznetsov M M and Smotrova L V 2013 *Moscow RegionState University. Series: Physics and Mathematics* (3) 66-73

[5] Kuznetsov M M, Kuleshova Ju D, Reshetnikova Yu G, Smotrova L V 2018 *Journal of Physics: Conference Series*, 996 012006.

[6] Oberai M M 1965 *The physics of fluids* (8) 826-833

[7] Oberai M M 1966 *The physics of fluids* (9) 16-37

[8] Kuznetsov M M, Kuleshova Ju D and Smotrova L V 2016 *Moscow RegionState University. Series: Physics and Mathematics* (3) 84-95

Semiclassical Qubits

T F Kamalov[1,2]
[1]Theoretical Physics Department, Moscow Region State University , Moscow Region, Mytishchi 141014, Russia

[2]Theoretical Physics Department, Moscow Institute of Physics and Technology (State University), Moscow Region, Dolgoprudny, 141700, Russia
E-mail: timkamalov@gmail.com

Abstract. The semiclassical approximation of quantum computing and quasi-qubits (s-bits) have been obtained by us as a result of our work over the past few years. This work can be conventionally divided into two parts. The first part, let's call it the programming model, contains a computer model of quasi-qubits and quantum computing. The second part, let's call it the microelectronic model, describes the microelectronic realization of qubits in the semiclassical approximation (quasi-qubits) and exists in the form of block diagrams, which are supposed to be easy to manufacture. How did we get the semiclassical approximation? The difficulty in solving such a problem was that microparticles in quantum mechanics are described in an infinite-dimensional Hilbert space. Classical models are much poorer in the number of variables; therefore, it is impossible to describe quantum mechanical objects by classical methods due to the small number of available parameters.

1. Introduction. Physics of Non-Inertial Reference Frames

Our way is to supplement the description of the classical model with nonlocal variables, which refine the behavior of qubits in the semiclassical approximation [1,2]. Dwelling on additional nonlocal variables in more detail, we note that they represent a decreasing series and therefore we can set any accuracy of modeling qubits in the semiclassical approximation based on the technical capabilities of the classical computer or microelectronic elements [3] used in the work for the second way of implementing this problem.

Changing the number of nonlocal variables depending on the required computational accuracy and based on our technical capabilities, we get a different computation speed. Note that it is rather difficult to distinguish the semiclassical approximation of qubits (quasi-qubits) from qubits, as well as their comparative speed of calculations, since for example, such differences described in Bell's theorem were experimentally obtained by A. Aspect, who received the Nobel Prize for these results. The maximum values of Bell's observable in these two cases differed by a factor of 1.4 and were difficult to detect due to the instability of the behavior of quantum objects, the states of which can change under any random small influences.

The main goal of this work was to create working semiclassical qubits, which at the first stage can replace a quantum computer with a computational speed comparable to that of a quantum computer.

Why do we need a semiclassical implementation of quantum computing and quantum cryptographic processes? To search for problems that can be solved using qubits. In solving a whole class of problems, the use of qubits will not give any advantages over classical methods. Finding problems that give a temporary gain when using qubits is now the most important task. The classical methods known today do not allow satisfactory modeling of quantum computing and quantum cryptographic processes due to the insufficient number of variables in computer experiments. Incomplete understanding of the work of qubits in a quantum computer prevents hopes for their wide application in practice, with the exception of a limited range of problems. The semiclassical realization of qubits in a quantum computer is necessary to construct the quantum algorithms needed for this. The semiclassical implementation allows us to take the next step towards understanding quantum calculations by working with the available classical elemental base, despite the existing limited resources available to us.

The proposed semiclassical implementation allows a more complete description of the operation of qubits through the use of additional non-local variables. Difficulties in understanding quantum processes are known to be associated with a small number of variables in the classical model compared to the quantum one. Supplementing the classical model with new variables in the form of nonlocal variables makes it possible to create a semiclassical implementation and understand the structure of qubits and their connections. Ultimately, this leads to an increase in the accuracy and speed of information processing and transmission, to the development of a promising element base for quantum computers and quantum-cryptographic communication systems.

Applying the WKB method to approximate qubits, you can get a semiclassical model with an accuracy that we can set ourselves based on our technical capabilities. As nonlocal variables complementing the description of qubits, we used the time derivatives of the generalized coordinates $q(t)$ i.e., $q^{(k)}(t)$, where k is the order of the derivative. In our case, these variables are non-local, that's why.

Modern physics is the physics of predominantly inertial reference systems. At the same time, it is impossible to find a non-inertial frame of reference in reality, because there are always small random fields and forces. Therefore, one can agree with the existence of inertial reference frames only in theory, extending the postulates of classical mechanics to the case of non-inertial reference frames. Such physics is the physics of non-inertial frames of reference. To do this, we introduce a coordinate transformation in such reference systems.

Conversion of coordinates of a point particle between two Non-Inertial Reference Frames [3] provided τ is a time interval for averaging, shall be expressed as

$$Q = q(t) + \dot{q}(t)\tau + \Delta q(t), \tag{1}$$

$$\Delta q(t) = \sum_{k=2}^{N} (-1)^k \frac{1}{k!} \tau^k q^{(k)}(t).$$

And same holds for momentum

$$P = p(t) + \dot{p}(t)\tau + \Delta p(t),$$

$$\Delta p(t) = \sum_{k=2}^{N} (-1)^k \frac{1}{k!} \tau^k p^{(k)}(t)$$

$$Q =< q(t) >= \frac{1}{2}[q(t+\tau) + q(t-\tau)].$$

Here, $\Delta q(t)$ and $\Delta p(t)$ are remainder terms of the Taylor expansion. The remainder terms $\Delta q(t)$ and $\Delta p(t)$ in Non-Inertial Reference Frame may be interpreted as uncertainties of coordinate and momentum of a point particle in this reference frame. In quantum mechanics, uncertainties of coordinate and momentum of a micro particle obey to the rule

$$\Delta q(t)\Delta p(t) \geq \hbar/2.$$

In Non-Inertial Physics can be introduced an General Uncertainty Relation, as there always exist random small fields and forces influencing either the very system to be described or an observer, that is

$$\left[\sum_{k=2}^{N}(-1)^k \frac{1}{k!}\tau^k q^{(k)}(t)\right] \left[\sum_{k=2}^{N}(-1)^k \frac{1}{k!}\tau \;\; p^{\;(k\;)}(t\;)\right] \geq \hbar/2$$

inertial one. In this case instability states of observing objects can be describe by

Quantum Theory with high-order derivatives $\psi(t) = |q, \dot{q}, \ddot{q}, \dddot{q}, \ldots, q^{(n)}\rangle = |Q(t)\rangle$. Note that here the time derivatives of the generalized coordinates q play the role of non-local variables, since they do not change from one point to another point in a non-inertial reference frame for a body of any mass.

2. Ostrogradsky Formalism and Wentzel–Kramers–Brillouin (WKB) approximation for Qubits

Classical Euler-Lagrange equations in Ostrogradsky formulation [5] accounting for random small influences in the form of higher derivatives will take on the form

$$\delta S = \delta \int L_r(q, \dot{q}, \ddot{q}, \dddot{q}, \ldots, \dot{q}^{(n)}, \ldots)dt = \int \sum_{n=0}^{N} (-1)^n \frac{d^n}{dt^n} \frac{dL_r}{drq^{(n)}} \delta rq^{(n)} dt = 0$$

or

$$\frac{\partial L_r}{\partial q}-\frac{d}{dt}\frac{\partial L_r}{\partial \dot{q}}+\frac{d^2}{dt^2}\frac{\partial L_r}{\partial \ddot{q}}-\frac{d^3}{dt^3}\frac{\partial L_r}{\partial \dddot{q}}+...+(-1)^n\frac{d^n}{dt^n}\frac{\partial L_r}{\partial q^{(n)}}+...=0 \qquad (2)$$

This equation with instability [6] of high-order derivatives coordinates on time can be rewritten as [7,8]

$$F-ma+\tau m\dot{a}-\frac{1}{2!}\tau^2 ma^{(2)}+....+(-1)^n\frac{1}{n!}\tau^n ma^{(n)}+...=0$$
$$\dot{a}(t_0)=0,\ddot{a}(t_0)=0,...,a^{(n)}(t_0)=0,... \qquad (3)$$

τ being the time interval, and a – acceleration, t_0 is the moment of the time of stability trajectory. The second Newton's law is a second-derivative differential equation describing stability dynamics

$$F-ma=0\,.$$

In mathematical physics WKB theory is a method for approximating the solution of a differential equation whose highest derivative is multiplied by a small parameter. Applying this method for formula (3), we obtain the solution with terms $S_n(t)$ in the expansion in $\delta \to 0$

$$\sum_{n=0}^{\infty}(-1)^n m_n(t)\frac{d^n}{dt^n}q=0$$
$$q(t)\sim exp\left[\frac{1}{\delta}\sum_{n=0}^{n=\infty}\delta^n S_n(t)\right].$$

3. Algorithm for semiclassical Qubits (s-bits)

Using a random number generator, the random process $\varphi(t)$ with $\langle\varphi(t)\rangle=0$ has been given the random phase on the angle 0÷2π. The algorithm of a random s-bits [8] is

$$f(t,\theta)=sign\{\varphi(t)+\theta\}$$

a being θ means measurement angle. Then $\langle f(\theta)\rangle=0$ and $f(\theta\pm\pi)=-f(\theta)$, means $f(\theta\pm\pi)$ and $f(\theta)$ are anticorrelated. Here we note that $f(\theta)$ and $f(\theta\pm\pi/2)$ are non-correlated s-bits.

It is very important that if, for example, the first s-bit is used for the process of s-bits initialization as signal s- bit, and during the computation process on the second s-bit the changes, then we must also change the state of the signal s-bit. Then the calculations on the second s-bit will affect the calculations of the others of the s-bits.

The implemented program of semiclassical computer with s-bits is posted on the website https://theorphys.org/.

4. Conclusion

Additional terms in the form of higher derivatives have non-local character, which enables their employment for description the non-locality of quantum mechanics. Random and instability variables in the form of high-order derivatives coordinates on the time can describe the effect of quantum correlations and non-locality of quantum states in Non-Inertial Reference Frames.

References

[1] Kamalov T F 2011 *Quantum Computers and Computing* 11 p 52 (*Preprint arXiv*: 1011.4256)
[2] Rybakov Y P, Kamalov T F 2016 *Int. J. of Theor. Phys.* **55** 9 4075
[3] García Zavala Y M et al 2011 *CIENCIA ergo sum* **18** 2 p 171
[4] Kamalov T F 2020 *J. Phys.: Conf. Ser.* 1557(1) 012003
[5] Ostrogradsky M V 1850 *Mem. Acad. St. Peterburg* **6** 4 p 385
[6] Woodard R P 2007 *Lecture Notes in Physics* **720** 403 (*Preprint arXiv:*astro-ph/0601672)
[7] Kamalov T F 2020 *Symmetry* **12** 1 p 63
[8] Kamalov T F, Rybakov Y P 2006 *Quantum Computers and Computing* **6** 1 p 125

Physics of nonlinear oscillations with nonlocal variables

O A Volkova[1], M H Khamis Hassan[1], T F Kamalov[1,2]

[1] Department of Theoretical Physics, Moscow Region State University, Moscow Region, Mytishchi 141014, Russia;
[2] Moscow Institute of Physics and Technology, Dolgoprudny, Moscow Region, Russia, 141701, Dolgoprudny, Russia

Abstract. In cases where physical processes cannot be described by linear equations, and nonlinear equations are difficult to solve mathematically, we have to use approximate solutions to such problems. One such example is the description of the Kapitsa pendulum, which is a pendulum with a vibrating suspension point. In contrast to the previously known methods of describing such a problem, in this paper we propose to use additional variables in the form of higher derivatives, which allows us to obtain corrections that give a more detailed contribution to the description of this problem.

1. Introduction

We will consider the physics of nonlinear oscillations on the example of the Kapitsa pendulum. As nonlocal variables, we will use them in the form of higher derivatives. A detailed experimental study of a pendulum with dynamic stability in an inverted position is described in the works of P. L. Kapitsa [1].

The Kapitsa pendulum is a system consisting of a body attached to a light, inextensible spoke that is attached to a vibrating suspension point. The pendulum named after P. L. Kapitsa, who published a description of such an oscillatory system in 1951 [1, 2].

When studying this type of pendulum, all attention was focused on the type of movement when the period of oscillation of the suspension T differed little from the period of oscillation of the pendulum itself τ. These studies were reduced to studying the properties of solutions to the Mathieu equation, which describes this motion at small oscillation amplitudes. The nature of the movement of the pendulum in this position and the degree of its stability at high frequencies of suspension vibrations remained completely unexplored.

For the problem of stabilization of a mathematical pendulum with a vibrating suspension point in an inclined position, the conditions for the existence of equilibrium are obtained [1,2]. It is shown that for any limited movement of the suspension, the exact equilibrium states of the pendulum can only be the vertical (upper and lower) positions. This result is generalized to the case of the laws of motion of the suspension, which have discontinuities in velocity, which mechanically means pulsed effects at the corresponding time points. For inclined positions, control actions are constructed that provide periodic vibration of the suspension point in time and harmonic oscillations of the pendulum relative to a fixed inclined position.

Therefore, a visual description of the movement of such pendulums was made with the help of differential equations of higher derivatives, which led to a formal solution. Ostrogradsky's formalism using the Lagrange function in such an oscillatory system leads to equations of motion that coincide with Newton's equations, which is the basis for the expression for the Lagrange function [1,2].

Let the axis of the pendulum suspension oscillates in the vertical direction, the angle between the spoke and the y-axis is denoted by φ, then the dependence of the coordinates of the body on time will have the form:

$$\begin{cases} x = l\sin\varphi \\ y = -(l\cos\varphi + A\cos\omega t) \end{cases} \quad (1)$$

where l – rod length, φ – the angle between the rod and the axis y, A - amplitude of the oscillations in the vertical direction, ω is the cyclic frequency of the oscillations, причем $A << l$.

A special case is also considered for the possibility of a theoretical study of the motion of the Kapitsa pendulum, which can be determined using the Euler-Lagrange equation of motion for the phase of the pendulum, limiting ourselves to the first derivative

$$\frac{\partial L}{\partial \varphi} = \frac{d}{dt}\frac{\partial L}{\partial \dot{\varphi}}$$

The sought differential equation describing the phase evolution of the pendulum is nonlinear due to the presence of the sin φ factor in it.

$$\ddot{\varphi} = -\frac{A\vartheta^2}{l}\cos\vartheta t\sin\varphi - \frac{g}{l}\sin\varphi.$$

Thus, we derive the equation of motion of the pendulum in time, which was solved numerically by the explicit Runge-Kutta method, and show that the Kapitsa pendulum has two stable positions, which is in agreement with the theory. Such an analytical method allows us to renounce the considered particular, narrower cases.

2. How to find the upward force for the stability of the Kapitsa pendulum?

The Ostrogradsky formalism [3] using the Lagrange function

$$L = L(q, \dot{q}, \ddot{q}, \dots q^{(n)}),$$

let's consider a vibrating reference frame, then

$$x = Asin\,\omega t.$$

The behavior of macroscopic mechanical systems in non-inertial reference frames can be described by higher-order differential equations. Here, we consider the case when the contribution of higher derivatives is small compared to lower ones. Therefore, at this stage, we restricted ourselves to only the third derivatives of the coordinates with respect to time. There are many examples of the description of mechanical systems in non-inertial reference frames due to the influence of the backgrounds of random fields and waves. Theoretical descriptions of such cases do not always fully describe the physical reality of processes occurring in this process. Such cases include Kapitsa's pendulum, the movement of bulk materials upwards, against the action of gravity. For describing vibrating mechanical systems, the principle of least action is traditionally used to obtain critical states of mechanical systems. All such cases are described by second-order differential equations. In this case, the direction of the resultant force remains uncertain. This is the main disadvantage of this method of description.

Here, we use a fourth-order differential equation. This allows to first obtain the correct direction of the resultant force. The differential equation of the Kapitsa pendulum is limited to the fourth-order of the time derivative of the coordinate [4]:

$$F - ma + \frac{mj}{\omega} - \frac{ms}{2\omega^2} = 0$$

where $a = \frac{d^2q}{dt^2}$ is the acceleration of the suspension coordinate, $j = \dot{a} = \frac{d^3q}{dt^3}$, $s = \ddot{a} = \frac{d^4q}{dt^4}$ is the rate of change in the acceleration of the coordinate q over time, called the jerk and impact s, где $\tau = 1/\omega$ this is the averaging time during the transition from the micro-to the macrocosm, opposite to the average cyclic frequency ω.

Then

$$F + m\omega^2 A(cos\omega t - \sin \omega t) = ma.$$

Ostrogradsky's formalism using Lagrange with high-order derivatives has a great advantage over the study. Namely, this approach makes it possible to consider the properties of the system in a large range of additional variables in the form of higher derivatives.

Let's designate

$$v_x = \frac{dx}{dt} = \frac{d}{dt}(l \sin \varphi) = l\dot{\varphi} \cos \varphi,$$

$$v_y = \frac{dy}{dt} = -\frac{d}{dt}(l \cos \varphi + A \cos \omega t) = l\dot{\varphi} \sin \varphi + A\omega \sin \omega t.$$

$$a_x = \frac{d^2x}{dt^2} = \frac{d^2}{dt^2}(l \sin \varphi) = \frac{d}{dt}(l\dot{\varphi} \cos \varphi) = l\ddot{\varphi} \cos \varphi - l\dot{\varphi}^2 \sin \varphi,$$

$$a_y = \frac{d^2y}{dt^2} = -\frac{d^2}{dt^2}(l \cos \varphi + A \cos \omega t) = \frac{d}{dt}(l\dot{\varphi} \sin \varphi + A\omega \sin \omega t) =$$
$$= l\ddot{\varphi} \sin \varphi + l\dot{\varphi}^2 \cos \varphi + A\omega^2 \cos \omega t.$$

$$\dot{a}_x = \frac{d^3x}{dt^3} = \frac{da_x}{dt} = \frac{d}{dt}(l\ddot{\varphi} \cos \varphi - l\dot{\varphi}^2 \sin \varphi) = l\dddot{\varphi} \cos \varphi - 3l\dot{\varphi}\ddot{\varphi} \sin \varphi - l\dot{\varphi}^3 \cos \varphi,$$

$$\dot{a}_y = \frac{d^3y}{dt^3} = \frac{da_y}{dt} = \frac{d}{dt}(l\ddot{\varphi} \sin \varphi + l\dot{\varphi}^2 \cos \varphi + A\vartheta^2 \cos \omega t) =$$
$$= l\dddot{\varphi} \sin \varphi + 3l\dot{\varphi}\ddot{\varphi} \cos \varphi - l\dot{\varphi}^3 \sin \varphi - A\omega^3 \sin \omega t.$$

The energy of the system is determined by the Lagrange function, namely, the difference between the kinetic and potential energy. Because of the symmetry of the Poincare group, the derivatives of the coordinates must be quadratic functions

$$L = -U + \frac{mv^2}{2} + \frac{ma^2}{2\omega^2}$$

or

$$L = -U + \frac{mv^2}{2} + \frac{ma^2}{2\omega^2} + \frac{mj^2}{4\omega^4}.$$

Substituting the values, we get

$$L = -U + \frac{m}{2}((l\dot{\varphi} \cos \varphi)^2 + (l\dot{\varphi} \sin \varphi + A\omega \sin \omega t)^2)$$
$$+ \frac{m}{2\omega^2}((l\ddot{\varphi} \cos \varphi - l\dot{\varphi}^2 \sin \varphi)^2 + (l\ddot{\varphi} \sin \varphi + l\dot{\varphi}^2 \cos \varphi + A\omega^2 \cos \omega t)^2)$$
$$+ \frac{m}{4\omega^4}((l\dddot{\varphi} \cos \varphi - 3l\dot{\varphi}\ddot{\varphi} \sin \varphi - l\dot{\varphi}^3 \cos \varphi)^2$$
$$+ (l\dddot{\varphi} \sin \varphi + 3l\dot{\varphi}\ddot{\varphi} \cos \varphi - l\dot{\varphi}^3 \sin \varphi - A\omega^3 \sin \omega t)^2)$$

After performing the transformations, the Lagrange function has the form:

$$L = -U + \frac{m}{2}l^2\dot{\varphi}^2 + \frac{3m}{4}A^2\omega^2 \sin^2 \omega t + ml\dot{\varphi} \sin \varphi A\omega \sin \omega t + \frac{m}{2\omega^2}l^2\dot{\varphi}^4 + \frac{m}{2\omega^2}l^2\ddot{\varphi}^2$$
$$+ \frac{m}{2}A^2\omega^2 \cos^2 \omega t + ml\ddot{\varphi} \sin \varphi A \cos \omega t + ml\dot{\varphi}^2 \cos \varphi A \cos \omega t + \frac{m}{4\omega^4}l^2\dddot{\varphi}^2$$
$$+ \frac{9m}{4\omega^4}l^2\dot{\varphi}^2\ddot{\varphi}^2 + \frac{m}{4\omega^4}l^2\dot{\varphi}^6 - \frac{m}{2\omega^4}l^2\dot{\varphi}^3\dddot{\varphi} - \frac{m}{2\omega}l\dddot{\varphi} \sin \varphi A \sin \omega t$$
$$- \frac{3m}{2\omega}l\dot{\varphi}\ddot{\varphi} \cos \varphi A \sin \omega t + \frac{m}{2\omega}l\dot{\varphi}^3 \sin \varphi A \sin \omega t.$$

The equations of motion of such a system satisfy the Euler-Lagrange equations. From here, we find the equation for the dependence of the pendulum phase φ on time and thus determine the position of the weight by the formula (1). The Euler-Lagrange equation for the pendulum phase is as follows:

$$\frac{\partial L}{\partial \varphi} - \frac{d}{dt}\frac{\partial L}{\partial \dot{\varphi}} + \frac{d^2}{dt^2}\frac{\partial L}{\partial \ddot{\varphi}} - \frac{d^3}{dt^3}\frac{\partial L}{\partial \dddot{\varphi}} = 0$$

Using the Lagrange formula, we calculate the derivatives that are used in this expression:

$$\frac{\partial L}{\partial \varphi} = -\frac{\partial U}{\partial \varphi} + ml\dot{\varphi}\cos\varphi\, A\omega\sin\omega t + ml\ddot{\varphi}\cos\varphi\, A\cos\omega t - ml\dot{\varphi}^2\sin\varphi\, A\cos\omega t$$
$$-\frac{m}{2\omega}l\dddot{\varphi}\cos\varphi\, A\sin\omega t + \frac{3m}{2\omega}l\dot{\varphi}\ddot{\varphi}\sin\varphi\, A\sin\omega t + \frac{m}{2\omega}l\dot{\varphi}^3\cos\varphi\, A\sin\omega t,$$

$$\frac{d}{dt}\frac{\partial L}{\partial \dot{\varphi}} = ml^2\ddot{\varphi} - ml\dot{\varphi}\cos\varphi\, A\omega\sin\omega t + ml\sin\varphi\, A\omega^2\cos\omega t + \frac{6m}{\omega^2}l^2\dot{\varphi}^2\ddot{\varphi} + \frac{m}{2}l\ddot{\varphi}\cos\varphi\, A\cos\omega t$$
$$-\frac{m}{2}l\dot{\varphi}^2\sin\varphi\, A\cos\omega t + \frac{9m}{\omega^4}l^2\dot{\varphi}\ddot{\varphi}^3 + \frac{27m}{2\omega^4}l^2\dot{\varphi}^2\ddot{\varphi}\dddot{\varphi} + \frac{15m}{2\omega^4}l^2\dot{\varphi}^4 - \frac{3m}{\omega}l^2\dot{\varphi}\ddot{\varphi}\dddot{\varphi}$$
$$-\frac{3m}{2\omega}l^2\dot{\varphi}^2\varphi^{(4)} - \frac{3m}{2\omega}l^2\dddot{\varphi}\cos\varphi\, A\sin\omega t + \frac{9m}{2\omega}l\dot{\varphi}\ddot{\varphi}\sin\varphi\, A\sin\omega t$$
$$+\frac{3m}{2\omega}l\dot{\varphi}^3\cos\varphi\, A\sin\omega t,$$

$$\frac{d^2}{dt^2}\frac{\partial L}{\partial \ddot{\varphi}} = \frac{m}{\omega^2}l^2\varphi^{(4)} - 2l\ddot{\varphi}\cos\varphi\, A\cos\omega t + 2ml\dot{\varphi}^2\sin\varphi\, A\cos\omega t + 2ml\dot{\varphi}\sin\varphi\, A\cos\omega t$$
$$-\frac{m}{2}l\dot{\varphi}\cos\varphi\, A\omega\sin\omega t - ml\sin\varphi\, A\omega^2\cos\omega t,$$

$$\frac{d^3}{dt^3}\frac{\partial L}{\partial \dddot{\varphi}} = \frac{m}{2\omega^2}l^2\varphi^{(6)} - \frac{3m}{\omega^4}l^2\ddot{\varphi}^3 - \frac{9m}{\omega^4}l^2\dot{\varphi}\ddot{\varphi}\dddot{\varphi} - \frac{3m}{2\omega^4}l^2\dot{\varphi}^2\varphi^{(4)} - \frac{m}{2\omega}l\dddot{\varphi}\cos\varphi\, A\sin\omega t$$
$$+\frac{3m}{2\omega}l\dot{\varphi}\ddot{\varphi}\sin\varphi\, A\sin\omega t - \frac{m}{2\omega}l\dddot{\varphi}\cos\varphi\, A\cos\omega t + \frac{m}{2\omega}l\dot{\varphi}^3\cos\varphi\, A\sin\omega t$$
$$+\frac{3m}{2}l\dot{\varphi}^2\sin\varphi\, A\cos\omega t - ml\ddot{\varphi}\cos\varphi\, A\cos\omega t + \frac{3m}{2}l\dot{\varphi}\cos\varphi\, A\omega\sin\omega t$$
$$+\frac{m}{2}l\sin\varphi\, A\omega^2\cos\omega t.$$

So

$$-\frac{\partial U}{\partial \varphi} + 2ml\dot{\varphi}^2\sin\varphi\, A\cos\omega t - \frac{3m}{2\omega}l\dot{\varphi}^3\cos\varphi\, A\sin\omega t - ml^2\ddot{\varphi} - \frac{5m}{2}l\sin\varphi\, A\omega^2\cos\omega t$$
$$-\frac{6m}{\omega^2}l^2\dot{\varphi}^2\ddot{\varphi} - \frac{m}{2}l\ddot{\varphi}\cos\varphi\, A\cos\omega t - \frac{9m}{\omega^4}l^2\dot{\varphi}\ddot{\varphi}^3 - \frac{27m}{2\omega^4}l^2\dot{\varphi}^2\ddot{\varphi}\dddot{\varphi} - \frac{15m}{2\omega^4}l^2\dot{\varphi}^4$$
$$+\frac{3m}{\omega}l^2\dot{\varphi}\ddot{\varphi}\dddot{\varphi} + \frac{3m}{2\omega}l^2\dot{\varphi}^2\varphi^{(4)} + \frac{3m}{2\omega}l^2\dddot{\varphi}\cos\varphi\, A\sin\omega t - \frac{9m}{2\omega}l\dot{\varphi}\ddot{\varphi}\sin\varphi\, A\sin\omega t$$
$$+\frac{m}{\omega^2}l^2\varphi^{(4)} - \frac{m}{2\omega^2}l^2\varphi^{(6)} + \frac{3m}{\omega^4}l^2\ddot{\varphi}^3 + \frac{9m}{\omega^4}l^2\dot{\varphi}\ddot{\varphi}\dddot{\varphi} + \frac{3m}{2\omega^4}l^2\dot{\varphi}^2\varphi^{(4)}$$
$$+\frac{m}{2\omega}l\dddot{\varphi}\cos\varphi\, A\cos\omega t = 0.$$

Getting rid of the terms that make a small contribution, we write

$$-\frac{\partial U}{\partial \varphi} - ml^2\ddot{\varphi} - \frac{m}{2}l\ddot{\varphi}\cos\varphi\, A\cos\omega t - \frac{5m}{2}l\sin\varphi\, A\omega^2\cos\omega t + \frac{3m}{2\omega}l^2\dddot{\varphi}\cos\varphi\, A\sin\omega t$$
$$+\frac{m}{2\omega}l\dddot{\varphi}\cos\varphi\, A\cos\omega t = 0.$$

or

$$ml^2\ddot{\varphi} + k\dot{\varphi} + ml(g + A\omega^2\sin\omega t)\sin\varphi = 0.$$

The upper position of the pendulum is stable when

$$\frac{mv^2}{2} = \frac{mA^2\omega^2}{2} > mgl$$

or

$$A\omega > \sqrt{2gl}.$$

3. Conclusion

The dynamics of mechanical systems in non-inertial reference frames can be described by differential equations above the second. Physics require the concept of inertial reference systems, the description of motion in nonlocal hidden variables in a more general description, and can be named non-inertial mechanics, complementing physics. In this particular case, for example, when using a fourth-order differential equation by the method of variable replacement, it can be represented by two second-order equations. In the general case, non-inertial dynamics can be described by high order differential equations.

The classical problem of the behavior of a pendulum with a vertically oscillating suspension axis is a special case of the physics of nonlinear oscillations. Nonlinear oscillations, which were considered on the example of the Kapitsa pendulum model, cover general patterns that can be described using high-order derivatives of all mechanical systems [5-8] and are more complete than the description with second-order derivatives.

References

[1] Kapitsa P L 1951 *ZhETF* **21** p 588
[2] Kapitsa P L 1951 *Physics-Uspekhi* **44** p 7
[3] Ostrogradsky M V 1850 *Mem. Acad. St. Petersburg* **6** 4 p 385
[4] Kamalov T F 2020 *Symmetry* 12 (1) p 63
[5] El-Nabulsi R A 2021 *Acta Mechanica* 232 (1) p 89
[6] El-Nabulsi R A 2017 *Mech. Res. Commun.* 86, p 18
[7] El-Nabulsi R A 2018 *Adv. Space Res.* 61 12 2914
[8] El-Nabulsi R A 2014 *Qual. Theory Dyn. Sys.* 13 149

The attack strategy for a quantum key distribution protocol based on Bell's theorem

Chan Myae Hein[1] **and T F Kamalov**[1,2]

[1]Department of Theoretical Physics, Moscow Institute of Physics and Technology (State University), 9 Institutskiy per., Dolgoprudny, Moscow Region, 141701, Russia

[2]Moscow Region State University MRSU, Very Voloshinoy str., 24, Mytishi, 141014, Russia

E-mail: chanmyaehein3@gmail.com

Abstract. A new eavesdropping strategy is proposed for the Quantum Key Distribution (QKD) protocol. This scheme represents a new kind of intercept/resend strategy based on Bell's theorem. Quantum key distribution (QKD) provides the foremost reliable form of secure key exchange, using only the input-output statistics of the devices to realize information-theoretic security. In this paper, we present an improved QKD protocol that can simultaneously distribute the quantum secret key. We are already using the QKD protocol with simulated results matched completely with the theoretical concepts.

1. Introduction

The security of quantum key distribution (QKD) is predicated on a fundamental characteristic of quantum mechanics: measuring a quantum system violates it. Thus, an interceptor attempting to intercept quantum exchange will inevitably leave visible traces. Legitimate exchangers can prefer to either discard the corrupted information or nullify the data available to the eavesdropper by extracting a shorter key. The first (QKD) protocol was proposed in 1984 by Charles H. Bennett and Gilles Brassard [1]. QKD protocols may be classified in keeping with the principle used. These are the Heisenberg uncertainty principle, quantum entanglement, and public and private key cryptography. The aim of quantum key distribution (QKD) and classical key distribution is that the same; the only difference is in how they're implemented. The quantum key distribution [2] relies on the principle of quantum physics, where both the persons are initially unaware of the keys before sharing, and the secret key's determined by matching the basis used at both ends, which is completely random in nature. A communication system will be implemented using quantum entanglement or quantum superposition and transmission of information in quantum states that detect eavesdropping. An important goal of this research is the analysis of the types of attacks that eavesdropper can make on quantum key distribution protocol (QKD).

2. The Heisenberg Uncertainty Principle

The recent results in (QKD) are related to the Heisenberg uncertainty principle in quantum mechanics and EPR effects. This principle plays a critical role in preventing the attempts of eavesdroppers in a cryptosystem based on quantum cryptography. For any two observable properties linked together like A and B, if the corresponding operators $\hat{A}$, $\hat{B}$ satisfy

$$[\hat{A}, \hat{B}] = \hat{A}\hat{B} - \hat{B}\hat{A} \tag{1}$$

$$\langle \Delta A \rangle \langle \Delta B \rangle \geq \frac{1}{2} \| \langle [A, B] \rangle \| \tag{2}$$

where $\langle \Delta A \rangle$ and $\langle \Delta B \rangle$ are average value of operators $\Delta\hat{A}$ and $\Delta\hat{B}$, respectively.

Thus, $\langle \Delta A \rangle$ and $\langle \Delta B \rangle$ are variances that measure the uncertainty of the observed. For observables A and B such that $[A, B] \neq 0$, decreasing the uncertainty $\langle \Delta A \rangle$ of A forces the uncertainty $\langle \Delta B \rangle$ of B to increase, and vice versa. Thus, observables A and B cannot be simultaneously measured with arbitrary precision. The Heisenberg Uncertainty Principle can be applied to design a fully secure channel in quantum cryptographic key distribution protocols that transmit random binary sequences (i.e. keys) with automatic interception detection. An example is the uncertainty in the polarization of light, which is widely used in a coding system. This coding system uses 4 non-orthogonal polarization states identified as horizontal, vertical, 45° and 135° . This system works with the sending side (Alice) sending polarized qubits to the receiving side (Bob) through the quantum channel. However, the 4 non-orthogonal states of polarization cannot be distinguished due to the limitation of the uncertainty principle. Rectilinear and diagonal polarizations are complementary properties in the sense that measuring one property necessarily randomizes another.

3. Quantum Key Distribution (QKD) based on Bell's theorem

EPR effect plays an important role in quantum information processing. This happens when a spherically symmetric atom emits two photons in opposite directions towards two observers, Alice and Bob. Two photons are produced in an initial state of indefinite polarization. For example, if Alice and Bob both measure rectilinear polarization, they are each equally likely to capture either 0 (horizontal polarization) or 1 (vertical polarization), but if Alice gets 0, Bob will get 1 and vice versa. Of course, the EPR effect may occur on various particles, and not only on photons. Einstein, Podolsky and Rosen (EPR) in their famous 1935 paper [3] challenged the foundations of quantum mechanics by pointing out a "paradox".

There are spatially separated pairs of particles, hereinafter referred to as EPR pairs. EPR pairs can be pairs of particles separated over long distances. In 1964 Bell [4, 5] obtained the means to actually test the theories of local hidden variables. He proved that all such theories of local hidden variables must satisfy Bell's inequality. Bell's theorem provides a method for testing eavesdropping. Below we provide brief overviews. For convenience, we explain Bell's theorem using pairs of particles with spin $\frac{1}{2}$. Consider two measurable quantities A and B and denote the (discrete) possible values of A and B by α_i and β_j, the corresponding unit vectors are a_i and b_j, respectively.

$$E(a_i, b_j) = P_{++}(a_i, b_j) + P - -(a_i, b_j) - P - +(a_i, b_j) - P + -(a_i, b_j) \tag{3}$$

where $P_{\pm\pm}(a_i, b_j)$, denotes the probability that the result ± 1 was obtained along a_i a and ± 1 along b_i. The correlation coefficient S is obtained [6]

$$S = \sum_{i,j} E(a_i, b_j) \tag{4}$$

According to the rules of quantum

$$E(a_i, b_j) = -a_i \cdot b_j \tag{5}$$

Thus, quantum mechanics requires

$$S = -2\sqrt{2}.$$

The eavesdropper intervention induces

$$S = \int \rho(n_a, n_b) dn_a dn_b [\sqrt{2} n_a \cdot n_b], \tag{6}$$

where n_a and n_b are two unit vectors (for particles a and b, respectively), $\rho(n_a, n_b)$ is the probability of intercepting the spin component along a given direction for a specific measurement. In this case,

$$-\sqrt{2} \leq S \leq 2$$

When Alice connects Bob and exchanges information, Eva intercepts the quantum bit sequences. For each quantum bit in the sequence that Eva intercepted. By the correlation coefficients, a legitimate user can detect an interceptor (Eva) and Eva cannot escape the detection of Alice and Bob. So Eva's work will be limited by the Bell's theorem, her measurement disturbs the quantum state. From this it follows that the QKD is quite reliable and the risks of its decryption by Eva are quite low.

4. New Attack on QKD

With the development of technology, the implementation of a single photon source and quantum memory is just around the corner. Eve can use these new technologies and the new attack to eavesdrop and receive additional information. An important goal of this research is to find upper bounds on the amount of secret key information that may be eavesdropped by Eve.

In the QKD protocol, there are four non-commuting. $|0\rangle, |\pi/2\rangle, |\pi/4\rangle, |3\pi/4\rangle$ Linearly polarized states $|0\rangle, |\pi/2\rangle$ and circular polarized states $|\pi/4\rangle, |3\pi/4\rangle$ are orthogonal respectively. The quantum states $|\pi\rangle$ and $|\pi/2\rangle$ are measured by the so-called rectilinear type of measurement. Representing this straight forward measuring type as L, we have:

$$L|0\rangle = \lambda_1 |0\rangle, \tag{7}$$

$$L|\pi/2\rangle = \lambda_2 |\pi/2\rangle, \tag{8}$$

where λ_i, $i = 1, 2$ are eigenvalues. Since the states form the base in Hilbert, an arbitrary quantum state can be expanded by this base, i.e.

$$|\psi\rangle = C_1|0\rangle + C_2|\pi/2\rangle, \tag{9}$$

$$|\pi/4\rangle = \frac{\sqrt{2}}{2}|0\rangle + \frac{\sqrt{2}}{2}|\pi/2\rangle, \tag{10}$$

$$|3\pi/4\rangle = \frac{\sqrt{2}}{2}|0\rangle - \frac{\sqrt{2}}{2}|\pi/2\rangle, \tag{11}$$

Own auxiliary quantum state has the form

$$|\alpha\rangle = \frac{\sqrt{3}}{2}|0\rangle + \frac{1}{2}|\pi/2\rangle, \tag{12}$$

The product of the auxiliary quantum state and basic quantum state gives

$$\langle\alpha|0\rangle = \frac{\sqrt{3}}{2} \to m_1 = \frac{3}{4} = 0.75, \tag{13}$$

$$\langle\alpha|\pi/2\rangle = \frac{1}{2} \rightarrow m_2 = \frac{1}{4} = 0.25, \tag{14}$$

$$\langle\alpha|\pi/4\rangle = \frac{\sqrt{6}+\sqrt{2}}{4} \rightarrow m_3 = \frac{\left(\sqrt{3}+\sqrt{1}\right)^2}{8} \approx 0.933, \tag{15}$$

$$\langle\alpha|3\pi/4\rangle = \frac{\sqrt{6}-\sqrt{2}}{4} \rightarrow m_4 = \frac{\left(\sqrt{3}-\sqrt{1}\right)^2}{8} \approx 0.067, \tag{16}$$

here m_J, $J = 1, 2, 3, 4$ corresponds only to the ground quantum state, $k = 1, 2, 3, 4$. It follows that the QKD protocol is sufficiently reliable and the risks of decryption by its Eve are rather low

Conclusions

Quantum Key Distribution protocol has been showed theoretically that it is possible to improve the Bell's theorem and QKD protocol is proven to be secure against eavesdropping attacks. The eavesdropper cannot obtain the information between the legitimate users without being detected. We used Bell's theorem to implement the quantum key distribution protocol. It can prevent impersonation and middle attack. Of course, this can also be realized with non-orthogonal quantum states with the Heisenberg uncertainty principle. In the future, QKD protocol can be performed using the latest technology on Bell's theorem to improve the accuracy of the results.

References

[1] C H Bennett and G Brassard 1984 *Quantum cryptography: Public key distribution and coin tossing Proceedings of IEEE International Conference on Computers Systems and Signal Processing*, Bangalore India (New York: IEEE,) pp 175-179.
[2] P W Shor and J Preskill 2000 *Phys. Rev. L.* **85** 441.
[3] A Einstein, B Podolsky and N Rosen 1935 *Phys. Rev. L.* **47**(777) 22
[4] J S Bell 1965 *Physics* (Long Island City, N.Y.) 1, 195.
[5] Yu P Rybakov and T F Kamalov 2016 *Journal of Theoretical Physics* **55**(9) 4075–4080
[6] Yu P Rybakov and T F Kamalov 2007 *Foundation of Probability and Physics* **889** 315-319

A software package for modeling the propagation of dynamic wave disturbances in heterogeneous multi-scale media

N. Khokhlov
National Research Nuclear University MEPhI (Moscow Engineering Physics Institute), Russia
E-mail: kl@lokip.ru

Abstract. The paper considers a software package designed to simulate the propagation of dynamic wave disturbances in heterogeneous media. One of the main features of the considered software package is numerical algorithms with an explicit selection of inhomogeneities. Within the framework of the work, such inhomogeneities as pores, fractures and interfaces between media (contact boundaries) are considered. The considered algorithms make it possible to perform calculations in different scale settings in micro and macro sizes. The mathematical model is based on the equations of the linear theory of elasticity. For the calculation, block structural meshes are used. The software package is parallelized using MPI and OpenMP technologies. Separate parts of the algorithm are parallelized using graphics accelerators such as GPGPU. The paper describes the features of the algorithms under consideration and provides examples of calculations that demonstrate the capabilities of the algorithm.

1. Introduction

Numerical modeling of the propagation of dynamic wave disturbances in getherogenious medium is used to solve a wide range of problems. These tasks include, among other things, the tasks of seismic [1], strength of materials, high-speed loads [2], and tasks of non-destructive testing. The role of numerical modeling in each of these areas is very important. Thus, the numerical modeling of the propagation of seismic waves is an essential part of the work in geological exploration in the oil industry. Mathematical modeling is carried out in various geological environments [3], including in layered environments and in environments with the presence of various inhomogeneities (for example, fractures or caverns) [4, 5]. Tasks of this kind seem to be very resource-intensive in terms of computational resources. The computation area is usually a seismic cube with an edge length of 1 km to 10 km. At the same time, irregularities can be several meters in size. When modeling seismic resistance problems, one also has to deal with many inhomogeneities. The parameters of buildings, such as the thickness of the walls, the dimensions of the openings, are much smaller than the dimensions of the computational regions, which sometimes include large rock masses, more than 10 km in size along one direction. Thus, the computational mesh must be detailed enough to be able to correctly select all inhomogeneities. To obtain a sufficient calculation accuracy and take into account a large number of inhomogeneities, the use of large computational grids is required; in real calculations, meshes with sizes up to several tens of billions of nodes are used. Another important task is the problem of non-destructive testing and ultrasonic diagnostics. To isolate all inhomogeneities and describe the complex structure of the considered media, the use of appropriate algorithms and methods is required [6].

2. Mathematical model

As mentioned earlier, the approaches under consideration are based on the equation of the linear theory of elasticity. According to [7], the state of a continuous linear elastic medium obeys the following equations:

$$\rho \mathbf{v_t} = (\mathbf{\Delta} \cdot \mathbf{T})^{\mathbf{T}}, \tag{1}$$

$$\mathbf{T} = \lambda(\Delta \cdot \mathbf{v})\mathbf{I} + \mu\Big(\Delta \otimes \mathbf{v} + (\Delta \otimes \mathbf{v})^T\Big), \tag{2}$$

where ρ is density, $\mathbf{v}$ is velosity, T is the stress tensor, and λ, μ are the Lame's parameters, characterizing the elastic properties of the medium.

The grid-characteristic method uses the characteristic properties of the systems of hyperbolic equations, describing the elastic wave propagation [8, 9]. The mathematical principles of the GCM based on representing the equations of motion of the linear-elastic medium in the following form:

$$\mathbf{q}_t + \mathbf{A}_1\mathbf{q}_x + \mathbf{A}_2\mathbf{q}_y + \mathbf{A}_3\mathbf{q}_z = 0. \tag{3}$$

In the last equation, $\mathbf{q}$ is a vector of unknown fields, having 9 components and equal to

$$\mathbf{q} = [v_1\ v_2\ v_3\ T_{11}\ T_{22}\ T_{33}\ T_{23}\ T_{13}\ T_{12}]^T. \tag{4}$$

Matrices $\mathbf{A_k}$, k=1,2,3, are the square 9×9 matrices. Matrix $\mathbf{A_1}$ is given by the following expression:

$$A_1 = \begin{bmatrix} 0 & 0 & 0 & \rho^{-1} & 0 & 0 & 0 & 0 & 0 \\ 0 & 0 & 0 & 0 & 0 & 0 & 0 & 0 & \rho^{-1} \\ 0 & 0 & 0 & 0 & 0 & 0 & 0 & \rho^{-1} & 0 \\ \lambda + 2\mu & 0 & 0 & 0 & 0 & 0 & 0 & 0 & 0 \\ \lambda & 0 & 0 & 0 & 0 & 0 & 0 & 0 & 0 \\ \lambda & 0 & 0 & 0 & 0 & 0 & 0 & 0 & 0 \\ 0 & 0 & 0 & 0 & 0 & 0 & 0 & 0 & 0 \\ 0 & 0 & \mu & 0 & 0 & 0 & 0 & 0 & 0 \\ 0 & \mu & 0 & 0 & 0 & 0 & 0 & 0 & 0 \end{bmatrix} \tag{5}$$

matrix $\mathbf{A_2}$ is given by the following expression:

$$A_2 = \begin{bmatrix} 0 & 0 & 0 & 0 & 0 & 0 & 0 & 0 & \rho^{-1} \\ 0 & 0 & 0 & 0 & \rho^{-1} & 0 & 0 & 0 & 0 \\ 0 & 0 & 0 & 0 & 0 & 0 & \rho^{-1} & 0 & 0 \\ 0 & \lambda & 0 & 0 & 0 & 0 & 0 & 0 & 0 \\ 0 & \lambda + 2\mu & 0 & 0 & 0 & 0 & 0 & 0 & 0 \\ 0 & \lambda & 0 & 0 & 0 & 0 & 0 & 0 & 0 \\ 0 & 0 & \mu & 0 & 0 & 0 & 0 & 0 & 0 \\ 0 & 0 & 0 & 0 & 0 & 0 & 0 & 0 & 0 \\ \mu & 0 & 0 & 0 & 0 & 0 & 0 & 0 & 0 \end{bmatrix} \tag{6}$$

and matrix $\mathbf{A_3}$ can be written as follows:

$$A_3 = \begin{bmatrix} 0 & 0 & 0 & 0 & 0 & 0 & 0 & \rho^{-1} & 0 \\ 0 & 0 & 0 & 0 & 0 & 0 & \rho^{-1} & 0 & 0 \\ 0 & 0 & 0 & 0 & 0 & \rho^{-1} & 0 & 0 & 0 \\ 0 & 0 & \lambda & 0 & 0 & 0 & 0 & 0 & 0 \\ 0 & 0 & \lambda & 0 & 0 & 0 & 0 & 0 & 0 \\ 0 & 0 & \lambda+2\mu & 0 & 0 & 0 & 0 & 0 & 0 \\ 0 & \lambda & 0 & 0 & 0 & 0 & 0 & 0 & 0 \\ 0 & \mu & 0 & 0 & 0 & 0 & 0 & 0 & 0 \\ \mu & 0 & 0 & 0 & 0 & 0 & 0 & 0 & 0 \\ 0 & 0 & 0 & 0 & 0 & 0 & 0 & 0 & 0 \end{bmatrix} \tag{7}$$

The product of matrix $\mathbf{A_k}$ and vector q can be calculated as follows:

$$A_k = \begin{bmatrix} \rho^{-1}(\mathbf{T}\cdot\mathbf{n}) \\ \lambda(\mathbf{v}\cdot\mathbf{n})\mathbf{I} + \mu\Big(\mathbf{n}\otimes\mathbf{v} + \mathbf{v}\otimes\mathbf{n}\Big) \end{bmatrix} \tag{8}$$

In the last equation n is a unit vector directed along the x, y, or z directions for matrices $\mathbf{A_1}$, $\mathbf{A_2}$, or $\mathbf{A_3}$, respectively.

As we discussed above, the GCM approach is based on representing the solutions of the acoustic and/or elastic wave equations at later time as a linear combination of the displaced at a certain spatial step solutions at some previous time moment. This representation can be used to construct a direct time-stepping iterative algorithm of computing the wave fields at any time moment from the initial and boundary conditions. In order to develop this time-stepping formula, we represent matrices $\mathbf{A_k}$ using their spectral decomposition. For example, for matrix $\mathbf{A_1}$ we have:

$$\mathbf{A}_1 = (\mathbf{\Omega}_1)^{-1}\mathbf{\Lambda}_1\mathbf{\Omega}_1, \tag{9}$$

where $\mathbf{\Lambda}_1$ is a 9×9 diagonal matrix, formed by the eigenvalues of matrix $\mathbf{A}_1$; and $\mathbf{\Omega}_1^{-1}$ is a 9×9 matrix formed by the corresponding eigenvectors. Note that, matrices $\mathbf{A}_1$, $\mathbf{A}_2$ and $\mathbf{A}_3$ have the same set of eigenvalues:

$$\{c_p, -c_p, c_s, -c_s, c_s, -c_s, 0, 0, 0\}. \tag{10}$$

In the last formula, c_p is a P-wave velocity being equal to $\Big(\rho^{-1}(\lambda+2\mu)\Big)^{1/2}$ and c_s is an S-wave velocity being equal to $(\rho^{-1}\mu)^{1/2}$.

Let us consider some direction x. We assume that the unit vector n is directed along this direction, while the unit vectors $\mathbf{n}_1$ and $\mathbf{n}_2$ form a Cartesian basis together with $\mathbf{n}$. We also introduce the following symmetric tensors of rank 2:

$$\mathbf{N}_i j = \frac{1}{2}(\mathbf{n}_i\otimes\mathbf{n}_j + \mathbf{n}_j\otimes\mathbf{n}_i), \tag{11}$$

where $\otimes$ denotes the tensor product of two vectors, the indices i and j vary from 0 to 2 in order to simplify the final formulas, and $\mathbf{n}_0 = \mathbf{n}$.

The solution of equation (5), vector **q**, at the x, y, and z directions can be written as follows:

$$q(t+\tau,x,y,z)=\sum_{j=1}^{J}[X_{1,j}q(t,x-\Lambda_{1,j}\tau,y,z)],$$
$$q(t+\tau,x,y,z)=\sum_{j=1}^{J}[X_{2,j}q(t,x,y-\Lambda_{2,j}\tau,z)],$$
$$q(t+\tau,x,y,z)=\sum_{j=1}^{J}[X_{3,j}q(t,x,y,z-\Lambda_{3,j}\tau)]. \quad (12)$$

3. The architecture of the software package

One of the results of this work is a software package that allows you to numerically solve the problems of wave propagation initiated by various dynamic loads on three-dimensional objects with a complex internal structure. The developed parallel version of the software package can significantly reduce the computation time, as well as increase the resolution of the models. The software package consists of three parts: a computational module based on the grid-characteristic method; input data preprocessing systems; post-processing systems for visualization and analysis of output data. Let's consider each part of the software package in more detail.

The computing module has the following characteristics:

- the ability to calculate on three-dimensional parallelepiped grids and curvilinear structural grids;
- calculation on block geometry from a combination of grids, at the points of contacts of the grids they must be consistent;
- calculation using chimeric (superimposed grids);
- includes monotone grid-characteristic schemes of 1-4 orders of accuracy;
- has the ability to calculate both in one thread and on multi-core computing systems and on computing systems with distributed memory.

The architecture of the core of the computing complex was developed taking into account further enhancements of the capabilities of the software package. Preprocessing includes the following stages: setting the computational geometry; setting boundary conditions; setting the parameters of the initial disturbance region; setting the conditions at the interface between the media and at the boundaries of neighboring blocks; setting the mechanical characteristics of the medium; setting a computational algorithm; setting save parameters. Design geometry can be specified in several ways. First of all, you can specify the parameters of all computational grids directly in the configuration file, namely, the number of nodes along all axes, steps of the grids along all axes and their position in space. Curved meshes can be loaded from files, at the moment it is supported to load such meshes from the text version of the VTK format.

Also, the computational geometry can be prepared using the free package for pre- and post-processing scientific data Gmsh. At this stage, the import of block geometry from parallelepiped meshes is supported, the import of curved meshes is under development. In fig. 1 shows the process of creating the geometry of a high-rise building located in the enclosing rock mass.

After creation, the computational module independently breaks the geometry into blocks and in each block builds a computational grid with a given step. Contact boundaries and stitching conditions are set automatically. Also, the Gmsh capabilities allow you to specify the types of boundary conditions at the boundaries of the computational domains, and to designate areas with the same mechanical parameters of the environment. The use of this tool made it

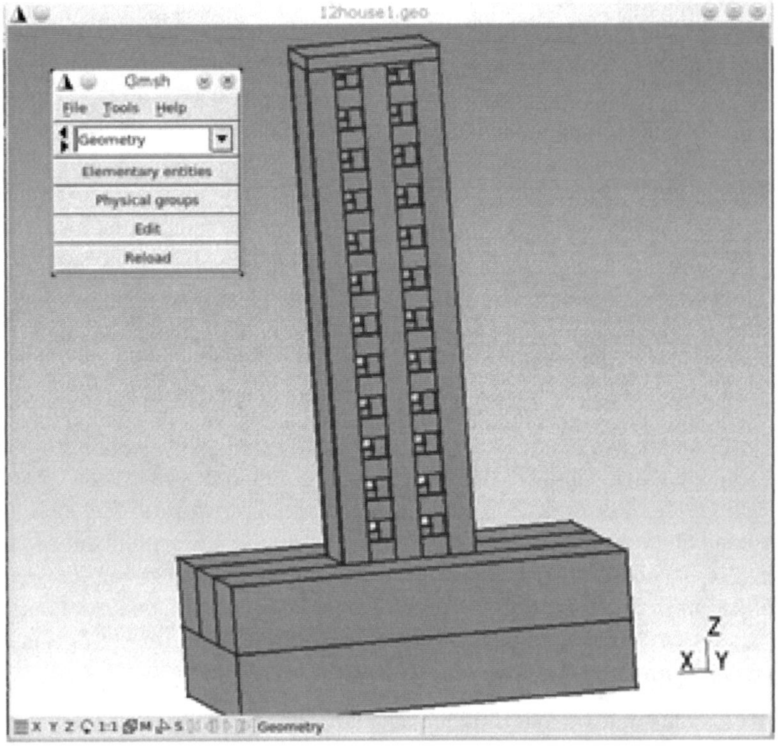

Figure 1. Creation of computational geometry in the Gmsh program

possible to significantly expand the possibilities of specifying design geometries and to carry out modeling on complex geometries. At the moment, the following types of boundary conditions are implemented in the software package: a constant external force at the boundary is set; free surface condition; the external speed is set at the border; symmetry conditions for longitudinal and transverse elastic waves; absorption condition; demolition condition, i.e. the derivatives of all the functions determining the behavior of the environment vanish along the corresponding coordinate directions. The initial disturbance can be specified both in the form of a load versus time, and in setting the values of the sought functions at all nodes of the computational grid at the beginning of the calculation. There is a set of built-in initial conditions with different parameters, but import from external files is also supported. The software package allows you to automatically determine the locations of contact boundaries on geometries and, depending on the characteristics of the medium in neighboring blocks, set the condition for contact or stitching of neighboring areas. Depending on the type of partition, the following conditions can be set: conditions of free sliding; condition of complete adhesion; condition for stitching adjacent computational blocks. The setting of the values of the mechanical characteristics of the medium (density, velocity of longitudinal and shear waves) is carried out as a function of coordinates.

At the post-processing stage, graphic visualization and analysis of the calculated data are performed. The main output data format is VTK, and free packages that support this data format are used for 3D post-processing. The current version of the program supports work with the packages ParaView and Mayavi. In fig. 2 shows the process of processing the calculated data in the ParaView program. Also, the software package allows you to save data for quick analysis in the form of tables and one-dimensional graphs. This greatly simplifies the process of building seismograms and quick analysis of calculated data.

Figure 2. Post-processing system based on the ParaView system

4. Parallelization

The considered software package is parallelized using MPI, OpenMP and CUDA technologies. Explicit grid methods are used in this work, i.e. the transition from one time step to the next occurs by traversing the entire computational grid and calculating the value at the new node based on the value at the original node and some of its vicinity. The mesh traversal is organized by two nested loops. The study of such an implementation has shown that with a sufficiently large size of the computational grid, a certain number of processor cache misses occurs, which leads to an increase in the program runtime. By changing the algorithm for traversing the computational grid, as well as through the use of additional memory arrays and forced data caching, it was possible to optimize the work with memory, which in some cases made it possible to speed up the operation of the algorithm on large computational grids [10].

To work in systems with distributed memory, the software package is parallelized using the MPI technology. When parallelizing, we used standard algorithms for decomposition of the computational domain and exchange of border cells. One of the main requirements for the algorithm is operation on a large number of computational cores (thousands) to provide an acceptable computation time. In Figure 3 shows the results of testing the scaling of the algorithm with an increase in the number of computational cores from 128 to 16384. In the test, a computational grid with a size of 1000× 1000× 1000 nodes was used [11].

The algorithm is also parallelized in shared memory systems using OpenMP technology. In this direction, work has just begun, and with an increase in the number of cores from 1 to 32, the algorithm shows an efficiency of about 60%. Such low rates are associated with insufficiently efficient work with memory, the computational cores have to access the memory areas of neighboring cores a lot, which slows down the calculation. Work is underway to improve the efficiency of work with the OpenMP technology [10].

The algorithm has also been parallelized on NVidia GPGPU using CUDA technology. A complete rewriting of a part of the computational module was required for this architecture. Currently, the speedup has been achieved about 80 times compared to a single CPU core using single precision numbers and 30 times using double precision numbers [12].

The software package also paid a lot of attention to parallelization in a distributed environment in the presence of many computational grids. Thus, an algorithm for the decomposition of multiblock meshes was proposed, which makes it possible to reduce the number of exchanges between processes [13].

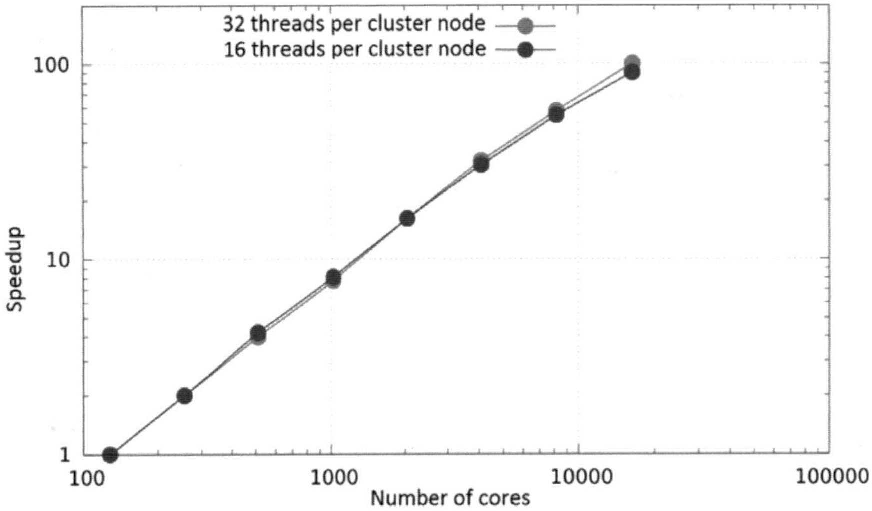

Figure 3. Acceleration of the parallel algorithm in the MPI environment

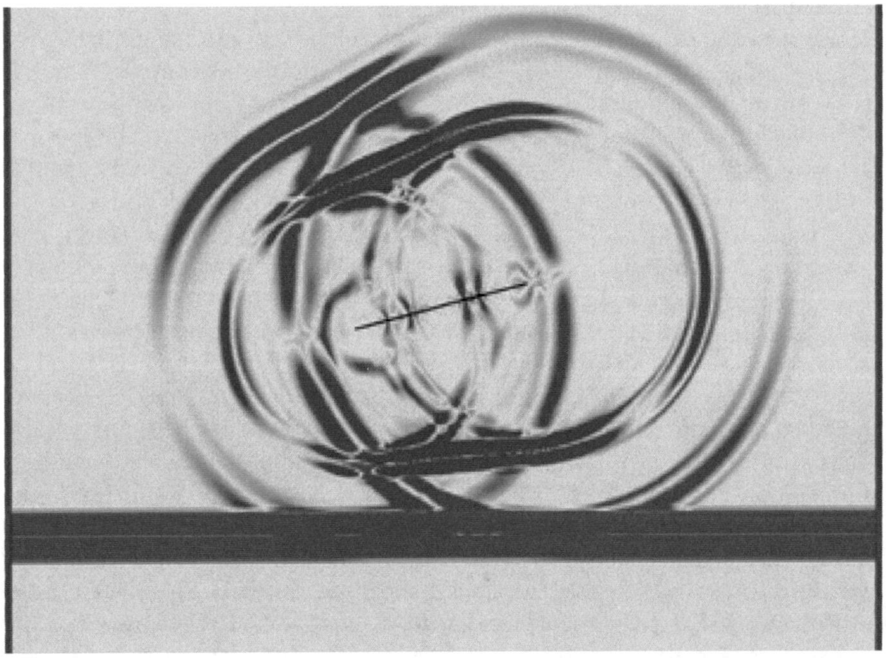

Figure 4. Passing of a plane longitudinal wave front through a fluid-saturated fracture

5. Computational examples

Next, we will consider a number of examples calculated using this software package and demonstrating its capabilities. Figure 4 shows an example of the passage of a plane longitudinal wave front through a fluid-saturated oblique fracture.

The color reflects the longitudinal and transverse components of the wave. The transmitted and incident longitudinal waves, reflected and diffracted longitudinal waves (red) are clearly visible. Diffracted and reflected shear waves (blue) are also formed.

In fig. 5 shows an example of the passage of a plane wave through a porous medium. A

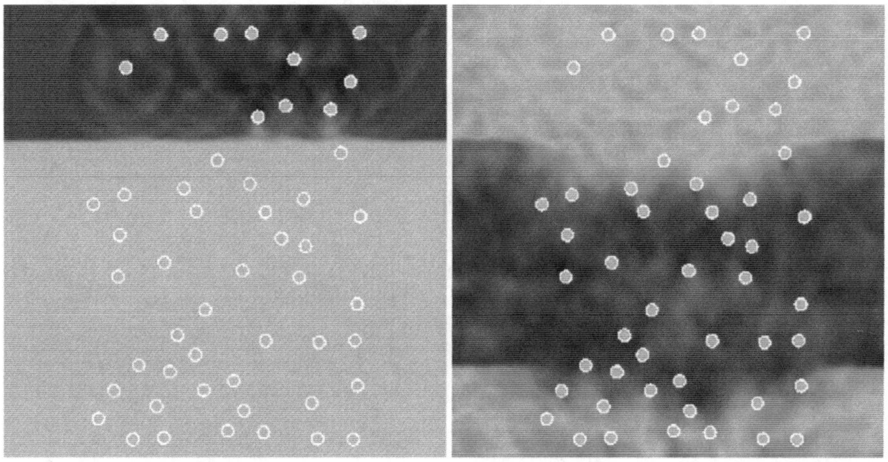

Figure 5. Passing of a plane wave front through a porous medium

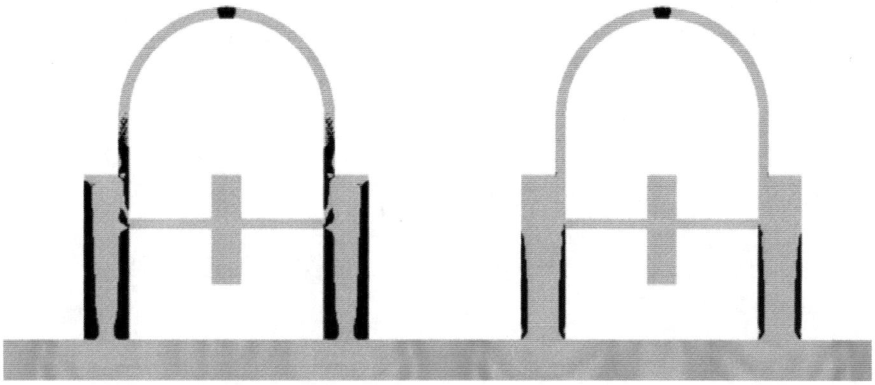

Figure 6. Seismic resistance of the dome structure at different concrete shear moduli. Destruction zones are shown in black

feature of the calculation is that each pore is clearly distinguished in the calculation algorithm.

The next task is to calculate the seismic resistance of ground structures (Fig. 6, 7). Previous calculations were carried out taking into account the formation of cracks, but now we present the results for the shear fracture model. An earthquake is described as a single plane wave train falling from the Earth's interior perpendicular to the earth's surface. The structures are presented as monolithic concrete structures. In the first case, the structure is a domed structure (NPP dome) made of concrete and has a complex structure. The reactor is located in the center and is mounted on beams, along which seismic waves practically do not reach the reactor. The roof of the station has a dome shape also for greater seismic resistance. The height of the station is 30 m, the thickness of the walls is 4 m. It can be seen from the calculations that the destruction does not reach the reactor, which confirms the correctness of the chosen design.

The following is an example of calculating the destruction of a two-story residential building with a strip foundation (three-dimensional calculation). The task was posed with the participation of Indian specialists from CDAC, studying the seismic resistance of low-rise buildings in the foothills of the Himalayas. Wall thickness 40 cm, earthquake magnitude 0.1 m/s, 5 times less than in the previous calculation. Two series of calculations were carried out for different strength moduli (1.0 MPa and 1.5 MPa). In fig. 7 shows a comparison of the fracture

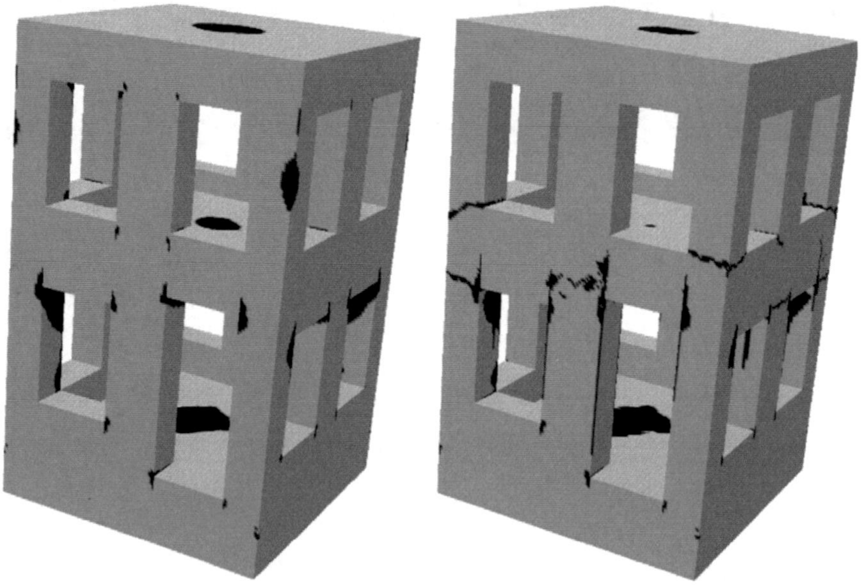

Figure 7. Seismic resistance of a two-story residential building

pattern without taking into account the corrections caused by the destroyed areas (left), and taking into account (right).

Conclusions
The paper describes a software package for modeling the propagation of dynamic wave disturbances in heterogeneous media. A feature of the algorithms is the explicit identification of inhomogeneities. The software package is parallelized in systems with shared and distributed memory, as well as on graphics coprocessors. Examples of calculations are given that show the broad capabilities of the developed software package.

Acknowledgements
This work was supported by RFBR (Russian Foundation for Basic Research), the Government of Krasnoyarsk krai and Krasnoyarsk Regional Fund for Supporting Scientific and Technological Activities Project № 20-48-242923 p_mk_Krasnoyarsk, Research of separation and purification of gas mixtures by methods of kinetic theory in new generation porous nanomaterials.

References
[1] Stognii P, Khokhlov N and Petrov I 2021 *Journal of Physics: Conference Series* **1715** 012058 DOI: 10.1088/1742-6596/1715/1/012058
[2] Kozhemyachenko A A, Petrov I B, Favorskaya A V and Khokhlov N I 2020 *Comput. Math. Math. Phys* **60** 1539–1554 DOI: 10.1134/S0965542520090110
[3] Golubev V I, Shevchenko A V and Petrov I B 2021 *Journal of Physics: Conference Series* **1715** 012056 DOI: 10.1088/1742-6596/1715/1/012056
[4] Stognii P V, Khokhlov N I and Petrov I B 2020 *Numer. Anal. Appl.* **13** 271–281 DOI: 10.1134/S1995423920030076

[5] Favorskaya A V and Khokhlov N I 2020 *Procedia Computer Science* **176** 2556–2565 ISSN 1877-0509 knowledge-Based and Intelligent Information Engineering Systems: Proceedings of the 24th International Conference KES2020 DOI: 10.1016/j.procs.2020.09.317

[6] Beklemysheva K A and Petrov I B 2020 *Lobachevskii J. Math.* **41** 2638–2647 DOI: 10.1134/S1995080220120082

[7] LeVeque R J 2002 *Finite Volume Methods for Hyperbolic Problems* Cambridge Texts in Applied Mathematics (Cambridge University Press) DOI: 10.1017/CBO9780511791253

[8] Favorskaya A V, Zhdanov M S, Khokhlov N I and Petrov I B 2018 *Geophysical Prospecting* **66** 1485–1502 DOI: 10.1111/1365-2478.12639

[9] Golubev V I, Petrov I B and Khokhlov N I 2013 *Comp. Math. and Math. Phys.* **53** 1523–1533

[10] Furgailo V, Ivanov A and Khokhlov N 2019 *2019 Ivannikov Memorial Workshop (IVMEM)* pp 79–85 DOI: 10.1109/IVMEM.2019.00019

[11] Ivanov A M and Khokhlov N I 2018 *Comput. Res. Model.* **10** 667–678 DOI: 10.20537/2076-7633-2018-10-5-667-678

[12] Khokhlov N, Ivanov A, Zhdanov M, Petrov I and Ryabinkin E 2016 *Distributed Computer and Communication Networks* ed Vishnevskiy V M, Samouylov K E and Kozyrev D V (Cham: Springer International Publishing) pp 577–588 ISBN 978-3-319-51917-3 DOI: 10.1007/978-3-319-51917-$3_4$9

[13] Fofanov V and Khokhlov N 2020 *Supercomputing* ed Voevodin V and Sobolev S (Cham: Springer International Publishing) pp 63–73 ISBN 978-3-030-64616-5 DOI: 10.1007/978-3-030-64616-5_6

Based on kinetic theory computer models of a molecular beam in a collimator

E Komarov [1], Yu Yu Kloss [1,2], F G Tcheremissine [1,3], I V Govorun [2]

[1] Moscow Institute of Physics and Technology, National Research University
[2] National Research Center "Kurchatov Institute"
[3] Federal Research Center "Computer Science and Control" Computing Center of RAS

Abstract. The paper studies a molecular jet that is formed during a gas outflow through a circular tube from a vessel and determines geometric characteristics of the vessel that reduce the jet slope. Developed software is applicable for studying gas flow in various geometries and uses the GPU for computation.

1. Introduction

Recent years have been marked by rapid development of computing technology: clusters of multiple computers, supercomputers, GPU computing technologies. All this creates the possibility of effective use of computational methods for which there were not enough resources before.

Finite-difference methods for solving differential equations can be applied for to study physical and mathematical problems without building simplifying models. One of these methods, namely the projection method [1, 2, 3], is used in this work to solve the Boltzmann equation [4]. To speed up computations, the Nvidia CUDA technology is used for computing on video cards and video card clusters [5, 6].

The main goal of this work is to study the formation of a molecular beam at low Knudsen numbers in a collimator with a circular tube. The study researches the dependence of the shape of this beam on geometric characteristics of the collimator - the length and the radius of the tube.

2. Statement of the problem

The paper studies the outflow of gas from a vessel through a circular tube of a finite length. Scheme of the installation is shown in Figure 1.

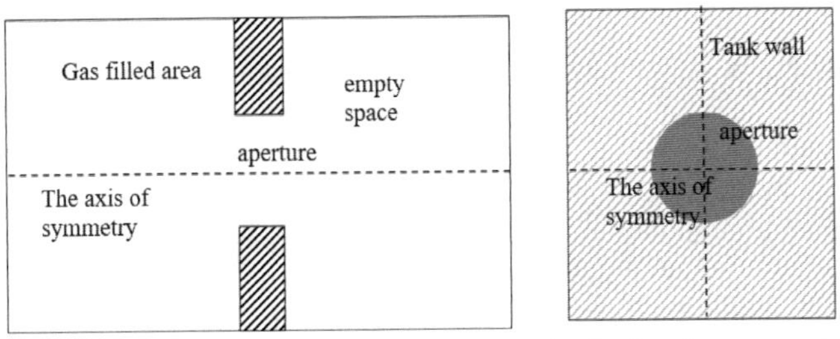

Figure 1. Scheme of the installation in longitudinal and cross sections

The vessel is initially completely filled with a gas having concentration n_0. Outside of the vessel, the gas concentration is many orders of magnitude lower, $n_1 = n_0 \cdot 10^{-8}$.

The gas flow through the left wall of the vessel is constant, which produces a steady flow through the tube. On the remaining walls of the vessel diffuse boundary conditions are posed. The dotted lines in the figure show the planes of symmetry, that makes it possible to reduce the required computation area by a factor of 4 applying peculiar reflective boundary condition at the indicated planes.

To describe the geometric characteristics of the jet, definitions from [7] are used. Let's designate $n_{axis}(x)$ - gas concentration on the axis of symmetry, and the cutoff concentration $n_{cut}(x) = 0.5 \cdot n_{axis}(x)$.

Definition 1. The beam radius is the distance from the symmetry axis at which the gas concentration reaches the cutoff concentration, i.e. $n(x, y, z) = n_{cut}(x)$.

Definition 2. Jet slope α is $tan^{-1}(dR/dx)$.

The dependence of the radius and divergence of the molecular beam on the geometric characteristics of the collimator are computed.

The jet slope is considered as the primary geometric characteristic of the jet. To calculate it, one has to find the gas concentration at all points in the computation area.

3. Numerical method

The Boltzmann equation is solved in Cartesian coordinates.

$$\frac{\partial f}{\partial t} + v_x \frac{\partial f}{\partial x} + v_y \frac{\partial f}{\partial y} + v_z \frac{\partial f}{\partial z} = I_f$$

The numerical solution is performed in two stages - solving the equation with zero right-hand side and calculating the contribution of the collision integral [8]. At the first stage, the finite-difference TVD scheme and the method of alternating directions are applied [9]. At the second one, the collision integral is taken into account using the method presented in [10].

The characteristic parameters of the gas T_0, n_0, m_0, d_0 are introduced: the initial temperature, the concentration of the gas in the vessel, the molecular weight of the gas and the diameter of the molecule. The characteristic values of velocity, distance and time are calculated using this parameters:

$$v_0 = \sqrt{\frac{T_0}{m_0}}, \lambda = \frac{1}{\sqrt{2}\pi n_0 d_0^2}, \tau = \frac{\lambda}{v_0}$$

The transition to dimensionless variables is carried out.

Introducing a finite time step, grids in the space of velocities and coordinates, and a grid function, we obtain the finite-difference scheme for solving the Boltzmann equation.

$$\frac{f_{i,j,k}^{l+1/3} - f_{i,j,k}^{l}}{\tau} + v_x \frac{\mathfrak{f}_{i+1/2,j,k}^{l} - \mathfrak{f}_{i-1/2,j,k}^{l}}{h_x} = 0$$

$$\frac{f_{i,j,k}^{l+2/3} - f_{i,j,k}^{l+1/3}}{\tau} + v_y \frac{\mathfrak{f}_{i,j+1/2,k}^{l+1/3} - \mathfrak{f}_{i,j-1/2,k}^{l+1/3}}{h_y} = 0$$

$$\frac{f_{i,j,k}^{l+1} - f_{i,j,k}^{l+2/3}}{\tau} + v_z \frac{\mathfrak{f}_{i,j,k+1/2}^{l+2/3} - \mathfrak{f}_{i,j,k-1/2}^{l+2/3}}{h_z} = 0$$

Where $\mathfrak{f}$ for each of the indices are defined as follows:

$$\mathfrak{f}_{i+1/2,j,k} = \begin{cases} f_{ij} + \frac{1-\gamma_x}{2}\phi(\theta_{i+1/2,j,k})(f_{i+1,j,k} - f_{i,j,k}), v_x > 0 \\ f_{i+1,j,k} - \frac{1+\gamma_x}{2}\phi(\theta_{i+1/2,j,k})(f_{i+1,j,k} - f_{i,j,k}), v_x < 0 \end{cases}$$

$$\gamma_x = \frac{v_x \cdot \tau}{h_x}$$

Then the obtained value of the distribution function is changed by including the collision integral.

The solution algorithm is described in more detail in [7].

4. Boundary conditions

Diffuse boundary conditions are imposed on the flat walls of the vessel. Taking into account the condition of no gas flow through the surface, the distribution function for reflected molecules takes the form:

$$f^{l+1}_{outi,j,k} = \frac{\sum_{v_x<0} v_x f^{l}_{inpi,j,k}}{\sum_{v_x>0} v_x e^{-v^2/2}} \cdot e^{-v^2/2}$$

For the left wall of the vessel and the rest of the boundaries of the counting area, the "zero boundary condition" is used. For example, on the left border:

$$\forall v_x > 0: f^{l+1}(x, v) = f^{l}(x, v) = f^{0}(x, v)$$

That is, the value of the distribution function on this surface does not change for velocities directed towards the interior of the computation area. This generates a constant incoming flux of particles, which allows the formation of a stationary molecular beam.

$$F = \sum_{v_x>0} v_x \cdot f = \sum_{v_x>0} v_x \cdot f_0 = const$$

The boundary conditions on the side surface of the tube in the collimator are somewhat difficult to pose. In the finite-difference expression, the cells have the shape of a cube. So, the surface of a circular cylinder needs to be replaced with a different surface. In order to bring the new surface as close as possible to the cylinder, we will replace it with a polyhedron of a special shape. Its structure is shown in Figure 2.

A normal and a tangent plane are drawn in each cell near the surface of the cylinder. These planes form a polyhedron, from which the gas passing through the tube is diffusely reflected.

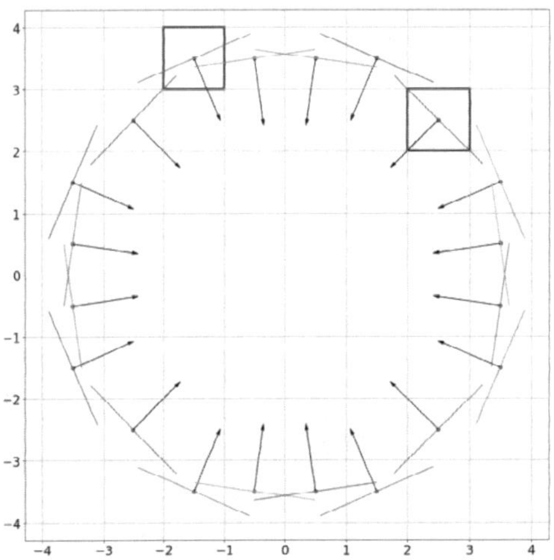

Figure 2. Circular hole approximation scheme

Diffuse boundary conditions will be applied for the values of ***v*** for which the condition $(v \cdot n) > 0$ is satisfied.

The expression for the reflected distribution function takes the form:

$$f^{l+1}_{outi,j,k} = \frac{\sum_{v\cdot n>0}(v \cdot n) f^{l}_{inpi,j,k}}{\sum_{v\cdot n<0}(v \cdot n) e^{-v^2/2}} \cdot e^{-v^2/2}$$

For the values of ***v*** that don't satisfy the condition, a zero boundary condition is used or a difference scheme is applied.

This method requires checking that the side surface of the tube really reflects all particles falling on it and does not adsorb its inside. This can be verified by counting the total number of molecules "inside the wall" which should not rise over time.

5. Impermeability check

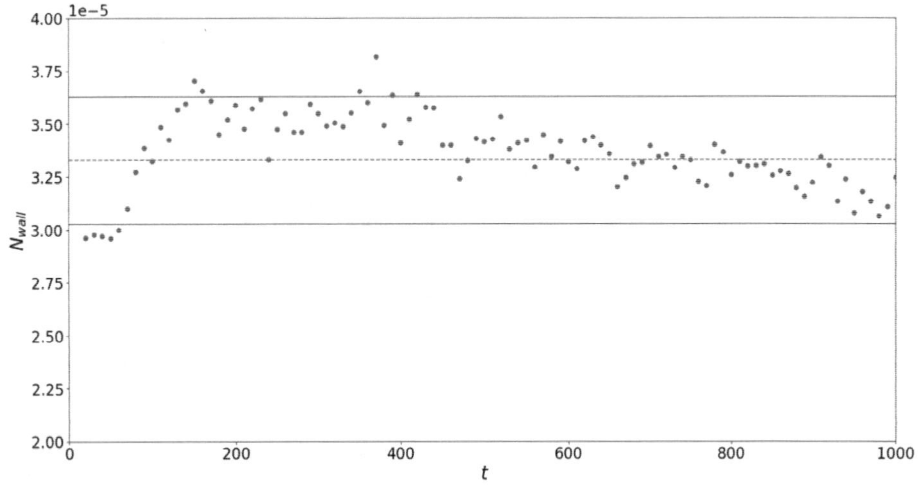

Figure 3. Dependence of the number of molecules inside the wall on time step

Using the least squares method, we approximate the number of molecules by a constant.

$$N_{wall} = C = (3.3 \pm 0.3) \cdot 10^{-5}$$

The number of molecules inside the wall during the flow remains at the level of measurement error, which shows that the side surface of the hole fully reflects the entire incident gas flow.

6. Reaching steady flow regime

In order to measure the geometric characteristics of the jet, it is required to continue the computation until a steady flow regime is formed. For this, the dependence of the gas flow through the plane perpendicular to the x axis is calculated:

$$F(x) = \int dy\, dz \int_{\forall v} d^3 v v_x \cdot f(x, y, z, v)$$

In finite-difference expression, this value takes the form:

$$F_i = \sum_{j,k} \sum_{|v|<v_{cut}} m \cdot \Delta v \cdot f_{i,j,k,m,n,p}$$

The flow is considered steady if the above value changes within the measurement error in 10 time steps. The natural metric on the function space is applied:

$$\rho(x, y) = max_j |y_j - x_j|$$

Let us introduce the value $\rho(x, y) = max_j |y_j - x_j|$ that is the distance between $F(t)$ and $F(t - \Delta t)$ in the sense of the above metric, $\Delta t = 10$. If the value is equal to zero within the measurement error, then it is accepted that the stationary flow has been established.

7. Simulation results

The simulation was carried out on a grid with the following parameters:

$\tau = 0.1$ is time step.

$h = h_x = h_y = h_z = 1$ is a step in coordinate space.

$\Delta v = \Delta v_x = \Delta v_y = \Delta v_z = 0.48$ is a step through the space of velocities.

$|v_{cut}| = 4.8$ is maximum molecular velocity.
$\Omega = [0; 100] \times [0; 25] \times [0; 25]$ is the size of the computation area.
$x_w = 25$ is location of the right side of the vessel.
$(y_h, z_h) = (12,12)$ is location of the center of the tube.
Additional parameters are determined for different configurations of tube radius and wall thickness.

Table 1. Configurations of the problem

Configuration	**a**	**b**	**c**	**d**	**e**
Tube radius r_h	4	4	4	2	6
Wall thickness d_w	2	5	10	4	4

For each configuration, values $\rho(t)$ were obtained to determine when the flow becomes stationary. They are shown in Figure 4.

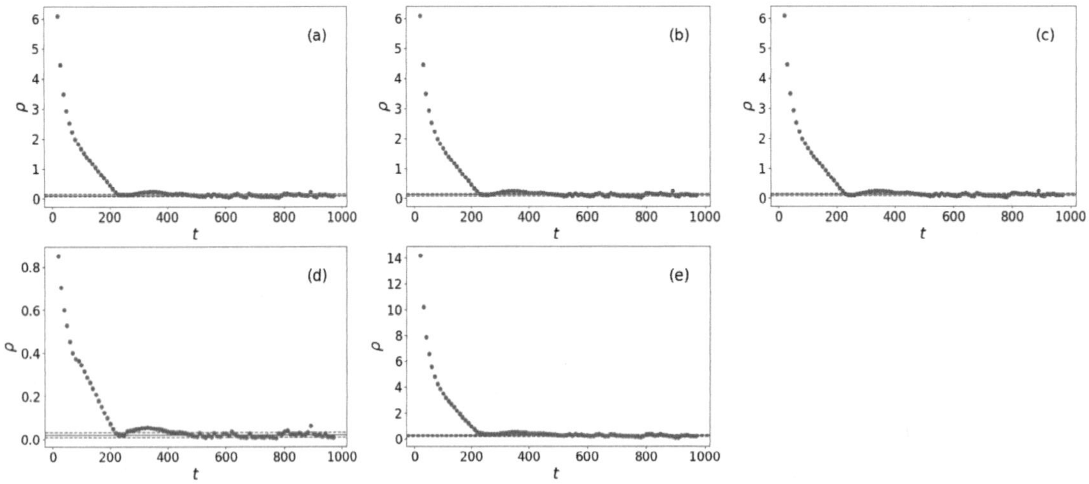

Figure 4. Transition to the steady state flow

One can see that for all configurations at t > 500, a steady flow is established.

The concentration distributions for different configurations in the longitudinal and cross sections are shown in Figure 5.

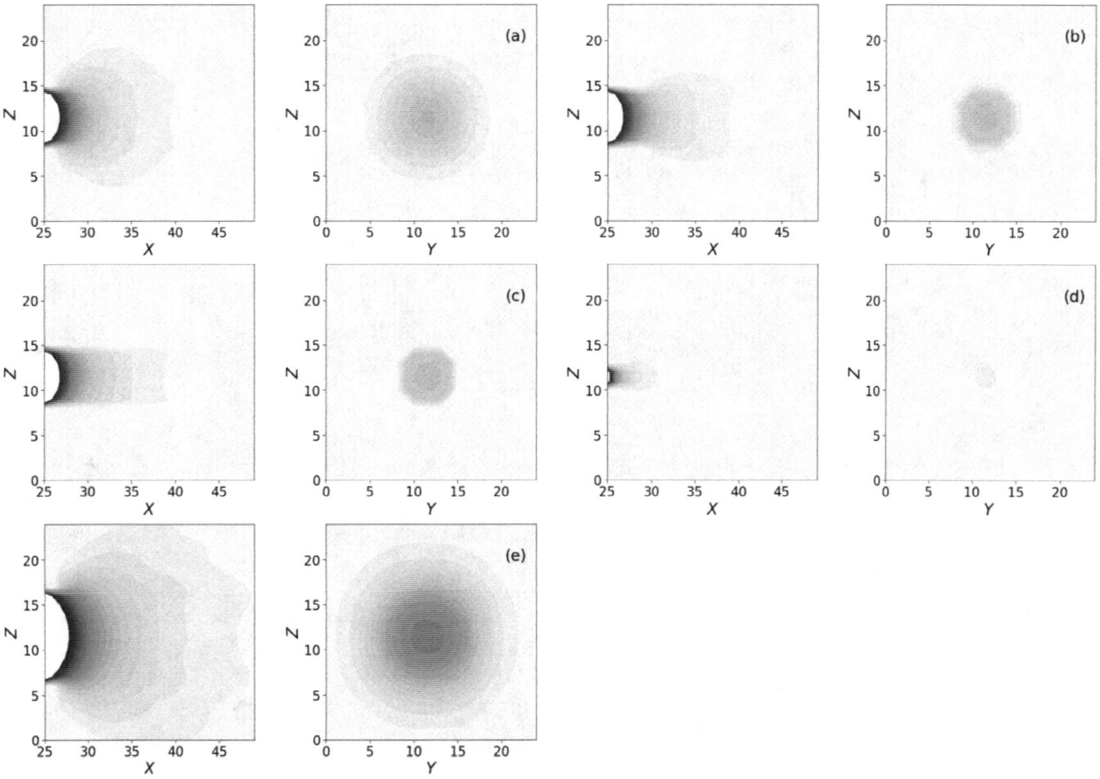

Figure 5. Distributions of concentration for stationary flow

The dependencies of the jet radius on the coordinate are shown in Figure 6.

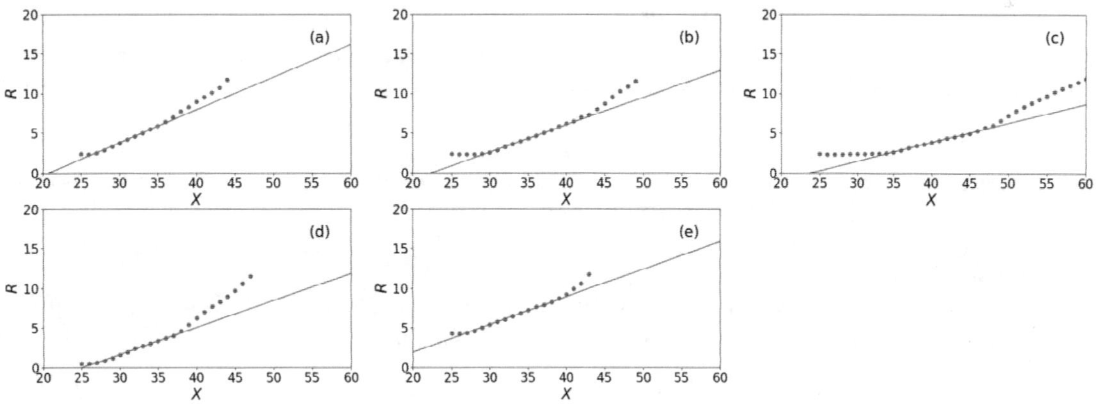

Figure 6. Jet radius versus distance from the jet exit

The R (x) dependence for each configuration is approximated by a straight line. For the approximation, not all points of R (x) are used, but only those for which the Knudsen number is not too large. This is explained by the fact that for a flow close to free molecular (that is, at high Knudsen numbers) a "ray effect" arises [11], which leads to distortion of the concentration values.

The dependence of the Knudsen number at the axis of symmetry on the x coordinate is shown in Figure 7.

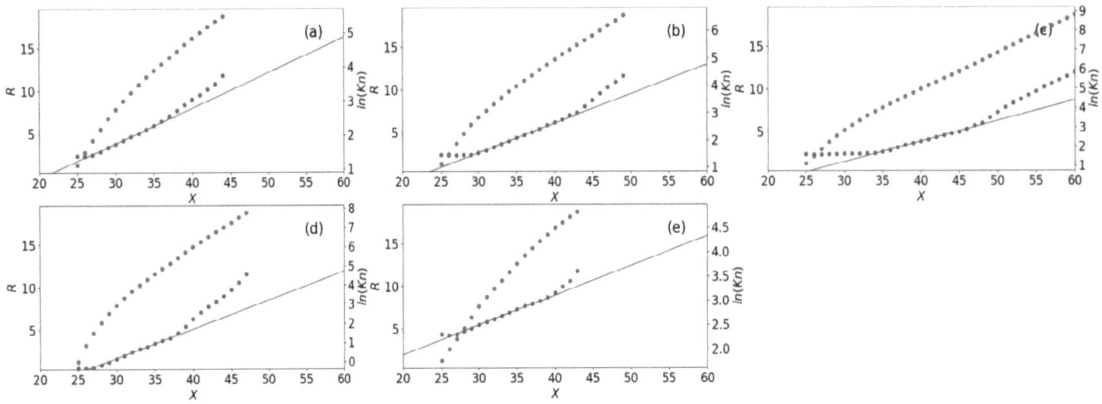

Figure 7. Jet radius (blue) and Knudsen number (red) versus distance from the jet exit

It is calculated that the value of the derivative changes sharply at $Kn \sim 10^2$. However, on the graph of the Knudsen number, there is no sharp change of the derivative at all.

The jet slope is calculated for each configuration, which is presented in Table 2.

Table 2. Jet slope values

Configuration [a]	$k = tan\alpha$	α – jet slope
a	0.414 ± 0.01	$22.5° \pm 0.5°$
b	0.355 ± 0.006	$20.3° \pm 0.4°$
c	0.240 ± 0.004	$13.5° \pm 0.3°$
d	0.33 ± 0.01	$18.8° \pm 0.6°$
e	0.34 ± 0.01	$19.2° \pm 0.6°$

[a] The parameters of each configuration are shown in the table 1

8. Conclusions

The article presents simulation of gas outflow from a vessel through a circular tube. It is shown that an increase in the length of the tube significantly affects the geometry of the molecular beam and decreases its slope. The radius of the exit hole have small effect on the beam geometry.

A method for modeling curved reflective surfaces is developed at an example of a circular tube.

The developed application uses computing technologies on GPUs, which can significantly speed up finite-difference solution of the Boltzmann equation.

9. Acknowledgements

The work was carried out with the support of RFBR projects (20-08-00634, 20-08-00448, 20-07-00135, 20-08-00211) and National Research Center Kurchatov Institute (order No. 1919 of 25.09.2020).

References

[1] Tcheremissine F G 1997 *Doklady RAN (in Russ.)* **357** 53

[2] Tcheremissine F G 2006 *Zhurnal Vychislitel'noi Matematiki i Matimaticheskoi Fiziki (in Russ.)* **46** 329

[3] Dodulad O I, Kloss Y Y, Potapov A P, Tcheremissine F G and Shuvalov P V 2016 *Zhurnal Vychislitel'noi Matematiki i Matimaticheskoi Fiziki (in Russ.)* **56** 89 DOI: 10.7868/S0044466916060119

[4] Tcheremissine F G 1970 *Zhurnal Vychislitel'noi Matematiki i Matimaticheskoi Fiziki (in Russ.)* **10** 654

[5] Suchard M A, Wang Q, Chan C, Frelinger J, Cron A, West M 2010 *J Computat. Graph. Statis.* **19** 419
[6] Preis T 2011 *The Eur. Phys. J Spec. Top.* **194** 87
[7] Komarov E, Kloss Yu Yu, Tcheremissine F G 2020 Simulation and analysis of physical characteristics of a molecular beam in a collimator using methods of kinetic theory *Journal of Physics: Conference Series* 1560 (2020) 012062 DOI:10.1088/1742-6596/1560/1/012062
[8] Aristov V V and Tcheremissine F G 1976 *Doklady AN SSSR (in Russ.)* **231** 49
[9] Samarsky A A and Babischevich P N 1967 *Additive schemes for ploblems of mathematical physics* (Moscow: Nauka)
[10] Dodulad O I, Kloss Y Y, Tcheremissine F G and Ryabchenkov V V 2011 *Vychislitel'nye metody i programmirovanie (in Russ.)* **12** 40
[11] John C Chai, HaeOk S Lee, Suhas V Patankar 1993 Ray effect and false scattering in the discrete ordinates method *Numerical Heat Transfer, Part B: Fundamentals* 24:4 p 373-389 DOI: 10.1080/10407799308955899

Intermolecular interactions of layers octa-phenyl-2,3-naphthalocyaninato zinc

D N Chausov[1,2], S Suleymanova[1], A V Kazak[1], A V Savin[1], I A Vasilyeva[1], M M Kuznetsov[1]

[1]Moscow Region State University MRSU, Very Voloshinoy str., 24, Mytishi, 141014, Russia
[2]Prokhorov General Physics Institute of the Russian Academy of Sciences, 119991, Russia, Moscow, 38 Vavilova St.

E-mail: d.chausov@yandex.ru

Abstract. The orientation of octa-phenyl-2,3-naphthalocyaninato zinc molecules is modeled by the atom-atom potential method. The dependence of the interaction energy of octa-phenyl-2,3-naphthalocyaninato zinc molecules on the angles characterizing their orientation relative to each other is established. It is shown that the obtained values of the interaction energy allow us to qualitatively explain the formation of molecules in a columnar structure.

1. Introduction

In contrast to the expensive and difficult-to-prepare silicon nanoelectronic elements, organic materials [1–3] can be deposited by cheaper wet methods and demonstrate controlled absorption over a wide optical range. Currently, phthalocyanines predominate in the field of modern organic materials. They are used or potentially suitable for creating polarizing optical structures in the form of thin-film anisotropic optical means of protection and identification of various industrial products. An important fundamental task of great practical importance is the control of the structure of such materials [4]. The control of the structure leads to the direct creation of thin-film nanomaterials with optimal physical and chemical properties, which is necessary for the creation of modern efficient nanoelectronic devices [5].

The intermolecular orientation of aliphatic hydrocarbon crystal compounds has been well studied [6–8]. At the same time, there are few papers devoted to the orientation of disc-shaped molecules with a macroheterocyclic nucleus.

The aim of this work is to study the intermolecular interactions of octa-phenyl-2,3-naphthalocyaninato zinc layers. Many phthalocyanine derivatives are known as substances that are used in organic nanoelectronics devices [4, 9].

2. Simulation of materials

The structural formula and model of octa-phenyl-2,3-naphthalocyaninato zinc are shown in Figure 1, the molecules were plotted from the coordinates of their atoms. The optimization of the atomic coordinates in the construction of the compound under study was carried out using the GROMACS physical and chemical modeling software package.

The mutual orientation of the molecules is determined by the angles θ, φ, ψ (Fig. 2), which varied from 0° to 180° in increments of 10°. Here X, Y, Z is the coordinate system associated

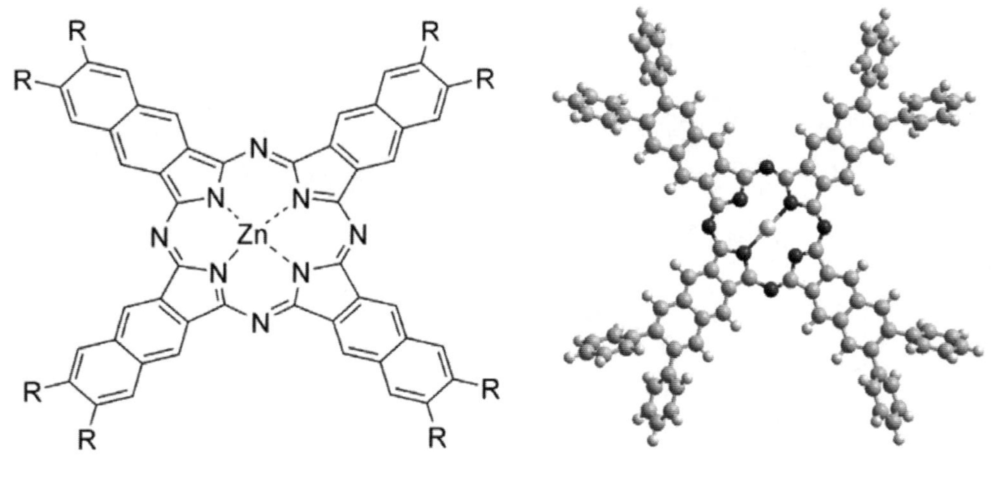

Figure 1. Structural formula of the studied octa-phenyl-2,3-naphthalocyaninato zinc

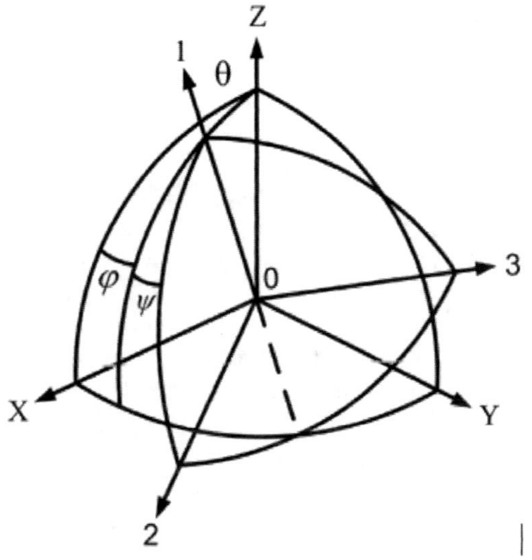

Figure 2. Mutual arrangement of the coordinates systems, connected with the molecule (1, 2, 3) and the surface (x, y, z). θ – the polar angle of axis 1 in the system x, y, z, φ – the azimuth of axis 1 (the angle of the surface xz with 1z)

with the crystal surface; 1,2,3 is the coordinate system associated with the macromolecule; θ is the polar angle of the axis 1 in the fixed coordinate system; is the azimuth of the axis 1 (the angle between the planes XZ and 1Z); ψ is the azimuth of the mobile system (the angle between the planes 1Z and 12).

The initial value of the distance a from the center of inertia of the molecule to the plane passing through the centers of inertia of the atoms composing the surface was taken to be 3 Å and the energy was minimized along the height a.

The calculations were carried out using atom–atom potentials method [10–12]. The interaction energy is shown as a sum of energy of interaction values between the atoms of the molecule and the atoms of the crystal surface, which is determined by the correlation

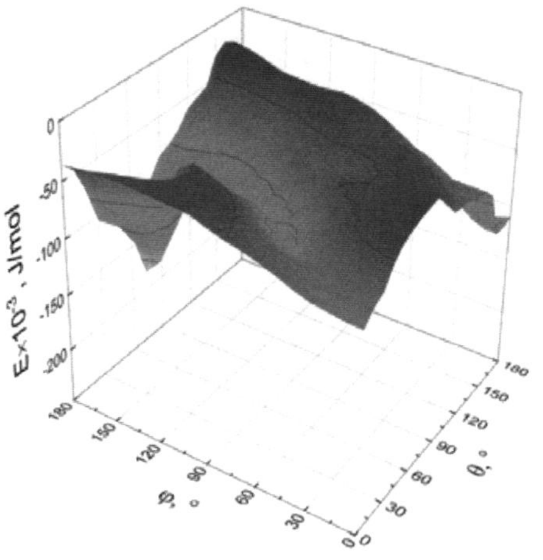

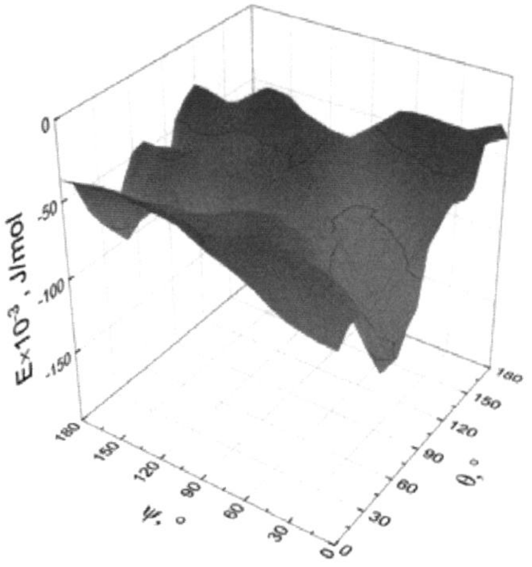

Figure 3. The dependence of the interaction energy of the octa-phenyl-2,3-naphthalocyaninato zince system on the orientation angles θ, φ, at ψ=30° (a), and the orientation angles θ, ψ, at φ =10° (b)

(Lennard–Jones potential)

$$E = \sum_{i,j} E_{i,j},$$

$$E_{i,j} = 4\varepsilon\left(\left(\frac{\sigma}{r_{i,j}}\right)^{12} - \left(\frac{\sigma}{r_{i,j}}\right)^{6}\right) \quad (1)$$

in which $\varepsilon > 0$ – depth parameter, σ – distance between certain atoms, at which $E_{ij} = 0$; r_{ij} – the distance between the atom surface i and the atom of the molecule j.

The interaction energy was determined by the ratio (1), where the values of the interaction parameters for the groups C'...C', N...N, H...H were taken from [13]. The interaction potentials, C'...H, and C'...N were calculated by us from the well-established principle of additivity of Van der Waals radii and the geometric mean rule, and are given in [12]. It can be noted that with these values of atom–atom potentials, the orientation values of liquid crystal octyl cyanobiphenyl molecules on the surface of graphite and polyethylene crystals were obtained [14–16], which have a good agreement with the experimental data [8, 13].

Conclusions

As a result, the dependence of the interaction energy of the octa-phenyl-2,3-naphthalocyaninato zinc molecules on the orientation angles θ, φ, ψ was obtained.

From figures 3, it can be seen that the energy of the intermolecular interaction of octa-phenyl-2,3-naphthalocyaninato zinc molecules significantly depends on the three orientation angles θ, φ, ψ, while there is a region of orientation states of naphthalocyanine molecules, the interaction energy of which is an order of magnitude greater than the energy of the other states.

Acknowledgements

The work was supported by the Ministry of Education and Science of the Russian Federation in the framework of the state task for Ivanovo State University (No. FZZM-2020-0006) RFBR and Ivanovo Region (Grant No. 20-47-370002)

References

[1] Usol'tseva N V, Smirnova I, Kazak V, Kovaleva I, Galanin N Shaposhnikov G P, Bodnarchuk V V and Yablonskii S V 2015 *Liq. Cryst. and their Appl. (in Russ.)* **15** 56 DOI: 10.18083/LCAppl.2015.4.56
[2] Yudin S G, Bodnarchuk V V, Lazarev V V, Smirnova A I and Yablonskii S V 2019 *Liq. Cryst. and their Appl. (in Russ.)* **19** 50–60 DOI: 10.18083/LCAppl.2019.4.50
[3] Kamanina N V, Studeonov V I and Tkachev A G 2020 *Liq. Cryst. and their Appl. (in Russ.)* **20** 78–83 DOI: 10.18083/LCAppl.2020.4.78
[4] Kazak V, Dubinina T V, Chausov D N, Kucherov R N, Kholodkov I V and Stsiapanau A A 2021 *Liq. Cryst. and their Appl. (in Russ.)* **21** 72–80 DOI: 10.18083/LCAppl.2021.1.72
[5] Smirnova A I, Giricheva N I, Soldatova K M, Ezhov A V, Glukhovskoy E G and Usol'tseva N V 2021 *Liq. Cryst. and their Appl. (in Russ.)* **21** 50–60 DOI: 10.18083/LCAppl.2021.1.50
[6] Foster J S and Frommer J E 1988 *Nature* **333** 542
[7] Hara M 1990 *Nature* **344** 228
[8] Poulin J C and Kagan H B 1991 *Acad. Sci., Ser.II: Mec., Phys., Chim., Sci. Terre Univers* **313** 1533
[9] Pavlova E I, Znoyko S A, Maizlish V E, Bumbina N V, Akopova O B and Usol'tseva N V 2020 *Liq. Cryst. and their Appl. (in Russ.)* **20** 65–71 DOI: 10.18083/LCAppl.2020.3.65
[10] Chausov D N, Belyaev V V and Dadivanyan A K 2014 *International Conference on Diamond and Carbon Materials* (Madrid, Spain) p 1.013
[11] Chausov D N, Dadivanyan A K and Belyaev V V 2015 *Mol. Cryst. & Liq. Cryst.* **611** 21 DOI: 10.1080/15421406.2015.1027991
[12] Chausov D N 2018 *J. Phys. Conf. Ser.* **996** 012019 DOI: 10.1088/1742-6596/996/1/012019
[13] Walba D, Stevens F, Parks D, Clark N and Wand M 1995 *Science* **267** 1144
[14] Dadivanyan A K, Chausov D N, Ryabova Y M, Belyaev V V and Solomatin A S 2010 *23rd International Liquid Crystals Conference.* (Krakov, Poland.) p 300
[15] Dadivanyan A K, Chausov D N, Pashinina Y M, Belyaev V V and Solomatin A S 2011 *Mol. Cryst. & Liq. Cryst.* **545** 159/[1383] DOI: 10.1080/15421406.2011.571998
[16] Dadivanyan A K, Noah O V, Chausov D N, Belyaev V V, Chigrinov V G and Pashinina Y M 2012 *Mol. Cryst. & Liq. Cryst.* **560** 108

On reconstruction of the coefficient in complex Helmholtz's equation

M Malovichko[1], A Orazbayev[2], Yu Kloss[2,3], N Khokhlov[2]
[1]Skolkovo Institute of Science and Technology, CDISE, Moscow, Russia
[2]Moscow Institute of Science and Technology, Dolgoprudny, Russia
[3]National Research Center Kurchatov Institute, Moscow, Russia

E-mail: kl@lokip.ru

Abstract. This note summarizes some preliminary results on the fast solution of the coefficient inverse problem for the Helmholtz equation, given measured pressure in a set of observation points. The Helmholtz equation is the model PDE for the harmonic problem of the linear theory of elasticity, and this work is a move in that direction. The problem has been the primary focus for several research areas, most notably seismic exploration. Still, practical problems are very challenging because they are non-linear and large. In this paper, we develop a novel numerical method for seismic full-waveform inversion based on Newton iterations. Its distinctive future is that it does not require the Jacobian of the target functional. Thus, in certain scenarios, it will perform only a fraction of computations comparing to the conventional Gauss-Newton algorithm. We present some early results on the Helmholtz equation in two dimensions.

1. Introduction

This paper is devoted to the coefficient inverse problem for the Helmholtz equation. In the geophysical community, it is known as the frequency-domain acoustic full-wave inversion (FWI) We refer to [1,2] for a general description.

Many algorithms to perform the FWI have been proposed in the literature, see [2] for a comprehensive review. Most of them fall into three categories: first-order methods, second-order methods (Newtonian), and quasi-Newtonian methods. The first-order methods use only objective function gradients. The most well-known first-order method is the nonlinear conjugate gradients (NLCG). NLCG computes the correction to a given distribution of P-wave velocity combining gradients of the objective function at the two consecutive iterations supplemented with the selection of optimal step size. The second-order methods use the second derivatives of the objective function. A typical representative of this family is the celebrated Gauss-Newton (GN) method. It computes the correction to the model by solving a normal system of linear equations, whose system matrix approximates the Hessian of the objective function. The quasi-Newton methods use information about second derivatives only partially, usually by approximating the Hessian with a diagonal matrix. Typical representatives of this class are L-BFGS and the variant of NLCG, in which a diagonal matrix scales gradients.

First-order and quasi-Newton methods have a common disadvantage. They do not sufficiently account for the nonlinearity of the coefficient inverse problem. In practice, the inversion may converge slowly, making hundreds of nonlinear iterations. The algorithm may break down at complicated velocity distributions. Newton's methods are mostly free from this drawback and often converge just in several iterations. The main difficulty is related to the fact that each Newton's step requires the solution of the normal system of linear algebraic equations, which has a large dense and ill-conditioned matrix. This system, due to its size, is solved by an internal iterative solver. No efficient preconditioner is known for it, and therefore each Newton update may require thousands of internal (linear) iterations. Thus, thousands of forward solves are potentially needed.

This paper proposes a new way of calculating Newtonian corrections. The normal system of equations with a dense matrix, which is difficult to precondition, is replaced by another system with a larger matrix. This matrix, however, is sparse and thus can be effectively preconditioned. Thus, the external (Newton) iterations coincide in both cases, but the procedure for calculating the corrections at each iteration may become much more economical for certain classes of problems. Here we present some early results which, though positive, should be considered as a proof of concept. The paper is organized as follows. In Section 2 we formulate the inverse problems, present the classical Gauss-Newton method and outline our approach. Section 3 is dedicated to a numerical experiment. Concluding remarks are given in Section 4.

2. Theory

Consider the forward problem

$$\begin{aligned} -\Delta U - \frac{\omega^2}{c^2}U = F, \ \ in \ \Omega, \\ U = 0, \ \ on \ \partial\Omega. \end{aligned} \tag{1}$$

Upon finite-difference discretization, it transforms to the following system of linear equations,

$$Au = f. \tag{2}$$

In practice, perfectly matched layers (PML) are added to (1) on all sides of Ω except the air-ground interface, for example, [3]. It leads to a complex-valued matrix A. Although in our numerical experiments we did not add PML layers, we will assume that A is Hermitian sparse matrix.

Let us by m denote a discretized inverse squared velocity. Matrix A depends on m, so we will sometimes write $A(m)$ to emphasize this dependency. We introduce the so-called forward operator $\mathcal{F}$, which computes data d (e.g. measured pressure at a set of receivers due to a given source) for a given m. Operator $\mathcal{F}$ is simply

$$\mathcal{F}(m) = QA(m)^{-1}.$$

The standard approach to the FWI (in fact, to any geophysical coefficient problem) is to minimize the following Tikhonov's functional

$$\underset{m}{\text{minimize}} \ \ \frac{1}{2}W\left(d - \mathcal{F}(m)\right)_2^2 + \frac{\varepsilon}{2}\left|\left|L^{1/2}m\right|\right|_2^2.$$

In the Gauss-Newton framework, m is iterated as $m_{n+1} = m_n + p$ with p being the update to m_n, satisfying the following minimization problem

$$\underset{p}{\text{minimize}} \ \ \frac{1}{2}W\left(r - Jp\right)_2^2 + \frac{\varepsilon}{2}\left|\left|L^{1/2}(mn + p)\right|\right|_2^2.$$

Here $r = d - \mathcal{F}(m_n)$ is data residual at n-th iteration. Matrix J is the matrix of Fréchet operator at m_n, $J = \left.\frac{\partial\mathcal{F}(m)}{\partial m}\right|_{m_n}$. Due to the fact that (1) is a linear PDE, it can be shown that the Jacobian is given by the following formula,

$$J = WQA^{-1}P,$$

where P is a diagonal matrix containing forward solution at m_n. In other words,

$$P = \omega^2 diag(u),$$

where u is the solution of the problem (2) with $m = m_n$. The gradient of the target functional with respect to m, which we will denote by g, can be expressed through the Jacobian,

$$g = -J^* W^T W r + \varepsilon L m_n.$$

These formulations are well known in the geophysical community, see [4].

In the standard (Gauss-Newton) approach, update p is the solution of the normal system of linear equations,

$$\left(J^* W^T W J + \varepsilon L\right) p = -g. \quad (3)$$

Matrix J is dense and so large that the use of direct factorization for 3D problems is untractable even on modern supercomputers. The largest singular value of J depends on the mesh step size through A. Thus, for a fixed ε, the condition number of the systems matrix in (3) grows rapidly with problem size, which is very unfavorable for iterative solvers.

One of the few practical options to mitigate computational burden is the use of "background" Jacobian. In this approach, one fixes a background velocity distribution, c_b, such that the forward problem is easier to compute. Then the system matrix in (3) is replaced with matrix

$$J_b^* W^T W J_b + \varepsilon L \quad (4)$$

with J_b being the Jacobian at the background velocity distribution c_b. The computation load decreases considerably when c_b is a constant value or a layered medium, for which the forward problem is fast. Unfortunately, as velocity distribution is updating, the efficiency of Newton's step deteriorates.

It can be shown that (3) is equivalent to the following system of linear equations,

$$\begin{bmatrix} \varepsilon L & -P^* & O \\ -P & O & A \\ O & A^* & Q^T W^T W Q \end{bmatrix} \begin{bmatrix} p \\ \lambda \\ v \end{bmatrix} = \begin{bmatrix} -\varepsilon L m_n \\ 0 \\ Q^T W^T W r \end{bmatrix}, \quad (5)$$

where matrix P corresponds to u computed for $m = m_n$. Lets us show that (5) is indeed equivalent to (3). First, we exclude λ using the first equation in (5). The second and the third equations become

$$\begin{aligned} -Pp + Av = 0, \\ \varepsilon A^* (P^*)^{-1} L (p + m_n) + Q^T W^T W Q v = Q^T W^T W r, \end{aligned} \quad (6)$$

Now, we express v from the first equation of (6) and substitute it to the second one,

$$\varepsilon L(p + m_n) + P^* (A^*)^{-1} Q^T W^T W Q A^{-1} P p = P^* (A^*)^{-1} Q^T W^T W r.$$

Using definition (4), we rewrite it as

$$\varepsilon L(p + m_n) + J^* J p = J^* W r,$$

which is (3).

System (5) is 3 times bigger than (3). It, however, has considerable advantages over the normal system: (a) it is sparse, and (b) its blocks correspond to the discretized PDE (Helmholtz's equation). Collectively, it means that an efficient iterative solver can be devised.

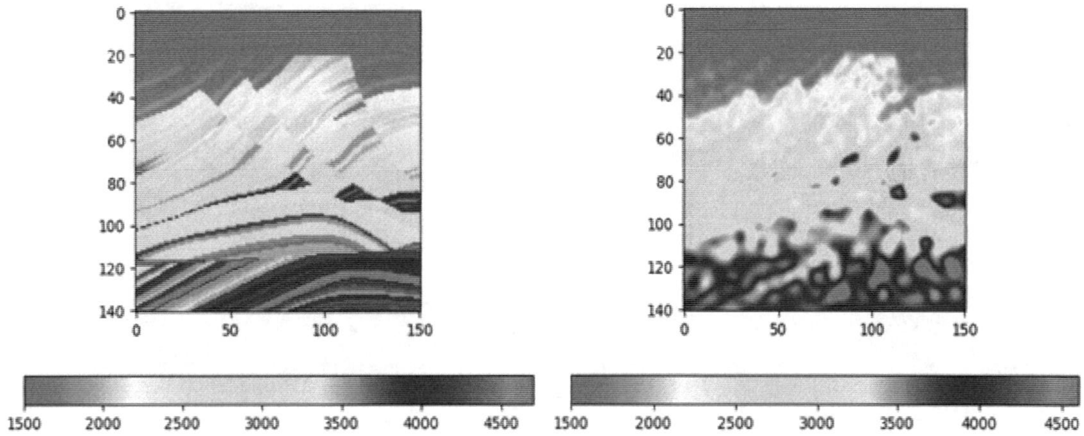

Figure 1. Initial velocity distribution (left) and the rusult of inversion (right) at a frequency of 16 Hz

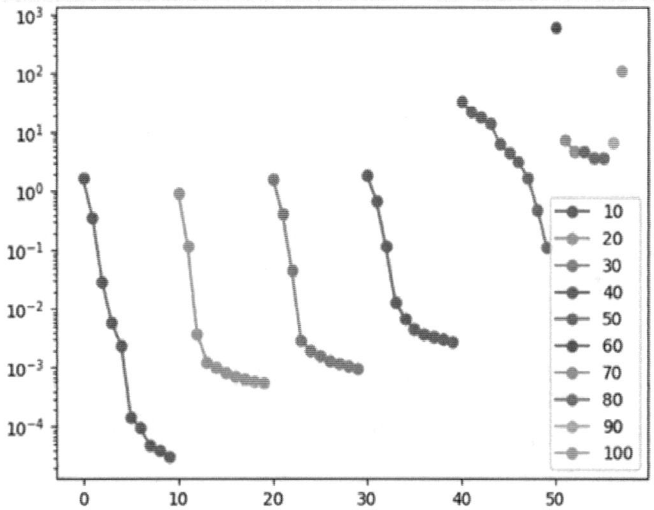

Figure 2. The history of outer iterations. The numbers denote angular frequency ω in rad/s.

3. Numerical experiments

To verify the presented formulas, we created a Newton-based inversion (3) in the two-dimensional case. Critical parts of the code, including the forward problem, were implemented in C++. The optimization part was programmed in Python. Here we solve the forward problem by a sparse direct solver.

A part of the Marmousi model has been taken as a true model; see Figure 1. We placed 40 sources and 80 receivers at a depth of 60 m. A uniform finite-difference grid with a step of 20 m was employed. The step of the computational grid was 20 m. We simulate input data in the frequency range from 1.6 Hz to 16 Hz and supply them to the inversion as input data.

We solve inverse problems progressively, from low to high frequencies. The algorithms tried to make as many updates at a single frequency as possible, switching to the next frequency when the misfit was below 1%. Figure 2 depicts history of outer (Newton) iterations. The final model, reconstructed by the Gauss-Newton method (3), is shown in Figure 1.

The two velocity models have a good match, which is expected since it was a purely synthetic

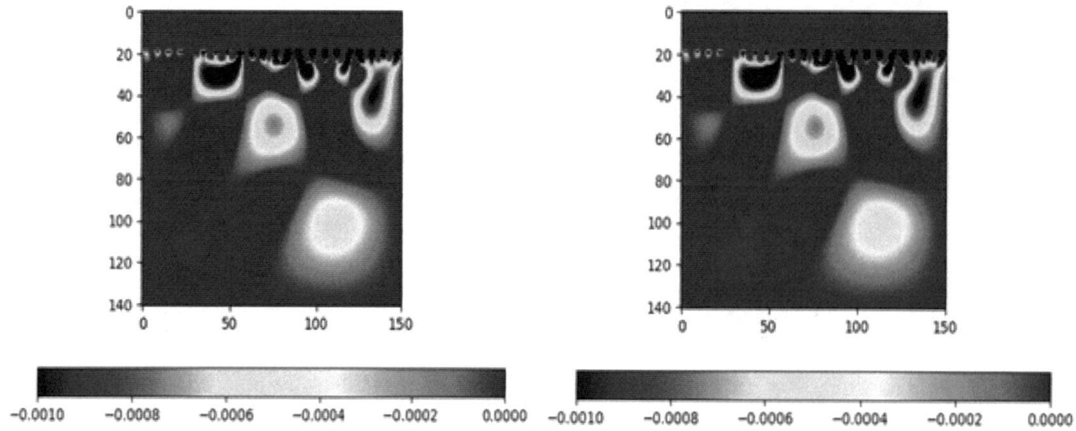

Figure 3. Comparison of Newton's update to c^{-2} for a single source computed with (3) (left) and with (5) (right). The source located at x=100 with the frequency 1.6 Hz.

example with precise data and with no noise. To verify formula (5), we compared the Newtonian steps calculated in two ways for individual sources. The comparison for a single source is shown in Figure 2. The two images are almost identical apart from a slight difference in amplitude visible in the bottom-left corner. This difference is likely to be related to the difference in the accuracy to which linear systems were solved.

Conclusions

We have proposed a new method for accelerating Newton seismic full-wave inversion. In contrast with the standard approach with a solution to the normal system, the problem is reduced to a 3×3 block sparse matrix, which can, at least in principle, be more easily treated with iterative solvers. Our numerical example demonstrated that the two approaches produce almost equivalent updates as predicted theoretically. Based on the results obtained, we are very optimistic that the new approach will accelerate Newton-based seismic data inversions.

Acknowledgements

This research was supported by the Russian Science Foundation, project no. 21-11-00139. The authors acknowledge computational resources granted by Complex for Simulation and Data Processing for Mega-Science Facilities at NRC "Kurchatov Institute", http://ckp.nrcki.ru/; and Skoltech CDISE's high-performance computing cluster, Zhores [5].

References

[1] J Virieux and S Operto 2009 *Geophysics* 74 WCC1–WCC26.
[2] L Metivier, R Brossier, Operto, and J Virieux 2017 *SIAM Review* **59**(1) 153–195
[3] N Yavich, N Khokhlov, M Malovichko, and M S Zhdanov 2021 *Computers and Mathematics with Applications* **86** 63–72.
[4] R E Plessix 2006 *Geophysical Journal International* **167** 495–503.
[5] I Zacharov, R Arslanov, M Gunin, D Stefonishin, A Bykov, S Pavlov, O Panarin, A Maliutin, S Rykovanov, and M Fedorov 2019 *Zhores—Petaflops supercomputer for data-driven modeling, machine learning and artificial intelligence installed in Skolkovo Institute of Science and Technology. Open Eng.* (9) 512–520

Molecular dynamics of DNA-binding protein and its 2D-crystals

E V Tereshkin, K B Tereshkina, and Y F Krupyanskii

Semenov Federal Research Center of Chemical Physics, RAS, 119991 Kosygina ul., 4, Moscow, Russia

E-mail: ramm@mail.ru

Abstract. In this work the dodecamers and the two-dimensional crystals of DNA-binding protein from starved cells (DPS) of *Escherichia coli* bacteria were investigated. The DPS monomer contains 167 amino acids residues. It can form dimers, trimers, and dodecamers. The versatility of the DPS protein structure can be used to design nanomaterials with structures and functions not found in living nature. The ability of this protein to self-assemble into complex shapes and structures defined on the nanometer scale can make them highly demanded for various technological applications. It was used all-atom classical molecular dynamics simulation on 0.1 microsecond scale to obtain the spatial and energy characteristics of the proteins and the components of the simulation box. The fluctuation mobility of DPS protein at various temperatures was discussed. The diffusion of ions in the presence of dodecamers and 2D crystals was compared. It has been shown that this protein retains its ability to accumulate ions in a wide range of biological temperatures from 277 to 369K. It also retains the mobility of key amino acid residues involved in the formation of nanocrystals and the transport of ions into the cavity, even at low physiological temperatures.

1. Introduction

Fabrication of biotemplated functional nanostructures is one of the important areas in nanotechnology [1]. The dimensions of biomolecules are nanometers. However, they can self-organize into subcellular structures, form cells and whole organisms, exceeding the size of the molecules themselves by several orders of magnitude. A detailed understanding of the processes of proteins functioning and the peculiarities of their structure made it possible to talk about proteins as natural nanosystems. These nanosystems have great potential in the field of nanomaterials, nanobiointerfaces, and nanomedicine. Understanding how these systems self-assemble, how they function during different assemblies, how structural features and other molecules and ions affect them is critical to the development of new synthetic materials. Each of the proteins has a unique structure. Moreover, the structure is repeated from molecule to molecule. Many of the proteins are capable of forming functional complexes or even crystals. In addition, protein molecules can interact with inorganic substances to form stable structures. One example of a protein that combines all of these characteristics is the DNA-binding protein from starved cells (DPS) [2]. This protein is the member of the ferritin superfamily and exists widely in bacteria and archaea. Thus, it can be easily and quickly produced by nanotechnology. DPS in the nature protects the cell from oxidative stress by inactivating the potentially dangerous Fe^{2+} ions and provides intracellular nucleoid crystallization. This protein is capable of accumulating up to 500 iron ions, which forms a ferrihydrite-like core. It protects the DNA under hazardous conditions. It was shown that its concentration reaches 180000 molecules per bacterial cell under stress. This quantity of protein is necessary for the cell to ensure biocrystallization: a special mechanism for the stabilization of the nucleoid, in which the nucleoid is packed with proteins into a compact, extremely stable structure. DPS can form of DPS–DNA liquid crystal mesophases that depends strongly on the protein dodecameric structure. In this work, we investigate the dynamics of protein at different temperatures. The migration of ions and the fluctuation subspecies of the protein at different temperatures are considered. The principal component analysis is used to assess the difference in protein dynamics in dodecameral and crystalline forms.

2. Materials and methods

2.1. Molecular structures of dodecamers and monolayer crystal of DPS protein

The crystal structure of Escherichia coli DPS was first obtained in 1998 by X-ray diffraction analysis with a resolution of 1.6 Å [3]. It was shown that the DPS protein forms spherical homododecamers with a cavity inside, similar to ferritin. The location of the subunits is such that the protein belongs to the tritetrahedral point symmetry group (the 23rd class, according to international classification). Later it was shown that the DPS protein can exist in the form of a monomer, dimer, trimer, or dodecamer (Fig. 1). The most interesting is the dodecamer form of the protein. In this form, the DPS is spherical with a cavity inside. The outer diameter of the sphere is 9 nm, the inner diameter of the cavity is 5 nm. DPS dodecamers can accumulate ions. Also, this form is capable of forming various crystal structures in the living cell and *in vitro*. The dodecamer of the DPS protein can be considered as a cube whose faces are formed by two subunits, between which there are two ferroxidase centers. Each of the centers is divided into two Fe^{2+} binding sites.

We obtained the molecular configuration of DPS monolayer from the RSCB Protein Database, entry 6gcm [4]. These molecular crystals were solved by our group at the European synchrotron radiation facility (ESRF). This is an x-ray structure of the DPS protein resolved at an average RMS of 2.45 Å, reconstituted from plasmid expressed in E. coli. Up to 13 N-terminal residues in each chain were unidentified in this experimental structure since they are disordered. These residues play an important role in DPS-DPS and DPS-DNA binding [5-7]. Therefore, they were added to the model using UCSF Chimera package [8].

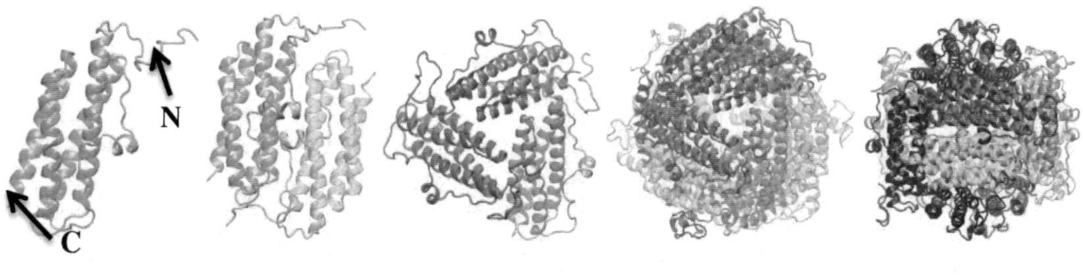

Figure 1. Structure of the DPS protein. From left to right: monomer, dimer, trimer, dodecamer (Fe-pore view), dodecamer (side view). Each subunit is coloured with its own colour. C and N letters correspond to C-terminus and N-terminus.

2.2. Molecular dynamics protocol

The systems were prepared for calculation in the Gromacs 2018 program [9]. The DPS dodecamer was placed in a cubic unit cell with an edge 14.7077 nm in length. DPS 2D-crystals consisted of 4 DPS dodecamers placed in the center of box modelling infinite single layer two-dimensional crystal. The box dimensions were 14.7 nm * 17.7 nm * 20.0 nm. To neutralize the charge of the system, 48 Na^+ ions were added to the cell as counter ions for each DPS dodecamer. In addition, other q^+, q^- and q^{2+} ions were placed to the box at low and physiological concentrations. To solvate the systems, the SPC/E water model was used. The calculations were carried out in the AMBER99-PARMBSC1 force field [10]. The energy minimization of the systems containing the DPS protein (due to its complex structure with an internal cavity) cannot be carried out with sufficient accuracy by the conjugate gradient method. Therefore, the fastest descent algorithm was used in all the systems. Then, sequential relaxation of the systems was carried out under conditions of constant volume and pressure of 100 ps duration. After relaxation, the dynamics of the systems was calculated. The calculations were carried out in an NPT ensemble. The constant temperature was maintained using a stochastic thermostat, the constant pressure was maintained using a Parrinello–Raman thermostat in isotropic (for the system of DPS dodecamers) or semi-isotropic manner (for DPS 2D-crystals), a time constant was 2 ps. Fast degrees of freedom were limited using the LINCS algorithm. The cut-off radii for the Coulomb and van der Waals interactions were taken as equal to 1.5 nm. The DPS dodecamer systems were calculated at temperatures from 277 to 369 K with a step of 4 K. The temperature of the 2D-crystal system was 300 K. The integration step was 2 fs, and the trajectory calculation time was 0.1 μs.

3. Results and discussion

The system of the DPS protein and its two-dimensional crystals is relatively large and requires trajectories of the order of 0.1 μs. Internal relaxation of systems, for which the parameters of the calculation box and the dynamic characteristics of protein molecules are balanced, occur at times up to 0.05 μs. We obtained molecular dynamics trajectories for dodecamers of the DPS protein at temperatures from 277 to 369 K and for two-dimensional crystals of this protein at 300 K. The Root mean square fluctuation (1) is shown as a bar plot of residue numbers on Fig. 2. The amino acid residues responsible for DPS binding for the formation of DPS crystals and co-crystals are active even at temperatures close to 0°C (yellow dots, 0-12 residues). This means that DPS crystals must maintain their stability over a wide temperature range. RMSF of several amino acid residues remain high. It is Arg70 in the inner cavity of the DPS, which is responsible for the coordination of ions. Lys105, which apparently stabilizes DPS molecules in crystals. And Lys140, which directs ions into the cavity.

$$RMSF = \left(\frac{1}{T}\sum_{t_j=1}^{T}\left(r_i(t_j) - r_{i,ref}\right)^2\right)^{\frac{1}{2}} \quad (1)$$

where T is the duration of the simulation (every time step), r_i (t_j) the coordinates of atom r_i at time t_j and $r_{i,ref}$ is reference position of atom r_i.

One of the capacious characteristics of biological polymers is the radius of gyration (2). For DPS, it increases slightly with increasing temperature (Fig. 3). The maximum values of the radius of gyration are achieved at 325K and 369K.

$$R_g = \left(\frac{\sum_i \|\boldsymbol{r}_i\|^2 m_i}{\sum_i m_i}\right)^{\frac{1}{2}} \quad (2)$$

where m_i is the mass of atom i and r_i the position of atom i with respect to the center of mass of the molecule.

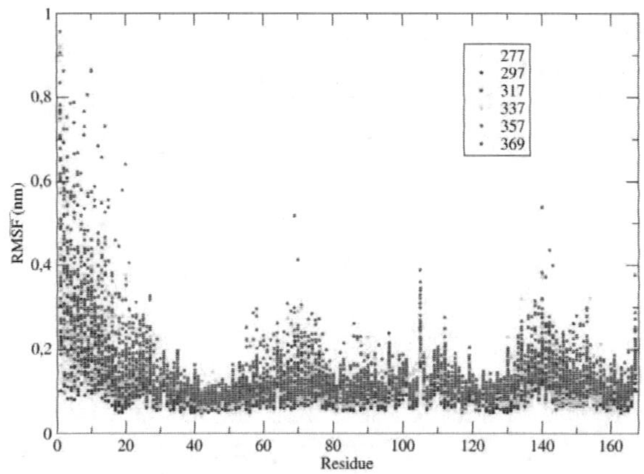

Figure 2. RMSF per residue of DPS monomers in DPS dodecamers at different temperatures. The legend shows the temperatures of simulation in K.

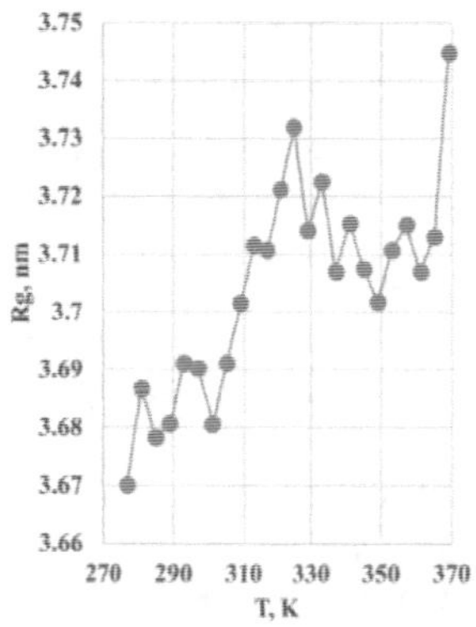

Figure 3. Radii of gyration for DPS protein at different temperatures.

Fig. 4 shows the self-diffusion coefficients of ions near DPS dodecamers calculated from the Einstein relation (3) based on the linear parts of the Mean square displacement (MSD) plots. Ion diffusion coefficient is small compared to the value for an aqueous solution [11-13], and practically does not change with increasing temperature.

$$\lim_{t\to\infty}(\|\boldsymbol{r}_i(t)-\boldsymbol{r}_i(0)\|^2) = 6Dt \qquad (3)$$

where $r_i(t)$ the coordinates of atom r_i at time t and $r_i(0)$ is the initial coordinate. For molecules consisting of more than one atom, r_i was taken as the center of mass positions of the molecules.

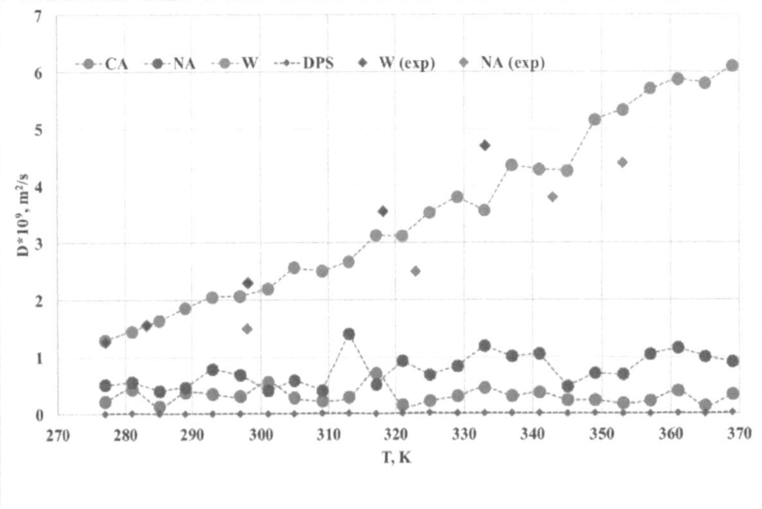

Figure 4. Calculated self-diffusion coefficients for water (grey circles), Na^+ ion (orange circles), Ca^{2+} ion (green circles) compared with experimental self-diffusion coefficient of water (blue diamonds) and inter-diffusion coefficient of NaCl-Water (orange diamonds). Red line with red diamonds corresponds to the DPS protein.

Bivalent ions have the lowest diffusion coefficient near two-dimensional DPS crystals, since they have a high affinity for the DPS surface. Monovalent positive ions accumulate in the DPS cavity, so their average diffusion coefficients are low. In this case, the diffusion coefficients of monovalent ions vary from 0 (for ions bound to the surface of the DPS) to 2.5 $10^{-9}m^2/s$, which corresponds to the diffusion of ions in aqueous solutions.

Table 1. Self diffusion coefficients of ions near 2D-crystals of DPS and lateral diffusion coefficients in the plane of 2D-crystal (10^{-9} m^2/s) at 300K.

	Na^+	K^+	Ca^{2+}	Cl^-
diffusion	0.50±0.04	0.46±0.07	0.24±0.15	0.81±0.19
lateral diffusion	0.63±0.06	0.65±0.04	0.34±0.23	1.10±0.25

DPS molecules inside crystals are packed very tightly. Crystals are formed in such a way that they provide the distribution of DPS molecules over only two angles of rotation. In this case, the rotational autocorrelation functions reveal an extremely high degree of correlation (more than 0.99) for DPS molecules both in space and relative to the normal axis to the 2D crystal. We observed a slowdown of protein rotational diffusion upon crowding conditions in 2D-crystals. In our previous studies, it was found that DPS molecules accumulate not only ions, but also small molecules (for example, 4-hexylresorcinol), as well as oligonucleotides. This property of DPS molecules with a strict orientation of DPS molecules inside the crystal can make it possible to create various ordered structures in the composition of co-crystals.

Conclusions

The results obtained show that DPS molecules are capable of accumulating various cations, ensuring their strong binding to the surface of the inner cavity at different temperatures. At the same time, DPS crystals remain active both in the accumulation of ions and in the binding of other molecules even at physiologically low temperatures close to 0°C. This property of DPS molecules can make it possible to create new functional materials in the inner cavity of a protein, as well as to obtain new crystalline materials like DPS – DNA liquid crystal mesophases.

Acknowledgements

This work was supported within frameworks of the state task #0082-2019-0015 for Semenov Federal Research Center for Chemical Physics, Russian Academy of Sciences (state registration #AAAA-A20-120031490003-7). The computations were carried out on MVS-10P at Joint Supercomputer Center of the Russian Academy of Sciences (JSCC RAS).

References

[1] Yamashita I 2016 *Surface Innovations* **4** (3) 111
[2] Karas V O, Westerlaken I, and Meyer A S 2015 *J. Bacteriol.* **197** 3206
[3] Grant R A, Filman D J, Finkel S E, et al. 1998 *Nat. Struct. Biol.* **5** 294
[4] Tereshkin E V, Tereshkina K B, Kovalenko V V, Loiko N G, and Krupyanskii Y F 2019 *Russian Journal of Physical Chemistry B* **13**(5) 769 DOI: 10.1134/S199079311905021X
[5] Haikarainen T, Thanassoulas A, Stavros P, et al. 2011 *J. Mol Biol.* **405**(2) 448
[6] Tereshkin E, Tereshkina K, Loiko N, Chulichkov A, Kovalenko V & Krupyanskii Y 2019 *J Biomol Struct Dyn.* **37**(10) 2600
[7] Moiseenko A, Loiko N, Tereshkina K, Danilova Y, Kovalenko V, Chertkov O, Feofanov A V, Krupyanskii Y F, Sokolova O S 2019 *BBRC* **517**(3) 463
[8] Pettersen E F, Goddard T D, Huang C C, Couch G S, Greenblatt D M, Meng E C, and Ferrin T E 2004 UCSF *Journal of Computational Chemistry* **25**(13) 1605
[9] Hess B, Kutzner ., van der Spoel D, and Lindahl E 2008 *J. Chem. Theory Comput.* **4** 435
[10] Ivani I, Dans P D, Noy A, et al. 2015 *Nat. Methods* **13** 55
[11] Harris K R and Woolf L A 1980 *J. Chem. Soc., Faraday Trans.* **76** 377
[12] Fell C J D, and Hutchison H P 1971 *Journal of Chemical & Engineering Data* **16**(4) 427
[13] Mills R 1955 *Journal of the American Chemical Society* **77**(23) 6116

Determination of Reynolds shear stress from the turbulent mean velocity profile and spectral characteristics of Orr-Sommerfeld -Squire equations

R. S. Selim[1]

[1] Moscow Institute of Physics and Technology, 9 Institutsky ave., Dolgoprudny, 141700, Moscow region, Russian Federation
selim.rs@phystech.edu

Abstract. Theoretically, based on a waveguide model, the expression of the tangential stress is formulated for steady, two-dimensional incompressible fluid flow over a flat plate in turbulent boundary layer. It is dependent on some factors, one of them, the behaviour of the last damping mode eigenvalues, and eigenfunctions, that are deduced from solution Orr-Sommerfeld equation by spectral Chebyshev collocation Method. Verification of the latter method is investigated by comparison the deduced formula of turbulent tangential stress with experimental data. In addition to, weight factors in this expression are connected to define the condition of dynamical system solution for multiple 3-wave resonance. This system is solved numerically, and the dynamic invariant is normalized to obtain the time average of the square modulus harmonic, and sub harmonics amplitudes by theorem Birkhoff-Khinchin. Comparison is made between the time-averaged and the phase average for the square modulus of harmonic, and sub harmonic amplitudes that defined on the unit sphere, in the state of multiple 3-wave resonance.

Keywords: Incompressible fluid flow – Orr-Sommerfeld eigenfunctions –Turbulent tangential stress

1. Introduction

In [1], based on Navier-Stokes equations, the validity of presentation of a developed turbulent field of fluctuations in the form of superposition of field coherent structures and stochastic component is considered. However, it is not explicitly shown that the mean field of the longitudinal velocity component has a known behaviour with the logarithmic part of the dependence on the transverse coordinate. In the presence of a given dependence of the tangential stresses on y , the problem is reduced to solve the Reynolds equations. Last works [2-7], studied the average longitudinal velocity profile, using different asymptotic methods with various auxiliary data borrowed from the experiment. Especially, in [3], an explicit expression for the average velocity profile as a function of the Reynolds number over the boundary layer thickness is obtained, which is heuristically compiled from the law of the wall behaviour and the behaviour in the region of the velocity defect. The models based on a dynamic waveguide representation by pulsation are considered for a turbulent boundary layer [8]. The solution of the Reynolds equations for a developed turbulent boundary layer on a plate in an incompressible fluid in the presence of specified tangential stresses is obtained by the method of asymptotic expansions [9]. In [10], an analytical method is presented for determining the Reynolds shear stress profile in steady, two-dimensional wall-bounded flows using the mean stream wise velocity. Since the near-wall region can generally be subdivided into viscous sub layer, buffer layer and fully turbulent layer for turbulent flow over a flat plate. Therefore, the meshed elements need to be carefully arranged to capture the velocity and turbulence in different layers. Moreover, in [11], the turbulent shear flow is generally divided into three layers (inner, meso, outer) with two overlap regions. The meso layer is the intermediate layer in between inner (Prandtl) limit and outer (Karman) limit in a turbulent shear flow which covers domain of buffer layer and traditional overlap region. In order to explain the theory of dynamic system more accurate, more exact numerical methods are achieved, which are requiring considerable expertise in modern mathematics [12]. The formula used in [13] can be used to formulate the well-known theorem of Birkhoff-Khinchin. In detail, more knowledge that under some conditions is proved that, the average time of a function is almost anywhere, and that is connected to the average space. The average time for nearly all initial points is the same, as a special class of Ergodic system [14]. Ergodic definitions and assumptions are fundamental to ergodic theory implementations. The underlying concept is that the average time of their properties for certain structures is equal to the average space [15]. Menyuk et al. [16] discussed ergodicity to some extent, clarifying the dependence on the values of the interaction coefficients, through the case of two or more triads that can lead to chaotic behavior. In this paper, we consider a model composed of incompressible fluid over a flat plate in the turbulent boundary layer. The objective of the present work is to compare the time-averaged, and the phase average for the squares modulus of harmonic, and sub harmonic amplitudes, that defined on the unit sphere, in the state of multiple 3-wave resonance. We compute theoretically the tangential stress expression, based on a waveguide model of turbulent boundary layer. The plan of this work is as follows: In the next section, a dynamical system of finite mode model is investigated. In the third section, the expression of the Reynolds shear stress is formulated to describe its behaviour. Finally, the salient results of our analysis are discussed in the conclusion.

2. Statement of the problem

To gain insight into the randomness of the turbulence, we consider the existence of random solutions for differential equations. We consider finite model consist of the main harmonic, and five sub harmonics, as a dynamical system for describing the coherent structure [17], for two-dimensional incompressible fluid flow, with zero pressure gradient, flowing over a flat plate.

$$\frac{da_1}{dt} = \sum_{s=1}^{n} \Lambda_1^s a_2^s a_3^s, \quad \frac{da_2^s}{dt} = \Lambda_2^s a_1 a_3^{s*}, \quad \frac{da_3^s}{dt} = \Lambda_3^s a_1 a_2^{s*}, \quad s = 1, 2, \cdots, n \quad \textbf{(1)}$$

The boundedness of system solution (1) is controlled by the positivity of the quantities q_2^s , q_3^s which based on the determination of coefficients $\Lambda_1^s, \Lambda_2^s, \Lambda_3^s$. The solution of the above system of differential equations is formulated with statistical initial conditions, which are distributed on the surface 6-dimensional ellipsoid. But a change of variables [1], in the model version, by $a_i(t_1,t_2) = \phi(t_2)\tilde{a}_i(\phi(t_2)\tilde{t}_1, t_2), \phi(t_2) = \sqrt{\lambda/\Phi}$ are introduced here to capture solutions of the dynamical system (1), on the unit sphere, and so on, the theory of Birkhoff-Khinchin is applied [13]. Explicit solutions for amplitudes are also found in classical textbooks [18], moreover, solutions for amplitudes and phases are found in modern form [19]. Based on the above explanation, solutions of the dynamical system (1) are represented in the form of dynamical invariant [17] as follow:

$$I^{(n)}(t_1) = |\tilde{a}_1|^2 + \sum_s q_{12}^s |\tilde{a}_2^s|^2 + q_{13}^s |\tilde{a}_3^s|^2 = 1, q_{12}^s = \frac{q_2^s}{q_1}, q_{13}^s = \frac{q_3^s}{q_1}, s = 1, 2, \ldots, n. \quad \textbf{(2)}$$

Here the sum is formulated by index s, for all sub harmonics in the case of multiple three wave resonance. It is found that, the dynamic invariant is defined on the unit sphere, and it is formulated by a quadratic form of the principal harmonic, and sub-harmonics amplitudes as illustrated in Eq. **(2).** Ergodicity of the dynamical system (1) is defined by the following relation [15] :

$$\lim_{T\to\infty} \overline{f(x)}^T = \frac{1}{\sigma(S)} \int_S f(x) d\sigma, \quad \textbf{(3)}$$

where $\sigma(S)$ is the area of the surface. It is clear that, as outlined in table.1, the time average over a single sub harmonic almost equal space overage over the appropriate sphere surface, for almost every sub harmonic.

Table.1. Comparison between the time average, and phase a average, for five sub harmonics

Number of sub harmonic	$\lim_{T\to\infty} \overline{f(x)}^{T}$	$\frac{1}{\sigma(S)}\int_s f(x)d\sigma$
first sub harmonic	0.0825247	0.0817011
second sub harmonic	0.0199271	0.0190730
third sub harmonic	0.0930313	0.0961254
four sub harmonic	0.0338027	0.0333333
five sub harmonic	0.0287763	0.0286095

3. Formulation of tangential stress

In this section, a wave guide theory of a developed turbulent boundary layer is presented, to relate the dynamics of Tollmien Schlichting waves to the turbulent tangential stresses. This expression is defined in terms of the last damping mode eigenvalue, and their corresponding eigenfunction, that are deduced from solving the spectral problem Orr-Sommerfeld equation by Chebyshev collocation method for a given mean velocity profile $U(y)$ [3]. Introducing the Fourier form of continuity, and the normal component of the vorticity equation, then the formula of tangential stresses, is defined by

$$u=\frac{i}{k^2}\left(\alpha\frac{d\,\mathrm{v}}{d\,y}-\beta\,\eta\right),\quad \eta=\frac{\beta\,\mathrm{v}}{-\omega\,(\mathbf{k})+\alpha\,U(y)}\frac{d\,U}{d\,y} \tag{4}$$

Here, α, β are the wave numbers of the oscillation in the stream-wise and span-wise directions respectively, and $\omega(\mathbf{k})$ is the dispersion relation of the spectral problem Orr-Sommerfeld equation, and v, η are the vertical components of velocity and vorticity that deduced from Orr-Sommerfeld, Squire, equations respectively. Substituting representation (4), in $\tau=-\langle \mathrm{u\ v}\rangle$, thus we have the following formula:

$$\tau=-\langle \mathrm{u\ v}\rangle=\frac{i}{k^2}\left(\alpha\left\langle \mathrm{v}.\frac{d\,\mathrm{v}}{d\,y}\right\rangle-\frac{\beta^2\langle \mathrm{v.\,v}\rangle}{-\omega+\alpha\,U(y)}\frac{d\,U}{d\,y}\right), \tag{5}$$

$$\begin{gathered}
\mathrm{v}_{\omega\mathbf{k}}(t_0,t_1)=A_{\omega\mathbf{k}}(t_1)\,\delta\left(\omega-\omega_{T-S}(\mathbf{k})\right)e^{i\,\omega\,t_0-i\,\mathbf{k}.\mathbf{r}}\varphi_0(\mathbf{k},y),\\
A_{\omega\,\mathbf{k}}(t_1)=a_1(t_1)\,\delta(\mathbf{k}-\mathbf{k}_1)+a_1^*(t_1)\,\delta(\mathbf{k}+\mathbf{k}_1)+\sum a_2(t_1)\,\delta(\mathbf{k}-\mathbf{k}_2)+a_2^*(t_1)\,\delta(\mathbf{k}+\mathbf{k}_2)+\\
\sum a_3(t_1)\,\delta(\mathbf{k}-\mathbf{k}_3)+a_3^*(t_1)\,\delta(\mathbf{k}+\mathbf{k}_3),\\
\langle \mathrm{u\ v}\rangle=\lim_{t_1\to\infty}\frac{1}{t_1}\int_0^{t_1}\left(\lim_{t_0\to\infty}\frac{1}{t_0}\int_0^{t_0}\mathrm{u\ v}\,d\,t_0\right)d\,t_1
\end{gathered} \tag{6}$$

Here $\boldsymbol{\delta}(\mathbf{k})$ is Delta-Dirac function, and $A_{\omega k}$ is the amplitude of the vertical velocity component in a developed turbulent boundary layer, summation is carried out for all sub harmonics of the lowest harmonic allowed by the dispersion equation and the condition of positive certainty of the wave invariant in the state of multiple three-wave resonance [17]. Averaging this expression over t_0, t_1 gives an expression for the turbulent tangential stress in the discrete representation of the coherent structure. Here, the index summation s, is simulated for the main harmonic, and five sub-harmonics, where $\left\langle|\tilde{a}_s|^2\right\rangle_{t_1}$ is the time average for harmonic or sub harmonic amplitude.

The value of parameter $\varepsilon=\varepsilon(\mathrm{Re}_{\delta^{**}})=0.033$ that deduced from Reynolds number, $\mathrm{Re}_{\delta^{**}}=7.57\times10^4$ is provided in [9], and also $\phi=0.417133$, in a stationary case [1] is calculated from formula, that defined above in section 2. Values of $q_{12}^{\ s}$, and $q_{13}^{\ s}$ are determined, to define the boundedness for solving the dynamical system (1), that based on coefficients $\left(\Lambda_1^s, \Lambda_2^s, \Lambda_3^s\right)$. It is clear that the mathematical expression of (7) represents structure of least damping mode eigenfunction $\varphi_0(\mathbf{k}, y)$ of the Orr-Sommerfeld equation, for different combination of the wave vectors. These vectors are especially located on the curve of three wave resonance, as displayed in Fig.1.

$$\tau = \frac{\varepsilon^2 \phi^2}{(2\pi)^4}\left(f_1(\mathbf{k}_1, y) + 2\sum_{s=1}^{5}\sum_{l=2}^{3} \frac{\left\langle \left|\tilde{a}_l^s\right|^2 \right\rangle_{t_1}}{q_{1,l}^{\ s}\left|\mathbf{k}_l^s\right|^2}\left(f_2\left(\mathbf{k}_l^s, y\right) + f_3\left(\mathbf{k}_l^s, y\right)\frac{\partial U(y)}{\partial y}\right)\right) \quad \textbf{(7)}$$

$$\langle \ldots \rangle_{t_1} = \frac{1}{t_1}\int_0^{t_1} \ldots d\,t_1, \quad q_{1,l}^{\ s} = q_{1l}^{\ s}$$

$$f_1(\mathbf{k}_1, y) = \frac{\left\langle \left| \tilde{a}_1 \right|^2 \right\rangle_{t_1} \alpha_1 \operatorname{Im}\left[\varphi_0^*(\mathbf{k}_1)\,\partial\varphi_0(\mathbf{k}_1)/\partial y\right]}{\left|\mathbf{k}_1\right|^2};$$

$$f_2\left(\mathbf{k}_l^s, y\right) = \alpha_l^s \operatorname{Im}\left[\varphi_0^*\left(\mathbf{k}_l^s\right)\partial\varphi_0\left(\mathbf{k}_l^s\right)/\partial y\right];$$

$$f_3\left(\mathbf{k}_l^s, y\right) = \frac{\left(\beta_l^s\right)^2 \left|\varphi_0\left(\mathbf{k}_l^s\right)\right|^2 \omega_i\left(\mathbf{k}_l^s\right)}{\omega_i^2\left(\mathbf{k}_l^s\right) + \left(\omega_r^2\left(\mathbf{k}_l^s\right) - \alpha_l^s\, U(y)\right)^2}.$$

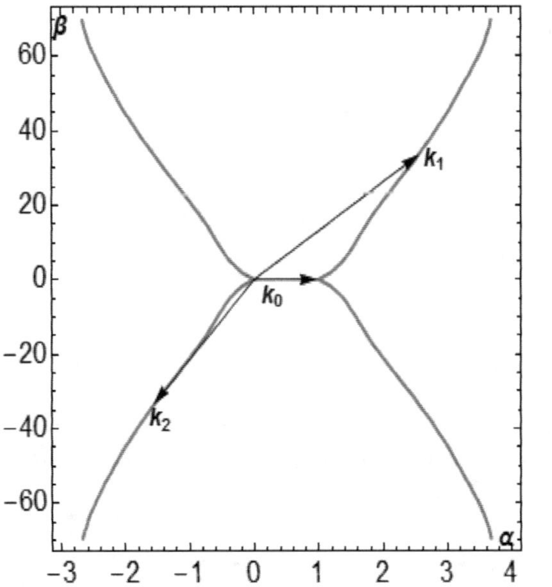

Fig.1 Curve three wave resonance, which defined the least damping mode of Musker Profile [3] for $\mathrm{Re} = 7.57\times10^4$.

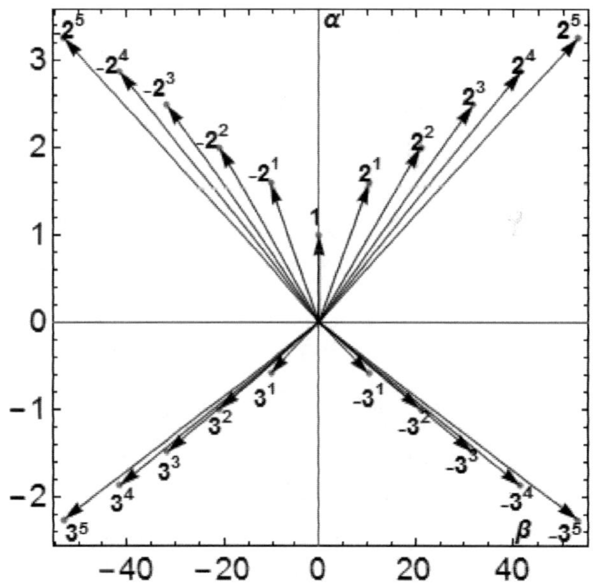

Fig.2 Structure of wave vectors which are located on three wave resonance curve, and formulated in Eq.(7)

As outlined in formula (7), the left side represent the tangential stress. The terms on the right hand side depend on some factors, like, parameter values $\left(\varepsilon, \phi, q_{12}^{\ s}, q_{13}^{\ s}\right)$, time averages of harmonic, and sub harmonics amplitudes $\left\langle \left|\tilde{a}_l^s\right|^2 \right\rangle_{t_1}$, the mean velocity profile Mucker $U(y)$ [3], eigenvalues, and eigenfunctions for the least damping mode of different wave vectors $\mathbf{k}_l^s = (\alpha_l^s, \beta_l^s)$. The real and imaginary parts of leading mode eigenvalues are outlined by $\omega_r = \mathrm{Re}[\omega]$, and $\omega_i = \mathrm{Im}[\omega]$ respectively. The asterisk for, $\varphi_0^*(\mathbf{k})$ denotes a complex conjugate of the least damping mode eigenfunction $\varphi_0(\mathbf{k})$. The index sum l is considered for two wave vectors $\mathbf{k}_2, \mathbf{k}_3$, which are satisfied the triad resonance with the main wave vector. The first part in the right, concerns only the leading mode eigenfunction, for the main wave harmonic $\mathbf{k}_1 = (1,0)$, as shown in Fig.2 with index number 1.

Table.2. Multiplier values, of five sub harmonics, are corresponding to five wave vectors

Number of sub harmonic	q_{12}^s	q_{13}^s
$s=1, \alpha_1 = 1.685, \beta_1 = 13.38$	0.156531	0.102806
$s=2, \alpha_2 = 2.313, \beta_2 = 27.84$	0.258746	0.22306
$s=3, \alpha_3 = 2.832, \beta_3 = 40.38$	0.296102	0.170738
$s=4, \alpha_4 = 3.308, \beta_4 = 54.84$	0.136389	0.07947605
$s=5, \alpha_5 = 3.633, \beta_5 = 69.78$	0.0849696	0.0216714

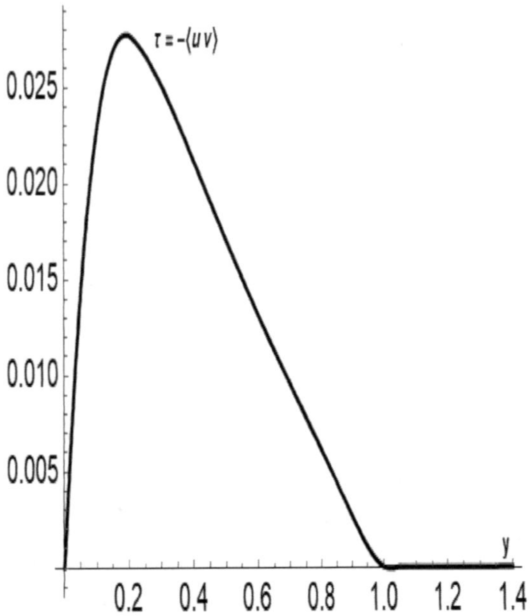

Fig.3. Behaviour of tangential stress for incompressible fluid flow over a flate plate in the turbulent boundary layer.

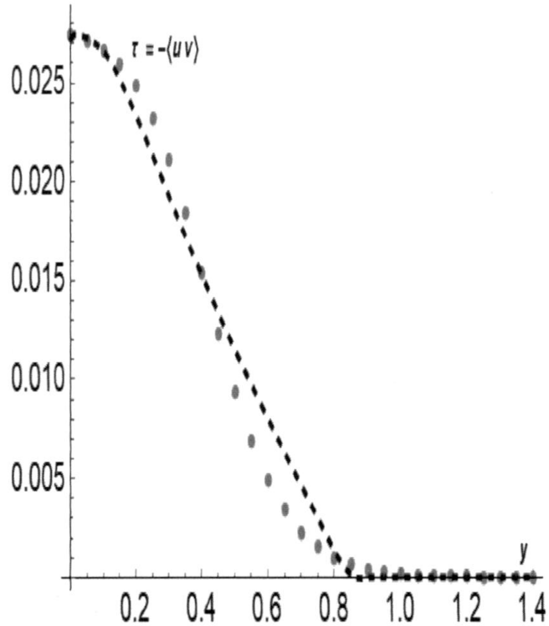

Fig.4. Comparison between tangential stress, which has a shift in the thickness sub layer, and experimental data, black points: theoretical expression, red points: Experimental data Klebanoff [20].

The second and third terms are represented leading mode eigenfunctions for five sub harmonics amplitudes, which are corresponding to five wave vectors $\mathbf{k}_2^s, \mathbf{k}_3^s$ respectively. In addition, these terms are multiplied by 2, due to the symmetry of the wave vector component β.

The wave vector $\mathbf{k}_2^s, s = 1, \ldots, 5$, are indicated in Fig.2 by index numbers, $(2^1, 2^2, 2^3, 2^4, 2^5)$, where the power is referred to the number of sub harmonic, that are corresponding to five multipliers values

$\left(q_{12}^{1}, q_{12}^{2}, q_{12}^{3}, q_{12}^{4}, q_{12}^{5}\right)$, respectively as outlined above in table 2, and the similar situation is also defined for $\mathbf{k}_3^s, s=1,.....5$, for $\left(q_{13}^{1}, q_{13}^{2}, q_{13}^{3}, q_{13}^{4}, q_{13}^{5}\right)$. Generally, the flow field with the normal coordinate y is divided into three sub domains, the inner, intermediate, and outer $y \sim \Delta$, $y \sim 1$, $y \sim 1/\varepsilon$ respectively as outlined in [9]. The behaviour of tangential stress with the normal coordinate y is completely defined in Fig.3. In addition, Fig. 4 is formulated to determine the behaviour of tangential stress, only in the external domain of the boundary layer. So that, a shift is applied to the thickness of the boundary layer, to focus the behaviour study of tangential stress on the outer region. It is found that, the behaviour of tangential stress has a qualitative and quantitative agreement with experimental data Klebanoff [20].

4. Conclusions

In this paper, we determine the expression of the tangential stress, in terms of the waveguide model in the turbulent boundary layer, for given mean velocity profile Mucker. Their behavior dependent mainly on the spectral characteristic of eigenvalues, and eigenfunction Orr-Sommerfeld equation, that obtained by using the spectral Chebyshev collocation method. Generation of eigenfunctions, for different combination of the wave vectors $\mathbf{k}_2^s, \mathbf{k}_3^s$ in the deduced formula of tangential stress are related to the structure for a set of sub harmonic amplitudes, in the case of multiple three wave resonance. Numerical solutions of the dynamical system are contributed to obtain agreement between the time, and space averages for the square of main harmonic, and five sub harmonics amplitudes. It is found that, the behaviour of turbulent tangential stress, with the shift in the normal coordinate y, has a good agreement with the experimental data Klebanoff.

References

[1] Zharov V A 2014 Model representation of a coherent structure in a developed turbulent boundary layer, *Scientific Notes of TsAGI*, XLV, **5**: 33-46.
[2] Kader B A and Yaglom A M 1980 *Laws of similarity in near-wall turbulent flows*. - M.: VINITI, 81-155
[3] Musker A J 1979 Explicit expression for the velocity distribution along a smooth wall in a turbulent boundary layer, *AIAA* J., **17** (**6**): 655-657.
[4] Sychev V V 1987 On the structure of a turbulent boundary layer, *PMM*, **51** (**4**): 593-599.
[5] Vigdorovich I I 1993 Asymptotic study at large Reynolds numbers of a turbulent boundary layer on a flat plate, *Izv. RAS. MZhG*, **4**:106-117.
[6] Mikhailov V V 2005 Universal law of velocity defect for turbulent boundary layer, *Izv. RAS. MZhG*, **2**:89-101.
[7] Mikhailov V V and Samoilova N V 2013 On the calculation of the friction coefficient of a turbulent boundary layer. Criterion of layer relamination, *Scientific Notes of TsAGI*, XLIV, **6**: 58-68.
[8] Landahl, M T 1967 A wave-guide model for turbulent shear flow, *J. Fluid Mech*, **29**: 441–459.
[9] Zharov V A and Le Van Ha 2016 Construction of the longitudinal velocity profile at the given tangential stress in the developed turbulent boundary layer on the plate, *Scientific Notes of TsAGI*, **47**(8):50-60.
[10] Ralph J V and Michael P S 2018 Determination of wall shear stress from mean velocity and Reynolds shear stress profiles, *Physical Review Fluids,* **3**, 034606.
[11] Afzal N and Seena A 2018 Mean velocity, Reynolds shear stress, fluctuations of velocity and pressure, due to log laws in a turbulent boundary layer and origin offset by Prandtl transposition theorem, *Journal of Fluids Engineering*, **47**(8):1-50.
[12] Katok A B and Hasselblat B 1999 *Introduction to the Modern Theory of Dynamical Systems*, Cambridge University Press, Moscow.
[13] Kolmogorov A N 1985 *Selected Works of Mathematics and Mechanics*. Nauka, Moscow.
[14] Joseph D Z 2002 A question of Priority and the Ergodic Theorems, *Historia Mathematica*, **29**:138–156
[15] Orszag S 1977 *Lectures on the Statistical Theory of Turbulence*, MIT, **31,** 216 p.
[16] Menyuk C R, Chen H H and Lee Y C 1982 restricted multiple three-wave interactions: Painleve analysis, *J. Mathematical Physics,* **24,** 1073.
[17] Zharov V A, Lipatov I I, and Selim R S 2020 Wave model of organized structures in a turbulent boundary layer on a plate with zero longitudinal pressure gradient, *Scientific Notes of TsAGI*, **51** (**6**): 51-59.

[18] Craik A D 1988 Wave Interactions and Fluid Flows, *Cambridge University Press*, 1986, XII, pp: 322, ISBN 0521 267404.
[19] Kartashova and Bustamante (2011) Resonance Clustering in Wave Turbulent Regimes: Integrable Dynamics, *Commun. Comput. Phys*. 10, 1211.
[20] Klebanoff P S 1955 Characteristics of turbulence in a boundary layer with zero pressure gradients. *NACA Rep*. **1247**.

The influence of Fermi surface anisotropy and the charge carrier surface scattering kinetics on the electrical conductivity of a thin metal film in the view of the quantum size effect

I A Kuznetsova, O V Savenko and D N Romanov*

Microelectronics and General Physics Department, P. G. Demidov Yaroslavl State University, 14 Sovetskaya Street, Yaroslavl 150000, Russia

*E-mail: romanov.yar357@mail.ru

Abstract. The electrical conductivity of a thin metal film in an alternating electric field is calculated considering the quantum size effect. The Fermi surface of the metal has the shape of an ellipsoid of rotation, the main axis of which is parallel to the plane of the film. The quantum kinetic equation obtained from the von Neumann equation (the Liouville quantum equation) is solved. The Soffer model is used as the boundary conditions for the distribution function. The dependence of the electrical conductivity on the film thickness is analyzed. A comparison is made with experimental data on the electrical conductivity of bismuth thin films.

1. Intoduction

Thin films are widely used as functional, reflective, conductive and dielectric materials in the formation of contacts, the manufacture of printed circuit boards, integrated circuit elements in microelectronics, the creation of light filters, the element base of optoelectronics, in modern lithographic processes. Modern technology makes it possible to produce objects (films, wires, dots) of several nanometers in size, in which electrons exhibit quantum properties. The reduction of the film thickness to the size of the order of the de Broglie wavelength of the charge carriers leads to the quantum size effect due to the quantization of the energy of the charge carriers. The dependence of the electrical conductivity on the thickness of the thin metal film is oscillating, which is shown in [1, 2].

There are several methods for calculating the electrical conductivity of thin metal films, taking into account the quantum size effect. In [3], the effect of surface scattering on the nature of the motion of charge carriers is calculated using the Kubo formalism using the classical kinetic equation. The contribution of impurity scattering is not considered. Surface scattering is described by the surface potential Hamiltonian. The conductivity in the film plane oscillates with a period equal to half the de Broglie wavelength of the electron. However, this method does not establish the possibility of switching from the quantum case to the classical one. This drawback was eliminated in [4] by introducing scattering potentials into the model Hamiltonian, which consider the scattering on the impurities and on both surfaces of the film. Another way to solve the problems was to directly calculate the integral of surface collisions [5]. The relaxation time is calculated considering the total volume and surface scattering and the quantum size effect.

The problem of the electrical conductivity of a thin metal film is solved considering the quantum theory of transport phenomena in this paper. The method of solving the problem is to find the diagonal elements of the density matrix (non-equilibrium distribution function) from the solution of the Liouville equation. The surface scattering of charge carriers is considered through the Soffer boundary conditions [6].

2. Solving method and mathematical calculations

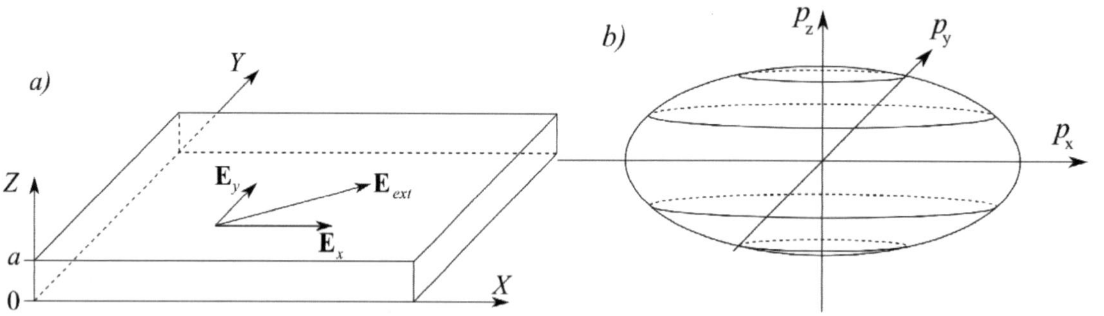

Figure 1. (a) Quantum metal film in a longitudinal electric field;
(b) surface of the Fermi metal film.

We consider a quantum infinite film of metal with thickness a, which is located in an external alternating electric field. We introduce a Cartesian coordinate system. The film plane lies in the XY plane (see figure 1 (a)). The origin is located on the lower surface of the film.

The electric field strength vector $\mathbf{E}_{\text{ext}} = (E_x, E_y, 0)$ is located in the film plane and varies according to the harmonic law:

$$\mathbf{E}_{\text{ext}} = \mathbf{E}_0 \exp(-i\omega t), \tag{1}$$

where ω is the angular frequency of the external field, $\mathbf{E}_0 = (E_{0x}, E_{0y}, 0)$ is the vector of the amplitude of the electric field strength. We assume that ω is much less than the frequency of the plasma resonance ω_p in the metal film (in the metal, ω_p lies in the ultraviolet region $\omega_p \approx 10^{16}\ \text{s}^{-1}$).. The thickness of the film is less than the depth of the skin layer (the skin effect is not considered). In this case, the specific electrical conductivity, considering the restriction on the plasma frequency, should be greater than $10^4\ \Omega^{-1}\text{m}^{-1}$. Since the electrical conductivity for most metals is in the range of values $10^6\ \Omega^{-1}\text{m}^{-1} < \sigma < 10^8\ \Omega^{-1}\text{m}^{-1}$, then this condition is met automatically.

The state of charge carriers in a metal is described by the energy, which can be expressed in terms of a quasi-pulse. Let the surface of the constant energy of the metal in the case of a thick film, in which there are no quantum dimensional effects, has the shape of an ellipsoid of rotation, the axis of rotation of which is directed along the axis of the projection of the quasi-pulse of the charge carriers on the X (p_x) axis (see figure 1 (b)). As the film thickness decreases, the energy of the charge carriers is quantized and takes a constant value on equidistant ellipses, the centers of which lie on the p_z axis, and the ellipses themselves are located on the ellipsoid of rotation (see figure 1 (b)). The energy of the electrons (holes) takes the form [7]:

$$\varepsilon_n = \frac{p_x^2}{2m_\parallel} + \frac{p_y^2}{2m_\perp} + \varepsilon_{zn};\ \varepsilon_{zn} = \frac{p_{zn}^2}{2m_\perp};\ n = \pm1, \pm2, \pm3, \dots, \pm N, \tag{2}$$

where $m_\perp$ and $m_\parallel$ are the transverse and longitudinal effective masses. Since the film thickness is comparable to the de Broglie wavelength of electrons, the projection of the electron momentum p_z takes a discrete series of values, and the projections p_x and p_y take a continuous series of values.

A metal film for free electrons (holes) can be represented as a potential well with infinitely high walls. For a given potential well, the quantization condition of the projection of the quasi-pulse on the Z-axis (p_{zn}) has the form

$$p_{zn} = \frac{\pi\hbar n}{a};\ n = \pm1, \pm2, \pm3, \dots \tag{3}$$

Since the isoenergetic surface is an ellipsoid of rotation, the specific conductivity of a thin film is a rank 2 tensor. If the main axis of the ellipsoid is parallel to the p_x axis (see figure 1 (b)), then the conductivity tensor is diagonal:

$$\tilde{\sigma} = \begin{pmatrix} \sigma_\parallel & 0 \\ 0 & \sigma_\perp \end{pmatrix}. \tag{4}$$

In the case of an arbitrary direction of the main axis of the ellipsoid in the plane (p_x, p_y), the expression for the conductivity tensor can be obtained using the rotation matrix around the axis p_z:

$$\sigma_{ij} = M_{ik}^{v_z}(\varphi_0)\tilde{\sigma}_{kl}M_{lj}^{v_z}(-\varphi_0), \tag{5}$$

where $M_{ik}^{v_z}(\varphi_0)$ are the components of the rotation matrix around the p_z axis by the angle φ_0:

$$M^{v_z}(\varphi_0) = \begin{pmatrix} \cos\varphi_0 & -\sin\varphi_0 \\ \sin\varphi_0 & \cos\varphi_0 \end{pmatrix}. \tag{6}$$

As a result, from expressions (4-6) we get the following form of the conductivity tensor:

$$\sigma = \begin{pmatrix} \sigma_\parallel \cos^2\varphi_0 + \sigma_\perp \sin^2\varphi_0 & (\sigma_\parallel - \sigma_\perp)\cos\varphi_0 \sin\varphi_0 \\ (\sigma_\parallel - \sigma_\perp)\cos\varphi_0 \sin\varphi_0 & \sigma_\parallel \sin^2\varphi_0 + \sigma_\perp \cos^2\varphi_0 \end{pmatrix}. \tag{7}$$

The components of the tensor (4) can be found using the kinetic equation. The kinetic equation for charge carriers in a crystal under the action of external fields is derived from the von Neumann equation (the Liouville quantum equation). The Liouville equation describes the evolution of mixed quantum states, which are characterized by a density matrix [8]:

$$\frac{\partial\hat{\rho}}{\partial t} = \frac{1}{i\hbar}[\hat{H}, \hat{\rho}], \tag{8}$$

where $\hat{\rho}$ is the density matrix operator, $\hat{H}$ is the Hamilton operator, $\hbar$ is the Planck constant, and the brackets denote the commutator.

In the framework of perturbation theory, the Hamiltonian of an electron (hole) in a crystal can be represented as

$$\hat{H} = \hat{H}_0 + \hat{U} + \hat{V}, \tag{9}$$

where $\hat{H}_0$ is the Hamiltonian of the free charge carrier in a vacuum, $\hat{U} = -\mathbf{Fr}$ is the potential energy of the external fields, and $\hat{V}$ is the potential energy of the impurity field in the crystal. Since the external fields and the field of impurity atoms are small perturbations, it is convenient to work in the basis of the eigenstates $\varphi_{\mathbf{p}}$ of the Hamiltonian $\widehat{H}_0$, that is, in the basis of plane waves.

Thus, the Liouville equation (8) in matrix form, taking into account (9), is as follows [8]

$$i\hbar\frac{d\rho_{\mathbf{p'p}}}{dt} = (\varepsilon_{n'} - \varepsilon_n)\rho_{\mathbf{p'p}} + \sum_{\mathbf{p''}}\left[(U+V)_{\mathbf{p'p''}}\rho_{\mathbf{p''p}} - \rho_{\mathbf{p'p''}}(U+V)_{\mathbf{p''p}}\right], \qquad (10)$$

The diagonal components of the density matrix $\rho_{\mathbf{pp}}$ correspond to the classical distribution function $f_{\mathbf{p}}$ [8]. Let's write equation (10) for $\mathbf{p'} = \mathbf{p}$:

$$i\hbar\frac{df_{\mathbf{p}}}{dt} = \sum_{\mathbf{p''}}\left[(U+V)_{\mathbf{pp''}}\rho_{\mathbf{p''p}} - \rho_{\mathbf{pp''}}(U+V)_{\mathbf{p''p}}\right]. \qquad (11)$$

Separately, it is worth noting that the function $f_{\mathbf{p}}$ is a set of distribution functions that describe the state of charge carriers in the plane (p_x, p_y) for each value of p_{zn} (n-th subzone): $f_{\mathbf{p}} = f_n(v_x, v_y)$. The distribution function f_n depends only on the coordinate along the film thickness (the z coordinate), since all points in the XY plane have the same properties: $f_n = f_n(z, v_x, v_y)$. The dependence of f_n on the z coordinate and on the velocity projections v_x and v_y does not contradict the Heisenberg uncertainty principle.

To unambiguously determine the nonequilibrium distribution function f_n from equation (11), it is necessary to set boundary conditions for it. As such, we use the Soffer model [6], which considers the dependence of the specularity coefficients of the lower and upper surfaces q_{1n} and q_{2n} on the roughness parameters g_1 and g_2 and the angle ϑ of the electron incidence on the film surface:

$$\begin{cases} f_n(0, v_x, v_y) = q_1(g_1, \vartheta_n) f_{-n}(0, v_x, v_y), \\ f_{-n}(a, v_x, v_y) = q_2(g_2, \vartheta_n) f_n(a, v_x, v_y), \end{cases} \qquad (12)$$

$$q_{1,2n}(g_1, \vartheta_n) = \exp\left(-\left[4\pi g_{1,2}\cos\vartheta_n\right]^2\right), \qquad g_{1,2} = g_{s1,2}/\lambda_F, \qquad (13)$$

where f_n and f_{-n} are nonequilibrium distribution functions of electrons (holes), respectively, with positive and negative velocity projections on the Z axis ($n = 1, 2, 3 \ldots$); ϑ_n is the angle between the charge carrier momentum and the p_z axis; g_{s1} and g_{s2} are the root – mean – square height of the surface relief for the lower and upper boundaries of the film; λ_F is the de Broglie wavelength of an electron with Fermi energy, which depends on the direction in the anisotropic crystal.

The current density $\mathbf{j}$ is defined as follows

$$\mathbf{j} = e\sum_n \iint \mathbf{v} f_n \frac{2dp_x dp_y}{(2\pi\hbar)^2}\left(\frac{p_{zn} - p_{z(n-1)}}{2\pi\hbar}\right), \qquad (14)$$

and the concentration of charge carriers does not depend on the film thickness and is defined as

$$n_c = \sum_n \iint f_n \frac{2dp_x dp_y}{(2\pi\hbar)^2}\left(\frac{p_{zn} - p_{z(n-1)}}{2\pi\hbar}\right). \qquad (15)$$

The average values (14) over the film thickness are usually found under experimental conditions

$$\langle \mathbf{j} \rangle = \frac{1}{a}\int_0^a \mathbf{j}\,dz, \tag{16}$$

Finally, the components of the tensor (4) are obtained from the formal local Ohm's law

$$\sigma_{\parallel} = \langle j_x \rangle / E_x, \qquad \sigma_{\perp} = \langle j_y \rangle / E_y. \tag{17}$$

The kinetic equation (11) is solved in the relaxation time approximation and in the linear approximation with respect to the external electric field. It is assumed that the scattering of charge carriers on impurity ions is much weaker than the scattering on neutral impurity atoms. This approximation allows us to neglect the dependence of the relaxation time on the energy of the conduction electrons [7].

We obtain the following expressions after performing a number of mathematical calculations for the components of the conductivity tensor (4):

$$\sigma_{\perp,\parallel} = \sigma_0 \Sigma_{\perp,\parallel}, \qquad \sigma_0 = \frac{n_c e^2 \tau_0}{m_0}, \tag{18}$$

$$\Sigma_{\perp,\parallel} = \frac{x_0}{z_{0\perp,\parallel} x_\lambda k_{m\perp,\parallel}} \left(1 - \frac{3\rho_\varepsilon}{64 x_0^4 z_{0\perp,\parallel} k_{m\perp}^{5/2}} \sum_{n=1}^{N} n \left(1 - \frac{n^2}{\rho_\varepsilon}\right) \times \right.$$

$$\left. \times \left(\frac{2 - q_{1n} - q_{2n} + (q_{1n} + q_{2n} - 2q_{1n}q_{2n})\exp(-\Omega_{\perp,\parallel n})}{1 - q_{1n}q_{2n}\exp(-2\Omega_{\perp,\parallel n})}\right)\left(1 - \exp(-\Omega_{\perp,\parallel n})\right)\right), \tag{19}$$

$$\Omega_{\perp,\parallel n} = \frac{2 z_{0\perp,\parallel} k_{m\perp} x_0}{n}, \quad q_{1,2n} = \exp\left(-\left[\frac{2\pi g_{s01,2}}{x_0} n\right]^2\right), \tag{20}$$

$$g_{s01,2} = \frac{g_{s1,2}}{\lambda_{F0}}, \quad z_{0\perp,\parallel} = \frac{\sqrt{\rho_\varepsilon}}{2\sqrt{k_{m\perp} x_\lambda k_{\tau\perp,\parallel}}} - i y_0, \quad x_0 = \frac{a}{\lambda_{F0}}, \quad x_\lambda = \frac{\lambda}{\lambda_{F0}}, \quad y_0 = \frac{a\omega}{v_0}, \tag{21}$$

$$\lambda = \tilde{v}_0 \tilde{\tau}_0 = v_0 \tau_0, \quad \tilde{v}_0 = \sqrt{\frac{2\varepsilon_F}{m_0}}, \quad v_0 = \sqrt{\frac{2\varepsilon_{FV}}{m_0}}, \quad p_0 = m_0 v_0, \quad \lambda_{F0} = \frac{2\pi\hbar}{p_0}, \tag{22}$$

$$m_0 = \sqrt[3]{m_\perp^2 m_\parallel}, \quad k_{m\perp} = \frac{m_\perp}{m_0}, \quad k_{m\parallel} = \frac{m_\parallel}{m_0} = \frac{1}{k_{m\perp}^2}, \tag{23}$$

$$\tau_0 = \sqrt[3]{\tau_{yy}^2 \tau_{xx}}, \qquad k_{\tau\perp} = \frac{\tau_{yy}}{\tau_0}, \qquad k_{\tau\parallel} = \frac{\tau_{xx}}{\tau_0} = \frac{1}{k_{\tau\perp}^2}, \tag{24}$$

where $\Sigma_{\parallel}$ and $\Sigma_{\perp}$ is the dimensionless longitudinal and transverse components of the tensor g_{s01} and g_{s02} the roughness parameters for the lower and upper bounds of the film, bezrazmernye on the characteristic wavelength of de Broglie λ_{F0}; $k_{m\parallel}, k_{m\perp}$ – dimensionless longitudinal and transverse effective masses, respectively; $k_{\tau\parallel}, k_{\tau\perp}$ – dimensionless longitudinal and transverse relaxation times of charge carriers, respectively; x_0 – dimensionless film thickness; y_0 is the dimensionless frequency of the external electric field; x_λ – dimensionless mean free path length; $\lambda = \tilde{v}_0\tilde{\tau}_0 = v_0\tau_0$ – characteristic free path length of electrons (holes) without taking into account surface scattering; $\tilde{v}_0$ and $\tilde{\tau}_0$ are the characteristic Fermi velocity and free path time of charge carriers in a quantum film; $\tilde{v}_0$ and $\tilde{\tau}_0$ are the characteristic Fermi velocity and free path time of charge carriers in a macroscopic sample. The free path of electrons (holes) Without taking into account the surface scattering, it does not depend on the film thickness, since this length is a characteristic of the crystal lattice.

3. Result analysis

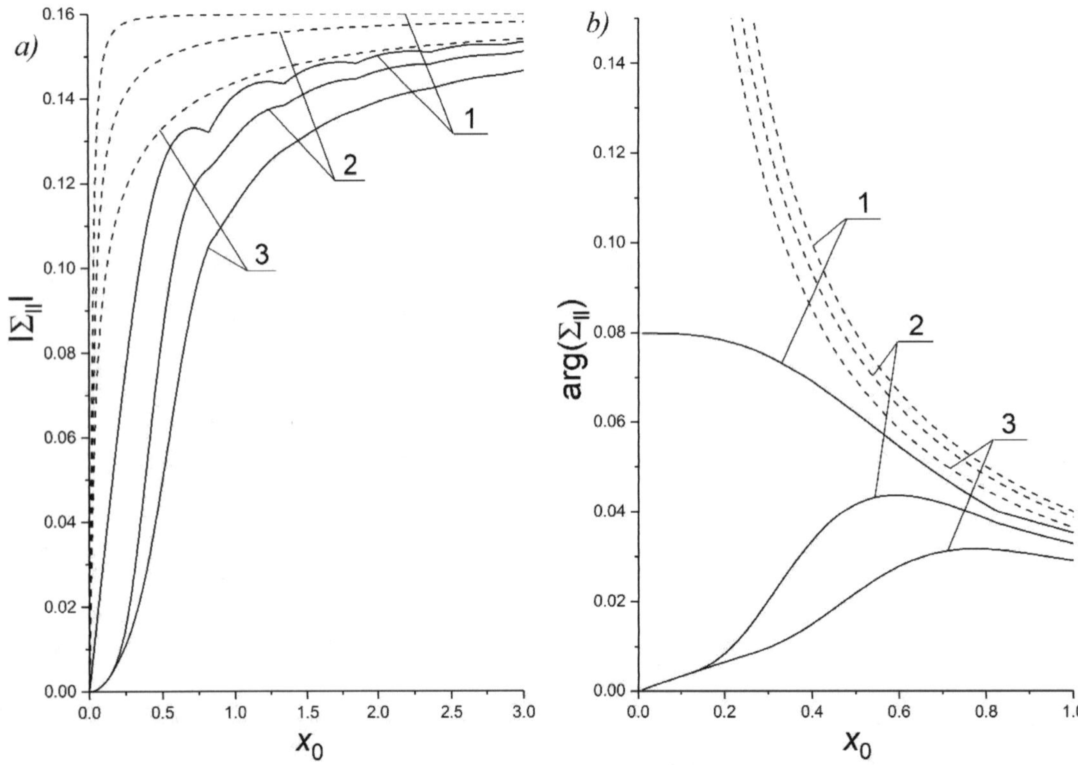

Figure 2. The dependence of the modulus (a) and argument (b) of $\Sigma_\parallel$ in the quantum case (19-21) (solid lines) and in the quasiclassical case [9] (dotted lines) on the dimensionless film thickness x_0 at $k_{m\perp} = 1, x_\lambda = 5, y_0 = 0.05, k_{\tau\perp} = 2.5$. Curves 1 – $g_{s01} = g_{s02} = 0$; curves 2 – $g_{s01} = g_{s02} = 0.05$; curves 3 – $g_{s01} = g_{s02} = 0.1$.

Figure 2 shows the dependences of the modulus and argument of the dimensionless longitudinal component of the integral conductivity tensor $\Sigma_\parallel$ on the dimensionless film thickness x_0. The solid curves are constructed considering the quantum theory of transport phenomena, and the dotted curves are constructed considering the quasi-classical theory. Considering the quantum size effect and the mirror boundaries, the dependence of the film conductivity on the thickness is oscillating. As the surface roughness increases, the amplitude of the oscillations decreases. The conductivity in the quasiclassical case is greater in modulus and in the argument of the conductivity in the quantum case and in the quasiclassical case, no oscillations are observed. This indicates that the oscillations are caused by quantization of the energy spectrum of charge carriers.

Figure 3 shows the dependences of theoretical and experimental data on the electrical conductivity of a quantum bismuth film [1]. The films were obtained by electron beam evaporation in a Bi target vacuum of 99.999% purity at a residual pressure of $\approx 10^{-6}$ Pa. The Fermi surface of electrons in bismuth consists of three symmetrically arranged ellipsoids, the axis of rotation of which is directed parallel to the plane (111). The angle between the axes of rotation is the same and is equal to $2\pi/3$. The agreement of theoretical calculations and experimental data is observed in the thickness range $a > 70$ nm. Within 10% of the relative error with a film thickness of less than 70 nm, there is agreement between the experimental and theoretical data. As the film thickness decreases, the difference between the theoretical and experimental values increases and becomes significant. The reasons for this difference are electron scattering at the boundaries of crystallites in a polycrystalline bismuth film, inter-valley scattering, changes in the structure of energy zones during the transition to two-dimensional materials (changes in the orientations of ellipsoids), etc.

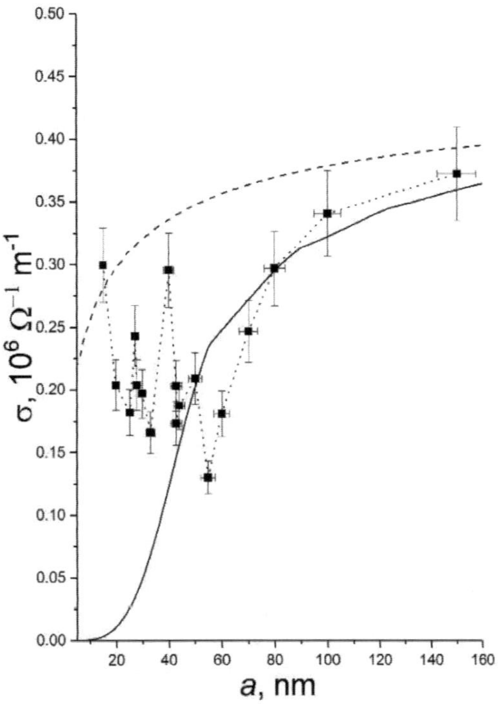

Figure 3. The dependence of the specific conductivity of a quantum bismuth film on the thickness. The dots represent the experimental data [1]. The solid curve is calculated in the quantum case (7, 18-24), and the dotted curve is calculated in the quasi – classical case [9]. The calculation is performed with the following parameters: $\sigma_0 = 2.1 \cdot 10^4\ \Omega^{-1}\text{m}^{-1}$, $\lambda = 400$ nm, $\lambda_{F0} = 27$ nm, $k_{m\perp} = 0.17$, $k_{\tau\perp} = 2.44$, $y_0 = 0$, $g_{s1} = g_{s2} = 1.6$ nm.

4. Conclusions

Theoretical model of the electrical conductivity of a thin metal film is constructed, considering the quantum size effect in this paper. The oscillatory dependence of the electrical conductivity on the thickness in the case of the mirror boundaries of the film is shown. As the roughness of the oscillation surfaces increases, they disappear. There is agreement between theory and experiment for the electrical conductivity of Bi thin films, however, in a limited range of film thicknesses.

Acknowledgments

The reported study was funded by RFBR, project number 19-32-90008.

References

[1] Demidov E V, Grabov V M, Komarov V A, Krushelnitckii A N, Suslov A V and Suslov M V

2019 *Semiconductors* **53** 729-31
[2] Tsysar K M, Smelova E M, Saletsky A M and Andreev V G 2020 *Thin Solid Films* **710** 138263
[3] Fishman G and Calecki D 1989 *Phys. Rev. Lett.* **62** 1302-5
[4] Sheng L, Xing D Y and Wang Z D 1995 *Phys. Rev. B: Condens. Matter* **51** 7325-8
[5] Meyerovich A E and Stepaniants A 2000 *J. Phys.: Condens. Matter.* **12** 5575
[6] Soffer S B 1967 *J. Appl. Phys.* **38** 1710
[7] Anselm A I 1978 *Introduction to semiconductor theory* (Nauka: Moscow)
[8] Blum K 1981 *Density matrix theory and applications* (London; New York)
[9] Kuznetsova I A, Romanov D N and Savenko O V 2021 *Phys. Scr.* **96** 045803

About boundary conditions for kinetic equations in metal

F Karimov[1], T N Lam[1] and A A Yushkanov[1]
[1] Moscow Region State University
10A, Radio str., Moscow, 105005, Russian Federation

faha_rtsu_2003@mail.ru, nhunglam279@gmail.com, yushkanov@inbox.ru

Abstract. Boundary conditions for kinetic equations describing the dynamics of electrons in the metal were analyzed. The boundary condition of the Fuchs and the boundary condition of Soffer were considered. The Andreev conditions for almost tangential moving electrons were taken into account. It is shown that the Soffer boundary condition does not satisfy this condition. The boundary condition was proposed that satisfies the Andreev condition. It is shown that this boundary condition passes in the limiting case into the mirror–diffuse Fuchs boundary condition.

1. Introduction

To describe the dynamics of electrons in the metal along with the kinetic equation one requires boundary conditions. These boundary conditions determine the nature of the interaction of electrons with the metal surface. Most often the mirror–diffuse Fuchs boundary condition are used. This condition implies that q–part of the electrons is reflected from the surface in the mirror manner. The remaining electrons are diffuse reflected. Then the electron distribution function f on the surface changes as follows [1]— [3]

$$f(\mathbf{v}) = qf(\mathbf{v}') + (1-q)f_0(v), \mathbf{v}' = \mathbf{v} - \mathbf{2n}(\mathbf{nv}). \quad (1)$$

Here $\mathbf{v}$ − the electron velocity before the collision with surface, $\mathbf{v}^0$ – the electron velocity after collision with the surface, $\mathbf{n}$ − a unit normal to the surface. The function $f_0(v)$ − the equilibrium distribution function. For the degenerate electron gas in the metal it has the following form

$$f_0(v) = \Theta(E_F - E), E = \frac{mv^2}{2}.$$

Here E – electron energy, E_F – the Fermi energy, $\Theta(x)$ − the Heaviside step function. It is equal to zero when $x < 0$. In other cases, it is equal to unit.

2. Kinetic equation

In the Fuchs boundary conditions (1), the value of q is considered constant. However, the coefficient of reflectivity q should depend on the angle of incidence of electrons on the metal surface.

In the work [4], it is shown that the reflectivity coefficient $q = q(\theta)$ (θ − the angle of incidence of the electron on the border) tends to one when the angle of incidence θ tends to $\pi/2$. From this it follows that the reflectivity coefficient $q = q(\theta)$ at $\theta \to \pi/2$ can be represented as the following decomposition

$$q(\theta) = 1 - a_1 cos\theta - a_2 cos^2\theta - a_3 cos^3\theta + \cdots. \quad (2)$$

Here a_n— some coefficients depending on the properties of metal surface.

It was proposed the model describing the dependence of the reflectivity coefficient on the angle of incidence of electrons on metal surface [5]

$$q(\theta) = \exp[-(-(4\pi G\cos\theta)^2)], G = \frac{h_s}{\lambda_F}. \quad (3)$$

In equation (3) value h_s – the mean-squared height of the surface relief, λ_F – the wavelength of an electron on the Fermi surface.

This dependence of the reflectivity coefficient on the incidence angle of electrons on the metal surface (3) has been used in several papers [6]– [8].

Let us consider the behavior of the reflectivity coefficient q with almost tangential incidence of the electron onto the metal surface in the model of Soffer (3). Then $\theta \to \pi/2$ and $\cos\theta \to 0$. Therefore

$$q(\theta) \simeq 1 - (4\pi G)^2 (\cos\theta)^2. \quad (4)$$

Hence $q(\theta) \sim 1 - A(\cos\theta)^2$ in this limit.

This contradicts to the Andreev condition (2). In the Soffer model the reflectivity coefficient q tends to unit too fast at $\theta \to \pi/2$.

Let us consider the model boundary conditions. These boundary conditions have to meet the Andreev condition (2). In addition they have under certain parameter values to go into Fuchs boundary conditions. And at certain angles of incidence of electrons on the metal surface with the appropriate parameters boundary conditions must reproduce the Soffer boundary conditions (3).

The following expression satisfies these conditions

$$q(\theta) = q_0 + (1 - q_0)\exp(-b_1 cos\,\theta - b_2\cos^2\theta). \quad (5)$$

In this expression there are 3 parameters: q_0, b_1, b_2. These parameters are non-negative.

Then, when $\theta \to \pi/2$ $(cos\,\theta) \to 0$ the value $q(\theta) \to 1$.

In the linear approximation for $\cos\theta$ we have

$$q(\theta) = 1 - (1 - q_0)b_1 cos\,\theta.$$

Therefore, there is the following relation with the expression (2)

$$a_1 = (1 - q_0)b_1.$$

The parameter b_2 is necessary to account for the Soffer boundary conditions [5], which can be implemented at intermediate values of the angle θ.

If the parameters b_1,b_2 are large, then when the angle θ not too close to $\pi/2$, the value $q(\theta)$ is almost constant and close to q_0. Then the case of an ordinary mirror–diffuse boundary conditions [1] is implemented.

In the case $b_1 = 0$ and the angles θ close to $\pi/2$ we get

$$q(\theta) \simeq 1 - b_2(\cos\theta)^2.$$

This relation coincides with the Soffer result (4) if $b_2 = (4\pi G)^2$.

For metal

$$cos\theta = \frac{|v_n|}{v_F}.$$

Here v_n – component of electron velocity perpendicular to the surface, v_F – the Fermi velocity. Then the expression (2) can be rewritten in the form

$$q(\theta) = q_0 + (1 - q_0)\exp(-\beta_1|v_z| - \beta_2|v_z|^2). \quad (6)$$

$$\beta_1 = \frac{b_1}{v_F}, \beta_2 = \frac{b_2}{v_F}.$$

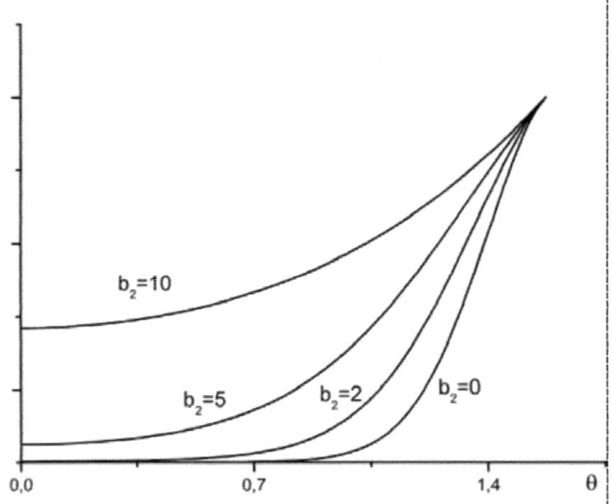

Fig. 1. The dependence of the reflectivity coefficient on the angle θ. Value q_0 = 0.5, and value b_1 = 1.

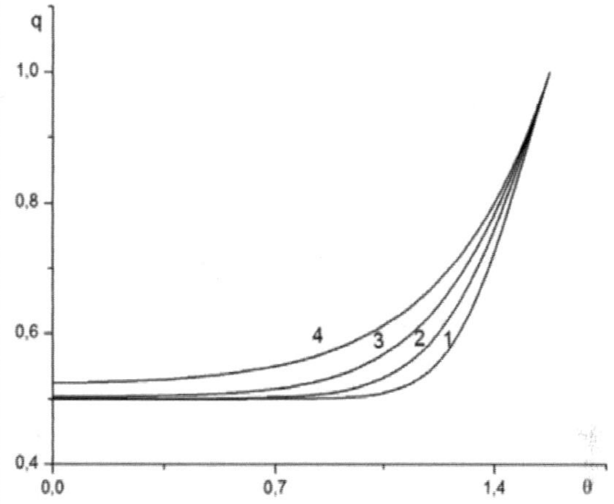

Fig. 2. The dependence of the reflectivity coefficient on the angle θ. Value q_0 = 0.5, and b_1 = 3. Curve 1 corresponds to the value of b_2 = 10. Curve 2 corresponds to the value of b_2 = 5. Curve 3 corresponds to the value of b_2 = 2. Curve 4 corresponds to the value of b_2 = 0.

From Fig. 1. and Fig. 2. we see that for large values of b_1 and b_2, the reflectivity coefficient q_0 for most angles of incidence remains constant. In this case the condition $q = q_0$ is satisfied. With the decrease of the coefficients b_1 and b_2 are deviations from the Fuchs boundary conditions (1) with constant reflectivity coefficient becomes evident. These deviations are particularly significant when the angle θ close to $\pi/2$.

3. Conclusion

The paper considers a boundary condition to the kinetic equation for the electrons in the metal. This boundary condition is a generalization of the Fuchs and Soffer boundary conditions. In the limit cases it goes into these boundary conditions. In addition it satisfies the Andreev condition.The Fuchs and Soffer boundary conditions this condition not satisfy. The considered boundary condition can be used to describe the electron kinetics in thin films and wires. It is possible to use this boundary condition for describing the kinetics of electrons in small metal particles.

References

[1] Fuchs K. 1938 *Math. Proceedings of the Cambridge Philosophical Society* **34** 100–108.
[2] Abrikosov A A 1998 *Fundamentals of the Theory of Metals* (North–Holland, Amsterdam) p 380.
[3] Sondheimer E H 2001 *Advances in Physics* **50** 499-537.
[4] Andreev A F 1971 *Physics-Uspekhi* **105** 114–123.
[5] Soffer S B 1967 *Journal of Applied Physics* **38**(4) 1710–1715.
[6] Dimmich and Warkusz F 1986 *Electrical conduct. of thin wires Active and Pass. Elec.* 103-109.

[7] Yushkanov A A, Savenko O V and Kuznetsova I A 2016 *Journal of Surface Investigation* **10** (3) 663671.
[8] Kuznetsova I A, Lebedev M E and Yushkanov A A 2014 *Condensed Matter Physics* **17**(1) 19.

Longitudinal Plasmons in a Thin Flat Conductive Film

A A Zotov and N V Zverev

Moscow Region State University,
Very Voloshinoi str. 24, 141014 Mytishchi, Moscow Region, Russian Federation

E-mail: aleksandr.zotov.99@mail.ru, zverev_nv@mail.ru

Abstract. The longitudinal plasmons in the plasma of conductivity electrons between the surfaces of a thin flat conductive film are investigated. It is shown that these plasmons lead to a resonant behaviour of the optical power coefficient of E-wave interaction with this film. The conditions for appearance of these plasmon resonances are found, and peculiarities of the dependence of the resonant frequencies on the film characteristics are revealed.

1. Introduction

A modern development of the nanotechnologies raise a question of a measurement of linear sizes of thin objects from 1 nanometer up to 1 micrometer with large enough precision. A perspective method for such measurements is a method of the plasmon resonances [1, 2].

This method is based on the longitudinal plasmon resonances appearing in thin layers of conductive matter like metals, semimetals and semiconductors, at the interaction of electromagnetic radiation with these layers.

This paper is devoted to investigation of the longitudinal plasmons in thin flat conductive film and an influence of these plasmons to the interaction of electromagnetic radiation with the film.

2. Longitudinal plasmons

Let us consider in more details the mechanism of appearance of the longitudinal plasmons [3]. These waves are the electromagnetic waves in the plasma matter when the electric field strength is co-directed with their propagation direction i.e. $\mathbf{E} \parallel \mathbf{k}$ where $\mathbf{k}$ is the wave vector. Electric field strength and electric displacement vectors of the longitudinal plasmon look as

$$\mathbf{E}(\mathbf{r},t) = \mathbf{E}(\omega,\mathbf{k})\, \mathrm{e}^{\mathrm{i}(\mathbf{kr}-\omega t)}, \qquad \mathbf{D}(\mathbf{r},t) = \mathbf{D}(\omega,\mathbf{k})\, \mathrm{e}^{\mathrm{i}(\mathbf{kr}-\omega t)}, \tag{1}$$

where ω is the cyclic frequency of the wave, and $\mathbf{E}(\omega,\mathbf{k})$ and $\mathbf{D}(\omega,\mathbf{k})$ are amplitudes of the longitudinal plasmons. Their components are related by the equation

$$D_a(\omega,\mathbf{k}) = \varepsilon_0 \sum_b \varepsilon_{ab}(\omega,\mathbf{k}) E_b(\omega,\mathbf{k}) \qquad (a,b=1,2,3). \tag{2}$$

Here ε_0 is the electric constant in SI units and $\varepsilon_{ab}(\omega, \mathbf{k})$ is the plasma dielectric permittivity tensor. In case of uniform isotropic plasma, the tensor looks as

$$\varepsilon_{ab}(\omega, \mathbf{k}) = \varepsilon_l(\omega, k)\frac{k_a k_b}{k^2} + \varepsilon_{tr}(\omega, k)\left(\delta_{ab} - \frac{k_a k_b}{k^2}\right), \tag{3}$$

where $\varepsilon_l(\omega, k)$ and $\varepsilon_{tr}(\omega, k)$ are respectively the longitudinal and the transverse plasma dielectric permittivity, $k = |\mathbf{k}|$ is the wave number, k_a are components of the $\mathbf{k}$ vector and δ_{ab} is the Kronecker symbol.

Substituting the right equation in (1) in the Maxwell equation

$$\nabla \cdot \mathbf{D}(\mathbf{r}, t) = 0$$

and using the equations (2) and (3), one gets the relation

$$\varepsilon_l(\omega, k)\mathbf{k}\mathbf{E}(\omega, \mathbf{k}) = 0.$$

Since $\mathbf{E} \parallel \mathbf{k}$ one arrives at the longitudinal plasmon dispersion law [3]:

$$\varepsilon_l(\omega, k) = 0. \tag{4}$$

It worth to note that the longitudinal plasmons arise in plasma owing to the dependence of the longitudinal permittivity ε_l not only on the frequency ω but also on the wave number k.

However, the longitudinal and transverse permittivities $\varepsilon_l(\omega, k)$ and ε_{tr} are the complex-valued functions [3]. This is caused by the interaction of plasma particles between each other and also with the radiation and with external particles. Such a feature of the ε_l leads from (4) to the complex-valued dependence $\omega = \omega(k)$ which shows that the longitudinal plasmons are the damped waves.

At the same time, these waves contribute to interaction of electromagnetic radiation with thin film of conductive matter. The longitudinal plasmons arise between borders of the film and hence for their appearance, the electric field strength $\mathbf{E}$ of the external electromagnetic wave has to possess a constituent which is orthogonal to the borders of the film. Therefore the longitudinal plasmons can arise only in the E-wave for which the electric field strength $\mathbf{E}$ lies in the incidence plane and which falls on the film surface under nonzero incidence angle.

3. Model of E-wave interaction with conductive film

Now we consider thin flat conductive film of the thickness d localized between two transparent isotropic dielectric media having positive dielectric constants ε_1 and ε_2. Let the plane E-wave falls on the surface of the film from the ε_1 medium under the incidence angle θ. Then the reflection and transmission optical power coefficients, R and T_r, for the wave look as [4]

$$R = \left|\frac{U^{(1)} + U^{(2)}}{V^{(1)} + V^{(2)}}\right|^2, \qquad T_r = \mathrm{Re}\left(\frac{\cos\theta'}{\cos\theta}\sqrt{\frac{\varepsilon_2}{\varepsilon_1}}\right)\left|\frac{U^{(1)}V^{(2)} - U^{(2)}V^{(1)}}{V^{(1)} + V^{(2)}}\right|^2, \tag{5}$$

where the values $U^{(j)}$ and $V^{(j)}$ $(j = 1, 2)$ are equal to

$$U^{(j)} = \frac{\cos\theta - Z^{(j)}\sqrt{\varepsilon_1}}{\cos\theta' + Z^{(j)}\sqrt{\varepsilon_2}}, \qquad U^{(j)} = \frac{\cos\theta + Z^{(j)}\sqrt{\varepsilon_1}}{\cos\theta' + Z^{(j)}\sqrt{\varepsilon_2}}. \tag{6}$$

In the equations (5) and (6), θ' is the refraction angle satisfying to the law

$$\sqrt{\varepsilon_1}\sin\theta = \sqrt{\varepsilon_2}\sin\theta',$$

and $Z^{(j)}$ is the surface inpedance of the conductive film for the E-wave. When one supposes the mirror reflection of the conductivity electrons from the borders of the film, the surface impedance is evaluated by the equation [5]

$$Z^{(j)} = \frac{2\mathrm{i}c\omega}{d}\sum_n \frac{1}{k_n^2}\left(\frac{k_x^2}{\omega^2\varepsilon_l(\omega,k_n)} + \frac{(\pi n/d)^2}{\omega^2\varepsilon_{tr}(\omega,k_n)-(ck_n)^2}\right). \tag{7}$$

Here ω is the E-wave frequency, c stands for the speed of light, the k_n and k_x values are equal to

$$k_n = \sqrt{(\pi n/d)^2 + k_x^2}, \qquad k_x = \frac{\omega}{c}\sqrt{\varepsilon_1}\sin\theta, \tag{8}$$

and the summation is performed over all odd $n = \pm 1, \pm 3, \pm 5, \ldots$ at $j = 1$ and over all even $n = 0, \pm 2, \pm 4, \ldots$ at $j = 2$.

The permittivities $\varepsilon_l(\omega,k)$ and $\varepsilon_{tr}(\omega,k)$ are known for the electron plasma with taking into account the wave properties of electrons [6, 7]. In case of the degenerate electron plasma in metals, the permittivities are evaluated by the equations

$$\varepsilon_l(\omega,k) = 1 + \frac{3}{4Q^2}\frac{(\Omega+\mathrm{i}\gamma)F(\Omega+\mathrm{i}\gamma,Q)F(0,Q)}{\Omega F(0,Q)+\mathrm{i}\gamma F(\Omega+\mathrm{i}\gamma,Q)}, \tag{9}$$

$$\varepsilon_{tr}(\omega,k) = 1 - \frac{1}{\Omega^2}\left(1 + \frac{\Omega G(\Omega+\mathrm{i}\gamma,Q)+\mathrm{i}\gamma G(0,Q)}{\Omega+\mathrm{i}\gamma}\right), \tag{10}$$

where

$$F(\Omega+\mathrm{i}\gamma,Q) = \frac{1}{r}\left[B_1(\Omega_+ +\mathrm{i}\gamma,Q) - B_1(\Omega_- +\mathrm{i}\gamma,Q)\right] + 2, \tag{11}$$

$$G(\Omega+\mathrm{i}\gamma,Q) = \frac{3}{16r}\left[B_2(\Omega_+ +\mathrm{i}\gamma,Q) - B_2(\Omega_- +\mathrm{i}\gamma,Q)\right] + \frac{9}{8}\left(\frac{\Omega+\mathrm{i}\gamma}{Q}\right)^2 + \frac{3}{32}Q^2r^2 - \frac{5}{8}, \tag{12}$$

$$B_\alpha(\Omega+\mathrm{i}\gamma,Q) = \frac{1}{Q^{2\alpha+1}}\left[(\Omega+\mathrm{i}\gamma)^2 - Q^2\right]^\alpha \ln\frac{\Omega+\mathrm{i}\gamma-Q}{\Omega+\mathrm{i}\gamma+Q}. \tag{13}$$

In case of the non-degenerate electron plasma in semimetals or semiconductors, these permittivities look as

$$\varepsilon_l(\omega,k) = 1 - \frac{2}{\tilde{Q}^2}\frac{(\Omega+\mathrm{i}\gamma)H(\Omega+\mathrm{i}\gamma,\tilde{Q})H(0,\tilde{Q})}{\Omega H(0,\tilde{Q})+\mathrm{i}\gamma H(\Omega+\mathrm{i}\gamma,\tilde{Q})}, \tag{14}$$

$$\varepsilon_{tr}(\omega,k) = 1 - \frac{1}{\Omega^2}\left(1 + \frac{\Omega H(\Omega+\mathrm{i}\gamma,\tilde{Q})+\mathrm{i}\gamma H(0,\tilde{Q})}{\Omega+\mathrm{i}\gamma}\right), \tag{15}$$

where

$$H(\Omega+\mathrm{i}\gamma,\tilde{Q}) = \frac{\tilde{Q}^2}{\sqrt{\pi}}\int_0^{+\infty}\frac{\left[(\Omega_+ +\mathrm{i}\gamma)(\Omega_- +\mathrm{i}\gamma)+(\tilde{Q}\xi)^2\right]\exp(-\xi^2)}{\left[(\Omega_+ +\mathrm{i}\gamma)^2-(\tilde{Q}\xi)^2\right]\left[(\Omega_- +\mathrm{i}\gamma)^2-(\tilde{Q}\xi)^2\right]}\,\mathrm{d}\xi. \tag{16}$$

In the equations (9) – (16), the following dimensionless values are introduced:

$$\Omega = \frac{\omega}{\omega_p}, \quad \gamma = \frac{\nu}{\omega_p}, \quad \Omega_\pm = \Omega \pm \frac{\hbar k^2}{2m_e\omega_p}, \quad Q = \frac{v_F k}{\omega_p}, \quad r = \frac{\hbar\omega_p}{m_e v_F^2}, \quad \tilde{Q} = \frac{v_T k}{\omega_p}. \tag{17}$$

Here ω_p, ν and m_e are respectively the plasma frequency, the collision frequency and the effective mass of the conductivity electrons, $\hbar$ is the Planck constant, v_F is the Fermi velocity of electrons in the degenerate electron plasma at zero absolute temperature $T = 0$ K, and v_T is the thermal velocity of electrons in non-degenerate electron plasma:

$$v_T = \sqrt{\frac{2k_B T}{m_e}}, \tag{18}$$

where k_B is the Boltzmann constant.

4. Results and discussion

By use of the equations (5) – (18) we study the reflection and transmission power coefficients R and T_r. For numerical simulations, we took the aluminium and graphite materials. The aluminium metal has the parameters [8] $\omega_p = 1.93 \cdot 10^{16}$ s^{-1}, $\nu = 10^{-3}\omega_p$, $v_F = 1.34 \cdot 10^6$ m/s, $m_e = 1.35 \cdot 10^{-30}$ kg. For graphite matter being a weak conductor we took the following parameters [9]: $\omega_p = 2.54 \cdot 10^{14}$ s^{-1}, $\nu = 2.273 \cdot 10^{11}$ Hz, $m_e = 9 \cdot 10^{-31}$ kg.

Comparing the equations (4) and (7), one sees that the longitudinal plasmons between film borders can contribute to the power coefficients (5) when the condition

$$\mathrm{Re}\,\varepsilon_l(\omega_{res}, k_n) = 0 \tag{19}$$

is fulfilled. Here ω_{res} are the resonant frequencies for which the power coefficients R and T_r have a critical behaviour. Note that it is obvious from (7) that the critical behaviour of power coefficients occurs then the incidence angle $\theta \neq 0$.

It was found that the resonant frequencies are compared with the plasma frequency according to the relations

$$\omega_{res} > \omega_p, \qquad \omega_{res} \sim \omega_p.$$

This is turn out that the ω_{res} values depend mostly on the ω_p, d, v_F and v_T parameters. This is obvious from the equations (9), (11), (13), (14), (16) and (17). They are almost independent of θ, ν, ε_1, ε_2 values when the conditions

$$d \ll \frac{c}{\omega_p}, \qquad d \gg \frac{\hbar}{m_e v} \tag{20}$$

are valid. Here $v = v_F$ for metals and $v = v_T$ for semimetals and semiconductors.

The left condition in (20) follows from (8) when one requires a negligibility of k_x in comparison with π/d at $\omega \sim \omega_p$. One explains this condition as the plasmon resonances in the film are observable when the film width is much less than the skin-depth length c/ω_p. But the right condition in (20) means in framework of classical physics that the measurable film thickness must be much large than the de Broglie wavelength. Then the relation

$$\frac{\hbar}{2m_e\omega_p}\left(\frac{\pi}{d}\right)^2 \ll 1$$

occurs since the $\hbar\omega_p \sim m_e v^2$ for metals and for semimetals and semiconductors at usual temperatures T. Independence of the ω_{res} on ν follows from the relation $\nu \ll \omega_p$.

In numerical studies, the d values satisfy to the (20) conditions.

Furthermore, investigations of R and T_r show that the ω_{res} values evaluated from (19), lead to critical behaviour of the power coefficients only in case of the **odd** numbers $n = 1, 3, 5, \ldots$ as it is demonstrated at the figures 1 and 2 (see also [1, 5]). Such a result can be explained as follows. In case of the even values n, the $n = 0$ term in (7) strongly prevails over all remaining

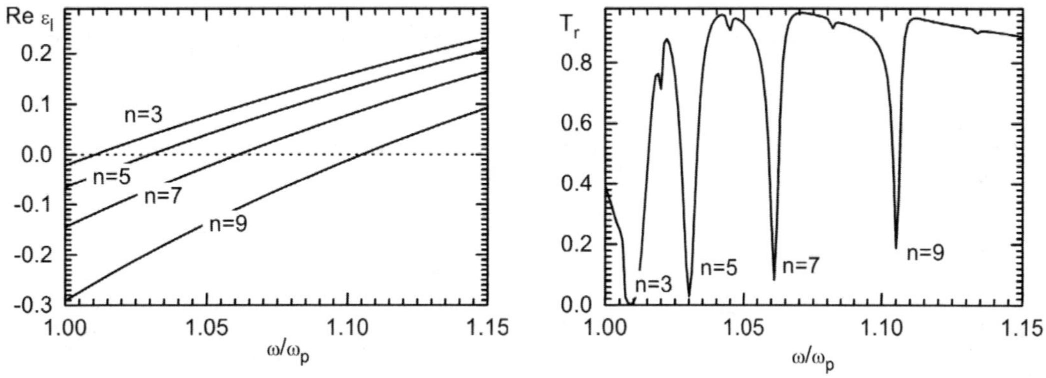

Figure 1. The real part of longitudinal permittivity $\mathrm{Re}\,\varepsilon_l(\omega, k_n)$ (left plot, various n) and transmission power coefficient T_r (right plot) as functions of the frequency ω for aluminium film. Values $d = 3.5$ nm, $\theta = 60°$, $\omega_p = 1.93 \cdot 10^{16}$ s^{-1}, $\varepsilon_1 = 1$ (air), $\varepsilon_2 = 2$ (quartz).

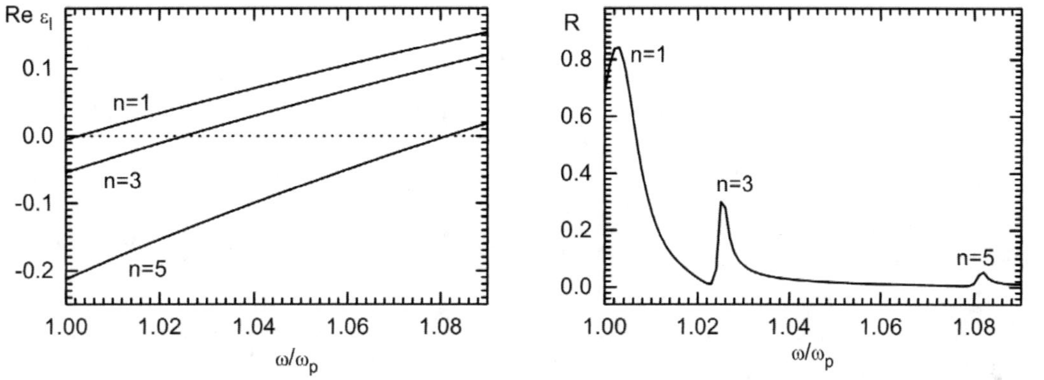

Figure 2. The real part of longitudinal permittivity $\mathrm{Re}\,\varepsilon_l(\omega, k_n)$ (left plot, various n) and reflection power coefficient R (right plot) as functions of the frequency ω for graphite film. Values $d = 20$ nm, $\theta = 60°$, $T = 294$ K, $\omega_p = 2.54 \cdot 10^{14}$ s^{-1}, $\varepsilon_1 = 1$ (air), $\varepsilon_2 = 2$ (quartz).

terms and critical behaviour of $Z^{(2)}$ is smeared. But in case of odd numbers n, the first positive and negative n terms in (7) have the same orders and the critical behaviour of $Z^{(1)}$ is visible for different n values.

The above analysis of ω_{res} leads to a conclusion that these resonant frequencies delivering a visible critical behaviour of the power coefficient, satisfy to the equation

$$\omega_{res} = \omega_p f\left(\frac{\pi n v}{\omega_p d}\right). \tag{21}$$

Here $n = 1, 3, 5, \ldots,$ $v = v_F$ for metals and $v = v_T$ for semimetals and semiconductors, and $f(\overline{Q})$ is a dimensionless function of dimensionless variable $\overline{Q}$ which is evaluated from (19) and from (9) for metals or from (14) for weak conductors. Of course, the d values should satisfy to the (20) conditions.

The dependence (21) is confirmed qualitatively by the numerical simulations of the coefficients (5) at various d and for weak conductor, at different T values (figures 3 and 4). These results show that since the resonant frequencies ω_{res} are measured with large enough precision, the relative errors of d values are less than 1% [1]. Hence the method of plasmon resonances is suitable for measurements of the thickness d of conductive film with a large precision.

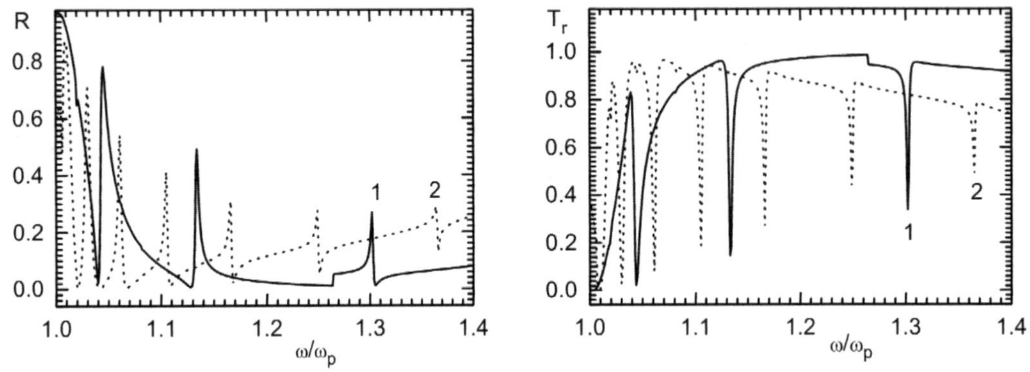

Figure 3. The reflection and transmission power coefficients R (left plot) and T_r (right plot) as functions of the frequency ω for aluminium film: $1 - d = 1.75$ nm (solid line), $2 - d = 3.5$ nm (dotted line). Values $\theta = 60°$, $\omega_p = 1.93 \cdot 10^{16}$ s^{-1}, $\varepsilon_1 = 1$ (air), $\varepsilon_2 = 2$ (quartz).

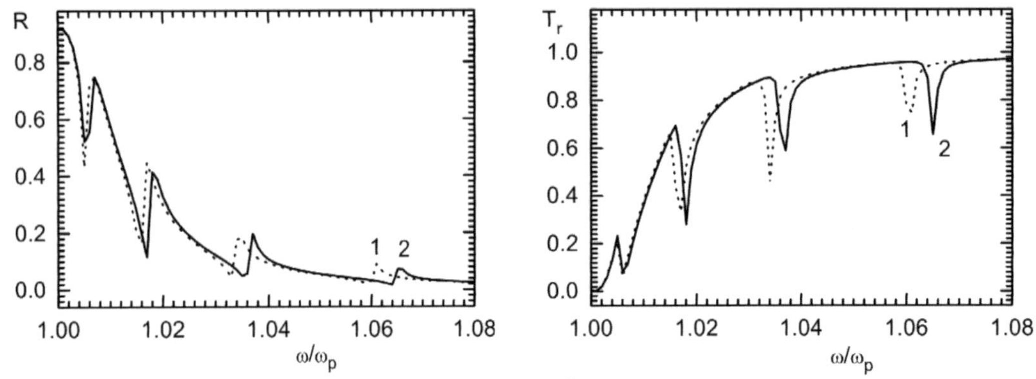

Figure 4. The reflection and transmission power coefficients R (left plot) and T_r (right plot) as functions of the frequency ω for graphite film: $1 - T = 283$ K (dotted line), $2 - T = 305$ K (solid line). Values $d = 40$ nm, $\theta = 60°$, $\omega_p = 2.54 \cdot 10^{14}$ s^{-1}, $\varepsilon_1 = 1$ (air), $\varepsilon_2 = 2$ (quartz).

Acknowledgements
The work is supported by the RBRF Grant *No 19-07-00537* and by the Moscow Region Governor Grant *No 18 on July 24, 2020.*

References

[1] Latyshev A V and Yushkanov A A 2015 *Quantum Electronics* **45**(3) 270–74
[2] Valyansky S I, Vinogradov S V, Kononov M A et al 2017 *Prikladnaya Fizika (Applied Physics)* **No 6** 103–08
[3] Lifshitz E M and Pitaevskii L P 1981 *Physical Kinetics. Vol 10 (1st ed.)* (Pergamon Press) p. 462
[4] Yushkanov A A and Zverev N V 2017 *Phys. Lett.* A **381** 679–84
[5] Jones W E, Kliewer K L and Fuchs R 1969 *Phys. Rev.* **178**(3) 1201–03
[6] Latyshev A V and Yushkanov A A 2013 *Theor. and Math. Phys.* **175**(1) 559–69
[7] Latyshev A V and Yushkanov A A 2014 *Theor. and Math. Phys.* **178**(1) 130–41
[8] Fuchs R and Kliewer K L 1969 *Phys. Rev.* **185**(3) 905–13
[9] Yushkanov A A and Zverev N V 2019 *J. Phys.: Conf. Ser.* **1309** 012011

Relaxation type Kinetic equation for electrons in polycrystalline metal

T N Lam[1], **F Karimov**[1] **and A A Yushkanov**[1]
[1] Moscow Region State University 105005, Moscow,
st. Radio, 10A, Russian Federation

nhunglam279@gmail.com, faha_rtsu_2003@mail.ru, yushkanov@inbox.ru

Abstract. The kinetic equation for electrons in a polycrystalline metal is considered. A kinetic equation is written that describes in a unified manner the scattering of conduction electrons both by impurities or phonons and by grain boundaries. This kinetic equation takes into account the scattering of electrons at the boundaries of crystallites of a polycrystalline metal. An expression is obtained for the bulk conductivity in the general case. Let us analyze the effect of electron scattering at grain boundaries on its electrical properties.

1. Introduction

In the granular conductor electron scattering at grain boundaries leads to that they are scattered mainly backward. Thus, some of the electrons acquire the velocity the opposite original one.

Up to now this was considered by a formal summation of the contribution to the electrical resistance of the conductor caused by electron scattering on the surface and by lattice defects (and phonons) and the contribution due to electron scattering at the grain boundaries [1]. Such an approach would not be quite satisfactory. It is obvious that the contributions of these processes in the electrical resistance of the conductor are not independent.

For correct accounting of contributions of these processes is necessary to use the unified kinetic approach to the problem. This approach must account for all phenomena in a unified way. Up to now such a kinetic approach was absent.

Scattered at the interface between the grains, the electrons will have a speed mainly directed in the opposite direction from the initial velocity. Therefore, the equilibrium electron velocity after the collision will be nonzero, because of the scattering on impurities and defects. And the sign of the average speed of the electrons after the collision with the grain boundaries will be the opposite of the original average speed of electrons.

2. Kinetic equation

Kinetic equation with the relaxation type collision integral for electrons will be as follows [2]— [4]

$$\frac{\partial f}{\partial \mathrm{t}} + \mathbf{v}\frac{\partial f}{\partial \mathbf{r}} + e\mathbf{E}\frac{\partial f}{\partial \mathbf{p}} = \nu\left(f_{eq} - f\right). \quad (1)$$

where $\mathbf{v}$, e — the velocity and charge of electrons, $\mathbf{E}$ — Electric field strength, ν — the rate of collision of electrons, the function f_{eq} - the equilibrium Fermi distribution for electrons in a solid plasma:

$$f_{eq} = \left[1 + \exp\frac{\mathcal{E} - \mu}{k_B T}\right]^{-1}, \tag{2}$$

here $\mathcal{E}$ – equilibrium electron energy after scattering, μ – chemical potential, T – temperature, k_B – Boltzmann's constant.

We assume that the conducting medium has a spherical symmetry. In this case for energy of electrons $\mathcal{E}_0$ we have:

$$\mathcal{E}_0 = \frac{m\mathbf{v}^2}{2}.$$

Here m — the effective mass of the electron.

However, in general, after scattering, the average speed of the electron will be nonzero. That is, the electron will "remember" the velocity, which had before the scattering. In the case when the electron–electron scattering is dominant, electron after scattering partially retains the original velocity.

The average speed of the electrons after scattering, we denote u. We assume that the magnitude of u is connected to the average electron velocity $\mathbf{u}_0$ by the following relation:

$$\mathbf{u} = \alpha\mathbf{u}_0.$$

The value of α took into account the fact that the scattering electrons partially retain their memory of their velocity before scattering. In this approach, the value of α is an empirical coefficient.

Then we get the following relationship:

$$\mathcal{E} = \frac{m(\mathbf{v} - \mathbf{u})^2}{2}. \tag{3}$$

A similar approach was previously used to take into account electron-electron collisions [5] — [7]. In the case of electron-electron collisions, the value of α is positive, since the scattering of electrons occurs predominantly forward. In the case of electron scattering at the crystallite boundary of a polycrystalline metal, the value of α is negative, that is, $\alpha < 0$. This is due to the fact that in this case the scattering of electrons occurs mainly in the direction opposite to the direction of the initial motion of electrons.

We assume that the velocity u is much less than the thermal velocity of electrons (or Fermi velocity for the case of degenerate Fermi–gas). Then the value of $\mathcal{E}$ (3) can be linearized:

$$\mathcal{E} \simeq \frac{m\mathbf{v}^2}{2} - m\mathbf{v}\mathbf{u} = \mathcal{E}_0 - m\mathbf{v}\mathbf{u}.$$

Through appropriate linearization for the locally equilibrium functions f_{eq}, we obtain the following expression:

$$f_{eq} = f_0 - \frac{\partial f_0}{\partial \mathcal{E}} m\mathbf{v}\mathbf{u}, \tag{4}$$

Similarly, in the linear case, the term with the electric field in the kinetic equation (1) takes the following form:

$$e\mathbf{E}\frac{\partial f}{\partial \mathbf{p}} \simeq e\mathbf{E}\mathbf{v}\frac{\partial f_0}{\partial \mathcal{E}}, \tag{5}$$

where $\mathbf{p} = m\mathbf{v}$ – electron momentum.

The magnitude of $\mathbf{u}_0$ is defined through the distribution function:

$$\mathbf{u}_0 = \frac{1}{n}\int f\mathbf{v}\frac{2d^3p}{(2\pi\hbar)^3}.$$

Here n – concentration of electrons:

$$n = \int f\frac{2d^3p}{(2\pi\hbar)^3}.$$

Note that the current density $\boldsymbol{j}$ can be represented in the form:

$$\boldsymbol{j} = ne\mathbf{u}_0.$$

The linearized distribution function has the form [2]:

$$f = f_0 + f_1.$$

For a function f_1 in the works [5] — [7] has been proposed model equation taking into account the contribution of electron-electron collisions in the kinetic processes. For a degenerate Fermi gas of electrons this equation has the following form:

$$\frac{\partial f_1}{\partial t} + \mathbf{v}\mathrm{grad} f_1 + e\mathbf{E}\mathbf{v}\frac{\partial f_0}{\partial \mathcal{E}} == -\nu\left(\frac{3m\alpha}{4\pi v_F^3}\mathbf{v}\frac{\partial f_0}{\partial \mathcal{E}}\int \mathbf{v}' f_1 d^3v' + f_1\right).$$

Here $\nu = {}^1/_\tau, \tau$ - the average time between two successive collisions of an electron. The integral term corresponds to the taking into account the value $\mathbf{u}_0$. v_F - the Fermi velocity, $v_F = {}^{p_F}/_m$, p_F - the Fermi momentum.

The function f_1 comfortably search in the form:

$$f_1 = -\frac{\partial f_0}{\partial \mathcal{E}}\psi. \tag{6}$$

Here ψ is a new unknown function. For an electron degenerate Fermi-gas we have:

$$\frac{\partial f_0}{\partial \mathcal{E}} = -\delta(\mathcal{E} - \mathcal{E}_F),$$

$$f_1 = \delta(\mathcal{E} - \mathcal{E}_F)\,\psi.$$

Here $\delta(x)$ - the Dirac Delta function. $\mathcal{E}_F$ - Fermi energy, $\mathcal{E}_F = {}^{mv_F^2}/_2$.

In this case

$$\mathbf{u}_0 = \frac{1}{n}\int f_1\mathbf{v}\frac{2d^3p}{(2\pi\hbar)^3}.$$

$$n = \int f_0\frac{2d^3p}{(2\pi\hbar)^3}.$$

Therefore

$$n = \frac{8\pi}{(2\pi\hbar)^3}\int_0^{p_F} p^2dp = \frac{8\pi p_F^3}{3(2\pi\hbar)^3}.$$

We have the following relation ${}^{p_F^2}/_m = \mathcal{E}_F$.

Then

$$\mathbf{u}_0 = \frac{1}{n}\int f_1\mathbf{v}\frac{2d^3p}{(2\pi\hbar)^3} == \frac{3(2\pi\hbar)^3}{8\pi p_F^3}\int f_1\mathbf{v}\frac{2d^3p}{(2\pi\hbar)^3} == \frac{3}{4\pi v_F^3}\int f_1\mathbf{v}d^3v. \tag{7}$$

We assume that the field $\mathbf{E} \sim \exp(-i\omega t)$. Then $\psi \sim \exp(-i\omega t)$.

Taking into account the relations (4), (5) and (6), the kinetic equation (1) can be written as follows for the function ψ [2]:

$$-i\omega\psi + \mathbf{v}\frac{\partial\psi}{\partial \boldsymbol{r}} - e\mathbf{E}\mathbf{v} = -\nu(\alpha m\mathbf{v}\mathbf{u}_0 - \psi). \tag{8}$$

The value u is related to the average velocity of the electron before collision. In the case of electron-electron collisions the electrons partially retain the initial impulse and the initial velocity. In this case, the coefficient α is positive, α > 0 . In the case of scattering by grain boundaries the situation is the opposite. Thus, the electrons after scattering by the grain boundaries have a velocity opposite to the initial one. Therefore, the sign α in the equation (8) is negative.

Equation (8) can be rewritten in the following form:

$$-i\omega\psi + \mathbf{v}\frac{\partial\psi}{\partial \boldsymbol{r}} + \nu\psi = \mathbf{v}(\,e\mathbf{E} + \nu\alpha m\mathbf{u}_0). \tag{9}$$

Consider the one-dimensional problem, when the function ψ depends only on the spatial variable x. Axis y will direct along a direction of the electric field E. Thus, $\mathbf{E} = \mathrm{E}_y\exp(-i\omega t)$. Then the equation (9) will be of the form:

$$-i\omega\psi + v_x\frac{\partial\psi}{\partial x} + \nu\psi = v_y\left(e\mathrm{E}_y + \nu\alpha m\mathbf{u}_{0y}\right). \tag{10}$$

We introduce a new function h by using the relation:

$$\psi = mv_yv_Fh. \tag{11}$$

The function h is dimensionless.

We calculate the value $\mathbf{u}_{0y}$:

$$\mathbf{u}_{0y} = \frac{3}{4\pi v_F^3}\int f_1 v_y d^3v = \frac{3}{4\pi v_F^3}\int \delta(\mathcal{E}-\mathcal{E}_F)\,\psi v_y d^3v =$$

$$= \frac{3}{4\pi v_F^3}\int \delta(\mathcal{E}-\mathcal{E}_F)\, m v_y v_F h v_y d^3v = \frac{3m}{4\pi v_F^2}\int \delta(\mathcal{E}-\mathcal{E}_F)\, v_y^2 h d^3v.$$

But

$$\delta(\mathcal{E}-\mathcal{E}_F) = \delta\left(\frac{mv^2}{2}-\frac{mv_F^2}{2}\right) = \delta\left(\frac{m(v-v_F)(v+v_F)}{2}\right) = \frac{1}{mv_F}\delta(v-v_F).$$

Then

$$\mathbf{u}_{0y} = \frac{3m}{4\pi v_F^2}\int \frac{1}{mv_F}\delta(v-v_F)\, v_y^2 h d^3v = \frac{3}{4\pi v_F^3}\int \delta(v-v_F)\, v_y^2 h d^3v.$$

Let's go over to a spherical frame of reference in the velocity space:

$$v_x = v\cos\theta, \quad v_y = v\sin\theta\cos x, \quad v_y^2 = v^2\sin^2\theta\cos^2 x.$$

Let's introduce the designation:

$$\mu = \cos\theta$$

Then

$$v_x = v\mu, \quad v_y^2 = v^2(1-\mu^2)\cos^2\varphi, \quad d^3v = v^2\sin\theta dv d\theta d\varphi = -v^2 dv d\mu d\varphi.$$

In this case:

$$\mathbf{u}_{0y} = \frac{3}{4\pi v_F^3}\int \delta(v-v_F)\, v_y^2 h d^3v =$$

$$= -\frac{3}{4\pi v_F^3}\int_0^\infty dv \int_1^{-1} d\mu \int_0^{2\pi} d\varphi h\delta(v-v_F)v^4(1-\mu^2)\cos^2\varphi =$$

$$= \frac{3}{4\pi v_F^3}\int_0^\infty dv \int_{-1}^{1} d\mu \int_0^{2\pi} d\varphi h\delta(v-v_F)v^4(1-\mu^2)\cos^2\varphi =$$

$$= \frac{3}{4\pi v_F^3}\pi v_F^4 \int_{-1}^{1}(1-\mu^2)h d\mu = \frac{3v_F}{4}\int_{-1}^{1}(1-\mu^2)h d\mu.$$

It is convenient to introduce the following dimensionless variables:

$$z_0 = \frac{a(\nu - \iota\omega)}{v_F}, \qquad \mu' = \frac{v_x}{v_F}, \qquad x' = \frac{x}{a}.$$

Then the kinetic equation takes the form:

$$\mu\frac{\partial h}{\partial x'} + z_0 h(x',\mu') = E_0 + \frac{3}{4}\beta\int_{-1}^{1}(1-\mu'^2)h(x',\mu')d\mu'. \tag{12}$$

Here

$$E_0 = \frac{eE_y a}{mv_F^2}, \qquad \beta = \frac{a\alpha\nu}{v_F}.$$

In the future, the variable x' will be written without dash for short.

3. Calculation of the conductivity coefficient

Let's calculate the current density $\boldsymbol{j}_y$:

$$\boldsymbol{j}_y = ne\mathbf{u}_{0y} = \frac{3v_F ne}{4}\int_{-1}^{1}(1-\mu^2)h d\mu.$$

Consider the case of not varying in space electric field. Then $\partial h/\partial x' = 0$.

Then the function $h(x,\mu')$ is a constant. We denote this using a constant H, i. e. $h(x,\mu') = H$. From equation (10) we find this constant:

$$z_0 H = E_0 + \frac{3}{4}\beta \int_{-1}^{1} (1 - \mu'^2) H d\mu'.$$

Or

$$z_0 H = E_0 + \beta H.$$

Therefore

$$H = \frac{E_0}{z_0 - \beta}$$

Calculate the current density $\boldsymbol{j}_y$:

$$\boldsymbol{j}_y = \frac{3 v_F n e}{4} \int_{-1}^{1} (1 - \mu^2) h d\mu = \frac{v_F n e E_0}{z_0 - \beta}$$

The expression for the electrical conductivity σ has the following form:

$$\sigma = \frac{\boldsymbol{j}_y}{E_y} = \frac{v_F n e}{z_0 - \beta}\frac{E_0}{E_y} = \frac{v_F n e}{\frac{a(\nu - \iota\omega)}{v_F} - \frac{a\alpha\nu}{v_F}} \frac{\frac{e E_y a}{m v_F^2}}{E_y} = \frac{n e^2}{m(\nu - \iota\omega - \alpha\nu)}.$$

When $\omega = 0, \alpha = 0$ we get the classical expression for static conductivity:

$$\sigma_0 = \frac{n e^2}{m\nu}.$$

In the general case for the electrical conductivity we have:

$$\sigma = \frac{\sigma_0}{1 - \alpha - \iota\omega/\nu}.$$

When $\omega = 0, \alpha \longrightarrow 1,$ the value $\sigma \longrightarrow \infty$. This case corresponds to the fact that in the scattering of electrons their momentum is conserved. So the friction of the electron gas on a lattice is missing.

When $\alpha \longrightarrow -\infty$, we get $\sigma \longrightarrow 0$. That is, in this case scattering at grain boundaries dominates. Moreover, the grain boundaries become impermeable for the electrons.

4. Conclusion

We have considered the kinetic equation for electrons in polycrystalline conductor (metal). The influence of scattering electrons at the boundaries of the crystallites on the kinetic the processes have been analyzed. We showed how these processes affect the value of electric conductivity. The considered kinetic equation can be used to study the electrical conductivity of thin metal polycrystalline films. It is possible to use this kinetic equation for the analysis of skin–effect in polycrystalline metal.

References

[1] Chawla J S, F Gstrein, O'Brien K P, Clarke J S and Gall D 2011 *Electron scattering at surfaces and grain boundaries in Cu thin films and wires* (*Physical Review B* vol 84) p 235423.

[2] Abrikosov A A 1998 *Fundamentals of the Theory of Metals* (North–Holland, Amsterdam) p 380.

[3] Sondheimer E H 2001 *The mean free path of electrons in metals (Advances in Physics* vol 50), chapter 6 pp 499-537.

[4] Lifshitz E M and Pitaevskii L P 1981 *Physical Kinetics: Volume 10* (Butterworth-Heinemann), p 466.

[5] Borchi E and De Gennaro S 1980 *On the low–temperature electrical resistivity of copper and gold* (*J. Phys. F* vol 10) pp 271–274.

[6] De Gennaro S and Rettory A 1984 *The low–temperature electrical resistivity of potassium: size effects and the role of normal electron–electron scattering* (*J. Phys. F* vol 14) pp 237–242.

[7] De Gennaro S and Rettory A 1985 *Normal electron–electron contribution to the anomalous surface impedance* (*J. Phys. F* vol 15) pp 227–230.

Method of the electron distribution function calculation in a problem of the kinetic rarefied plasma plume modeling

A M Bishaev[1], M V Abgaryan[1,2]

[1]Research Institute of Applied Mechanics and Electrodynamics of the Moscow Aviation Institute
Leningradskoye shosse 5, Moscow, 125080, POB 43, Russia
[2]Moscow Aviation Institute (National Research University)
Volokolamskoe shosse 4, Moscow, 125993, Russia.

E-mail: bishaev@bk.ru

Abstract. A problem related to the rarefied plasma plume of the stationary plasma thruster (SPT) is considered in the paper. The consideration is conducted fully in terms of kinetics, namely, distribution functions are introduced to describe motion of every plasma component. The system of kinetics equations for the distribution functions should be solved in combination with the Maxwell's equations. The paper discusses methods for solving the stated problem.

Introduction

The most comprehensive solution of the SPT plume dynamics problem was considered in [1] and [2]. The geometry of such problem statement is shown in the figure 1.

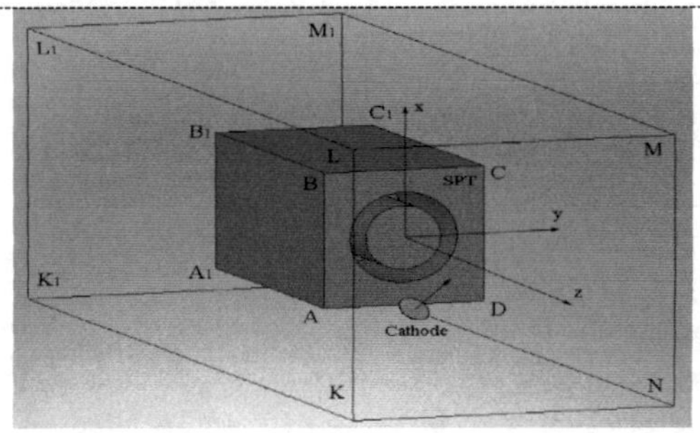

Figure 1. The geometry of the SPT plasma plume problem

A thruster, out of the ring opening of which the ions and neutral particles are flowing into the environment, is modeled with an $ABCDA_1B_1C_1D_1$ parallelepiped. The electrons are emitted by the cathode, which is shown in the figure 1. The papers mentioned above include the results of previously made estimations that show, that the flowing out medium represents the quazineutral plasma, which requires a kinetic approach for the adequate description of its dynamics.

In the papers [1] and [2], the kinetic approach was used to describe motion of ions and neutrals, i.e. the kinetic equations for the presented distribution functions of the ions and neutrals definition were formulated. To describe the electron component, its motion equations were used, from which the expression for electric field was derived via the A.I. Morozov's hypothesis of 'thermalized potential' [3]. This allowed completing the kinetic equations system for ions and neutrals.

Paper [1] presents also the development of a numerical method for solving the resultant kinetic equations system, and results of the conducted calculations are presented in [2]. The parallelepiped $KLMNK_1 L_1 M_1N_1$ (figure 1) represents the computational domain, where all calculations were made. It is obvious that they were made both upstream and downstream.

Here we propose to reject the 'thermalized potential' hypothesis and to describe the electron component in terms of kinetics.

1. The problem statement

Let's assume that $f_i = f_i(t, \vec{x}, \vec{\xi})$ is the ion distribution function, where $\vec{\xi} \in D_i$ is the velocity space for ions; and $f_e = f_e(t, \vec{x}, \vec{\mathrm{v}})$ is the electron distribution function, where $\vec{\mathrm{v}} \in D_e$ is the velocity space for electrons. Therefore, $f_n = f_n(t, \vec{x}, \vec{\mathrm{w}})$ is the neutral particle distribution function, where $\vec{w} \in D_n$ is the velocity space for neutrals. It is assumed, that the introduced distribution functions obey the following kinetic equations system:

$$\frac{\partial f_i}{\partial t} + \xi^k \frac{\partial f_i}{\partial x^k} + \frac{e}{m_i} E^k \frac{\partial f_i}{\partial \xi^k} + \frac{e}{m_i c} \varepsilon_{kjl} \xi^j H^l \frac{\partial f_i}{\partial \xi^k} = J_{in}$$

$$\frac{\partial f_e}{\partial t} + \mathrm{v}^i \frac{\partial f_e}{\partial x^i} - \frac{e}{m_e} E^i \frac{\partial f_e}{\partial \mathrm{v}^i} - \frac{e}{m_e c} \varepsilon_{ijl} \mathrm{v}^j H^l \frac{\partial f_e}{\partial \mathrm{v}^i} = 0 \qquad (1)$$

$$\frac{\partial f_n}{\partial t} + w^i \frac{\partial f_n}{\partial x^i} = J_{ni}$$

The equation (1) utilizes the following nomenclature: $\vec{H} = \{H^i\}, i = 1,2,3$ is the given time-independent magnetic field. Then the electric field $\vec{E} = \{E^i\}$ is to be a conservative one with the intensity $E^i = -\frac{\partial \varphi}{\partial x^i}$, where $\varphi = \varphi(t, \vec{x})$ is the electrostatic potential, which is defined by the Maxwell's equation:

$$\Delta\varphi = 4\pi e(n^e - n^i)\ \Delta = \frac{\partial^2}{\partial x^2} + \frac{\partial^2}{\partial y^2} + \frac{\partial^2}{\partial z^2} \tag{2}$$

In both equations (1, 2), e is an absolute value of the electron charge, c is the light velocity, m_i and m_e are the ion and electron masses, respectively, ε_{ijl} is the Levi-Civita symbol. The values n^e, n^i in equation (2) are the electron and ion densities, respectively, which are defined by the following formulas: $n^e = \int_{D_e} f_e d\,\mathrm{v},\ n^i = \int_{D_i} f_i d\xi$. The current density is defined as $\vec{j} = e(\vec{j}_i - \vec{j}_e)$, where $\vec{j}_i = \int_{D_i} \vec{\xi} f_i d\vec{\xi},\ \vec{j}_e = \int_{D_e} \vec{\mathrm{v}} f_e d\vec{\mathrm{v}}$ $, \vec{u}_i = \frac{\vec{j}_i}{n_i}, \vec{u}_e = \frac{\vec{j}_e}{n_e}$.

The expressions $\frac{3}{2}kT^i = \frac{m_i}{2}\int_{D_i} (\vec{\xi} - \vec{u}_i)^2 f_i d\vec{\xi}, \frac{3}{2}kT^e = \frac{m_e}{2}\int_{D_e} (\vec{\mathrm{v}} - \vec{u}_e)^2 f_e d\vec{\mathrm{v}}$ define the ion and electron temperatures.

The right sides of the first and of the third equations of (1) are the integrals of collisions of ions and neutrals with neutrals and ions, correspondingly. They were defined by the following formulas: $J_{in} = \nu_{in} f_n - \nu_{ni} f_i, J_{ni} = \nu_{ni} f_i - \nu_{in} f_n$, here ν_{in}, ν_{ni} are the frequencies of ion-neutral and neutral-ion collisions, respectively. The corresponding expressions for these values in a case of resonant charge-exchange reaction were derived in [2].

It is assumed, that the electrons in the thruster plume are moving in collisionless mode. Note that the formulas considered below summarize by repetitive indices. The indices *i, e, n* are introduced only to indicate that the corresponding value relates to ions, electrons or neutrals, regardless of whether the index is superscript or suffix. A.A. Vlasov proposed similar approach to plasma describing in [4].

To obtain the macroscopic equations for ions and electrons, it is needed to integrate the first and the second equations of system (1), each over its own velocity space, and then to multiply the first one by $\vec{\xi}$, and the second one – by $\vec{v}$; once done, they both are integrated again over their velocity spaces. These steps resulted in:

$$\frac{\partial n_i}{\partial t} + \frac{\partial j_k^i}{\partial x^k} = 0, \frac{\partial n^e}{\partial t} + \frac{\partial j_k^e}{\partial x^k} = 0,$$

$$\frac{\partial (j^s)^i}{\partial t} + \frac{\partial (j^s u_k)^i}{\partial x^k} + \frac{\partial P_{sk}^i}{\partial x^k} - \frac{e}{m^i} n^i E^s - \frac{e}{m^i c}\varepsilon_{skl} j_k^i H^l = \left\langle \sigma_{in} g^{in} \right\rangle \left(j^n - j^i \right)^s \tag{3}$$

$$\frac{\partial\left(j^s\right)^e}{\partial t}+\frac{\partial\left(j^s u_k\right)^e}{\partial x^k}+\frac{\partial P_{sk}^e}{\partial x^k}+\frac{e}{m^e}n^e E^s+\frac{e}{m^e c}\varepsilon_{skl}j_k^e H^l=0, s=1,2,3$$

Let's make equations (1) and (2) dimensionless. The length scale a is taken to be equal to the half-width of the ABCD square (figure 1). The scale values of the velocity spaces are taken to be equal to the following values: $\xi_0=\sqrt{2kT_0^i/m_i}, v_0=\sqrt{2kT_0^e/m_e}, w_0=\sqrt{2kT_0^n/m_n}$, where T_0^i, T_0^e, T_0^n are the representative values of temperatures of respective components. Since $m_e \prec\prec m_i$, it is commonly fair that $v_0>\xi_0$, that is why one can choose $t_0=a/v_0$ as the time scale. The self-consistent electric field in the plasma is defined by the motion of its components. And the scale value of the field is taken to be kT_0^e/e.

Generally, the values $f_i^0=n_0/\xi_0^3, f_e^0=n_0/v_0^3, f_n^0=n_0/w_0^3$ are accepted to be the representative values of the corresponding distribution functions. Here, n_0 is the scale value of ion and electron densities. It is the same for both components. The system (1) in the dimensionless form is as follows:

$$Sh_i\frac{\partial f_i}{\partial t}+\xi^k\frac{\partial f_i}{\partial x^k}-B\frac{\partial\varphi}{\partial x^k}\frac{\partial f_i}{\partial\xi^k}+\frac{1}{R_i}\varepsilon_{kjl}\xi^j H^l\frac{\partial f_i}{\partial\xi^k}=\frac{1}{Kn_{in}}J_{in}$$

$$\frac{\partial f_e}{\partial t}+\mathrm{v}^k\frac{\partial f_e}{\partial x^k}+\frac{\partial\varphi}{\partial x^k}\frac{\partial f_i}{\partial \mathrm{v}^k}-\frac{1}{R_i}\varepsilon_{kjl}v^j H^l\frac{\partial f_e}{\partial v^k}=0 \qquad (4)$$

$$Sh_n\frac{\partial f_n}{\partial t}+w^k\frac{\partial f_n}{\partial x^k}=\frac{1}{Kn_{ni}}J_{ni}$$

In (4), $Sh_i=\frac{\mathrm{v}_0}{\xi_0}=\sqrt{\frac{m_i T_0^e}{m_e T_0^i}}>1$ is an analogue of the Strouchal number for ions, $Sh_n=\frac{\mathrm{v}_0}{w_0}$ is the same for neutrals, $B=\frac{T_0^e}{T_0^i}\geq 1, R=r_l^s/L$, here $r_l^s=m_s c_s c/eH_0, s=i,e; c_i=\xi_0, c_e=v_0$ are the Larmor radius for ions and electrons, respectively. In the plume problem, the electron Larmor radius is much less than that of the ion, therefore, it is possible to ignore the magnetic field component in the equation for ion Kn_{in}, Kn_{ni} are the Knudsen numbers for ion-neutral and neutral-ion interactions, and the components in front of them are the dimensionless values of collision integrals. They are defined in [1] and [2].

Equation for the electric field potential is as follows

$$\varepsilon^2\Delta\varphi=(n_e-n_i) \qquad (5)$$

The formula (5) is the Maxwell's equation in a dimensionless form. Here $\varepsilon = r_d / L,$ with $r_d = \sqrt{\dfrac{kT_0^e}{4\pi e^2 n_0}}$ being the Debye length. In the plume problem, its magnitude is of the order of 10^{-3}, i.e. it is very small.

2. Solution method

The small value of the Debye length stipulates difficulties in finding numerical solutions for the equations (4), (5). Indeed, assuming that the electric field in (4) in the previous time step is known (as it is in [1]), it becomes possible to define n^i, n^e by solving the system (4), and typically $n^i \neq n^e$. By substituting the found values in the right side of (5), we obtain very high magnitudes of electric field, which divides the charges in the next step of time in a way that $n^i - n^e \approx 0$. As the new electric field is weak, this process will be repeated. Hence, the developed numerical scheme reproduces the Langmuir oscillations, which aren't observed it the plume. That is why it is needed to develop such scheme, which would include the possibility for plasma oscillations damping.

Another, perhaps, fundamental difficulty in obtaining the solution of the previously presented problem concerns the absence of boundary conditions for the equation (5). To overcome the considered difficulties, the following scheme is proposed. Let's set that $N = \left(n^i - n^e\right)$ and subtract the electron continuity equation from the similar equation for ions. The result is $\dfrac{\partial N}{\partial t} + \dfrac{\partial j^k}{\partial x^k} = 0$.

Then, we subtract the momentum conservation equation for electrons from the corresponding equation for ions. After the appropriate manipulation the following equation is derived:

$$\frac{\partial j^s}{\partial t} + \left\langle \sigma_{in} g^{in} \right\rangle j^s + e\frac{n^e}{m^e}\frac{\partial \varphi}{\partial x^s} = R^s, s = 1,2,3 \tag{6}$$

The value R^s comprises all components needed for the left hand side of (6) to be as it appeared after all the described processing. An immediate deriving of this value is not needed further. The value $\dfrac{n^e}{m^e}$ in (6) exists, if in an accurate deriving $m_e \prec\prec m_i$ is accounted. By taking the divergence of left and right hand sides of (6) with due regard for (2) and transposing some components to the right-hand side, the following equation may be derived:

$$\frac{\partial^2 N}{\partial t^2} + \left\langle \sigma_{in} g^{in} \right\rangle \frac{\partial N}{\partial t} + \frac{n^e}{m^e} e^2 N = Q \tag{7}$$

Here $Q = div\vec{R} - (grad\left\langle \sigma_{in} g^{in} \right\rangle \vec{j}) - \left(grade\dfrac{\dot{n}}{m^e}\, grad\varphi \right)$.

It is clear, that the formula (7) was deduced with direct utilizing of the continuity equations. Let us make the equation (7) dimensionless by using the previously introduced formulas. The following is obtained:

$$\frac{\partial^2 N}{\partial t^2}+\frac{\nu(t,\vec{x})}{2\sqrt{\pi}\varepsilon K_{ni}}\frac{\partial N}{\partial t}+\frac{n^e}{\varepsilon^2}N=Q \tag{8}$$

The presence of the small parameter ε in (8) indicates that the motion of the charged plasma components would proceed in a two-scale mode, and to examine it we employ the asymptotic methods, which were developed in [5].

Let us assume that $N(t_0)\neq 0$ Then during the plasma evolution in the time interval $\left[t_{0,}t_{0,}+\Delta t\right]$, such time interval $O(\varepsilon)$ exists, when the quasi-neutrality is violated. Let's introduce 'fast' time $\tau=\frac{t}{\varepsilon}$. The solution of (8) is to be found in the form of asymptotic series expansion $N(\tau,\vec{x})=\bar{N}(\tau,\vec{x})++\varepsilon N_1(\tau,\vec{x})+\ldots$. The equation for $\bar{N}(\tau,\vec{x})$ will be as follows:

$$\frac{\partial^2 \bar{N}}{\partial \tau^2}+\frac{\nu(t_0,,\vec{x})}{2\sqrt{\pi}K_{ni}}\frac{\partial \bar{N}}{\partial \tau}+n^e(t_0,,\vec{x})\bar{N}=\varepsilon^2\bar{Q}(t_0,,\vec{x}), \tag{9}$$

Let us set $\alpha=\nu(t_0,\vec{x})/2\sqrt{\pi}K_{ni},\omega=\sqrt{n^e-\alpha^2/4}$.

Under $n^e-\frac{\alpha^2}{4}>0$, the solution of (9) is written as:

$$\bar{N}(\tau,\vec{x})=Ae^{-\alpha\tau}\cos(\omega\tau+\varphi)+\frac{\varepsilon^2}{n^e(t_0,,\vec{x})}Q(t_0,,\vec{x}) \tag{10}$$

The formula (10) describes the damping of plasma (Langmuir) oscillations in quick time mode. From the equation (10) the following may be derived: $\bar{N}(t_{0,}+\Delta t,\vec{x})=Ae^{-\alpha\frac{\Delta t}{\varepsilon}}\cdot\cos\left(\omega\frac{\Delta t}{\varepsilon}+\varphi\right)+\frac{\varepsilon^2}{n^e(t_0,,\vec{x})}Q(t_0,\vec{x}).$

When $\frac{\Delta t}{\varepsilon}>1$, the oscillations will decay. And by the time moment $t_0+\Delta t$ the quasi-neutrality sustains. Then, for the definition of the electrostatic potential, the following equation could be utilized: $\varphi(t_0+\Delta t,\vec{x})=-Q(t_0,\vec{x})$. In this equation the small parameter is missing. The issue would concern the boundary conditions set. Let's write the solution of the above equation in the following form:

$$\varphi(t_0 + \Delta t, \vec{x}) = \frac{1}{4\pi} \int \frac{Q(t_0, \vec{x})}{\sqrt{(x_1 - y_1)^2 + (x_2 - y_2)^2 + (x_3 - y_3)^2}} d\vec{y} + \varphi_{ext} \quad (11)$$

The integration in (11) is performed over the whole computational domain. The first summand in (11) is a Newtonian potential (specific solution of Poisson equation), and the second one is the known preassigned potential. It is defined as $\varphi_{ext} = 0$. As a result, the following solution scheme for the stated problem is proposed. The electrostatic potential at time $t_0 + \Delta t$ is defined via formula (11) based on the known values of the macro-variables at time moment t_0. The obtained value is plugged into the kinetic equations system, from the solution of which all macro-variables at time moment $t_0 + \Delta t$ are defined, and then the process is repeated. The idea of utilizing the conservation equations to solve the kinetic equations has been proposed earlier in [6].

3. Conclusion

A method is proposed, which allows to calculate the electron distribution function in the problem of plasma dynamics calculation as applied to the stationary plasma thruster plume based on describing the motion for every plasma particle with kinetic equations.

The application of a fully kinetic approach to the study of the plasma jet made it possible to obtain a system of conservation equations that govern this motion. The presence of distribution functions made it possible to consider this system as closed, which made it possible, by modernizing the method used in [7], to study qualitatively and quantitatively the process of establishing quasineutrality in a plasma medium, and to use the Maxwell equation directly to determine the electric field.

The principal point in the proposed work of the scheme is the presence of a system of kinetic equations describing the motion of charged and neutral components.

In the case of considering the problem of plasma motion in the thruster channel. The motion of the electrons will no longer be collisionless. We need to simulate the ionization process. There will also be a question about the boundary conditions for the electron distribution function on the walls of the thruster channel.

References

[1] M V Abgaryan, A M Bishaev 2018 Zhurnal vychislitelnoi matematiki i matematicheskoi fiziki **58** (7) 1132-46

[2] A M Bishaev, M V Abgaryan , V Kim, D V Merkur'ev, E P Ivanova, A I Fomenkov, M Wartelski and C Theroude 2018 *Plasma Physics Reports* **44**(2) 278-88

[3] P I Zhevandrov, A I Morozov and S A Yakunin 1984 *Plasma Physics Reports* **10**(2) 353-65

[4] A A Vlasov 1978 *Non-local statistical mechanics* ed. Fedyanin V K (Moscow: Nauka), chapter 8, 172-84

[5] J D Cole 1972 *Petrurbation Methods in Applied Mathematics* Moscow: MIR

[6] A M Bishaev and V A Rykov 1975 *USSR Computational Mathematics and Mathematical Physics* **15**(1) 166-76

[7] E M Lifshits, L P Pitaevskiy 1987 *Physical Kinetics* **X** «Nauka» 527.

Moving an atom by swapping places across a crystal lattice without vacancies

Askerova V I, Kalashnikov E V

Moscow Region State University, Radio street, 10A, 105005, Moscow, Russia.

e-mail address: ekevkalashnikov1@gmail.com; vera_askerova@mail.ru

Abstract. The mechanism of moving an atom through a crystal lattice without vacancies is considered. This situation can be observed when all sites of the lattice are occupied by atoms of the same type, or when some of them are occupied by isotopes of the same atoms. It is shown that in a three-dimensional crystal lattice constructed from identical atoms without vacancies, one-dimensional motion of the own atom along the lattice chain is possible. This movement is realized as a Frenkel-Kontorova soliton.

1. Introduction

One of the most common types of diffusion is self-diffusion. – A situation in which atoms of the same type (or substance) move along a lattice made up of (own) atoms of the same type (all sites are occupied by atoms). In addition, this situation is similar to that in which the sites of such a crystal can contain isotopes of atoms of the same type. When all the sites (nodes) are occupied, it is difficult the atom to move. The realization of self-diffusion is facilitated if Frenkel's defects are included (when the atom passes into the interstitial) or Schottky defects (when a vacancy occurs and it moves sequentially to the interface) [1]. On the other hand in the literature [1-3] models are proposed that allow us to take into account the possibility of moving an atom to a nearby occupied place by exchanging the places of the nearest atoms. This can be taken into account through the probability that the atom will move to a neighboring position, provided that the neighboring atom has left its place in the opposite direction. At the same time, the elementary mechanism of such a transition remains unclear. And, in addition, in the listed variants of self-diffusion implementation, it is not taken into account that the atom, moving along the lattice, participates in a collective movement, involving the entire environment in the act of its movement. Here we propose an approach for describing the motion of a own atom along a lattice without vacancies, taking into account the elementary mechanism of the exchange of places of the nearest atoms and taking into account the collective mechanism of such movement.

2. Model

Each atom in the lattice experiences deviations from its equilibrium position, thereby involving neighboring atoms in its displacements. We write down the Lagrangian for this behavior of atoms. To do this, select the two nearest atoms (Fig. 1).

$$\mathcal{L} = {m_1\dot{\boldsymbol{\Gamma}}^2}/{2} + {m_2\dot{\boldsymbol{D}}^2}/{2} - A_1\big(1 - cos(\boldsymbol{\Gamma}\cdot\boldsymbol{n})\big) - A_2\big(1 - cos(\boldsymbol{D}\cdot\boldsymbol{n})\big) -$$

$$- {\alpha_{jm}(\boldsymbol{\Gamma} - \boldsymbol{D})^2}/{2} - ({1}/{2})\textstyle\sum_j \mathcal{U}(\boldsymbol{\Gamma} - \boldsymbol{D}, \boldsymbol{\varepsilon}_j) +$$

$$+({m}/{2})\textstyle\sum_j \dot{\boldsymbol{\varepsilon}}_j^{\,2} - ({\alpha}/{2})\textstyle\sum_j (\boldsymbol{\varepsilon}_j - \boldsymbol{\varepsilon}_{j+1})^2 \quad (1)$$

Here, $\boldsymbol{\Gamma}$ *and* $\boldsymbol{D}$ – are the displacement vectors of the first and second separated atoms (Fig. 1). The first line in (1) expresses the Lagrangian for atoms with mass m_1 and m_2 as independently moving in the periodic field of the lattice for each with amplitude $A_1$and A_2. In this case, each displacement of the selected atom will experience an "elastic" interaction with its neighbor and with the rest of the lattice atoms. The second line of the Lagrangian corresponds to this situation. The third line expresses the behavior of the lattice in the harmonic approximation. m is the mass of the atoms in the lattice; $\boldsymbol{\varepsilon}_j$ is the displacement vector of the $\boldsymbol{j}$ -th atom of the rest of the lattice; vector $\boldsymbol{n}$ is the direction of motion.

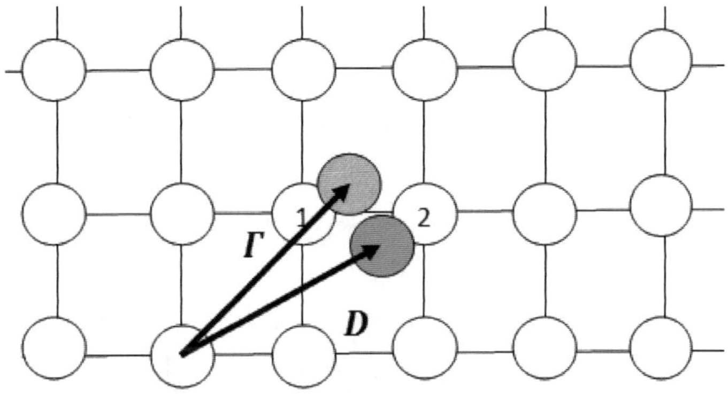

Fig. 1. Diagram of the nearest neighbor offsets "1" and " 2 " in the lattice.

The atom " 1 " has shifted to the position given by the vector $\boldsymbol{\Gamma}$. The atom " 2 " has shifted to the position set by the vector $\boldsymbol{D}$. This formulation of the problem turns out to be similar to the problem describing the movement of a nitrogen molecule along the core of a helical dislocation [4]. The local chain approximation [4] allows us to decompose the potential energy of the interaction of the selected atoms into the sum of the products of the differences between the deviations of the displacement of the selected atom and the displacements of the nearest neighbors in each local chain. The resulting equations of motion for the selected pair of atoms from (1) look like this:

$$m_1\, {d^2\boldsymbol{\Gamma}}/{dt^2} = \alpha_{jm}(\boldsymbol{\Gamma} - \boldsymbol{D}) + \alpha_1(\boldsymbol{\Gamma} - \boldsymbol{\varepsilon}) + A_1({2\pi}/{a})\sin(\boldsymbol{\Gamma}\cdot\boldsymbol{n}) \quad (2)$$

and

$$m_2\, {d^2\boldsymbol{D}}/{dt^2} = \alpha_{jm}(\boldsymbol{D} - \boldsymbol{\Gamma}) + \alpha_2(\boldsymbol{D} - \boldsymbol{\varepsilon}) + A_2({2\pi}/{a})\sin(\boldsymbol{D}\cdot\boldsymbol{n}) \quad (3)$$

Adding up these two equations and going to the center of mass

$$R = {(m_1\boldsymbol{\Gamma} + m_2\boldsymbol{D})}/{(m_1 + m_2)}$$

and taking into account the relative position $\boldsymbol{d} = \boldsymbol{\Gamma} - \boldsymbol{D}$ of the isolated atoms, (provided that all the masses in the crystal are equal to each other $m_1 = m_2 = m$ and are also equal to each other $\alpha_1 = \alpha_2 = \alpha_{jm} = \alpha$) we obtain:

$$m\, {d^2\boldsymbol{R}}/_{dt^2} = Z\alpha\left({d^2\boldsymbol{R}}/_{dt^2}\right)\cdot\tau^2 + A\left({2\pi}/_{a}\right)\cos\left(\boldsymbol{n}\cdot{\boldsymbol{d}}/_{2}\right)\sin(\boldsymbol{n}\cdot\boldsymbol{R}) \quad (4)$$

here Z is the number of nearest neighbors in the lattice, τ – parameter (time) during which the neighboring deviations do not have time to change [4]. Equation (4) can be rewritten to take into account the effective mass $m^* = m - Z\alpha\cdot\tau^2$ and $A\left({2\pi}/_{a}\right)\cos\left(\boldsymbol{n}\cdot{\boldsymbol{d}}/_{2}\right) = \mathcal{A}(\Theta)$ in a more compact form:

$$m^*\, {d^2\boldsymbol{R}}/_{dt^2} = \mathcal{A}(\Theta)\sin(\boldsymbol{n}\cdot\boldsymbol{R}) \qquad (5)$$

here Θ is the angle between the direction $\boldsymbol{n}$ and ${\boldsymbol{d}}/_{2}$ lies either in the plane oxz or oyz. The solution to this equation is the Frenkel-Kontorova soliton:

$$R_{n,z} = \left({2a}/_{\pi}\right) arctg\left[\pm 2\pi\left(-{\mathcal{A}(\Theta)}/_{m^*}\right)^{1/2} {t}/_{a}\right] \qquad (6)$$

In this case, the Lagrangian is transformed to the form:

$$\mathcal{L} = {m_1\dot{\boldsymbol{R}}_n^{\,2}}/_{2} - {\alpha(\boldsymbol{R}_n - \boldsymbol{R}_{n-1})^2}/_{2} - \mathcal{A}(\Theta)\left(1 - cos(\boldsymbol{R}_n\cdot\boldsymbol{n})\right) \qquad (7)$$

here $\mathcal{A}(\Theta)$ from (6) describes the rotation of the selected pair of atoms relative to the center of mass. This means that the movement of its own atom along the lattice without vacancies will be carried out from site to site by mutual "rolling" of the atom located in position $n - 1$ and the atom moving towards it from position n relative to the center of mass of this pair. The very center of mass of such a "rolling" formation moves like a Freenkel-Kontorova soliton.

Conclusion

It is shown that the movement of an atom along a one-dimensional chain without vacancies, constructed from the same atoms, is realized as the movement of a pair of atoms, the center of mass of which moves like a Frenkel-Kontorova soliton, and its "core" is a pair of atoms rolling over each other. Such a one-dimensional situation is possible if the possible connections with the nearest neighbors in the three-dimensional crystal are taken into account.

References

[1] Mehrer H 2007 *Diffusion in Solids* Springer-Verlag. Berlin, Heidelberg 651p.
[2] Girifalco L A 1964 *Atomic Migration in Crystals* Blaisdell Pablishing Company NY 162 p.
[3] Manning J R 1968 *Diffusion Kinetics for atoms in crystals*. D. Van Nostrand Company, INC. Princeton, Toronto 275 p.
[4] Kalashnikov E V, Klyavin O V, Titarenko I G 2013 *Technical Physics* **58**(4) 546-549.

MATLAB Modelling electron energy spectrum in one-dimensional quantum well with infinitely high walls for GaAs solid solution

E R Kozhanova[1], I M Tkachenko[2]

[1]Yuri Gagarin State Technical University of Saratov, Saratov, Russia
[2] École de Technologie Superieure, Montreal, Canada

E-mail: kozhanovaer@gmail.com

Abstract. The paper presents calculations of electron states in semiconductor nanostructures Al_xGa_{1-x} As (x=0) and simulates the energy spectrum of the electron in a one-dimensional quantum well with infinitely high walls of a given width (from 10 to 30 atomic monolayers) using MATLAB application. The data is visualized using a wavelet transform with different wavelet functions.

1. Introduction

Calculations of electronic states in semiconductor nanostructures, performed in the effective mass method, are based on the solution of the stationary Schrödinger equation, which is one-dimensional for the motion of electrons in the perpendicular plane of the layers (excluding cases where the potential relief is a function of time) [1 - 5]:

$$-\frac{\hbar^2}{2m}\cdot\frac{\partial^2\psi(z,E)}{\partial^2 z}+U(z)\cdot\psi(z,E)=E\cdot\psi(z,E), \tag{1}$$

where m is the effective mass of the electron, E is its total energy, U(z) is the potential relief for the electron along the z axis directed perpendicular to the plane of the layers.

A solution of this Schrödinger equation is the z - component of the envelope of the wave function $\psi(z,E)$ which characterises the motion of electrons in the perpendicular plane of the layers and determines the probability of finding an electron with energy E of motion along the Oz axis at a point with coordinate z within the normalisation [1 - 2].

2. Modelling the energy spectrum of an electron in a one-dimensional quantum well

Consider simulation of the energy spectrum of an electron in a one-dimensional quantum well.

If a thin layer of narrow band material is placed between two quite thick layers of wide-band material, a potential well can be formed for an electron moving in a direction transverse to the planes of the layers (Fig. 1) (the energy is counted from the bottom of the conduction band of the well material). This kind of relief is called a rectangular potential well. For modeling we will consider a symmetric potential well (Fig. 1, b) [1-2].

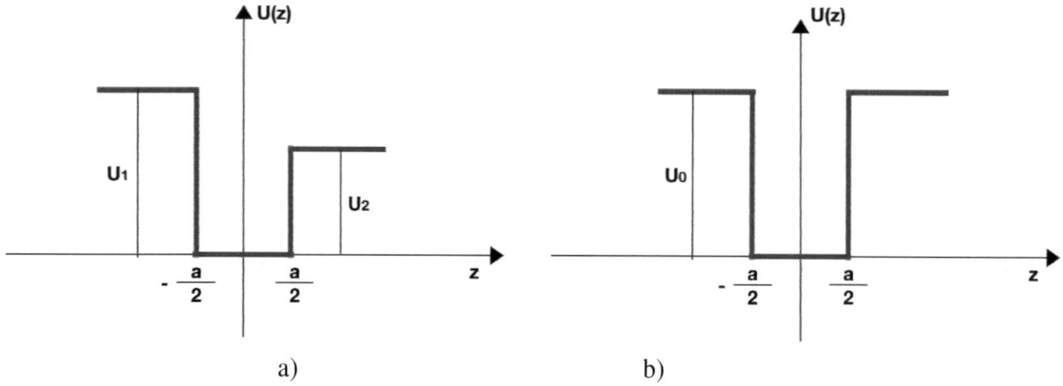

Figure 1. Energy diagram of a rectangular potential well:
a - asymmetrical, b – symmetrical [1]

Consider the idealized case of a potential well with infinitely high walls, in which case the envelope wave function $\psi(z,E)$ outside the well is identically equal to zero and inside the well satisfies the Schrödinger equation (1) with boundary conditions [1-5]:

$$\psi\left(\pm\frac{a}{2},E\right)=0 \tag{2}$$

where a is the width of the potential well (thickness of the middle layer of the three-layer structure). Then $\pm a/2$ are coordinates of interfaces between the layers.

The solution at discrete energy E - eigenvalues of this equation (1) [6] has the form:

$$E=\frac{\hbar^2}{2m}\cdot\left(\frac{\nu\pi}{a}\right)^2 \tag{3}$$

where ν is the number of the energy level.

The system has symmetry along the Oz axis with respect to the origin, so the set of eigen solutions of the Schrödinger equation is divided into subgroups of even and odd solutions [1, 3-5]:

$$\psi(z,E)=\begin{cases}\cos\left(C\cdot\frac{\nu\pi}{a}\cdot z\right) & at \quad \nu=1,3,5... \\ \sin\left(C\cdot\frac{\nu\pi}{a}\cdot z\right) & at \quad \nu=2,4,6...\end{cases} \tag{4}$$

where C is the normalisation factor, depending on ν.

Plots of the envelope wave functions of electrons use the square of the modulus of expression (4) (Fig. 3):

$$|\psi(z,E)|^2 \tag{5}$$

This task can be solved for the solid solution $Al_xGa_{1-x}As$, the band structure for any x is shown in figure 2.

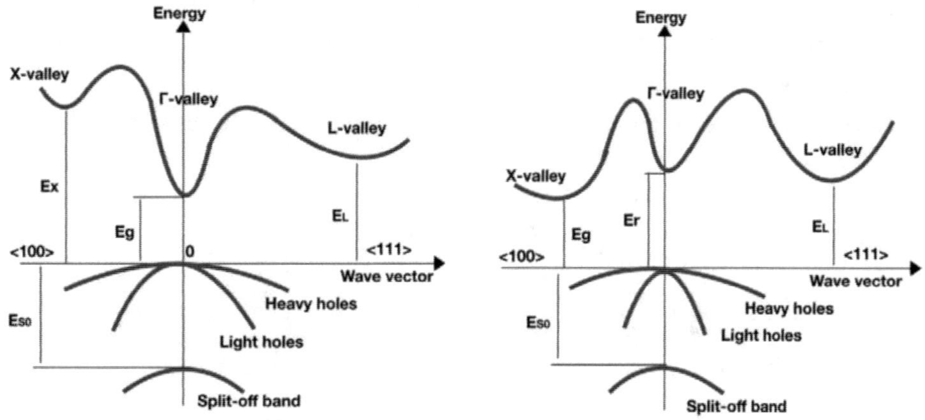

a) b)

Figure 2. Zone structure of $Al_xGa_{1-x}As$ solid solution: a - $x < 0.45$, b - $x > 0.45$ [1]

3. Modeling of electron energy spectrum in a one-dimensional quantum well with infinitely high walls of solid solution $Al_xGa_{1-x}As$

As a test case, the authors considered a quantum well with infinitely high walls of 20 atomic monolayers wide (11.3 nm) for the first four allowed energy levels (Fig.3, a), the energy spectrum of the electron in such a well shown in Fig.3, b [1]. The well material is GaAs ($x = 0$). The graphs are plotted on the basis of calculations in MathCad.

The next step was written a program in the MATLAB mathematical package to simulate the energy spectrum of the electron in a one-dimensional quantum well with infinitely high walls of solid solution $Al_xGa_{1-x}As$ for the well material is GaAs ($x = 0$) (Fig. 3, c and d).

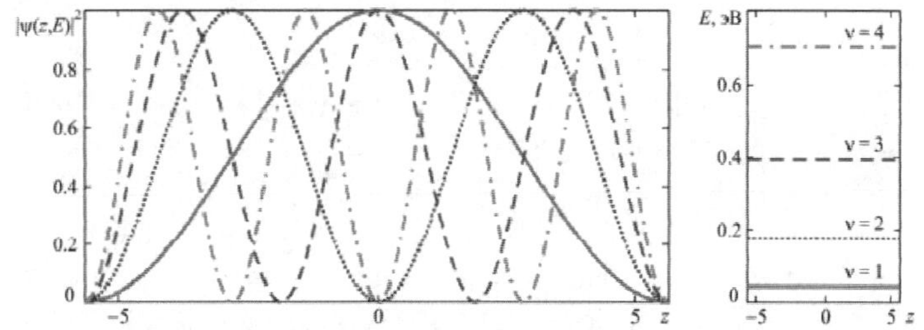

a) b)

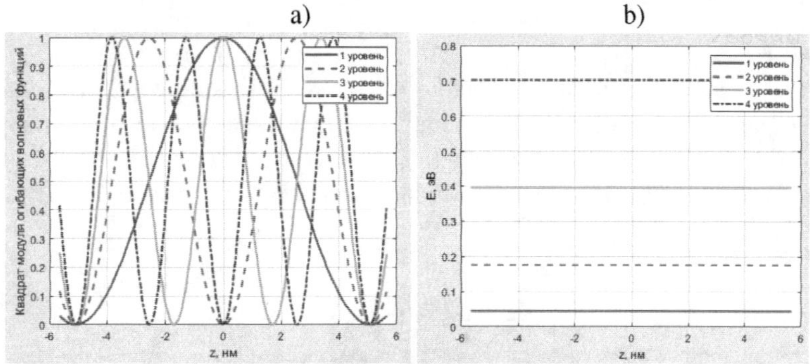

c) d)

Figure 3. Wave function envelopes (a, c) and quantised energy levels (b, d) of an electron in a rectangular quantum well

First, consider simulations of GaAs ($x = 0$) at different values of the number of atomic monolayers *NN* for the first four resolved energy levels (Table 1, Figure 4).

Table 1

Number of atomic monolayers *NN*	Width of the quantum well, *a*, nm	Quantisation levels in the well			
		1	2	3	4
10	5.6533	0.0281 e-36	0.1125 e-36	0.2532e-36	0.4502 e-36
15	8.4800	0.0125 e-36	0.0500 e-36	0.1125e-36	0.2001 e-36
20	11.3066	0.0070 e-36	0.0281 e-36	0.0633e-36	0.1125 e-36
25	14.1333	0.0450 e-37	0.1801 e-37	0.4052e-37	0.7203 e-37
30	16.9599	0.0313 e-37	0.1250 e-37	0.2814e-37	0.5002 e-37

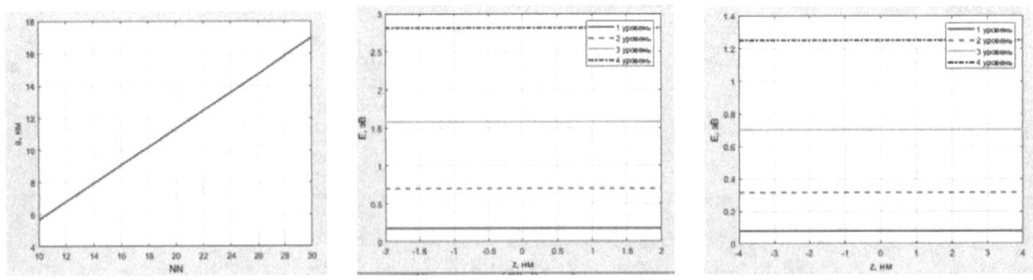

a) b) NN = 10 c) NN=15

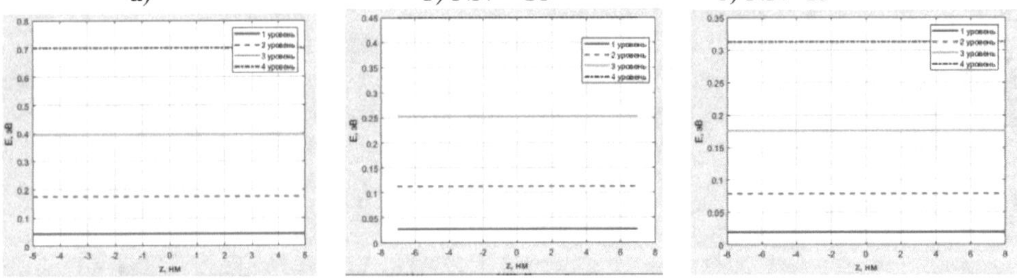

d) NN = 20 e) NN=25 f) NN = 30

Figure 4. Simulation results of electron energy spectrum in one-dimensional GaAs quantum well with infinitely high walls ($x = 0$) at different values of number of atomic monolayers *NN* for first four resolved energy levels

4. Wavelet transform

We performed a continuous wavelet transform [7-9] using the following wavelet functions: «Haar» (column 2 of Table 2), capturing inflection points, and «Mexican Hat», capturing points of extremum (column 3 of Table 2), as well as a *wtmm* function from the Matlab library [10] (in 0.01 steps; number of samples 1131) for each envelope function of quantised energy levels (Fig. 3,c).

Table 2

Energy level	Wavelet «Haar» / *wtmm*	Wavelet «Mexican Hat» / *wtmm*
1	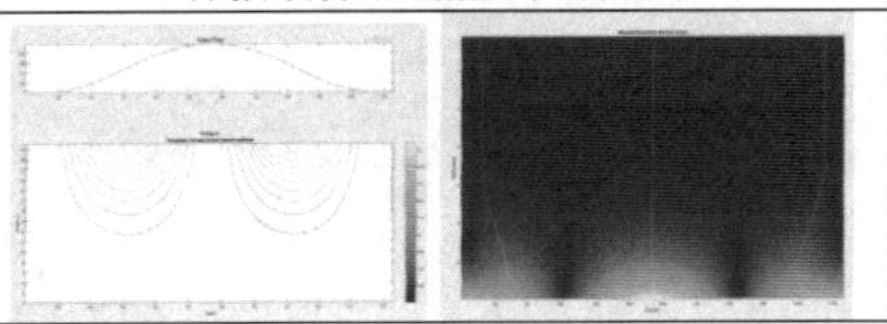	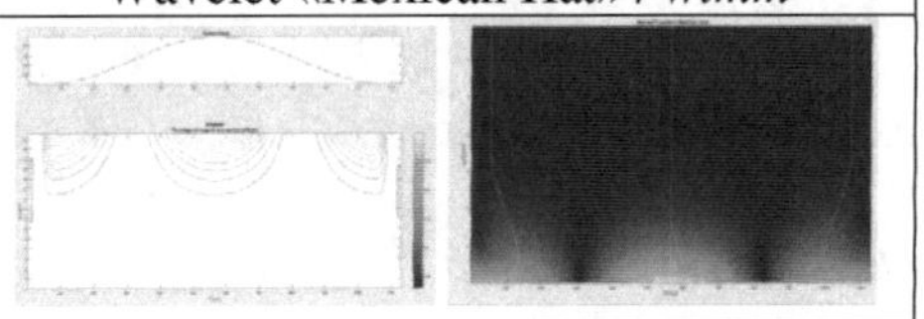

2	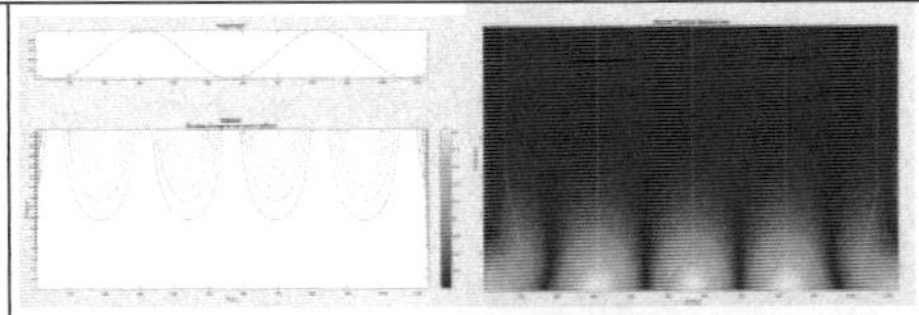	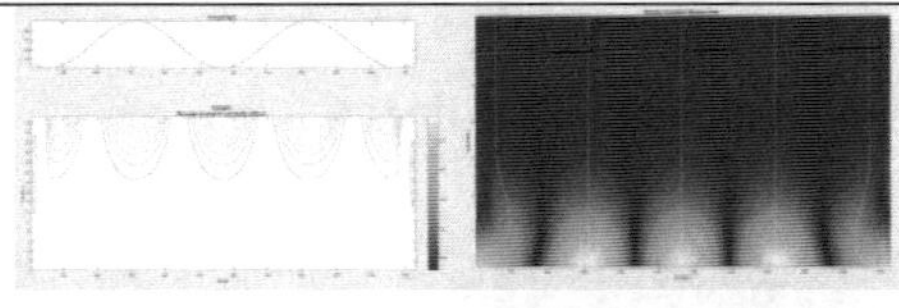
3	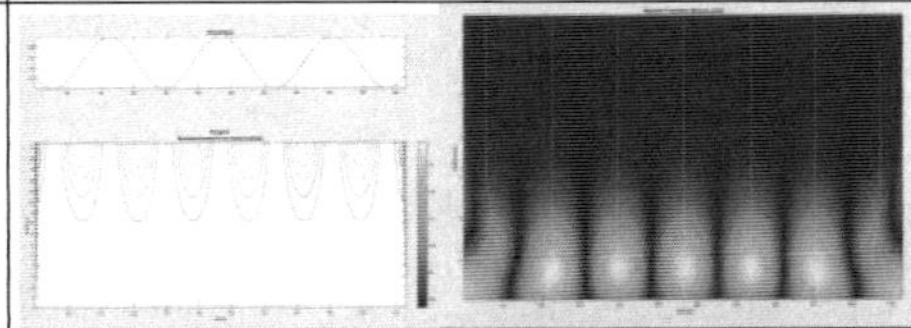	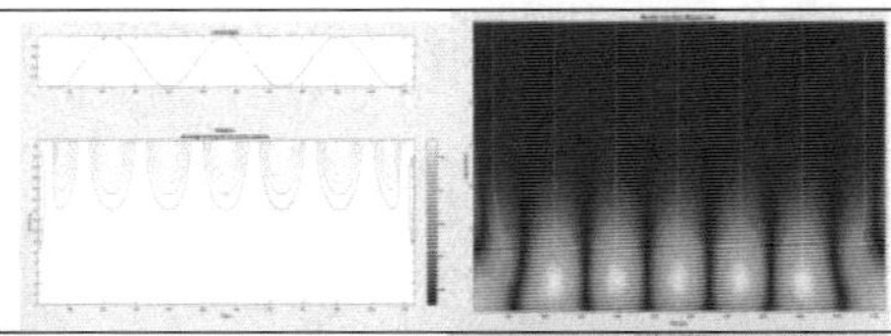
4	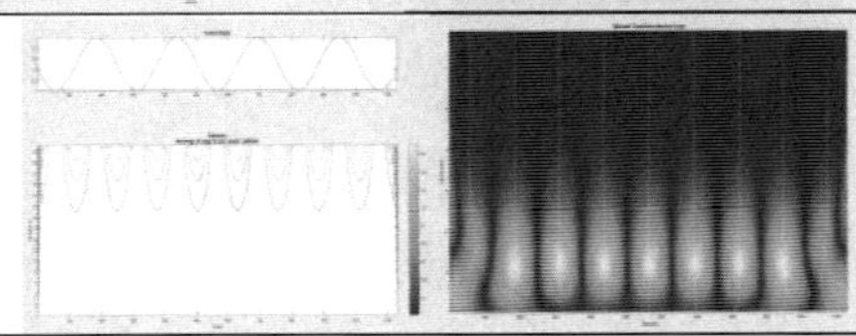	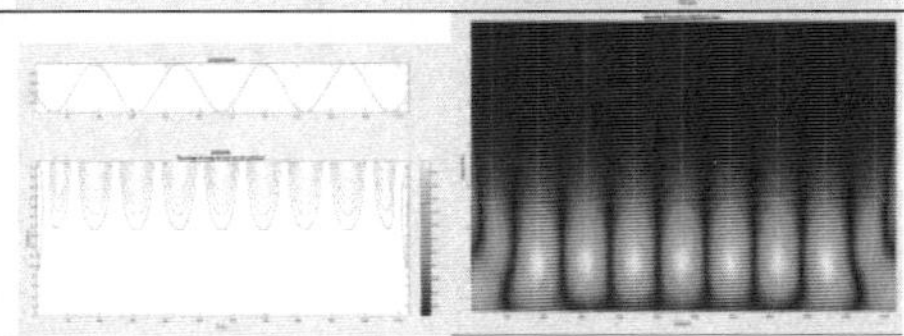

If we leave the Mexican Hat wavelet function [7-9] for the study and set the step to 0.001 (number of counts 11307), the appearance of the plots will change (Table 3).

Table 3

Energy level	Wavelet «Mexican Hat» / *wtmm*	
	step = 0,01	step = 0,001
1	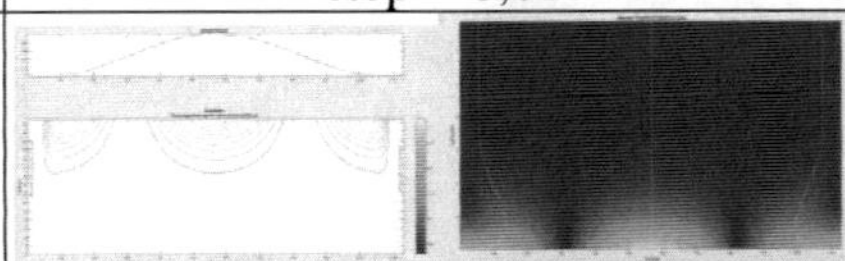	
2	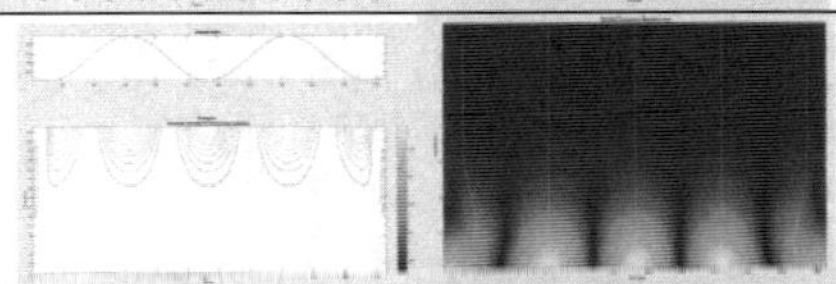	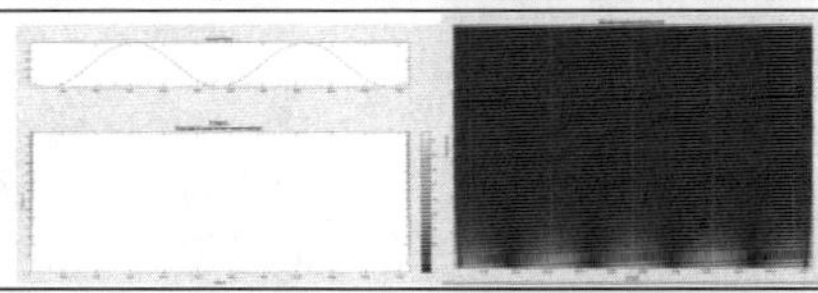
3	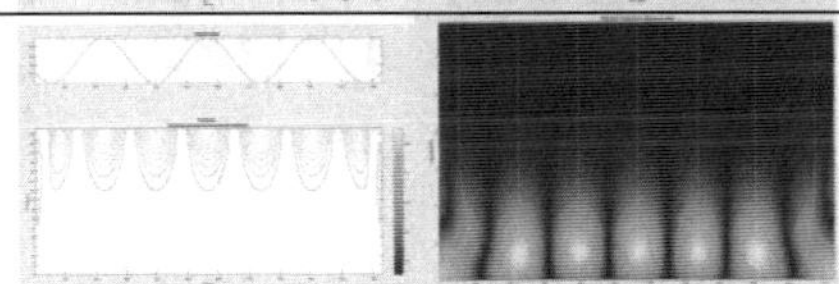	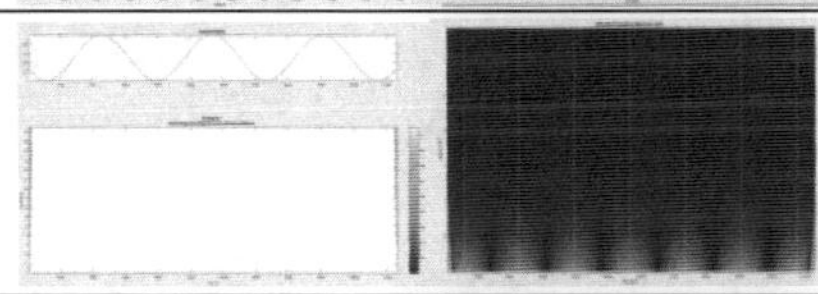
4	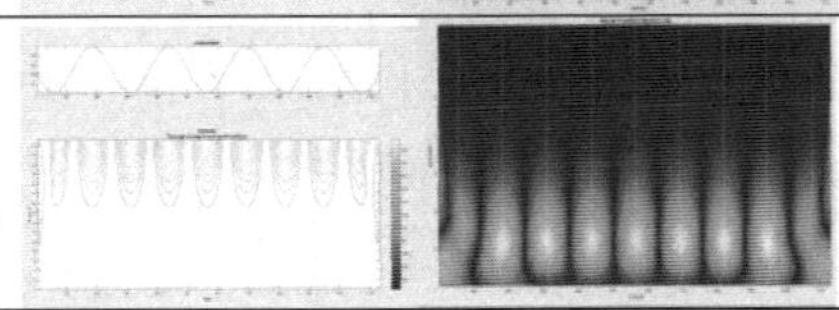	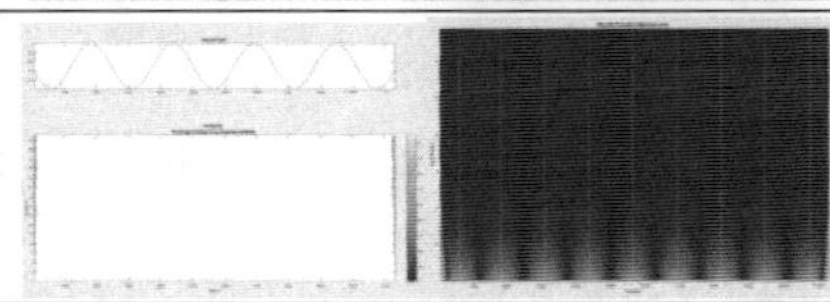

All wavelet transform results are plotted with respect to the number of reference points with a given step, so the task arises of translating the number of reference points with respect to the coordinate z defining the width of the quantum well from $-a/2$ to $a/2$ (Fig. 5).

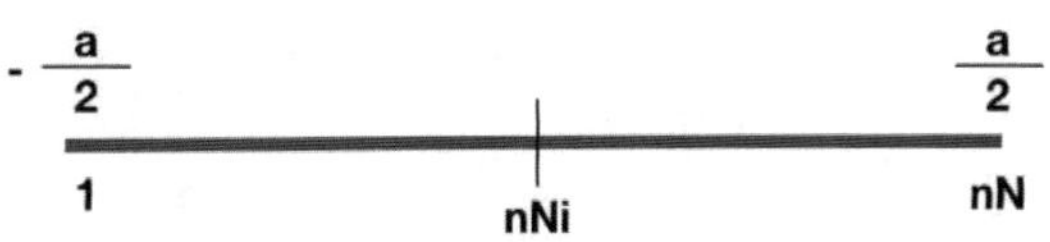

Figure 5. Conversion from one scale to the other

A formula for conversion from one scale to the other can be derived from this:

$$z_i^* = -a/2 + shag*(nN_i - 1) \tag{6}$$

where *shag* is the step, nN_i is the number of counts of a given point.

Consider first the errors that arise when converting from one scale to another by summarising the data in Table 4.

Table 4

Parameters	*shag* = 0,01	shag =0,001
a	11.3066	11.3066
a/2	5.6533	5.6533
nN	1131	11307
Error Δ_l	0,0066	6e-4

Substitute *nN* and shag from Table 4, calculate the last scale value and compare it with the value of *a/2*, finding the scale conversion error.

- for step *shag* = 0.01:

$$z1 = -5{,}6533 + 0{,}01*(1131 - 1) = 5{,}6467$$
$$\Delta_1 = 5{,}6533 - 5{,}6467 = 0{,}0066$$

- for step *shag* = 0.001:

$$z2 = -5{,}6533 + 0{,}001*(11307 - 1) = 5{,}6527$$
$$\Delta_2 = 5{,}6533 - 5{,}6527 = 6e - 4$$

Compare the results obtained with the *wtmm* function for different values of the *shag* step (Fig.6).

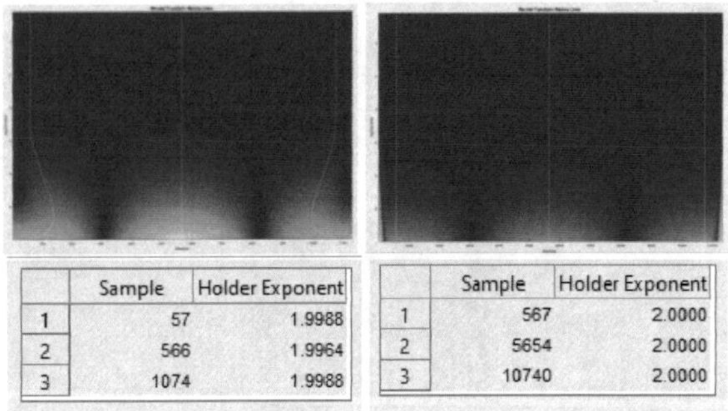

a) *shag* = 0,01 b) *shag*=0.001

Figure 6. Results obtained with the wtmm function

Use formula (6) to find the values (Table 5).

Table 5

Extremes (Fig.6)	*shag* =0,01		*shag* =0,001	
	sample (Fig.6,a)	calculation z_i	sample (Fig. 6,b)	calculation z_i
1	57	-5.0933	567	-5.0873
2	566	-0.0033	5654	-3.0000e-04
3	1074	5.0767	10740	5.0857

5. Conclusion

The authors present the results of a program written using the MATLAB application for solving the task of modeling the energy spectrum of an electron in a one-dimensional quantum well with infinitely high walls of the $Al_xGa_{1-x}As$ (x=0) solid solution. As a result of the wavelet transform the wave function envelopes of quantised energy levels, wavelet spectrograms were obtained using the wavelet functions «Haar» and «Mexican Hat» at different values of the number of atomic monolayers (from 10 to 30). The authors have proposed a formula (6) to convert the number of nN_i counts obtained by dividing the width of quantum well by the calculation step *shag*, into a z scale with a range from $-a/2$ to $a/2$. In further research, the authors plan to use a function *wtmm*, which captures points of extremum and outputs a table with their values.

6. Acknowledgement

The paper is prepared under partial support of Russian Foundation for Basic Researches (RFBR) (project No. 19-07-00602_a).

References

[1] Usanov, D.A., Skripal, A.V., Skripal, An.V., Abramov, A.V. 2013 Textbook for students of nano- and biomedical technologies. Saratov: SGU. 100 p.
[2] Usanov, D.A., Skripal, A.V. 2013 Textbook for students of nano- and biomedical technologies. Saratov, 128 p.
[3] Physics of Quantum Electron Devices 1990 Berlin: Springer.
[4] Brown, E.R. 1991 Hot Electrons in Semiconductor Heterostructures / Ed. By J.Shah. Boston: Academic Press.
[5] Ozbay E., Bloom, D.M., Diamond, S.K. 1991 Phys. and Appl. New York: Plenum.
[6] Korn, G. 2003 Handbook of mathematics for scientists and engineers: definitions, theorems, formulas 6th edition, stereotyp. SPb. Lan. - 832 p.
[7] Tkachenko, I.M., Kozhanova, E.R., Belyaev, V.V., Yazbeck, H. 2019 JPConfSer DOI:10.1088/1742-6596/1309/1/012020.
[8] Tkachenko, I.M., Kozhanova, E.R. 2018 Volga Region Scientific and Technical Bulletin **11**. P. 261-264.
[9] Smolentsev, N. K. Fundamentals of wavelet theory. Wavelets in MATLAB. Moscow: DMK Press, 2005. 304 p.
[10] Matlab Documentation.URL: https://www.mathworks.com/solutions.html?s_tid=gn_sol (Date of request 01.04.2021г.).

MATLAB Wavelet analysis of electron energy spectrum in one-dimensional quantum well with infinitely high walls for Al-Ga-As system (0 < x < 1)

E R Kozhanova[1], I M Tkachenko[2], V V Belyaev[3], S Maignan[3]

[1]Yuri Gagarin State Technical University of Saratov, Saratov, Russia
[2] École de Technologie Superieure, Montreal, Canada
[3] Moscow Region State University «MRSU», Very Voloshinoy str., 24, Mytishi, 141014, Russia

E-mail: kozhanovaer@gmail.com

Abstract. This paper presents calculations of electronic states in $Al_xGa_{1-x}As$ semiconductor nanostructures and simulates the envelope wave functions of quantum energy levels in a one-dimensional quantum well with infinitely high walls of a given width at various values of *x*. For the analysis of results the authors choose the function *wtmm* from the Matlab library that fixes the extremums and which is a characteristic of the fractality of the envelope wave functions of quantum energy levels.

1. Introduction

Nowadays, when new nanostructures are created, such as quantum dots or superlattices, the issue of studying $Al_xGa_{1-x}As$ semiconductor nanostructures remains relevant, where *x* is the fraction of gallium atoms substituted by aluminum atoms, it usually varies from 0.15 to 0.35.

Figure 1 shows the band structure of $Al_xGa_{1-x}As$ at different contents of *x* [1-2].

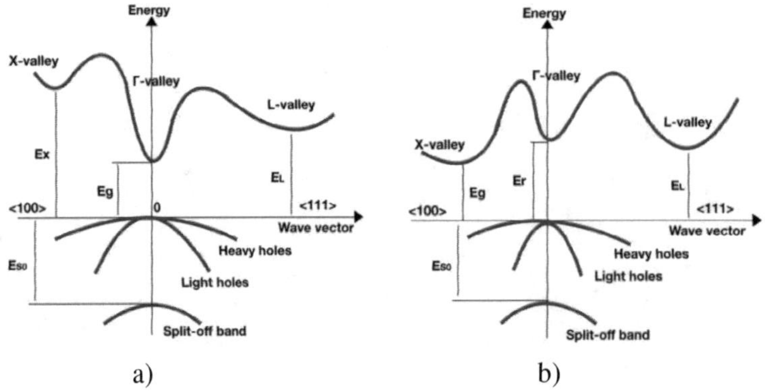

Figure 1. Zone structure of $Al_xGa_{1-x}As$ solid solution: a - $x < 0.45$, b - $x > 0.45$ [1]

The electronic states in semiconductor nanostructures are determined by the effective mass method, based on the solution of the stationary Schrödinger equation, which is one-dimensional for the motion of electrons in the perpendicular plane of the layers (excluding cases where the potential relief is a function of time) [1 - 2]:

$$-\frac{\hbar^2}{2m}\cdot\frac{\partial^2\psi(z,E)}{\partial^2 z}+U(z)\cdot\psi(z,E)=E\cdot\psi(z,E), \tag{1}$$

where m is the effective mass of the electron, E is its total energy, $U(z)$ is the potential relief for the electron along the z axis directed perpendicular to the plane of the layers.

A solution of this Schrödinger equation is the z - component of the envelope of the wave function $\psi(z,E)$ which characterises the motion of electrons in the perpendicular plane of the layers and determines the probability of finding an electron with energy E of motion along the Oz axis at a point with coordinate z within the normalisation [1 - 2].

Consider simulation of the energy spectrum of an electron in a one-dimensional quantum well.

If a thin layer of narrow band material is placed between two quite thick layers of wide-band material, a potential well can be formed for an electron moving in a direction transverse to the planes of the layers (Fig. 2) (the energy is counted from the bottom of the conduction band of the well material). This kind of relief is called a rectangular potential well. For modeling we will consider a symmetric potential well (Fig. 2, b) [1-2].

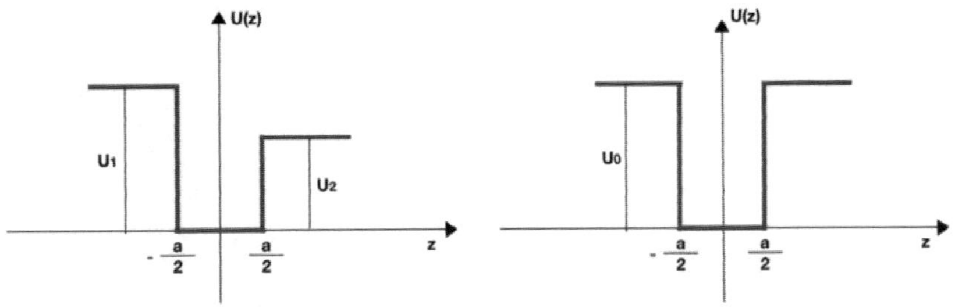

a) b)

Figure 2. Energy diagram of a rectangular potential well:
a - asymmetrical, b – symmetrical [1]

Consider the idealized case of a potential well with infinitely high walls, in which case the envelope wave function $\psi(z,E)$ outside the well is identically equal to zero and inside the well satisfies the Schrödinger equation (1) with boundary conditions [1 - 2]:

$$\psi\left(\pm\frac{a}{2},E\right)=0 \tag{2}$$

where a is the width of the potential well (thickness of the middle layer of the three-layer structure). Then $\pm a/2$ are coordinates of interfaces between the layers.

The solution at discrete energy E - eigenvalues of this equation (1) [3] has the form:

$$E=\frac{\hbar^2}{2m}\cdot\left(\frac{\nu\pi}{a}\right)^2 \tag{3}$$

where ν is the number of the energy level.

The system has symmetry along the Oz axis with respect to the origin, so the set of eigen solutions of the Schrödinger equation is divided into subgroups of even and odd solutions [1, 3-5]:

$$\psi(z,E)=\begin{cases}\cos\left(C\cdot\frac{\nu\pi}{a}\cdot z\right) & at \quad \nu=1,3,5...\\ \sin\left(C\cdot\frac{\nu\pi}{a}\cdot z\right) & at \quad \nu=2,4,6...\end{cases} \tag{4}$$

where C is the normalisation factor, depending on ν.

2. Modelling the energy spectrum of an electron in a one-dimensional quantum well at different values of *x*

The following equations were proposed to describe the $Al_xGa_{1-x}As$ structure [1]:

- effective mass of the electrons in the Γ-minimum:

$$m_n=(0.067+0.083\cdot x)\cdot m_e \tag{5}$$

- lattice constant:

$$a_0=(0.56533+0.00078\cdot x)\cdot 10^{-9} \tag{6}$$

- bandgap energy (eV):

$$Eg=\begin{cases}(1.424+1.247\cdot x)q_e & at\ x\ <0.45\\ \left(1.9+0.125\cdot x+0.143\cdot x^2\right)q_e & at\ x\ \geq 0.45\end{cases} \tag{7}$$

- width of the quantum well, which depends on the number of atomic layers:

$$a=NN\cdot a_0\cdot 10^{+9} \tag{8}$$

where $m_e = 9.1093897\cdot 10^{-31}$ - electron mass, q_e - electron charge, NN - number of monolayers.

For modelling purposes, the following parameters are assumed constant: number of monolayers $NN = 20$ and the first four allowed energy levels – $\nu = 4$. The results of the simulations at different values of x are shown in Table 1 and Figure 3.

Table 1

Structure description parameters	Fraction of gallium atoms substituted by aluminum atoms, x					
	0	0.2	0.4	0.6	0.8	1.0
m_n	6.1033e-32	7.6154e-32	9.1276e-32	1.0640e-31	1.2152e-31	1.3664e-31
a_0	1.4240	1.6734	1.9228	2.0265	2.0915	2.1680
Eg	5.6533e-10	5.6549e-10	5.6564e-10	5.6580e-10	5.6595e-10	5.6611e-10
a	11.3066	11.3097	11.3128	11.3160	11.3191	11.3222

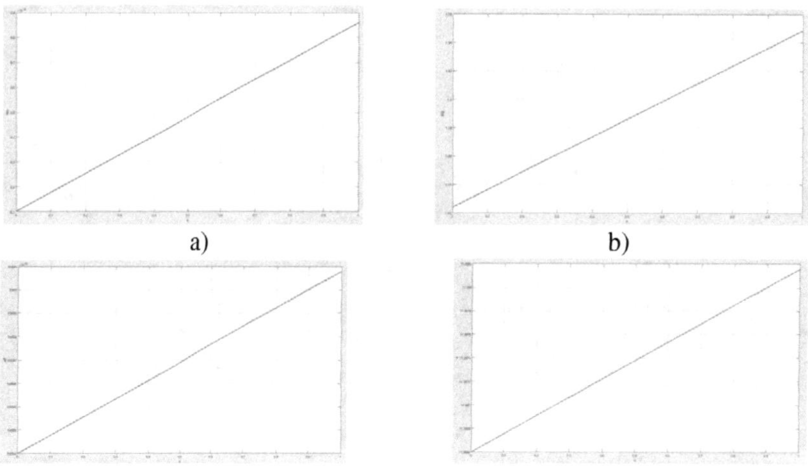

a) b) c) d)

Figure 3.Graphs of dependence on structure description parameters от x:

a) effective mass of the electrons in the Γ-minimum m_n, b) lattice constant a_0, c) bandgap energy Eg, d) width of the quantum well, which depends on the number of atomic layers a

Analysis of the graphs shows that all dependences are linear, which simplifies the finding of the structure parameters of Al_xGa_{1-x} As by formulas (5 - 8). Only one of these parameters depends on the parameter - number of monolayers *NN*.

Figure 4 shows the dependence of the quantum well width depending on the number of atomic layers a at the parameter number of monolayers $NN = 10, 20, 30$. The graphs coincide at $NN = 10 = 30$.

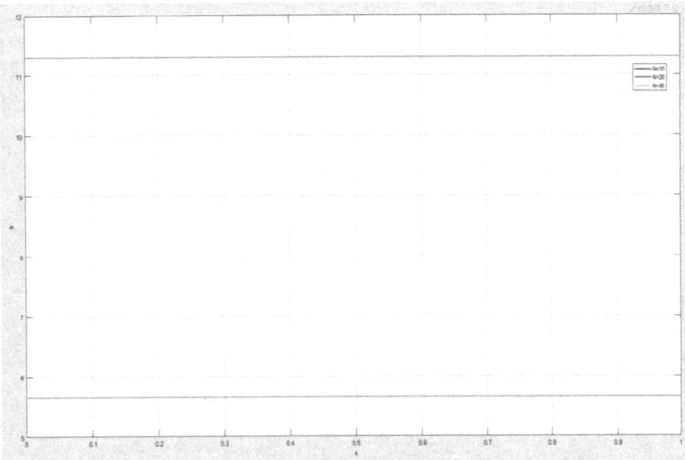

Figure 4. Dependence of quantum well width depending on the number of atomic layers a at the parameter number of monolayers $NN = 10, 20, 30$

3. Holder exponent (Lipschitz exponent and scaling exponent) as a tool for fractal analysis

The Holder exponent (Lipschitz exponent and scaling exponent) is used for signal analysis tasks in purpose of revealing their regular changes or features on the processing interval and is a tool for fractal signal analysis [4-6].

The signal *f(t),* satisfies the Holder condition Hol, so there is a constant C_H:

$$|f(\tau)-f(t)| \le C_H\, |\tau-t|^{Hol}, \quad 0<Hol<1 \tag{9}$$

it can be represented as the exponent value raised to the power, so the difference in the time argument will be proportional to the difference in the values of the signal f(t) at a given time interval [6].

If we perform a wavelet transform of this signal and write it in terms of scaling (the scale parameter tends to zero), the transform is [4-6]:

$$E\left\langle \left|W_\varphi(t,A,\phi)\right|^2 \right\rangle \approx |A|^{2Hol+1}, \quad A\to 0 \tag{10}$$

where A is the scale parameter of the wavelet transform, E<>- averaging sign.

The analysis of expression (10) shows that various kinds of signal regularities are clearly visible in the area of small values of the scaling factor [6].

In Matlab, this function is implemented by the *wtmm* function, which finds the maximum value of the wavelet transform modulus. By default, the wavelet transform of the original signal is performed by a second-order Gaussian wavelet, which differs from the "Mexican Hat" wavelet by the correction normalizing coefficient.

The authors chose the following syntax for this function [7], which draws lines of wavelet maxima and displays them as a table on the right side of the Holder local exponent estimation graph by which we can judge the fractality of the signal:

wtmm(signal, 'ScalingExponent', 'local')

4. Analysis of results

For modeling, we take the values of x from 0 to 1 in steps of 0.1.

1) Consider the dependence of quantised energy levels of an electron in a rectangular quantum well on different values of x (Figure 5). Analysis of the graphs showed that an increase in the value of x leads to a decrease in the energy levels.

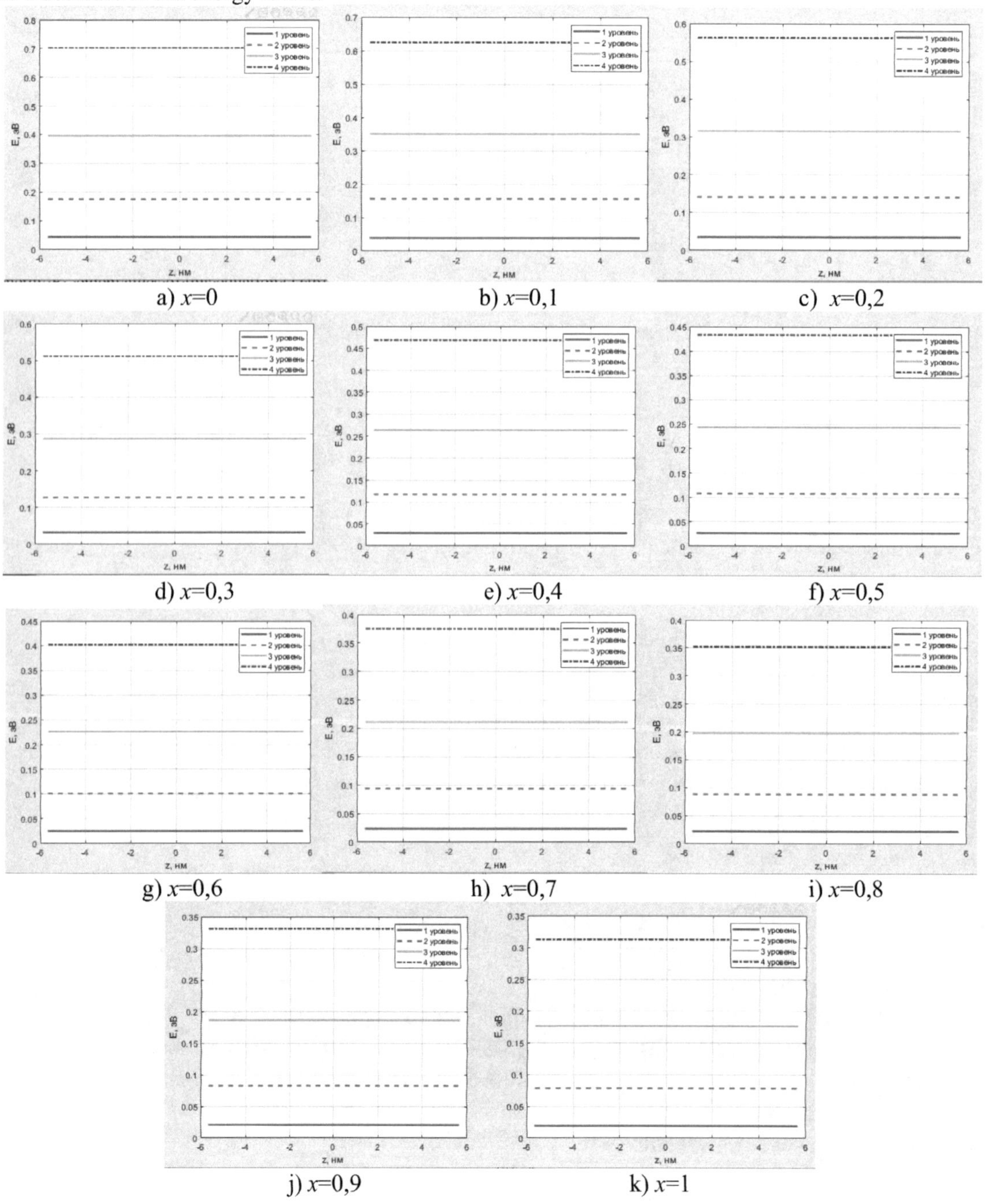

a) x=0 b) x=0,1 c) x=0,2

d) x=0,3 e) x=0,4 f) x=0,5

g) x=0,6 h) x=0,7 i) x=0,8

j) x=0,9 k) x=1

Figure 5. The dependence of quantised energy levels of an electron in a rectangular quantum well with values of *x*

2) The wavelet analysis with the *wtmm* function will be performed in the range of 0.1 to 0.4, assuming that the fraction of gallium atoms substituted by aluminum atoms *x* usually varies between 0.15 and 0.35 [1 - 2]. The visual representation of the plots of the local Holder exponent estimation is shown only in Figure 5b, since in the other cases they are very similar because of the small deviations of the lines of maximums for each envelope function.

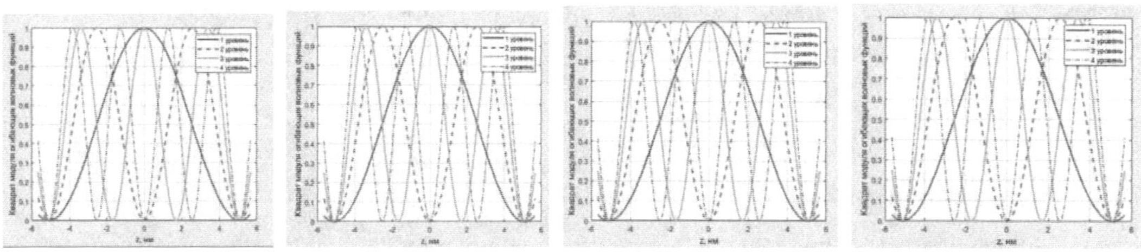

a) x=0.1 b) x=0.2 c) x=0.3 d) x=0.4

Figure 6. Wave function envelopes of an electron in the rectangular quantum well at different values of *x*

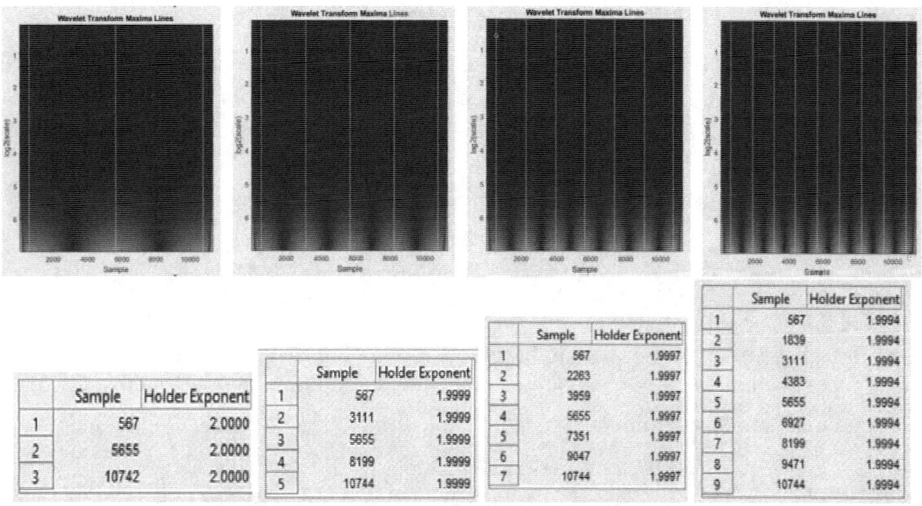

	Sample	Holder Exponent
1	567	2.0000
2	5655	2.0000
3	10742	2.0000

	Sample	Holder Exponent
1	567	1.9999
2	3111	1.9999
3	5655	1.9999
4	8199	1.9999
5	10744	1.9999

	Sample	Holder Exponent
1	567	1.9997
2	2263	1.9997
3	3959	1.9997
4	5655	1.9997
5	7351	1.9997
6	9047	1.9997
7	10744	1.9997

	Sample	Holder Exponent
1	567	1.9994
2	1839	1.9994
3	3111	1.9994
4	4383	1.9994
5	5655	1.9994
6	6927	1.9994
7	8199	1.9994
8	9471	1.9994
9	10744	1.9994

Figure 7. Results of wavelet analysis in the form of local Holder exponent estimation plots at four wave function envelopes (x=0.1)

	Sample	Holder Exponent
1	567	2.0000
2	5656	2.0000
3	10745	2.0000

	Sample	Holder Exponent
1	567	1.9999
2	3112	1.9999
3	5655	1.9999
4	8200	1.9999
5	10745	1.9999

	Sample	Holder Exponent
1	567	1.9997
2	2262	1.9997
3	3959	1.9997
4	5656	1.9997
5	7353	1.9997
6	9048	1.9997
7	10745	1.9997

	Sample	Holder Exponent
1	567	1.9994
2	1838	1.9994
3	3111	1.9994
4	4384	1.9994
5	5656	1.9994
6	6928	1.9994
7	8201	1.9994
8	9472	1.9994
9	10745	1.9994

Figure 8. Results of wavelet analysis in the form of local Holder exponent estimation plots at four wave function envelopes (x=0.2)

	Sample	Holder Exponent
1	565	2.0000
2	5656	2.0000
3	10747	2.0000

	Sample	Holder Exponent
1	565	1.9999
2	3112	1.9999
3	5657	1.9999
4	8202	1.9999
5	10747	1.9999

	Sample	Holder Exponent
1	565	1.9997
2	2263	1.9997
3	3960	1.9997
4	5657	1.9997
5	7353	1.9997
6	9050	1.9997
7	10747	1.9997

	Sample	Holder Exponent
1	565	1.9994
2	1838	1.9994
3	3111	1.9994
4	4384	1.9994
5	5657	1.9994
6	6929	1.9994
7	8202	1.9994
8	9475	1.9994
9	10747	1.9994

Figure 9. Results of wavelet analysis in the form of local Holder exponent estimation plots at four wave function envelopes (x=0.3)

	Sample	Holder Exponent
1	565	2.0000
2	5657	2.0000
3	10750	2.0000

	Sample	Holder Exponent
1	565	1.9999
2	3112	1.9999
3	5657	1.9999
4	8203	1.9999
5	10750	1.9999

	Sample	Holder Exponent
1	565	1.9997
2	2262	1.9997
3	3960	1.9997
4	5657	1.9997
5	7355	1.9997
6	9053	1.9997
7	10750	1.9997

	Sample	Holder Exponent
1	565	1.9994
2	1838	1.9994
3	3111	1.9994
4	4384	1.9994
5	5657	1.9994
6	6931	1.9994
7	8204	1.9994
8	9477	1.9994
9	10750	1.9994

Figure10. Results of wavelet analysis in the form of local Holder exponent estimation plots at four wave function envelopes (x=0.4)

Analysis of the results of Figures 7-10 shows that the values of local Holder exponents for the four wave function envelopes coincide, which indicates the equality of extremum values for each envelope function for different values of x. Depending on the value of x, the coordinates of points of extremum change by insignificant value, for example, for the first envelope function the second coordinate of extremum (in units of counts) is 5655 (x=0.1) – 5656 (x=0.2) – 5656 (x=0.3) – 5657 (x=0.4).

A formula for conversion from sample to z can be derived from this:

$$z_i^* = -a/2 + shag*(nN_i - 1) \qquad (11)$$

where $shag$ is the step, nN_i is the number of counts of a given point (sample).

3) Wavelet analysis with the wtmm function at x=0 for different numbers of monolayers: 10 monolayers, 20 monolayers, and 30 monolayers (Table 3). Table 2 shows quantised energy levels for the considered numbers of monolayers.

Table 2

Number of monolayers, NN	Quantised energy levels			
	1	2	3	4
10	0.0281e-36	0.1125e-36	0.2532e-36	0.4502e-36
20	0.0070e-36	0.0281e-36	0.0633e-36	0.1125e-36
30	0.0313e-37	0.1250e-37	0.2814 e-37	0.5002e-37

Table 3

Wave function envelopes and quantised energy levels (x=0)	Estimation graphs of local Holder exponents (wtmm function) with step $shag$ = 1.0000e-03
1	2

10 monolayers

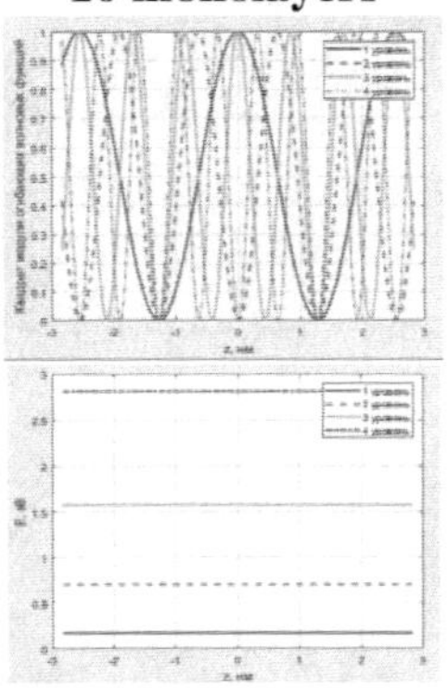

a = 5.6533,
number of counts nN = 5654

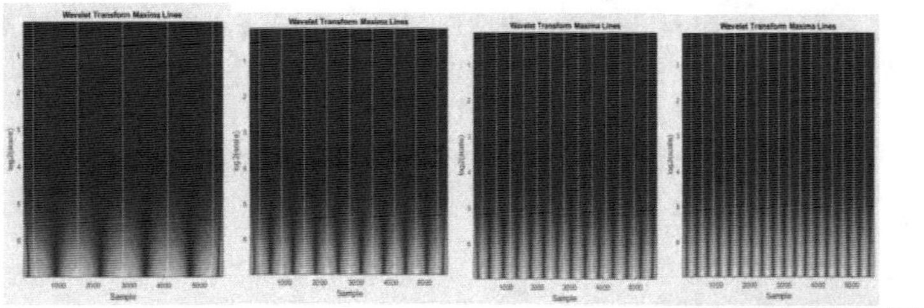

	Sample	Holder Exponent
1	285	1.9994
2	1556	1.9994
3	2828	1.9994
4	4100	1.9994
5	5371	1.9994

	Sample	Holder Exponent
1	285	1.9977
2	920	1.9977
3	1556	1.9977
4	2192	1.9977
5	2828	1.9977
6	3463	1.9977
7	4099	1.9977
8	4735	1.9977
9	5371	1.9977

	Sample	Holder Exponent
1	285	1.9948
2	708	1.9948
3	1132	1.9948
4	1556	1.9948
5	1980	1.9948
6	2404	1.9948
7	2828	1.9948
8	3252	1.9948
9	3675	1.9948
10	4099	1.9948
11	4523	1.9948
12	4947	1.9948
13	5371	1.9948

	Sample	Holder Exponent
1	284	1.9907
2	602	1.9907
3	920	1.9907
4	1238	1.9907
5	1556	1.9907
6	1874	1.9907
7	2192	1.9907
8	2510	1.9907
9	2828	1.9907
10	3146	1.9907
11	3463	1.9907
12	3781	1.9907
13	4099	1.9907
14	4417	1.9907
15	4735	1.9907
16	5053	1.9907
17	5371	1.9907

1 *2*

20 monolayers

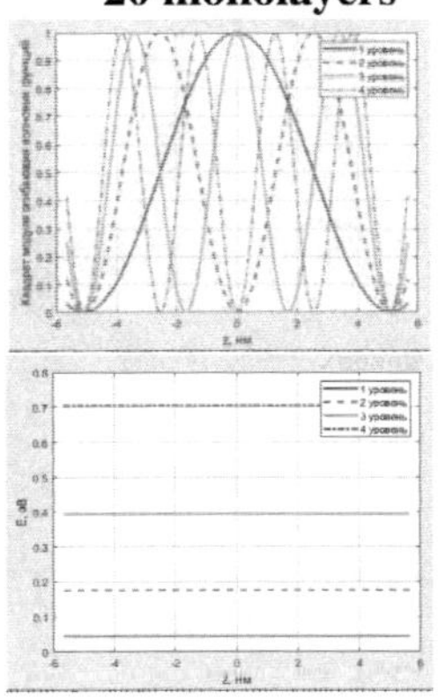

a = 11.3066,
number of counts nN = 11307

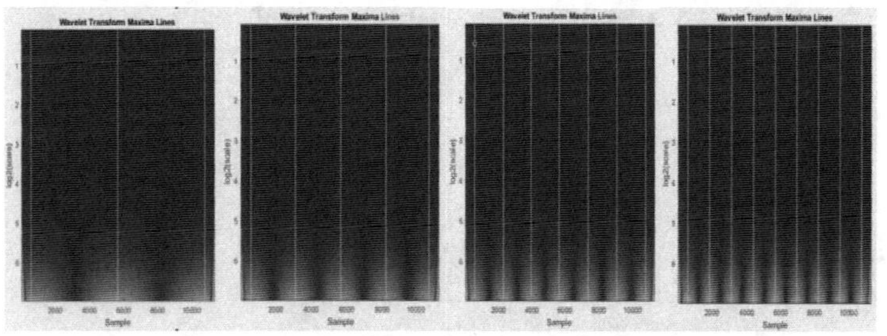

	Sample	Holder Exponent
1	567	2.0000
2	5654	2.0000
3	10740	2.0000

	Sample	Holder Exponent
1	567	1.9999
2	3110	1.9999
3	5654	1.9999
4	8197	1.9999
5	10740	1.9999

	Sample	Holder Exponent
1	567	1.9997
2	2264	1.9997
3	[illegible]	1.9997
4	5654	1.9997
5	7350	1.9997
6	9045	1.9997
7	10740	1.9997

	Sample	Holder Exponent
1	567	1.9994
2	1840	1.9994
3	3111	1.9994
4	4383	1.9994
5	5654	1.9994
6	6926	1.9994
7	8198	1.9994
8	9469	1.9994
9	10740	1.9994

30 monolayers

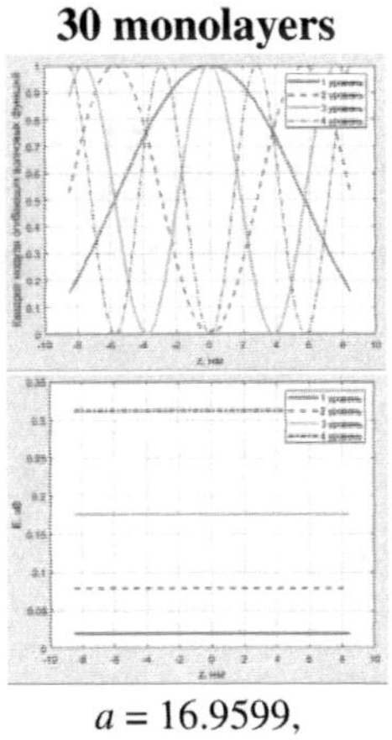

a = 16.9599,
number of counts nN = 16960

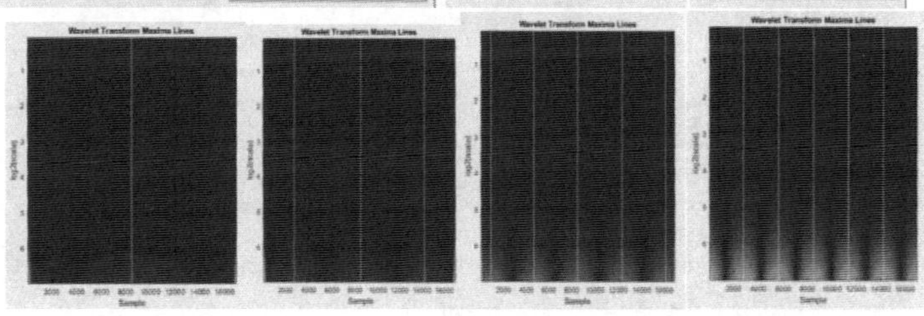

	Sample	Holder Exponent
1	8481	2.0000

	Sample	Holder Exponent
1	2759	2.0000
2	8480	2.0000
3	14203	2.0000

	Sample	Holder Exponent
1	851	1.9999
2	4666	1.9999
3	8481	1.9999
4	12296	1.9999
5	16111	1.9999

	Sample	Holder Exponent
1	2759	1.9999
2	5620	1.9999
3	8481	1.9999
4	11342	1.9999
5	14203	1.9999

5. Conclusion

This paper presents the results of a wavelet analysis of the electron energy spectrum in a one-dimensional quantum well with infinitely high walls Al_xGa_{1-x} As at different values of x (x values from

0 to 1) and at different numbers of monolayers (*NN* = 10, 20, 30), using the *wtmm* function calculating the Holder exponent to determine the coordinates of extrema and analyze the fractality of envelope functions. The obtained values can be used to determine the coordinates of extremums of envelope functions.

6. Acknowledgement
The paper is prepared under partial support of Russian Foundation for Basic Researches (RFBR) (project No. 19-07-00602_a).

References

[1] Usanov D A, Skripal A V, Skripal An V, Abramov A V 2013 *Textbook for students of nano- and biomedical technologies* Saratov: SGU. 100 p.
[2] Usanov D A, Skripal A V 2013 *Textbook for students of nano- and biomedical technologies* Saratov 128 p.
[3] Korn G 2003 *Handbook of mathematics for scientists and engineers: definitions, theorems, formulas 6th edition, stereotyp.* SPb. Lan, 832 p.
[4] Mallat S, Hwang W L 1992 *IEEE Transactions on Information Theory* **38**(2) 617 DOI: 10.1109/18.119727
[5] Rioul O, Flandrin P 1992 *IEEE Transactions on Signal Processing* **40**(7) 1746 DOI:10.1109/78.143446
[6] Dvornikov S V 2019 *Proceedings of educational institutions of communication* **5**(2) 117 DOI :10.31854/1813 – 324X-2019-5-2-117-125
[7] Matlab Documentation URL: https://www.mathworks.com/solutions.html?s_tid=gn_sol (Date of request 01.04.2021r.).
[8] Tkachenko I M, Kozhanova E R, Belyaev V V, Yazbeck H 2019 *J P.Conf.Ser.* DOI:10.1088/1742-6596/1309/1/012020.
[9] Tkachenko I M, Kozhanova E R 2018 *Volga Region Scientific and Technical Bulletin.* **11** 261.
[10] Smolentsev N K 2005 *Fundamentals of wavelet theory. Wavelets in MATLAB.* Moscow: DMK Press. 304 p.

Light-reflecting characteristics of optically rewritable electronic paper

A A Kudreyko[1], **V G Chigrinov**[2,3]
[1]Bashkir State Medical University, Department of Medical Physics and Informatics, 450008 Ufa, Russia
[2]Moscow Region State University, Department of Theoretical Physics, 141014 Mytishi, Russia
[3]Nanjing Nanhui Intelligent Optical Sensing and Manipulation Research Institute Co. Ltd, 210093 Nanjing, China

E-mail: akudreyko@bashgmu.ru

Abstract. Light-reflecting characteristics play a crucial role in optically rewritable liquid crystal display (electronic paper). By using a special optimizer software, light-reflecting characteristics of the twisted nematic cell were calculated. The achieved values of the normalized reflection coefficients range within $\sim 0.03 - 0.43$, which is comparable or better than other commercial technologies. The calculated contrast ratio of the device is 10 : 1. A 2D image, which can be displayed on the electronic paper was obtained. Comparison of the simulated background color and contrast ratio is very similar with the results, which were obtained in experimental cells of other studies. Our calculations indicate that angular dependence of the reflectance coefficient is almost uniform within a wide range of viewing angles.

1. Introduction

Optical properties of liquid crystal (LC) displays can be easily tuned if its layer is placed between electrodes. The consequent response to the electric field definitely leads to the change of reflective or transmittive characteristics of the display. There also exists another way to control reflectance/transmittance of the display: exposing azo dye coated substrate surface to ultraviolet light. This idea is the basis for light-printed images, which is the concept of optically rewritable electronic paper (ORW e-paper).

There exist many other technologies of e-paper fabrication, where high reflection coefficient (%) is crucial. For example, electrophoretic (E-Ink, ~ 40%), electrowetting (Liquavista, ~ 50%), cholesteric liquid crystal (Kent Displays Inc., ~30%), electrochromic (NTerra Inc., ~45%, DIC, ~65%), micro-electromechanical interference (Qualcomm Inc., ~25%) and liquid powder (Bridgestone, ~40%). However, reflection characteristics of all these technologies are lower than those of a typical white paper (over than 80%) [1].

The function of ORW e-paper is to store and display graphical information, generated by polarized light on thin flexible paper-like carrier with good reflective characteristics (preferably, $\gtrsim$ 40%) [2]. Optically rewritable technology represents a modified method of azo dye photoalignment with high azimuthal anchoring energy and reversible in-plane alignment. Optically rewritable LC cells consist of two substrates with different properties [3]. One substrate is optically passive, while another one is optically active, i.e., by manipulating with the polarization plane and the absorbed dose of the incident radiation, different alignment directions can be achieved on the same cell. Here, the key element in the ORW e-paper is photosensitive azo dye SD1. This compound is sensitive to ultraviolet light, and if one of the glass plates is coated with SD1, the cell performs new optical properties [4]. High thermal stability of SD1 makes this compound promising for many applications in optical devices. In particular, azo dye SD1-coated substrate produces continuous anchoring energy, which is proportional to the expose energy. Polar and azimuthal anchoring strengths between SD1 molecules and LC boundary are about 10^{-4} J/m^2, which is comparable with the rubbing technique (e.g., [5]).

In order to observe a reflected image, a reflective polarizer film is placed below the photo-stable aligning film [6]. A number of reflective characteristics of the ORW e-paper have not received proper consideration. This enables us to formulate the problem of estimation of ORW e-paper reflective characteristics.

Optimization of optical characteristics in LC cells has always been in the center of interest of liquid crystal studies [7, 8, 9, 10]. Indeed, during almost five decades of development, LC models have been transformed from lattice [11] to atomistic with predictive calculations[12, 13].

In this article we aim to show potential characteristics of e-paper by using universal Modelling and Optimization System of Liquid Crystal Devices (MOUSE-LCD, developed by the Hong Kong University of Science and Technology and the Saratov State University) [7, 14, 15]. In particular, we simulate image formation by changing boundary conditions on SD1-coated substrate. Then we introduce 8×8 transfer matrix description for computation of reflective characteristics and contrast ratio. The obtained characteristics are consistent with other e-paper technologies [1, 16].

2. Model

Schematic cross-sectional view of the twisted nematic (TN) cell is shown in figure 1 (a), where the right substrate is a reflective polarizer film. This cell will have high optical performance in the reflective mode if the phase retardation is Δnd, where Δn is the birefringence and $d = 6.85\,\mu$m is the cell gap. As it was mentioned above, flexible (polyethersulfone, PES) substrates are indispensable in the ORW concept. Meanwhile, we consider only flat surfaces of dielectric media.

Schematic representation of the cross-sectional view of ORW e-paper is depicted in figure 1 (b). In order to obtain more realistic results, we have introduced dispersion in the model. Let the refractive indexes along and perpendicular to the long molecular axes of LCs range within $n_{\|} = 1.642\ldots1.695$ and $n_{\perp} = 1.495\ldots1.515$ for all wavelengths of the visible spectrum. The following elastic constants were used $K_1 = 1.3 \cdot 10^{-6}, K_2 = 6.5 \cdot 10^{-7}, K_3 = 1.95 \cdot 10^{-6}$ dyn in the model. Knowing these parameters, the static director distribution in the LC layer can be calculated.

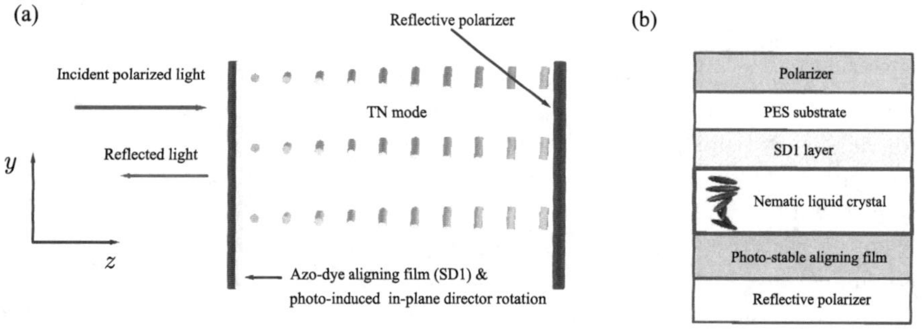

Figure 1. (a) Schematic representation of nematic liquid crystal cell after reorientation process occurs; (b) Device structure of ORW e-paper

When the incident polarized light beam propagates through the azo dye layer, its photoactive molecules change the alignment direction on the top substrate. Consequently, the twist angle can be changed from 0 to 90°, and grey levels can be continuously generated. The reorientation time is a very specific characteristic of the azo dye aligning film, and ranges within 5 – 20 s for intensities from 100 – 20 W/cm^2, respectively.

3. Simulation of performance characteristics

Reflective displays must be able to provide different grey scale levels and have more or less uniform reflectance characteristics versus different viewing angles. This suggests that viewing direction dependence must be described as the reflectance coefficient vs. viewing direction. Optics of multi-layered LC displays involves multiple reflections. It seemed earlier that a 4×4 differential-matrix technique [17] can solve the problem of multiple reflections. One of the drawbacks of this method is the necessity of spectral averaging, which involves time consuming numerical algorithms. Accurate and fast calculation of the reflective characteristics can be done by using 8×8 transfer matrix description for one-dimensional layered structures [14, 18].

By changing the twist angle in the boundary conditions (i.e., simulation of photo-induced in-plane director rotation) on the photo active substrate (left substrate in Figure 1 (a)), different chromatic colours can be generated. Thus, figure 2 depicts saturation trend of the normalized reflectance-twist angle curve. It is also possible to generate an image, which would appear on the e-paper. The image in the inset of figure 2 shows different grey levels, which we could achieve for different twist angles. One can see that the image quality is very close to the previously reported experimental results [19].

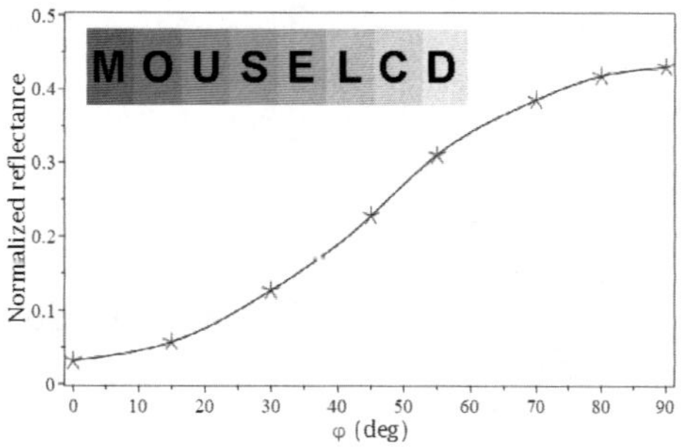

Figure 2. Normalized reflectance of ORW e-paper. Insert: computer generated-image with gray scales for different twist angles.

Normalized reflectance versus viewing angles can also be calculated. The isolines in Figures 3 show almost symmetric dependencies of the reflective coefficients for different twist angles. It can be clearly seen that higher twist angles increase the normalized reflectance coefficient. Similar simulations of the contrast ratio (the twist angle is 80°) show that it achieves 10 : 1 and reduces to 7 : 1 for viewing angles higher than $\sim 60°$, which is acceptable result for e-paper displays [20].

These results show that when the reflection coefficient increases, the dark state is no longer dark. Consequently, the contrast ratio decreases.

4. Discussion

In order to model a standard set of electrooptical characteristics of LC displays, three fundamental computational problems must be sequentially solved.

(a) (b)

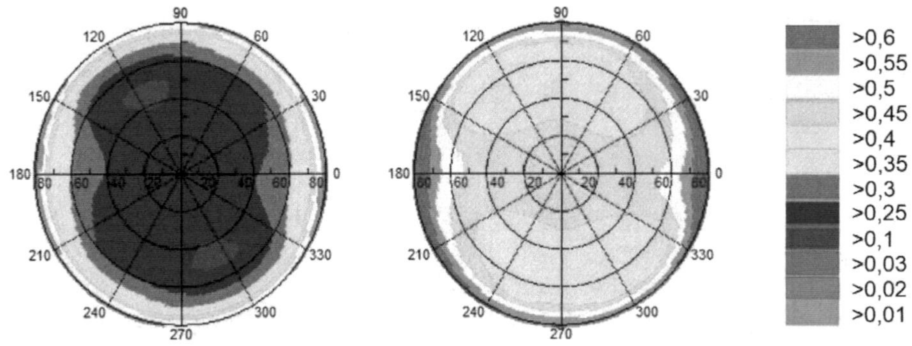

Figure 3. Angular dependence of the ORW reflectance coefficient: twist angle 45° (a) and 80° (b). The legend is relevant to both figures.

(i) Computation of the equilibrium state of the director configuration in the LC layer for the determined boundary conditions and the applied voltage (if it is exists).

(ii) Computation of dynamics of the director field (if the external voltage is applied).

(iii) Simulation of optical characteristics of the display with the given LC director field configuration.

The director field in LC layer was calculated by using numerical integration of Euler equations for the LC layer with the free energy minimization problem. The presented work does not involve computation of temporal characteristics, however the algorithm of MOUSE-LCD uses numerical integration of Ericksen-Leslie equations, which were solved by using the modified Eulers method.

It is clear that the reflective characteristics can be obtained by using using 4×4-transfer matrix technique. Meanwhile, this technique requires computation of the spectral averaging, which is time consuming. Application of transfer 8×8-matrix method provides a compromise between the accuracy and comutational costs. Transfer 8×8 matrix method was developed to simulate optics in LC displays considering propagation of a quasi-monochromatic plane wave in a special one dimensional non-uniform layered structure. This method was derived from the theory of computation of optical characteristics in anisotropic media including partial coherence. Optimization tests with other cell thicknesses and LC compounds indicate no significant change in light reflectance and contrast ratio due to interference losses.

5. Conclusions

We have studied a TN configuration for ORW e-paper displays. Optical characteristics were calculated by using MOUSE-LCD optimizer. A 2D image, which would appear on e-paper was simulated. The achieved values of the normalized reflectance coefficients range within ~0.03 to 0.43, which is higher or comparable with other commercial technologies. The calculated contrast ratio achieved 10 : 1, which is also comparable with other studies. A significant advantage of ORW e-paper is that it does not require formation of static charges and electrodes. The calculations also indicate that angular dependence of the reflectance coefficient is almost uniform within a wide range of viewing angles.

Acknowledgments

This study was supported by the Russian Science Foundation (project no. 20-19-00201).

References

[1] Heikenfeld J, Zhou K, Kreit E, Raj B, Yang S, Sun B, Milarcik A, Clapp L and Schwartz R 2009 *Nature Photonics* **3** 292–296
[2] Heikenfeld J, Drzaic P, Yeo J S and Koch T 2011 *Journal of the Society for Information Display* **19** 129–156
[3] Muravsky A, Murauski A, Chigrinov V and Kwok H S 2008 *Japanese journal of applied physics* **47** 6347
[4] Chigrinov V, Muravski A, Kwok H S, Takada H, Akiyama H and Takatsu H 2003 *Physical Review E* **68** 061702
[5] Seo D, Matsuda H, Oh-Ide T and Kobayashi S 1993 *Molecular Crystals and Liquid Crystals Science and Technology. Section A. Molecular Crystals and Liquid Crystals* **224** 13–31
[6] Muravsky A, Murauski A, Li X, Chigrinov V and Kwok H S 2007 *Journal of the Society for Information Display* **15** 267–273
[7] Chigrinov V G and Yakovlev D A 2006 *Molecular Crystals and Liquid Crystals* **453** 107–121
[8] MacDonald C S, Mackenzie J A and Ramage A 2020 *Journal of Computational Physics: X* **8** 100065
[9] Zannoni C 2018 *Liquid Crystals* **45** 1880–1893
[10] Allen M P and Wilson M R 1989 *Journal of computer-aided molecular design* **3** 335–353
[11] Lebwohl P A and Lasher G 1972 *Physical Review A* **6** 426
[12] LISA C, Curteanu S, Bulacovschi V and Apreutesei D 2008 *Revue Roumaine de Chimie* **53** 283–290
[13] Watanabe G and Yoshida J 2016 *The Journal of Physical Chemistry B* **120** 6858–6864
[14] Yakovlev D A, Simonenko G V, Tsoi V I, Chigrinov V G, Khokhlov N A and Podyachev Y B 2002 *Saratov Fall Meeting 2001: Coherent Optics of Ordered and Random Media II* vol 4705 (International Society for Optics and Photonics) pp 255–263
[15] Yakovlev D A, Chigrinov V G and Kwok H S 2015 *Modeling and optimization of LCD optical performance* (John Wiley & Sons)
[16] Kim M J, Sung G, Sun J Y *et al.* 2019 *Nano Convergence* **6** 1–24
[17] Berreman D W 1972 *Josa* **62** 502–510
[18] Yakovlev D A, Chigrinov V G and Kwok H S 2001 *Molecular Crystals and Liquid Crystals Science and Technology. Section A. Molecular Crystals and Liquid Crystals* **366** 327–335
[19] Geng Y and Yao L 2020 *Liquid Crystals* 1–7
[20] Gao H Y, Yao Q X, Liu P, Zheng Z Q, Liu J C, Zheng H D, Zeng C, Yu Y J, Sun T and Zeng Z X 2016 *Chinese Physics B* **25** 094203

Dependence of the integral characteristics of a nematic light modulator on the physical parameters of a liquid crystal

G V Simonenko
Saratov National Research State University named after N. G. Chernyshevsky, Astrakhanskaya st., 83, 410012, Saratov, Russia

Corresponding author: G.V. Simonenko
Saratov National Research State University named after N. G. Chernyshevsky, Astrakhanskaya st., 83, 410012, Saratov, Russia, e-mail: gvsim1960@hotmail.com

Abstract. In the present work, the effect of physical constants of an LC material on the integral characteristics of a modulator operating in a waveguide mode in an LC structure with a twist angle of 270 and antisymmetric boundary conditions in the working cell was studied using the method of computer simulation. It is shown that, under antisymmetric boundary conditions, the small value of the total response time of the LC modulator is due to the absence of a "backflow" in the LC cell and the role of only half of the LC working gap thickness in the dynamics of switching between two states in such a device. In addition, it was found that in order to simultaneously achieve high contrast ratios and a short total response time of the device, it is necessary to choose an LC material with a high value of its dielectric anisotropy. In this case, there is an optimal set of elasticity constants for the LC material, which simultaneously realizes a high contrast ratio and a short total response time of the modulator.

1.Introduction

Currently, liquid crystals (LC) have found wide application in various systems for displaying and converting information [1]. The most well-known areas of use of LCD systems are all kinds of indicator and display systems [1 - 4], as well as modulators of optical radiation [3 - 7]. In connection with the development of virtual reality systems, a class of LC modulators called active 3D glasses is of particular interest [4, 8]. Usually, a high-speed LC modulator is based on various designs π - cells [9], in which LC structures with a zero or 180-degree twist angle are used, however, as recent works have shown, the LC shutter, in which an LC with an angle swirls 270° and with antisymmetric boundary conditions on orienting surfaces [10]. Such a design of an LC shutter has been known for a long time [11], but how the physical parameters of an LCD affect its characteristics at high control voltages has not yet been clarified. This work is aimed at filling this gap and presents the results of computer simulation of the characteristics of LC modulators of optical radiation.

2.Integral characteristics of LC information processing device and their modeling

Integral characteristics of LC devices can be divided into three groups: optical; electro-optical; dynamic. The integral optical characteristics of an LCD device are the characteristics averaged over the wavelength of the optical range of the signal being processed: the spectrum average transmission, the spectrum average contrast ratio (or contrast), the angular dependences of the spectrum average contrast ratio and the transmission spectrum average, color coordinates and achromaticity of the open or closed states of the modulator. The dynamic characteristics of the LC modulator include the response (τ_{on}) and relaxation (τ_{off}) times when the control voltage is switched from the "off" state to the "on" state and back. To select the optimal design of the LC modulator, it is sufficient to use a set of only four integral optical characteristics [12]:

• average over the spectrum transmission of the LC device in the "open" or "off" state. Usually this condition occurs when there is no voltage at the gate electrodes. Therefore, the spectrum average transmission of the LC modulator is denoted as T_{off};
• color coordinates in the color triangle and achromaticity of the image. By achromaticity H we mean the distance of the current color point of the device on the color triangle from the white point, for example, the source D_{65}. This characteristic describes the distance of a given point from a white point on a color triangle;
• average over the spectrum contrast ratio C. Its value is determined as follows:

$$C = \frac{T_{off}}{T_{on}},$$

where T_{on} is the average over the spectrum transmission of the device in the "closed" or "on" state when voltage is applied to the control electrodes;
• total response time of the LC device (τ) ($\tau = \tau_{on} + \tau_{off}$).

Usually, an LC device for converting information is considered promising for further operation if these four integral characteristics satisfy the following conditions:

$$T_{off} \geq 35\%, C \geq 100{:}\ 1, H \leq 0.05, \tau \leq 6 \text{ ms}.$$

Due to the large number of design and technological parameters that determine the characteristics of the LC shutter, for engineering practice, an idea of the features of these dependencies, as well as the search for the optimal design of such a device, can be obtained only by computer simulation [13]. In this regard, we used the MOUSE-LCD computer simulation system [14] to study the characteristics and search for the optimal parameters of the LC device. This software package makes it possible to simulate the characteristics of various LC devices with an accuracy within 10%, which usually does not exceed the experimental error, and the simulation error is mainly determined by the accuracy of specifying the physical and design parameters of the modulator.

3.Results and discussion

Earlier it was shown that at high values of control voltages the minimum total response time has a device based on an LC structure with a twist angle $\Phi_T = 270^o$ under antisymmetric boundary conditions [10]. The device contains an input polarizer, an LC cell, a phase compensator and an output polarizer located in series on one optical axis. The input polarizer with its maximum transmittance is oriented in the direction of rubbing on the input substrate, and the output polarizer - in the direction of rubbing on the output substrate. The thickness of the d_{LC} LC layer in the working cell is 3.5 μm. This thickness of the LC working layer was chosen because a decrease in d_{LC} leads to a violation of the waveguide regime of the LC structure and greatly deteriorates the optical characteristics of the LC modulator, and an increase in d_{LC} results in an increase in the response time of the device. Antisymmetric boundary conditions on the orienting cell substrates are due to the standard technology of rubbing the cell substrates, which provides the pre-tilt angle of the liquid crystal molecules on the input surface equal to 2°, and on the output surface – (-2)°. The phase compensator is a uniaxial phase plate oriented at an angle of 28° to the rubbing direction on the first substrate of the LC cell and having an optical thickness of 0.0345 μm. The described modulator device has the following values of integral characteristics: contrast ratio reaches 350: 1; achromaticity of the device in the "off" state $H <$ 0.05; spectrum average transmission $T_{off} > 38\%$; total response time of the device $\tau \sim 6$ ms. These data indicate good optical characteristics of such an LC device for use as an LC shutter.

Let us discuss the dynamics of switching the LC modulator at different control voltages. The response time of the LC modulator is the sum of the on time and the off time. The turn-on time is the time required for the LC modulator to go from a state with zero control voltage to a state with a high control voltage. The turn-off time is the time required for the LC modulator to make a reverse transition when the control voltage is switched from high to zero. In this case, the short response times of this LC modulator are achieved due to three factors: 1) high values of control voltages; 2) antisymmetric boundary conditions in the structure; 3) small thickness of the LC cell.

A typical dependence of the total response time (τ) of such an LC modulator on the control voltage U is a decreasing function. Analysis of this dependence showed that it can be approximated with an error of 15% by a simple function [15]:

$$\tau = c + \frac{a}{U^2 - b}, \qquad (1)$$

where the constants *a*, *b*, and *c* depend on the physical constants of the LC and the boundary conditions in the LC cell (the twist angle of the structure, the pre-tilt angles of the LC molecules on the substrates). Both experimental and theoretical description of back flow effects in nematic and cholesteric liquid crystals is presented in [17-19].

A few words should be said about the effect of the “backflow” on the relaxation time, which significantly affects the dynamics of switching off the LC modulator under symmetric boundary conditions [13]. We carried out a computer simulation of the switching dynamics of an LC modulator with antisymmetric boundary conditions, and as a result, it was found that the relaxation time of such a device is the same with and without taking into account the "backflow". It can be assumed that the short response time of the LC modulator with antisymmetric boundary conditions can be explained by two factors: 1) antisymmetric boundary conditions in the LC cell, in which only half of the working gap thickness plays a role [16]; 2) in the dynamics of switching between the two states in an LC modulator with antisymmetric boundary conditions, there is no influence of the “backflow”. The latter statement was confirmed by us by computer simulation of the dynamics of switching the LCD shutter with and without the “backflow”, which showed the same relaxation times for both cases.

Let us consider the influence of the physical parameters of an LC substance (anisotropy of the dielectric constant and elastic constants) on the characteristics of an LC modulator, the design and physical parameters of which are described in the previous section. The following values were chosen as operating voltages: U_{on} = 8V, and U_{off} = 0V.

Two integral characteristics of the modulator (average transmission in the off state T_{off} and achromaticity in this state H_{off}) do not depend on the physical parameters of the LC substance, since this state is observed at zero control voltage and does not depend on all possible parameters and conditions. At the same time, the conditions of prospects for further use in data processing systems are met, which are superimposed on the optical integral characteristics ($T_{off} \geq 35\%$, $H \leq 0.05$).

In what follows, we will consider only the dependences of the spectrum-averaged contrast ratio C and the total response time тора of the modulator on the physical constants of the LC material. In addition, the influence of the physical constants of the LC material on the voltage of the optical threshold of the electro-optical effect U_{opt}, on the basis of which the LC modulator operates, should be considered. The optical threshold voltage is the value of the control voltage in the LC cell at which the optical response changes when a certain electro-optical effect is observed. In the case of the effect of electric field-controlled birefringence in an LC, the voltage of the optical threshold coincides with the voltage of the threshold of the Fredericksz effect. In our case, the voltage of the optical threshold of the waveguide effect is significantly higher than the voltage of the Freedericksz threshold, since the observation of this effect requires a significant deformation of the LC layer in the working cell [13]. Therefore, the magnitude of the voltage of the optical threshold characterizes the degree of deformation of the LC layer in the working cell, which affects the total response time of the LC modulator.

In fig. 1 shows the dependences of C and τ on the dielectric anisotropy of the LC material. As can be seen from this figure, the dependence εΔ(τ=τ) with an error of 15% is described by linear regression. At the same time, an increase in the anisotropy of the dielectric constant of the LC leads to a decrease in the total response time of the modulator due to the fact that the degree of deformation of the LC layer increases. The dependence C = C (Δε) has the shape of a curve with a minimum. Thus, in the LCD modulator described above, in order to simultaneously achieve high contrast ratio values and a short total response time of the device, it is necessary to choose an LCD material with a high value of its dielectric anisotropy.

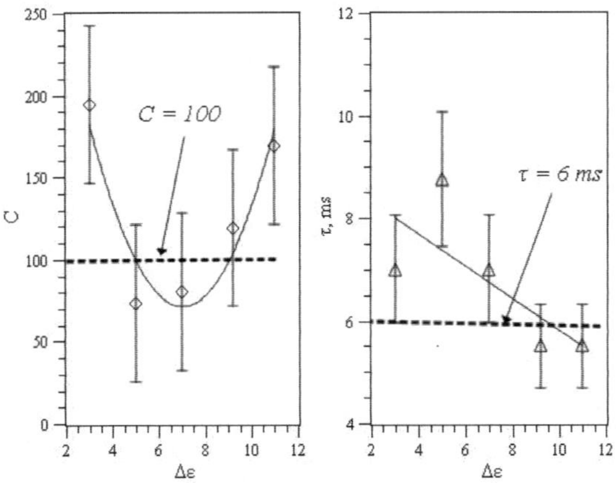

Figure 1. Dependence of characteristics (contrast ratio C and total response time τ) of the modulator on the anisotropy of the dielectric constant of the liquid crystal Δε

Let us consider the influence of the elasticity coefficients of the liquid crystal on the characteristics of the liquid liquid crystal. Let $\omega_1 = k_{33}/k_{11}$ and $\omega_2 = k_{33}/k_{22}$. Analysis of the data shows that the behavior of the dependence $U_{opt}=U_{opt}(\omega_1)$ is opposite to the behavior of $U_{opt}=U_{opt}(\omega_2)$. In this case, the dependence $U_{opt}=U_{opt}(\omega_1)$ with an error of 10% is a linear regression, and the dependence $U_{opt}=U_{opt}(\omega_2)$ is a linear progression with the same error. This behavior of the dependencies $U_{opt}=U_{opt}(\omega_1)$ and $U_{opt}=U_{opt}(\omega_2)$ is explained as follows. The magnitude of the voltage of the Fredericksz junction depends on these parameters as follows:

$$U_{Fred} \sim (k_{11} \cdot (1 + C_a \cdot \omega_1 \cdot (1 - 2/\omega_2 C_b)))^{1/2}, \quad (2)$$

where C_a and C_b are constants determined by the LC material.

The ratio of the LC elasticity constants ω_1 increased due to a decrease in k_{11}, while the parameter ω_2 remained constant. As can be seen from expression (2), with an increase in ω_1 U_{Fred} decreases, which leads to a decrease in U_{opt}. In turn, the ratio of LC elasticity constants ω_2 increased due to a decrease in k_{22}, while the parameter ω_1 remained constant. In accordance with expression (2), this leads to an increase in U_{Fred}, and, accordingly, U_{opt}.

The dependences of C and τ on the ratio of the elastic constants k_{33}/k_{11} and k_{33}/k_{22} of the LC material are shown in Fig. 2. As can be seen from this figure, the behavior of the dependences $C = C(\omega_1)$ and $\tau = \tau(\omega_1)$ is opposite to the behavior of the functions $C = C(\omega_2)$ and $\tau = \tau(\omega_2)$. In addition, from Fig. 2 it follows that the dependences of the contrast ratio of the total response time of the LC modulator on the ratio of the elasticity coefficients of the LC ω_1 and ω_2 indicate the presence of optimal values of these parameters at which the maximum possible value of the contrast ratio is reached and, simultaneously, the minimum value of the total response time of the device. The opposite behavior of the dependences $C = C(\omega_1)$ and $C = C(\omega_2)$ is explained by the fact that, in the opposite way, the values of the optical threshold voltages, which characterize the degree of distortion of the LC layer in the working cell, depend on these parameters. In this case, for a fixed value of the control voltage, the contrast ratio is the higher, the lower the voltage of the optical threshold, that is, the more the LC layer is deformed. The opposite behavior of the curves $\tau = \tau(\omega_1)$ and $\tau = \tau(\omega_2)$ is similarly explained. The nonlinear nature of the dependences $\tau = \tau(\omega_1)$ and $\tau = \tau(\omega_2)$ can be understood on the basis of the complex influence on the dynamics of switching parameters ω_1 and ω_2, which does not allow obtaining simple expressions for these functions. All these dependences indicate the presence of an optimal set of elasticity constants of the LC material, at which a high contrast ratio and a short total response time of the modulator are simultaneously realized. For this, it is necessary to choose an LC material for which the value of the parameter ω_1 is close to unity, and the value of the parameter ω_2 is about 3.

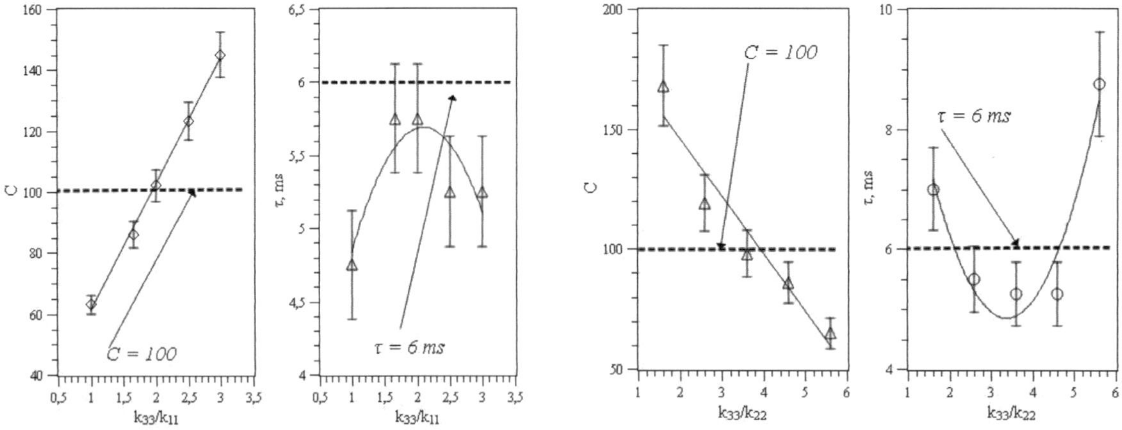

a b

Figure 2. Dependence of the characteristics (contrast ratio *C*, and total response time τ) of the modulator on the ratio of LC elasticity coefficients (a) k_{33} / k_{11} and (b) k_{33} / k_{22}.

It follows from the previous results that the values of the characteristics of the LC modulator *C*, () and U_{opt} should depend on the absolute value of the elasticity coefficients of the LC at constant values of the parameters ω_1 and ω_2. Let K be the value of the average coefficient of elasticity of the liquid crystal K ($K = (k_{11} + k_{22} + k_{33})/3$). Our calculations show that the dependence $U_{opt} = U_{opt}(K)$ is a linear progression with an error of 10%, which is easy to understand using expression (2). And the dependence $C = C(K)$ has the shape of a curve with a minimum, which indicates the presence of an optimal value of the average coefficient of elasticity of the liquid crystal.

The dependence $\tau = \tau(K)$, which is a monotonically increasing function, requires separate consideration. Indeed, it is known that the reaction and relaxation times are inversely proportional to the value of the LC elasticity coefficient [13]. However, we must remember that we are in the mode of a fixed value of the control voltage for the "on" state. In this mode, with an increase in the parameter K, the value of the threshold voltage also increases. This leads to the fact that the degree of deformation of the LC layer decreases, which, as is known, leads to an increase in the total reaction time of the modulator. This conclusion is confirmed by our calculations when fixing not the absolute value of the control voltage in the “on” state, but when fixing the voltage of the Fredericksz transition $U_{nor} = U_{on}/U_{Fred}$ normalized to the threshold value. In this case, at the same U_{nor} value, the full response times of the modulator are the same. The latter circumstance suggests that each LC material has its own modulator control mode. The magnitude of the control voltage and the method of switching between the "on" and "off" states should be selected based on the physical constants of the LCD and the design parameters of the device.

4. Conclusion

In the present work, the influence of physical constants of the LC material on the integral characteristics of the modulator was studied using the method of computer simulation. The device operates in a waveguide mode in an LC structure with a twist angle of 270° and antisymmetric boundary conditions in the working cell. Shown, that:

1) under antisymmetric boundary conditions, the short total response time of the LC modulator is due to the absence of a "backflow" in the LC cell and the fact that only half of the working gap thickness plays a role in the dynamics of switching between two states in such an LC modulator;

2) for the simultaneous achievement of high contrast ratio values and a short total response time of the device, it is necessary to choose an LC material with a high value of its dielectric anisotropy;

3) there is an optimal set of elasticity constants of the LC material, at which a high contrast ratio and a short total response time of the modulator are simultaneously realized; for this, it is necessary to choose an LC material for which the value of the parameter ω_1 is close to unity, and the value of the parameter ω_2 is about 3;

4) for each LC material there is its own modulator control mode, and the control voltage value and the method of switching between the "on" and "off" states should be selected based on the physical constants of the LCD and the design parameters of the device.

References

1. Jones Cliff J. 2018 Liquid Crystals Today **27** (3) 44
2. Khokhlov A. R., Emel'yanenko A. V. 2018 *Beilstein J. Nanotechnol.* **9** 2644
3. Belyaev V. V., Ostrovsky B. I., Pikina E. S. 2018 *Liq. Cryst. and their Appl.* **18** (1) 84
4. Minko A.A. 2017 *Physics of liquid crystals: a tutorial.* (Minsk: BSU) p 111
5. Qifeng Niu, Chunyang Wang, and Hongwei Shi 2019 *Review of Scientific Instruments* **90** 055001
6. Santec Corporation. Spatial Light Modulator https://www.santec.com/en/products/components/slm?gclid=EAIaIQobChMIzMDt4fiP6gIVAcKyCh2oCwvOEAAYAiAAEgKp6PD_BwE
7. Meadowlark Optics, Inc. Spatial Light Modulator https://www.meadowlark.com/?gclid=EAIaIQobChMIzMDt4fiP6gIVAcKyCh2oCwvOEAMYASAAEgLinfD_BwE
8. Belyaev V.V. 2015 *Liq. Cryst. and their Appl.*, **15**(3) 7
9. Studentsov S.A., Ezhov V.A. 2011 Russian patent No. RU 2488150
10. Simonenko G. V. 2020 Journal of Physics: Conference Series 1560 012028
11. Lipton L., Tilton M. 1994 Patent US 5327269
12. Simonenko G.V. 2018 *Computer simulation of characteristics of high-speed classical modulators based on liquid crystals* (Saratov: From Saratov University) p 134
13. Sukharier A.S. 1991 *Liquid crystal indicators* (Moscow: Radio I svyaz) p 256
14. Chigrinov V.G., Simonenko G.V., Yakovlev D.A., Podjachev Yu.B. 2000 *Mol. Crys. Liq. Crys.* **351** 17
15. Simonenko G.V. 2021 *Civil Eng. Res. J.* **11**(3) 555811
16. Simonenko G.V. 2018 Liq. *Cryst. and their Appl.* **18** (3) 26
17. V Belyaev, V Chigrinov 1977 *Sov Phys Crystallography* 22 344
18. V Belyaev, V Chigrinov 1978 *Sov Phys Crystallography* 23 454
19. V V Belyaev, S V Belyaev, V G Chigrinov, M F Grebenkin 1980 «*Advances in Liquid Crystal Research & Applications*» Oxford-Akademia Kiado, Budapest 673–691.

Mathematical Modeling of Survivability Function for Thermoelectric Module

Sh Sattar[1], A Osipkov[2], V V Belyaev[1,3]

[1]RUDN University (People's Friendship University of Russia), 117198, Miklukho-Maklaya, Moscow, Russia

[2]Bauman Moscow State Technical University, Moscow, Department of Rocket and Space Engineering Technologies, 2nd Baumanskaya Street, 105005 Moscow, Russia

[3]Moscow Region State University MRSU, Very Voloshinoy str., 24, Mytishi, 141014, Russia

E-mail: sattar_sh@pfur.ru

Abstract. Developing an optimized reliability model for thermoelectric module at the stress where the probability of module to functions without abruptive failure is a challenging aspect. One of the major reasons is the mismatch of thermal expansion coefficient, which has severe effects on segmented moduli compared to unsegmented moduli. The likelihood of a thermoelectric module to survive at certain level of thermo-mechanical stresses varies by varying number of component (layers) in thermoelectric leg. On another hand, selection of an adequate distribution model to predict reliability and sustainability of the thermoelectric module requires development of new optimized stress-strength-based model. In this paper the predictive reliability model for high temperature segmented module is derived from parametric Lognormal mean residual life and nonparametric Lognormal-kernel survival function to measure probability of module to survive at certain thermo-mechanical stress. A comprehensive comparative discussion has been done to illustrate the maximum likelihood based on Bayesian nonparametric lognormal-Kernel inference method regarding to Monte Carlo simulation, Weibull's distribution, and Lognormal mean residual life for various shapes for the survival function. It has been demonstrated that nonparametric lognormal-kernel survival function has high ratio of probability to predict the survival of module at higher discrete thermo-mechanical stress data.

1. Introduction

Among the other feasible technologies for this purpose, thermoelectric (TE) energy is an interesting viable alternative because the TE module (shown in Figure 1) can convert given heat, through different sources, into electric power using the Seebeck affect without requiring moving components [1]. Although, this particular character provides a reliable source of electrical power for space applications but terrestrial applications, heat recovery and automobile industry, confront limitations. For terrestrial application the major drawbacks for TE module is the fragile reliability [2]. The major reason behind this vary fact is dynamic operational condition that subject module thermomechanical fatigue and structural failure [3]. Thermomechanical fatigue is a product of thermomechanical stress produced due to thermal expansion mismatch between different TE materials [4]. Another significant challenge is increase of temperature difference between thermoelements in order to increase the conversion efficiency. The increase in the temperature gradient intensifies operating conditions, which cause chemical and mechanical instability in TE devices [5].

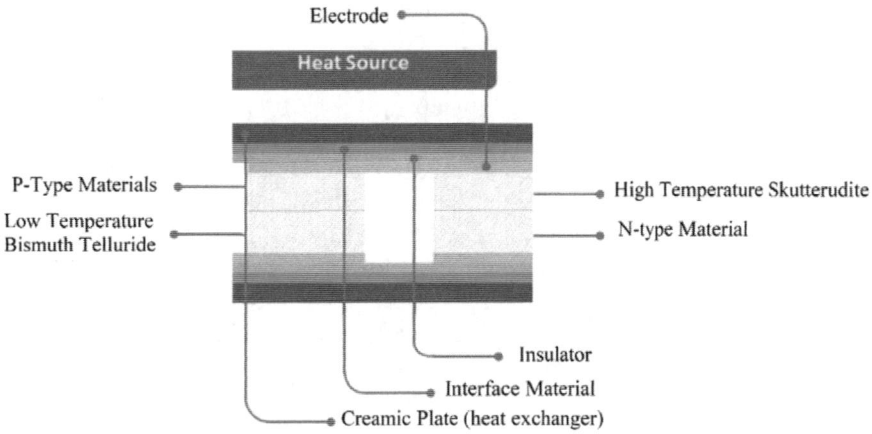

Figure 1. Simple Structure of segmented TE module

Potential candidacy of TE module to be environmentally friendly technology have led to intensive work over the past 15-20 years [6]. Series of studies have been conducted to produce mechanically reliable high performance TE modules by evaluating and limiting the operational stresses. These studies indicated that high level of thermomechanical stresses comes from material's properties, mechanical durability, and module's geometry. Hee Seok Kim at el. [7] studied cumulative temperature dependence to reduce the gap between material and device by engineering thermal conductivity. The paper suggests that an accurate evaluation of p-n configuration and mechanical characteristics of interfacial structures can solve thermomechanical reliability of the module. Moo-Yeon Lee at el. [8] paper did a comprehensive discussion on leg's geometry and different material's compatibility with structure. Investigation suggests that cylindrical shape segmented leg provides 13.6% efficiency at thermal stress of 0.69 GPa. Whereas Naveen K. Karri at el. [3] and Suhir's [9] mathematical model, on TE legs, gives insight to understand effect of shear stress between the layers and at the boundaries. Malzbender's [10] work demonstrates that the stiffness, thickness and thermal expansion of a material are active variable to influence overall stress level of the module. Z.-H. Jin [11] continued Malzbender's model and predicted that failure can occur in multilayer leg, if length to thickness ratio increases. G. Nikolova [12] developed a comprehensive mathematical model to study thermal and mechanical behavior of bonded layers. She has demonstrated that debonding (failure) of layers happen when module reaches at its critical shearing stress.

On another hand, modelling probabilistic reliability for TE module entails many challenges. Much of the existing experimental studies apply accelerated testing method to calculate mean-time-between-failure (MTBF) based on number of cycles to failures [13]. Whereas, the ceramics and glasses experimental studies apply Weibull's distribution for reliability calculation [14]. Naveen K. Karri [3] and Andrew A. Wereszczak [15] studied structural reliability to evaluate thermomechanical stress using brittle material failure theory based on Weibull distribution. The paper described significant gain in reliabilities on small reduction in maximum principle stress. Consequently, the shape of TE leg has less effect on estimated probability of failure. Alternative to this notion, we published a paper [16], presenting simulative results, where lognormal distribution is used to calculate failure rate. Generally lognormal distribution is used to measure the rate of failure for micro-electric devices at high temperature stress. The distribution is based on the multiple failure model, which mean that at given temperature range the TE legs in a the given module undergoes a random increase of degradation, interlayer diffusion, sublimation or oxidation, leading to complete abruptive failure [17-19]. Therefore, the use of lognormal distribution is mostly used to model components or devices that fail primarily due to stress or fatigue.

In this paper a comprehensive discussion is done around the fact that the module has survived some certain temperature gradient without any abruptive failure. An optimized survival function is presented for a segmented module that has survived a specific temperature range, to obtain mean residual life (MRL) for measuring the reliability of high temperature TE modules. This measure contains two aspects of information, the lifetime of a module and the temperature at which the module has ability to work operate without any failure.

A stress-strength model-based lognormal MRL function is obtained to differentiate various characteristics of segmented TE device at higher thermal stresses. This new model assists us to determine the temperature range at which the reliability of segmented device could be achieved above 90. Additionally, an optimized Bayesian nonparametric survival function is derived to measure the probability of segmented module to survive beyond the interface stress (σ_y).

2. Analytical model

For a given set of operating conditions, the reliability is defined as the probability that a system survives for some specific period of time. In our case, we drive our survival function on random variable "thermal stress (σ)" at specific temperature "T". In order to operate efficiently, the generated thermal stress in TE module must be lower than the strength (s) of the overall module. The condition of reliability is that the TE module survives within modeled strength. This condition can be mathematical expressed as $S = P(s > \sigma) = \int_{\sigma_0}^{\infty} f(s)ds$. Here σ_0 is initial mechanical stress before the temperature effect and S is survival function. The survival function (S) of any TE module, with a probability that stress is less than the strength, where stress and strength are two independent variables, for all the possible values of the strength can be computed as

$$S = \int_0^{\infty} f_S(s)\Big[\int_0^{S} f_\sigma(\sigma)d\sigma\Big]ds. \tag{1}$$

Then the Failure function can be given as case that the value of stress is higher than the strength of the device, that is, $F = 1 - S = P(s \leq \sigma)$. And for independent values of stress (σ) and strength (s) variables, the failure function is given as

$$F = \int_0^{\infty} F(s) \cdot f_\sigma(\sigma)d\sigma. \tag{2}$$

Since TE module is temperature dependent device, we assume here interface random real number (y) which is interim thermal stress at which module operates without any interruption, that is, $y = s - \sigma$. If we assume that σ and s are non-negative independent random variables then the ability to operate without interruption at interface temperature is given by S $(y > 0)$ and the P.D.F of the failure function $(y = 0)$ can be determined through following solution

$$F_y(y)dy = f_s(s)ds \int_{\sigma}^{\infty} f_\sigma(\sigma)d\sigma. \tag{3}$$

In equation (3) the value of strength is a fixed variable whereas the value of Stress has a random magnitude. In this scenario the sustainability of the device is assumed that all possible value of σ are less than the value of s i.e. $(s > \sigma)$. By assuming the range of σ from $0 \to \infty$, the equation (3) can be solved as,

$$f_y(y)dy = f_S(s)ds \int_0^\infty f_S(y+\sigma) \cdot f_\sigma(\sigma)d\sigma. \tag{4}$$

Based on equation (4) reliability of the TE module can measured through survival function, given as $S = \int_0^\infty f_y(y)dy$, and the probability of failure can be written as

$$F = 1 - S = \int_{-\infty}^{0} \int_0^\infty f_S(y+\sigma) \cdot f_\sigma(\sigma)d\sigma dy. \tag{5}$$

When we consider the reliability of segmented Module, the survival function of a system with multiple layers with multiple possible failure mode,

$$S = 1 - \prod_{i=1}^{\varphi} F_{n_i} \cdot F_{p_i}. \tag{6}$$

Here, S, φ and $F_{n_i} \cdot F_{p_i}$ present over all module survival function of TE module, number of significant failure mode and significant failure mode of n and p type TE leg with i number of layers respectively.

As elaborated above that we study here segmented TE modules for high temperature operating atmosphere, we take lognormal distribution, instead of Weibull distribution. The strength-stress density function for lognormal distribution is derived [20] as

$$f_y = \frac{1}{y\delta\sqrt{2\pi}} \exp\Big[-\frac{1}{2\delta^2}(\ln y - \mu)^2\Big]. \tag{7}$$

Here, $y > 0$, and the parameters μ and δ are the mean and the standard deviation, respectively, of the variable $\ln y$. If we take log of interference condition, i.e. $\ln y = \ln s - \ln \sigma$, we get mean log values to get probability plot. The condition is $\overline{y} = \frac{\overline{S}}{\overline{\sigma}}$, where bar shows average log value for thermal stress and strength. Here the mean log value for strength is taken as ultimate stress at which the component materials in form of segmentation can survive at modeled temperature. The system is reliable if the probability of survival function is equal to 1 (see Figure 2) and probability of module to survive can be given as

$$S = P\Big(\frac{s}{\sigma} > 1\Big) = P(y > 1) = \int_1^0 f_y(y)dy \tag{8}$$

The lognormal distribution provides, as shown in Figure 2, an Upside-Down Bathtub shape for MRL and survival function.

3. Mean Residual Life for Lognormal distribution

In reliability analysis, lifetimes are mostly taken as random variables when probability distributions are considered. In thermoelectric field most famous distribution used in reliability analysis is Weibull distribution, which is the most suitable model for modules operating $T >$

50°C [21]. But rising demand of high temperature modules limits Weibull's distribution and requires new methods to predict accurately operating life of TE modules, especially segmented modules. In order to develop a mathematical model, there are five main characteristics which must be define by the reliability model and those are PDF, CDF, failure rate, survival function and MRL function. In this paper we focus on MRL function in order to develop a condensed information to measure precise reliability of the segmented TE module. The main purpose of MRL is to measure at which thermal stress value TE component works before the module faces abruptive failure. This give us two aspects of module, its lifetime and fact that module has been working to a certain thermal stress (σ) without halt.

Suppose σ is a continuous non-negative random variable with CDF $F(\sigma_y)$, PDF $f(\sigma_y)$ and survival $S(\sigma_y) = 1 - F(\sigma_y)$. To define residual life random variable at threshold thermal stress (σ_{Th}), the life expectancy would be $\sigma_{Th} = \sigma - \sigma_y|\sigma > \sigma_y$ [22]. This implies that mean residual life (MRL) can be calculate as

$$M(\sigma_y) = E(\sigma - \sigma_y|\sigma > \sigma_y) = \frac{1}{S_{\sigma_y}} \int_{\sigma_y}^{\infty} S(y)dy, \quad \sigma_y \geq 0 \tag{9}$$

The lognormal distribution belongs to those distribution which have no closed form of survival function, so will use equation (9) to obtain the MRL function for lognormal distribution. In this regard, the PDF and CDF of a lognormal are given as

$$f(\sigma) = \frac{1}{\sigma\sqrt{2\pi\delta^2}} \exp\left[-\frac{1}{2}\left[\frac{\ln\sigma_y - \mu}{\delta}\right]^2\right] \tag{10}$$

$$F(\sigma_y) = \int_0^{\sigma_y} \frac{1}{\sigma_y\sqrt{2\pi\delta^2}} \exp\left[-\frac{1}{2}\left[\frac{\ln\sigma_y - \mu}{\delta}\right]^2\right] \partial\sigma_y \tag{11}$$

Therefore, the survival function TE module can be written as

$$S(\sigma) = 1 - \varphi \cdot \frac{\ln\sigma - \mu}{\delta} \tag{12}$$

Let $z_\sigma = \dfrac{\ln\sigma - \mu}{\delta}$, then $\sigma = \exp[z_\sigma\delta + \mu]$ and $\partial\sigma = \sigma\exp[z_\sigma\delta + \mu]dz_\sigma$.
Main residual life for lognormal distribution will become [23]

$$M(\sigma) = \exp\frac{\mu + \delta^2}{2} \cdot \frac{\left[1 - \varphi \cdot \dfrac{\ln\sigma_y - (\mu + \delta^2)}{\delta}\right]}{1 - \varphi \cdot \dfrac{\ln\sigma_y - \mu}{\delta}} - \sigma \tag{13}$$

Figure 2 provides PDF (top left), CDF (bottom left), Survival function (Top right) and MRL function (bottom right). MRL function demonstrates reported number of TE modules that survive at specific thermal stress. As compare to Weibull survival function, where no transformation is necessary, both lognormal MRL and survival functions shows fitting graph for higher value of scale (δ) and location (μ) parameters. The domains of the survival functions, based on equation (13) for $\sigma \leq \sigma_y$ can take on values from 0 to infinity, however for $1 < \sigma_y < 0$ goes from 0 to σ_{Th}/σ.

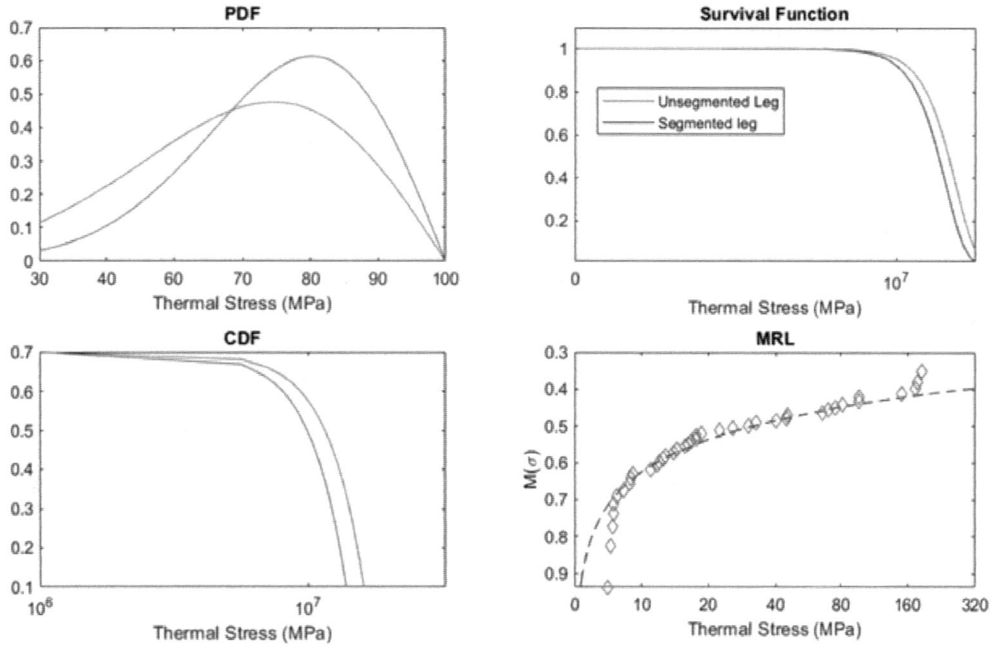

Figure 2. Mean residual life of segmented and unsegmented TE leg

4. Non-Parametric Lognormal Survival Function

Nonparametric models, compared to parametric modelling, are mostly used to analyze failure data to estimate the MRL function [24]. Under extreme right censorship, the nonparametric model can provide an innovative method to deal with discrete and nonlinear censoring data. The relationship between nonparametric MRL and failure rate function can assist us to develop a better reliability model and effective system. Consequently, nonparametric Survival function aims to measure sustainability of a device based on their failure behaviors.

We used a Dirichlet process for obtaining a common nonparametric lognormal survival function [25]. In this regard we take density of survival function ($\acute{S}$) at threshold stress (σ_{Th}) with the respect of interference and strength stress. i.e.

$$\acute{S}(\sigma_y, G) = \int K(\sigma_y, \sigma_S) dG(\sigma_{Th}) \tag{14}$$

Here K is a lognormal distribution for the kernel, $G(\sigma_{Th})$ is nonparametric vector defined at threshold stress where we can still find our TE modules operating, the moment before module confronts abruptive failure. The nonparametric lognormal-Kernel distribution, for the discrete data based on stress-strength data distribution, can be given as

$$K\Big(\sigma_y; \sigma_S = (\mu, \delta^2)\Big) = \frac{1}{\sigma_y\sqrt{2\pi\delta^2}} \exp\left[-\frac{1}{2}\cdot\left[\frac{\ln\sigma_y - \mu}{\delta}\right]^2\right] \tag{15}$$

And nonparametric vector for a module to operate at threshold stress (surviving stress) can be defined as

$$G(\sigma_{Th}) = \sum_{\sigma=1}^{\infty} \omega_\sigma \delta_{f(\sigma)}(\sigma_{Th}) \tag{16}$$

Here $f(\sigma)$ presents independent stress distributed identically over the module. And $\delta_{f(\sigma)} = \{1, 2 \ldots\}$ is the parametric indicative function of stress, which is the baseline distribution. ω_σ

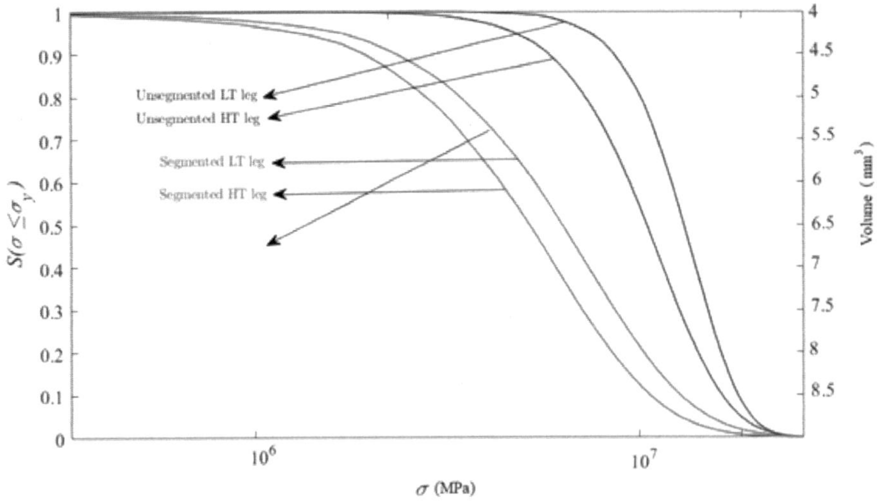

Figure 3. Comparative Nonparametric Survival Function graph for Segmented and Unsegmented

present stick-breaking of discreteness and explicitly used to give random probability of discrete distribution. Since the distribution is random itself, its precision is derived from location parameter, define as $\{\delta_\sigma\}_{\sigma=1}^{\infty}$ and shape of survival $S(\sigma_y, G)$. The expansion of equation (13), we can locate the survival function curve, at σ_y and $G(\sigma_{Th})$

$$\acute{S}(\sigma_y G(\sigma_{Th}) = \sum_{l=1}^{N} \rho\sigma_{Th} LN(\sigma_y; \mu, \delta^2) \tag{17}$$

Here ρ comes from Bayesian inference to include posterior probability, which presents maximum likelihood of survival function derived from failure rate data. It is given as $\rho = E_{v_i} \prod_{v=i}^{i-1}(1 - E_{\sigma_y})$, where E_{v_i} volumetric elastic constant of ith layer at σ_y and $v_{\sigma_y} = \beta(1, \alpha)$. β and α are two positive shapes defined as per Dirichlet process [26] and i is the total number of components in the model. The nonparametric lognormal distribution mixed with Kernel distribution as per Dirichlet process model, for positive real number, is transformed as $Y = \log(\sigma) - \log(\sigma_y)$. The nonparametric survival probability for unsegmented module, to survival within domain of threshold stress but higher than interference stress, can be given as

$$\acute{S}(\sigma_y \leq \sigma \leq \sigma_{Th}; G) = Pr(\sigma \leq e^Y; G) \tag{18}$$

Respectively for segmented TE modules, the nonparametric probability can be given as

$$\acute{S}(\sigma_y \leq \sigma \leq \sigma_{Th}; G) = Pr(\sigma \leq e^Y; G) \tag{19}$$

$$\acute{S}(\sigma_y \leq \sigma \leq \sigma_{Th}; G) = \sum_{i \geq 1}^{i} \rho\varphi \cdot \frac{Y - \mu_i}{\delta_i} \tag{20}$$

Figure 3 shows the evaluation of survival function based obtained results for segmented and unsegmented TE legs. TE module working on high and low temperature result different ratio of reliability. Low temperature unsegmented TE modules demonstrated higher rate of survival on given standard (4 mm^3) leg volume. One of the major reasons is that unsegmented TE module has low adverse effect from thermal expansion compare to segmented TE module. At given leg volume unsegmented modules has 95% survival rate within range of $10 \leq \sigma < 60$ (MPa). Whereas the 95% of survival rate for low temperature segmented TE module exists only between $0 \leq \sigma < 30$ (MPa), due to mismatch of thermal expansion between different materials. The main difference between simple MRL and Survival function and nonparametric survival function is rapid decline in curve of Figure 3 compare to Figure 2. MRL has definite method to predict reliability of module based on failure rate data whereas nonparametric survival function can give us precise dimensions (volume) and thermally induce stress to produce reliable characteristics for segmented and unsegmented TE modules. Hence more intuitive conditions (especially boundary conditions of TE leg) are produced as alternatives on given survival curves. This makes easy for us to obtain graphical MRL function-based reliability to analyze survival function on sufficient conditions.

5. Comparative Discussion

The influence of temperature on thermoelectric module and its material properties plays a significant role to predict reliability as outlined in the introduction. The analytical model corresponds to stress- strength covariance to obtain probability of survival and MRL of the module. The temperature gradient along the TE leg, between cold and hot sides, leads significant changes by influencing thermal conductivity. Two main aspects that govern the reliability of the module are (i) amount of heat absorbed and (ii) material property changes due to heat distribution along the module. In our previous paper [16] we illustrated the linear relationship between temperature distribution and thermal stress generation, here we have notice that there is minor bow, shown in figure4 (a) in the temperature-stress relationship (in segmented cases). The inclusion of Thomson and joule heat [37] in temperature distribution profile leads to this bow in graph (shown in figure 4).

The Figure 4 (b) corresponding to temperature profiles by considering contribution of boundary conditions at each end (hot and cold) for the emergence of deviation from linear temperature profile at maximum efficiency. The simulated results show the temperature profile dependence on material properties, which contribute nonconstancy, leading temperature line bending, especially on hot side. Thomson heat generation (or absorption) gradient, in double integration $\left(k(T)\frac{\partial^2 T}{\partial y^2}\right)$, fills the loss of heat, causing a bow on the gradient line [27].

For visualization, the effect of Thomson heat generation (absorption) on stress-strength covariance profile, based on temperature gradient, is obtained for nonparametric lognormal model and compared with obtain simple Monte Carlo simulation (see Figure 5).

The simulation results demonstrate that bending with torsion and phase displacement, varying with length to thickness ratio, have influence over the maximum thermal shearing and mechanical shearing stresses. Figure 5(a) shows positions of maximum covariance between stress (shearing) and strength for a stable (segmented and unsegmented) TE module using general stress-strength relationship through Monte Carlo simulation. Whereas Figure 5 (b) presents the covariance profile for high and low temperature segmented TE modules under highly induced thermal stresses, where the maximum covariance corresponds to the critical position. For the ratio of thermal shearing stresses to the strength, at significant bending stresses, the plane of maximum covariance occurs from 300 MPa onwards (specific case of segmentation using bismuth telluride material). The plane of maximum covariance varies from material to material and module to module. The resemblance between graph (a) and (b) shows the relevancy of nonparametric lognormal distribution to obtain critical stress for segmented TE module through covariance plane.

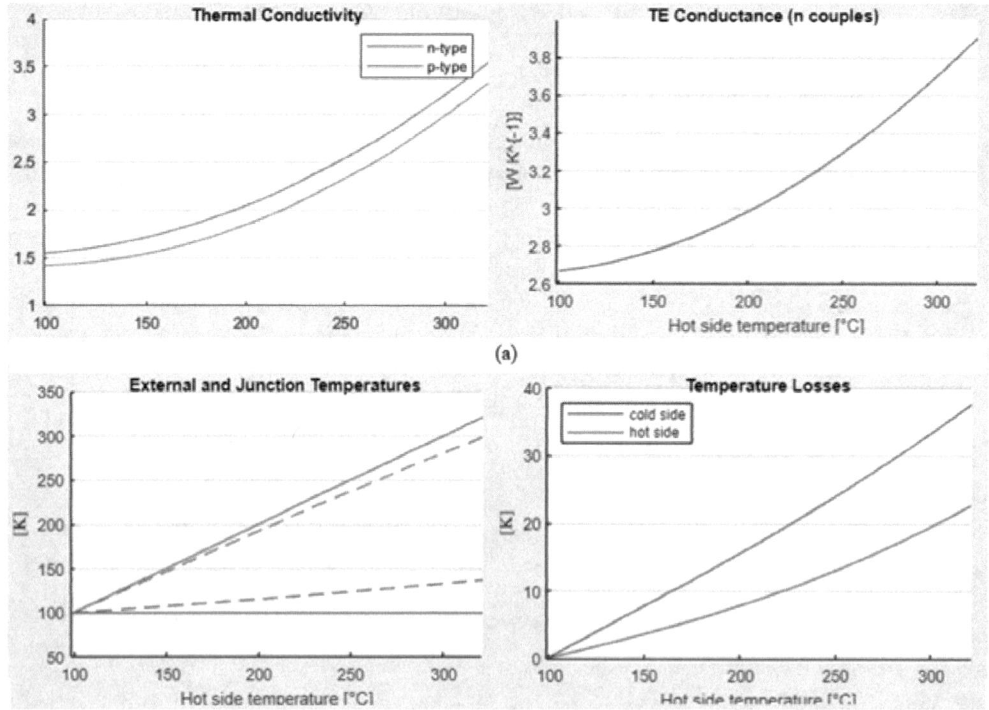

Figure 4. (a) & (b) Temperature Profile of Segmented TE module

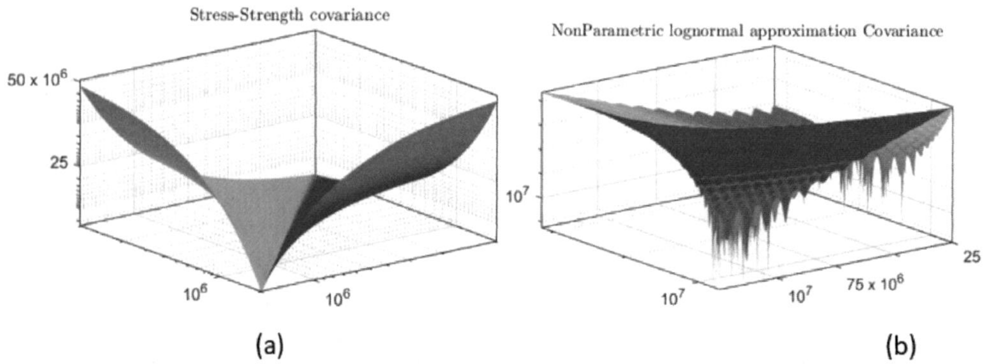

Figure 5. Stress Strength Covariance

On further investigation, the demonstrated Figure 6 shows relevance vector regression on stress-strength covariance. The optimized nonparametric lognormal distribution model, developed to understand stress-strength covariance, was simulated under Fast Multi-output relevance vector regression (MRVR) [28] in MATLAB. Figure 6 illustrate the distribution of Survival function in domain of threshold stress. A noteworthy observation is that the probability of survival function encompasses characteristics of upside-down bathtub shape for individual module threshold stress discrete analysis. The distribution of failure data has same characteristics for shape of survival function as Figure 3. The deviations are different when compared with high temperature segmented (SKD) and relatively decreases after 300 MPa when compare to low temperature TE modules (bismuth telluride). This has significant implications on TE system design and optimization of discrete data distribution to apply nonparametric

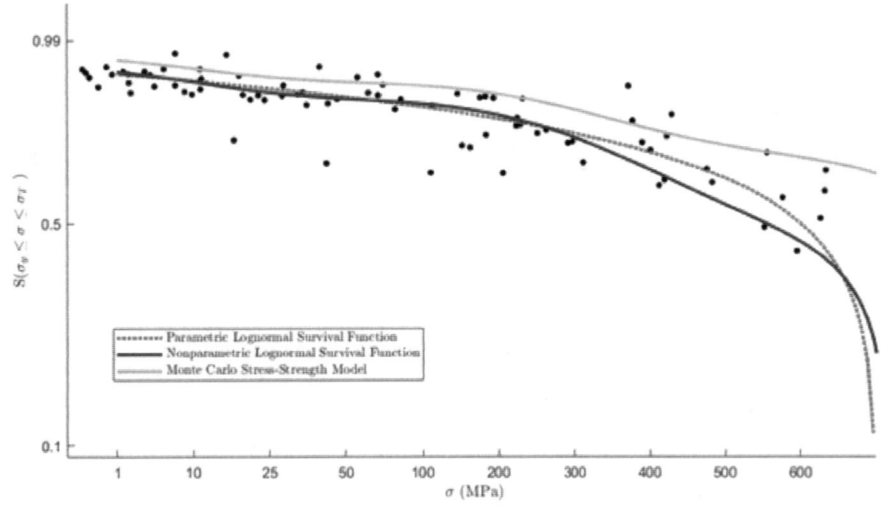

Figure 6. Parametric and Nonparametric lognormal survival function.

Figure 7 illustrates the nonparametric model, at 200, 300,500, 700 and 900 MPa stress data to approximate the appropriate priors for segmented and unsegmented modules. We used interference random variable y, under the random stress σ. The range of the unsegmented high temperature TE module on the log scale was (3.69356, 8.2986), extending the prior variance about 0.85. The shape parameter was set 2 so that the unsegmented high temperature module could have finite variance. Whereas for segmented high temperature module, following the same approach, the range on the log scale was set on (4.646774, 6.350078) so the expansion of prior variance is 0.55. Number of components varies from 4-9, depending on the module.

Posterior estimates for the densities for the segmented and unsegmented (low and high temperature) are shown in the Figure 7. Nonparametric lognormal function is able to describe the peaks and valleys that the parametric model can't. There is a slight discrepancy from the point of estimate and the density of the data around 900-1000 (MPa).

But nevertheless, the data density remains within the estimated interval for given model. By comparing the densities regarding to different temperatures under nonparametric lognormal function gives us a clear insight to understand mean residual life. Figure 8 in that respect points interval estimates of the posterior density for survival function for segmented (Low and High Temperature) and unsegmented (high temperature).

Looking at the estimated densities we can see that segmented high temperature module has lowest surviving life compared to segmented low temperature module. The survival function estimates show that after 400 MPa the survival curve monotonically decreases. Subsequently the underlying explanation is that failure rate function at thermal stress ($\sigma > \sigma_y$) is finite for nonparametric survival model which distinguishes it from other existing models with infinite initial failure rates. If modules are stressed to failure lower than σ_y, then it is difficult to distinguish between lognormal and Weibull distribution. If Its above then σ_y, then the probability of making a correct choice is fair, especially for small sizes samples and nonparametric lognormal module becomes quite good. The subsequent a comparative investigation was conducted for the probability of failure for various thickness of segmented TE leg. The probability of failure for Weibull's distribution, Monte Carlo simulation and nonparametric survival function, for both low and high temperature modules are shown in Figure 9. The Monte Carlo simulation-based technique assists to distinguish different distributions and predict

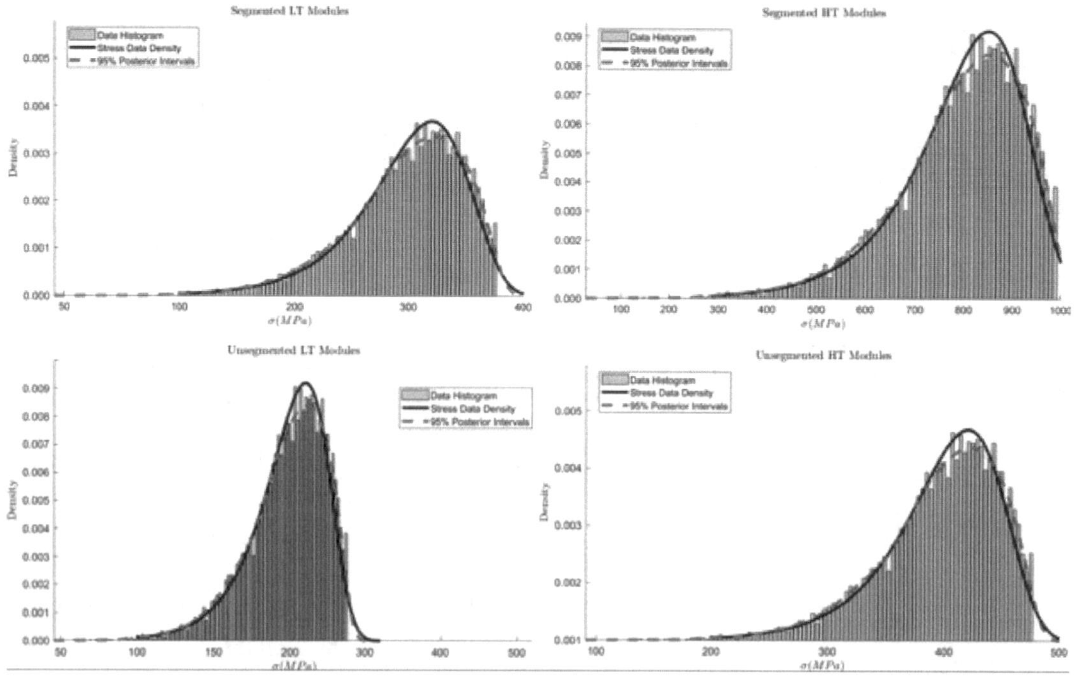

Figure 7. Relative frequency histogram and densities of survival function for segmented and unsegmented modules under the Nonparametric model

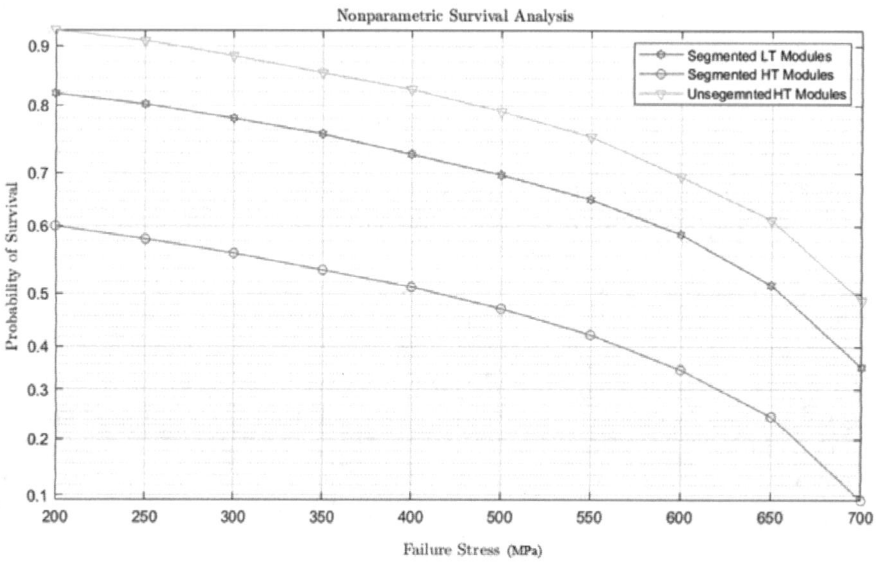

Figure 8. Probability of Survival for segmented (Low and High Temperature) and unsegmented (high temperature)

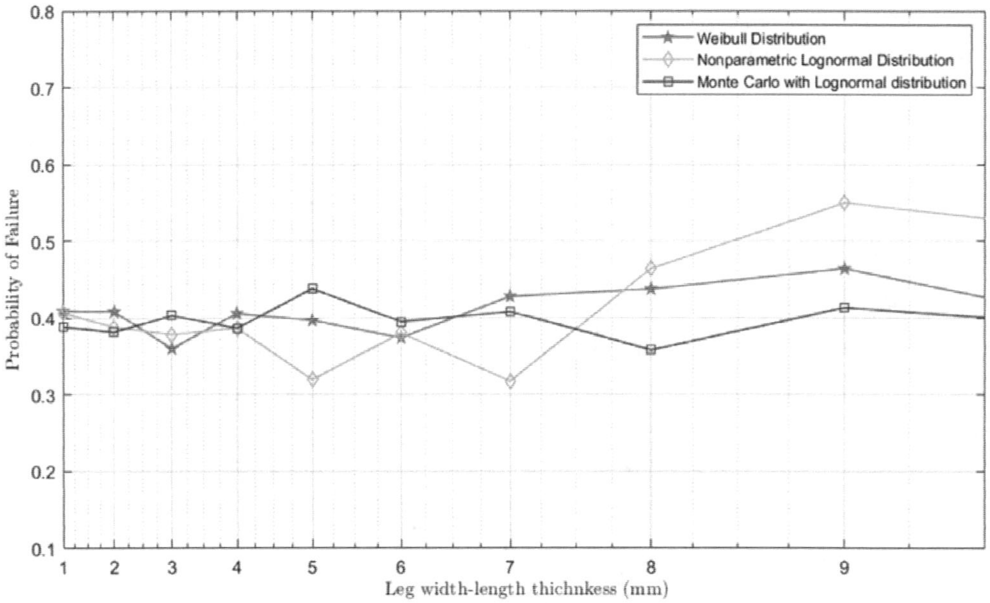

Figure 9. Comparative Probability Analysis between Weibull, Monte Carlo and Nonparametric distributions

reliability based on empirically determined failure data [29]. Especially distinguishing between nonparametric lognormal and Weibull distribution is of interest here because both are used to model probability of failure. The changing the in thickness corresponds to temperature very closely, producing varying probability map.

We used the posterior predictive survival approach, introduced by Gelfand and Gosh [30] to compare Weibull's model to the nonparametric lognormal mixture model. The comparative method is used to minimize expectation of specified survival function under posterior predictive distribution model to replicate response observed in data. For both segmented HT and LT, the nonparametric lognormal mixture model performs significantly better than the Weibull's model. The comparison of distribution regarding to thickness of the thermoelectric leg provides us insight for modeling devices. This aspect ultimately changes the prospect of sizing the legs. We have noticed that unsegmented TE legs can survive between 4-5 (mm) with probability of failure of 30% , whereas the segmented TE legs can survive between 6-7 (mm) with probability of failure of 35%. As Shown in graph that onwards, the probability of failure enhances up to 60% for nonparametric lognormal distribution whereas 45% for Weibull distribution. The result of the comparison supports our earlier argument [16] that nonparametric lognormal survival function is indeed a better model for these circumstances compared to Weibull model.

Conclusions

The article presents some basic properties and characteristics of nonparametric survival function, taking MRL function to define it. The provided method for obtain survival stress compared to different distribution allow us to study various factors to find definite shape of survival function. It has been shown that the optimal survival ability of a segmented module depends on stress-strength density function. The survival function is obtained through non-parametric lognormal distribution and an appropriate model for thermal stress, caused by change in temperature gradient, entails 80% accuracy to predict life of module for both high and low temperature. It has been shown that when using non-parametric lognormal distribution to extrapolate beyond the range of critical thermal stress, the non-parametric lognormal survival function can predict, based on average failure rates, operating life of TE module better than parametric lognormal and Weibull distribution.

Consequently, the comparative discussion between non-parametric model, Weibull distribution and Monte Carlo simulation based on module area gives a meaningful insight over sustainability of module on different temperature gradient and thickness of each layer. The average thickness of the leg is 5 to 7 mm, the range between which leg can survive at critical thermal stress without failing.

Acknowledgements

This research was partially supported by RFBR, Research Grant No. 19-07-00602 A and 19-57-45011 Ind_a.

References

[1] D Thesis 2009 *Modelling of thermoelectric devices for electric power generation.*
[2] S Baschel, E Koubli, J Roy, and R Gottschalg 2018 *Impact of Component Reliability on Large Scale Photovoltaic Systems' Performance* doi: 10.3390/en11061579.
[3] N K Karri and C Mo 2018 *Metallization and Processing Temperatures* **47**(10) doi: 10.1007/s11664-018-6505-1.
[4] G Li et al. 2016 *Acta Materialia Atomistic explanation of brittle failure of thermoelectric* **103** 775–780 doi: 10.1016/j.actamat.2015.11.021.
[5] S Twaha, J Zhu, Y Yan, and B Li 2016 *Renew. Sustain. Energy Rev.* **65** 698–726 doi: 10.1016/j.rser.2016.07.034.
[6] C Gayner and K K Kar 2016 *Prog. Mater. Sci.* **83** 330–382 doi: 10.1016/j.pmatsci.2016.07.002.
[7] H S Kim, W Liu, and Z Ren 2017 *Energy Environ. Sci.* **10**(1) 69–85 doi: 10.1039/c6ee02488b.
[8] M Y Lee, J H Seo, H S Lee, and K S Garud 2020 *Symmetry (Basel)* **12**(5) doi: 10.3390/SYM12050786.
[9] A Ziabari, E Suhir, and A Shakouri 2014 *Microelectronics J.* **45**(5) 547–553 doi: 10.1016/j.mejo.2013.12.004.
[10] J Malzbender 2004 *J. Appl. Phys.* **95**(4) 1780–1782 doi: 10.1063/1.1642289.
[11] Z H Jin 2014 *Microelectron. Reliab.* **54**(6–7) 1363–1368 doi: 10.1016/j.microrel.2014.02.028.
[12] N Geargana and I Jordika 2013 *Structure Subjected To Monotanically Increasing* 937–947.
[13] A Note 2004 *Reliability and the Electronic Engineer* **1** 1–4.
[14] Z Bertalan, A Shekhawat, J P Sethna, and S Zapperi 2014 *Fracture strength: Stress concentration, extreme value statistics and the fate of the Weibull distribution* 1–8.
[15] O M Jadaan and A A Wereszczak 2009 *Probabilistic Mechanical Reliability Prediction of Thermoelectric Legs Under contract* DE-AC05-00OR22725.
[16] S Sattar 2020 *J. Phys. Conf. Ser.* **1560**(1) doi: 10.1088/1742-6596/1560/1/012025.
[17] T Torstensson 2004 *Reliability in fatigue.*
[18] A D Telang and V Mariappan 2008 *Hazard Rate of Lognormal Distribution: An Investigation* **4**(2) 103–108.
[19] E Bridget and M Abiodun 2017 *Inference on Stress-Strength Reliability for Log-Normal Distribution based on Lower Record Values* **22** 77–97.
[20] M Baro-tijerina and G Duran-medrano 2020 March.
[21] B Basu, D Tiwari, D Kundu, and R Prasad *Distribution for Brittle Materials* 1–25.
[22] D Banjevic 2009 *Metrika* **69**(2–3) 337–349 doi: 10.1007/s00184-008-0220-5.
[23] R C Gupta and S Lvin 2005 *Math. Comput. Model.* **42**(9–10) 939–946 doi: 10.1016/j.mcm.2005.06.005.
[24] V Poynor and A Kottas 2019 *Biostatistics* **20** (2) 240–255 doi: 10.1093/biostatistics/kxx075.
[25] G S Mudholkar, D K Srivastava, and M Freimer 1995 *Technometrics* **37**(4) 436–445 doi: 10.1080/00401706.1995.10484376.
[26] J Sethuraman 1994 *Statistica Sinica* **4** 639–650 [Online] http://www3.stat.sinica.edu.tw
[27] P Ponnusamy, J de Boor, and E Müller 2020 *Appl. Energy* **262** doi: 10.1016/j.apenergy.2020.114587.
[28] Y Ha and H Zhang 2019 *Econ. Model.* **81** 217–230 doi: 10.1016/j.econmod.2019.04.007.
[29] E Environ, M Zebarjadi, K Esfarjani, M S Dresselhaus, Z F Ren, and G Chen 2012 *Environmental Science Perspectives on thermoelectrics: from fundamentals to device applications* pp. 5147–5162 doi: 10.1039/c1ee02497c.
[30] A E Gelfand and S K Ghosh 1998 *Biometrika* **85**(1) 1–11 doi: 10.1093/biomet/85.1.1.

Diamagnetic anisotropy and rotational viscosity of 6CHBT nematic liquid crystal

A D Kurilov[1,2], I S Chekulaev[1], S M Shammai Irani[1], N S Parashchuk[1], V S Bezruchenko[3], V V Belyaev[1] and D N Chausov[1,2]

[1]Moscow Region State University, 24 Very Voloshinoy St., Mytishchi, 141014, Russia
[2]Prokhorov General Physics Institute of the Russian Academy of Sciences, 119991, Russia, Moscow, 38 Vavilova St.
[3]Institute of Chemistry of New Materials NAS of Belarus, 36 F. Skoriny St., Minsk, 220141, Belarus

E-mail: ad.kurilov@gmail.com

Abstract. Nematic liquid crystal with polar terminal NCS group was studied by using acoustic and electro-optical methods. Experimental results on rotational viscosity have been obtained by two methods. The anisotropy of diamagnetic susceptibility of the studied LC with temperature variation was determined based on the comparison of these methods. The present results may be useful for testing physical models of the orientation dynamics of nematic liquid crystals in alternating electric and magnetic fields.

1. Introduction

Nematic liquid crystals (NLC) with high birefringence Δn and low viscosity are particularly attractive for displays, infrared and microwave applications [1–7]. The birefringence of NLC is mainly determined by the length of π-conjugation, molecular shape, functional and terminal groups. The presence of benzene rings as well as double and triple bonds lead to highly conjugated compounds [8]. The main drawbacks of this highly conjugated structures for practical applications are usually high melting point and strong smectogenic character. The addition of a terminal nitrogen-containing group make it possible to achieve ultra-high birefringence. Isothiocyanates (compounds with polar terminal NCS groups) have the highest birefringence among such materials [9]. Isothiocyanates, as compounds with high Δn, seem to be more suitable for application than other polar ones due to their lower bulk and rotational viscosities.The latter is lower by up to 35% than for structurally comparable CN compounds [10].

It is also known that the influence of ultraviolet (UV) radiation adversely affects operational parameters of LC materials: clearing point, birefringence, dielectric permittivity, as well as viscous and elastic constants. However, the addition of nitrogen containing groups increases UV resistance, and isothiocyanates have the highest stability among appropriate structures of other polar groups [11].

Experimental studies of the influence of alternating electric and magnetic fields on the orientation structure of liquid crystals allow us to obtain information about the relaxation, elastic and viscous properties [12]. In the present work the rotational viscosity of NLC with

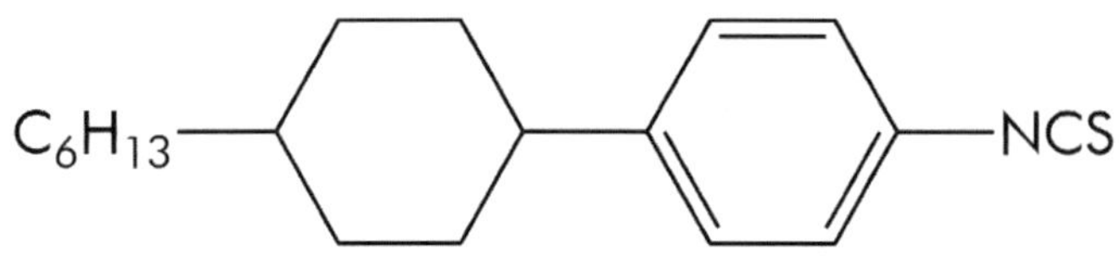

Figure 1. The structural formula of 6CHBT

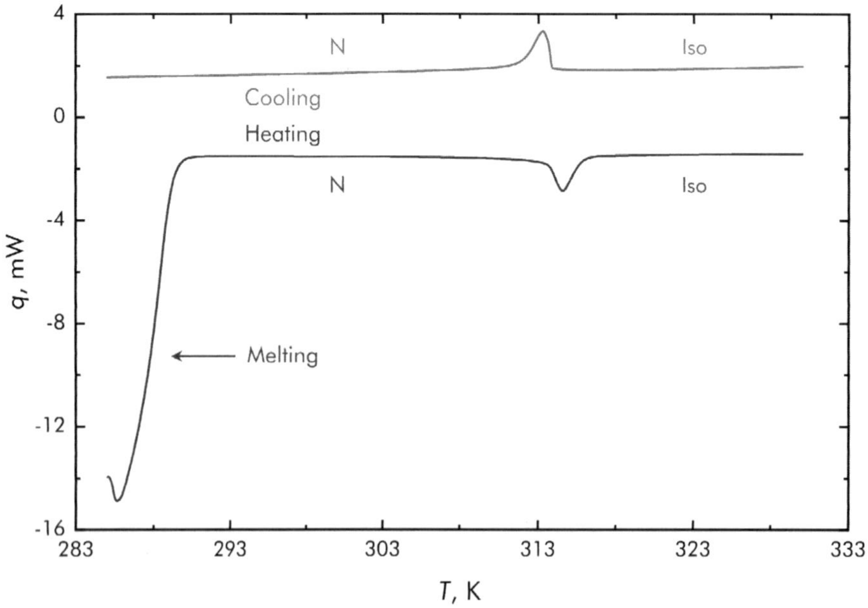

Figure 2. Thermogram of 6CHBT

polar terminal NCS group is measured using the Bogdanov method [13] and electro-optical method [14, 15] with temperature variation. The approach used, based on the comparison of the two methods, allows us to determine the anisotropy of diamagnetic susceptibility of the studied LC.

2. Materials and Methods

2.1. Materials

In this paper we present the results of rheological measurements of nematic liquid crystal *1-(4-hexyl-cyclohexyl)-4-isothiocyanato-benzene* (6CHBT, Military Technical University, Warsaw, Poland) with polar terminal NCS group. The structural formula of 6CHBT is shown in Fig. 1.

According to literature data 6CHBT has a melting point of $T_{\mathrm{Cr-N}} = 285.7$ K and a clering point of $T_{\mathrm{N-Iso}} = 316.7$ K [16]. It was found, using differential scanning calorimetry (Mettler Toledo DSC 3) and polarizing light microscopy (Altami Polar 3), that phase transition temperatures of the studied LC correspond to these data. The measurements were carried out using standard aluminum crucibles with a volume of 40 μl in the heating and cooling mode from 183 to 383 K with a rate of 33 mK/s. The temperature of the sample was kept at a constant temperature for at least 30 minutes for every measurement. The resulting thermogram is shown in Fig. 2.

2.2. Electro-optical response

The electro-optical properties of 6CHBT were studied using a standard technique based on the determination of the volt-contrast performance of a planar NLC cell [17, 18]. A continuous sinusoidal signal with a frequency of 5 kHz was applied to the cell. The signal cycle in this case is much less than both the director reorientation time and the relaxation time of ionic charges. Under these conditions, the director does not respond to the fluctuations of the electric field, but is sensitive to its RMS value. The amplitude of the alternating signal ranged from 0 to 80 V. A linearly polarized helium-neon laser ($\lambda = 632.8$ nm) was chosen as a light source. A planar LC cell is placed between crossed polarizers so that the angle between the director and the allowed orientation of each of the polarizers is $\pm 45°$. After passing through the cell, the optical signal reaches a photodiode; it is amplified and registered by an automatic recorder. Together with the intensity of the transmitted laser radiation, the recorder simultaneously registers the signal being supplied from the generator to the cell.

The dynamic parameters of the electro-optical response were studied using the S-Recorder L recording voltmeter with a sampling rate of 100 kHz. Instead of a continuous signal, an amplitude-modulated signal of the same frequency with an interval between bursts of 500 ms and a duration of 100 ms is supplied to the cell. The amplitude of the test signal was 5 V. The rising τ_{on} and the decay τ_{off} times were determined analyzing the change in the phase delay $\Delta\Phi$ from 10% to 90% of the maximum value for a specified voltage and from 90% to 10%, respectively.

Rotational viscosity is calculated from the experimental data of the decay time according to the formula [19, 20]

$$\gamma_1 = \frac{\tau_{\text{off}} U_{\text{th}}^2 \varepsilon_0 \Delta\varepsilon}{d^2}, \tag{1}$$

where U_{th} is a threshold voltage of the Freedericksz transition, $\Delta\varepsilon$ is a dielectric permittivity anisotropy, ε_0 is an electric constant, d is a thickness of the liquid-crystal layer.

Experimental data on the temperature dependence of the dielectric permittivity anisotropy of 6CHBT were taken from [21].

2.3. Acoustic implementation of the rotating magnetic field method

The acoustic method of Bogdanov [13] was developed to measure the rotational viscosity of nematic liquid crystals in a rotating magnetic field by the phase diagram of the angular dependence of the ultrasonic absorption. This method clearly shows the existence of rotational viscosity without additional processing of experimental data [22–26].

The pulse ultrasonic method with a fixed distance between piezoelectric transducers is used to determine the acoustic properties of LC. The rotating magnetic field is formed by a permanent magnet, which is fixed on a rotating platform. The platform rotation frequency can be varied in the range of $f_H = 0.056 - 280$ mHz. The magnetic field strength is 110 kA/m. The acoustic chamber is fixed between the poles of a permament magnet. The chamber is a nonmagnetic tin bronze cylinder with a length of $l = 11$ mm and a diameter of $D = 11$ mm. Piezoelectric transducers from lead zirconate titanate (PZT) with a resonant frequency of 4.2 MHz are glued to the ends of this cylinder.

The measurements of the amplitude of the probe pulse were carried out using an automatic recorder S-Recorder L, while the second channel of the device received control marks from the photodiode every 90° during the rotation of the platform. The initial displacement of the acoustic chamber ψ_0 as well as the lag angle of the director ψ is determined by the control marks from the photodiode. Then the angular dependence of the ultrasonic absorption coefficient changes $\Delta\alpha(\varphi) = \alpha(\varphi) - \alpha(90°)$ is determined. The signal frequency was set to $f_0 = 4.2$ MHz and corresponded to the main resonance of the piezoelectric transducer. The amplitude of the voltage applied was 15 V.

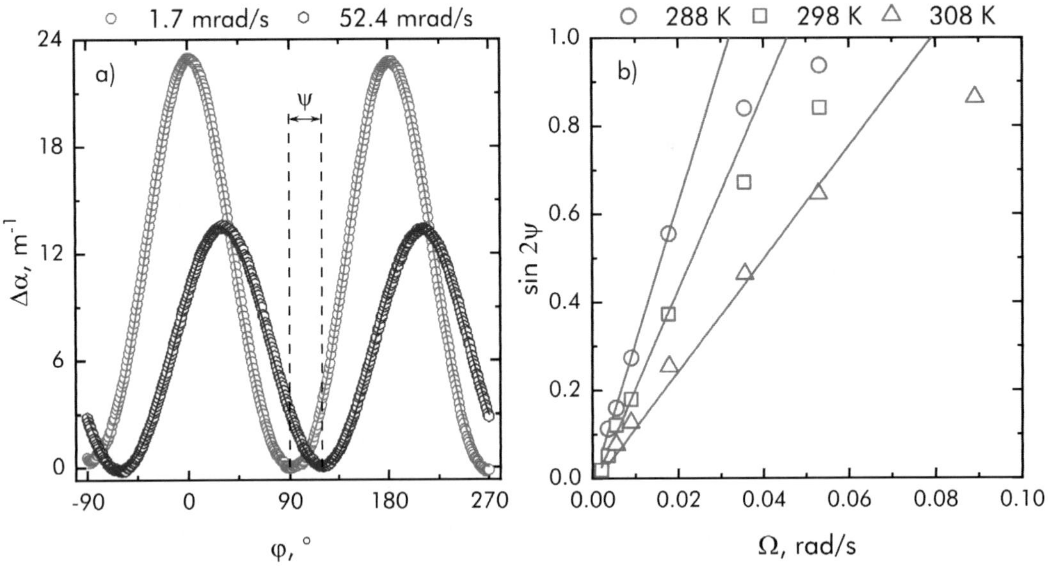

Figure 3. Demonstration of the phase lag of the director: a) the dependence of the ultrasonic attenuation coefficient changes of the 6CHBT on the angle between the external magnetic field and the wave vector of the ultrasonic wave b) the dependence of the $\sin 2\psi$ on the rotation frequency of the magnetic field

3. Results and discussion

The expression for $\Delta\alpha(\varphi)$ in the case of longitudinal waves propagating in the NLC at an angle φ to the director according to the Ericksen-Leslie theory has the form [27–29]

$$\Delta\alpha = \frac{\omega^2}{2\rho c_0^3}\left(a\cos^2\varphi + b\cos^4\varphi\right), \tag{2}$$

where a, b are numerical coefficients, that have dimension of viscosity, c_0 is a low-frequency limit of the speed of sound, ρ is a density of the NLC, ω is a circular frequency of the ultrasonic wave.

Figure 3a shows the phase lag ψ of the director and the magnetic field strength vector at $\Omega = 52.4$ rad/s and $T = 295$ K. The solid lines are best fits to expression (2). The phase difference ψ between the external field and director orientation is a consequence of the presence of a torque acting on the molecular director in a rotating magnetic field. The torque experienced by the liquid crystal sample in this case is $M = \Delta\chi\mu_0 H^2 V \sin 2\psi/2$, where μ_0 is a vacuum permeability.

The equality of the moment of friction and the magnetic moment corresponds to a steady rotation

$$\gamma_1 V\Omega = \frac{\Delta\chi\mu_0 H^2 V \sin 2\psi}{2}. \tag{3}$$

A steady rotation of the liquid crystal is possible only at frequencies

$$\Omega \leq \frac{\Delta\chi\mu_0 H^2}{2\gamma_1} = \Omega_0. \tag{4}$$

Figure 3b shows the experimental curves $\sin 2\psi$ vs Ω, which are satisfactorily described by expressions (3) and (4). The reason for the deviation of the experimental data from the theoretical predictions at large Ω is the inhomogeneities in the orientation of the sample caused by vortices. Thus, reliable data can only be obtained for sufficiently small Ω.

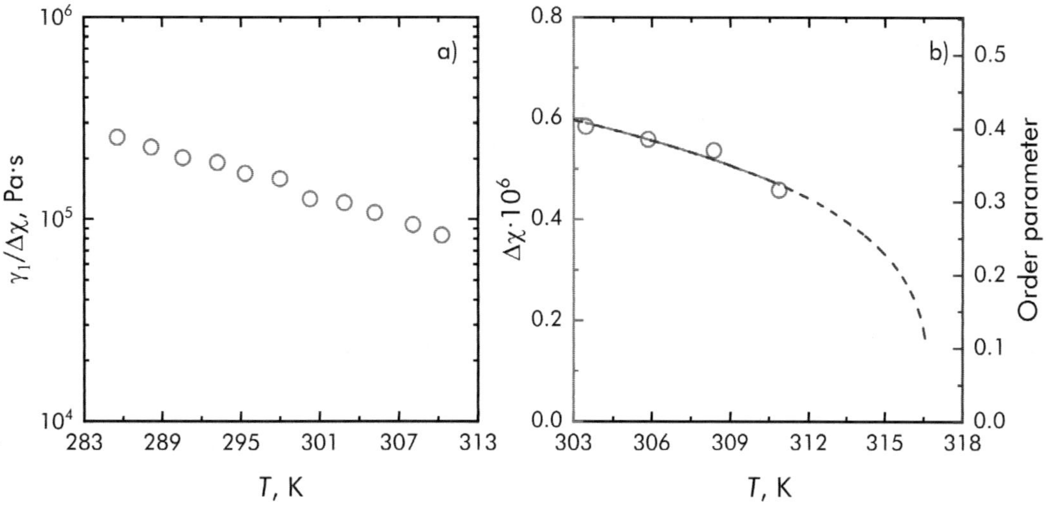

Figure 4. Temperature dependence of a) the $\gamma_1/\Delta\chi$ ratio and b) the orientation order parameter S

The critical frequency Ω_0 of the synchronous mode is determined by the experimental values of the $\sin 2\psi$. Then, the ratio of the rotational viscosity to diamagnetic susceptibility anisotropy $\gamma_1/\Delta\chi$ is calculated according to the expression

$$\frac{\gamma_1}{\Delta\chi} = \frac{\mu_0 H^2}{2\Omega_0}. \tag{5}$$

The temperature dependence $\gamma_1/\Delta\chi$ shown in Figure 4a can be used to determine the anisotropy of diamagnetic susceptibility. For this purpose, the electro-optical response of a thin layer of 6CHBT liquid crystal was studied. It is possible to calculate the temperature dependence of the anisotropy of diamagnetic susceptibility (Figure 4b) by comparing the experimental data obtained by electro-optical and acoustic methods.

Diamagnetic anisotropy of liquid crystals is determined by two factors: the structure and orientation of the molecules [30]. It is known that the magnetic anisotropy of aromatic compounds is mainly determined by the anisotropy of the benzene rings. Despite the fact that a liquid crystal can be considered macroscopically homogeneous under experimental conditions, the axes of its molecules deviate from the equilibrium position at significant angles due to thermal fluctuations. Therefore, diamagnetic anisotropy decreases with increasing temperature. The orientation order parameter is defined by the degree of alignment of the molecules $S = \Delta\chi/\Delta\chi_0$, where $\Delta\chi_0$ is a diamagnetic anisotropy at $T = 0$ K [31]. It can be determined from the experimental dependence $\Delta\chi(T)$ as follows [32, 33]

$$\Delta\chi = \Delta\chi_0 S = \Delta\chi_0 \left(1 - \frac{T}{T_{\text{N-Iso}}}\right)^{\beta}, \tag{6}$$

where β is a numerical parameter.

The solid lines in Figure 4b stand for the fitted curves using (6). The values of the order parameter obtained using the Bogdanov method and dielectric spectroscopy [21] are in good agreement. The behavior of the temperature dependence of the order parameter is well described by the expression (6).

Conclusions

The combination of the well-known electro-optical method with the rotating magnetic field method provides extensive information on the rheological, magnetic and orientational properties of nematic liquid crystals. The temperature dependences of the rotational viscosity, the diamagnetic susceptibility anisotropy, and the orientation order parameter of NLC with polar terminal NCS group were obtained. The results obtained are in good agreement with the literature data. The acoustic method of Bogdanov may be of interest for setting up experiments with complex configurations of alternating magnetic and electric fields, as well as for studying the structural and relaxation features of liquid crystals under high pressure.

Acknowledgements

The reported study was funded by RFBR and BRFBR, projects No. 19-32-90032 and 20-57-04003.

References

[1] Galeeva A I, Selivanova N M and Galyametdinov Y G 2021 *Liquid Crystals and their Application* **21** 23–33 DOI: 10.18083/LCAppl.2021.1.23
[2] Reuter M, Vieweg N, Fischer B M, Mikulicz M, Koch M, Garbat K and Dabrowski R 2013 *APL Materials* **1** 012107 DOI: 10.1063/1.4808244
[3] Averyanov E M 2019 *Liquid Crystals and their Application* **19** 42–51 DOI: 10.18083/LCAppl.2019.1.42
[4] Arakawa Y, Kang S, Tsuji H, Watanabe J and Konishi G i 2016 *RSC advances* **6** 92845–92851 DOI: 10.1039/C6RA14093A
[5] Kamanina N V, Zubtcova Y A, Kuzhakov P V, Zak A and Tenne R 2020 *Liquid Crystals and their Application* **20** 41–48 DOI: 10.18083/LCAppl.2020.3.41
[6] Mrukiewicz M, Perkowski P, Garbat K, Dabrowski R and Parka J 2013 *Acta Physica Polonica A* **124**
[7] Osipova V V, Kurilov A D, Galyametdinov Y G, Muravsky A A, Kumar S and Chausov D N 2020 *Liquid Crystals and their Application* **20** 84–92 DOI: 10.18083/LCAppl.2020.4.84
[8] Parish A, Gauza S, Wu S T, Dziaduszek J and Dabrowski R 2008 *Molecular Crystals and Liquid Crystals* **489** 22–348 DOI: 10.1080/15421400802241316
[9] Guan X L, Zhang L Y, Zhang Z L, Shen Z, Chen X F, Fan X H and Zhou Q F 2009 *Tetrahedron* **65** 3728–3732 DOI: 10.1016/j.tet.2009.02.080
[10] Dabrowski R, Dziaduszek J, Ziółek A, Stolarz Z, Sasnouski G, Bezborodov V, Lapanik W, Gauza S and Wu S T 2007 *Opto-Electronics Review* **15** 47–51 DOI: 10.2478/s11772-006-0055-4
[11] Wen C H, Gauza S and Wu S T 2004 *Liquid crystals* **31** 1479–1485 DOI: 10.1080/02678290412331298076
[12] Shmeliova D V, Kharlamov S S, Semina O A, Bluzhin A V and Pasechnik S V 2020 *Liquid Crystals and their Application* **20** 49–56 DOI: 10.18083/LCAppl.2020.3.49
[13] Bogdanov D L, Gevorkyan E V and Lagunov A S 1980 *Acoustic Journal* **26** 28–34
[14] Wu S T and Wu C S 1990 *Physical Review A* **42** 2219 DOI: 10.1103/PhysRevA.42.2219
[15] Wu S T and Wu C S 1990 *Liquid Crystals* **8** 171–182 DOI: 10.1080/02678299008047339
[16] Bogoslovov R B, Roland C M, Czub J and Urban S 2008 *The Journal of Physical Chemistry B* **112** 16008–16011 DOI: 10.1021/jp805421g
[17] Pasechnik S V, Chigrinov V G and Shmeliova D V 2009 *Liquid crystals: viscous and elastic properties in theory and applications* (John Wiley & Sons)
[18] Belyaev V V, Solomatin A S and Chausov D N 2013 *Applied Optics* **52** 3012–3019 DOI: 10.1364/AO.52.003012
[19] Belyaev V V 2009 *Viscosity of nematic liquid crystals* (Cambridge International Science Pub.)
[20] Belyaev V V 2001 *Physics-Uspekhi* **44** 255 DOI: 10.1070/PU2001v044n03ABEH000831
[21] Kurilov A D and Volosnikova N I 2019 *Bulletin of the Moscow State Regional University. Series: Physics-Mathematics* 83–96 DOI: 10.18384/2310-7251-2019-1-83-96
[22] Tsvetkov V N 1939 *Journal of Experimental and Theoretical Physics* **9** 602–615
[23] Gasparoux H and Prost J 1971 *Journal de Physique* **32** 953–962 DOI: 10.1051/jphys:019710032011-12095300
[24] Meiboom S and Hewitt R C 1973 *Physical Review Letters* **30** 261 DOI: 10.1103/PhysRevLett.30.261
[25] Prost J and Gasparoux H 1971 *Physics Letters A* **36** 245–246 DOI: 10.1016/0375-9601(71)90443-9
[26] Kneppe H, Schneider F and Sharma N K 1982 *The Journal of Chemical Physics* **77** 3203–3208 DOI: 10.1063/1.444195
[27] Miyano K and Ketterson J B 1975 *Physical Review A* **12** 615 DOI: 10.1103/PhysRevA.12.615
[28] Herba H and Drzymała A 1990 *Liquid Crystals* **8** 819–823 DOI: 10.1080/02678299008047392

[29] Kemp K A and Letcher S V 1971 *Physical Review Letters* **27** 1634 DOI: 10.1103/PhysRevLett.27.1634
[30] Tsvetkov V and Sosnovsky A 1943 *Journal of Experimental and Theoretical Physics* **13** 353–360
[31] Buka A and De Jeu W H 1982 *Journal de Physique* **43** 361–367 DOI: 10.1051/jphys:01982004302036100
[32] Zywucki B J and Kuczynski W 2001 *IEEE Transactions on Dielectrics and Electrical Insulation* **8** 512–515 DOI: 10.1109/94.933375
[33] Simões M and Simeão D S 2006 *Physical Review E* **74** 051701 DOI: 10.1103/PhysRevE.74.051701

Optical rotation dispersion of cholesteric-nematic mixture

E N Vasilchikova[1], A D Dmitrieva[1], A V Kondakova[1], A D Kurilov[1], V V Usachev[2], A A Muravsky[3], D N Chausov[1,4]

[1]Moscow Region State University MRSU, Very Voloshinoy str., 24, Mytishi, 141014, Russia
[2]RUDN University, 6 Miklukho-Maklay St., Moscow, 117198, Russia
[3]Institute of Chemistry of New Materials NAS of Belarus, 36 F. Skoriny St., Minsk, 220141, Belarus
[4]Prokhorov General Physics Institute of the RAS, 119991, Russia, Moscow, 38 Vavilova St.

E-mail: d.chausov@yandex.ru

Abstract. The dispersion of the specific rotation constant of a cholesteric-nematic mixture of 5CB and Ch17 at a low concentration was studied by optical spectroscopy. The dispersion of the anisotropy of the refractive index as a function of the wavelength is studied, and the helical pitch of the cholesteric-nematic mixture is calculated.

1. Introduction

The widespread use of cholesteric liquid crystals mixed with nematic materials in optics and photonics is due to the wide orientation-structural optical properties, in particular, such systems were used as resonators for the creation of microlasers [1]. The use of a cholesteric-nematic mixture allows us to significantly expand the range of existence of the mesomorphic state to the most commonly used temperatures (from 20°C to 90°C) [2, 3]. In addition, by varying the chemical composition of LC mixtures, it becomes possible to fine-tune their physical and chemical properties, which makes it possible to improve the operational parameters of existing LC devices, as well as to find new applications for liquid crystals in science and technology [4,5]. For example, the use of mixtures of two-frequency liquid crystals can significantly speed up the response of LC devices [6–8].

Cholesteric mixtures are now used to create tunable diffraction optics [9, 10], the possibility of electrical adjustment of selective reflection is found [11–13]. The use of CLC together with photoactive polymers makes it possible to create both transmitting and reflecting holographic lattices [14]. Recently, it has been proposed to use a cholesteric-nematic mixture as an optical biosensor for the determination of amino acids [15].

In this paper, the contribution of a cholesteric additive to the dispersion of the anisotropy of the refractive index of nematic liquid crystals is investigated. The data on the optical properties of the cholesteric-nematic mixture are compared with the resulting helix pitch. As you know [16], the continuous change in the step of the cholesteric helix is carried out in limitless samples. In cells whose size is comparable to the step of the undisturbed structure of the cholesteric, the external fields induce a stepwise unwinding of the helix [17]. The influence of various factors on the helix pitch in CLC droplets has not been sufficiently studied, as a rule, the helix pitch, and,

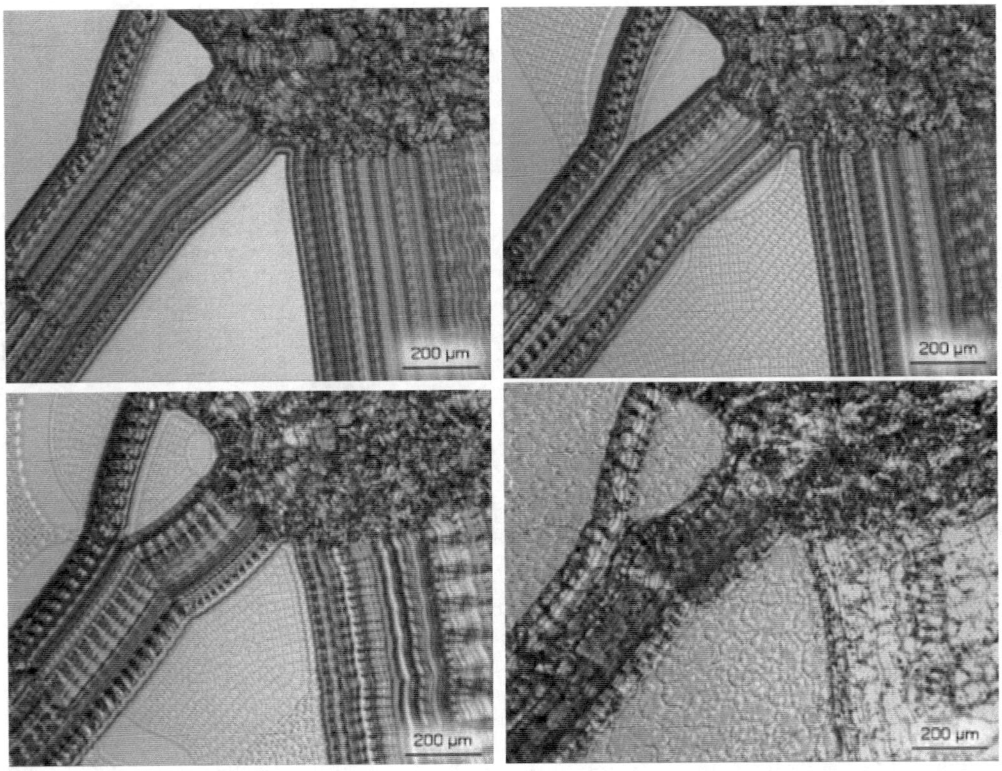

Figure 1. Phase cholesteric-nematic transitions. a – 25°C, b - formation of the "grid" texture 33°C, c - texture formation at 36°C, d - 37°C

as a result, the characteristics of the CLC material can be controlled by temperature [18]. The results obtained can be used to develop LC devices that operate in narrow ranges of the optical spectrum (for example, in the near IR), but have improved characteristics in comparison with classical solutions. Physical properties of the CLC in different Grandjean zones of wedge cells are described in [19, 20].

2. The object of the study and the method of the experiment

The object of the study is a cholesteric-nematic mixture of nematic (5CB) and cholesteric LC (Ch17) with a concentration of 7% mass. Cholesteric liquid crystals are mixed with nematics and have their own packaging of molecules, twisted relative to each other. The prepared mixture is filled with a cell with a thickness of d=100 microns and a planar orientation, after which it is uniformly heated to a temperature above 87°C (isotropic liquid). Next, this cell cools down to room temperature.

The non-oriented cholesteric liquid crystal has a confocal texture. A planar-oriented mixture under the influence of temperature, the cholesteric-nematic transition in the structure usually has three phases (fig. 1), which depend on the step of the helix. The grid (fig. 1b) is a two-dimensional spatial periodic deformation of a planar texture.

The study of the dispersion of the specific constant of rotation of the mixture can be carried out by the method of "grooved" spectrum using a monochromator. At the beginning of the experiment, the calibration curve of the monochromator is plotted using the known mercury spectrum.

A diagram of the experimental setup is shown in fig. 2. Cell 1 is placed between the crossed

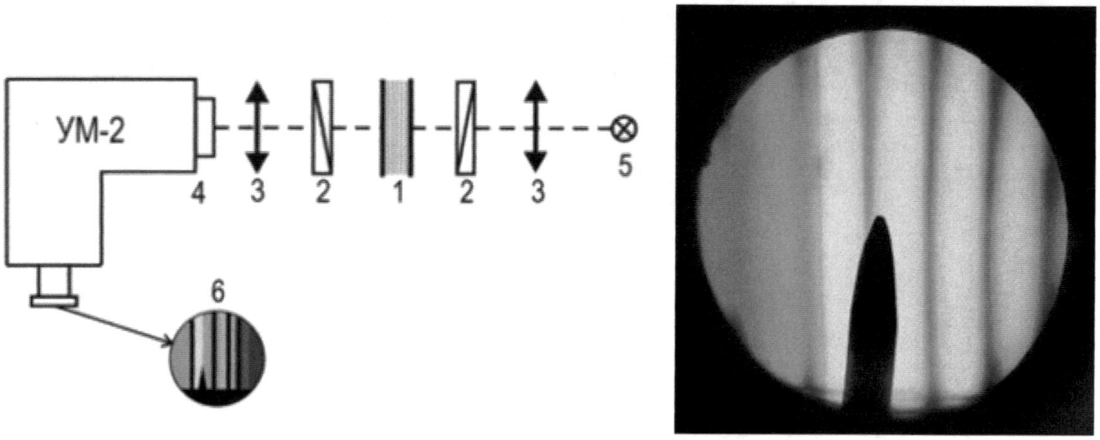

Figure 2. (a) Diagram of the experimental setup, (b) the fading lines in the blue part of the spectrum

polarizers 2 at an angle of 45° relative to the allowed directions of the polarizer. The lenses 3 focus the light on the input slit of the monochromator 4. The light source 6 is an incandescent lamp that emits white light. The radiation spectrum 6 is observed using the spectral nozzle of a monochromator. In the field of view, vertical black bands are clearly visible against the background of the spectrum (fig. 2b).

The spectrum has extinction bands for those wavelengths for which the sample is extinguished in crossed polaroids. The intensity of the light passing through the sample will be zero when the plane of light polarization rotates during the passage of the sample by an angle that is a multiple of π. That is, when moving from one fade band to another, the angle changes by the value of π.

Specific rotation of the mixture helix [11] is calculated according to the equation

$$\rho = -\frac{2\pi}{P} \cdot \frac{\delta^2}{8\lambda'^2\left(1 - \lambda'^2\right)}, \tag{1}$$

P – helix step, $\lambda' = \lambda/(\sqrt{\varepsilon}P)$, $\delta = (\varepsilon_2 - \varepsilon_1)/(\varepsilon_2 + \varepsilon_1)$.

If $\lambda' \ll 1$, i.e. $\lambda \ll nP$, the formula (1) is simplified:

$$\rho = -\frac{\pi P(\Delta n)^2}{4\lambda^2}. \tag{2}$$

Here Δn is the optical anisotropy of the layer, λ is the wavelength of the incident light, P is the pitch of the helix. The pitch of the P helix depends on the nature of the molecules and external influences.

The concentration of cholesteric is small, so the condition $\lambda < P$ is feasible, and the value of Δn inside the cholesteric layer can be replaced by Δn of the nematic component.

As can be seen from (2), the rotation angle of the polarization plane is inversely proportional to λ^2. Therefore, we can assume that the minimum for λ=668 nm corresponds to the rotation angle $\varphi = \pi + \varphi_0$, where φ_0 is some indefinite angle that will be constant for all other minima (fig. 3a), so it can be ignored

The deviation of the experimental curve from the straight line is due to the variance $\Delta n = \Delta n(\lambda)$ NLC (fig. 3b). To account for this variance, another planar cell filled with pure nematic 5CB was made. The cell thickness d_1=109 μm. This cell was used to obtain the

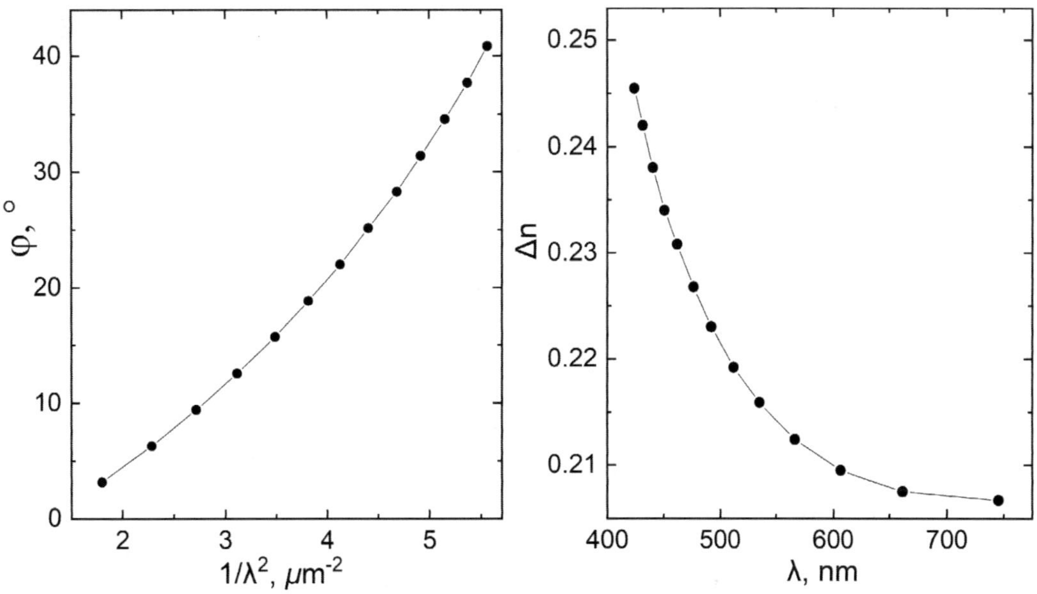

Figure 3. (a) The dependence of the angle of rotation of the plane of polarization of the NLC 5CB on $1/\lambda^2$, (b) the dispersion of the anisotropy of the refractive index as a function of the wavelength

dispersion curve using the transmission spectrum on the spectrophotometer. Based on the measurement results, the dispersion curves are plotted (fig. 3a). The values of Δn corresponding to the wavelengths of the extinction lines (fig. 3b).

Plot of the dependence of the rotation angle of the polarization plane on the value $(\Delta n/\lambda)^2$ is represented by fig. 4. As follows from the theory of (2), the graph can be approximated with a sufficient degree of accuracy by a straight line, the angular coefficient of which makes it possible to calculate the step of the cholesteric helix.

The step of the helix, this is the period at which the director makes a complete revolution around the cholesteric axis, respectively, can be calculated from

$$P = \frac{4}{\pi d} \cdot \frac{\Delta y}{\Delta x} = 2.07 \mu m$$

The equation of the approximation line $y = -12.844 + 16202.4x$ determined from the data on fig. 4.

Knowing the pitch of the helix, we can calculate (1) the value of the specific rotation of the helix of the cholesteric-nematic mixture

$$\lambda' = \lambda/\sqrt{\varepsilon}P = \lambda/3.2,$$

$$\sqrt{\varepsilon} = n, \quad \Delta\varepsilon/2\sqrt{\varepsilon} = \Delta n,$$

$$\delta = \Delta n/n \approx 0.125.$$

The error in measuring the pitch of the cholesteric helix is due, on the one hand, to the error in determining the optical anisotropy and the accuracy of determining the coordinates of the wavelengths of the extinction lines, which is associated with the visual method of reference. In addition, the error in determining the cell thickness is taken into account

$$\varepsilon_P = 2\varepsilon_{\Delta n} + 2\varepsilon_\lambda + 2\varepsilon_d \cong 0.07.$$

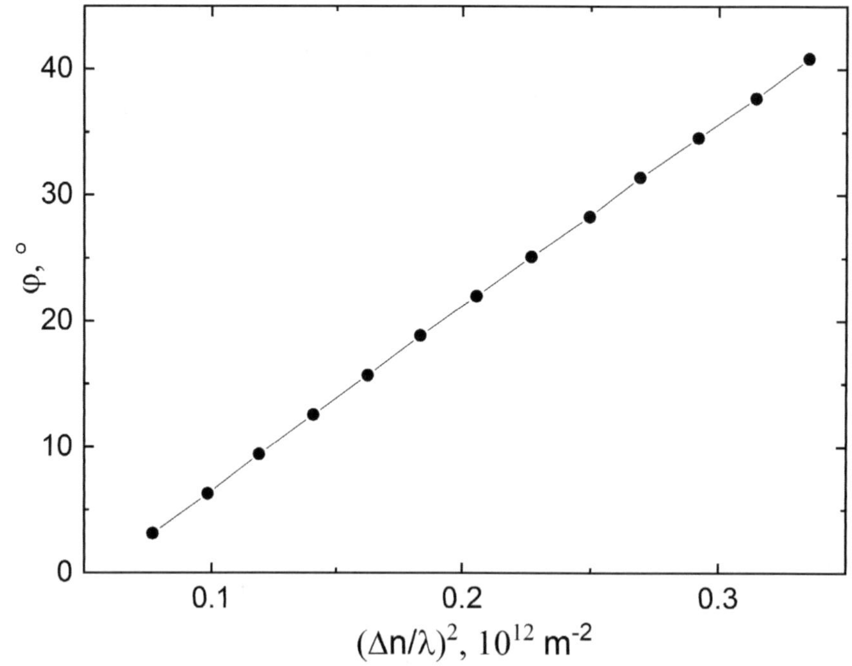

Figure 4. Dependences of the rotation angle of the polarization plane taking into account optical anisotropy

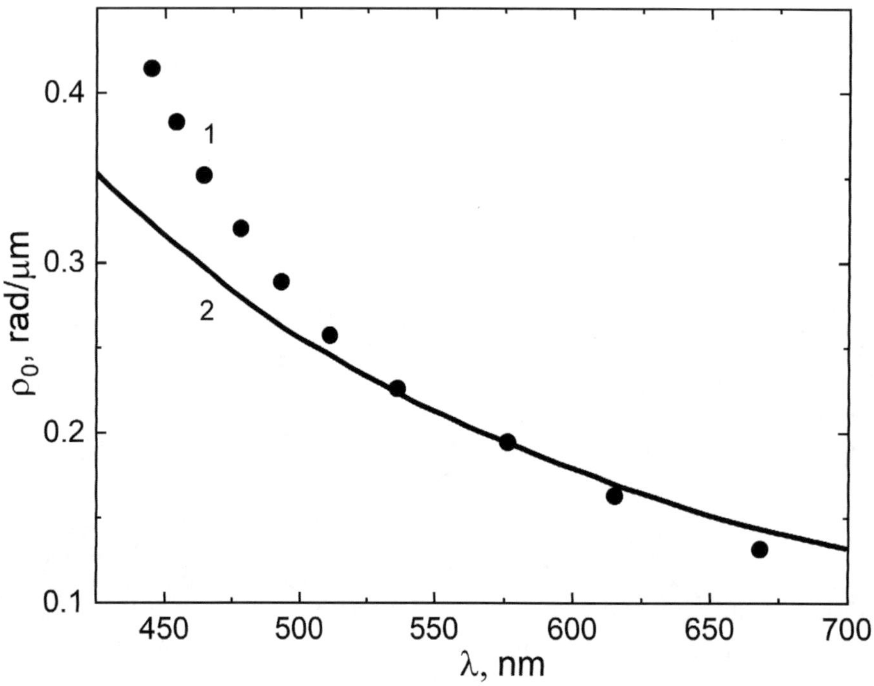

Figure 5. Dispersion of the specific rotation constant of a mixture of cholesteric-nematic mixture (1) experimental data, (2) theoretical curve

On fig. 5, the red curve is the theoretical dependence $\rho = \rho(\lambda)$, the black dots are the experimental values (P=2 microns). The discrepancy between the experimental data and the theory is probably due to the large variance of $\Delta n(\lambda)$ in the region of short wavelengths.

Conclusions

In the course of the work, the optical properties of the cholesteric-nematic mixture were investigated, various methods for determining the step of the cholesteric helix were described, and various factors on which this value depends were identified. The step of the cholesteric-nematic mixture was measured by the method of the grooved spectrum. From the experiment, the numerical parameters of the cholesteric-nematic mixture were determined, which are in good agreement with the theory.

Acknowledgements

This research was supported by RFBR, Research Grant No. 20-57-00012 Bel-a and Presidential Grant No. MD-2272.2021.1.6.

References

[1] Belyaev V V, Ostrovsky B I and Pikina E S 2018 *Liq. Cryst. and their Appl.* **18** 84–94 DOI: 10.18083/LCAppl.2018.1.84
[2] Wang J, Bergquist L, Hwang J I, Kim K J, Lee J H, Hegmann T and Jákli A 2018 *Liquid Crystals* **45** 333–340 DOI: 10.1080/02678292.2017.1369594
[3] Kushwaha J K, Malik P, Raina K K and Arora V P 2011 *Molecular Crystals and Liquid Crystals* **541** 236/[474]–242/[480] DOI: 10.1080/15421406.2011.569244
[4] Weil C, Müller S, Scheele P, Best P, Lüssem G and Jakoby R 2003 *Electronics Letters* **39**(24) 1732–1734(2) ISSN 0013-5194 DOI: 10.1049/el:20031150
[5] Park N H, Noh S C, Nayek P, Lee M H, Kim M S, Chien L C, Lee J H, Kim B K and Lee S H 2015 *Liquid Crystals* **42** 530–536 DOI: 10.1080/02678292.2015.1006698
[6] Xianyu H, Wu S T and Lin C L 2009 *Liquid Crystals* **36** 717–726 DOI: 10.1080/02678290902755598
[7] Sun J, Xianyu H, Gauza S and Wu S T 2009 *Liquid Crystals* **36** 1401–1408 DOI: 10.1080/02678290903242968
[8] Duan W, Chen P, Ge S J, Liang X and Hu W 2019 *Crystals* **9** ISSN 2073-4352 DOI: 10.3390/cryst9020111
[9] Ryabchun A and Bobrovsky A 2018 *Advanced Optical Materials* **6** 1800335 DOI: https://doi.org/10.1002/adom.201800335
[10] Ryabchun A, Yakovlev D, Bobrovsky A and Katsonis N 2019 *ACS Applied Materials & Interfaces* **11** 10895–10904 DOI: 10.1021/acsami.8b22465
[11] Simdyankin I V, Geivandov A R, Barnik M I, Palto V S and Palto S P 2019 *Liq. Cryst. and their Appl.* **19** 48–56 DOI: 10.18083/LCAppl.2019.2.48
[12] Mrukiewicz M, Iadlovska O S, Babakhanova G, Siemianowski S, Shiyanovskii S V and Lavrentovich O D 2019 *Liquid Crystals* **46** 1544–1550 DOI: 10.1080/02678292.2019.1587528
[13] Xiang J, Li Y, Li Q, Paterson D A, Storey J M D, Imrie C T and Lavrentovich O D 2015 *Advanced Materials* **27** 3014–3018 DOI: https://doi.org/10.1002/adma.201500340
[14] Ryabchun A, Sakhno O, Stumpe J and Bobrovsky A 2017 *Advanced Optical Materials* **5** 1700314 DOI: https://doi.org/10.1002/adom.201700314
[15] Vistak M, Mykytyuk Z, Vezyr F and Polishchuk V 2018 *Molecular Crystals and Liquid Crystals* **672** 67–72 DOI: 10.1080/15421406.2018.1542108
[16] Kuvshinov G V, Potemkina O V, Kuvshinova S A and Koifman O I 2017 *Liq. Cryst. and their Appl.* **17** 43–55 DOI: 10.18083/LCAppl.2017.1.43
[17] Pozhidaev E P, Barbashov V A, Kesaev V V, Pogonin V I, Samagin S A, Kotova S P, Torgova S I and Chigrinov V G 2017 *Liq. Cryst. and their Appl.* **17** 90–96 DOI: 10.18083/LCAppl.2017.4.90
[18] Kuvshinova S A, Gorshkova N M, Novikov I V and Litov K M 2017 *Liq. Cryst. and their Appl.* **17** 31–40 DOI: 10.18083/LCAppl.2017.4.31
[19] Chigrinov V G, Belyaev V V and Belyaev S V 1999 *JETP* **50** 994
[20] Belyaev S V, Rumyantsev V G and Belyaev V V 1977 *JETP* **46** 337

Special features of short-duration processes in condensed media

Tatyana A. Khantuleva and Victor M. Kats
St Petersburg University, 7/9 Universitetskaya Emb., St Petersburg 199034, Russia
E-mail: khan47@mail.ru

Abstract. The problem of the short-duration processes is considered on the base of the non-local theory of non-equilibrium transport, taking into account inertial effects. The system temporal evolution out of equilibrium connected to the dynamic structure transition described by the Speed-Gradient principle (SG-principle or SGP) developed in control theory and cybernetic physics. In the manuscript, we show that retardation of the system response to the short-duration loading due to inertial effects influences on the system evolution and can change its direction. The response to the shock loading of condensed matter is compared to quasi-stationary loading in a wide range of conditions. The short duration loading can lead the system into the structure unstable state and even give rise to self-organization of turbulent structures in the medium. The use of SGP for the modelling of such processes opens new possibilities to control them.

1. Introduction

Unlike long-duration processes, fast, highly non-equilibrium processes cannot be described by models of continuum mechanics. Therefore, the responses of the medium to shock and long-term loading, determined by different stages of relaxation, differ from each other [1]. At the first stage, elastic waves induced by an external action interfere and scatter on inhomogeneities of the real medium, forming mesoscopic wave packets [2]. At the second stage, a transition to plastic deformation occurs, which is accompanied by the self-organization of various turbulent effects. The last stage is a hydrodynamic flow. The initial state of the medium determines the stage at which relaxation begins, and the final state depends on the energy received by the system from the outside.

The mathematical model of the formation and propagation of elastic-plastic waves [3, 4], developed using the nonlocal theory of non-equilibrium transport processes [5], allowed us to explicitly demonstrate the effect of relaxation and inertia in a wide range of loading conditions for condensed media. The obtained relationships between stress and deformation, taking into account the duration and the strain-rate of loading, showed that relaxation and inertia play a very important role in fast processes. It has been found that in fast processes, the effects of self-organization of turbulent motions are possible, as a result of which the irreversible part of the mesoscopic structures can remain in the material after shock loading [3, 4, 6, 7, 8].

2. The integral model of the short-duration processes with retardation effects

A problem on the shock-induced elastic-plastic wave propagating in condensed matter has been solved by Meshcheryakov and Khantuleva [3, 4]. The conception of the shock-induced elastic-plastic transition based on the nonlocal theory of nonequilibrium transport processes [5] radically differs from the conventional one. Within the nonlocal theory based on nonequilibrium statistical mechanics [9] and cybernetical physics [10], in the case of the plane shock loading when the induced wave propagates along the x-axis, the integral relationship between longitudinal stress component $J(x,t)$ and the strain-rate $\dot{e} = -\partial v/\partial x$ (v is the mass velocity) takes a form

$$J(x,t) = -\rho_0 C^2 t_r \int\limits_0^{\omega(t)} \frac{dt'}{t_r} \int\limits_0^{\Omega(t)} \frac{dx'}{l_r} R \frac{\partial v}{\partial x'} \tag{1}$$

$$R = R(t,t';x,x';t_r,l_r)$$

$$\omega(t) = \begin{cases} t, & t < t_R \\ r_R, & t \geq r_R \end{cases}$$

$$\Omega(t) = \begin{cases} Ct, & Ct < L \\ \Gamma, & Ct \geq L \end{cases}$$

Here ρ_0 is initial unperturbed density, C is longitudinal sound velocity defined by the relationship $\rho_0 C^2 = K + 4G/3$, where K, G are volume compression and shear elastic moduli. t_R is typical loading time (force acts over a period t_R) not included in conventional deformation models, L is a typical distance from the impact surface (target thickness). The correlation function $R(t,t';x,x')$ describes the impulse relaxation, inertial and collective effects. R depends on the parameters: typical shear relaxation time t_r and length $l_r = Ct_r$, and on retardation time t_m. Relaxation of shear degrees of freedom considers being much faster than volume relaxation. For the short duration processes, we can assume the volume relaxation frozen. Unlike differential wave models, the integral relationship (1) determines the stress component $J(x,t)$ by the strain-rate history all over the wave period. Due to the inertial and relaxation effects in a dense medium, the wave duration can essentially exceed the loading time t_R. During the loading, impulse accumulates in the medium and relaxes after the loading.

In the reference connected to the elastic precursor running at the constant longitudinal sound velocity C, $\zeta = (t - x/C)/t_R$ $\xi = x/L$ an essential simplification of the non-local model for the impulse transport is gained on the condition $\tau\partial/\partial\zeta \gg \epsilon\partial/\partial\xi$ resulted from the experimentally tested evaluation $\epsilon/\tau = Ct_R/L \ll 1$. It must be noticed that the division of the typical parameters into fast and slow ones is the necessary condition for self-organization effects. Here $\tau = t_r/t_R$, $\theta = t_m/t_R$, $\epsilon = Ct_r/L$ are normalized parameters of relaxation, retardation and nonlocality. Due to the variables division we can consider the wave propagation quasi-stationary and depending on the distance x only by means of the parameters.

The momentum transport equation with the terms of the order $\epsilon/\tau \ll 1$ neglected results in an equation for the mass velocity in the wave with the model integral kernel depending on the parameters τ, θ.

$$v = \int\limits_0^{\omega} d\zeta' \exp\left\{-\frac{\pi(\zeta - \zeta' - \theta)^2}{\tau^2}\right\} \frac{\partial v}{\partial \zeta'} \tag{2}$$

Here the velocity v is normalized by the impact velocity V_0. The stress component $J(\zeta,t) = = \rho_0 C V_0 v = \rho_0 C^2 e$ corresponds to the elastic stress with the mass velocity satisfying to (2). However, unlike the elastic stress, v in equation (2) depends on the relaxation and delay effects. The equation (2) describes the waveform evolution accompanied by relaxation of the elastic precursor and retardation of the plastic front in terms of the parameters τ, θ depending on the distance x/L traveled by the wave. In the elastic limit $\tau \to \infty, \zeta < 1$ (2) becomes an identity.

Due to the parameters τ, θ the velocity $v(\zeta,\tau,\theta)$ in the left side of (2) differs from that under the integral in the right side $v(\zeta_0,\tau_0,\theta_0)$, where parameters τ_0,θ_0 correspond to an initial waveform. It means that stress and strain relate to different spatial points due to the nonlocal and delay effects. So, the integral operator in (2) plays a role of an evolution operator that describes the waveform evolution during its propagation. Substituting an acceleration induced by the loading under the integral in eq. (2), we can get the medium response at a distance from the loading surface.

3. The shock-induced and quasi-stationary loading of condensed matter

An approximate solution to (2) during the loading interval t_R at the constant acceleration $\partial v/\partial\zeta = 1$ has an explicit form [3, 4]

$$v(\zeta,\tau,\theta) = \frac{\tau}{2}\left(\operatorname{erf}\frac{\sqrt{\pi}(\zeta-\theta)}{\tau} + \operatorname{erf}\frac{\sqrt{\pi}\theta}{\tau}\right), \quad \zeta \le 1. \tag{3}$$

The solution (3) describes the waveform evolution during the loading.

At the initial stage when $t \le t_R \ll t_r$ the parameter $\tau = t_r/t_R \gg 1$ is large. Taking into account $\partial v/\partial\zeta = 1$, in the limit $\tau \to \infty$ we get the initial wave front $v(\zeta) = \zeta$ without relaxation. It means that at small times during the loading, the strain $e = (V_0/C)v(\zeta)$ and the stress component $J_1 = \rho_0 C V_0 v(\zeta)$ are linearly connected in the same spatiotemporal point as in the elastic case. The relaxation has not begun and considered frozen. The smaller duration of the shock, the closer is the medium response of solids to the elastic one. With large τ the medium response is always elastic.

During long loading $t_r \ll t \le t_R$ with small τ, the shear relaxation generates a viscous flow of the Newtonian fluid. In the limit $\tau \to 0$, the shear relaxation is entirely completed

$$J_1 = \rho_0 C V_0 v(\zeta) \xrightarrow[\tau\to 0]{} \rho_0 C V_0 \left(\frac{C_0^2}{C^2}v + \frac{C_G^2}{C^2}\tau\frac{\partial v}{\partial\zeta}\right) \to \rho_0 C V_0 \frac{C_0^2}{C^2} v$$

During the transition region between the two limits the shear elastic module is changing due to the shear relaxation. In the general case when the loading is more intensive, both modules can change due to beginning of the volume relaxation.

After the short-duration loading $t_R \le t \ll t_r$ the solution to the equation (2) is different

$$v(\zeta,\tau,\theta) = \frac{\tau}{2}\left(\operatorname{erf}\frac{\sqrt{\pi}(\zeta-\theta)}{\tau} + \operatorname{erf}\frac{\sqrt{\pi}(1-\zeta+\theta)}{\tau}\right), \quad \zeta > 1. \tag{4}$$

It describes the shear relaxation as post-shock effect that accompanies the wave propagation. The post-shock effect can be considered as a non-monotone relaxation or as an inertial effect of a dense medium. In the limit $\tau \to \infty$, the waveform and the stress preserve their amplitude.

$$J_1 = \rho_0 C V_0 v(\zeta) \xrightarrow[\tau\to 0]{} \rho_0 C V_0 \left(\frac{C_0^2}{C^2} + \frac{C_G^2}{C^2}\right) \to \rho_0 C V_0$$

Without the memory effects after the loading the shear stress disappears

$$J_1 = \rho_0 C V_0 v(\zeta) \xrightarrow[\tau\to 0]{} \rho_0 C V_0 \left(\frac{C_0^2}{C^2} + \frac{C_G^2}{C^2}\tau\right) \to \rho_0 C V_0 \frac{C_0^2}{C^2} = \rho_0 C_0^2 e.$$

Let us consider as an example the behavior of the stress in a wave induced in a condensed medium by loading and subsequent unloading under various conditions

$$\frac{J_1}{\rho_0 C V_0} = \frac{C_0^2}{C^2}v(\zeta) + \frac{C_G^2}{C^2}\Pi(\zeta;\tau,\theta).$$

Here we take into account only the effects of shear relaxation and assume that the averaged loading force causes a constant acceleration $\partial v/\partial\zeta = 1$, and during unloading, respectively, acceleration in the opposite direction $\partial v/\partial\zeta = -1$.

$$\Pi_{g1}(\zeta,\tau,\theta) = \frac{\tau}{2}\left(\operatorname{erf}\frac{\sqrt{\pi}(\zeta-\theta)}{\tau} + \operatorname{erf}\frac{\sqrt{\pi}\theta}{\tau}\right),\ \zeta \le 1.$$

$$\Pi_{g2}(\zeta,\tau,\theta) = \frac{\tau}{2}\left(\operatorname{erf}\frac{\sqrt{\pi}(\zeta-\theta)}{\tau} + \operatorname{erf}\frac{\sqrt{\pi}(1-\zeta+\theta)}{\tau}\right) -$$
$$-\frac{\tau}{2}\left(\operatorname{erf}\frac{\sqrt{\pi}(\zeta-1-\theta)}{\tau} + \operatorname{erf}\frac{\sqrt{\pi}\theta}{\tau}\right),\ 1 < \zeta \le 2.$$

$$\Pi_{g3}(\zeta,\tau,\theta) = \frac{\tau}{2}\left(\operatorname{erf}\frac{\sqrt{\pi}(\zeta-\theta)}{\tau} + \operatorname{erf}\frac{\sqrt{\pi}(1-\zeta+\theta)}{\tau}\right) -$$
$$-\frac{\tau}{2}\left(\operatorname{erf}\frac{\sqrt{\pi}(\zeta-1-\theta)}{\tau} + \operatorname{erf}\frac{\sqrt{\pi}(2-\zeta+\theta)}{\tau}\right),\ \zeta > 2.$$

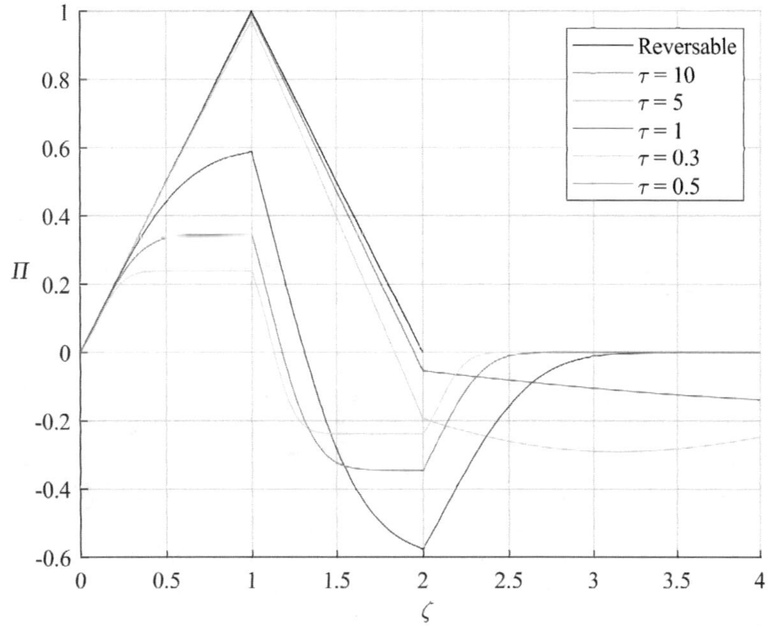

Figure 1. Behavior of the shear stress during relaxation at the retardation parameter $\theta = 0.1$

It can be seen from the Figures 1, 2 that with an increase in the loading duration, which corresponds to a decrease in the parameters τ, plastic effects arise, which eventually turn into hydrodynamic flows. In this case, with an increase in the delay θ, which can be caused by the inertia of the condensed medium, the maximum stresses are reached already during unloading, after which areas of significant negative stresses are formed. We see that the model of elastic solid is valid at large τ and low θ whereas the Newtonian model of liquid is valid at low values

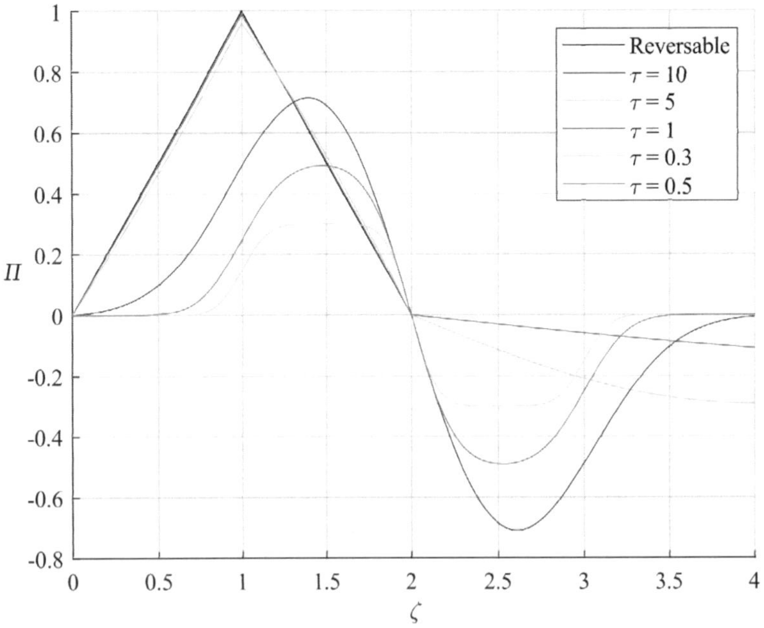

Figure 2. Behavior of the shear stress during relaxation at the retardation parameter $\theta = 1$

of both τ and θ. The most deviation of the system response to loading is observed in the intermediate region $\tau \sim \theta \sim 1$ with inertial effects increasing. This is highly non-equilibrium region where the models of continuum mechanics become unsuitable [8, 11, 12].

Just the same value of the stress $J_1 = \rho_0 C V_0$ can be achieved by two different ways. One way is stationary long-duration loading at low strain-rate, the other is the shock loading when during the very short-duration loading, the medium acquires a large impulse inducing an elastic-plastic wave. The retardation effects play an important role in the elastic-plastic waveform.

During the time interval θ, the shear relaxation is already completed, the wave amplitude reaches its maximal value and remains constant on the waveform plateau as far as the volume relaxation is frozen. The waveform amplitude is defined by the shock velocity but not by the plastic medium flow. Unlike the conventional approaches to the shock-induced processes, the plastic front is forming by the shear relaxation after the force of the shock proportional to the medium acceleration (or strain-rate) stops acting [3, 4]. So, the plastic front should be considered as the relaxation front resulted from the retarding response of condensed matter to the high-rate loading. First, this point of view has been presented in the paper [13]. As it was shown in [3, 4], the shock-induced waveforms given by the solutions to the integral equation (2) adequately describe all experimentally obtained dependencies beyond the concepts of continuum mechanics.

Figure 3 shows the fundamental difference between quasistatic and shock loading. According to the solution (3)-(4), during the shock loading, the elastic precursor reaches its the maximal amplitude at the time $t = t_R$ and after that, the shear relaxation forms the plastic front. During the slow continuous loading at the small constant strain-rate no two-wave front forms, one front arises and beyond the elastic limit becomes plastic. Comparison of the two type loading at the same maximal amplitude shows that the strain-rate during the continuous loading should be approximately 14 times lower than during the shock.

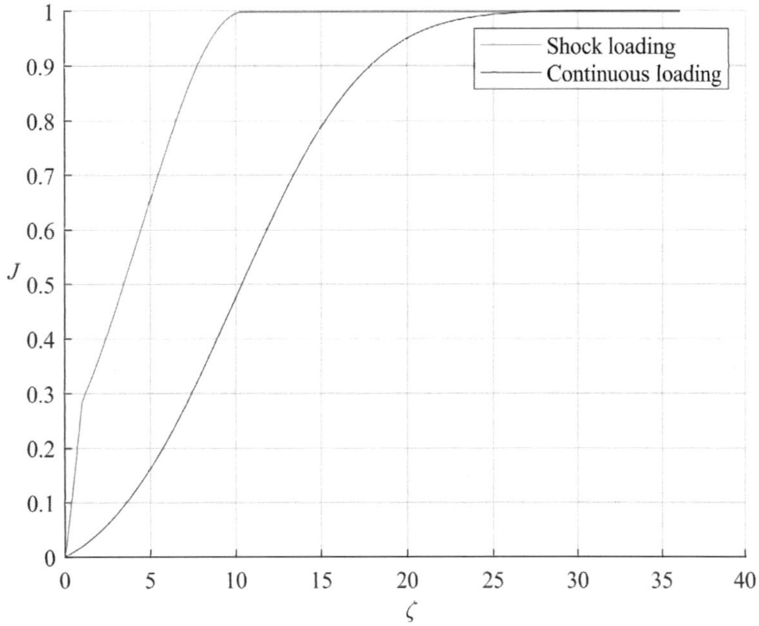

Figure 3. Stress growth under shock and continuous loading at $\tau = 15$, $\theta = 10$ and maximal amplitude

4. Temporal evolution of the traveling waveform via SG-principle

High-rate processes out of equilibrium are followed by multi-scale and multi-stage effects that make the system response nonlinear, probabilistic and introduce a feedback between the system structure and its response to an external loading [3, 5, 7, 8]. The system becomes unstable and evolves in the direction of the more equilibrium states under the imposed conditions [14]. The temporal evolution of the system is governed by an internal control with thermodynamically determined goal function and internal control parameters. Maximum entropy principle by Jaynes [14] defines the goal function indicating the direction of the system evolution. Speed gradient principle developed in control theory designed for the control problems of continuous-time systems [10] can be applied to construct an algorithm describing the fastest way to reach the goal.

Unlike conventional continuum mechanics, far from equilibrium, as it was derived in non-equilibrium statistical mechanics, only total entropy generation [15] in the system meets the second law of thermodynamics [9]. Therefore, we choose the total entropy generation as a control goal function that reaches its maximal value in a steady state under the imposed conditions. In the paper, we describe the temporal evolution of the condensed medium response to various loading in a wide range conditions.

The local entropy production σ, the total entropy generation S and the rate of the entropy generation $dS/d\xi$ (integral entropy production) are defined as follows [16]

$$\sigma(\zeta,\tau,\theta) = J_1\frac{\partial v}{\partial \zeta},\ S(\xi) = \int_0^{\xi} d\xi \int_0^{p} d\zeta\sigma(\zeta,\tau,\theta),\ \Sigma = \frac{dS}{d\xi}\left(\tau(\xi),\theta(\xi)\right) = \int_0^{p} d\xi J_1\frac{\partial v}{\partial \zeta}, \tag{5}$$

where p is duration of a loading process.According to the SG-principle, the integral entropy production tends to its minimum [1, 16].

$$\Sigma(\xi)=\int_0^p d\zeta\frac{\partial v}{\partial\zeta}\int_0^\omega d\zeta'\exp\left\{-\frac{\pi(\zeta-\zeta'-\theta(\xi))^2}{\tau^2(\xi)}\right\}\frac{\partial v}{\partial\zeta'} \tag{6}$$

For the processes presented in Figure 1 and Figure 2 the integral entropy production takes a form

$$\Sigma(\tau(\xi),\theta(\xi))=\int_0^1 d\zeta\Pi_1(\zeta;\tau(\xi),\theta(\xi))-\int_1^2 d\zeta\Pi_2(\zeta;\tau(\xi),\theta(\xi))$$

The surface Σ constructed over the plane of the parameters $\tau(\xi),\theta(\xi)$ is presented in Figure 4.

In Figure 4 we can see an elevation and a recess on the surface of the integral entropy production. The elevation corresponds to the zone of dissipation located close to local equilibrium. In the recess zone the values of the integral entropy production become negative. This part of the surface corresponds to the case in Figure 1 — 2 when the positive stress values retain during unloading due to the inertial effects. In non-equilibrium statistical mechanics, it was shown that such case is possible far from equilibrium [9].

In high-rate processes induced by short-duration loading when the wave propagation is followed by self-organization of new internal structures, the generalized integral entropy production becomes negative due to the energy consumption for the formation of internal structures [6, 7]. For example, in experimental research on shock loading of solid materials growth of the mass velocity dispersion indicating to the turbulent velocity pulsations were observed inside the waveform in real-time. After the wave propagation, an irreversible part of the pulsations was found out in a material in the form of mesoscopic rotational structures. As it was revealed the turbulent structures may change macroscopic properties of the material, including its strength.

According to the SG-principle [10] the slow temporal evolution of the control parameters τ, σ, connected to the relaxation and retardation effects is a gradient descent on the surface. The finite form of the SG algorithm in this case coincides to the gradient descent algorithm with the gain parameter $g>0$.

$$\frac{d\tau}{d\xi}=-g_\tau\frac{\partial\Sigma}{\partial\tau},\ \frac{d\theta}{d\xi}=-g_\theta\frac{\partial\Sigma}{\partial\theta}. \tag{7}$$

Coefficients $g_\tau>0$, $g_\theta>0$ are empiric constants characterizing inertial properties of the system internal structure.

Depending on the initial state of the medium structure, the descent path along the surface can lead to quite different final states. The descent from the hill to the origin means the full system relaxation to equilibrium. The descent in the opposite direction can lead to the recess and irreversible self-organization of shear and rotational structures in the medium. Otherwise, the medium response becomes reversible elastic waves. On the top of the hill there is a meta-stable point. Even a slight influence can direct the system evolution to the desired final state.

The control parameters during the descent change, the stress due to the feedback changes too. Through the entropy production rate in the right parts of the (7) when the surface evolves, a feedback between the medium response and turbulent structure evolution is introduced. The formulation of the control problem to describe the system evolution becomes closed and self-consistent. The control close-loops are known to make systems more stable than rigid program control.

The constructed integral mathematical model indicates that only with a short loading regime such structurally unstable states of the system are achievable. The medium properties can

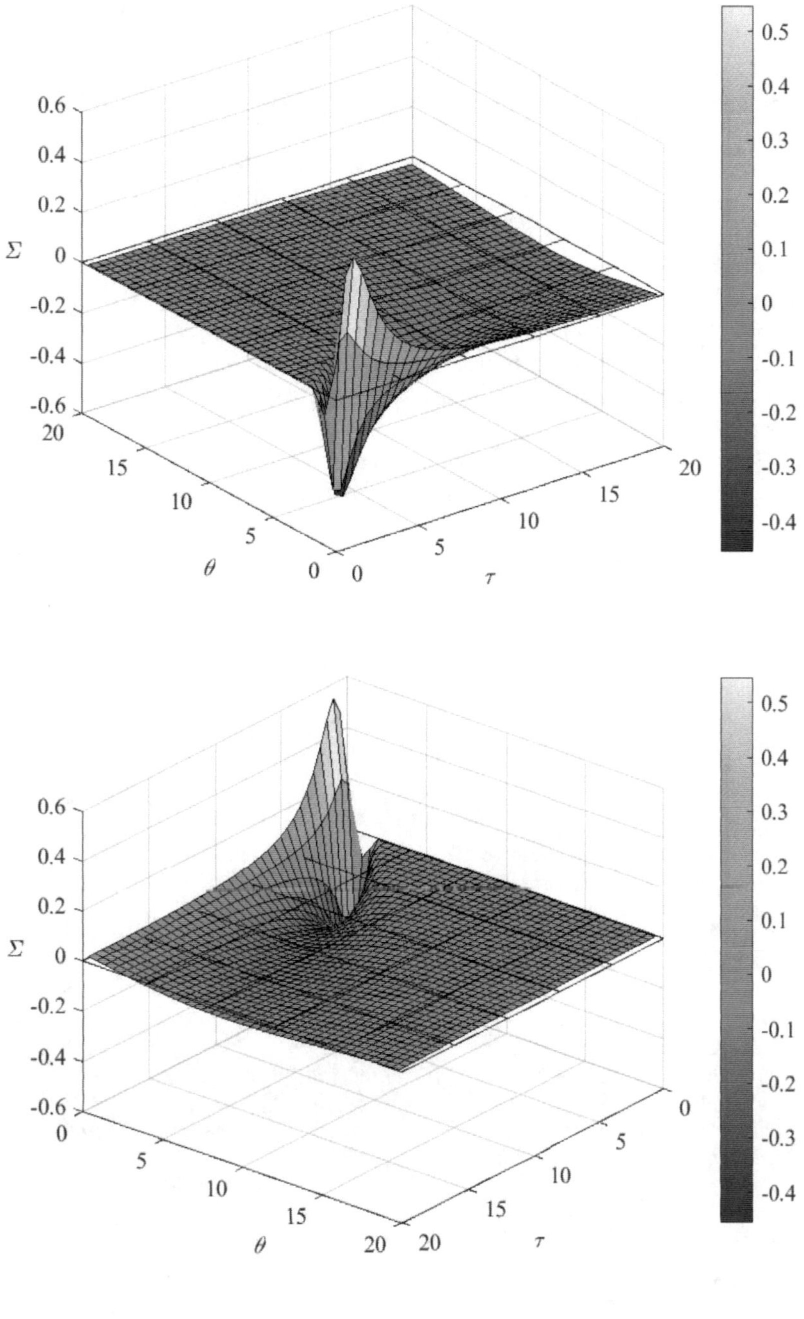

Figure 4. The surface $\Sigma(\tau(\xi), \theta(\xi))$ at $p = 1$

change because of the internal feedback between the stress in the medium and its deformation, which, in turn, can lead to changes of the deformation mechanisms and the medium internal structures. The controlled external influence at the system in the structure unstable state can be used to construct materials with given properties.

5. Conclusion

The proposed approach to modeling high-strain-rate processes based on the nonlocal theory of non-equilibrium transport [5] and cybernetic physics [10] allows an adequate description in a wide range of the loading conditions including transitions between different modes of deformation. Unlike conventional models within continuous mechanics, the integral model relationships between stress, strain and strain-rate depend on the loading conditions and take into account aftereffects arising due to inertial effects of condensed matter. The integral models do not require the separation of stress and strain components into elastic and plastic parts a priori.

Methods of cybernetic physics describe the time evolution of non-equilibrium systems under imposed conditions. The comparison of the shock-induced waveform evolution during its propagation along condensed matter in the framework of SG-principle with the results obtained in experimental study of shock loading materials [3, 4, 7] for different target widths shows that all experimental points on the plane of the control parameters lay down the theoretical straight lines derived for the integral model [17].

The evolution paths allow us to predict the macroscopic properties of the system after loading in the future. For the short-duration loading, such structure unstable states are available in which even a small influence can lead to a radical change of the system properties and behavior. So, the use of the SG-principle for modeling the short-duration processes opens new possibilities to control them.

References

[1] Khantuleva T A 2018 *AIP Conf. Proc.* **1959** 100003-1
[2] Khantuleva T A and Kats V M 2020 *Particles* **3** 562
[3] Mescheryakov Yu I and Khantuleva T A 2015 *Phys. Mesomech.* **18** 228
[4] Mescheryakov Yu I and Khantuleva T A 2015 *Materials Physics and Mechanics* **22** 136
[5] Khantuleva T A 2013 *Nonlocal Theory of Nonequilibrium Transport Processes* (StPetersburg: St-Petersburg University Publishing)
[6] Mescheryakov Yu I and Khantuleva T A 2016 *Phys. Mesomech.* **19(1)** 69
[7] Meshcheryakov Yu I, Divakov A K, Zhigacheva N I, Makarevich I P and Barakhtin B K 2008 *Phys. Rev.* B **78** 64301
[8] Khantuleva T A 2003 *High Pressure Shock Compression of Solids VI. Old Paradigms and New Challenges* vol VI, ed Y Horie, L Davison et al (New York: Springer-Verlag) pp 215–254
[9] Zubarev D N 1974 *Non-Equilibrium Statistical Thermodynamics* (New York: Springer-Verlag)
[10] Fradkov A L 2007 *Cybernetical Physics. From Control of haos to Quantum Control* (Berlin Heidelberg: Springer-Verlag)
[11] Mescheryakov Yu I and Khantuleva T A 2017 *Phys. Mesomech.* **20** 417
[12] Khantuleva T and Shalymov D 2020 *J. Phys.: Conf. Series* **1560** 012057
[13] Gilman J J 2003 *High Pressure Shock Compression of Solids VI. Old Paradigms and New Challenges* vol VI, ed Y Horie, L Davison et al (New York: Springer-Verlag) pp 279–296
[14] Jaynes E 1957 *Phys. Rev.* **106** 620
[15] Lucia U 2013 *Chem. Phys. Lett.* **583** 209
[16] Khantuleva T A and Shalymov D S 2017 *Phil. Trans. R. Soc.* A **375** 20160220
[17] Fradkov A L and Khantuleva T A 2016 *Procedia Structural Integrity* **2** 994

Resistance investigation of diamond-like carbon coatings to cyclic temperature changes

M O Makeev[1], A S Osipkov[1], V I Batshev[1,3], P A Mikhalev[1], B A Parshin[1], C V Suasnavas[2], A B Kozlov[3] and A S Machikhin[3]

[1] Bauman Moscow State Technical University, Moscow, Russia
[2] Peoples' Friendship University of Russia (RUDN University), Moscow, Russia
[3] Scientific and Technological Center of Unique Instrumentation, Russian Academy of Science, Moscow, Russia

E-mail: m.makeev@bmstu.ru

Abstract. Optical elements used in outer space must be designed considering the effects of such factors as space vacuum, atomic oxygen in low Earth orbit, solar and space radiation, large temperature drops, gas release of spacecraft materials and structural elements, space dust and debris. In order to harden and protect mirror surfaces of optical elements from external factors, it has been promisingly applied diamond-like carbon coatings on their surface. These coatings are characterized by high strength and wear-resistant properties, in particular, high hardness, low friction coefficient, high wear resistance and chemical inertness. This leads to their widespread use in various fields of science and technology, including optical instrumentation. This paper presents the results of testing an aluminum mirror with a diamond-like carbon coating under the effect of cyclic temperature changes for determining their ability to withstand a rapid cyclic ambient temperature change, and specifically, to maintain optical and mechanical properties.

Keywords: diamond-like carbon coating, temperature cycling, optical properties, mechanical properties, hardness, reflection coefficient, roughness, shape deviation.

1. Introduction

Optical elements used in outer space must be designed to withstand such factors as space vacuum, atomic oxygen in low Earth orbit, solar and space radiation, large temperature drops, gas release of spacecraft materials and structural elements, space dust and debris [1].

Thermal cycling is one of the main factors affecting the performance of the spacecraft optical elements. The temperature of optical surfaces in space, depending on the field of application, can vary from -150 °C (and even lower in devices for deep space exploration) to several hundred degrees. These temperature fluctuations can cause a shift in spectral lines at lower temperatures, which can be compensated by installing heaters, and also can cause a change in the geometric dimensions of the deposited layers, which will lead to a change in the refractive index. In addition, the difference in the coefficients of thermal expansion of the deposited layers materials on other layers will produce an increase in internal stresses, causing the destruction of the applied coating. In this regard, the choice of deposition technological modes or, if necessary, the application of intermediate subcoatings play a decisive role in ensuring the coating durability [1].

A promising option for protecting reflective mirror surfaces operating in the visible and infrared ranges is the deposition of diamond-like carbon (DLC) coatings on their surface. This type of nanostructured coatings, consisting of carbon atoms with both diamond (sp^3) and graphite-like (sp^2) bonds, is characterized by high strength and wear-resistant properties, in particular, high hardness from 5000 to 10000 Hv, low friction coefficient from 0.15 up to 0.08, high wear resistance and chemical inertness. This leads to their wide application in various fields of science and technology, including optical instrumentation [2-5]. At the same time, the refractive index of diamond-like carbon coatings is in the range from 1.55 to 2.20 [6-8], depending on the manufacturing technology, and the absorption coefficient is close to zero in a wide range of IR wavelengths [9-11]. Such unique properties make it possible to improve the reliability indicators (durability and trouble-free operation) of optical elements while maintaining the performance indicators (transmission or reflection coefficients).

This paper presents the results of testing mirrors with a diamond-like carbon coating under the effect of cyclic temperature changes for determining their ability to withstand ambient temperature changes, and specifically, to maintain optical and mechanical properties.

2. Methods

The procedure for testing mirrors with a diamond-like carbon coating under the effect of cyclic temperature changes was developed on the basis of a series of the Russian Federation State Standards (GOST R) [12-14]. The tests were carried out in 3 stages:

- preliminary measurements;
- impact of temperature cycling;
- final measurements.

Preliminary and final measurements included:

- measurements of light reflectance;
- measurements of surface roughness;
- measurements of shape deviations;
- measurements of hardness and Young's modulus.

The test procedure under the effect of temperature cycling included the following main stages:

1. The test sample, which has the same temperature as the laboratory air, is immersed in a cold bath with a liquid temperature of -196 °C.
2. The sample is kept in the cold bath for 18 minutes.
3. Then, the sample is removed from the cold bath, kept in normal atmospheric conditions for 2-10 minutes, and transferred to the heat chamber.
4. The heat chamber must be kept at a high temperature of 100 °C for 2 minutes.
5. During the next cycle, the sample is kept in normal atmospheric conditions for 2-10 minutes and transferred to the cold chamber.
6. One cycle includes two dwell times and two transfer times.
7. The sample is exposed to at least three cycles.
8. At the end of the last cycle, the sample is kept under normal atmospheric test conditions for the time necessary to achieve thermal equilibrium.

An experimental diagnostic stand (MKF-3M) was used as a heat chamber to study the degradation of materials under the influence of the outer space factors. The holding chamber at a low temperature was a Dewar vessel with liquid nitrogen and a boiling point of -196 °C. The test samples were always immersed in liquid.

The testing objects were samples of flat mirrors made of silicon carbide SiC with an aluminum coating deposited on them and a protective diamond-like carbon coating of 50 nm thickness (figure 1). The aluminum layer was deposited 200 nm by magnetron sputtering. The diamond-like carbon film was obtained by laser-triggered cathodic arc deposition and plasma separation [15].

Figure 1. Test sample.

The test sample was subjected to temperature cycling in accordance with the developed methodology. Three cycles were carried out.

The procedure described in GOST R 54164-2010 was taken as the reference for assessing the reflection coefficient. The measurements of the spectral reflectance before and after the tests were performed in the wavelength range from 780 to 1700 nm with a step of 5 nm on a PHOTON RT spectrophotometer.

The procedure for measuring hardness and Young's modulus is based on GOST R 8.748-2011 (ISO 14577-1: 2002) "Metals and alloys. Measurement of hardness and other characteristics of materials under instrumental indentation. Part 1. Test method". The procedure consists in introducing a geometrically and physically certified pyramid (Berkovich pyramid with an apex angle of 65.3° and a radius of curvature of 200 nm) into the material and determining with high accuracy the load-deformation relationship (indentation depth). The hardness before and after testing was measured using a Nanovea nanohardness tester according to the Oliver-Pharr method [16].

The roughness was measured in accordance with [17] by the method of coherent correlation interferometry using a TALYSURF CCI 6000 profilometer, which belongs to the type of non-contact optical interference devices. In this case, 10 sectors of the mirror surface with an area of 0.35 × 0.35 mm were selected for measurements.

The measurement of shape deviations of the samples was carried out using a Fizeau interferometer [18], designed to control the shape parameters of flat optical surfaces. The controlled parameter was the standard deviation of the mirror shape from the plane.

3. Results

It was determined by means of Raman spectroscopy that the diamond-like carbon coating of studied sample has sp^3 fraction equal 8% (figure 2).

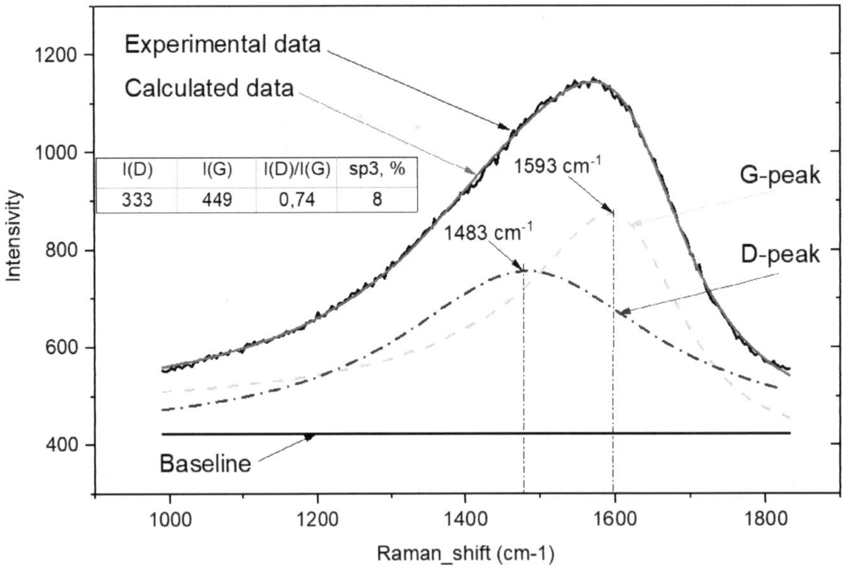

Figure 2. Study results of the DLC coating by the Raman method.

The measurement results of the reflection coefficients before and after thermal tests are shown in figure 3.

The averaged values of the spectral reflection coefficient of the sample ρ (λ) in the wavelength range from 780 to 1700 nm, as well as the coating hardness, roughness and shape deviation before and after testing are given in table 1.

The differences obtained in the investigated parameters of the tested mirror are small and commensurate with the instrumental and measurement methods errors.

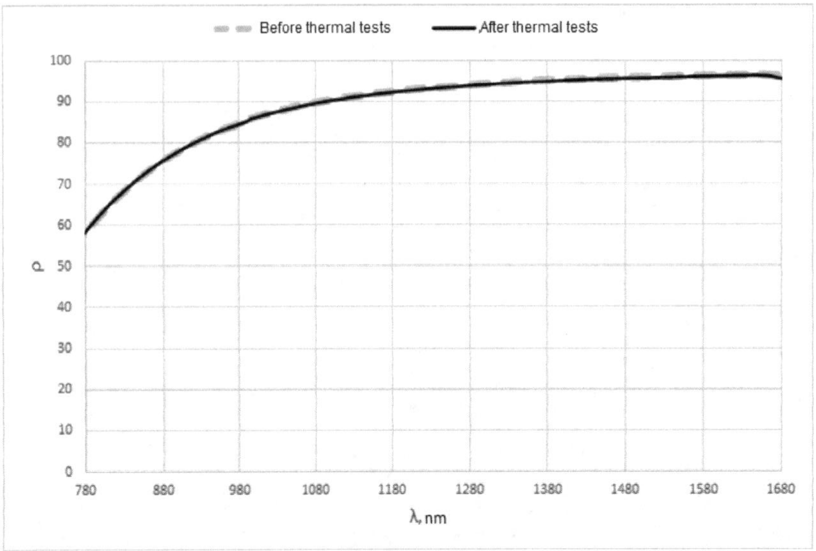

Figure 3. Light reflectance of the sample before and after thermal tests.

Table 1. Investigated parameters before and after testing.

Parameter	Before test value	After test value
Integral reflectance, %	89.21	89.19
Hardness, GPa	2.39	2.41
Elastic modulus, GPa	157	181
Roughness Ra, nm	2.12	1.79
RMS flatness deviation, nm	26.08	25.78

4. Conclusions

The tests were carried out on mirrors with a diamond-like carbon coating under the effect of temperature cycling (3 cycles) in the range from -196 to 100 °C. It was also measured the optical (reflection coefficient in the wavelength range from 380 to 1700 nm), mechanical (hardness and elastic modulus) and geometric (roughness and shape deviation) characteristics of the mirror with a diamond-like carbon coating. The results of the investigation allow to conclude that mirrors with a diamond-like carbon coating are suitable to withstand a rapid cyclic change of ambient temperature and to maintain optical and mechanical properties.

Acknowledgments

This work was supported by the Ministry of Science and High Education of the Russian Federation (state task No. 0705-2020-0032).

References

[1] Wernham D and Piegari A 2018 Optical coatings in the space environment *Optical Thin Films and Coatings* ed A Piegari and F Flory (Cambridge: Woodhead Publishing) chapter 22 pp 789–811

[2] Bewilogua K and Hofmann D 2014 *Surf. Coat. Technol.* **242** pp 214–225

[3] Macleod H 2010 *Thin-Film Optical Filters* (Boca Raton, Florida: CRC Press) p 782

[4] Osipkov A, Bashkov V, Belyaeva A, Stepanov R, Mironov Y and Galinovsky A 2015 Surface hardening of optic materials by deposition of diamond like carbon coatings from separated plasma of arc discharge *Proc. IOP Conf. Ser.: Materials Science and Engineering (St. Asaph)* vol 74 p 012013

[5] Batshev V, Kozlov A, Machikhin A, Makeev M, Osipkov A, Bulatov M, Kinzhagulov I and Stepanova K 2019 *Optics and Spectroscopy* **127** pp 634–638

[6] Suasnavas C, Makeev M, Osipkov A, Solano N, Shupenev A and Mikhalev P 2019 Relationship between hardness and optical properties of diamond-like carbon coatings *Proc. IOP Conf. Ser.: Materials Science and Engineering (Moscow)* vol 675 p 012051

[7] Mironov Y, Stepanov R, Osipkov A, Mironova A, Makeev M, Mikhalev P and Sedih N 2015 Optical and mechanical properties of diamond-like carbon coatings deposited by filtered cathodic vacuum arc deposition *Proc. 5th Int. Workshop on Computer Science and Engineering: Information Processing and Control Engineering (Moscow)* pp 295–300

[8] Klyui N, Litovchenko V, Lukyanov A, Neselevska L, Osovskiy V, Yaroschuk O and Dolgov L 2006 *Ukr. Phys. J.* **51** pp 710–714

[9] Sizov F, Klyuj N, Lukyanov A, Savkina R, Smirnov A and Evmenova A 2008 *Pis'ma v zhurnal tekhnicheskoj fiziki* **34** pp 32–40

[10] Robertson J 2002 *Materials Science and Engineering* (Amsterdam: Elsevier Science B.V.) pp 129–281
[11] Makeev M, Zhukova E, Mikhalev P, Osipkov A and Mironov Y 2015 Physical chemical and protective properties of the diamond-like carbon coatings synthesized from separated plasma of electric arc *Proc. 5th Int. Workshop on Computer Science and Engineering: Information Processing and Control Engineering (Moscow)* pp 255–259
[12] IEC 2006 GOST 28209-89 *Basic environmental testing procedures part 2: tests. Test N: change of temperature* (Moscow: Standartinform)
[13] IEC 2006 GOST 28209-89 *Basic environmental testing procedures. Part 2: Tests. Test B. Dry heat* (Moscow: Standartinform)
[14] IEC 2006 GOST 28209-89 *Basic environmental testing procedures. Part 2: Tests. Test A: Cold* (Moscow: Standartinform)
[15] Osipkov A, Makeev M, Mikhalev P, Machikhin A, Batshev V, Shiriaev P and Shishov K 2020 Diamond-Like Carbon Coatings to Protect the Optical Surfaces of Orbital Telescopes from the Outer Space Factors *Proc. Int. Conf. on Advances in Astronautical Sciences (Moscow)* vol 170 (California: Univelt Inc.) pp 665–670
[16] Oliver W and Pharr W 1992 An improved technique for determining hardness and elastic modulus using load and displacement sensing indentation experiments. *J. Mater. Res.* 7 pp 1564–83
[17] 2019 GOST 8.296-2015 *State system for ensuring the uniformity of measurements. State verification schedule for instruments measuring the surface roughness parameters Rmax, Rz in the range from 0,001 to 3000 μm and Ra in the range from 0,001 to 750 μm* (Moscow: Standartinform)
[18] Malacara D 1985 *Opticheskij proizvodstvennyj kontrol'* (Moscow: Mashinostroenie) p 400

The formation of strip-like vortex motion by surface gravity waves

A V Poplevin[1,2], S V Filatov[1,2] and A A Levchenko[1,2]

[1]Osipyan Institute of Solid State Physics, Russian Academy of Sciences, 2 Academician Osipyan str., 142432 Chernogolovka, Russia

[2]Landau Institute for Theoretical Physics, Russian Academy of Sciences, 1-A Akademika Semenova av., 142432 Chernogolovka, Russia

Abstract. We studied experimentally the generation of vortex flow by non-collinear gravity waves with a frequency of 2.34 Hz. The vortices formed on the water surface have the form of stripes, the width L =π/(2k sin θ) of which is determined by the wave vector k and the angle between them, and the length is determined by the size of the system. We demonstrate that the measured dependence Ω(t) can be described within the recently developed model that considers the Eulerian contribution to the generated vortex flow and the effect of surface contamination.

Introduction

Waves on the water surface have been the subject of research for quite some time. Detecting particles are used to study experimentally horizontal transport. The Lagrangian velocity for several surface waves propagating in arbitrary directions [1] has nonzero vertical vorticity $\Omega = \partial_x V_y - \partial_y V_x$, which permits creating horizontal drift flows of various spatial structures [2-4].

If traveling and standing waves excited on the liquid surface intersect at a small angle, a vortex flow is formed in the form of stripes. The length of the strips is close to the length of the experimental cell, and the characteristic distance between the stripes is determined by the length of pump waves and the angle between them. Vertical vorticity generated by the flow on the liquid surface can be described by the sum of two terms [5,6]:

$$\Omega = \Omega_E + \Omega_S. \quad (1)$$

The first term, the Eulerian contribution, represents an average velocity of the liquid, while the second term represents the Stokes drift in an ideal liquid [7] in the case of non-collinear waves. These two contributions have different stationary amplitudes and characteristic time scales. The situation was analyzed in detail theoretically in [8], and it was found for two surface waves propagating on the surface of infinitely deep water at an angle of 2θ to each other that the ratio of steady-state amplitudes is $\Omega_E/\Omega_S = 1/\sin\theta$. The characteristic Stokes drift time T_S can be estimated as half of the wave decay time since the Stokes drift is quadratic in amplitude. The characteristic time of the Eulerian contribution can be estimated as $T_E = L^2/(\pi^2 \nu)$, where ν is the kinematic viscosity of the liquid, L is the horizontal dimensions of the slow flow, and k is the wavenumber of excited surface waves.

For typical experimental conditions, $T_E \gg T_S$ due to additional dissipation of wave motion near the system boundaries [6, 9] and surface contamination [5, 10]. The latter factor also significantly affects the amplitude of the Eulerian contribution Ω_E on the liquid surface [5, 6], but in all cases, the theory predicts its universal dependence on time:

$$\Omega_E(t) \propto erf\left[\sqrt{t/T_E}\right], \quad (2)$$

where erf is the error function. Expression (2) is valid for a liquid of finite depth $\gg 1/k$, but the applicability is limited in time $t \ll d^2/\nu$, during which the Eulerian flow does not have time to penetrate to the bottom due to viscous diffusion.

The experiments aimed to observe the formation of strip-like vortex flows, study the development of vorticity on the water surface at a pump frequency of 2.34 Hz, as well as study the formation of the Eulerian vorticity Ω_E on the liquid surface [5].

1. Experimental Methods

The investigations were performed on an experimental setup, the scheme of which is presented in Figure 1. The setup consists of a bath a length of 70 cm, a width of 70 cm, and a height of 25 cm; it was made from 10-mm-thick glass. Distilled water was poured into the bath to the level of 10 cm. / The bath was filled with distilled water to the level of 10 cm.The bath was placed on a Standa vibration isolation table with an air suspension.

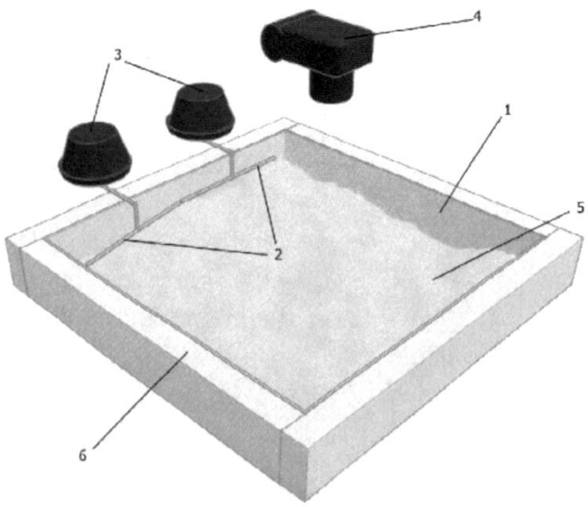

Figure 1. Scheme of the experimental setup: (1) – bath, (2) – plungers, (3) – plunger drives, (4) – camera, (5) – waves on the surface, (6) – heat insulation.

Surface gravity waves were excited by wavemakers consisting of drives (Pioneer TS-W254R subwoofers with a nominal power of 250 W) and plungers (stainless steel tube 1 cm in diameter and 30 cm in length). One of the plungers was placed at an angle of 19 ° to the bath wall, the second was located parallel to the bath wall at a distance of 1 cm from it. Sinusoidal signals with a frequency of 2.34 Hz were generated by a dual-channel generator, amplified, and supplied to the subwoofers. The phase difference between the signals was $\psi = \pi/2$. The surface was recorded using a Canon EOS 70D camera at 24 fps. The vortex motion was recorded by decorating the surface with white polyamide PA12 particles with an average diameter of about 30 μm, that were illuminated by an LED strip located along the bath perimeter. The PIVLab code [11] for MATLAB permitted obtaining the field of horizontal velocity associated with the motion of the detecting particles, after which the vertical vorticity Ω was calculated. At the same time, vertical oscillations of the liquid surface were recorded using a recently developed technique [12], that is based on restoring the surface curvature by analyzing optical distortions of a contrast image at the bottom of the bath.

The distribution of vertical vorticity on the liquid surface and the spectrum of waves in the k-space were found by processing the obtained data by the Fourier transform with the Hann window function. The Hann function rolls off at the edges, so its application allows us to weaken the finite size effects and achieve periodic boundary conditions for the signals. Then we calculated the wave elevation and vorticity amplitudes based on the maximum values of corresponding peaks in the k-space. The registration method and data processing algorithms were discussed in detail earlier, see [13].

2. Results

Figure 2 (a) shows the tracks of polyamide particles 100 seconds after the start of pumping. The vortices have an elongated shape, and their inclination to the X axis is $\theta = 18°$. The width of These three vortices is L = 23 cm, and the length is limited by the size of the system. Figure 2 (b) illustrates the corresponding distribution of vertical vorticity on the water surface. The middle stripe has vorticity of the opposite sign with respect to the other two stripes. The data presented in Figure 2 correspond to the time instant t = 1000 seconds after the start of pumping, and the amplitude of the wave propagating along the X axis in the stationary mode is 0.43 mm.

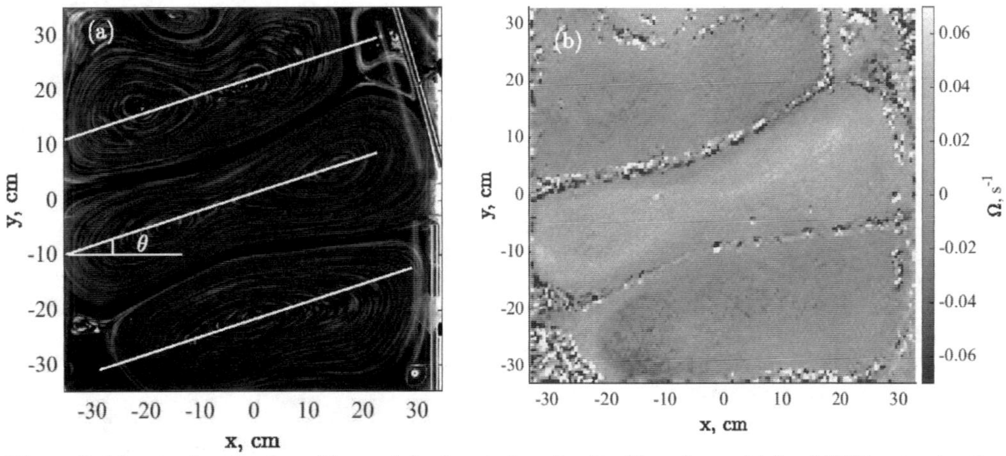

Figure 2. (a) – tracks of polyamide particles located on the liquid surface, obtained 1000 seconds after the pumping was switched on. (b) – the corresponding distribution of vertical vorticity on the water surface. The amplitude of the wave propagating along the X axis in the stationary regime is 0.43 mm.

Figure 3a shows the distribution of the energy of vertical oscillations at a frequency of 2.34 Hz in the k-space at the time t = 1000 seconds. On the water surface, two standing waves are excited, the energies of which differ by several times, and the angle between the waves is $2\theta \approx 36°$ to the X axis. The wavenumber of both waves is $k \approx 0.22\ cm^{-1}$. Note that $kd \approx 2.2$, therefore, we can assume that the waves propagate in deep water.

Figure 3b illustrates the distribution of vertical vorticity in the wave vector space. Two bright peaks correspond to an elongated vortex structure inclined at an angle $\theta \approx 18°$ to the X axis and shown earlier in Figure 2a. The width of the most intense vortex structure should be equal to L $= \pi/(2k \sin \theta) \approx 23$ cm, which is in good agreement with the results presented in Figure 2.

Next, the amplitude of the rise of waves H_1 and H_2 and the amplitude of vorticity were calculated; they corresponded to the most intense vortex structure in the k-space. Let H_1 – be the amplitude of the wave propagating in the X direction, and H_2 – be the amplitude of the wave propagating at an angle of $2\theta \approx 36°$.

Figure 4a shows the time dependence of H_1H_2 for different pump intensities. The nonmonotonicity of this dependence is due to a small discrepancy between the pump frequency and the frequency of the resonant mode. The beating of the oscillation amplitude completely dies out after about 100 seconds, after which the product of the amplitudes becomes constant.

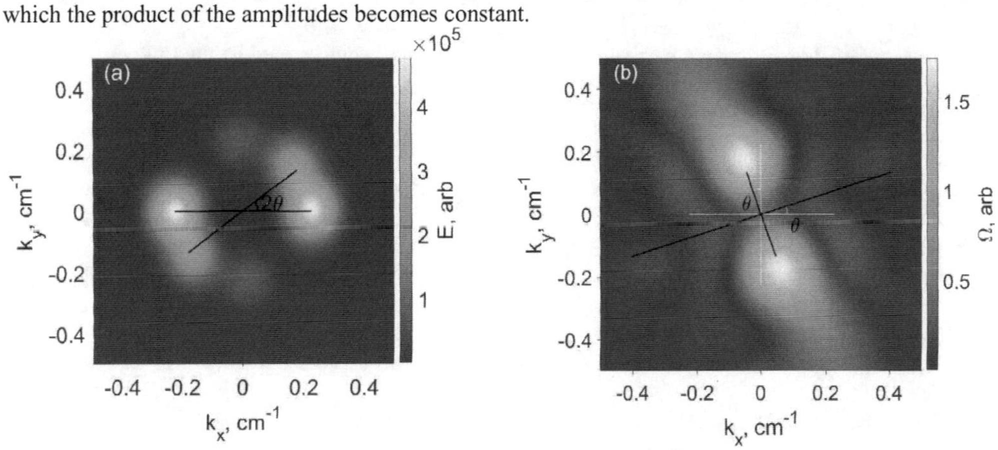

Figure 3. Distribution of (a) wave energy and (b) vertical vorticity in the k-space obtained 1000 seconds after the start of pumping. The amplitude of the wave propagating along the X axis in the stationary mode is 0.43 mm. The ends of the black lines indicate the positions of the theoretical

peaks using the dispersion law $\omega^2 = gk\tanh(kd)$.

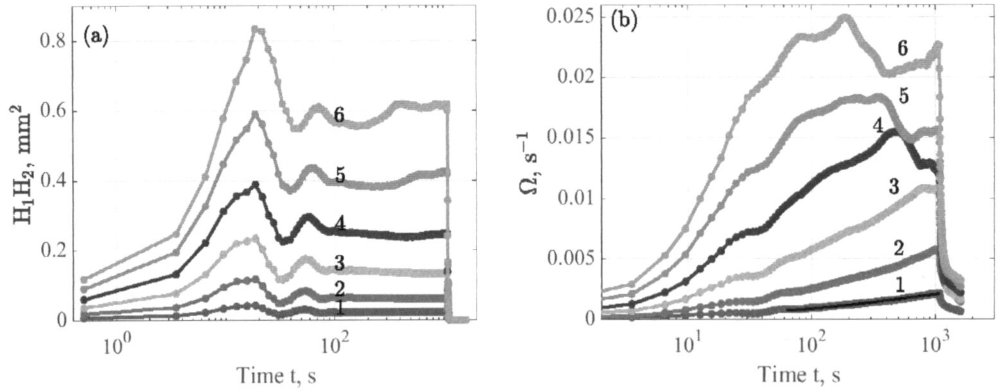

Figure 4. Time dependence of the product of amplitudes $H_1H_2(t)$ (a) and dependence of the amplitude of the vertical vorticity $\Omega(t)$ on the liquid surface (b). Different curves correspond to different values of the amplitude H_1 of the wave propagating along the X axis in the stationary mode: 1 – 0.17 mm, 2 – 0.29 mm, 3 - 0.43 mm, 4 – 0.57 mm, 5 – 0.72 mm, 6 – 0.85 mm.

Figure 4b shows the dependence $\Omega(t)$. At times shorter than 100 seconds, it follows changes in the product $H_1H_2(t)$ of the wave amplitudes. At longer times, the wave amplitude does not change, and the vorticity $\Omega(t)$ continues to increase. Thus, we observe the birth of the Eulerian contribution $\Omega_E(t)$ since= the Stokes drift contribution ΩSt remains at a constant level.

3. Discussion

Let us direct the X' axis in the direction of the stripes shown in Figure 2 and direct the Y' axis - perpendicular to it. In this coordinate system, the surface elevation can be written as:

$$h(t,x',y') = H_1\cos(\omega t)\cos(kx'cos\theta - ky'sin\theta) + H_2\cos(\omega t + \psi)\cos(kx'cos\theta + ky'sin\theta), \quad (3)$$

where ψ is the phase shift between the excited waves. In these experiments, the value of ψ was close to -π/2.

Next, consider the vertical vorticity corresponding to the two brightest peaks in Figure 3b. The contribution associated with the Stokes drift is:

$$\Omega_S = \omega k^2 H_1 H_2 \sin 2\theta\, cos^2\theta \cos[2ky'sin\theta], \quad (4)$$

see [8, Eq. E2]. The Eulerian contribution to vertical vorticity is determined by the following expression:

$$\Omega_E = \left(2 + \frac{\varepsilon^2/cos^2\theta}{2\sqrt{2}\gamma(\varepsilon^2 - \varepsilon\sqrt{2} + 1)}\right)\omega k^2 H_1 H_2\, cos^3\theta \cos[2ky'sin\theta]\, erf\left[\sqrt{t/T_E}\right]. \quad (5)$$

Here we introduced $\gamma = \sqrt{\nu k^2/\omega}$ and assumed that the liquid surface might be covered with a thin insoluble liquid film due to contamination [10]. The presence of such a film on the liquid surface can increase significantly the intensity of a vortex flow. The properties of this film can be described by the dimensionless compression modulus $\varepsilon \geq 0$ introduced in [5] and characterizing the properties of the film.

To explain the measured dependence Ω (t) we use equation (1), substituting relations (4) and (5) into it. The result is shown in Figure 4b with a black curve. The optimal ratio corresponds to the film compression modulus $\varepsilon = 0.37$ and $T_E = 1000$ seconds.

Let us estimate the relative contributions of the Stokes drift and the Eulerian term. At T = 100 seconds and with an amplitude of $H_1 = 0.43$ mm, the Eulerian contribution is $\Omega_E = 5 \times 10^{-3}$ s^{-1}, while the Stokes drift is $\Omega_S = 5 \times 10^{-4}$ s^{-1}. At T = 1000 seconds and the same wave amplitude, the Eulerian contribution is $\Omega_E = 10^{-2}$ s^{-1}, and the Stokes drift remains the same. It should be noted that the ratio of the Stokes and Euler contributions to vertical vorticity does not depend on the wave amplitude if the Reynolds number is not very high since both terms are proportional to the product $H_1H_2(t)$, see (4) and (5).

4. Conclusion

It has been shown experimentally that non-collinear waves propagating at an angle of 2θ on the water surface form strip-like vortex flows. The width of the stripes can be estimated as $L = \pi/(2k \sin\theta)$, and their length is limited by the size of the system.

The Eulerian contribution has relatively slow kinetics compared to the Stokes drift. An increase in the Eulerian contribution is well described by dependence (2). This contribution leads to an increase in the total vorticity $\Omega(t)$, which can be seen in Figure 4b. Quantitative agreement with the theory was achieved only when possible contamination of the liquid surface was taken into account [5].

Acknowledgements

The work was carried out according to the state assignment of ISSP RAS.

References

1. J A Nicolás and J M Vega, 2003 *Fluid Dynamics Research* **32**, 119.
2. S V Filatov, M Y Brazhnikov, and A A Levchenko, 2015 *JETP letters* **102**, 432.
3. S V Filatov, V M Parfenyev, S S Vergeles, 2016 *Phys. Rev. Lett.* **116**, 054501.
4. A P Abella and M N Soriano, 2020 *JETP* **130**, 452.
5. V M Parfenyev and S S Vergeles, 2018 *Phys. Rev. Fluids* **3**, 064702
6. V M Parfenyev, S V Filatov, M Y Brazhnikov, 2019 *Phys. Rev. Fluids* **4**, 114701.
7. G G Stokes, 1847 Trans. Cambridge Philos. Soc. 8, 441.
8. V M Parfenyev 2020 *Phys. Rev. Fluids* **5**, 094702.
9. L D Landau and E M Lifshitz, 1987 Course of theoretical physics. Vol. 6 (London).
10. A Campagne, R Hassaini, I Redor, J Sommeria, 2018 *Phys. Rev. Fluids* **3**, 044801.
11. W Thielicke and E Stamhuis, 2014 *Journal of Open Research Software* 2.
12. S Filatov, A Levchenko, M Y Brazhnikov, 2018 *Inst. and Exp. Tech.* **61**, 757
13. S V Filatov, S A Aliev, A A Levchenko, and D A Khramov, 2016 *JETP Letters* **104**, 702.

The formation of super-long nano-chains during evaporation of metals with a help of intensive impacts

N V Klassen, P V Provotorov
Institute of Solid State Physics, Russian Academy of Sciences, Chernogolovka, Russia
E-mail: klassen@issp.ac.ru

Abstract. Extended nanostructures, in particular, chains of nanoparticles, are of great interest for microelectronics, biomedicine, and energy. But until now, it was possible to obtain such structures with the lengths not more than a few millimeters. We have developed several techniques based on laser and electric evaporation of metals, which provide the formation of nanochains with the lengths up to 60 cm. A significant factor in the formation of such structures is the constructive participation of turbulent vortices, which create the necessary compression and subsequent agglomeration of the vapors of metals and oxides subjected to the intensive electrical or laser impacts. Research and production of the nano-structures with large surfaces and lengths provide promising opportunities for the direct converters of solar energy and ionizing radiation to electricity, development of new optoelectronic and microelectronic devices as well as highly sensitive biosensors.

1. Introduction

The influence of various nano-dimensional solids on deeper understanding of the mechanisms of natural processes in materials and development of new unique technologies and devices based on nanostructures is expanded monotonously. Nanostructures are applied more and more intensively in biomedical techniques and materials, new methods of generation, storage and transformation of energy, electronic and optical devices with record parameters, information technologies, constructions for aviation and space applications, etc [1 – 5]. Separate nano-particles and extended structures from nanoparticles of various solids including nano-chains and nano-grids belong to the most popular nano-dimensional objects either in scientific research or in practical development of technologies and installations. So, the amounts of publications devoted to the development and applications of various nanostructures is increasing rather quickly (e.g., see [6 – 32]). It should be emphasized that preparation of these kinds of nanostructures with necessary structures and morphologies is rather a difficult problem. In spite of the formation of the initial nano-particles with definite parameters it is rather problematic to prevent them from spontaneous non-controlled self-coagulation on the one hand and to provide their binding to required nano-chains of nano-grids on the other hand. Big amounts of chemical, electrical, optical, thermo-mechanical methods were studied for these tasks [5 – 21]. Our paper describes several promising results of the formation of chains and laces from oxide and metal nano-particles obtained with applications of electrical and laser impacts on metal raw materials. It is worth to emphasize that we have obtained super-long nano-chains with the record length being about 30 times longer than nanochains described in other publications (50 cm and more in comparison with one centimeter).

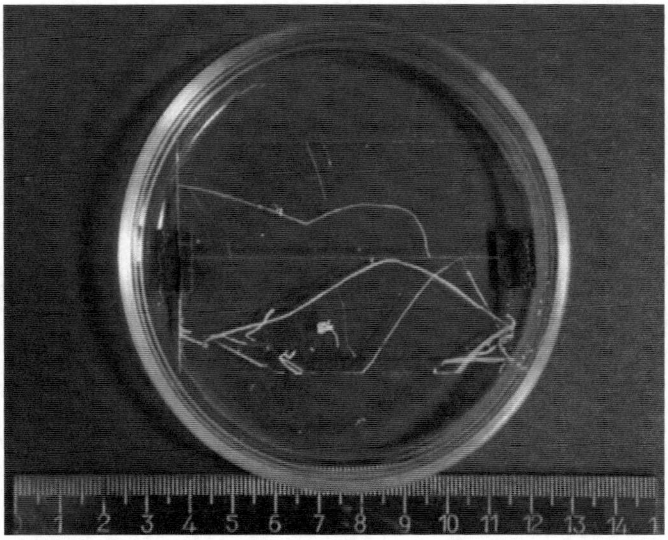

Figure 1. A photograph of the thread-like formations of MoO3 lying on glass substrates in the Petri dish

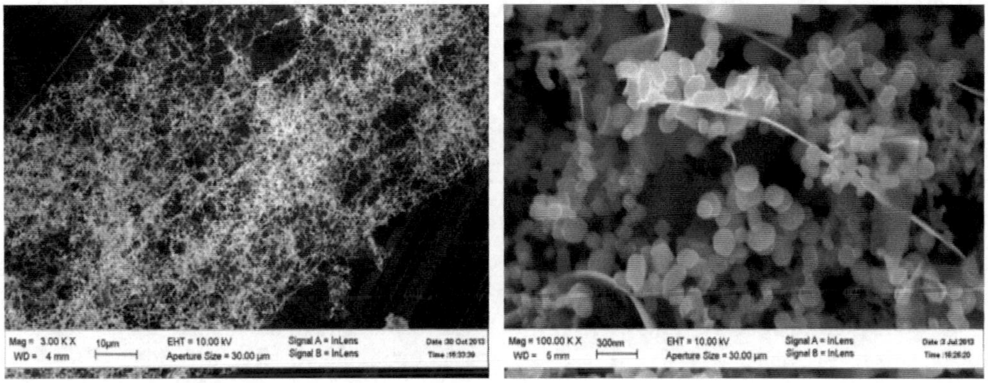

Figure 2. Scanning electron microscopy of the thread-like formations of MoO3. It can be clearly seen chains of nanoparticles and nanoribbons

2. Experimental techniques and results

2.1. Preparation of chains of nanoparticles via electrical explosion of metal wires and electrical discharge heating

For the experiments with a pulsed electrical discharge, we used a setup consisting of a large-capacity capacitor, a high-voltage power supply for charging it, a circuit that controls the charge, and a high-current switch for supplying an electric pulse from the capacitor. To study in detail the mechanism of the formation of nanostructures formed during the explosion of metal wires, two chambers were used: the first with a volume of 1000 liters for the statistical analysis of the formed structures, and the second with a volume of 5 liters to determine the dependence of the morphology of the forming nanostructures on the composition of the ambient atmosphere. Thus, experiments were carried out on the electric explosion of thin metal wires.

During the electric explosion of a molybdenum wire, the formation of thread-like formations (Fig.1) with a lace-like nano-scale morphology was observed (Fig.2).

Similar structures were also obtained in an experiment with a high-frequency generator (33

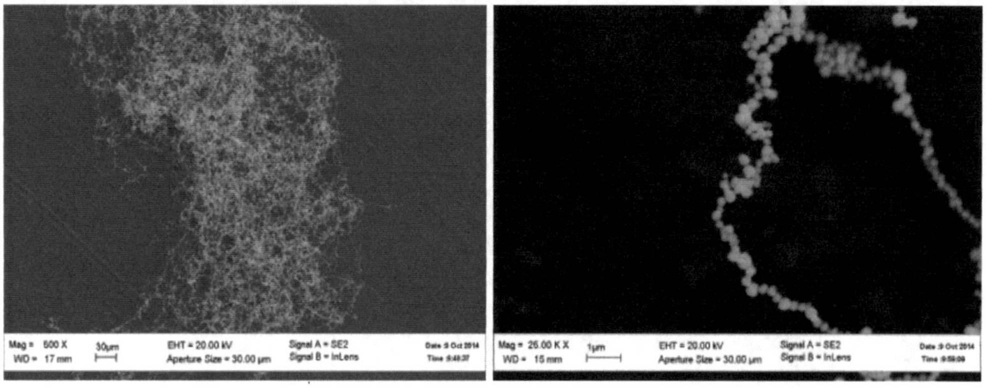

Figure 3. Scanning electron microscopy of the lace-like complex structures of PbO

Megahertz). A high-frequency electric discharge melted a metal wire, the melt drops then under the action of the gravity force were falling down, forming chains of nanoparticles along the trajectory of the fall. Rows of nanoparticles parallel to the axis of the structures are observed in the morphology, closed rings also prevail. The most interesting fact is that these agglomerates have macroscopic dimensions in the form of lace ribbons with a length of up to 60 cm and a width of up to 1 mm, which are formed in a fraction of a second in a sphere with a radius of the order of a meter from the exploded wire. Their weight was very light, so they were floating freely in the air for a long time and the rate of their lowering speed did not exceed 1 cm / sec. It was found that these structures were not formed during the explosion of the metal wires in an oxygen-less environment.

The formation of such lace-like nanostructures can be explained by the processes as follows: since the boiling point of molybdenum is much lower than the temperature of the molten drops themselves, the formed oxide became overheated essentially. This overheating results in abrupt evaporation. Due to the generated high pressure, the vapors are pushed out from the surface of the drop at high speed. As it is well-known, the formation of turbulent vortices is possible at high velocities of gas flows [27 – 31]. The formed turbulent vortices are pushed aside by the oncoming air flow into the rear hemisphere of the drop, where they cool down, forming a lace nanostructure. Since the flow of oxide vapor from the front surface of the droplet continues with the formation of turbulent vortices with solidifying at the rear surface of the droplet, the chain grows and lengthens. This process continues till all of the molten metal turns into a lace ribbon of oxide. It should be noted that for the formation of a continuous chain, the flight speed of the drop should not exceed the rate of the formation near the rear hemisphere of the drop, otherwise the solidified oxide particles do not have time to bind to each other; therefore, the nanochain formation is observed only when the drop is decelerated to a sufficiently low velocity.

2.2. Production of chains of nanoparticles via laser heating and evaporation

In this series of experiments we used a setup that consisted of a 100-watt and 35 watt infrared lasers and focusing lenses. We found out that under focused laser irradiation with an optical power of 35 watts of fusible indium and lead separately, neither in the case of indium nor in the case of lead, the formations of nano-chains were not observed. But upon irradiation of lead-indium alloys with comparable concentrations of components, stable generation of a thin parallel jets of smoke were observed. In these jets we found nanochains (Fig.3) with a lace-like complex structure consisting of nanoparticles of lead and its oxide.

At the same time, the formation of a tube-like agglomerate (Fig. 4) with an outer diameter of up to 4 mm was observed on the substrate.

Figure 4. The tube-like agglomerate on the substrate during laser heating and evaporation of lead-indium alloy

It is known that the melting and boiling points of indium oxide are 1637 °C and 3027 K, respectively, and the melting and boiling points of lead oxide are 613 K and 1262 K, respectively.

These data provided the explanation of the formation of nanochains and structures on a substrate as follows: when the lead-indium alloy is heated, the vapors of these metals are transformed into the corresponding oxides, but due to the difference in the melting and boiling points of the resulting substances, indium immediately crystallizes, forming a tube-like structure on the substrate, and lead oxide vapor propagating through this tube, forms the turbulent vortices, due to which nanochains are formed by the solidification of the vapor.

2.3. The evaporation of lead with 100-watt laser

In this experiment a piece of lead was irradiated with focused 100-watt laser beam. At the initial moment the metal melted, and then intense evaporation began at the point of the focusing of the laser beam. It turned out that the evaporation process induced by the continuous-wave laser had pulsed character with the intervals between moments of the intense emissions of the vapor from 7 to 20 s. An aluminum plate with a hole for the beam passage was installed above the crucible with the lead sample. The evaporated material was deposited on this substrate (Fig.5) and lead oxide prevailed in the precipitation layer. Outgoing smoke practically did not rise upward presumably due to the large mass of particles composing it (lead and lead oxide).

Detailed electron microscopic analysis showed that the deposition we obtained is a developed with lace-like structure of nano-chains of lead oxide nanoparticles (Fig.6).

The data obtained make us to suppose that turbulent vortexes are directly involved in the formation of the resulting deposition. In addition, it can be assumed that due to the accumulated vapors of the material near its surface, the laser beam focuses itself, which leads to even more intense evaporation. The layer of the formed lead oxide forms a film that prevents the formed vapors from leaving the surface, which leads to their accumulation in this bubble of oxide film. At the moment when the bubble bursts, an intense release of the accumulated vapor occurs, in which turbulent vortices are formed, contributing to the formation of the resulting developed structure of lead oxide.

2.4. Laser evaporation of copper

In this experiment laser evaporation of copper nanoparticles and sawdust was carried out.

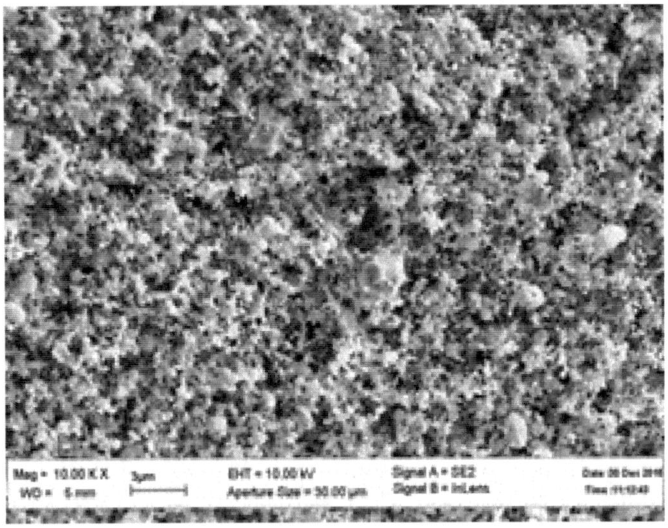

Figure 5. Scanning electron microscopy of the deposition of PbO formed during laser evaporation of lead

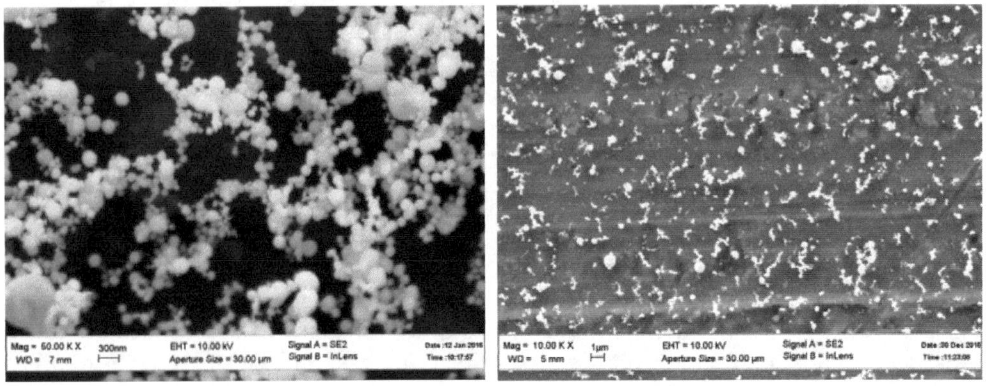

Figure 6. Scanning electron microscopy of lace-like structures in the deposition of PbO

Due to the excellent ability of metallic copper to reflect laser radiation in this range of laser wavelength (10.6 μm), as well as the high thermal conductivity of the material, fine fractions of the metal were selected for evaporation under the laser in order to improve the conditions for their heating.

When using nanoparticles, the metal easily evaporated under the laser. Vapors of the resulting oxides were deposited on a copper plate mounted above a ceramic crucible with an evaporated substance.

As a result, an oxide deposition was formed on the plate with a developed dendrite-like morphology (Fig.7) consisting of chains of nanoparticles (Fig.8).

In addition to the structure described, large spheres (Fig.9) with a diameter of up to 100 microns are also observed on the plate. The surface of the spheres either consists of individual spheres with a diameter of 0.5 to 5 microns, or it is covered with formations with a morphology similar to the described spraying.

When using copper sawdust, under the laser, they melted into a single drop, boiling took place throughout the entire volume without the formation of an intense steam flow. Following

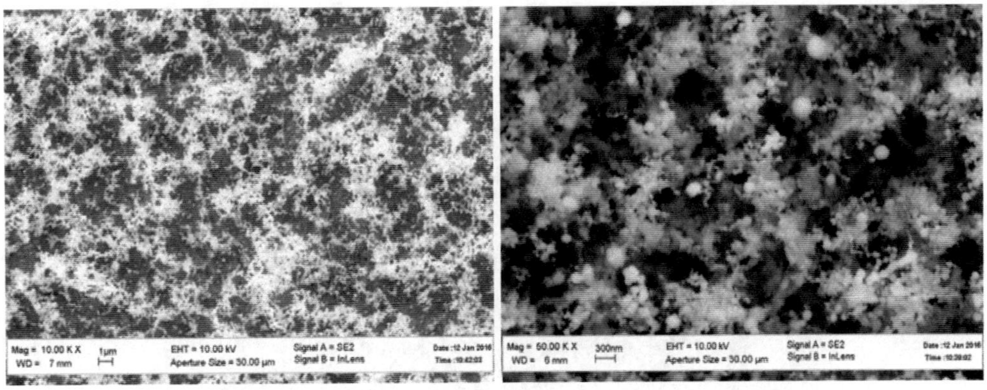

Figure 7. Scanning electron microscopy of copper oxide dendrite-like deposition

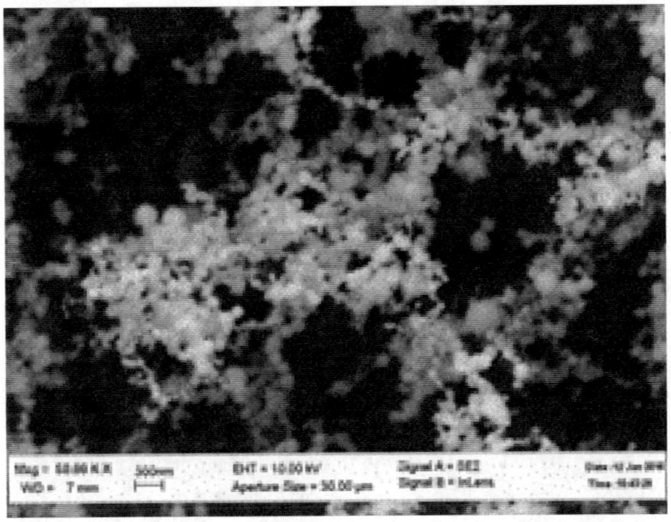

Figure 8. Scanning electron microscopy of chains of nanoparticles in copper oxide deposition

from these facts, it can be assumed that, due to the small size, as well as the oxide layer, which effectively absorbs laser radiation of this range, the nanoparticles instantly evaporated at the initial moment of time, and the resulting melt (Fig.10) was smaller in volume compared to melt from copper sawdust, which led to its greater heating, as a result, more intense evaporation and the formation of dense streams of steam.

Due to the lower atomic weight in comparison with lead, the vapor did not accumulate at the metal surface and pulsations of steam flows were not observed.

3. The formation of extended structures in cryogenic liquids

Fig. 11 presents scanning electron microscopy of the lace-like nanostructure of lead produced by irradiation of a sample of lead immersed into liquid nitrogen with pulsed Nd:YaG laser. It is worth to emphasize the distinct difference between the nanostructures produced by laser evaporation of lead at open air (room temperature) and in liquid nitrogen (77 K temperature). In the last case the long thread-like chains are not observed as well as the evaporation has not the pulsed character being continuous. The lace-like structure reveals the presence of turbulences in the vapor flows. But it can be supposed that the lengths of the vortices in this case are not

Figure 9. Scanning electron microscopy of copper oxide spheres with a dendrite-like surface consisting of nanoparticles

Figure 10. A photograph of copper in the crucible vaporized under the continuous CO_2 laser

big enough for the formation of the long nano-chains. So, they are not observed here (e.g., due to much stronger mechanical resistance of the liquid nitrogen and much faster cooling of the vapor).

4. Discussion of the importance of auto-generation of pulsed evaporation and long turbulent vortices for creation of extended nano-structures

Various kinds of extended nano-chains are applied in new types of electronic and optical devices, bio-medical sensors, energy generation and storage, etc. [22 – 26]. Due to the results of the experiments, it was found that an essential role for the formation of thread-like and lace-like structures from nanoparticles belongs to the turbulent vortices created by the intense evaporation of materials. In addition, we find confirmation of this assumption in the papers of the team of authoritative Russian scientist E. Gordon ("Formation of metal nanowires using laser ablation in

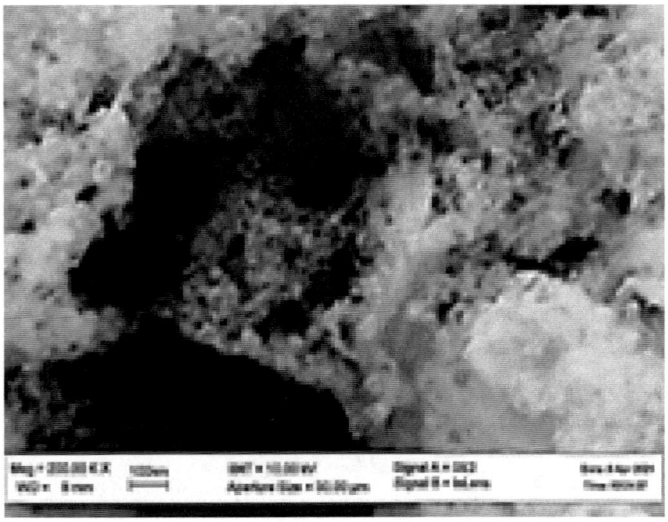

Figure 11. Scanning electron microscopy of the lace-like nanostructure of lead produced by irradiation of a sample of lead immersed into liquid nitrogen

liquid helium" and more late their publications [16, 18, 19, 21]. In these papers the phenomenon of the formation of nano-chains with the morphologies close to the structures obtained in our work was described (Figs. 7.8 in [16]). But in our case the nano-chains were longer significantly (up to 60 cm instead of 1 cm). This difference can be attributed to much better conditions for propagation of long vortices in the open air in comparison with cryogenic liquids. One more factor providing the creation of the longer vortices and subsequent super-long nano-chains for the case of continuous laser irradiation is connected with the auto-generation of the severe pulsations of the evaporation process [27]. The interruption of the evaporation is induced by accumulation of the vapor of irradiated heavy metal (like lead as it was in our cases) in the of the irradiated surface. When the pressure of the accumulated vapor achieves a certain threshold the breakthrough and intensive jet of the vapor is emitted. Due to its high speed the Reynolds number is rather large providing the generation and propagation of the long vortices. The essential amount of experimental and theoretical papers show that this kind of the turbulence facilitates effectively the coagulation of the vapor with formation of extended structures from nano-particles [28-32].

Conclusions

We have developed five new techniques of the formation of extended nano-structures from various metals subjected to electrical and laser impacts: electrical explosion of thin metal wires, melting of metal wires by high frequency electrical discharges, evaporation of solids by continuous-wave infrared lasers with 10.6 μm and 0.9 μm lengths of radiation, evaporation of solids immersed into liquid nitrogen by pulsed infrared laser. The chemical compositions of the nanostructures obtained by these means are regulated by variations of the content of the ambient atmosphere (air, argon, liquid and gaseous nitrogen). The analysis of the experimental observations manifests active participation of turbulent vortices in the formation of these nanostructure for all of the methods described above.

Acknowledgements

The work was carried out according to the state assignment of ISSP RAS.

References

[1] S Soni, A Salhotra, M Suar, 2015, *Engineering Science Reference* (USA), 1950 p.
[2] Journal of Pharmacevtics and Nanotecnology, 2016, 4, Special Issue: Reviews on Pharmaceutics and Nanotechnology
[3] A Neyes, R Saletti, 2018 *Nanoparticles in Life Sciences and Biomedicine*, 435 p.
[4] W Mattheu, B Noiurnikan 2021 *Emerging Nanotechnologies for Renewable Energy* 624 p.
[5] T Chen 2018, *Semicomd. Nanocryst. and Metal Nanopart.: Phys. Propert. and Device Applic.* 425
[6] F Christian 2013, *Frontiers in Energy* **7**, 5 – 18.
[7] C Huttei 2018, *Gas-Phase Synthesis of Nanoparticles*, Wiley (USA), 314 p.
[8] J P Merlin, S S Rajan, 2020 *J Cancer Sci. Treatment* 2, 87-89
[9] F E Kruis, H Fissan, and A Peled, 1998 *J of Aerosol Science* **29**, 511-535.
[10] K K Harish, V Nagasamy, B Himangshu, K. Anuttam 2018, Biomed J Sci & Tech Res **4**(2).
[11] M Kim, S Osone, T Kim, H Higashi, T Seto, 2017, *KONA Powder and Particle Journ* **34**, 80 – 90.
[12] Boguslavsky L.I. 2010, *Fine Chemical Technologies* **5**(5) 3-12 (In Russ.)
[13] M Yadi, et.al., 2018 *Artif Cells Nanomed Biotechnol* **46**(3) 336-S343.
[14] D Sharma, et.al., 2019 *Arabian Journ. Chemistry* **12** 3576-3600.
[15] A Elias, M Saravanakumar 2017, *IOP Conf. series: Materials, Science and Engineering* **263** 032019
[16] E Gordon, et.al. 2011 *Journal of Low Temperature Physics* **165**(3) 166-176
[17] N V Klassen, P V Provotorov, A E Ershov, V N Kurlov, V V Solodovnikov 2015 A device for the formation of superlong nanochains, Patent of Russia №153513 from July 20,
[18] E Gordon, et.al. 2017, *J. Phys. Chem. A* **121**(48) 9185–9190
[19] E B Gordon, A V Karabulin, M I Kulish, V I Matyushenko, M E Stepanov, 2017, *Mendeleev Communications* **27** 484-386
[20] S V Stovbun, A A Skoblin, 2019 *Phys.Chem. Chem. Phys.* **21** 5771-5779
[21] E Gordon, et.al. 2020 *High Energy Chemistry* **54** 164-169
[22] C Fang, et.al. 2008 *Applied Physics Letters* **92** 263108
[23] X Yuan, et. al. 2017 *ACS Energy Letters* 7b00223
[24] A Ruiz-Sanchez, et.al. 2017 *Journ. Mater. Chem B* **5** 7262
[25] V T Tran, et.al. 2014 *Sensors and Actuators B: Chemical* **203** 817-823
[26] N V Klassen, P V Provotorov, A E Ershov, 2015 A device for the direct transformation of ionizing radiation to electricity , Patent of Russia №154180 from July 21
[27] S Bugaychuk, 2014, *OSA Advanced Photonics Congress, Nonlinear Photonics* 27-31 July, Barcelona, Spain, JM5A.26
[28] S Friedlander, 2000 Smoke, dust and haze: fundamentals of aerosol dynamics, Oxford University Press (USA), 430 p.
[29] S Garrik, 2011, *Aerosol Science and Technology* **45**(10) 1272-1285.
[30] S Garrik, 2015 *Journ. Nanotechnology* 2015, 642014
[31] J Lin et.al. 2016 *Appl. Mat. Mech.* **37**(10) 1275-1288
[32] M Shigeta, 2020 *Plasma Chemistry and Plasma Processing* **40** 775-794

Diagnostics of atomic- molecular interactions in nano-compositions organics-inorganics by means of infrared spectroscopy

A S Azhgalieva[1], N V Klassen[2] , A M Likhter[1], A P Pokidov[2], T N Fursova[2], I S Tsebruk[2]

[1]Astrakhan State University, Astrakhan, Russia
[2]Institute of Solid State Physics, Russian Academy of Sciences, Chernogolovka, Russia

E-mail: klassen@issp.ac.ru

Abstract. This paper prolongs our studies of the formation and properties of the organic-inorganic scintillating compositions from activated polystyrene and inorganic nanoparticles from cesium iodid and sulfate. The compositions from the solution of polystyrene in benzene and nano- or micro-particles of cesium iodide were solidified in the form of thin films. Then some of the films were subjected to rolling. The variations of the infrared transmission spectra of the compositions were measured either during their solidification or after the solidification and rolling. The new absorption band was revealed in the process of the solidification and disappeared after it. The fractal-like structures of parallel rows were formed in the films. Quasi-periodical oscillations of the transmission spectra were found in several infrared regions ascribed to the interaction of the infrared radiation with the fractal super-structure of the films. The modifications of the absorption spectra found in our experiments are attributed to the interactions of polystyrene molecules with benzene molecules and nanoparticles of cesium iodide.

1. Introduction

This paper is devoted to development and studies of compositions from nano- and micro-particles of cesium iodide immersed into solid matrices from polystyrene. This work is a part of highly actual studies of physics and technology of composites organics-inorganics. The actuality of these studies is determined by quickly expanding applications of these materials in new trends of micro-electronics, photonics, generation and storage of energy, bio-medicine, etc. [1.2]. On the other hand various hybrids from nano-dimensional organic and inorganic components are of great interest for fundamental research of new features of condensed matters because the amount of the atoms placed at the interfaces between the regions with structures and properties differing significantly from each other is comparable with their total amount. So it is natural to expect revealing of qualitatively new kinds of intermediate substances with unusual properties. Our previous publications devoted to the development and studies of new scintillators composed from nanoparticles of heavy inorganic materials and matrices from scintillating polymer phosphors confirm the reality of these expectations [3 - 7]. It is worth to emphasize that rather promising organic-inorganic composites were obtained when biological polymers like proteins etc. were used as the organic components [8]. As usual any successful kind of composite materials should connect advantages of the initial constituents and eliminate their disadvantages. Our composites scintillators got from inorganic nanoparticles their high density and corresponding stopping power for ionizing radiation combined with good radiation hardness. On the other hand organic phosphors gave excellent temporal resolution and high light yield of scintillations. This work prolongs our RD on composite nano-materials organic-inorganic with record sets of scintillation parameters. We have obtained the composite films with new kinds of the internal morphology and unusual optical properties which are presented below. For example the composite films with polystyrene matrices and fillers from nano-particles of cesium iodide acquired complicated fractal-like internal morphology and interesting peculiarities of optical properties resulting from this morphology. The structural transformations of the composites were produced by solidification of the solutions of polystyrene in benzene or toluene with cesium iodide particles inserted into the solution as well as by plastic deformation of the solid composite films with the help of rolling. The fractal-like super-structures observed in our samples are attributed to self-organized crazing and deformation processes taking place in polymers subjected to severe deviations from equilibrium [9 – 12]. The main portion of the information about modifications of atomic-molecular bonds induced by these structural transformations was obtained by means of Fourier Transformed Infrared Spectroscopy technique [13,14,15] which is well known and effective method to reveal these modifications due to high sensitivity of atomic and molecular vibrations to structural disturbances.

2. Experimental materials and methods

The samples of polystyrene-cesium iodide composites were prepared in the form of films with the thicknesses from 300 to 600 micrometers. These films were formed by means of pouring of the solutions of polystyrene in benzene onto the substrates from glass-slides and plates of potassium chloride single crystals. The fillers from the particles of cesium iodide were inserted into the solution before its solidification. The content of the filler was regulated from 0 to 30 volume percentage. The particles of cesium iodide were prepared by 3 techniques: milling of the bulk crystals in the ball mill with steel or zirconia balls, chemical synthesis by condensation of cesium iodide from water solutions and laser evaporation of the powders of cesium iodide [16]. The first two methods produced microparticles with dimensions of grains from 10 to 50 micrometers, whereas the particles obtained by the laser evaporation had nanoscopic dimensions from 30 to 200 nanometers. Some films were subjected to plastic deformation by rolling in the jewelry rolls either in one or in two mutually perpendicular directions at temperatures about 70°C. The internal structures of the films were investigated by means of polarization optical microscopy and scanning electron microscopy (electron microscope Supra 50 VP with microanalysis system INCA Energy +). Infrared spectra were measured by means of Fourier spectrometer Vertex 80 v with infrared microscope Hyperion. The measurements of the infrared transmission were made in the spectral range from 350 to 7000 cm^{-1} with the spectral resolution 2 cm^{-1}. These measurements were made in the infrared microscope Hyperion with the spatial resolution of about 50 micrometers. To determine the normalized transmission spectra, the results of the transmission measurements were divided by the results of the analogous measurements of the transmission of the plate from single crystalline potassium chloride.

3. Experimental results and discussion

Fig.1 presents the scanning electron microscopy of the composite films of polystyrene with microparticles of CsI subjected to rolling in the direction making up the angle about 45° with the horizontal line. The range of the magnification at these figures is varied by about 40 times from Fig. 1a to Fig. 1c.

It is worth to emphasize that in spite of the significant variations of the scales the appearances of the systems of the parallel rows observed at these photos look likely. The invariance of

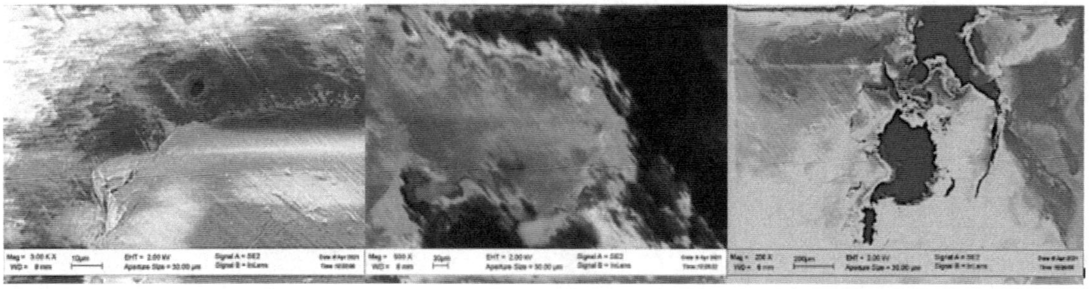

Figure 1. The sequence of electron-microscopy photos of internal morphologies of the composite films polystyrene – cesium iodide with varied magnifications manifesting their fractal-like structure.

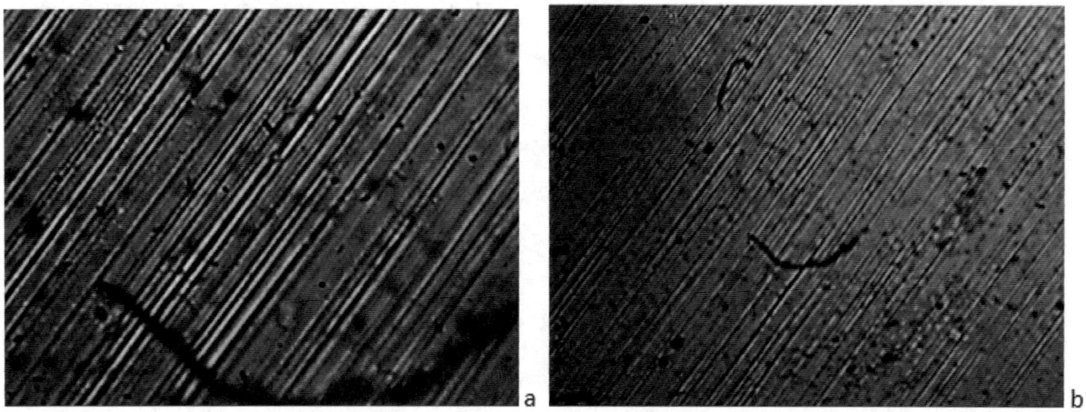

Figure 2. The optical microscopy photos of the polystyrene film solidified from the solution of polystyrene in benzene. The magnification at the left picture is 8 times higher than at the right one

the spatial distributions of the structural elements being independent on the scales of the observations of a given structure is appropriate to fractal structures [17]. Various structures of fractal types are formed in condensed matter subjected to any severe deviations from equilibrium conditions [17,18]. The transformations of the matters producing structures close to fractal kinds can be attributed to a wide variety of self-organization processes [19]. We have found that the severe plastic deformation by rolling or other methods is not the only way to obtain fractal-like structures in polystyrene films. It turned out that the process of the solidification of the solution of polystyrene in benzene proceeds with distinct signs of self-organization and formation of fractal-like internal structures. This is confirmed by the photos of the polystyrene film obtained by this way made with various magnifications (Fig. 2).

The difference in magnifications of Fig. 2a and 2b is about 8 times. But nevertheless the sets of the distances between the parallel rows, created by coagulation of structural defects formed in the film during its solidification are close to each other. The formation of regular systems of structural defects during the solidification of polymers is well known as crazing [10]. The observations of the fractal-like behavior of these defects delivers additional information useful for further development of technologies of polymer systems and their practical applications.

It turned out that the fractal-like superstructures in the polystyrene films studied in this paper produce several new features of the infrared spectra of these samples. The influence of the structural non-homogeneities described above is influenced essentially by the residual

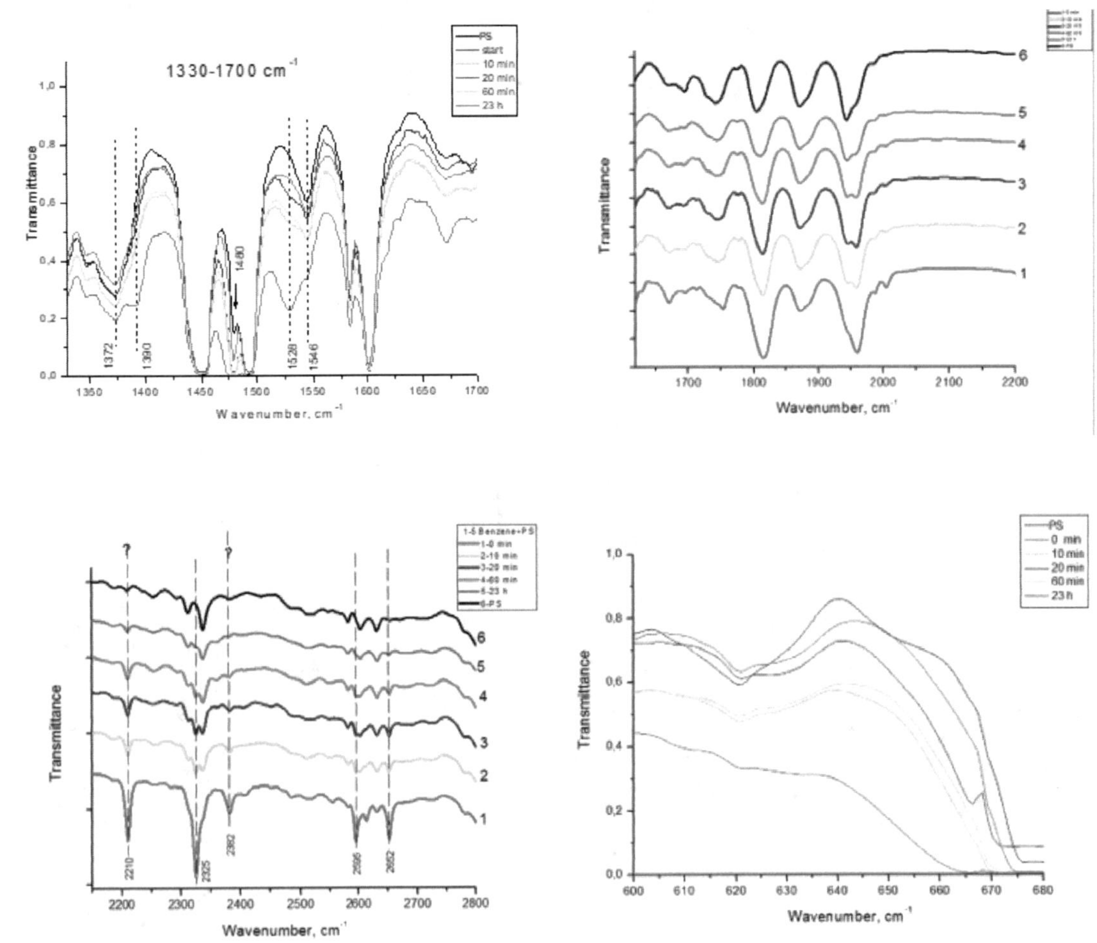

Figure 3. Dynamics of variations of the transmission spectra in various IR ranges of the polystyrene film during its solidification from the solution of polystyrene in benzene

benzene which is conserved inside the films for rather a long period of time (not less than for one week). The dynamics of the evaporation of benzene molecules from the solidified solution of polystyr ene was studied by means of the Fourier spectroscopy via the microscope Hyperion. The duration of one measurement of the FTIR spectrum in the full range from 350 to 7000 cm^{-1} did not exceed 30 seconds. The intervals in the sequence of these measurements was about 10 minutes. The results of these measurements for various spectral ranges are presented at Fig.3. The total duration of the measurements was about 60 minutes. The control polystyrene film for comparison of the spectra at various moments of the solidification was obtained by the same technique 7 days before this experiment.

Comparison of the "dynamic" spectra at Figs. 3 with the spectrum of the control film induced the conclusion that in spite of relatively fast evaporation of the solvent residual benzene was present inside the films even after 7 days from the creation of the film. This conclusion is proved by the residual picks of the infrared absorption observed at the well-known benzene frequencies (674, 1030, 1180, 1480, 1815, 1959 cm^{-1}). We assume that this residual benzene was accumulated inside the nano-scopic cracks (crazers [10]) produced during the process of the solidification of polystyrene. The spectrum in the range 600 – 680 cm^{-1} (right at the bottom) presents one more new spectral absorption band which has been induced during the solidification (the band at about 665.99 cm^{-1}, the blue curve). The particularity of this band is connected with its dynamics: it was absent at the beginning of the process, then it began to appear after 10 minutes of the solidification (a weak kink at about 665.99 cm^{-1}, green curve). Up to the 20-th minute this kink was transformed to a quite distinct Lorentz-like band at 665-665.99 cm^{-1} (the blue curve). But then the sequence of the curves measured after 60 minutes and 23 hours from the beginning demonstrated almost complete elimination of this band. This non-monotonous behavior of the new band at 665.99 cm^{-1} makes us to assume that its existence is induced by the interaction between the molecules of benzene and polystyrene. The necessary conditions for this interaction includes the existence of the optimal distances between these molecules. Just at the beginning the concentration of polystyrene in the solution does not exceed 15 %. So the distances between these molecules are too large for the active inter molecular interaction. After 20 minutes of the evaporation of benzene the concentration of the solution becomes close to the optimum resulting in active interaction between these molecules. Later the concentration of benzene becomes too low interrupting this cooperative process. The additional argument for the validity of this assumption is connected with the values of the resonance frequencies of polystyrene and benzene close to 665.99 cm^{-1}. One of the resonance frequencies of the bending vibrations of the benzene rings is 674 cm^{-1} [20]. One of the frequencies of the vibrations of polystyrene is 670 cm^{-1} [20]. In the first approximation the interaction of the polystyrene and benzene oscillators placed close enough to each other arranges various combinations of these frequencies. Up to now we do not know any possible details of this interaction, but one of these combinations (the "mirror' reflection of 674 cm^{-1} with respect to 670 cm^{-1}) gives 665.99 cm^{-1} quite exactly.

It is well known that the propagation of electromagnetic waves through 3-D fractal structures produces spatial, temporal and spectral oscillations of the distribution of their intensity [21]. Using the infrared microscope for the measurements of the transmission of the infrared radiation through the films with fractal-like internal super-structure we have observed indeed the quasi-periodical spectral oscillations in various parts of the spectrum. The oscillations were observed the most clearly at the long wavelength edge with the approximate periods about 20 – 25 cm^{-1} (Figs. 4a, 4b). In the interval 350 – 500 cm^{-1} these oscillations demonstrate distinct sinus-like character. These oscillations are observed the most clearly for the films rolled in one direction. In the longer interval (600 – 700 cm^{-1}) the oscillations are observed as quasi-periodical distribution of kinks at the transmission curves. At higher wave vectors these oscillations return sometimes to the sinus-like character (Figs. 5a, b, c). These oscillations are observed with relatively small amplitudes for the films subjected to the rolling in one direction – for the films either with or without CsI fillers. For the films rolled in two mutually perpendicular directions these oscillations are much weaker. At the wave-vectors higher than 4000 cm^{-1} these oscillations were not noticed.

It is worth to emphasize that the one-dimensional rolling of the polystyrene films doped with cesium iodide particles results in splitting of the particles to nano- dimensions. The electron microscopy of one these films is presented at Fig. 6a and its transmission spectrum presented at Fig. 6b.

It turned out that the combined action of the doping and the one-directional rolling resulted in severe enhancement of the absorption band with the center at about 540 cm^{-1}. This band of polystyrene is attributed to vibrations of its phenyl-groups (modified benzene rings). Usually it is connected with symmetrical vibrations, so the probability of its presence in the infrared spectra is low [13, 20]. Its appearance in this spectrum can be attributed to local disturbances of the symmetry of the vibrations of the phenyl groups. Hence the introduction of the cesium iodide particles combined with the one-directional rolling should enhance these disturbances. For example this process can take place when one of the edges of the phenyl

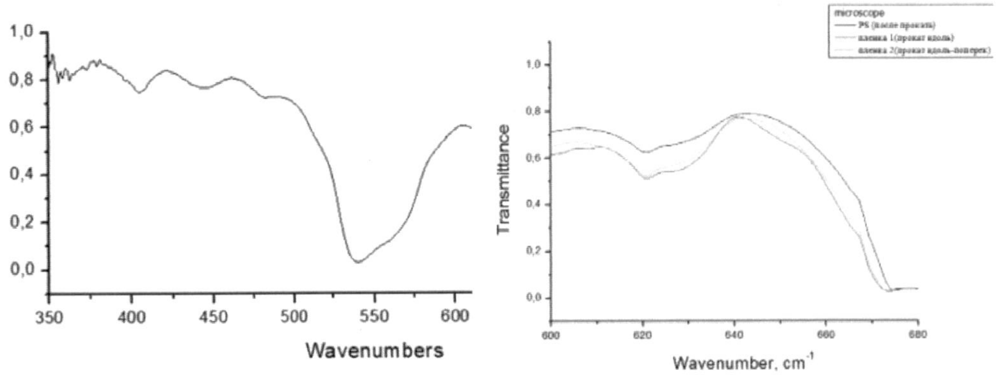

Figure 4. Quasi-periodical oscillations of the infrared transmission spectra of the composite films from polystyrene and particles of cesium iodide in the spectral range 350-680 cm^{-1}

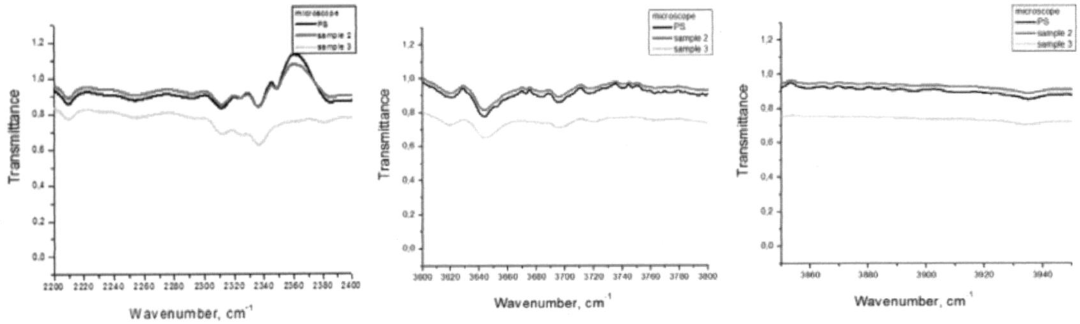

Figure 5. The oscillations of the transmission spectra of the composite films from polystyrene and cesium iodide particles in the spectral range from 2200 to 3950 cm^{-1}

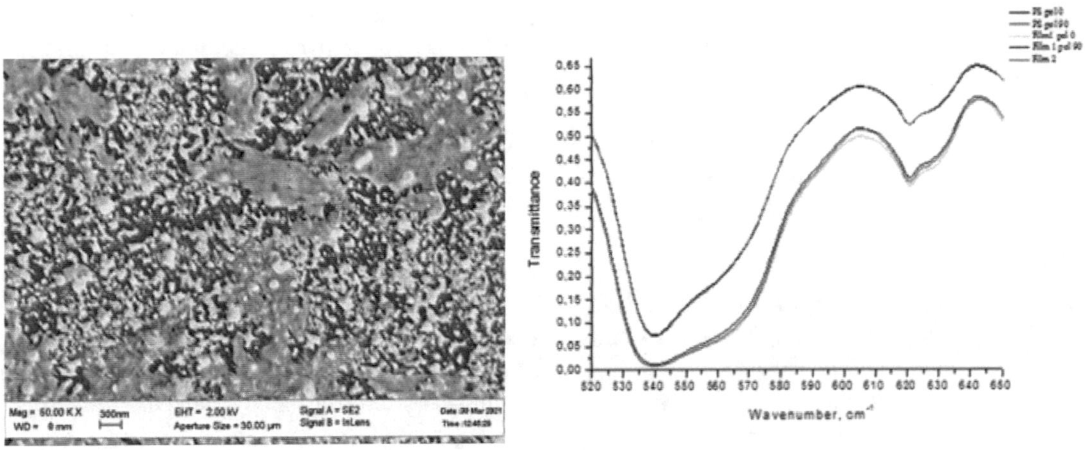

Figure 6. The electron microscopy image of the composite film from polystyrene and cesium iodide particles (6a, left) and its transmission spectra in the range 520 – 680 cm^{-1}

group becomes attached to the surface of the particle. In order to get the information about the asymmetry of the internal structure of the rolled films we studied the transmission spectra for two polarizations of the infrared radiation: parallel and perpendicular to the direction of the rolling and corresponding rows of the structural defects produced by this rolling. But the spectra at Fig. 6b did not reveal any polarization anisotropy in the region around 540 cm^{-1}. From the first impression this fact is in the direct contradiction with quite clear structural anisotropy of the films rolled in one direction shown at Fig. 1. But the electron microscopy with higher spatial resolution of the same film (Fig. 6a) helps to resolve this misunderstanding. The cesium iodide particles are splitted by the rolling to nanoparticles with mean dimensions less than 100 nm. The local compression applied to the films during the rolling provides effective mixing of these nanoparticles and polystyrene molecules. In this situation wrapping of cesium iodide nanoparticles with polystyrene molecules becomes probable. So the asymmetrical disturbances of the rings of the phenyl groups proceed in a wide set of angles. The local asymmetrical disturbances for selected rings are present but the integration along the polystyrene chain wrapped around the nanoparticle eliminates this asymmetry. It is worth noting that the local asymmetry of the contacts of definite phenyl rings can be found for example by polarized excitation of the luminescence combined with analysis of the polarization anisotropy of the corresponding photoluminescence.

Conclusions

It was found that organic-inorganic compositions from polystyrene, benzene and particles of cesium iodide subjected to severe deviations from equilibrium conditions by solidification and plastic deformation demonstrate essential transformations of their atomic-molecular structure and morphology.

Regular systems of internal non-homogeneities with fractal - like behavior are formed inside the films from polystyrene and filling particles of cesium iodide.

The new band in the infrared absorption spectrum of polystyrene films appears during the formation of the films and disappears up to the finishing of the solidification of polystyrene. It is explained by the activation of the quasi-resonance interaction of the benzene rings of the solvent and the phenyl rings of polystyrene when the concentrations of these organic molecules become comparable.

Quasi-periodical oscillations of the spectra of the infrared transmission spectra of the composite films from polystyrene and cesium iodide are observed in several spectral regions. They are explained by the interaction of the infrared radiation propagating through the films with their fractal-like internal structure.

Significant enhancement of the absorption band of polystyrene at 540 cm-1 is found in the films of polystyrene with cesium iodide fillers subjected to unidirectional plastic deformation by rolling. This enhancement is explained by the activation of asymmetrical interactions of the polystyrene phenyl rings with the nanoparticle of cesium iodide. The absence of the polarization anisotropy of this enhanced band is explained by the circular wrapping of the nanoparticles by polystyrene molecules.

Acknowledgements

The work was carried out according to the state assignment of ISSP RAS.

References

[1] R J Martin-Palma, J M Martinez-Duart, 2017, *Nanotechnology for Microelectronics and Photonics* (Cambridge MA, USA, Elsevier)

[2] R Mobin, et. al., 2019, Organic-Inorganic Hybrid Materials and Their Applications, in Functional Polymers and Polymeric Composites, Springer Nature, Switzerland.

[3] S Z Shmurak, V V Kedrov, N V Klassen, O A Shakhrai 2012, *Phys Solid State* **54**, 2266-2276
[4] A P Pokidov, N V Klassen, V V Kedrov, I M Shmytko, A V Bazhenov, T N Fursova 2016 *Inorganic Materials: Applied Research* **7** 85 – 90
[5] N V Klassen, A P Pokidov, V V Kedrov 2018, *Inorganic Materials: Applied Research* **9**(1) 141–146
[6] I S Cebruk, A P Pokidov, V V Kedrov, N V Klassen 2019, *J. Phys.: Conf. Ser.* 1309 012002
[7] I S Tsebruk, N V Klassen, A P Pokidov 2020, *J. Phys.: Conf. Series* 1560, 012043.
[8] N V Klassen, et.al. 2014, *Inorganic Materials: Applied Research* **5** 312–314
[9] M Antonietti 2003, *Nature materials* **2** 9-10, www.nature.com/naturematerials
[10] H -H Kausch, G Michler 2005, *Adv. Polym. Sci.* **187** 1-33
[11] R Mukerjee, A Sharma 2015, *Roy. Soc. Chem. Soft Matter Review* 282343500
[12] C Yuan, et.al. 2017, *ACS Appl. Mater. Interfaces* **9**(17) 14700–14708
[13] S Thomas, D Rouxel, D Ponnamma 2016, *Spetroscopy of polymers*, Elesevier.
[14] M Catauro, S Picconella, C Leonelli 2020, *Applied Sciences*, **10** 1180
[15] T Runka 2017, *Physical Sciences Reviews* 20170025.
[16] H Naser, et.al. 2019, *J. Nanopart Res.* **21** 249
[17] G-F. Gouyet 1995, *Physics and Fractal Stuctures, Paris, Ecole Polytechnique* 247.
[18] A Fedosimova et.al. 2018 *J Physics: Conf. Series* 1141 012004
[19] M Condepudi, I Prigogine 2002, *Modern Thermodinamics*, Wiley and Sons
[20] A H Koupsov, G N Zhizhin 1998, Handbook of Fourier Transform Raman and Infrared Spectra of Polymers, Elsevier
[21] C M Sorensen 2001, *Aerosol Science & Technology* **35**(2) 48-687

Particularities of photonic and structural transformations in the polystyrene film during pulsed infrared laser irradiation

I S Tsebruk, P.V. Provotorov, N V Klassen

Institute of Solid State Physics, Russian Academy of Sciences, Chernogolovka, Russia

E-mail: `cebruk@issp.ac.ru; klassen@issp.ac.ru`

Abstract. Two new kinds of the structural in-homogeneities in the polystyrene films were created by the irradiation of the film with nanosecond pulses of the infrared laser with 1064 nm wavelength. At the beginning of the irradiation dark micro-inclusions having unusual close to rectangular geometries appeared. After about 30 seconds of the irradiation one more kind structural distortions appeared having the geometries of the systems of the concentric micro-rings. Simultaneously with the creation of these concentric micro-rings the film started to emit bright blue luminescence all over its area and significant focusing of the laser beam during its propagation through the film was observed resulting in the generation of the bright white flashes at the screen installed behind the film. The creation of the dark rectangular spots as well as of the concentric micro-rings are ascribed to the non-linear interactions between the laser radiation and the structural distortions produced in the film due to photo-thermal processes excited in the film.

1. Introduction

This paper concerns the development of new techniques for creation of regular microscopic and nanoscopic superstructures in organic polymers by laser treatment. This topic is actual either from practical or fundamental considerations. Organic polymers like polystyrene and many others are highly prospective materials for new kinds of electronic and optoelectronic devices. In comparison with traditional semiconducting materials organic polymers possess a wide set of advantages [1-4]. Their production is cost-effective, the materials have low weight, mechanical flexibility, chemical resistivity, etc. The morphologies, chemical compositions and technical parameters of polymer products can be modified rather easily. On the other hand, the understanding of the nature of optical and electron phenomena in organic polymers as well as techniques of governing of their properties is important for applications in various optical and electronic devices. But it is far from completion in comparison with inorganic materials. For example, the science and technology of creation of periodical micro- and nano-structures from inorganic materials by means of self-organization is rather well investigated [5 – 7], providing an effective basement for modern cost-effective technologies. On the other hand, studies of self-organization processes in organic polymers have not achieved the inorganic level yet [8]. This is explained by essentially younger of synthetic polymer history on the one hand and significantly more complicated structures of these substances on the other hand. It is worth to emphasize that modification of microscopic and nanoscopic morphologies of polymer surfaces by especially regulated laser irradiation is rather well developed [9 – 12]. But this development concerns preparations of ripple – like structure from parallel lines by means of interference of two or more laser beams. The preparation of more sophisticated morphologies inside the volume of polymer materials is problematic up to now. In the studies described in this paper we have found that irradiation of polystyrene films by infrared laser nanosecond pulses being applied with proper parameters produces inside the volume of the material periodical microstructures from concentric rings possessing interesting optical properties for practical applications. Besides self-organization of the circular structure and its influence on the interaction of laser beam with the irradiated material the originality of the results presented here is connected with the creation of the modifications of the internal structure of the polystyrene by the flow of photons with the energies several times lower than the energy necessary for direct excitation of polystyrene interatomic bonds. Our studies of the nature of the self-organized structural transformations induced in the polystyrene films by nanosecond infrared radiation showed that the main reason of these processes is connected with the positive non-linear feedback between the local structural transformations of the material and the redistributions of the laser beam intensity.

2. Methods of experiments

The polystyrene films were prepared by solidification of the solution of polystyrene in benzene poured onto optically polished glass slides (usually applied in optical microscopes). Mean thickness of the films was about 500 micrometers. The luminescence ability of polystyrene was activated with molecules of 2.5 - diphenyloxazole (PPO) and 2,2- p-phenylene-bis(5-phenyloxazole) (POPOP) [13]. The film was irradiated with pulsed infrared laser (wavelength 1.064 μm, pulse duration about 20 ns, mean energy in the pulse 0.03 Joule). Irradiation could be arranged by single pulses or in periodical regime with the frequency from 6 to 30 pulses per second. The process of laser treatment of the films was recorded by video-camera registering the spatial and spectral characteristics of the luminescence. The modifications of the film structures were analyzed by optical and scanning electron microscopes. The laser irradiation of the films has been made at open air and in liquid nitrogen.

3. Experimental results and discussion

Irradiation of the films of polystyrene, activated with PPO and POPOP, with 20 ns pulses of 1.064 μm laser described above resulted in the generation of the light flux in violet and blue spectral regions (Figs. 1 and 2). It should be emphasized that at the beginning of the laser irradiation the violet luminescence is observed only, whereas after 20 – 30 seconds of the irradiation the blue luminescence appears. The spatial distributions of these two kinds of the luminescence are quite different. The violet emission inside the film is induced in the region where the film is crossed by the laser beam. Its angular distribution does not have noticeable anisotropy (Fig. 1a, left). On the other hand, the blue luminescence is seen across the whole film area (Fig.1b, right).

On the other hand, the blue luminescence is seen across the whole film area (Fig. 1b). It is important to emphasize that simultaneously with the generation of the blue luminescence the bright white emission appears at the screen installed behind the film/ at the distance about 4 cm. The reasons of these transformations of the luminescence of the film will be discussed lower in correlation with the transformations if the film morphology. In addition, it should be noted that the dependencies of the violet and blue light emissions on the intensity of the laser pulses demonstrate distinct super-linear character with the sharp threshold of the intensity. The luminescence is not observed when the intensity of the laser irradiation becomes lower than this threshold.

We analyzed the transformations of the morphology of the film at each stage of the spectral and spatial distributions of the luminescence described above. At the first stage when the

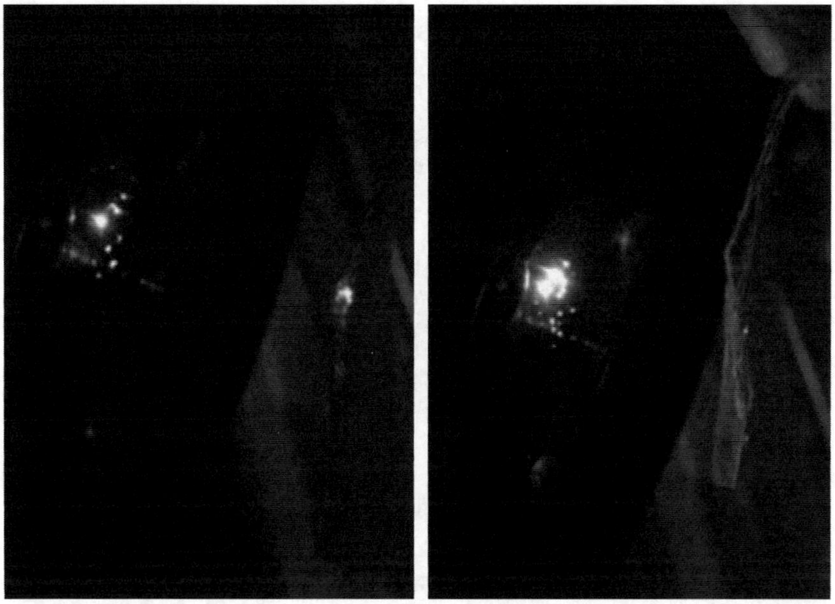

Figure 1. (a) (left) and 1b(right). Fig. 1a presents the violet emission of the polystyrene film excited by the infrared laser pulses (the right edge of the photo) and the moderate white emission at the screen (the left edge of the photo). (b) presents blue luminescence all over the same polystyrene film and simultaneously excited bright flame at the screen

violet luminescence is observed small dark spots (with dimensions of several μ m) appear in the irradiated region (Fig. 2).

It is worth to emphasize that these dark spots in the regions of the films with any structural in-homogeneities demonstrate the anisotropy of their geometry. Say, some polystyrene films acquired the superstructure of parallel straight lines during their solidification at the glass slides, revealed clearly with the optical microscopy (Fig. 3).

It turned out that in the regions where this superstructure prevails the dark spots demonstrate the rectangular geometry close to squares with the sides of these squares being parallel to the lines of the super-structure. On the other hand, when the blue light emission of the films appears the new regular super-structures inside the films are observed (Fig.4).

These super-structures have the geometry of the concentric rings. The distances between the neighbor rings revealed by the optical microscopy are about several tens of micrometers. It is worth to note that the internal appearances of the systems of the concentric rings created in the different regions of the films look like each other. This means that the concentric rings produced in different regions of the same film are characterized by the analogous internal morphologies of the rings concerning the microscopic scale.

We suppose that either the rectangular dark spots formed at the initial stage of the laser irradiation or the concentric rings formed at the next stage appeared as the result of the mutual interactions with the positive feedback between the laser radiation propagating through the film and the structural in-homogeneities created inside the film by this radiation. This means that the fluctuations of the spatial distribution in the films morphology (and the local optical parameters correspondingly) and the corresponding distortions of the spatial distribution in the laser radiation flow enhance mutually each other. And these mutual enhancements result in the creation of the two rather unusual kinds of the in-homogeneities of the films: the rectangular dark spots and the concentric systems of the rings. Concerning the creation of the dark spots

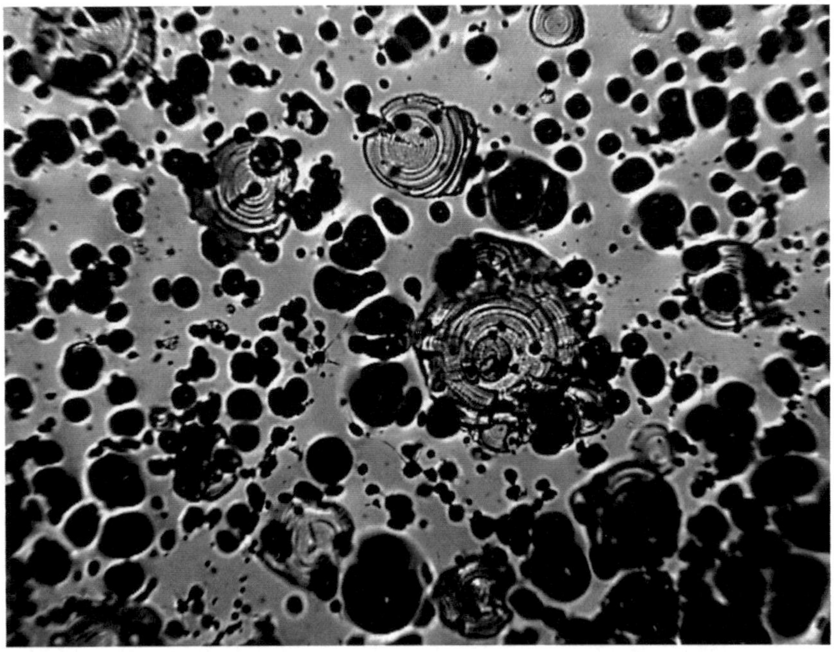

Figure 2. Two new kinds of the self-organized structural distortions in the polystyrene film produced by irradiation of nanosecond pulses of 1.064 μm infrared laser (rectangular dark spots and concentric rings)

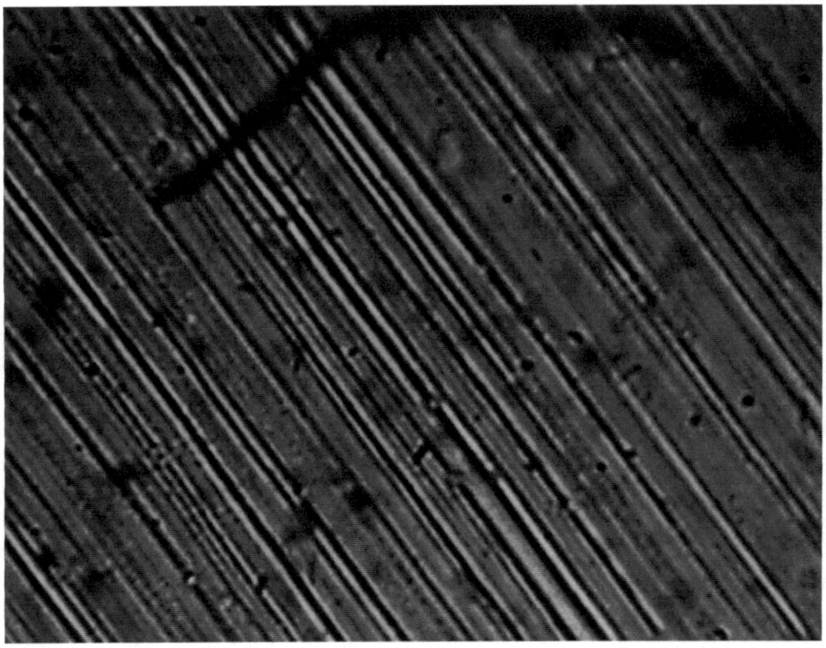

Figure 3. Self-organized linear structure of the polystyrene film solidified from the solution

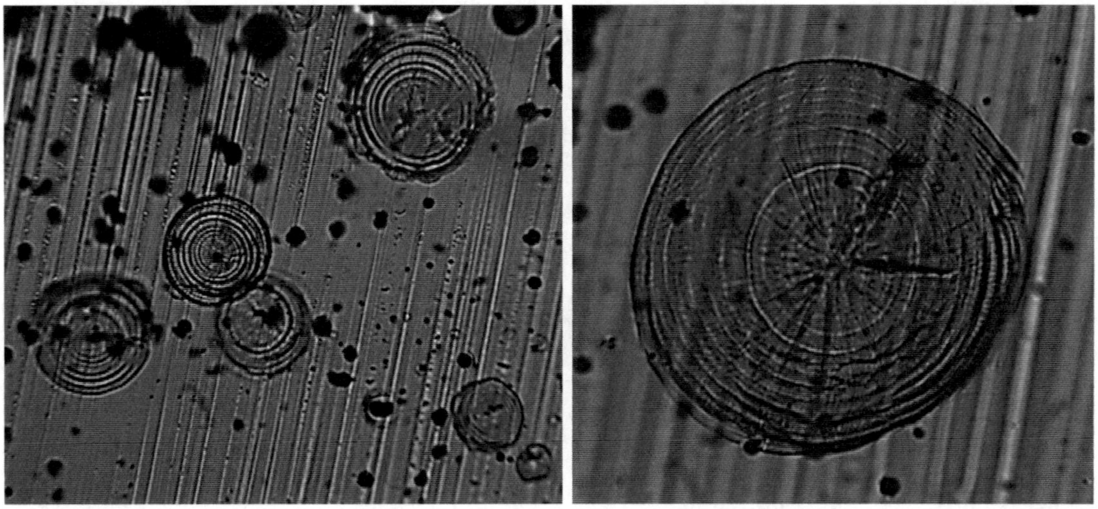

Figure 4. The systems of concetric rings produced by nanosecond laser irradiation of the polystyrene film. The horizonytal dimension of the left photo is 500 μm, of the right photo 120 μm

we suppose that when the laser beam passes through the polystyrene in the vicinity of one of the linear in-homogeneities its trajectory is distorted due to the gradient of the refractive index induced by this structural in-homogeneity. Due to the local bending of the light flow the distribution of its intensity will be varied. This variation will induce the further structural and refraction variations enhancing the bending of the trajectories of the beams, etc. In the local regions of the increase of the light intensity the refraction index will be changed due to the optical non-linearity of the substance inducing the further local enhancement of the intensity. Hence the fluctuations of the local structure of the film will induce the local enhancement of the absorption of the light energy. Due to the increased energy absorption the enhanced local chemical decomposition of the polystyrene will start with the subsequent creating of the dark spot (e.g. its darkness can result from the local carbonization of polystyrene induced by the combined actions of light and local heating [14]). The local increase of the refraction index will induce the attraction of the light flow to the place of this increase. Hence the light flow will be redistributed in the direction parallel to the initial linear superstructure creating the rectangular dark spots with limited dimensions along and perpendicular to the initial linear super-structure. The detailed quantitative analysis of these processes will be published later.

The systems of the concentric rings created by the laser irradiation of polystyrene films with 20 nanoseconds light pulses and observed in the transmission light manifest the corresponding local modifications of the refractive index as well. The positive feedback between local disturbances of the refractive index and the light flow redistribution will be able to produce the essential enhancement of the light intensity at the screen behind the film with corresponding generation of the bright flame. In this case the self-generated structural and refraction in-homogeneities will work as the local Fresnel focusing phase lattice. On the other hand, the part of the incident laser flow will be deviated by this phase lattice in the directions along the film exciting the blue luminescence across the whole film area. The quantitative analysis of these processes which we plan to finish soon will take into consideration the instabilities of the morphology, chemical composition and corresponding optical parameters of the film combined with the spatial redistributions of the delivered and absorbed light beam energy. The mutual positive influence of these factors is capable to create new interesting kinds of the combined waves of the transformed

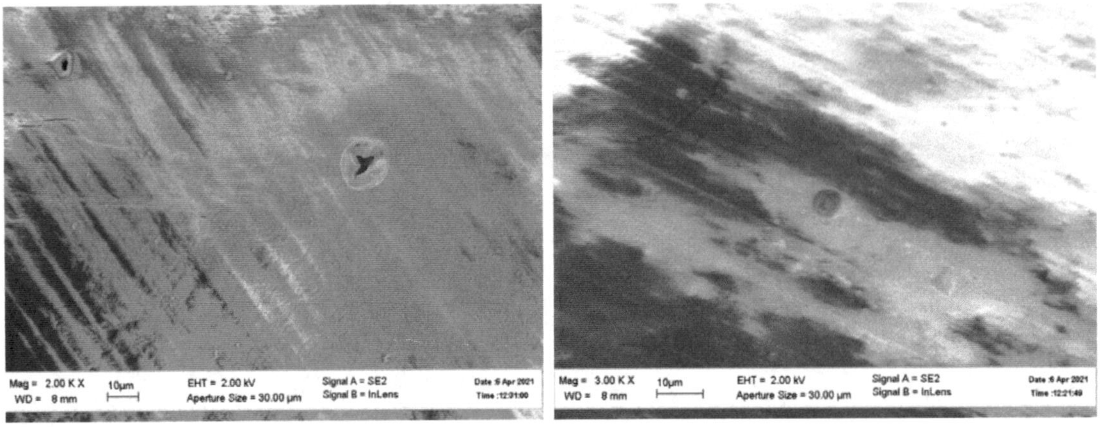

Figure 5. Scanning electron microscopy of the blisters produced in the polystyrene film by nanosecond pulses of the infrared laser

photon flows and structural defects (see, for example, the theoretical papers of the team of Prof. V. Emelyanov [15,16]).

The scanning electron microscopy of the films at the stage of the concentric rings revealed blisters with the micrometer diameters (Fig. 5).

Some of these blisters had anisotropic crack-like holes in their centers and manifest severe local over-heating. We suppose that these blisters work as the sources of the initial scattering of the laser light inside the film resulting in the self-organizing creation of the concentric rings as well as of the excitation of the blue luminescence all over the film. Generally speaking, the both new super-structural components found in our experiments (the rectangular dark micro-spots and the circular systems of the concentric rings) were formed due to essentially non-linear behaviors of the absorption and refraction of the polystyrene films as functions of the light intensity. The analytical estimations of their creation which we are going to make later will consider the in-homogeneities of the local temperature and of the intermolecular and interatomic bonds as well as the redistributions of the energy supply by the laser beam and the outflow of the energy by means of the convection and thermal conductivity. Our experiments on the laser irradiation of the same films immersed in liquid nitrogen showed that the balance between the inflow and outflow of the energy is one of the main factors of regulation of these super-structures. It turned out that the behavior of the films immersed into liquid nitrogen has changed radically in comparison with the open air. We have not observed in this case either the creation of the dark spots or the formation of the concentric rings. Correspondingly the excitation of the blue luminescence all over the film was not observed as well. We ascribe these changes to the severe enhancement of the outflow of the light energy via liquid nitrogen. Instead of the concentric rings the star-like systems of the micro-cracks induced by the laser irradiation were found (Fig. 6).

These facts can be attributed to the significant enhancement of the energy introduced into the film by the laser via liquid nitrogen, so the photo-thermal destruction of the polystyrene structure induced by the laser beam at the room temperature could not start. On the other hand, the thermo-elastic stresses induced by the local thermal expansion of the polymer at the point of the strong multiphoton absorption created tensile stresses around this point. Due to the decrease of the plasticity of the polymer these tensile stresses could not relax by the deformation mass-transfer and created the star-like system of the radial cracks.

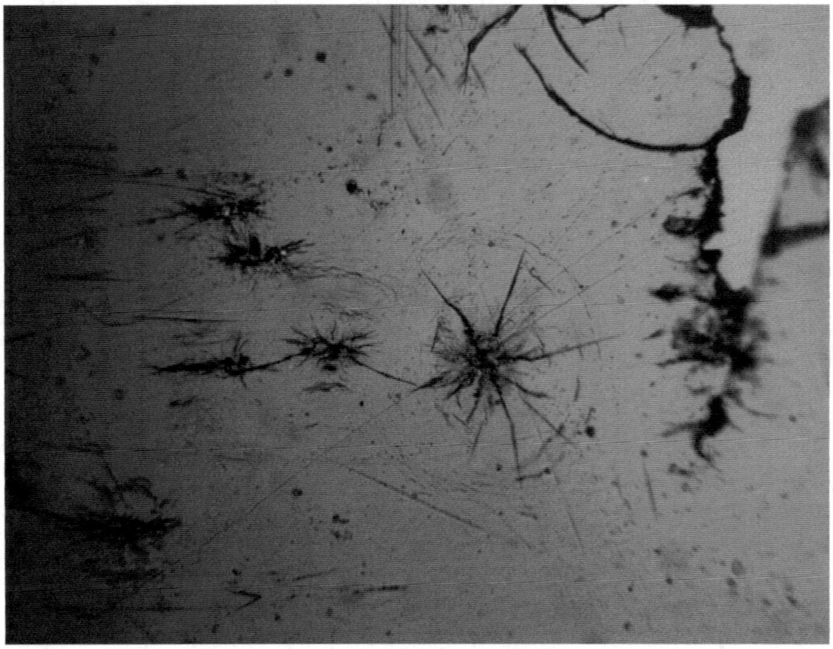

Figure 6. Optical microscopy of the polystyrene film irradiated by nanosecond laser pulses in liquid nitrogen

Conclusions

We found two new kinds of self-organized super-structures produced in the polystyrene films by the multiphoton absorption of the infrared 1.064 μm laser nanosecond pulses: the non-transparent micro-inclusions with unusual rectangular geometry and the systems of coaxial micro-rings. The appearances of both of these in-homogeneities are ascribed to the active interactions between the embryo of the structural distortions and the redistributions of the laser light flow induced by these distortions resulting to their mutual enhancement.

The systems of the coaxial micro-rings created generate severe redistributions of the laser flow propagating through the polystyrene film: the effective focusing of the light flow behind the film inducing the laser avalanche and the precise partial scattering of the laser flow inside the film resulting in its overall blue luminescence.

The creation of the rectangular and the coaxial structural distortions produced in the polystyrene films by infrared laser pulses are ascribed to the effective mutual interactions with the positive feedback between the emerging structural in-homogeneities and redistributions of the laser flow induced by these in-homogeneities.

Acknowledgements

The work was carried out according to the state assignment of ISSP RAS.

References

[1] S Sun, R Dalton (Ed.) 2017 *Introduction to Organic Electrinic and Optoelectronic Materials and Devices* CRC Press, USA 1170

[2] T Nakanishi (Ed.) 2019 *Supramolecular Soft Matter. Applications in Materials and Organic Electronics* John Wiley & Sons, USA 490 p.

[3] S Kalimtech, *Polymers in Organic Electronics* 2020, ChemTech pUblishing, Toronto 612

[4] O Ostroverkhova (Ed.) 2013 *Handbbook of Organic Materials for Electronic and Optoelectronic Devices* Woodhead Publishing Ltd., USA 824

[5] M Condepudi, I Prigogine 2002 *Modern Thermodinamics* Wiley & Sons 435
[6] G-F Gouyet 1995 *Physics and Fractal Stuctures, Paris, Ecole Polytechnique* 247
[7] N V Klassen, M M Myshliaev 2019 *J. Phys.: Conf. Ser.* **1309** 012005
[8] A L Volynskiy, N F Bakeev 2005 *Structural Self-Organization of Amorphous Polymers* Moscow, PhysMathLit 232 (in Russian)
[9] S Kozuukhin et.al. 2019 *Optics and Laser Technology* **113** 87 – 94
[10] K Ding et.al. 2020 *Optical Materials* **109** 110284
[11] S Graf 2020 *Adv. Opt. Techn.* **9**(1-2), 11-39
[12] E Rebollar et.al. 2015 *European Polymer Journal* **73** 162-174
[13] N V Klassen, A P Pokidov, V V Kedrov 2018 *Inorg.Mater.: Applied Research* **9** 141–146
[14] M Biyurin et, al. 2003 *Chem. Rev.* **103**(2) 519 – 552
[15] V I Emelyanov et.al 2000 *Laser Physics* **10** 733-740
[16] V I Emelyanov, A V Rogacheva 2002 *Proc. SPIE* **4748** 283 - 292

Structural transformations in the system lanthanum bromide – water – electrical field

S A Vinokurov, N V Klassen, A D Orlov

Institute of Solid State Physics, Russian Academy of Sciences, Chernogolovka, Russia

E-mail: sergvin@issp.ac.ru

Abstract. We found several manifestations of severe influence of relatively week direct electrical fields on interactions of water with lanthanum bromide. The rate of dissolution of these crystals in water at the negative electrode exceeds this value at the positive electrode by several times. Pulse-like growth of solid micro-rods at the negative electrode from the solution of lanthanum bromide subjected to direct electrical field was observed. Auto-oscillations between the contents of water bound in hydrated lanthanum bromide and free water were found. These oscillations are attributed to severe difference of the electrical polarizabilities of free and bound water.

1. Introduction

The actuality of studies presented in this paper is determined by several reasons. Single crystals of lanthanum bromide belong to the best scintillators required for medical diagnostics, radiation monitoring etc. [1-3]. They are highly hydroscopic and inclusions of water deteriorate their technical parameters significantly. On the other hand lanthanum bromide belongs to highly brittle materials. This property induces severe problems for the processes of preparation of scintillators from this crystal. Penetration of water produces structural non-homogeneities and internal stresses between the hydrated and non-hydrated regions and can be one of the reasons of this brittleness. Lanthanum bromide has ionic interatomic bonds . So its structural non-homogeneities can create local electrical fields. Due to high electrical moment of water molecules these fields are capable to rearrange their distribution enhancing the internal stresses. So studies of behavior of various combinations from lanthanum bromide, water and electrical field are important for practical applications. E.g., developments of any ways of protection of these crystals against penetration of water molecules or of their removal are actual. On the other hand these studies will give new information about details of structure of water which is vague up to now [4, 5]. The spatial distribution of molecules in liquid water is not chaotic. They form well arranged clusters including several tens or even hundreds of elementary water. Generally speaking the interaction of water with other substances should be dependent strongly on structures of these clusters. So systematical studies of these interactions with materials having well known atomic structures can give important information about arrangements of water clusters. Anyway more than 70% of living bio-systems including humans consist from water solutions. So collecting detailed information of behavior of these solutions in various situations is necessary for deeper understanding of functional activities of human organisms. Water solutions of salts of inorganic metals and electrical fields play important roles in these functions, so our work can be interesting for biologists. Earlier we published a paper about particularities of interactions of water with micro-capillaries of plants in weak electrical fields [6]. Comparison of results of the former paper and of this one can be useful for deeper understanding of interactions of water with inorganic and organic systems [4, 5].

2. Experimental techniques and materials

We used pieces of lanthanum bromide with irregular initial forms obtained from cracked single crystals grown from the melt [7]. They were kept in hermetic vessel but nevertheless were hydrated partially with molecules of water [8 - 12]. For studies of transformations of these samples in water they were placed at the glass slide of the optical microscope between two aluminum electrodes made from parallel plates having 1 mm thickness and fixed to the glass hermetically with 3 mm distance between them. Drops of distilled water were poured between the electrodes. The transformations were registered with the video-camera via the microscope. Direct electrical field was applied to the electrodes with the values varied from zero to 22 volts.

3. Experimental results and discussion

Generation of significant asymmetry of the dissolution rates of lanthanum bromide samples in water was found during application of the direct electrical field which was rather weak (not more than 100 V/cm) – Figs. 1, 2. It turns out that the dissolution rate in the region of the solution close to the negative electrode exceeds the rate in the region close to the positive electrode by several times (the ration of these rates can achieve even 10 times). Fig. 1 (a,b,c,d) demonstrates the sequence of the states of the lanthanum bromide crystals at various moments of time with the temporal intervals between them of about 20 seconds.

The negative electrode is place at the right side, the positive one – at the left. It is seen clearly that the extreme right piece is dissolved faster significantly than the pieces with more left positions.

Fig. 2 demonstrates two stages of dissolution of the pair of $LaBr_3$ pieces. The left piece is placed at the positive electrode whereas the right one – at the negative. The time interval between the photos is 20 seconds. The comparison of the changes of the samples manifests the difference in the dissolution rates being not less than 10 times.

Fig. 3 reveal the distinct difference of the chemical compositions of the solution at the left and the right sides of the sample of lanthanum bromide. The solution at the left side is much more red than the solution at the right side. It is well known that bromide agglomerations in water have red coloration [13]. During the dissolution of lanthanum bromide in the direct electrical field the negative ions of bromine should be transported preferentially to the positive electrode whereas the positive ions of lanthanum should be attracted in the opposite direction. The bromine ions are coagulated in the red agglomerations. These agglomerations deteriorate the removal of the dissolution products by diffusion. Hence the dissolution at the left side of the sample should be retarded. This explanation is in good correspondence with the experimental results.

One more "geometrical" experimental particularities of the dissolution processes of lanthanum bromide is not connected immediately with the external electrical field. Some pieces demonstrated significant anisotropy of the dissolution rate (Fig.4). This means that this rate in a certain direction was bigger than the rate in perpendicular direction by several times. Optical microscopy of the dissolution of the samples with this kind of anisotropy revealed quasi-periodical superstructure of parallel lines with the distances between these lines being several tens of micrometers. The maximal dissolution rate was observed in the direction perpendicular to these lines. This fact is confirmed by the Figs. 1 and 4. Fig. 5 shows one more particularity of the dissolution induced by the direct electrical field. As it was emphasized earlier the dissolution rate at the right side of the solution (close to the negative electrode) was higher significantly

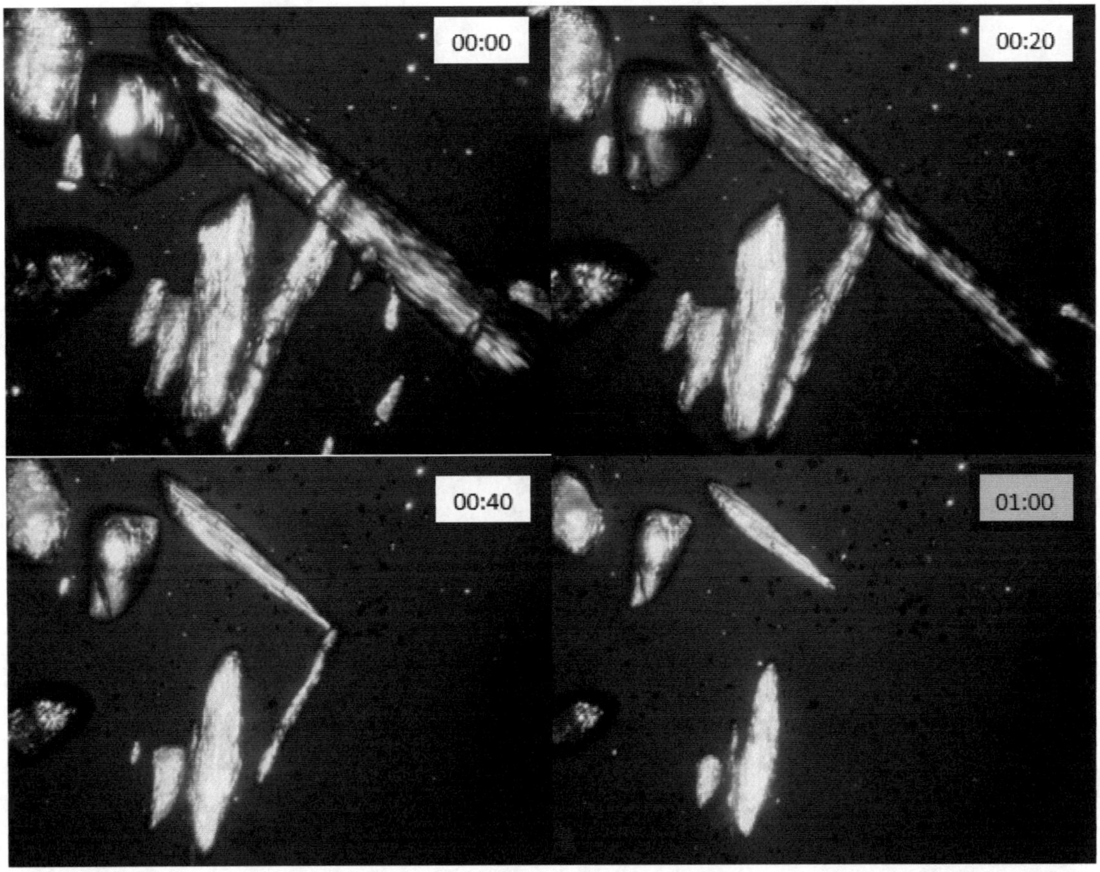

Figure 1. a,b,c, d. Severe asymmetry of the dissolution of lanthanum bromide crystals in the direct electrical field. The negative electrode is at the right side

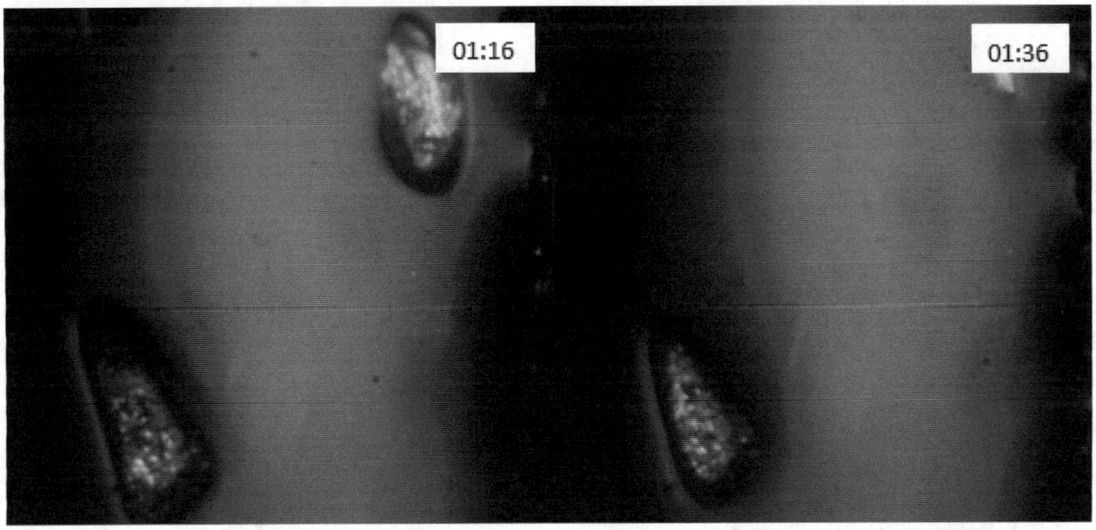

Figure 2. The asymmetry of the dissolution rates of lanthanum bromide crystals at the left (positive) and the right (negative) electrodes. The temporal interval between the photos is 20 seconds

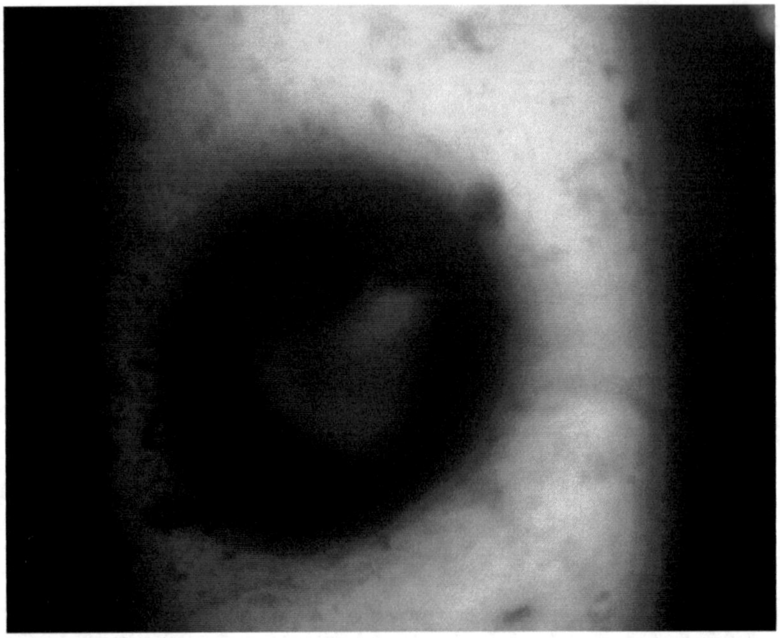

Figure 3. The asymmetry of the coloration of the water solution of lanthanum bromide in the direct electrical field. The negative electrode is at the right

than at the left side. This asymmetry at Fig, 5 was observed when the field was 10 V/cm. But when the external field has been increased abruptly to 100 V/cm, the dissolution has been changed by local growth of several rod-like ledges, starting from the negative electrode and separated from each other by distances about 0.5 millimeters. Figs. 5a, 5b and 5c were made with the time intervals of 1 second. So the growth rate of the upper ledge is about 0.5 mm/sec. When this growing ledge achieved almost the middle of the inter-electrode space the longitudinal growth has been changed by the transverse dissolution (Fig. 5c). The thickness of the ledge was decreased by about 3 times in one second. The analogous change from the longitudinal growth to the transverse dissolution was observed at the lower ledge.

Fig. 6 shows one more version of the anisotropy of the dissolution induced by the combined influence of external electrical field and water. The initial state of .a piece of lanthanum bromide in the direct field of 100 V/cm but without water is presented at Fig.6a. The sample has an oval geometry and does not demonstrate internal anisotropy. The next Fig.6b shows the same piece after pouring water between the electrodes. In the regions contacting with neighbor pieces of lanthanum bromide some features of anisotropy appeared (e.g., a beak-like ledge at the left side of the upper part of the sample). The next photos are made with the time intervals of one second. Fig.6c shows slight rotation of the sample counterclockwise. Fig. 6d demonstrates further rotation and the formation of a triangle at the left side and a thin horizontal line going from the sharp left end of the triangle towards the negative electrode. The creation of the beak-like ledge and the subsequent triangle geometry resulted from the anisotropy of the dissolution is induced by combined action of water. electrical field and point-like contacts with the neighbors. For example relatively sharp angle between the edges of the sample under consideration and its upper neighbor produces significant local enhancement of the electrical field due to anomalously high electrical permittivity of water [9]. This enhanced electrical field induces modification of internal structure of the crystals and subsequent anisotropic acceleration of the dissolution. This anisotropy produced a thin ledge at the left side of the sample (Fig.6e).

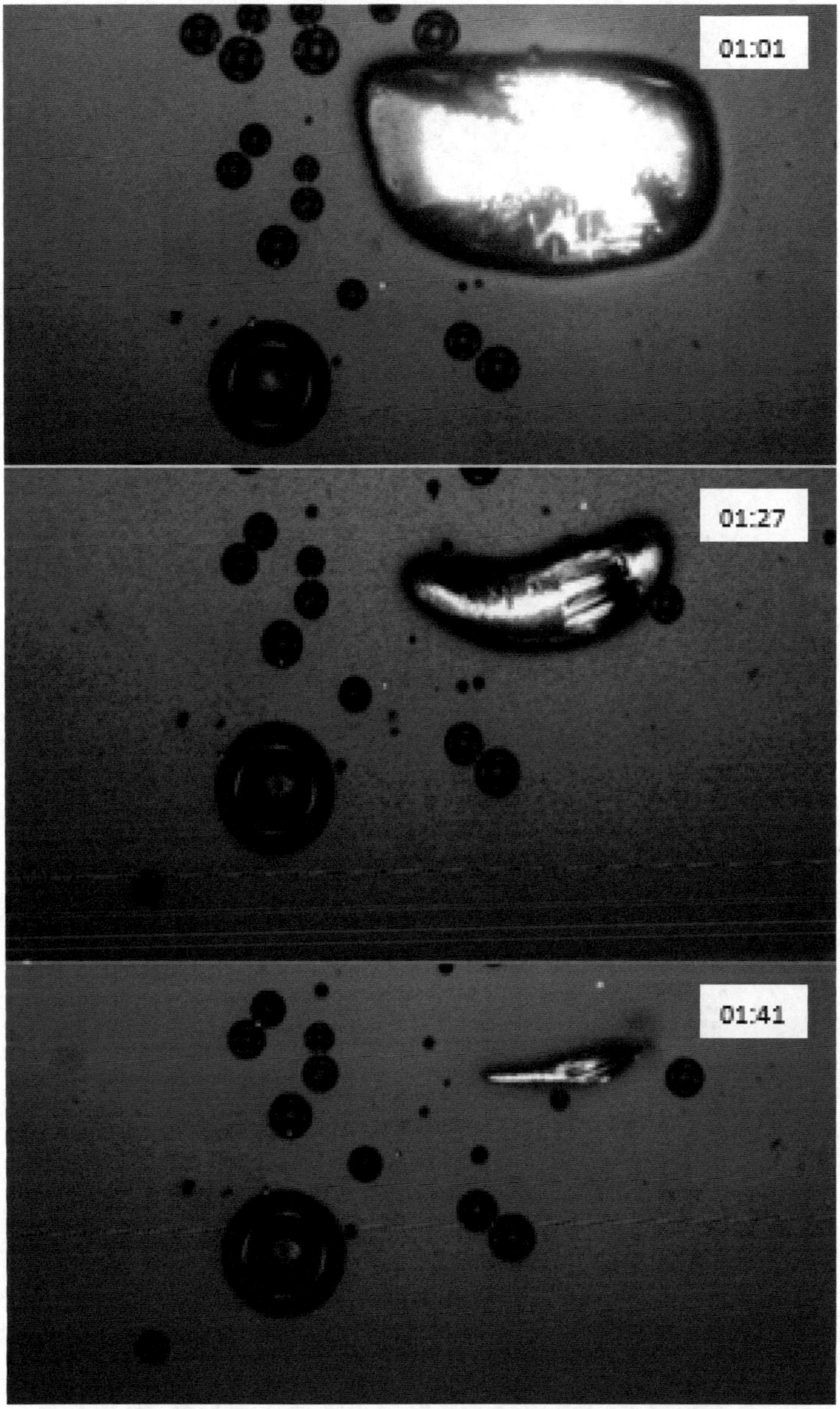

Figure 4. The anisotropy of the dissolution of lanthanum bromide crystals. The time of the photos is shown at the left of the bottom line.

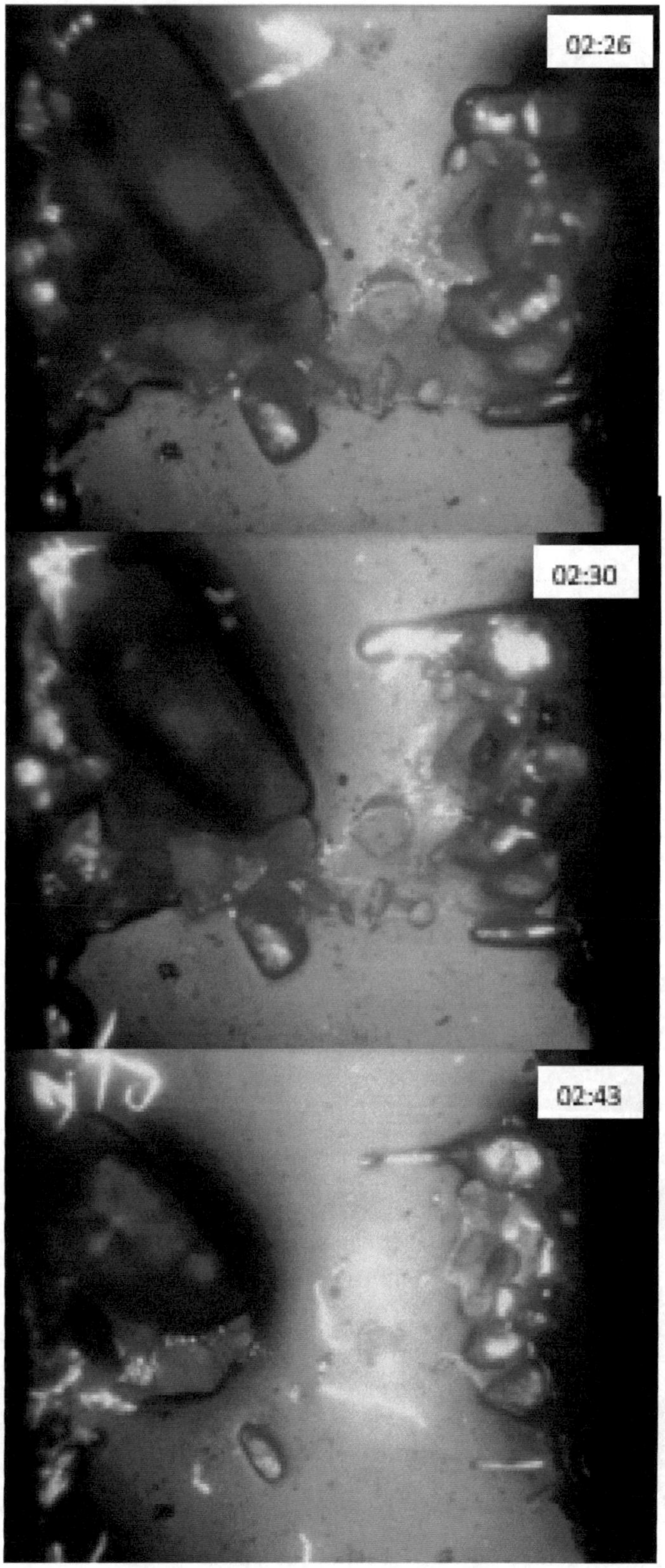

Figure 5. The asymmetry of lanthanum bromide grains in water solution at the right (negative) and the left (positive) electrodes.

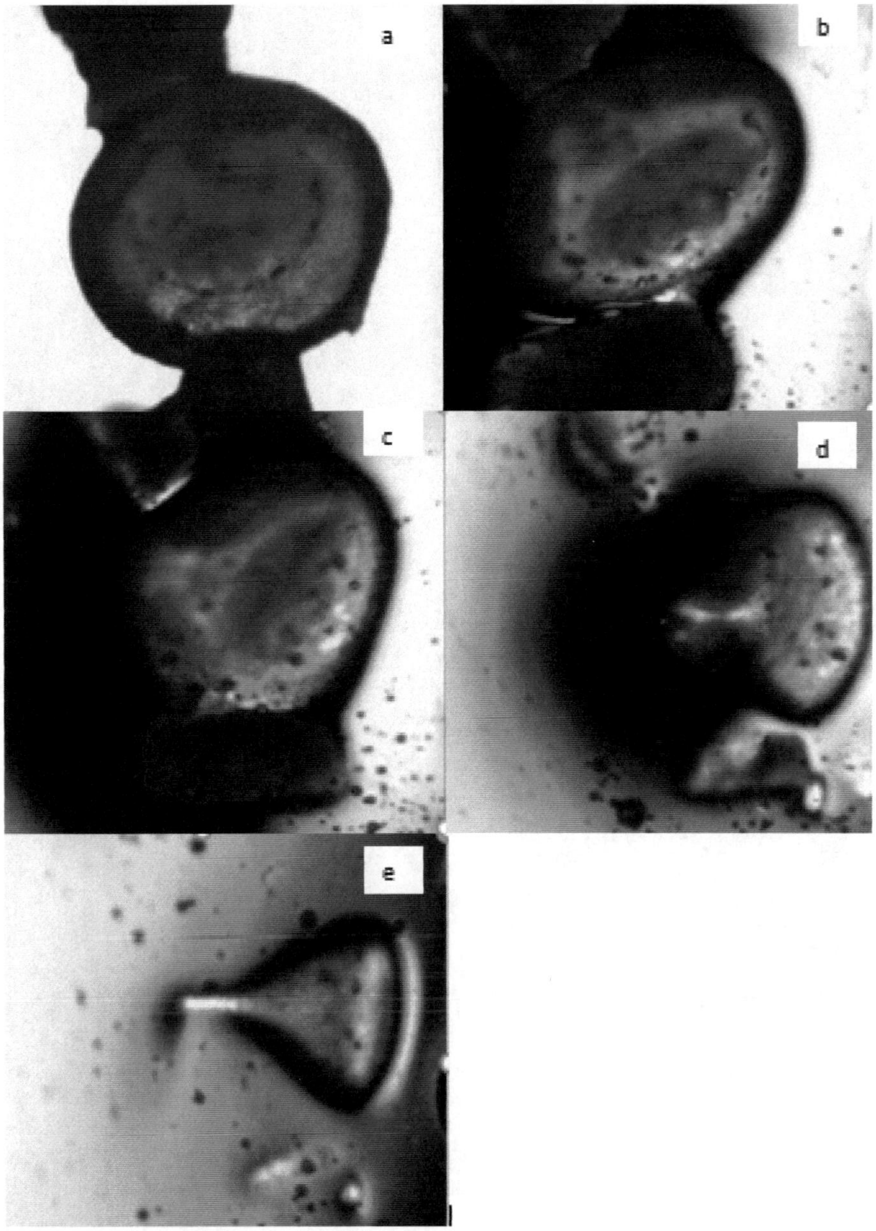

Figure 6. a,b,c,d,e. Dissolution anisotropy caused by the combined action of the external electrical field and water

The counterclockwise rotation of the sample (Fig.6c, 6d) and the final fixation of the thin needle-like ledge of the sample in the direction parallel to the external electrical field make us to assume that the initial vertex of the triangle and the resulting ledge are charged negatively. So they are repelled by the negative electrode. The creation of this negative charge can be attributed to accelerated dissolution of the sample surface and fast removal of positive ions of lanthanum.

One more manifestation of the influence of the electrical field on the system lanthanum - water was found when the direct filed was applied to the thin polycrystalline layer of lanthanum

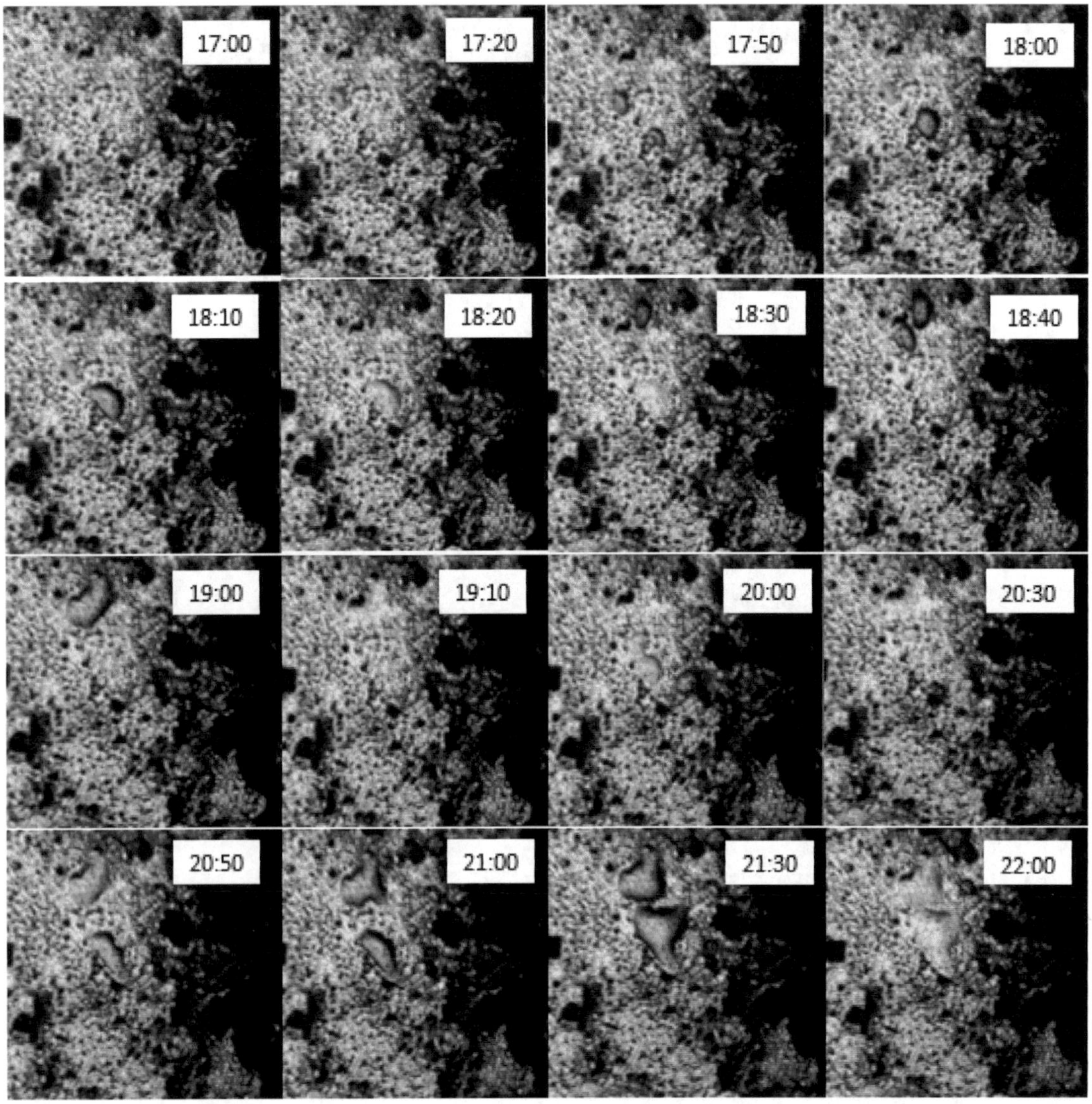

Figure 7. Auto-oscillating appearances and disappearances of water drops at the surface of the polycrystalline lanthanum bromide precipitation layer subjected to the direct electrical field

bromide precipitated from water solution on the galss slid eof the microscope (Fig. 7). The first photo presents this layer at the innitial moment (17.00). The transverse dimensions of the photo are 0.5 mm. After 20 seconds of observations two small drops of water appeared at the left half of the top region (17.20). In the next 30 seconds one more drop appeared in the middle of the sample (17.50). Theh the drops at the left part disaapeared whereas the middle drops was growing during 20 seconds (18.00 and 18.10). At the next stage the middle drop began to diappeare whereas at the top region of the sample a small drop appeared once more (18.20). The photos at 18.30 and 18.40 demonstrate creation of two more drops at the top left part and subsequent connection of these two drops (19.00). Then the upper drop disappeared (19.10). During the next intervals of 30 seconds two new drops appeared in the middle part (20.00) and disappeared (20.30). During the next 120 seconds two new drops appeared, grew significantly achieving the biggest dimensions and disppeared again in the middle region .

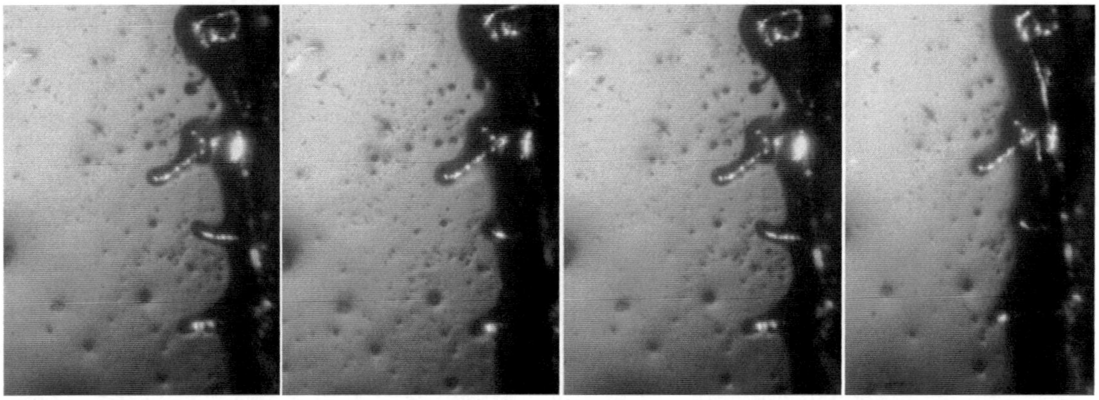

Figure 8. Auto-oscillations of the water meniscus level at the negative electrode in the direct electrical field. The interval between the photos is about 10 seconds

The experimental observations presented at Fig. 7 reveal oscillating behavior in the system water – lanthanum bromide induced by the expernal electrical field. These oscillations manifest themselves in the appearance of water drops at a certain place of polycrystalline lanthanum bromide, subsequent their disappearance, next appearance in another point, etc. The typical time intervals of these oscillations are of the order of 10-20 seconds.

Fig 8 presents another manifestation of the oscillating extraction of water from polycrystalline lanthanum bromide precipitation induced by the direct electrical field of the amplitude about 100 V/cm. Here the modifications of the meniscus of water adjusted to the negative electrode from aluminum is presented. The spatial amplitude of the oscillations of the edge of the water meniscus is about 100 micrometers. The time intervals between the neighbor photos are about 10 second. The sequence of the positions of the meniscus corresponds to oscillations of the amount of free water extracted from the hydrated lanthanum bromide when the electrical field is applied.

Conclusions

- Interaction of water with crystals of lanthanum bromide is regulated in a wide range by relatively low electrical field either increasing or decreasing significantly the rate of dissolving of these crystals by water.
- Essential asymmetry of dissolving of lanthanum bromide grains has been found in direct electrical field: at the negative electrode the rate of the dissolving exceeds the rate at the positive electrode by several times. This is explained by significant difference in the mobilities of the lanthanum and bromine ions dissolved in water.
- Oscillations of hygroscopicity of lanthanum bromide crystals in weak direct electrical field with periods of several seconds has been found. This phenomenon was expressed in abrupt extraction of micro-drops of water at the surface of solid polycrystalline lanthanum bromide precipitated at glass slide of the and subsequent their disappearance. These oscillations are explained by strong difference of the electrical polarizabilities of free and bound water molecules resulting in local variations of the internal electrical fields and the mechanical stresses induced by hydration and dehydration of lanthanum bromide.
- Pulse forming of needle – like microcrystals in water solutions subjected to pulses of direct electrical field has been found . These pulsed forming is explained by high electrical polarizability of water-less lanthanum compounds in comparison with the hydrated compounds.

- Strong anisotropy of the dissolution rate of some grains of lanthanum bromide in water has been found even without the external electrical field. Ths grains differ from others by the presence of microscopic systems of periodical parallel rows. The dissolution rate was bigger significantly in the direction perpendicular to these rows. The resulting forms of such grains before their complete dissolution became needle-like.
- The facts of asymmetry and anisotropy of dissolution of lanthanum bromide in water, auto-oscillations of local content of water and presence of layer-like superstructures are important for the development of optimized techniques of processing of as-grown lanthanum bromide crystals.

Acknowledgements
The work was carried out according to the state assignment of ISSP RAS.

References

[1] E V D Van Loef, P Dorenbos, C W E Van Eijk, K Krämer, H U Güdel 2001 *Appl. Phys. Lett.* **79**(10) 1573–1575. doi:10.1063/1.1385342
[2] *A Performance Summary of Lanthanum Bromide Scintillators*, https://www.azom.com/article.aspx?ArticleID=18025
[3] G Hull, et.al., 2019 *Nucl. Inst. and Meth. in Phys. Research* **A 925** 70-75
[4] X F Pang 2014, *Water: Molecular Structure and Properties*, World Scientific Publishing Co. Pte Ltd, 491 p.
[5] I Olovsson 2018 *Wonders of Water: The Hydrogen Bond in Action*, World Scientific Publishing Co. Pte. Ltd., 140 p
[6] S A Vinokurov, I S Tsebruk, T D Betenina, N V Klassen 2020 *J of Phys. C* 1560,01 2042
[7] A V Berdnikova, A D Orlov, S Z Shmurak, N V Klassen, et.al. 2017 *Instruments and Experimental Techniques* (2) 29-34
[8] A F Wells 1975 *Structural Inorganic Chemistry* Oxford University Press, Ely House, London, 1127 p
[9] D Brown et.al., 2017 *Cryst. Growth. Des.,* **17**(8) 4347-4364.
[10] T J Boyle, et.al. 2010 *Polyhedron* May, 10, **29**(7), 1784 – 1795
[11] K Kiyohara, R Minami 2018 *J. Chem. Phys.* **149** 014705
[12] A S Larsen, et.al. 2017 *J. Pharmats. Sciences* **106** 148 – 155
[13] M H William, ed., 2011 *CRC Handbook of Chemistry and Physics* (92nd ed.). Boca Raton, FL: CRC Press. p. 4.121

Spatial and spectral features of the luminescence of liquid nitrogen stimulated by infrared lasers

N V Klassen, P V Provotorov

Institute of Solid State Physics, Russian Academy of Sciences, Chernogolovka, Russia

E-mail: klassen@issp.ac.ru

Abstract. Spectral features of cryogenic liquids, especially nitrogen, seem to be clearly investigated due to the wide application in science. On the other hand, material properties can really be different under intensive laser irradiation. During irradiation of liquid nitrogen with pulsed (pulse duration 20 ns) YaG: Nd 1064 nm laser with average power of 0.3 W we have found bright luminescence in the visible region with five narrow spectral lines. This phenomenon can be explained by the multiphoton excitation of nitrogen molecules and ions with infrared photons. Its spatial and spectral characteristics are attributed to Amplified Spontaneous Emission. The main features include super-linear dependence of the emission intensity on the intensity of the excitation radiation, limited amount of narrow spectral bands and conical geometry of the emission with the clear separation of the cones in correspondence with the spectral lines of the emission. This phenomenon manifests liquid nitrogen as the convenient substance of Kerr type for studies of various non-linear optical processes by means of nanosecond lasers instead of femtosecond lasers applied usually for these purposes.

1. Introduction

This paper is dedicated to one of the fields of light-matter interactions. R&D on the interactions of light and matter have the vast history starting from ancient centuries. At that time the great Greece scientist Aristotle became one of the first persons, who tried to apply transformations of light in practice. The legend tells that Aristotle destructed Persian ships attacking Greece with the beams of focused Sun light. During many hundreds of years which have passed from that moment the science and applications of the light matter interactions expanded fantastically and penetrated into all the fields of humanity activities [1-7]. Diagnostics of a wide variety of materials including inorganics, organics and bio-organics by means of combinations of optical microscopy and optical spectroscopy belong to the most popular branches of this work. Studies of photoelectrical and photochemical processes inducing transformations in electron and atomic-molecular systems in various materials form the next step. Development of lasers opened the new and highly productive branch of the light-matter interactions [8 – 12] including various kinds of laser technologies (cutting, drilling, welding, hardening, forming of nanoparticles by means of laser ablation and sublimation, etc.). Active interactions of laser beams propagating through various suspensions of nanoparticles in colloidal water solutions with these nanoparticles created wide sets of rather complicated nanostructures [9]. Irradiation of materials (solids, liquids, gasses) with femtosecond laser pulses opened a new branch of the light-matter interactions [13]. Several principally new phenomena have been discovered: filamentation of laser beams, conical emission of the luminescence light and ultrafast generation of the emission pulses with supercontinuum optical spectrum. This our paper presents a set of new results concerning unusual features of the spatial and spectral distributions of the visible light emission of liquid nitrogen irradiated with nanosecond pulses of the YAG:Nd infrared laser with 1.064 μm wavelength. On the one hand the spatial distribution of this emission reminds the patterns observed in the cases of femtosecond irradiation. But on the other hand, the spectral features of these emissions demonstrate essential differences.

2. Experimental techniques

The infrared laser emitted 1.064 μm radiation by means of excitation of YAG:Nd crystal with the pulses of light. The mean optical power of the laser was about 0.3 W. Due to the optical amplifier the laser could be used in two kinds of the pulsed emission: millisecond and nanosecond (the duration of the pulses was about 20 nanoseconds). The laser beam had the transverse dimension about 5 mm. Several kinds of lenses were used for focusing of the laser beam (the focal lengths of the lenses were in the range 30 – 150 mm). Liquid nitrogen used in our experiments had the purity of the technical level and contained about 0.5 % inclusions of oxygen and argon. Three kinds of dewars for the laser irradiation of the liquid nitrogen were used in our experiments: the glass dewar with optically transparent walls, external diameter about 12 cm, the molten quartz dewar, external diameter about 4 cm and the metal dewar with two pairs of mutually perpendicular optical windows.

The observations of the propagation of the laser beam through the liquid nitrogen was made using the two schemes: the beam entered the glass or fused quartz dewar vertically top down. The propagation of the beam inside the nitrogen was observed in the horizontal direction via the transparent walls of the dewars. In the second scheme the beam entered the dewars horizontally via either through the transparent walls of the glass or silica dewars or through one of pairs of the windows of the metal dewar. The observations of the propagations of the beam were made in the horizontal direction perpendicular to the entering laser beam. Before the entering the dewars the beam could be focused with one of the lenses. In addition, the pattern of the radiation exiting the dewars was fixed and photographed at the screen with the plane perpendicular to the entering beam.

The spectra of the radiation emitted from the liquid nitrogen were registered by means of the spectrometer with the spectral resolution about 0.1 nm. The dynamics of the emission was registered by the photodiode and oscilloscope with the temporal resolution of about 1 ms. The dynamics of the patterns of the nitrogen emission was registered by the video camera with the temporal resolution of about 3 milliseconds.

3. Experimental results

During laser irradiation of liquid nitrogen in dewar with YAG:Nd pulsed laser (pulse duration 20 ns, average power 0.3 W) with the beam focused by the lenses in the both schemes of the irradiation we have revealed the bright emission of the liquid nitrogen in the visible spectral region. A bright white spot of the emission is seen clearly at the sub-surface region of the liquid. In the lower regions of the liquid the red color of the emission prevailed.

The spatial distributions of the emission were observed rather clearly when the laser beam entered the liquid nitrogen horizontally. The photos of these patterns are presented at Fig.1 and 3 (the beam propagates from the right side to the left).

In the situation presented at Fig.1 the laser beam was focused with the lens having the focal length of 7 cm. The focus point is placed at about 1/3 of the width of the picture from its right edge. Several features observed at this photo are as follows:

- Several small whitish spots are observed in the vicinity the focal point.

Figure 1. Luminescence of liquid nitrogen in glass dewar under horizontal irradiation of pulsed infrared laser. Laser beam goes from right to left

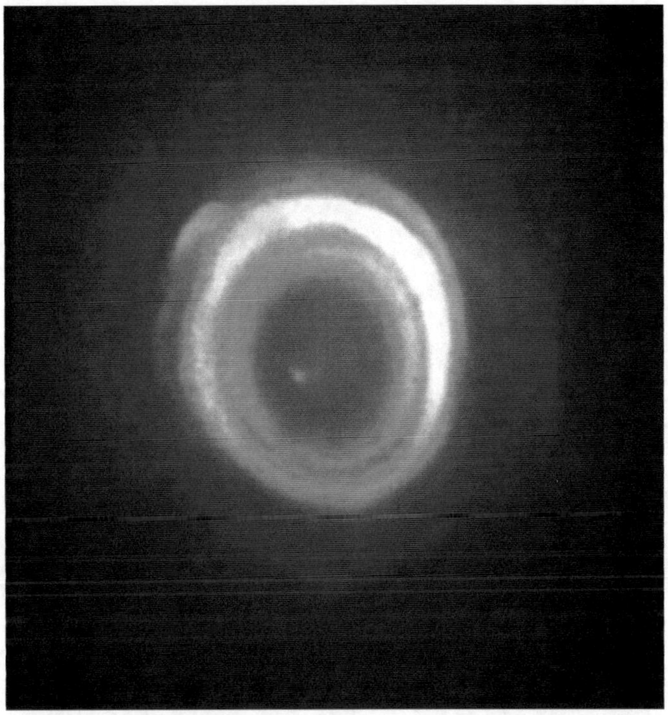

Figure 2. The concentric series of the colored rings at the perpendicular projection of the emission exiting the dewar with liquid nitrogen

- The reddish emission is seen clearly at the left side of the beam trajectory whereas the emission is not observed at the right side of the beam.
- The reddish emission at the left side has a form close to the narrow cone.

When we put a screen at the left side of the beam prolongation 5 cm apart from the exit of the beam from the dewar we observed that the cross-section of the visible radiation emitted from the liquid nitrogen looked like series of concentric circles corresponding to different wavelengths. Spectral analysis showed that in the center of this series the laser beam is situated with a wavelength of 1064 nm. Then the sequence of the rings from the center to the edge corresponds to the wavelengths of 852 nm, 711.7 nm, 610.9 nm, 610.3 nm and 534.7 nm respectively. The borders between these rings look like dark circles and can be seen clearly (Fig.2).

When we used the focusing lens with the focal length of 3.5 cm (instead of 7.0 cm in the

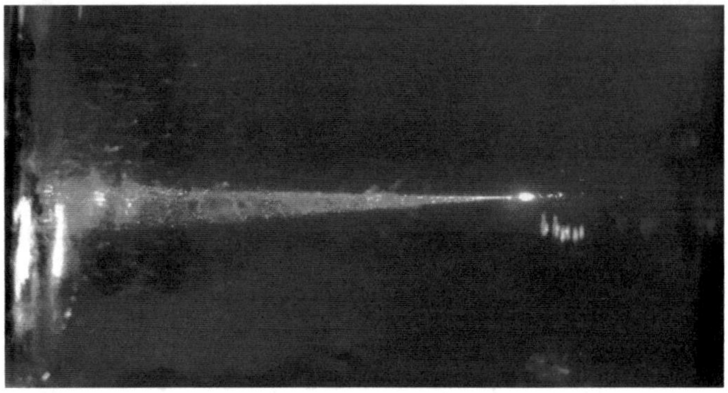

Figure 3. Photograph of the light trace in liquid nitrogen during laser irradiation through a short throw lens. Laser beam goes from right to left

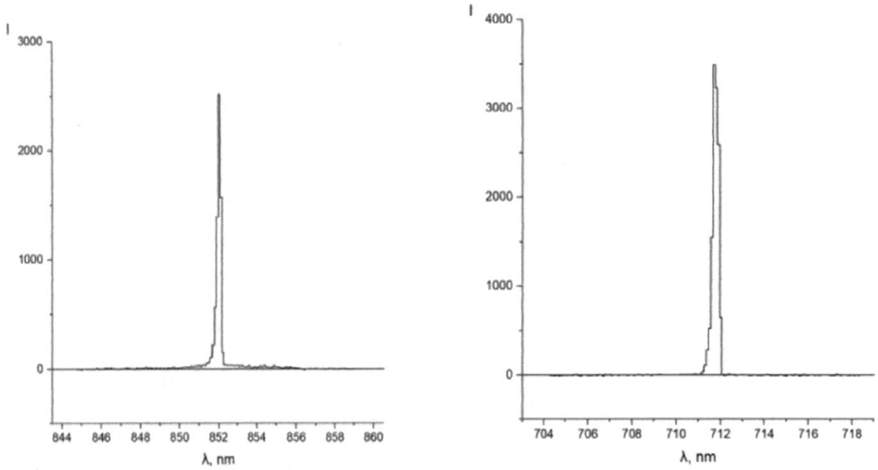

Figure 4. Photoluminescence spectra of liquid nitrogen under a pulsed laser 1064 nm

previous experiment) the brightness of the emission became higher significantly (Fig.3). But the angle at the apex of the cone remained the same approximately. A wider dimension of the red spot at the point of the crossing of the exit radiation and the wall of the dewar is explained by the shift of the focal spot to the right side along the beam.

The study of the spectral composition of the radiation showed the presence of 5 narrow spectral lines with wavelengths at 852 nm, 711.7 nm, 610.9 nm, 610.3 nm, 534.7 nm, the accuracy of the measurements of the wavelengths is $\pm$ 0.5nm. Most notably, the half-width of these spectral lines is about 0.7 nm and the line broadening is quite small.

The measurements of the emission spectra of the bright flashes of the whitish spots appearing spontaneously in the trajectory of the laser beam showed that these spectra consist from several components: the broadened line of the emission of argon neutral atoms and of ionized argon atoms as well as the quasi-continuum plasma-like emission.

One more rather important experimental fact consists in the threshold-like and super-linear dependence of the light emission from liquid nitrogen on the intensity of the exciting laser irradiation. The nanosecond pulses do excite this emission whereas the millisecond pulses of the same wavelength and the same mean power do not. On the other hand, the dependence of the emission intensity on the excitation intensity is essentially super-linear and in the process of gradual decreasing of the excitation intensity the emission disappears abruptly below a certain

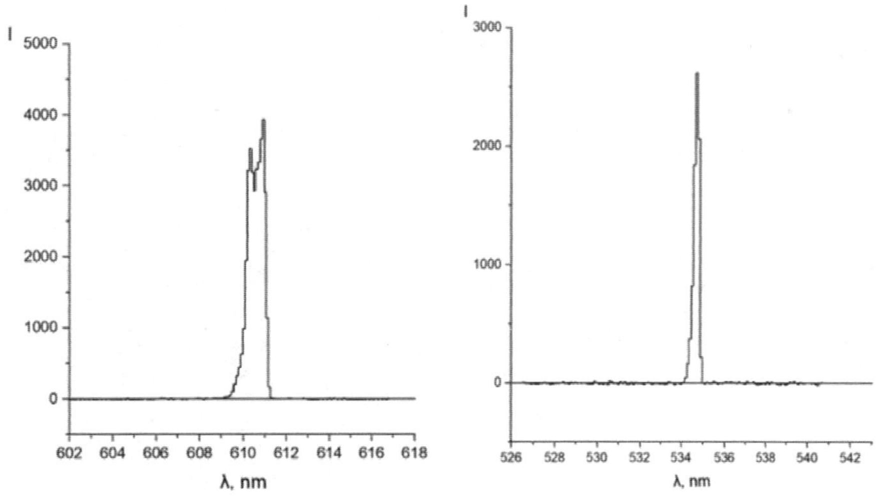

Figure 5. Photoluminescence spectra of liquid nitrogen under a pulsed laser 1064 nm

excitation level.

4. Discussion of the experimental results

The spatial and spectral distributions of the light emission from liquid nitrogen described above demonstrate several unusual features. First of all, the energies of the exciting photons (1.064 μm wavelength) are lower significantly than the energies of the emitted photons (say, for 534 nm emission line this deficiency is two times). At least two possible reasons of this deficiency are described in publications [14,15,16] – collapses of the bubbles of gaseous nitrogen produced by the intense laser irradiation and multiphoton excitation. The formation of the gaseous nitrogen bubbles during laser irradiation of liquid nitrogen manifests itself in the boiling-like behavior of the liquid nitrogen subjected to laser irradiation. Fig. 6 demonstrates the light scattering at the sphere from the gaseous nitrogen micro-bubbles in liquid nitrogen formed during its irradiation with the focused beam of continuous wave infrared CO_2 laser (the wavelength 10.6 μm, the optical power 70 W).

The bubbles formed in the liquid phase below its boiling temperature are non-equilibrium and should collapse. The difference of the internal energies between the gaseous and liquid phases is emitted partially via infrared or visible photons. When these non-equilibrium bubbles are formed by acoustical irradiation of liquids the light emission is called sono-luminescence [14]. In another case connected with the emission of photons during the fast structural transformation of materials from the phase with higher internal energy to the phase with lower energy is known as Perelman – Tatartchenko effect (PeTa-effect [15]). But, the spectral widths of the emission bands observed in the both of these cases are higher significantly than in our experiments. So, we assume that the excitation of the visible emission of liquid nitrogen excited by infrared lasers is produced by multi-photon mechanisms. Say, the emission lines presented at Figs. 4 and 5 can be excited by simultaneous absorption of two or three 1.064 μm photons [16]. The assumption about the multiphoton excitation explains the super-linear and threshold-like dependence of the liquid nitrogen emission. Moreover, it can be attributed to the explanation of the non-symmetrical distribution of the red emission along the laser beam (Figs.1 and 3). The red emission is observed at the left side from the focal point only. At the right side it is not observed, although the distributions of the exciting beam intensities before the focal region (the right side) and after it (the left side) should be symmetrical. The asymmetrical distribution of

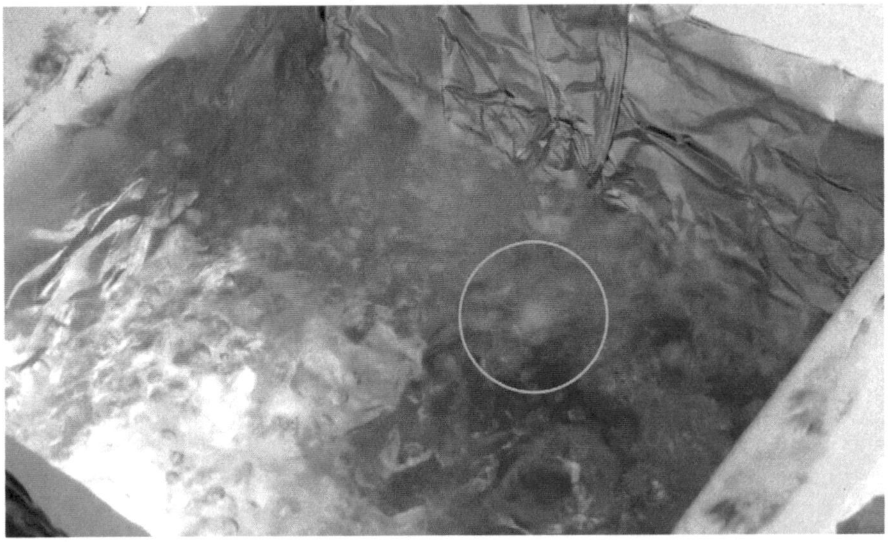

Figure 6. Photograph of the formation of a structure of microbubbles in the form of a cloud at the site of incidence of the CW CO_2 70W laser beam

the red emission manifests the necessity of achieving the threshold level of the exciting intensity for the start of the nitrogen emission. Naturally, this threshold level is achieved in the focal region first of all.

But the assumption about the multiphoton excitation is not sufficient for complete explanation of the spatial distribution of the emission concerning its narrow conical geometry and distinct separations of the cones corresponding to different emission lines (Fig.2). To resolve these problems, we attract the mechanisms of the Amplified Spontaneous Emission (ASE [17]) and of the filamentation of the intense laser beams propagating through optical refractive media [18]. ASE between two levels of any electron system appears when the density of excitations at the upper level becomes higher than the density at the lower level (the so called inverse distribution [17]). The filamentation occurs due to the positive feedback between the intensity of the light and non-homogeneity of the refraction index of the medium induced by this light. The phases of the electron excitations along this filament are determined by the propagating beam. If these excitations are transformed to the self-stimulated emission of the photons, the wave vectors of the emitted light waves should be in the spatial resonance with the double wave vector of the exciting photons. This means that due to the rule of conservation of the impulse the projections of the wave vectors of the emitted photons onto the axis of the exciting beam should be equal to the double wave vector of the exciting photons (for the two-photon excitation). So, the directions of the self-stimulated emissions of the separate spectral lines form discrete cone surface along the exciting beam. Due to the fact that the wave vectors of light waves are increasing with the decreasing of the wavelength the cones of the shorter lines are situated at bigger angles (Fig.2). This explanation of the conical emission observed in our experiments corresponds to the model of Cherenkov radiation excited in a certain substance by particles moving with the speeds higher than the speed of light in this substance [19].

The assumption of ASE explains the narrow spectral distributions of the emitted light and relatively small amount of the spectral lines observed in spite of the fact that the total amount of the emission lines of nitrogen molecules, atoms and ions is bigger significantly [20,21]. The self-stimulated emission (the so called super-luminescence) selects usually the most profitable ways of the radiative recombination.

Our preliminary studies of the spectral distributions of the white flashes emission show that these flashes are produced by the micro-inclusions of argon, which is present in ordinary air and in liquid nitrogen of the technical purity. In more detail the emission of impurities present in liquid nitrogen will be described later.

Conclusions

- We found the generation of the visible light emission in liquid nitrogen by nanosecond pulsed radiation of the infrared laser (the wavelength 1.064 μm). This generation is ascribed to the multiphoton excitation of nitrogen molecules and ions by the infrared radiation.
- This visible light emission of liquid nitrogen demonstrates several signs of the Amplified Spontaneous Emission: highly anisotropic spatial concentration of the emitted light flow around the direction of the exciting laser beam, the narrow widths of the spectral lines of the emission and severely limited amount of several spectral lines of the emission among rather a wide set of various possible channels of the radiative recombination of nitrogen, the super-linear dependence of the intensity of the emission on the intensity of the laser beam with the threshold intensity of the start of the emission.
- The visible light emission of liquid nitrogen is concentrated in the conical surfaces along the exciting beam having the general apex at the focal region of the laser beam. The cones corresponding to different wavelengths of the emitted light are separated clearly from each other following the rule of the phase synchronism: the projection of the wave vector of the emitted light onto the direction of the laser beam should be in resonance with the wave vector of the exciting radiation.
- The experimental results presented in this paper reveal liquid nitrogen as the flexible and convenient Kerr substance for generation of various non-linear optical phenomena by means of nanosecond lasers which usually are arranged only by means of femtosecond lasers.

Acknowledgements
The work was carried out according to the state assignment of ISSP RAS.

References

[1] R Tilley 2011 *Colour and the Optical Properties of Materials* Wiley 510 p.
[2] J Singh (Ed.) 2006 *Optical properties of condensed Matter and Applications* Wiley 435 p.
[3] R A Ganeev 2013 *Nonlinear Optical Properties of Materials* Springer 250 p.
[4] J Z Zhang 2009 *Optical Properties and Spectroscopy of Nanomaterials* World Scientific Publishing 395 p.
[5] M Kira, S W Koch 2006 *Progress in Quantum Electronics* **30** 155 – 296
[6] Yu Huakang et. al. 2019 *Nature Partner Jour., Computational Materials* 45
[7] S Brem et.al. 2018 *Scientific Reports* **8** 8238
[8] Y Zhang et.al. 2019 *Phys. Rev. B* **100** 075435
[9] S A Vinokurov, N V Klassen 2019 *Journal of Physics C* **1309** 112006
[10] K Jachymcki, A Negretti 2020 *Phys. Rev. Res.* **2** 033326
[11] Y Lu et.al. 2021 *Nature Communs* **12** 3135
[12] P Schaaf 2010 *Laser Processing of Materials* Springer, 240 p.
[13] A Dubietis, A Couairon 2019 *Ultrafast Supercontinuum Generation in Transparent Solid State Media* Springer 125 p.
[14] F R Young 2004 *Sonoluminescence, CRC Press* 227 p.
[15] V A Tatartchenko 2019 *Optics and Photonics Journal* **9** 189-217
[16] Y Toyozawa 2003 *Optical processes in solids* New York, Cambridge University Press 423 p.
[17] R Paschotta 2008 *Field Guide to Lasers* SPIE Press, 140 p.
[18] M R Vaziri 2013 *Laser Physics* **23**(10) 153401
[19] G N Afanasiev 2004 *Vavilov-Cherenkov and Synchrotron Radiation Foundations and Applications* Kluwer Academic Publishers, 503 p.

[20] S Remillard 2018 *Optical Spectroscopy of Diatomic Nitrogen: A Tutorial for Work in the Hope College Microwave Group* https://www.researchgate.net/publication/325157257
[21] A Lofthus, P Krupenle 2009 *Journ. of Chemical Reference Data* **6** 113, published online Oct 15 https//doi.org/10.1063/1565546

Journal of Physics: Conference Series **2056** (2021) 012039 doi:10.1088/1742-6596/2056/1/012039

Self- concentrated mass-transfer during deformation treatments of organic-inorganic compositions

I S Tsebruk, A P Pokidov, V V Kedrov, N V Klassen

Institute of Solid State Physics, Russian Academy of Sciences, Chernogolovka, Russia

E-mail: cebruk@issp.ac.ru; klassen@issp.ac.ru

Abstract. This paper prolongs the series of our previous papers where we found super-fast and super-deep introduction of foreign substances in crystalline materials by means of the ball rolling. A set of new experimental results was used to justify the new version of the mechanism of this introduction with the record speed and depth. The main process which determines this phenomena is connected with the sequence of openings and closings of nanocracks at the surface subjected to the rolling and the capture of the substance introduced from the surface by these cracks. The process of this introduction with the record parameters is supported by the intense chemical interactions between the matrix and the substance being introduced. This chemical interaction is intensified by several times with the deformation treatments. The analogous super-fast mass transfer is observed in the situation of the pulling out of the polystyrene fibers from the solution of polystyrene in benzene when the interaction of the organic components with cesium iodide nanoparticles was activated by the deformation treatment of the solution during its pulling out resulting in the formation of big amounts of nano-channels promising for effective utilization of hazardous radioactive wastes.

1. Introduction

This paper is a prolongation of our works on the scintillating radiation detectors and other profitable useful compositions of inorganic and organic raw materials [1-4]. R&D studies on various compositions from organics and inorganics are developing rather quickly resulting in new materials for electronics, nano-optics, biomedicine with sufficiently improved properties [5, 6]. Our works on the composite scintillators confirm the productivity of assembling in the unified material of the advantages of organics (fast light emission and high light yield) with high stopping power and good radiation hardness of inorganics achieving the scintillating material with record combination of the technical properties and the low-cost and convenient production technology [7, 8]. One of the main characteristics of the composite scintillators is the atomic-electron structure of the bonds between the organic and inorganic constituents. This structure determines the quality of the transfers of the electron excitations and atomic-molecular vibrations between the components which are responsible for the whole set of the scintillation parameters of the material. Our former papers [1-4] and another our paper of this issue are devoted to the studies of the formation, infrared and visible optical spectra of the composites which delivered the useful information about the behavior of the bonds organic-inorganic in various conditions. Rare earth fluorides and cesium sulfate were used as inorganic fillers whereas the activated polystyrene phosphor was applied as the binding matrix of the composites. In this paper we changed the inorganic filler and took cesium iodide crystals instead of the fluorides and cesium sulfate. The choice of cesium iodide is justified by three reasons. This material is well known for specialists in nuclear physics and nuclear energy because on the one hand CsI crystals are the most common material for radiation detectors and their interaction with ionizing radiation has been studied in details [7]. On the other hand, radioactive cesium and iodine belong to the most common waste materials of nuclear fission. So their utilization by means of the interactions of these substances with various organic and inorganic materials have been studied during many years [9–14]. The active interactions of cesium iodide and its components with organic materials belong to one the promising results of these studies for the applications of cesium iodide in the compositions for the studies in this paper. [12-14]. For example, cesium is capable to create strong bonds with the carbon atoms of the benzene rings or phenyl groups, whereas iodine forms complexes with the same rings and is capable to form bound states with hydrogen. So it is natural to expect any signs of these interactions in the compositions of cesium iodide with polystyrene and benzene. And several manifestations of these interactions have been found here indeed.

We used two techniques of preparation of these compositions which are quite different from each other [3, 4]. The first one used introduction of any foreign substance into the sample of solid cesium iodide by means of the ball rolling of the upper surface of the crystal with this foreign substance deposited onto the surface. As the second method we used the solidification of the solutions of polystyrene in benzene with various contents of cesium iodide in the forms of micro- or nano-particles inserted into the solution. The solidification was made by means of pulling of a jet of the solution with a thin glass stick touching the surface of the solution before the pulling. The speed of the pulling out was regulated from 0.5 cm/sec to 20 cm/sec. In spite of the severe differences in the methods of the preparation of the composites several analogies in the structures and properties of these compositions have been found. They are described below.

2. The results of the experiments and their discussion

In the previous paper [4] we paid attention to two types of unusual geometries of the trajectories of the polystyrene molecules introduced into single crystalline cesium iodide by means of the ball rolling: the dagger-like regions going preferentially in perpendicular to the surface subjected to the rolling and thin and long filaments oriented analogously. The formation of the "daggers" could be attributed to the creation of a new kind of a mixed material from cesium iodide and polystyrene appeared due to the active interaction between these substances stimulated by the intense deformation treatment. The 'dagger' formed by this way could be harder significantly with respect to cesium iodide. So the compressive stresses from the ball taking place at some positions of the ball were capable to drive this "dagger" as a nail into the crystal. In order to clarify this question, we studied the dynamics of the introduction of polystyrene by means of the optical microscopy. It turned out that the sharp apexes of these "daggers" did not create significant concentrations of the internal stresses which should be present in the case of the "driving in". Moreover, the introduction of thin and long threads cannot be explained by the "driving in" process due to the fact that these threads are too flexible.

Detailed examination of the dynamics of the introduction of the polystyrene molecules as well as of the carbon particles induced the alternative assumption about the mechanism of the super-fast and super-deep introduction of foreign substances by the ball rolling. This process was registered by the polarization-optical microscope equipped with the video-recorder. The optical axis of the microscope during the observations was oriented along the surface subjected to the ball rolling and perpendicular to the direction of the introduction of polystyrene [4]. The spatial resolution of the optical microscopy varied from 2 to 10 micrometers by changing the objectives. Figs. 1 – 4 demonstrate the stages of introduction of polystyrene to cesium iodide. The optical microscopy photos at Fig. 1. show the initial stage of the rolling. The long dark threads with the thicknesses of several micrometers penetrated into the subsurface layer rather deeply (the depth of their penetration achieves 300 micrometers). It is seen clearly that the

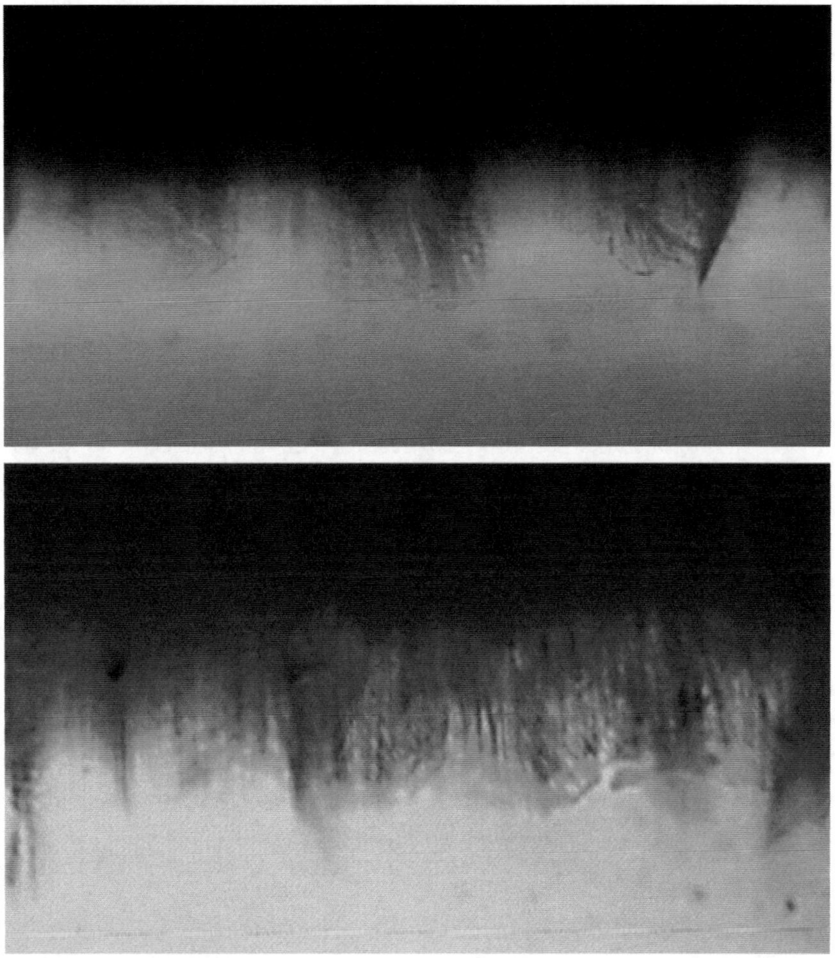

Figure 1. The optical microscopy of the initial stage of the introduction of polystyrene molecules into cesium iodide by means of the ball rolling. The horizontal dimension of the region shown at the photo is 2 mm

penetration is not connected with the "driving in" or "pressing in" of any hard daggers. One of the daggers is emerging at the right edge of the upper photo. At least 4 emerging daggers are seen at the next photo. The patterns of their formations lead to the assumption that these daggers result from the deep penetration of polystyrene in the forms of threads but not vice versa. Rather a wide variety of the geometries of these threads is observed. Some of them are oriented at the angle of 45° with respect to the surface, i.e. in the most profitable direction for the plastic deformation. Other straight threads are perpendicular to the surface. But on the other hand several threads look like rather curved arks. These observations make us to conclude that the geometry of these threads do not determine the directions of the mass transfer.

Three polarization-optical micro-photos of the morphologies of the subsurface layer of the cesium iodide crystal at the next stage of the introduction of polystyrene are presented at Fig. 2. The upper photo made in the transmitted light through the crossed polarizers shows several daggers formed by the rolling under the surface, placed at the upper part. It is worth to emphasize that no signs of the production of any internal stresses are seen at these photos. This is one more evidence that the "daggers' were the results of the deformation mass-transfer but they were not the actual participants of this process and were not pressed into the crystal.

The third photo presented at Fig. 2 shows the polarization - optical microscopy in the transmitted light of the region presented at the second photo but in the perpendicular direction. The third photo presents the appearance of the rolled cesium iodide crystal when the polarized light propagated in the direction parallel to the daggers shown at the middle photo. The bright spots in the middle part of this photo correspond to the points of the entrances of these daggers into the crystal. The brightness of these spots in the crossed polarizers manifests severe optical anisotropy of the internal structure of these daggers. This information is important for understanding of the mechanism of the formation of these daggers during the rolling of cesium iodide surface with polystyrene solution deposited onto the surface subjected to the rolling.

Our microscopical observations of the dynamicsof the rolling processes of the cesium iodide crystals with polystyren deposited preliminary onto their surface revealed that these "daggers" are created by accumulation of the thin threads produced by the rolling at its initial stages. The emerging of these accumulations were shown at Fig. 1. The micro-photo at Fig. 3 demonstrates this process of the emerging of the "daggers" more clearly. In these cases the molecules of polystyrene were introduced into cesium iodide by the ball rolling. All long enough threads at this photo demonstrate mutual attractions resulting in the formation of the initial stages of the "daggers" with the geometry of sharp apexes.

The information from publications about the active interactions between cesium and iodine with various organic substances including polymers and benzene was presented in the Introduction.

This information combined with the experimental facts described above made us to prepare the new version of the mechanism of the super-fast and super-deep introduction of organic substances into cesium iodide. The last photo (Fig. 2) of the surface of the cesium iodide sample subjected to the ball rolling besides the optical anisotropy demonstrates the traces of the micro-cracks produced by the motions of the ball. We suppose that these cracks are the main participants in the mass-transfer process. The scheme of this process is presented at Fig. 4.

The ball 1 is rolling back and forth along the surface of the solid sample with the substance deposited onto this surface (blue small balls). When the tensile stress induced by the ball in the subsurface layer stimulates the creation of the crack, some portion of the substance is inserted into the crack by the ball and by the natural attraction of this substance to the fresh surface of cesium iodide. The capability of the substance for the active interaction with the solid facilitates the fast penetration of the substance into the crack. The return movement of the ball tries to close the crack, but the compound formed by the surface substance with the walls of the crack create essential resistance against this process. Due to the rule of the lever the compressive stress created by the ball at the top of the crack is transformed to the tensile strass at its mouth. If the stress is sufficient the crack below the internal substance opens wider and the compressive stress pushes the introduced substance deeper. This version is capable to explain either superfast and super-deep introduction of foreign substances by the ball rolling or the formation of super-thin and super-long filaments of foreign substances inside the subsurface layers of solid subjected to the ball rolling. This means that the threads decorate the edges of the cracks when they do not move.

It is necessary to emphasize that the interactions between cesium iodide with polystyrene or carbon in ordinary conditions is not observed. But when the crystal is subjected to the deformation treatment the interaction is enhanced by many times. The reasons of this enhancement are connected with the modifications of electron states of cesium, iodine, polystyrene and carbon subjected to mechanical loads [16]. Detailed studies of the influence of the deformations on the electron states of these substances will be studied later.

Anyway the experiments described above manifest the ability of various combinations from

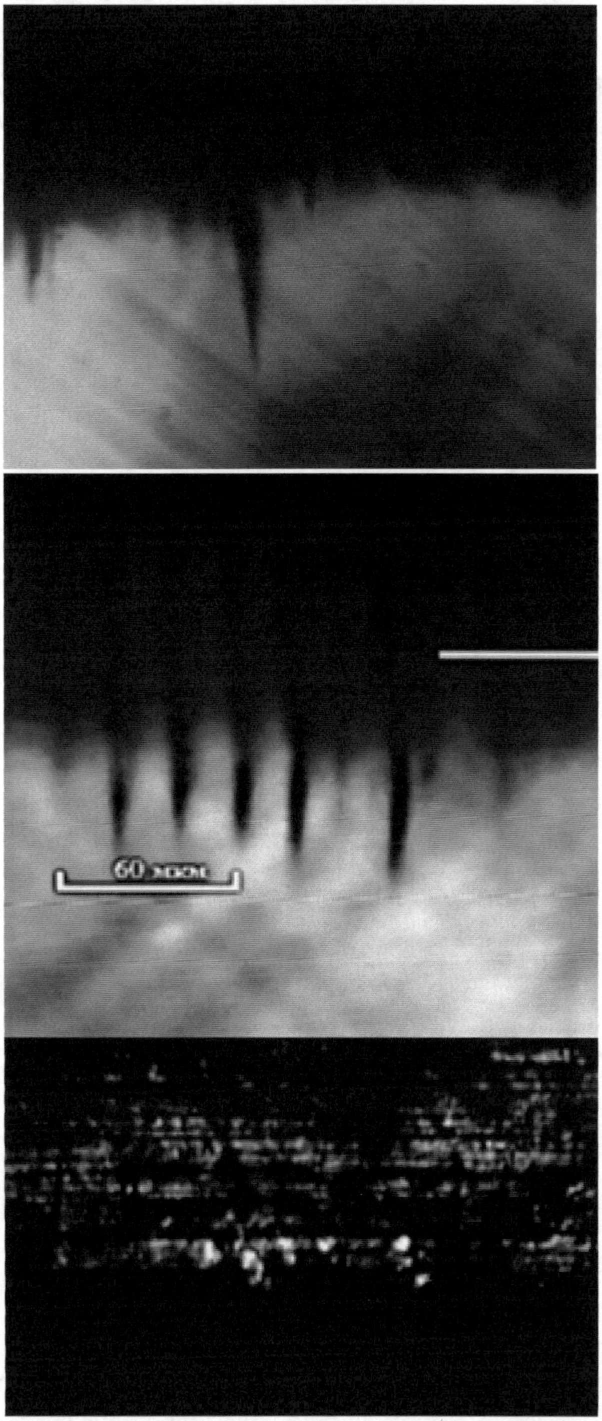

Figure 2. The optical microscopy of the subsurface layer with the "daggers" from polystyrene molecules introduced by the ball rolling (two upper photos). The third photo shows the transmission optical microscopy of the "daggers" introduced by the ball rolling showing that the daggers have optically anisotropic internal structures

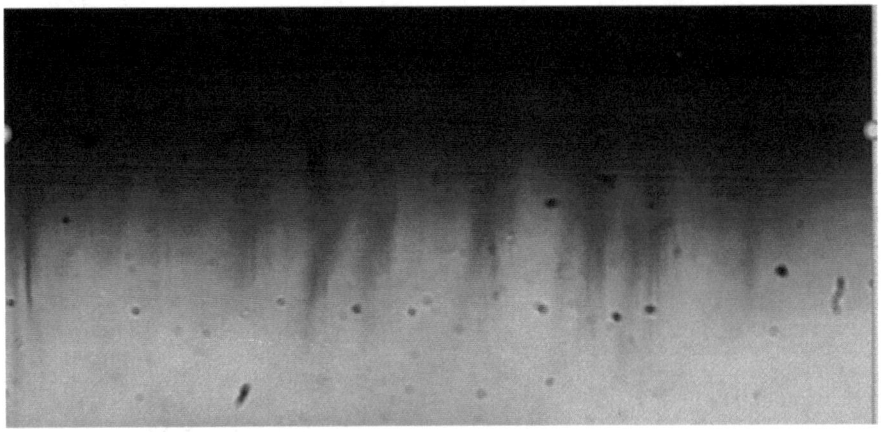

Figure 3. The optical microphoto of the super-long thin threads produced from polystyrene and introduced into cesium iodide by the ball rolling which demonstrates the start of the accumulation of the threads into "daggers"

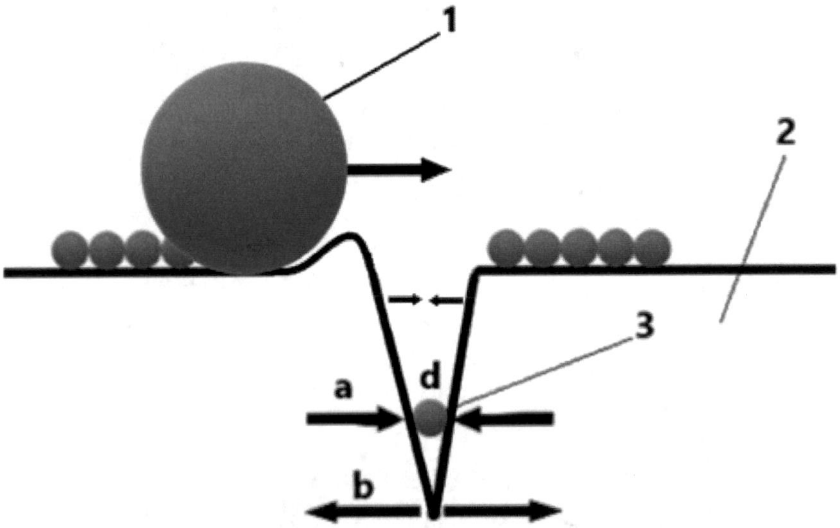

Figure 4. The scheme of the super-fast and super-long introduction of foreign substances into solids by means of the ball rolling of their surface with these substances deposited onto it

cesium iodide and its constituents, polystyrene and carbon being subjected to some kinds of the deformation treatment to form new phases and new compounds. Hence these interactions turn out to be profitable from thermodynamics point of view, facilitating creation of various super-structures by means of self-organization [17, 18]. Some signs of these manifestations of the self-organization can be noticed at the Figures presented above. For example, the spatial distributions of the "daggers", micro-threads and some other traces of the rolling introduction demonstrate signs of periodical or fractal super-structures. The quasi-periodical distributions of the micro-threads parallel to the directions of the dislocation glide bands of cesium iodide at the initial stage of the rolling and analogous quasi-periodical distributions of the "daggers" at the later stages result from usual competitions of the internal energy factors trying to place the

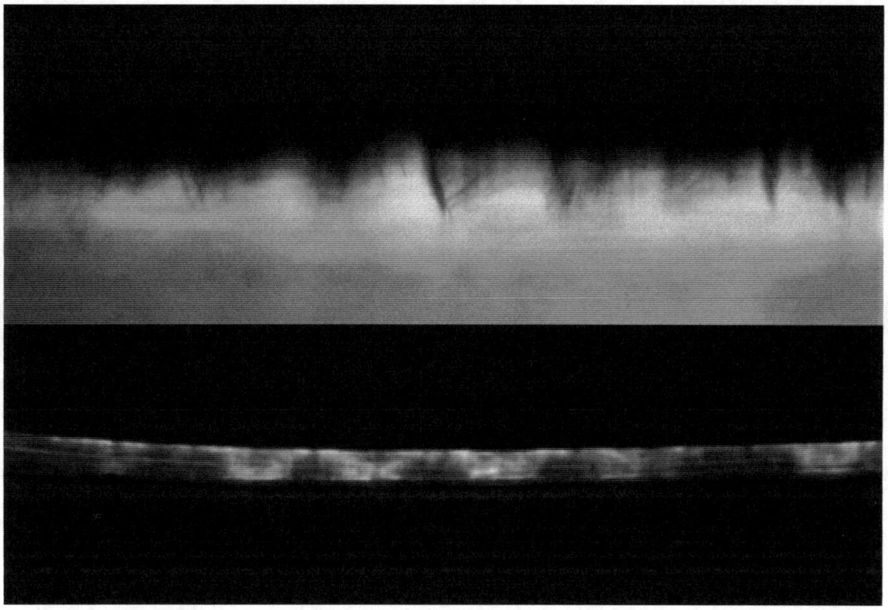

Figure 5. The transmission optical microscopy of the spatial distributions of the foreign substances introduced into the basic matrices: Top – polystyrene into cesium iodide matrix by means of the ball rolling. Bottom: cesium iodide particles into polystyrene matrix by means of pulling out the fibers from the solutions doped with inorganic particles

interacting substances as close to each other as possible and the diffusion and entropy factors trying to disturb these signs of self-organization [17, 18].

In spite of significant differences between the processes of preparation of the compositions organic – inorganic by means of the ball rolling and by means of the pulling out the fibers from solutions some results obtained in these two kinds of the studies are close to each other. For example, the spatial distributions of the polystyrene molecules introduced into cesium iodide by rolling and of the cesium iodide micro-particles introduced into the polystyrene fibers by the preliminary introduction of these particles into the solution of polystyrene in benzene and subsequent pulling out the composite fibers resemble each other (Fig. 5).

The photos presented at Fig. 5 manifest the obvious likeliness in the distributions which are "vice-versa" mutually. In the both cases the degree of the periodicity is determined by the ordering attraction between cesium iodide and polystyrene components and disordering factors induced by thermal a diffusion processes. Concerning the fibers of polystyrene doped with cesium iodide particles the attracting interactions organic-inorganic between these two substances resulting in super-fast and super-deep introductions in the case of the ball rolling gave a new interesting structural feature in the case of the fibers. When the speed of pulling out of the fibers was low enough the morphology of the fibers with microscopic diameters turned out to be rather interesting (Fig. 6). In addition to the transversal nano-folds created by the transversal gradients of the speed of the solidification of the fibers the walls of the fibers demonstrated the presence of the nano-channels directed from the surface into the depth of the fiber. The diameters of the channels did not exceed 100 nm. The surface density of the channels was rather high – up to 10^9 of the channels per square cm of the lateral surface of the fiber.

This new fact of creation of nano-channels in the composite fibers from polystyrene impregnated with particles of cesium iodide is interesting from at least two sides. The first one concerns the practical needs of effective utilization of radio-active cesium and iodine residual

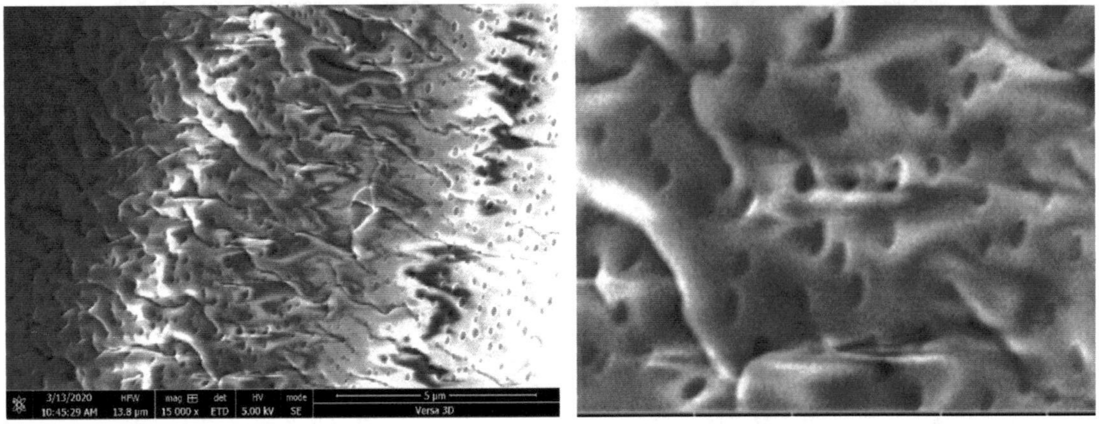

Figure 6. The electron microscopy of the polystyrene fiber pulled out from the solution of polystyrene in benzene doped with nanoparticles of cesium iodide. The nano channels formed during the pulling out the fiber are seen clearly

materials which are present usually in the cooling water of the nuclear plants. In the case of any accidents at these plants these radioactive remnants penetrate into the environment, so their utilization is sharply actual ecological problem [9–14]. Our polystyrene fibers impregnated with cesium iodide promise to be highly effective absorbers of radioactive cesium and iodine due to their active interactions with these materials and on the other hand due to high density of nano-channels providing big effective area of the active surface. It is worth to emphasize that this way of creation of nano-poured polymer materials is much more economic in comparison with well known method of preparation of nano-membranes by means of nuclear irradiation [19, 20].

On the other hand, understanding of the mechanism of this new phenomenon of auto-generation of the nano-channels in compositions organic- inorganic in addition to economic profits is capable for the generation of new knowledge about the interactions between these two kinds of materials. We assume that the main factor of this process is connected with the deformation activated creation of new kinds of interatomic bonds and corresponding atomic-molecular structures from cesium iodide and its constituents separately as well as of carbon and hydrogen. For example, deformation stimulated synthesis of hydrogen iodide can create nano-channels produced by the gaseous HJ exiting from the internal regions of the fibers. So two deformation activation factors can work in the case of the fibers being pulled out: the surface tension when the solution is transformed from the bulk state to the thin jet and the pressure from the bubbles of the gaseous hydrogen iodide. The joint action of these two factors is capable to induce the avalanche – like process of the deformation – chemical transformations.

Conclusions

A set of the experimental facts is presented justifying the version that the super-fast and super-deep introduction of foreign substances into crystals by the ball rolling proceeds by means of the sequences of openings and closings of the nano-cracks capturing the introducing substance at the stage of opening.

The formation of nanostructured channels in the fiber compositions from polystyrene and cesium iodide nanoparticles has been found during pulling out the fibers from solutions of polystyrene in organic solvents doped with nanoparticles.

The both of the processes described above are ascribed to the deformation activation of the intense chemical interactions between cesium iodide and the components of the organic substances. The possibility of the preparation of new kinds of materials (inorganic crystals with containing dopants from organic molecules immersed into them) has been confirmed experimentally.

The new cheep technology has been demonstrated of the preparation of the nets of polymer fibers with high density of nanoscopic channels capable for effective adsorption of hazardous impurities (e.g. radio-active ions of cesium and iodine in wastes of nuclear fission).

Acknowledgements
The work was carried out according to the state assignment of ISSP RAS.

References

[1] S Z Shmurak, V V Kedrov, N V Klassen and O A Shakhrai 2012 *Phys. Solid State* **54**(11) 2266–2276.
[2] N V Klassen, A P Pokidov, V V Kedrov 2018 *Inorganic Materials: Applied Research* **9**(1) 141–146
[3] I S Cebruk, A P Pokidov, V V Kedrov, N V Klassen 2019 *J. Phys.: Conf. Ser.* **1309** 012002
[4] I S Tsebruk, A P Pokidov, V V Kedrov, N V Klassen 2020 *J. Phys.: Conf. Ser.* **1560** 012043
[5] V Mittol (Ed.) 2010 *Optimization of Polymer Nanocomposite Properties* (WILEY) 420 p.
[6] R J Martin-Palma, J M Martinez-Duart 2017 *Nanotechnology for Microelectronics and Photonic* (USA, Elsevier), 336 p.
[7] M E Globus and B V Grinev 2000 *Neorganicheskie stsintillyatory* (Inorganic Scintillators), Kharkov: Akta,
[8] B V Grinev and V G Senchishin 2003 *Plastmassovye stsintillyatory* (Plastic Scintillators), Kharkov: Akta,.
[9] R Lorentz et.al. 1976 *Behavior of Iodine, Methyl Iodide, Cesium Oxide, and Cesium Iodide in Steam and Argon* Oak Ridge National Laboratory 47 p.
[10] E Beahm, C Weber, T Kress 1991 *Iodine Chemical Forms in LWR Severe Accidents* NUREG/CR-5782, ORNL/TM-11861 Oak Ridge National Laboratory,
[11] J Collins, N Osborne, R Lorentz 1984 *Behavior of Ce, I, and Te in the Fission Product* Oak Ridge National Laboratory,
[12] K Vats et.al. 2020 *ACS Nano* **14** 4665-4635
[13] T Kobayashi, K Yokozawa 2016 *Journ. Nucl. Sci. Technol.* **53**(10) 1489-1453
[14] T Petrovich, et.al. 2013 *Nature Communs.* **14** 2772
[15] J Murphy, D Baker 1965 *Canad. Journ. Chem* 1271-1278
[16] Y Simon, S Craig 2018 *Mechanochemystry in Materials, Roy.Soc. Chem.* 195
[17] M Condepudi, I Prigogine 2002 *Modern Thermodynamics* J. Wiley and Sons
[18] N V Klassen, M M Myshliaev 2019 *J. Phys.: Conf. Ser.* **1309** 012005
[19] T Brock 1987 *Membrane Filtration* 464 p.
[20] P Rauwell, E Rauwell 2019 *Nanomaterials* **9** 682

Specificity of the second sound standing waves behavior in a medium with nanoparticles

A A Esina, V B Efimov

Institute of Solid State Physics, Russian Academy of Sciences, Chernogolovka, Russia

E-mail: efimov@issp.ac.ru; orlova_lkk@issp.ac.ru

Abstract. The experimental studies of the second sound waves propagation in a resonator with a deuterium-helium gel were carried out. The latest experimental results, combined with those obtained in earlier experiments, have shown that the propagation of the second sound waves in gels leads to their significant attenuation and a decrease in the propagation velocity. This behavior differs from the case of the propagation of the sound waves of a two-component system with a strongly slow normal component and may indicate changes in the properties of superfluidity under confined geometry conditions.

1. Introduction

Impurity-helium gels have been studied since the 1960s and are still attracting the interest of researchers from different fields of science. Impurity-helium gels are a loose semitransparent substance that reminds slightly of “jelly”. It consists of clusters of impurity atoms surrounded by localized He^4 atoms. The attractive forces are determined by the induced polarization of neutral He^4 atoms by an impurity due to Van der Waals interaction. The sizes of nanoclusters formed depend on the impurity nature and can vary from several Å to several μm. The investigation of the properties of these gels is of interest for scientists in different areas of fundamental researches, for example, in astrophysics, since the mechanism of nanocluster formation can take place in an interstellar gas, the temperature of which is slightly below 3 K.

Since the pores of impurity nanoclusters in helium are soaked with a superfluid, one can observe interesting phenomena in these systems. For example, when the counterflow of normal and superfluid He^4 components (the second sound or heat flows) is excited, the case of the strong retardation of a normal component against the nanopore walls is realized, whereas a superfluid component passes through them freely. That is, this is the case of the second sound passing to the fourth sound. Another unusual and poorly studied phenomenon in these structures is bounded expanse superfluidity that, in particular, manifests itself in a decrease in the temperature of superfluid He^4 transition when He^4 contains nanoparticles. One can observe this change, for example, having measured the velocity of the second sound waves propagation in a resonator with a gel at different temperatures or measured the change of the resonant mode amplitudes at a fixed temperature. In the present work, we will describe the results obtained using the second method.

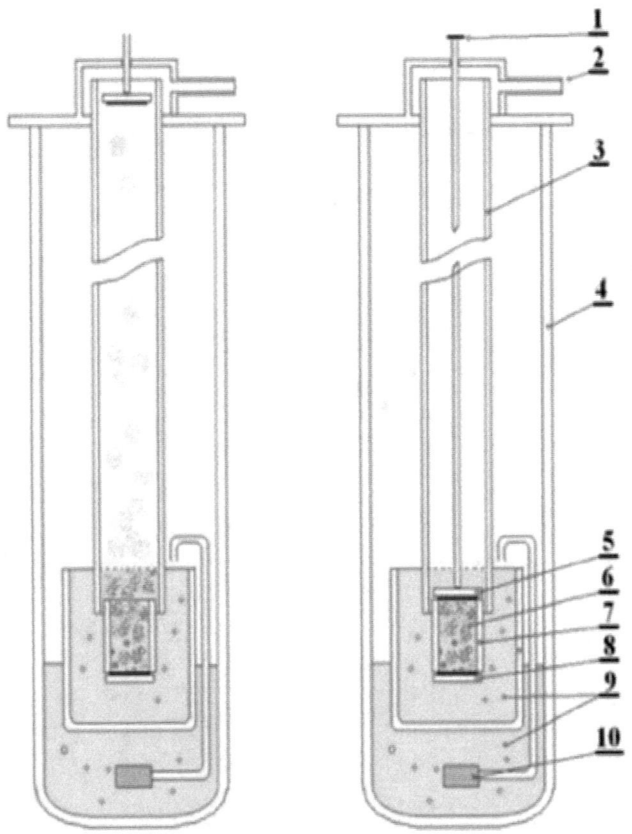

Figure 1. The left side is the device in the gel condensation regime (the mixture flux passes through filling pipe (3)). The right side is in the regime of measuring the second sound waves propagation in the resonator. (1) is a lift rod with heater (5) fixed at its end; (2) is a branch pipe for supplying a He^4 + (2÷ 4%) D2 mixture; (4) is a helium dewar with superfluid helium (9); (7) is a quartz resonator with gel (6); (8) is a superconducting bolometer; (10) is a cryopump

2. Experimental technique and method

In the present work, deuterium was used as an impurity. A deuterium-helium mixture was prepared in a tank with a capacity of 50 L and pressure slightly below the atmospheric one. The concentration of the impurity was about 2-4 %. An experimental setup was a helium cryostat in which there were a cylindrical resonator with the length L≈30 mm and diameter D≈15 mm, placed into a special cup, and a filling pipe through which the mixture was supplied laminarly. The upper end of the resonator was placed into the lower end of the pipe, which allowed all condensed gel to be gathered in the resonator. A superconducting bolometer was placed on the lower end of the resonator. It was a tin bronze thin film sputtered onto a quartz plate. The superconducting transition of the bolometer was shifted to the desired temperature range using a magnetic field generated by a superconducting solenoid placed under the bolometer. The upper end of the resonator was closed with a cover using a special rod after filling with a gel. A heater was placed on the inside of the cover. As the heater, a nonsuperconducting tin bronze film was used on which current strips were drawn in the form of a meander. This created a quasi-one-dimensional heat flow in the resonator. The operating capacity of this construction was tested by measuring the resonator characteristics after each opening and closure of the top cover.

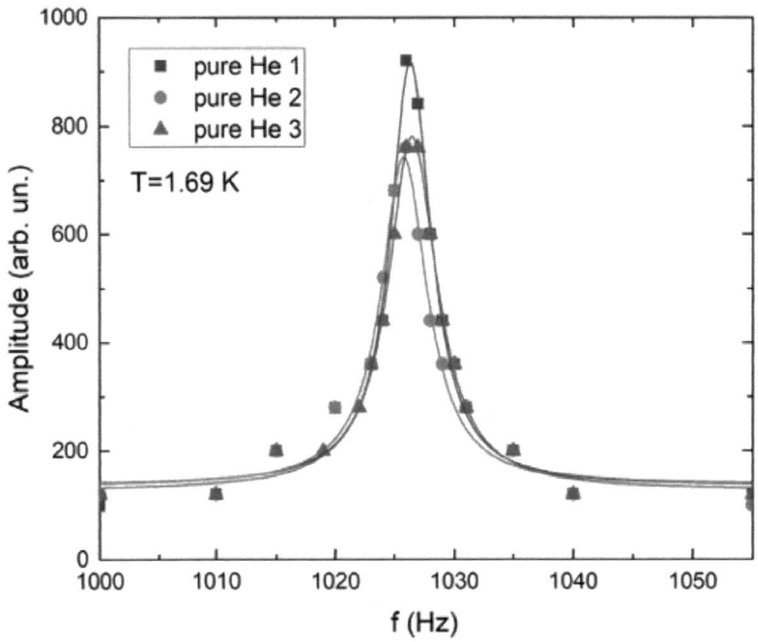

Figure 2. Three successive measurements of the Q-factor and resonant frequency of the resonator with pure He^4 after the closure of the top cover of the resonator. The 6-th resonance, T = 1.69 K, U_G = 5 V

3. Results

Before measuring the second sound waves attenuation in the resonator with a gel, the modes of the resonator with pure He were measured. The results are shown in Fig. 3.

The measurements were taken at the 6-th resonance and T = 1.65 K. In the vicinity of this temperature, the second sound velocity changed slightly with a temperature change and was about 20 m/s. Correspondingly, the resonant frequencies did not change either at this temperature.

After filling the resonator with a deuterium-helium gel, it was attempted to measure the Q-factor of the 6-th resonance. In this experiment, however, the obtained gel apparently turned out to be too non-uniform, which led to the very strong attenuation of the second sound waves. At that, the higher the number of the resonance was, the more significant the attenuation was. As a result, we did not succeed in measuring adequately the amplitudes of the resonances and comparing their attenuation degree.

However, the situation arisen contributed to us estimating well enough how much the velocity of the second sound decreases when passing through a dense deuterium-helium gel from a change of the frequency of the resonant modes. The results obtained are depicted in Fig. 4.

One can see in the plot that the frequencies (and, correspondingly, wave velocities) of each corresponding mode decreased markedly.

The second sound waves propagation in porous media was studied earlier [1, 2]. If a small pore size is present, the retardation of a normal component occurs, whereas a superfluid component passes through a porous medium without friction. At that, the velocity of heat waves increases and becomes close to the first sound velocity ($V_1 \sim 200$ m/s) at 1.5 K. In our experiments, the situation is opposite: the velocity of heat waves propagation decreases. The explanation can be related to the peculiarities of He^4 transition to the superfluid state in confined geometry. In this case [3], the transition to the superfluid state occurs at lower temperatures. Then, the

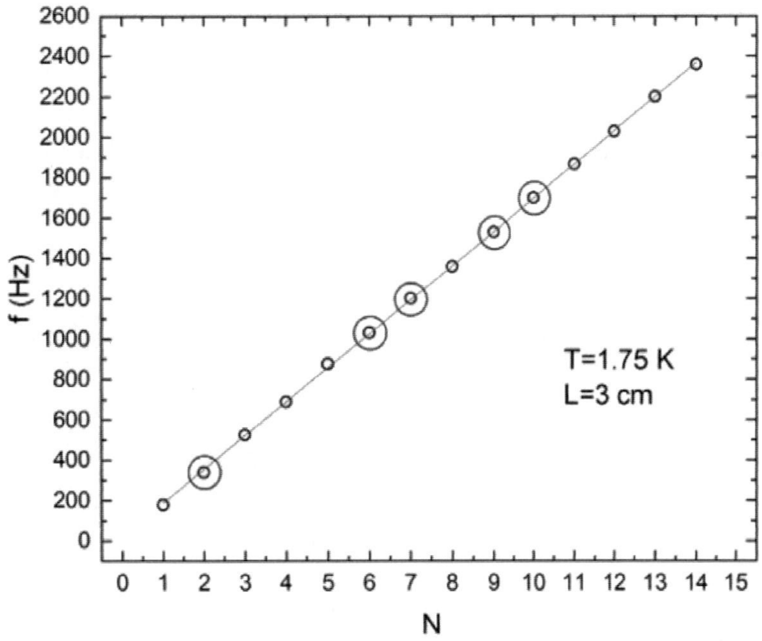

Figure 3. Resonant frequencies for the first 14 longitudinal resonances in pure He^4. The resonances with the highest Q-factor are marked with blue circles. T = 1.75 K

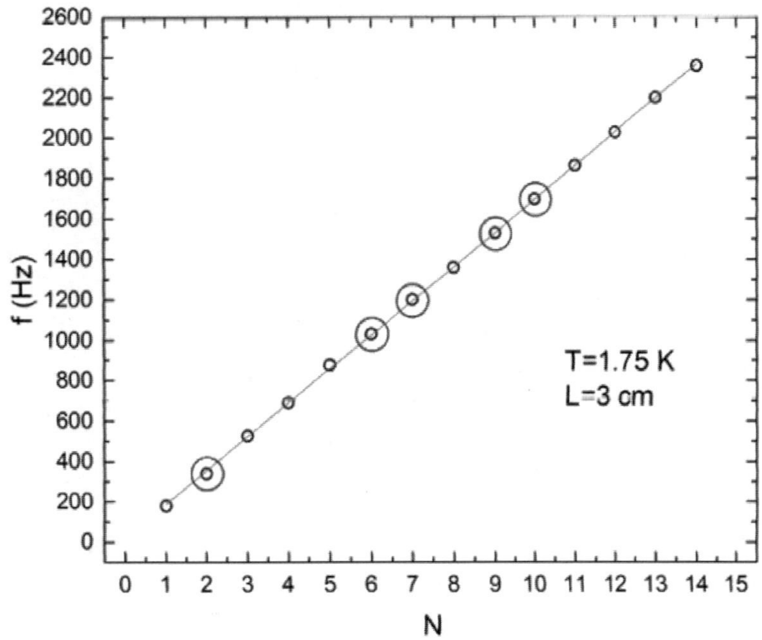

Figure 4. Comparison of the frequencies of the resonator with pure He^4 (blue squares) with the frequencies measured in the resonator with a gel (red squares). T = 1.69 K, U_G = 5 V

dependence $V_2(T)$ shifts to lower temperatures, which was observed in the experiment.

Conclusions

The experiments carried out have shown that, when introducing an impurity-helium gel (a deuterium-helium gel) to the resonator bulk, the second sound waves attenuate strongly, as might be expected when introducing a porous medium to the bulk. The result with a decrease in the velocity of the second sound waves propagation has turned out to be unexpected. The results obtained can be explained as follows. In the bulk of an individual nanopore, the fraction of a superfluid component turns out to be lower compared with the case of an unbounded liquid since the atoms on the nanocluster walls are in a "normal" state. Since the fraction of the surface is high at such small volumes, correspondingly, a normal component of a liquid occupies a higher fraction than in the absence of constraints. Because of this, bounded expanse superfluidity emerges at lower temperatures, which directly affects the value of the velocity of the second sound waves propagation. This situation differs from the case of the strong retardation of a normal component against the pore walls in that in the latter case the second sound velocity and the resonance Q-factor would have to increase, and the second sound would pass to the fourth sound. However, in this case, questions arise: how mobile is a normal component in the structure having a very large surface area and what has a decisive influence on mobility? Experiments with gels of other impurities, for example, nitrogen, can answer these questions. The structure of the frame of this impurity will be more rigid and similar to the structure of an aerogel for which the passages of the second sound to the fourth sound have been observed.

Acknowledgements

The work was carried out according to the state assignment of ISSP RAS.

References

[1] V Kiryukhin, B Kremer, R E Boltnev, V V Khmelenko, E B Gordon, 1997 *Phys. Rev. Lett.* **79** 9
[2] E P Bernard, R E Boltnev, V V Khmelenko, V Kiryukhin, S I Kiselev, D M Lee, 2004 *JLTP* **134** 133-143
[3] J D Reppy 1992 *JLTP* **87** 205-245
[4] V B Efimov, A N Izotov, L P Mezhov-Deglin 2013 *Bulletin of the russian academy of sciences: Physics* **77**(1) 48 DOI: 10.7868/S0367676513010080

Broad temperature range ferrielectric liquid crystal with subwavelength helix pitch

A V Kuznetsov[1,2], M V Minchenko[1], T P Tkachenko[1], E P Pozhidaev[1,2]

[1] Lebedev Institute of Physics, Russian Academy of Sciences, Moscow 119991, Russia
[2] Moscow Aviation Institute (National Research University), Moscow 125993, Russia

E-mail: kuznetsovav@lebedev.ru

Abstract. The ferrielectric liquid crystal (FerriLC) phase with the helix pitch p_o of about 100 nm was revealed in a broad temperature range from -3°C to + 35°C in a mixture containing an achiral smectic-C (SmC) biphenylpyrimidine and two non-mesogenic chiral diesters of 4,4"-terphenyl dicarboxylic acid.

1. Introduction

The phenomenon of ferielectricity began to be considered more than sixty years ago as an inherent property of a specific phase state of polar crystals [1]. To date, it has been well proven that a common property of ferrielectric phases, which are intermediate between ferroelectric and antiferroelectric phases, is the presence of nonequivalent subsystems of oppositely directed dipoles [2]. The same statement is also true for the ferrielectric phases of polar smectic liquid crystals [3, 4].

Until now, ferrielectric phases in polar liquid crystals have been observed and investigated experimentally mainly in the chemical class of (R)- and (S)-1-Methylalkyl 4'-(4"-n-Alkoxybenzoyloxy) biphenyl-4-carboxylates [3 - 7]. Ferrielectric phases were observed at temperatures above 60°C, and their existence temperature range has been varied from 2°C [6] to 12°C [4], depending on the molecular structure. The main efforts of researchers were devoted to proving the existence [3 - 6] and the classification of ferrielectric phases in liquid crystals on the basis of various theoretical approaches [6 - 8].

We have developed a multicomponent chiral smectic liquid crystal with a rather broad temperature range of the ferrielectric phase, from -3°C to + 36°C. Below is a description of this result, including data on the chemical composition of the mixture and evidence of the presence of a broad-temperature range ferrielectric phase.

2. Results and discussions

The chemical structure of the developed three-component mixture, called as FerriLC-1, in which the ferrielectric phase is observed in the temperature range from -3C to + 36C, is shown in Table 1.

Table 1. Chemical structure of the FerriLC-1 mixture.

Entry	Liquid crystal components	Concentration, % weight
1		49.7
2	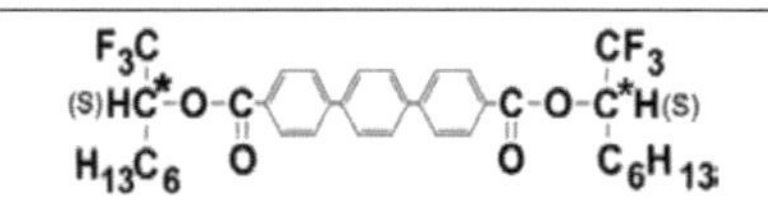	38.9
3	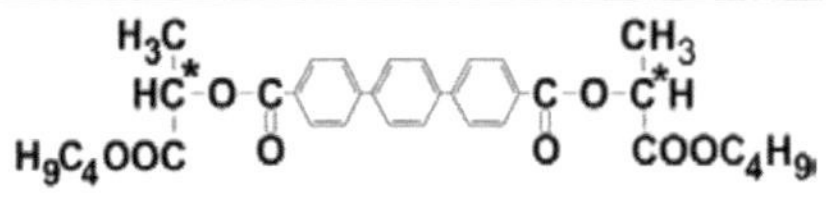	11.4

The FerriLC-1 mixture contains an achiral smectic-C biphenylpyrimidine [9] as the SmC host (see Table 1, entry 1) and two chiral non-mesogenic dopants shown in Table 1, in entry 2 (description of this structure syntheses was published in [10]) and in entry 3 (the structure was published in [11, 12]). A mixture of the same three components that are shown in Table 1 but with a different percentage of these components has already been investigated earlier in work [13], where it was shown that with a weight ratio of the components of 52%, 39%, 9% in entries 1 - 3 of Table 1, respectively, the mixture was a smectic C* ferroelectric liquid crystal.

Now we claim that at the concentrations of the mixture components indicated in Table 1, a broad-temperature ferrielectric phase appears in the mixture. This statement is confirmed by the shape of the hysteresis loop, typical for ferrielectrics (see Figure 1). The hysteresis loop shape shown in Figure 1 is observed over a temperature range of -3°C to + 36°C, which can be seen in the inset to Figure 1.

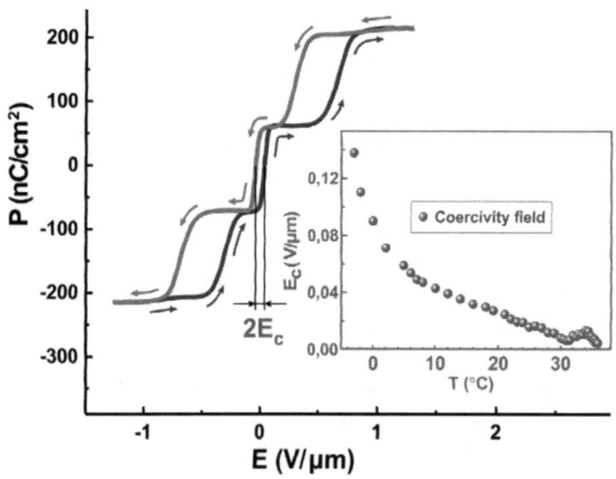

Figure 1. Macroscopic polarization ***P*** of a 107 μm FerriLC-1 based cell versus triangular electric field ***E*** of $2{\cdot}10^{-2}$ Hz frequency applied to the cell; the insert shows the temperature dependence of coercive field E_c.

Static hysteresis loops of FerriLC-1 based cells were measured at the applied triangular voltage with frequency $2 \cdot 10^{-2}$ Hz to exclude the influence of viscous torque on the voltage coercivity and on the loop shape [14 - 16]. The FerriLC-1 layer thickness in electro-optical cells was around 107 μm, and the hysteresis loops measurements were carried out via measurements of polarization reversal currents and their subsequent integration according to the well-known method described in detail [16].

The pitch p_0 of the FerriLC-1 helix measured by the known method [17], turned out to be less than 120 nm in the entire temperature range of the ferroelectric phase existence (see Figure 2), that is, several times less than any wavelength in the visible spectral range. The spontaneous polarization $\boldsymbol{P_s}$ of the FerriLC-1 mixture is observed up to the temperature about of 110°C, Figure 2. It means that within the temperature interval 36°C < T < 110°C other polar phases exist: either smectic C* phase or other ferrielectric phases predicted in [7, 8]. However, we have not yet completed the identification of these phases by now.

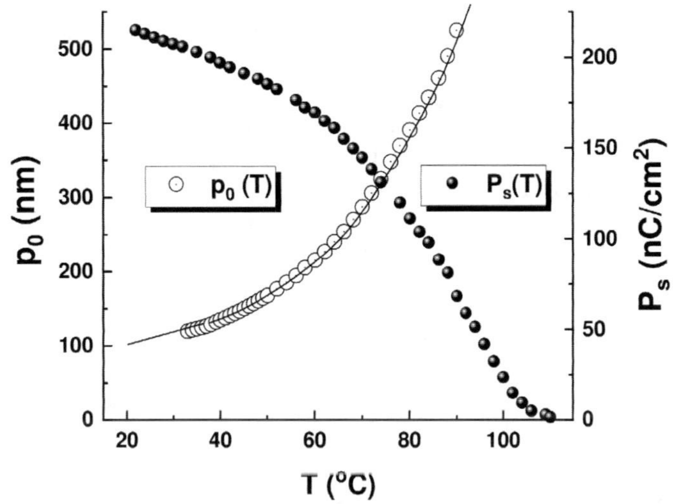

Figure 2. Temperature dependencies of the helix pitch and spontaneous polarization of the FerriLC-1 mixture.

3. Conclusion

A ferrielectric phase with a subwavelength helix pitch (p_0 < 120 nm) was discovered in the temperature range from -3C to + 36C in the mixture of achiral smectic C biphenylpyrimidine with two non-mesogenic chiral compounds.

Acknowledgments

This work was partially supported by the following RFBR grants: № 19-52-06005 MNTI_a and № 20-02-00746.

References

[1] Pulvari C F 1960 *Physical Review* **120**, 1670
[2] Fu Z, Chen X, Li Z, Hu T, Zhang L, Lu P, Zhang S, Wang G, Dong X, Xu F 2020 *Nat. Commun.* **11**, 3809

[3] Takezoe H, Lee J, Ouchi Y, Fukuda A 1991 *Mol. Cryst. Liq. Cryst.* **202**, 85-90
[4] Goodby J W, Patel J S, Chin E 1992 *J. Mater. Chem.* **2**(2), 197-207
[5] Isozaki T, Fujikawa T, Takezoe H, Fukuda A 1993 *Physical Review B* **48**(18), 13439 – 13450
[6] Muševič I, Škarabot M S 2001 *Physical Review E* **64**, 051706
[7] Emelyanenko A V, Pozhidaev E P, Shtykov N M, Molkin V E 2008 *Journal of the SID*, **16**(8), 811- 817
[8] Dolganov P V, Zhilin V M, Dolganov V K, Kats E I 2008 *Jetp Lett.* **87**, 253–257
[9] Terashima K, Ichihashi M, Takeshita F, Kikuchi M, Furukawa K 1987 *EP* **0293763 A2**
[10] Pozhidaev E P, Torgova S I, Molkin V E, Minchenko M V, Vashchenko V V, Krivoshey A I, Strigazzi A 2009 *Mol. Cryst. Liq. Cryst.* **509**, 1042–1050
[11] Rabinovich A Z, Loseva M V, Chernova N I, Pozhidaev E P, Petrachevich O S, Narkevich J S 1989 *Liq. Cryst.* **6**, 533–543
[12] Pozhidaev E P, Ganzke D, Zyryanov V Ya, Smorgon S L, Haase W 2002 *Liq. Cryst.* **29**(10), 1305-1310
[13] Pozhidaev Evgeny P, Srivastava Abhishek Kumar, Kiselev Alexei D, Chigrinov Vladimir G, Vashchenko Valery V, Krivoshey Alexander I, Minchenko Maxim V, Kwok Hoi-Sing 2014 *Optics Letters* **39**(10), 2900-2903
[14] Dahl I, Lagerwall S T, Skarp K 1987 *Phys. Rev. A* **36**, 4380
[15] Reynaerts C, De Vos A 1991 *Ferroelectrics* **113**, 439
[16] Kaznacheev Anatoly, Pozhidaev Evgeny, Rudyak Vladimir, Emelyanenko Alexander V, Khokhlov Alexei 2018 *Phys. Rev. E* **97**, 042703
[17] Berreman D W 1973 *Mol. Cryst. Liq. Cryst.* **22**, 175–184

Effect of aluminum alloying on the structure and properties of rapidly quenched TiNiCu alloy

A V Shelyakov[1], N N Sitnikov[1,2], I A Zaletova[1,2], S A Eroshenkov[1] and O N Sevryukov[1]

[1]National Research Nuclear University MEPhI (Moscow Engineering Physics Institute), Kashirskoe sh. 31, 115409 Moscow, Russia
[2]JSC State Scientific Center of The Russian Federation "Keldysh Research Center", Onezhskaya St. 8, 125438 Moscow, Russia

E-mail: AVShelyakov@mephi.ru

Abstract. The efficiency of shape memory alloys for the MEMS technology has been recently demonstrated. Quasibinary intermetallic TiNi-TiCu alloys produced by rapid quenching from
liquid phase in the form of thin (about 40 μm) ribbons are an attractive material for the fabrication of micro-actuators due to their narrow temperature hysteresis of the shape memory effect (SME) and relatively large recoverable strain. In order to broaden the functionality of SME microdevices, in this work we have alloyed TiNiCu containing 25 at.% copper with aluminum.
The results have shown that alloying with 0.6 at.% Al increases the cast characteristics of the composition and favors its amorphization. Upon crystallization by isothermal annealing or electropulse treatment the resultant microstructure and SME properties of the Al containing alloy change but slightly in comparison with the original alloy however there is a significant shift (by more than 15°C) of the SME temperature range toward lower temperatures.

1. Introduction

Shape memory alloys (SMAs) are a bright example of the so-called smart functional materials [1, 2]. The ability of SMAs to retain their unique properties on a microscopic scale allows fabricating high-efficiency devices on their basis for microelectromechanical systems (MEMS) [3-5] which is one of the most rapidly developing innovative technologies today. Alloys of the TiNi-TiCu quasibinary intermetallic system produced by rapid quenching from the liquid phase in the form of thin (about 40 μm) ribbons are an attractive material for the fabrication of micro-actuators due to their narrow temperature hysteresis of the martensitic transformation (MT) which is the basis of the shape memory effect (SME) and relatively large recoverable strain [6-10]. We recently developed a series of microtweezers – a micromechanical tool for gripping and holding of micro- and nanoscale objects for their further spatial manipulation and preparation [11-13]. Thereby we experimentally showed the possibility of fabricating composite amorphous-crystalline micrograppers with SME using the pulse laser treatment technique and demonstrated the operation of the devices and their compatibility with most of the existing micro- and nanopositioning systems.

An efficient way to broaden the functionality of SME microdevices is to provide new alloys with modified properties. It is well-known that even a slight change in the chemical composition of TiNi-based alloys causes major changes in their thermomechanical properties [14]. For example, an increase in the Ni content in binary TiNi alloys by only 1 at.% relative to the equiatomic composition leads to a considerable decrease in the MT critical temperatures and a strong change in the alloy structure (phase composition). On the other hand, addition of third component, e.g. hafnium, to TiNi alloys allows shifting the MT temperature range by decades of degrees toward higher temperatures.

Research is currently underway into the effect of alloying of ternary TiNiCu alloys with fourth component (aluminum) on the alloy structure, phase MT characteristics and SME parameters.

2. Experimental

The test material was the quasibinary TiNi–TiCu alloy containing 25 at.% copper (hereinafter referred to as 25Cu) and the 25Cu alloy with 0.6 at.% Al addition (denoted as 0.6Al). The alloy compositions and notations are summarized in Table 1.

Table 1. Compositions of the studied alloys and their designation.

Alloy	25Cu	0.6Al
Composition (at.%)	$Ti_{50.2}Ni_{24.8}Cu_{25}$	$Ti_{50.2}Ni_{24.8}Cu_{24.4}Al_{0.6}$

The alloys were produced by rapid quenching from the liquid phase (the planar flow casting technique) [15, 16] in the form of 30–50 μm thick ribbons at a melt cooling rate of 10^6 K/s which was maintained constant for both the alloys.

The alloys were heat treated by two methods: isothermal annealing at 500°C for 300 s and electropulse method, i.e., passing a short electric current pulse with preset duration and amplitude through the specimen for specimen heating to the specified temperature due to Joule heat release. To calculate the required parameters of electrolpulse crystallization of the amorphous specimens we used the current density to electric pulse duration relationship derived by us earlier [8].

The structure of the alloys was studied by metallography, electron microscopy and X-ray diffraction. For the metallogra[hic study of ribbon specimens we prepared their cross-sections using Buehler equipment. Final polishing was conducted using an acid-containing mixed suspension with an abrasive grain size of 50 nm. The cross-section microstructure of the specimens was studied under a FEI Quanta 600 FEG scanning electron microscope (SEM). X-ray diffraction analysis was carried out with Bragg-Brentano focusing and a hybrid monochromator on a PANalytical Empyrean X-ray diffractometer in Cu-K_α radiation.

The characteristic temperatures and enthalpy of the MT in the alloys were determined using differential scanning calorimetry (DSC) on an STA 449 F1 Jupiter calorimeter by heating and cooling cycles in the 20 to 100°C range with 5°C/min steps.

The SME in the alloys was studied by bend testing [17]. A ribbon specimen with thickness d having a preset straight shape memory was bent by 180° in the martensitic state and placed between a stable and a movable pressure plates. The distance D between the plates was set so the ribbon acquire the preset initial strain $\xi_i = d/D$ providing for the maximum specimen surface strain. After the removal of the load and subsequent heating to above the reverse MT finish temperature A_f the SME occurred in the specimen to recover the preset straight shape, either completely or partially, thus returning part ξ_{sme} of the accumulated strain which characterizes the SME magnitude. As shown earlier [17], ξ_{sme} is determined by the relationship $\xi_{sme} = \xi_i\ (180°-\alpha)/180°$, where the angle α is in degrees. The changes in the specimen shape and the angle α were traced using a special video control system and data processing software.

3. Results and Discussion

The initial structure of the rapidly quenched ribbons of the test alloys was studied using X-ray phase analysis both at the contact (oriented towards the quenching wheel) and the noncontact (free) surfaces of the ribbon (Fig. 1). The room-temperature X-ray diffraction patterns for the ribbon free sides of the 25Cu alloy contained clearly expressed diffraction peaks of both the austenitic phase $B2$ (CsCl type bcc lattice) and the martensitic phase *B*19 (AuCd type orthorhombic lattice) whereas those patterns for the 0.6Al alloy contained a smeared halo near 2θ = 42 deg indicating its amorphous state. Electron microscopic studies of the alloy microstructure for the ribbon cross-sections showed that this difference between the alloys is caused by the formation of a ~3 μm thick surface crystalline layer at the 25Cu alloy ribbon free side during quenching (Fig. 2d) whereas such a crystalline layer did not form on 0.6Al ribbons (Fig. 2a). The non-typical positions of the most intense reflections in the X-ray diffraction patterns of the B2 and B19 phases are accounted for by the formation of textures on the formed crystalline layer. It should be noted that the contact surfaces of both alloy ribbons had neither crystalline layers nor structure-related peaks in the X-ray diffraction patterns. Thus study of the 0.6Al alloy showed that additional aluminum alloying of the 25Cu alloy leads to its complete amorphization.

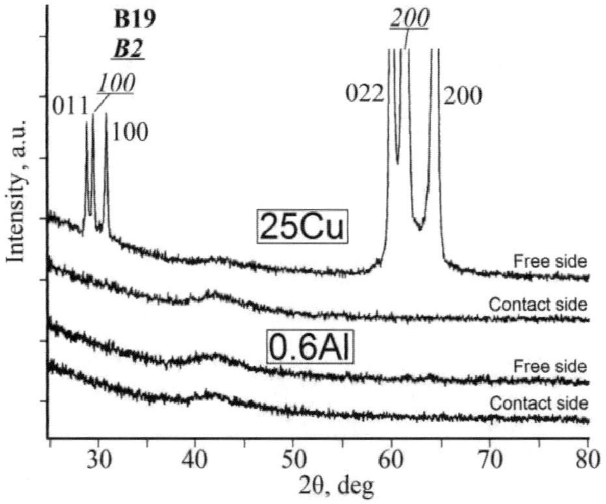

Figure 1. X-ray diffraction patterns for contact and free surfaces of 25 Cu and 0.6Al alloys in as-quenched state (at room temperature).

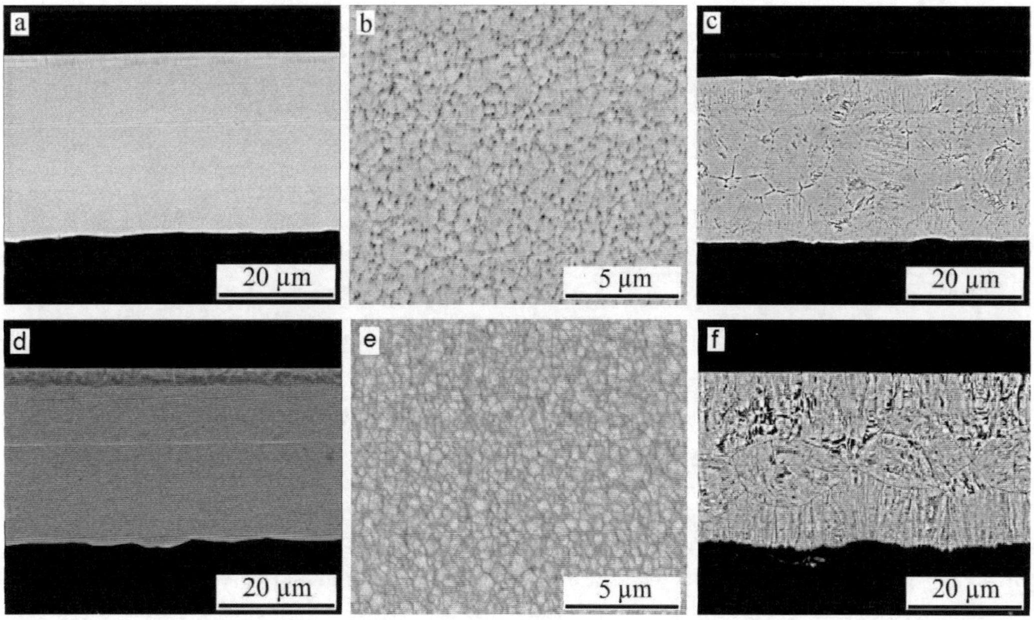

Figure 2. Cross-sectional SEM images of rapidly quenched ribbons of 0.6Al (a, b, c) and 25Cu (d, e, f) alloys, in the initial state (a and d), after isothermal crystallization (b, e) and after electropulse crystallization with 5 ms pulse duration (c, f)

Analysis of the cross-sectional SEM images of the ribbons after isothermal crystallization from the amorphous state showed that a submicron crystalline structure forms both in the 25Cu and in the 0.6Al alloys with an average grain size of 0.5-0.7 μm (Fig. 2b, e). Noteworthy aluminum alloying of the 25Cu alloy changes its structural morphology but slightly. X-ray study revealed that aluminum alloying does not either considerably affect the phase composition of the 25Cu alloy (Fig. 3a). It can be seen that the 0.6Al alloy is completely in the *B*19 martensitic state at room temperature which is confirmed by the presence of *B*19 phase reflections and the absence of strong peaks of other phases in the X-ray diffraction patterns.

The DSC curves of the 0.6Al alloy specimens after isothermal crystallization contained typical heat absorption peaks (during heating) and heat emission peaks (during cooling) in the 20 to 70°C range (Fig. 3b), indicating the single-stage polymorphic MT *B*2↔*B*19. This suggests that the crystallization leads to the formation of a homogeneous single-phase structure. A similar MT behavior for the 25Cu alloy was reported in a number of earlier works [8, 16, 18]. However according to the DSC curves of the 0.6Al alloy its MT temperature ranges are noticeably shifted (by more than 15°C) toward lower temperatures in the 30 to 45°C range. This fact can be obviously used for broadening the functionality of the TiNiCu alloy with 25 at.% Cu for the fabrication of SME devices.

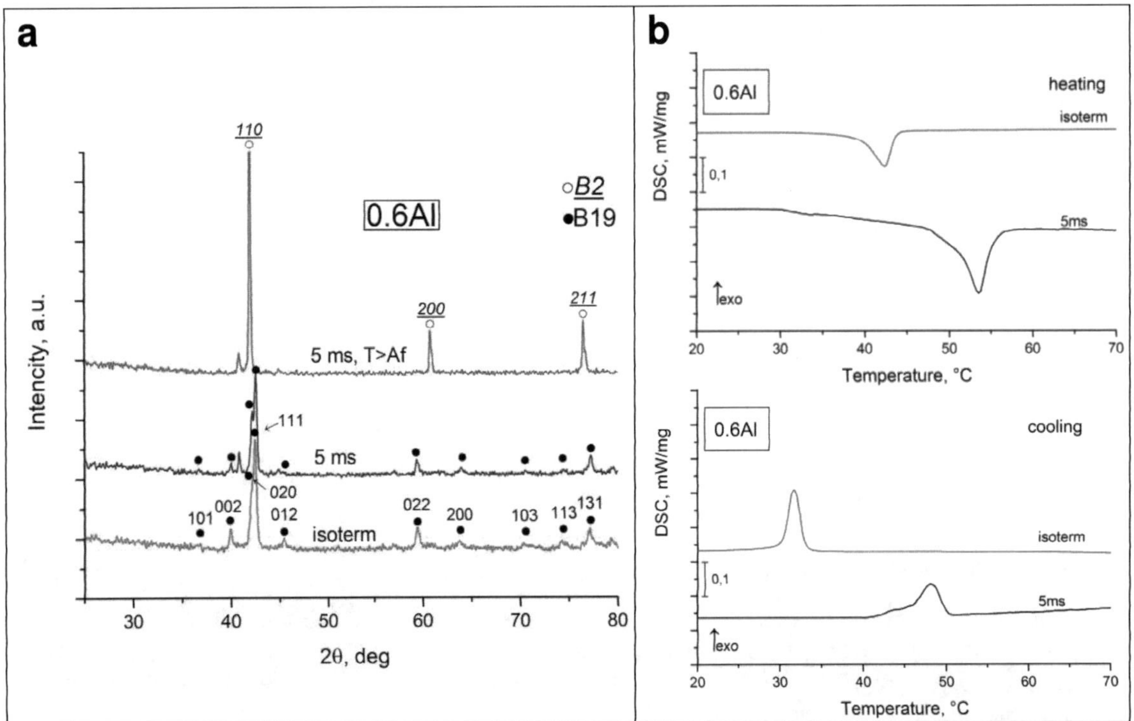

Figure 3. X-ray diffraction patterns (a) and heating and cooling DSC curves (b) of rapidly quenched 0.6Al alloy after isothermal annealing and electropulse crystallization with 5 ms pulse duration.

Electropulse (dynamic) heat treatment with 5 ms pulse duration dramatically changed the microstructure of the alloys as compared with isothermal annealing (Fig. 2c, f). The cross-sectional SEM images of the dynamically crystallized ribbons show an inhomogeneous distribution of crystals in the ribbon thickness: a columnar crystal structure forms near the ribbon surfaces whereas discrete or agglomerated large grains with typical sizes of 3 to 12 μm are observed in the ribbon bulk. The columnar crystals observed near the surface migrate inward to the crystals forming in the ribbon bulk. The transverse sizes of the columnar crystals at the contact and free sides of the ribbons are nearly the same, i.e., 0.5 to 2.0 μm, their heights ranging between 2 and 5 μm. It is well-known that annealing with an high-rate heating rate favors a decrease in the size of the forming crystalline features while retaining the same morphology and crystalline structure and should therefore lead to the formation of finer structures in comparison with conventional isothermal furnace annealing. As we noted earlier [19, 20], this considerable increase in the grain size for dynamically crystallized TiNiCu alloys can be caused by a decrease in the number of crystallization centers in the amorphous matrix with an increase in the copper content in the alloys or by the lack of a regular structure of the "frozen-in" crystallization centers.

The X-ray diffraction patterns show that by analogy with isothermal annealing, dynamically crystallized 25Cu and 0.6Al alloys are completely in the martensitic state at room temperature with the *B*19 structure (Fig. 3a). Heating of the 0.6Al alloy to 70°C (above the temperature A_f) leads to the elimination of the *B*19 phase peaks, and only the *B*2 phase reflections remain, i.e., the alloy transits to the completely martensitic state as a result of the MT $B19 \leftrightarrow B2$. It should be noted that the diffraction pattern of the 0.6Al alloy in the martensitic state contains quite an intense peak near $2\theta = 40.8$ deg which is not eliminated upon the transition to the austenite state. This peak can be identified as the (110) reflection of the *B*11 phase but one cannot affirm safely the presence of this phase in the alloy structure because the diffraction patterns do not contain visible reflections from other crystal planes of this phase.

The experimentally shown structural properties of the dynamically crystallized specimens have a significant effect on the DSC curve patterns of the alloys (Fig. 3b), i.e., splitting of the heat emission and absorption peaks. This is likely to be associated with the bimodal structure of the alloys which contain a grain structure in the ribbon bulk and columnar crystals near the ribbon surfaces (Fig. 2c, f). These structures may have different critical MT temperatures resulting in superposition of peaks from different structures and splitting of the common peak. Furthermore electropulse crystallization appreciably increases the MT critical temperatures (by more than 10°C) for the 0.6Al alloy; the same phenomenon was earlier found in other TiNiCu alloys [20].

Then we determined the SME parameters of the alloys by recording curves of the SME recovered strain ξ_{sme} vs the initial strain ξ_i. The results showed that after isothermal annealing ξ_{sme} increases in an almost linear manner with an increase in the strain ξ_i, whereas at $\xi_i > 5\%$ this growth becomes considerably slower which seems to be caused by the exceeding of the pseudoplastic strain for the TiNi-TiCu system alloys [10, 21] and the generation of a tangible plastic strain. The maximum measured deformation $\xi_{sme} \approx 5.1\%$ is almost the same for the 25Cu and 0.6Al alloys. However the dynamic crystallization of the specimens considerably increases their maximum strain ξ_{sme} (to 6.4%). Thus aluminum alloying of the 25Cu alloy (0.6 at.%) does not compromise its SME parameters.

4. Conclusions

A quasibinary TiNi-TiCu system alloy with a copper content of 25 at.% in the form of a layered amorphous-crystalline ribbon was produced by rapid quenching from the melt (the planar flow casting technique). We showed that aluminum alloying to 0.6 at.% causes complete amorphization of the alloy during quenching.

Isothermal crystallization of both the initial and the alloyed alloys produces a submicron crystalline structure of orthorhombic martensite *B*19 with an average grain size of 0.6-0.7 μm which transforms to the austenitic phase *B*2 during heating as a result of the polymorphic martensitic transformation $B2 \leftrightarrow B19$. However aluminum alloying significantly (by more than 15°C) reduces the critical MT temperatures and hence shifts the SME temperature range toward lower temperatures. The maximum SME-recovered bending strain is 5.1% regardless of alloying.

High-rate electropulse crystallization with 5 ms duration dramatically changes the microstructure of the alloys as compared with isothermal annealing and increases the SME magnitude to 6.4% but also noticeably increases (by about 10°C) the critical MT temperatures.

Thus aluminum alloying of ternary TiNiCu alloys with SME allows one to broaden their functionality by varying the SME temperatures while retaining the high mechanical parameters of the SME.

Acknowledgments

The study is supported by Russian Science Foundation (project No. 19-12-00327).

References

[1] *Advances in shape memory materials* 2017 ed Q-P Sun, R Matsui, K Takeda and E A Pieczyska (Cham: Springer).
[2] Rao A, Srinivasa A R and Reddy J N 2015 *Design of Shape Memory Alloy (SMA) actuators*. (Cham: Springer).
[3] Stachiv I, Alarcon E and Lamac M 2021 *Metals* **11** 415.
[4] Kohl M, Ossmer H, Gueltig M and Megnin C 2018 *Shape Mem. Superelasticity* **4** 127.
[5] Jani J M, Leary M, Subic A and Gibson M A 2014 *Mater. Design* **56** 1078.
[6] Nespoli A, Besseghini S and Pittaccio S, Villa E and Viscuso S 2010 *Sensors Actuators A Phys* **158** 149.
[7] Morgiel J, Cesari E, Pons J, Pasko A and Dutkiewicz J J 2002 *Mater. Sci.* **37** 5319.
[8] Shelyakov A V, Sitnikov N N, Menushenkov A P, Koledov V V and Irjak A I 2011 *Thin Solid Films* **519** 5314.
[9] Senkovskiy B V, Usachov D Yu, Fedorov A V, Vilkov O Yu, Shelyakov A V and Adamchuk V K 2012 *J. Alloys Compd.* **537** 190.
[10] Kang S W, Lim Y M, Lee Y H, Moon H J, Kim J W and Nam T H 2010 *Scr. Mater.* **62** 71.
[11] Shelyakov A, Sitnikov N, Borodako K, Koledov V, Khabibullina I and von Gratowski S 2020 *J Micro-Bio Robot* **16** 43.
[12] Shelyakov A, Sitnikov N, Menushenkov A and Fominski V 2018 *Acta Physica Polonica* **134(3)** 708.
[13] Irzhak A, Zakharov D, Koledov V et al. 2014 *J. Alloys Compd.* **586** 464.
[14] Otsuka K and Ren X 2005 *Prog Mater Sci* **50** 511.
[15] Shelyakov A V, Sevryukov O N, Sitnikov N N, Borodako K A and Khabibullina I A 2020 *J. Phys.: Conf. Ser.* **1686** 012056.
[16] Shelyakov A V, Sitnikov N N, Menushenkov A P, Korneev A A, Rizakhanov R N and Sokolova N A 2013 *J. Alloys Compd.* **577** S251.
[17] Shelyakov A V, Sitnikov N N, Borodako K A and Sevryukov O N 2020 *Russ. Metall.* 4 345
[18] Chang S H, Wu S K and Kimura H 2007 *Intermetallics* 15 233.
[19] Chernysheva O, Shelyakov A, Sitnikov N, Veligzhanin A, Borodako K and Sundeev R 2021 *Mater. Lett.* **285** 129104.
[20] Shelyakov A, Sitnikov N, Khabibullina I, Tabachkova N, Fominski V and Andreev N 2019 *Mater. Lett.* **248** 48.
[21] Glezer A, Sitnikov N, Sundeev R, Shelyakov A, Khabibullina I 2019 *Materials* **12** 2670.

Giant spin-valve effect in Al/$Y_3Fe_5O_{12}$ heterostructures

O V Skryabina and L S Uspenskaya

Institute of Solid State Physics RAS, Chernogolovka, 142432 Russia

uspenska@issp.ac.ru

Abstract. Superconducting phase transition in the aluminium stripes fabricated on yttrium-iron garnet and on oxidized silicon substrates is studied. Reduction of critical temperature, widening of superconducting transition depending upon the mutual current and magnetization orientations as well as on current strength are observed in aluminium on garnet comparing with aluminium on silicon. The proximity and triplet proximity effects, Andreev reflection, exchange interaction, spin-orbit coupling and self YIG magnetic field impacts on the observed effects are discussed. We show that only accounting for the spin-orbit coupling and self-magnetic field of YIG is allowed to explain the observed phenomenon of asymmetric change in the transition under the variation of the current and magnetization alignment. Based on the obtained results, we suggest a new geometry of spin-valve with hundreds percent variation of the resistance controlled by the current.

1. Introduction

Intensive development of cryoelectronics and the prospects for creating a cryogenic quantum computer [1-9] imply the low temperature circuits and elements with the ability to be switched in ultra-short times with low energy losses. These are elements of static and dynamic memory, diodes, transistors, switches and much more. In particular, to create magnetoresistive memory and spin valves, it is proposed to use a superconductor (S) - ferromagnet (F) heterostructure [10-14]. The operation of such devices is based on their ability to change the superconducting properties of the S-layer by reversing the magnetization of the F-layer. Two-layer SF structures and three-layer FSF, SFS, SF_1F_2 structures have been studied. Each of them has its own advantages and disadvantages. For example, Nb/FeNi is capable of switching in a field of only a few Oersted, but the magnetoresistive effect in it is small [15]. Nd/PdFe/Nb switches in the weakest magnetic field of the order of 1 Oe, but its high resistive state turns out to be ambiguous and is realized in a very narrow range of the field [16,17]. The magnetoresistance in such structures is observed when an inhomogeneity of magnetic state appears, e.g. the domain walls in ferromagnet are formed, or vice-versa, when the stray fields vanishes due to demagnetization of the ferromagnetic layer. Resistance of trilayer spin-valve based on two magnetic layers with different coercivity depend upon the magnetization alignment in the ferromagnetic layers [18-20]. The resistance of such structures is low when the magnetization in ferromagnetic layers is aligned. It becomes higher when the magnetization in the layers is misaligned. Advantage of such devices is relatively low switching magnetic field, stability of low and high resistive states in wide enough range of fields determined by the difference in coercivity of the layers. Disadvantage is not high variation of the resistivity. However, not studied till now is the switching rate of SF_1F_2, which could be too low because of exponential growth of magnetic viscosity with temperature decrease [21-24].

In this work, we suggest a new type of bilayer spin-valve controlled by an alignment of the current and magnetization and report on hundreds percent resistance variation.

2. Experimental

We investigated the resistive properties of aluminum/yttrium iron garnet structures (Al/YIG). Aluminum was chosen as a technologically advanced superconductor with a large coherence length ξ about of 130 nm in 100 nm thick films, which facilitates the fabrication of structures. YIG is a magnetic dielectric with a record low attenuation, with gigahertz rates of magnetization switching by weak magnetic field or light. These properties of the materials determine their choice.

The experiments were carried out on Al stripes 1 μm wide and 80 nm thick, fabricated by magnetron sputtering on relatively large YIG plates with the size about of 5 mm × 5 mm and thickness of 60 μm. Pairs of identical stripes located over the same magnetic domain were made. One of the stripes of each pair was aligned with the direction of magnetization in the domain, while another goes perpendicular to the magnetization. The stripes were located hundreds micrometer away from magnetic domain walls. The location was verified by direct observation in a polarizing optical microscope in transmitted polarized light. In this way we took advantage of YIG's characteristics: transparency, large Faraday rotation and birefringence [25].

Reference Al stripes were fabricated on oxidized Si simultaneously with the samples under the study to illuminate the influence of the YIG vicinity on the Al properties.

The resistance of Al-structures was measured by standard 4-probe method. The distance between potential contacts was about of 100 μm. The current contacts were located outside the potential ones on the distance about of 100 μm from them, The current with the strength 1 - 100 μA was applied during the measurements; the smaller value was determined by the sensitivity of our set-up, the upper value was limited by significant Al overheating appeared at higher current.

3. Results

The temperature dependence of the resistance measured on the Al stripe fabricated on oxidized silicon substrate is shown in figures 1a-1c. Superconducting phase transition occurs at $T_c \sim 1.264$ K when 1 μA current is applied (figure 1a, curve R). The transition takes place at the same T_c when 10 μA current is used (figure 1b, curve R). The transition shifts down to T_c=1.258 K under the current of 100 μA (figure 1c, curve R). The 0.006 K shift we ascribe to overheating of the Al under the current.

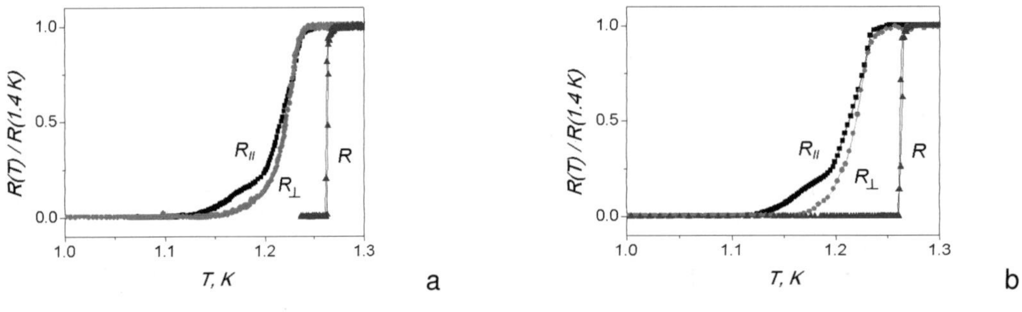

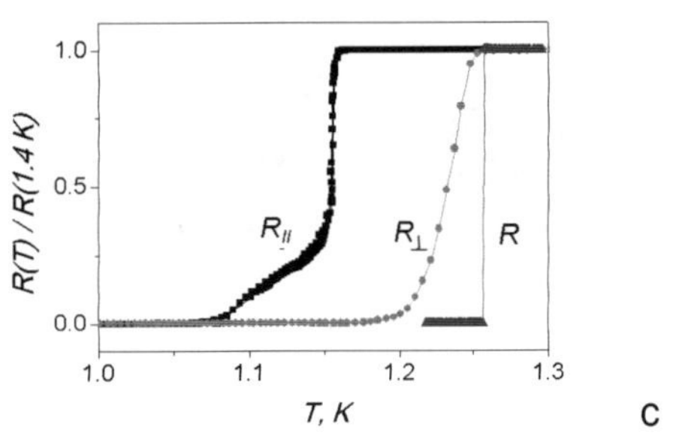

Figure 1. *R(T)* measured on Al stripes which are parallel with YIG magnetization ($R_{\|}$) and perpendicular to it ($R_{\perp}$) and *R(T)* on Al/Si (*R*).

The dependences measured under the current about of 1, 10 and 100 μA are shown in figure a, b and c, correspondingly.

Temperature dependences of the resistance measured on identical Al stripe fabricated on YIG differs remarkably from those on Al/Si. Four specific features can be seen from figures 1a-1c, in which $R_{\parallel}$ marks the curves showing the resistance of the stripe aligned with magnetization and $R_{\perp}$ is used as notation of the resistance of perpendicular stripe. First, the phase transition begins at lower temperature. Second, the transition becomes much wider. Next, the orientation of Al stripe relative to magnetization of YIG turns out to be important: the transition widening is larger when the current flow parallel to magnetization of YIG than in perpendicular direction. Further, temperature of the transition depends upon the strength of the current. We see that reduction of the beginning of the transition: measured at level $R = 0.9\ R(1.4K)$, it occurs at 1.236 K for 1 μA current, 1.234 K for 10 μA current, and 1.157 K or 1.247 K for 100 μA current of perpendicular directions. The transition temperature measured at level $R = 0.1\ R(1.4K)$ is 1.163 K and 1.192 K for 1 μA, 1.160 K and 1.192 K for 10 μA, 1.10 K and 1.212 K for 100 μA currents. It is definitely seen that the difference between $R_{\parallel}(T)$ and $R_{\perp}(T)$ increases with the current strength. Besides, the transition temperature increases with current strength when the current is directed perpendicular to magnetization and decreases when the current flow parallel with it. As a result, large interval of temperature appears in which the resistance depends upon mutual orientation of the current and magnetization. Difference in resistivity between $R_{\parallel}(T)$ and $R_{\perp}(T)$ measured under the 100 μA current, $\Delta R = R_{\parallel}(T) - R_{\perp}(T)$, corresponds now to the difference in the Al resistivity in the superconducting state and in the normal state, figure 2. Determined as $\Delta R / R_{\perp}(T)$, it turns to be infinitely large.

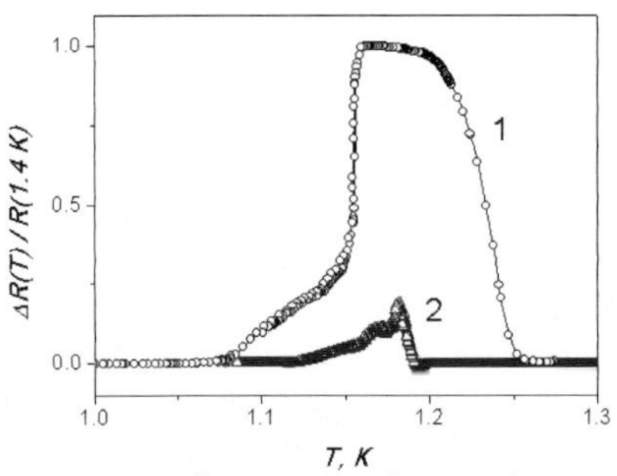

Figure 2. Relative difference in the resistance $(R_{\parallel}(T) - R_{\perp}(T)) / R$ (1.4K) in Al stripes with the width of 1 μm (curve 1) and 3 μm (curve 2). Measurements are performed under 100 μA current.

Listed above features were observed in all studied Al/YIG samples, although some details in $R(T)$ differ, e.g. the value of the reduction of T_c and the value of ΔR could be a little different.

We also found that the effect depends upon the width of Al stripe. Comparing the results obtained under the same applied current for stripes with the width of 1 and 3 μm, we have established that in wider stripes the effects are less pronounced. For example, the ΔR changes by only 0.2 from R(1.4K), the temperature range with the effect shrinks from 0.15 K down to 0.07 K (compare curves in figure 2).

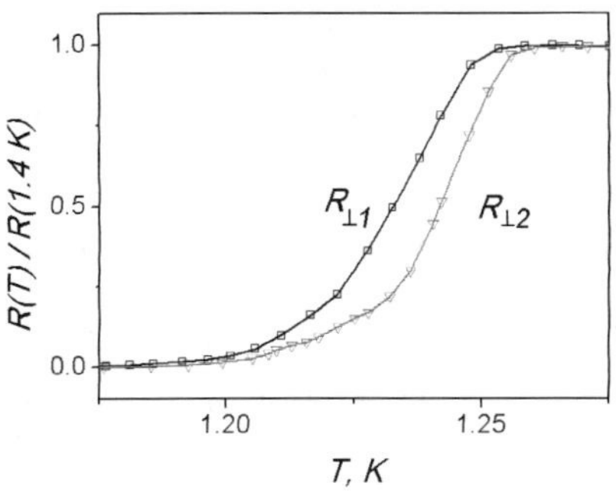

Figure 3. $R_{\perp}(T)$ for oppositely directed 100 μA current.

One more very important feature was found in the manifestations of the influence of the perpendicular current on the superconducting transition. The switching the current on 180 degrees led to different shift of the transition, figure 3. This showed that the resistance in Al/YIG can be switched using just the current inversion. Unfortunately, the resistance variation in this case was less than when the current was turned by 90 degrees, and took place in a narrower temperature range.

4. Discussion

Reduction of superconducting transition temperature in superconductor-ferromagnet structures can be caused by many reasons. The proximity effect [9,26,27], i.e. the decay of Cooper pairs penetrating into ferromagnet, does not matter in our case, because we are using a ferromagnetic insulator. In the same time, the decay of Cooper pairs due to the Andreev reflection [28,29] from SF interface could contribute to the observed superconductivity suppression. However, it cannot explain the observed dependence of the superconducting transition from the alignment of magnetization and the current.

The effect of triplet proximity in SF structures is sensitive to gradients of magnetic induction [11,12,30,31] but it is invariant of current-magnetization disorientation. We did not see magnetic inhomogeneity in YIG. Differently from [32], we do not have any domain walls under the Al-structure; we paid special attention to fabricate the Al-structures over the domain wall-free region in YIG. Therefore, we could not refer to appearance of triplet states in the Al.

Exchange interaction [33,34] is considered as an important factor, which could influence the T_c, but exchange is isotropic, so it cannot be responsible for the difference in $R_{\parallel}(T)$ and $R_{\perp}(T)$ seen in figures 1a-1c and 3.

Spin-orbit coupling (SOC) [35,36] in superconductor-magnetic isolator structures was supposed to be confirmed in the experiments [37-39]. They observed the resistance variation under the field rotation from perpendicular to planar orientation and even observed the anisotropy in the effect, which was strictly correlated with crystallographic anisotropy of the ferromagnet. However, the change in the direction of magnetization in these experiments was accompanied by a change in the demagnetizing factor and a complex transformation of the magnetic domain structure. These effects, in turn, could change the interaction of the superconductor and the ferromagnet. We were changed the current-magnetization off-orientation in the plane of the ferromagnet and observed the effect. However the SOC should give larger influence on the resistance of that Al stripe which is perpendicular to the magnetization because of non-zero vector product of $\boldsymbol{n} \times \boldsymbol{j} \times \boldsymbol{m}$, where $\boldsymbol{n}$ is the normal to the interface, $\boldsymbol{j}$ - current direction and $\boldsymbol{m}$ is magnetization direction. We get quite opposite; the effect was larger in the stripe with aligned $\boldsymbol{j}$ and $\boldsymbol{m}$. However, the amplification of the current applied to the structure made it possible to notice the difference in the shift of $R_{\perp}(T)$ for oppositely directed current, figure3, and the shift of $R_{\perp}(T)$ to higher temperature, while $R_{\parallel}(T)$ was shifted to lower temperature, figure2c. Therefore, we believe that we have some signatures of the SOC of superconducting Al with YIG.

However the effect is too large to be attributed to SOC only. Large demagnetizing field H_m always exists in and around the YIG plate. If YIG has an in-plane magnetization, the out-of-sample field H_m has maximum on YIG surface decreasing slowly with distance from it. The Al layer fabricated on YIG is very thin relative to the thickness of YIG plate. Therefore, H_m penetrates the whole Al layer without changing its strength. The estimations show that for YIG with magnetization value $4\pi M_s = 2400$ Oe, which is table low temperature value, the demagnetizing field H_m is ~ 8 Oe [34]. Table value of critical field for Al film is H_c~ 100 Oe. This field completely destroys the superconductivity. The H_m is smaller than H_c. Therefore it does not destroy superconductivity but reduces critical temperature T_c by appropriate value. Roughly estimated [40] temperature reduction ΔT is about of ~ 0.05 K, which is in reasonable agreement with our experiments. Further, the current flowing through the Al stripe produces some field H_J, which is summarised with H_m; the H_m slightly deviates from stripe direction by H_J for parallel with magnetization stripe and the strength of summarised field varies between $H_1 = H_J + H_m$ and $H_2 = H_J - H_m$ for perpendicular stripe, namely, the field is increased on one stripe surface and decreased on opposite one. Thus, self-field of the YIG together with current field could bring asymmetric responses in differently oriented Al stripes upon variations of an applied current.

The widening of superconducting transition in SF structures is ascribed usually to the enhanced flux flow states which are described in many publications, e.g. in [41-44]. Different from our observations (compare curves 1 with 2,3 in figure 2), the expected broadening is independent upon mutual orientation of the current and magnetization. The spatial fluctuations of superconductivity or normal phase can be differently oriented in parallel and perpendicular stripes because of the interaction with considered above field H_m and these would cause their different drift (in directions and velocity). The field H_m can introduce vortices in Al stripe, orientation of which and as a consequence motion under the current flow would be different; as a consequence, the energy losses in two types of stripes as well as *R(T)* would be different. Unfortunately, we cannot define exactly what impact gives each of discussed above mechanisms in large orientation dependent variation of *R(T)* that we observe. New experiments are needed.

Finally, we suggest the using of the observed effects in some low-temperature applications. For example, the resistance of a structure consisting of two crossed stripes or squared shaped can be switched between low and high resistance by changing the strength or direction of the current. Otherwise, relatively weak rotating magnetic field, below 10 Oe, can be applied for switching of the resistance of Al/YIG. We believe that similar effects can take place in any SF structures where the equilibrium magnetization of the ferromagnet is oriented in the structure plane.

5. Conclusions

We compared the superconducting phase transitions in Al stripes fabricated on plates of oxidized silicon and yttrium-iron garnet which is ferrimagnet dielectric. In accordance with the known literature data, we observed a decrease in the temperature of the superconducting transition and its widening due to the proximity of the magnetic dielectric. Studying the influence of orientation of the magnetization in the plane of the structure, we discovered unexpected phenomena. We found that the widening of the superconducting phase transition largely depended on the mutual orientation of the magnetization $\boldsymbol{m}$ and testing current $\boldsymbol{j}$. The transition was remarkably wider at parallel orientation, i.e when $[\boldsymbol{j} \times \boldsymbol{m}] = 0$. We also found that the transition depends both on the strength of the current and on the direction of vector product $[\boldsymbol{j} \times \boldsymbol{m}]$, being shifted towards higher or lower temperature when the product change the sign in the case of $[\boldsymbol{j} \times \boldsymbol{m}] \neq 0$. Considering the generally accepted mechanisms of the influence of the proximity of a ferromagnet on the properties of a superconductor, we came to the conclusion that the interface spin-orbit coupling along with the influence of self-field of ferromagnet on the flux-flow state and on the nucleation and drift of vortices could explain the found asymmetry. We believe that similar effects can be observed in any SF structures. Finally, we suggest the circuits of low-temperature resistive devices, the resistance of which can be switched by changing the strength or direction of the current or otherwise by the rotation of magnetic field.

6. Acknowledgments

The authors acknowledge the financial support of the state task of ISSP RAS. We thank V.S. Stolyarov for samples fabrication, N. Bebenin, V. Edelstein, A. Mel'nikov, Yu. Proshin, N. Pugach for useful discussions of the experimental results and O. Tikhomirov for the discussion of Zeeman field influence.

References

[1] Heikkilä T T, Silaev M, Virtanen P and Bergeret F S, 2019 *Progr. in Surf. Science* **94** 100540
[2] Bergeret F S, Silaev M, Virtanen P and Heikkilä T T, 2018 *Rev. Mod. Phys.* **90** 041001
[3] Kushnir V N, Sidorenko A, Tagirov L R, Kupriyanov M Y 2018 *Basic Superconducting Spin Valves.* In: Sidorenko A. (eds) *Functional Nanostructures and Metamaterials for Superconducting Spintronics.* NanoScience and Technology. Springer, Cham. https://doi.org/10.1007/978-3-319-90481-8_1

[4] Eschrig M, 2015 *Rep. Prog. Phys.* **78** 104501
[5] Blamire M G and Robinson J W A 2014 *J. Phys.: Condens. Matter* **26** 453201
[6] Ryazanov V V, Oboznov V A, Bol'ginov V V 2004 *Physics-Uspekhi* **47** 732
[7] Izyumov Yu A, Proshin Yu N and Khusainov M G 2002 *Physics-Uspekhi* **45** 109
[8] Ryazanov V V 1999 *Physics-Uspekhi* **42** 825
[9] Buzdin A I, Bujicic B and Kupriyanov M Y, 1992 *Sov. Phys. JETP* **74** 124
[10] Hauser J J 1969 *Phys. Rev. Lett.* **23** 374
[11] Halterman K and Alidoust M 2018 *Phys. Rev. B* **97** 064517
[12] Halterman K and Alidoust M 2016 *Phys. Rev. B* **94** 064503
[13] Zdravkov V I, Kehrle J, Obermeier G, Lenk D, Krug von Nidda H-A, Müller C, Kupriyanov M Yu, Sidorenko A S, Horn S, Tidecks R. and Tagirov L R 2013 *Phys. Rev. B* **87** 144507
[14] Sangjun Oh, Youm D. and Beasley M R 1997 *Appl. Phys. Lett.* **71** 2376
[15] Rusanov A Yu, Golikova T E and Egorov S V 2008 *JETP Letters* **87**175
[16] Bol'ginov V V, Stolyarov V S, Sobanin D S, Karpovich A L and Ryazanov V V 2012 *JETP Lett.* **95** 366
[17] Karelina L N, Bolginov V V, Erkenov Sh A, Egorov S V, Golovchanskiy I A, Chichkov V I, Hamida A B and Ryazanov V V 2020 *JETP Lett.* **112** 705
[18] Deutscher G and Meunier F 1969 *Phys. Rev. Lett.* **22** 395
[19] Gu J Y, You S Y, Jiang J S, Pearson J, Bazaliy Ya B and Bader S D 2002 *Phys. Rev. Lett.* **89** 267001
[20] Moraru I C, Pratt W P and Birge N O 2006 *Phys. Rev. Lett.* **96** 037004
[21] Uspenskaya L S 2010 *Bulletin RAS: Physics* **74** 711
[22] Uspenskaya L S, Nurgaliev T and Miteva S 2010 *Acta Phys Pol A* **117** 207
[23] Egorov S V and Uspenskaya L S 2016 *JETP Letters* **103** 272
[24] Uspenskaya L S and Tikhomirov O A 2019 *Journ. Exper. Theoret. Phys.* **128** 624
[25] Vlasko-Vlasov V K, Dedukh L M and Nikitenko V I 1976 *Zh. Eksp. Teor. Fiz.* **71** 2291
[26] Petrashov V T, Sosnin I.A, Cox I, Parsons A and Troadec C 1999 *Phys. Rev. Lett.* **83** 3281
[27] Vas'ko V A, Larkin V A, Kraus P A, K. Nikolaev R, Grupp D E, Nordman C A and Goldman A M 1997 *Phys. Rev. Lett.* **78** 1134
[28] Upadhyay SK, Palanisami A., Louie R N, Buhrman R A 1998 *Phys. Rev. Lett.* **81** 3247
[29] Beckmann D, Weber H B and v. Löhneysen H 2004 *Phys Rev Lett.* **93** 197003
[30] Fominov Ya V, Golubov A A and Kupriyanov M Yu 2003 *JETP Letters* **77** 510
[31] Fominov Ya V, Golubov A A, Karminskaya T Yu, Kupriyanov M Yu, Deminov R G and Tagirov L R 2010 *JETP Letters*, **91** 308
[32] Aladushkin A Yu, Fritzshe J and Moshchalkov V V 2009 *Appl. Phys. Lett.* **94** 222503
[33] Li Bin, Roschewsky N, Assaf B A, Eich M, Epstein-Martin M, Heiman Don, Münzenberg M and Moodera Ja S 2013 *PhysRevLett.* **110** 097001
[34] Xiong Y M, Stadler S, Adams P W and Catelani G 2011 *PhysRevLett* **106** 247001
[35] Edelstein V M 1990 *Solid State Comm.* **73** 233
[36] Edelstein V M 1996 *J. Phys.: Condens. Matter.* **8** 339
[37] Banerjee N, Ouassou J A, Zhu Y, Stelmashenko N A, Linder J and Blamire M G 2018 *Phys. Rev. B* **97** 184521
[38] González-Ruano C, Johnsen L G, Caso D,Tiusan C, Hehn M, Banerjee N, Linder J and Aliev F G 2020 *Phys. Rev. B* **102** 020405
[39] Johnsen L G, Svalland K and Linder J 2020 *Phys.Rev.Lett.* **125** 107002
[40] Tikhomirov O A, Skryabina O V, Uspenskaya L S 2021 Journ. Magn. Magn. Mater. **535** 168071
[41] Fulde P and Ferrell R 1964 *Phys. Rev.* **135** 550
[42] Larkin I and Ovchinnikov Yu N 1964 *Zh. Eksp. Teor. Fiz.* **47** 1136.
[43] Proshin Yu N and Khusainov M G 1997 *JETP Lett.* **66** 562
[44] Bobkov A M and Bobkova I V 2014 *JETP Lett.* **99** 333–339

The concept of biaxial surface potential and polymer aligning layers in electro-optical cells based on helical nanostructures of ferroelectric liquid crystal

T P Tkachenko[1], **A A Zhukov**[2], **E P Pozhidaev**[1,2]

[1] Lebedev Institute of Physics, Russian Academy of Sciences, Moscow 119991, Russia
[2] Moscow Aviation Institute (National Research University), Moscow 125993, Russia

E-mail: tptkachenko@lebedev.ru

Abstract. The paper considers the possibility of controlling the alignment quality of helical nanostructures of ferroelectric liquid crystals (FLCs) within the concept of biaxial surface potential due to variation the FLCs helical pitch p_0 and polymer aligning layers structures.

1. Introduction

The anchoring energy of liquid crystals (LCs) with bounding surfaces plays an important role in the analysis of the LCs orientational structure including the FLCs ones. At the moment, the anchoring energy coefficient W_n is the parameter of well-known uniaxial Rapini potential Φ_n [1], which can be written also for FLCs as [2]:

$$\Phi_n = -\frac{W_n}{2} sin^2 \psi \qquad (1)$$

where W_n – is the anchoring energy between the director $\boldsymbol{n}$ of the liquid crystal and the easy alignment axis $\boldsymbol{R}$, and ψ-angle is illustrated in Figure 1.

Commonly, the potential (1) corresponds to a monodomain LC structure. However, within this approach, it is hardly possible to obtain an exhaustive description of multi-domain over-molecular helical nanostructures of FLCs that arise most often in reality [3]. To solve this problem, the Rapini potential was extended to a more general case of a biaxial object [2], which is any FLC. It was proposed for this purpose to characterize a spatial position of the biaxial FLC structure not only by the director $\boldsymbol{n}$, as is done for uniaxial molecular structures of liquid crystals, which are equivalent in symmetry to cylinders of rotation but also by a molecular core plane (MCP) that allows introducing additionally to the ordinary $\boldsymbol{n}$-director a second component of the FLC director, so-called $\boldsymbol{m}$-director, which is perpendicular to the MCP (Figure 1).

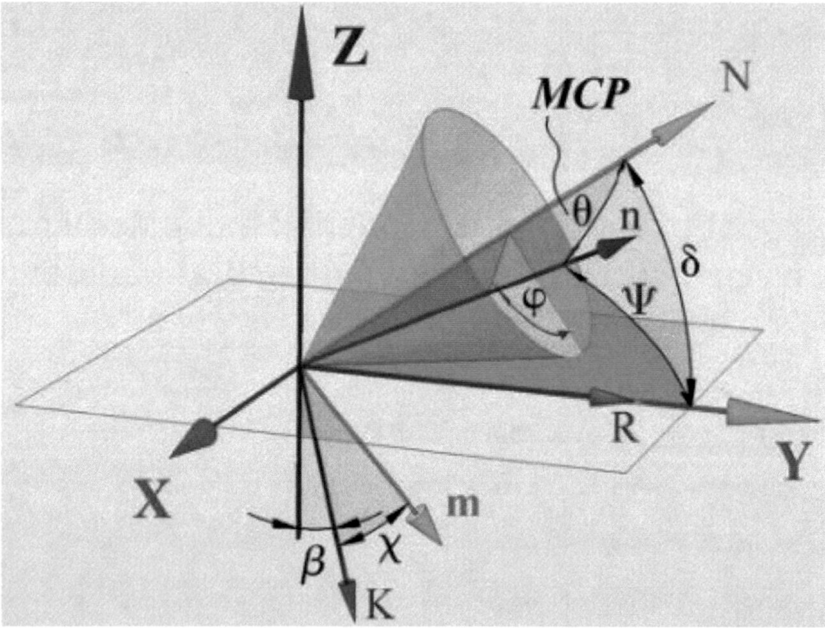

Figure 1. The designation of the structure of the helix-free (HF) FLC layers in the concept of biaxial surface potential, where $\boldsymbol{N}$ – is the normal to the smectic layers, δ - is the angle between the normal to the plane of the smectic layer $\boldsymbol{N}$ and the $\boldsymbol{R}$, φ – is the azimuthal angle of the position of the director $\boldsymbol{n}$ in the plane of the smectic layer, θ – is the angle of inclination of the molecules in the smectic layers, ψ – is the angle between $\boldsymbol{n}$ and the easy alignment axis $\boldsymbol{R}$, the unit $\boldsymbol{K}$ shows the direction of easy alignment of the $\boldsymbol{m}$ (normal to the MCP). The direction of the $\boldsymbol{K}$ can be, in general, tilted by an angle β relative to the normal to the of the solid boundary surface, χ the angle between $\boldsymbol{K}$ and $\boldsymbol{m}$ [2].

Given this addition, the biaxial surface potential Φ_s is written in the following form [2]:

$$\Phi_s = -\frac{W_n}{2}\left[(\sin\delta\cos\varphi\sin\theta + \cos\delta\cos\theta)^2 + \widetilde{W}_1 sin^2\varphi\, cos^2(\delta-\beta) \pm \widetilde{W}_2 \sin\varphi\cos(\delta-\beta)\right] \quad (2)$$

where $\widetilde{W}_1 = W_m/W_n$, $\widetilde{W}_2 = W_p/W_n$ - are dimensionless anchoring energies, and the signs + or − correspond to the upper and lower boundaries of the HF FLC cell, respectively, W_m - is the anchoring energy coefficient of the $\boldsymbol{m}$ director with respect to the easy alignment axis $\boldsymbol{K}$, W_p - is the polar part of the anchoring energy of vector $\boldsymbol{p_1}$ (collinear to vector $\boldsymbol{m}$) with the direction of easy orientation $\boldsymbol{K}$.

By solving equation (2), for certain values $\widetilde{W}_1$ and $\widetilde{W}_2$, it is possible to have several energy minima (Fig. 2.). One minimum (Fig.2. a) corresponds to the ordinary uniaxial Rapini potential that provides a monodomain texture (Fig.2. c). The presence of three energy minima (Fig.2. b) indicates the three-domain texture of the FLC (Fig.2. d), which often observed in FLC cells.

In this paper, we consider the possibility of controlling the alignment quality of FLC helical nanostructures by transforming the biaxial potential of the FLC to an effectively uniaxial potential due to the diminishing of the FLC helix pitch.

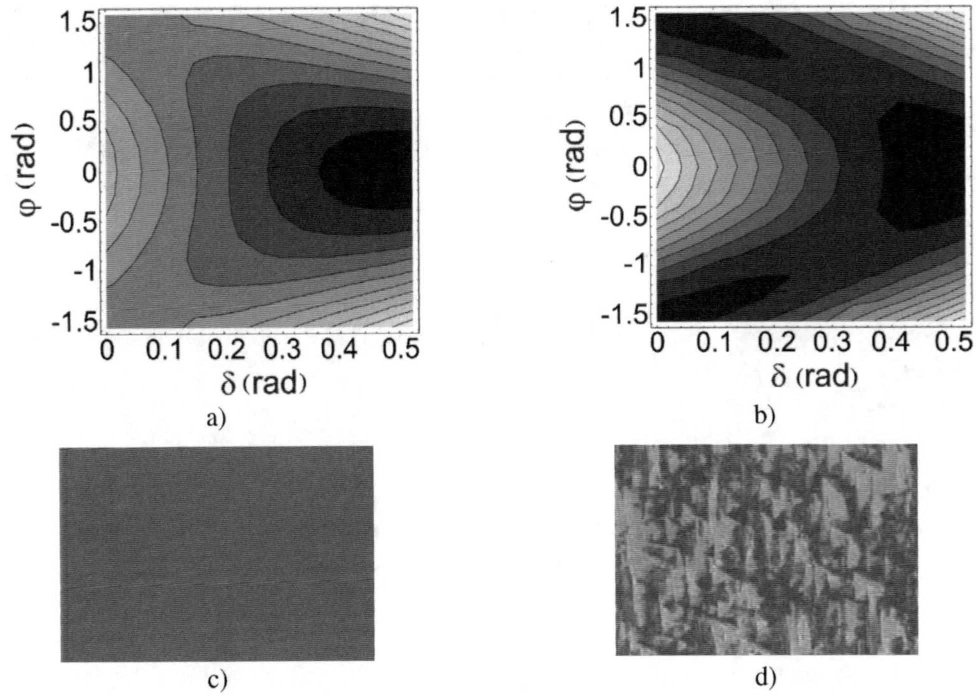

Figure 2. Lines of the surface potential level and the positions of its minima and textures of HF FLC cells. a) One minimum of surface potential, b) Three minima of surface potential, c) a monodomain structure at the one minimum of surface potential, d) three types of domains corresponding to three minima of surface potential [2].

2. Experiment

Continuous films and films with an island structure (Fig.3) made of PMDA-ODA polyimide layers were used as aligning layers. Liquid crystal mixtures of FLC-661/7 series with different helix pitch p_0 [4] were used to check the proposed concept of controlling the alignment quality of FLCs helical nanostructures. Images of the FLCs textures were taken using the Olympus BX 53 polarizing microscope, and the optical contrast ratio (CR) of electro-optical cells placed between crossed polarizers was measured in white light according to equation (3):

$$CR = \frac{T_{max}}{T_{min}} \tag{3}$$

where T_{min} – minimum light transmittance of the FLC cell, when the helix axis is oriented along the polarizer plane, T_{max} - maximum light transmittance of the FLC cell, when the helix axis is oriented at an angle of 45 degrees to the polarizer plane.

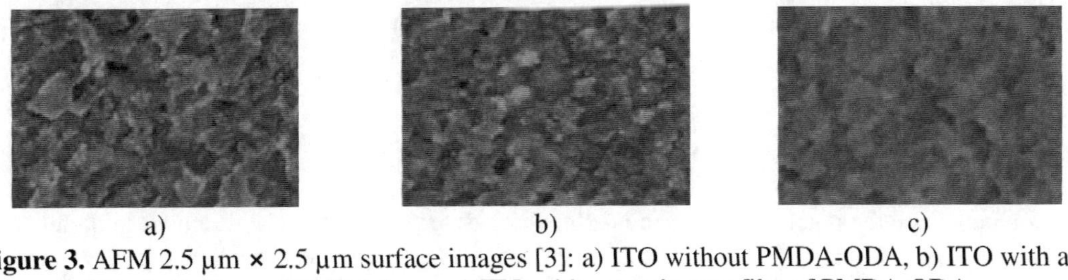

Figure 3. AFM 2.5 μm × 2.5 μm surface images [3]: a) ITO without PMDA-ODA, b) ITO with an island structure of PMDA layer, c) ITO with a continuous film of PMDA-ODA.

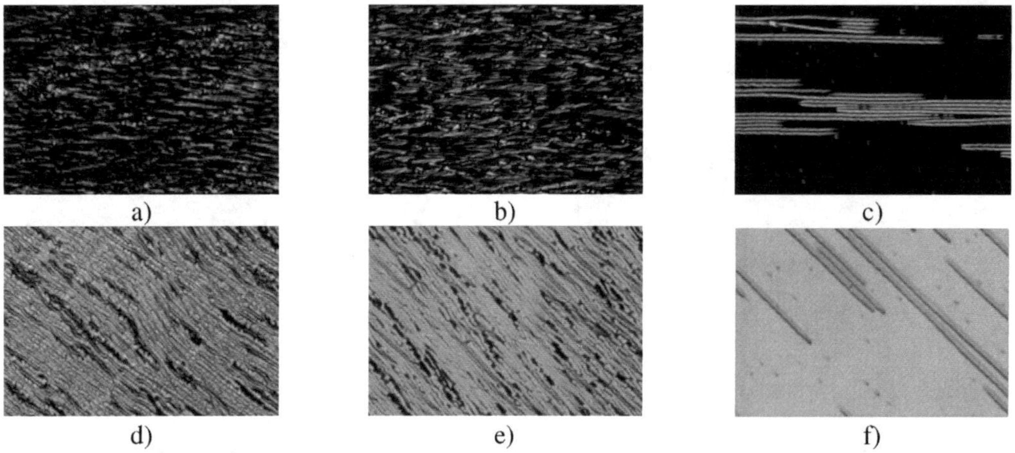

Figure 4. The FLC-661/7 series textures of electrooptical cells with continuous aligning layers at FLC layer thickness of about 3.0 μm. Textures a), b), c) are observed when the helix axis is oriented along the polarizer plane, and textures are d), e), f) when the helix axis is perpendicular to the polarizer plane; a), d) the FLC helix pitch $p_0 = 143$ nm, $CR = 4$; b), e) $p_0 = 112$ nm, $CR = 10$; c), f) $p_0 = 70$ nm, $CR = 28$. Size of images is 250 μm × 200 μm.

In Fig.4. The FLC-661/7 series textures of electrooptical cells with continuous aligning layers and with different helix pitch are shown. A decrease in the helix pitch allows one to achieve a significant decrease in the dislocations density or, in other words, a significant increase in the texture homogeneity. The presence of several colors of birefringence (Fig. 4 a, d) indicates the presence of several types of domains, which appear, probably, due to several potential minima of the biaxial surface potential, which it could be true also for FLCs with a sufficiently large helix pitch (p_0 = 143 nm) the same manner as it is valid for the HF FLC. As the helix pitch decreases down to 70 nm (Fig. 4 c, f), the dislocations density sharply decreases, the texture becomes more uniform and with alone birefringence color, which may indicate a gradual transformation of the biaxial surface potential into a uniaxial one. The improvement in optical quality of FLC cells due to diminishing of the p_0 magnitude is confirmed by the measured contrast ratio values.

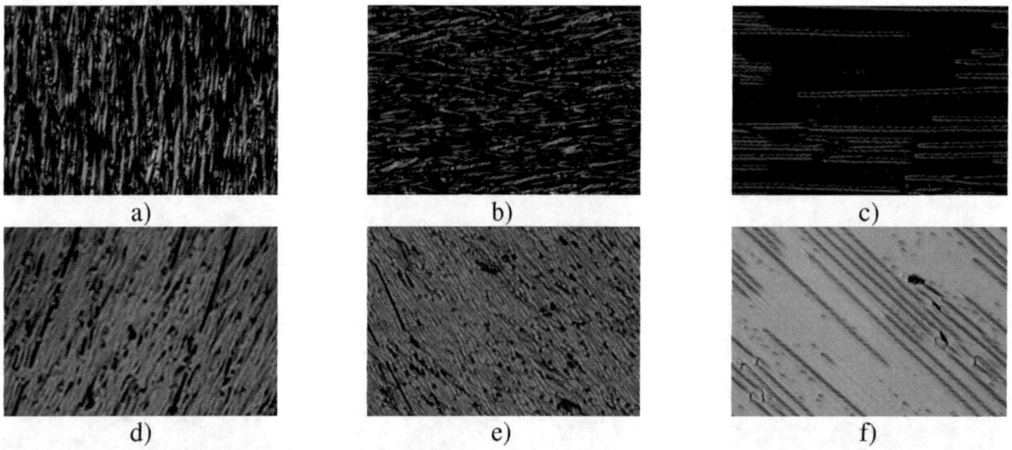

Figure 5. The FLC-661/7 series textures of electrooptical cells with an island structure of aligning layers at FLC layer thickness of about 3.0 μm. Textures a), b), c) are observed when the helix axis is oriented along the polarizer plane, and textures are d), e), f) when the helix axis is perpendicular to the polarizer plane; a), d) $p_0 = 112$ nm, $CR = 7$; b), e) $p_0 = 70$ nm, $CR = 14$; c), f) $p_0 =$ 50 nm, $CR = 22$; in the dark and light state, respectively. Size of images is 250 μm × 200 μm.

In Fig.5. The FLC-661/7 series textures of electrooptical cells with an island structure of aligning layers are shown. The deterioration of the optical quality of the FLC with p_0 = 70 nm when replacing the continuous aligning layer with the island one is quite obvious when comparing Figs 4 c, f and 5 b, e. However, a further decrease in p_0 to 50 nm makes it possible to diminish significantly the dislocation density and improve the optical quality of FLC layers even with the island structure of aligning layers, see Figs. 5 c, f.

Thus, an improvement in the optical quality of FLC layers with a decrease in the helix pitch is observed both for continuous and island structures of aligning layers. A decrease in the helix pitch obviously leads to a more isotropic distribution of vectors ***m*** in space that means a transition to an effectively uniaxial surface potential, which provides a monodomain FLC structure.

3. Conclusion

It is shown experimentally that a decrease in the helix pitch contributes to the FLC monodomain structure formation, in other words to the optical quality of FLC cells improvement independently on aligning layer's structure. At the qualitative level, the observed phenomenon can be interpreted in terms of the transition from a biaxial to an effectively uniaxial surface potential as the helix pitch decreases.

Acknowledgments

The work was partially supported by the following RFBR grants: № 19-52-06005 MNTI_a and № 20-02-00746.

References

[1] Rapini A and Papoular M J 1969 *J. Phys. (Paris) Colloq.* **30** C4-54
[2] Kaznacheev A V, Pozhidaev E P, Rudyak V, Emelyanenko A V, Khokhlov A R 2018 *Physical Review E* **97** 042703
[3] Mikhailenko V, Krivoshey A, Pozhidaev E, Popova E, Fedoryako A, Gamzaeva S, Barbashov V, Srivastava A K, Kwok H S, Vashchenko V 2019 *Journal of Molecular Liquids* **281** 186–195
[4] Huang D D, Pozhidaev E P, Chigrinov V G, Cheung H L, Ho Y L, Kwok H S 2004 *Displays* **25** 21 - 29
[5] Pozhidaev E P, Torgova S I, Budynina E M, Tkachenko T P, Kuznetsov A V, Barbashov V A 2020 *Liq. Cryst. and their Appl.* **20** (3) 26–33

Influence of the temperature field in the energy-saving carbon-graphite thermal unit on configuration of the solidification front under different conditions of growing fluoride single crystals by the HDS method

V A Yusim[1], S E Sarkisov[1], Y Y Kloss[3], F A Yusim[1], L V Ivanova[2]

[1]*NRC Kurchatov Institute, 1, Akademika Kurchatova pl., Moscow, 123182, Russia*
[2]*MGIMO University, 76, Prospect Vernadskogo Moscow, Russia, 119454*
[3]*MIPT, 9, Institutskiy per., Dolgoprudny, Moscow Region, 141701, Russian Federation*

E-mail: nrcki@nrcki.ru, intoff@mipt.ru, international@inno.mgimo.ru

Abstract. Mathematical modeling of the influence of the temperature field created by the ECU on the position and on configuration of the solidification front under different conditions of growing single crystals of fluorides by the HDS method has been carried out. Data were obtained for different growing conditions for single crystals of fluorides and their effect on configuration of the solidification front in various modifications of the ECU of the first and second types. From a comparison of mathematical calculations, a conclusion was made about the most suitable conditions for the synthesis of these single crystals.

1.Introduction

The synthesis of single crystals from a melt is a complex physicochemical process in which the most important, often decisive, role is played by heat and mass transfer. Strictly specified temperature conditions must be observed to obtain structurally perfect single crystals of high optical quality. It is also necessary to maintain constant temperature gradients at the growth front of the single crystal and high stability of the process. The accuracy of maintaining the temperature for the synthesis of most single crystals of fluorides with the fluorite structure at a level of 1500° C reaches several fractions of a degree [1-3]. Disturbance of these growth conditions can lead to the appearance of various types of defects and high residual stresses in single crystals.

One of the most important tasks in the synthesis of single crystals is to control the shape of the solidification front, which is set by the thermal and kinetic growth conditions [4]. Thus, the mathematical calculation of the temperature field of the energy-saving carbon-graphite thermal unit (ECU) and its influence on the position and configuration of the solidification front is a non-trivial problem. This task requires strict knowledge of the boundary conditions of crystallization and the thermophysical properties of the system, including its aggregate states. The problems of the dependences of the elastic modulus of steels and alloys depending on high temperatures were discussed in detail in the articles [5-10].

The goal of this work is to study the processes of heat and mass transfer in various modifications of the ECU and their influence on configuration of the solidification front [11]. The possibility of varying the value of the temperature gradient in the zone of the crystallization front is shown by letting gas flow directly into this zone. The effect of the admitted gas flow on the shape of the solidification front and on the axial temperature gradient of the ECU has been studied. The most suitable flow rates of the admitted gas have been determined to achieve optimal conditions for the synthesis of a single crystal. The effect of the created temperature field in various modifications of carbon graphite heating units under different growing conditions on configuration of the solidification front of the synthesized single crystal and the temperature distribution in it is investigated. This work is a further development [12] on growing single crystals with a fluorite structure by the HDS method in a carbon-graphite thermal unit.

2.Experimental part

In addition to the basic modification of the carbon-graphite growth ECU described and calculated in [12], two additional modifications of the structures of growth units for growing fluoride crystals were developed.

Figure 1 shows a modification of the ECU of the first type. In this design, there is no system of replaceable heat shields of the external heat-shielding circuit on all three modules, instead of which thermal insulation made of graphite felt, such as a "sandwich panel", is used. Thermal insulation is laid in several layers tightly adjacent to each other. Thus, the finished structure is a total thermal insulation volume like a muffle. The heating block of the central module also has additional thermal protection in the form of "sandwich panels", made up according to the same principle into a common heat-insulating muffle. This modification of the ECU provides low-gradient zone (soft gradient zones) in the crystal growth zone and in the crystal cooling zone due to more efficient energy saving. This modification of the ECU allows to reduce the thermal inertia of the entire system, to reduce the thermal load on the inner walls of the crystallization apparatus (at the boundary of the furnace atmosphere and the channel heat exchanger). The power values on the heaters also decrease at the same melt temperatures in the crystal growth zone. The applied thermal insulation made of graphite felt, of the "sandwich panel" type, is a low-density carbon-carbon composite material based on carbon felt and carbon fiber, molded individually or in combinations in the form of layered structures compacted with pyrolytic carbon (SGL Carbon SIGRAFLEX, SIGRASIC, SIGRATHERM etc.).

A modification of the ECU of the second type is described [11] (Figure 2). In this modification, changes are present both in the design of the growth unit and in the crystallization apparatus. An additional unit for gas puffing is introduced into the design of the central chamber of the crystallization apparatus. This unit consists of a vacuum needle leak mounted on the outside of the central chamber of the crystallization apparatus, transition sleeves and nozzles made of graphite. In the design of the apparatus in the central module, the upper part of the inner contour of the right block has a central insulated gas supply channel with outlets into the holes of which nozzles are inserted. This adit has a single sealed flow channel with a vacuum needle leak. The tight connection of graphite couplings, branch pipes with each other, with the steel parts of the needle leak, as well as with the adit, is carried out through the use of liquid coal tar pitch as a binding material at the joints and seals. This compound hardens during the start of operation of the furnace (that is, with an increase in temperature) [11]. During furnace operation (during crystal synthesis), an inert gas, Ar, N_2 or He, is continuously supplied through the gas supply channel directly to the crystal growth zone, namely, to the receiving area located immediately after the diaphragm. This contributes to a sharp change in the axial temperature gradient, while the temperature gradient is controlled directly by changing the flow rate of the supplied gas and the temperature on the heaters. Thus, this modification makes it possible to vary the value of the temperature gradient in the zone of the solidification front during the crystal synthesis process.

For each study, a semitransparent CaF_2 single crystal with dimensions of *300mm x 198mm x 40mm*, placed in a graphite crucible, which, in turn, was mounted on a tungsten drag, was considered in the calculation model. The thermal process was considered as quasi-stationary due to the low crystallization rates from 1 to 5 mm / h, [13, 14]. Thus, the crucible with the synthesized crystal was installed motionlessly in the central module through the gradient field from the melting zone (through the heaters) to the cold annealing zone, thereby undergoing asymmetric heating.

It is noteworthy that a distinctive feature of the HDS method from other melt ones is that the height of the melt in it is much less than its surface area, so the investigated crystal melt made up 2/3 of the crucible boat. So in this method, with a small height of the melt, large length and large surface area of the melt open in relation to the heater, temperature conditions of soft natural convection [2] of a laminar nature arise in the bottom part and along the surface of the melt without vortex flows. This fact excludes the formation of intense hydrodynamic and convective flows, which affect the degree of stability of the crystallization process and the morphological stability of the crystallization front. Thus, the velocity of the flow inside the melt can be neglected to simplify the problem being solved. And as a consequence, the investigated single crystal is considered in one state of aggregation - solid.

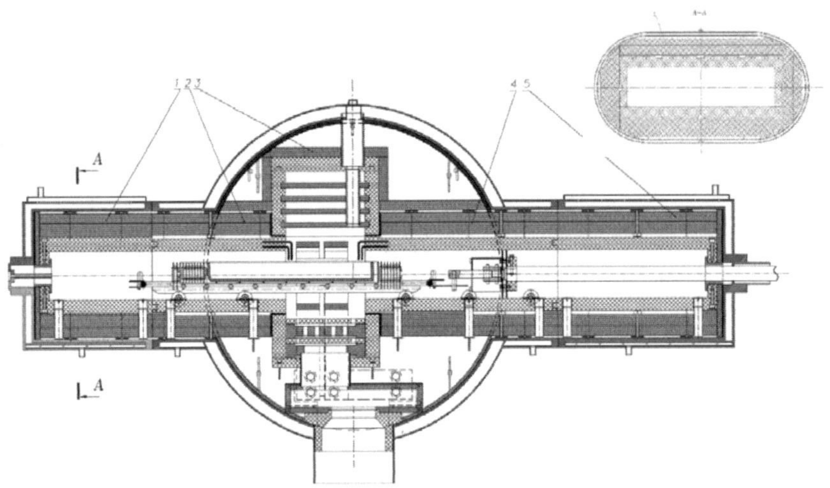

Figure 1. Layout diagram of the modified construction of the ECU of the first type placed in the crystallization apparatus "Rubitek - 1", where 1 - thermal insulation made of graphite felt, such as a "sandwich panel" of the loading module; 2.4 - thermal insulation made of graphite felt, such as a "sandwich panel" of the central module corridor blocks; 3 - additional external thermal insulation of the central module; 5 - thermal insulation made of graphite felt, of the "sandwich-panel" type of the receiving module;

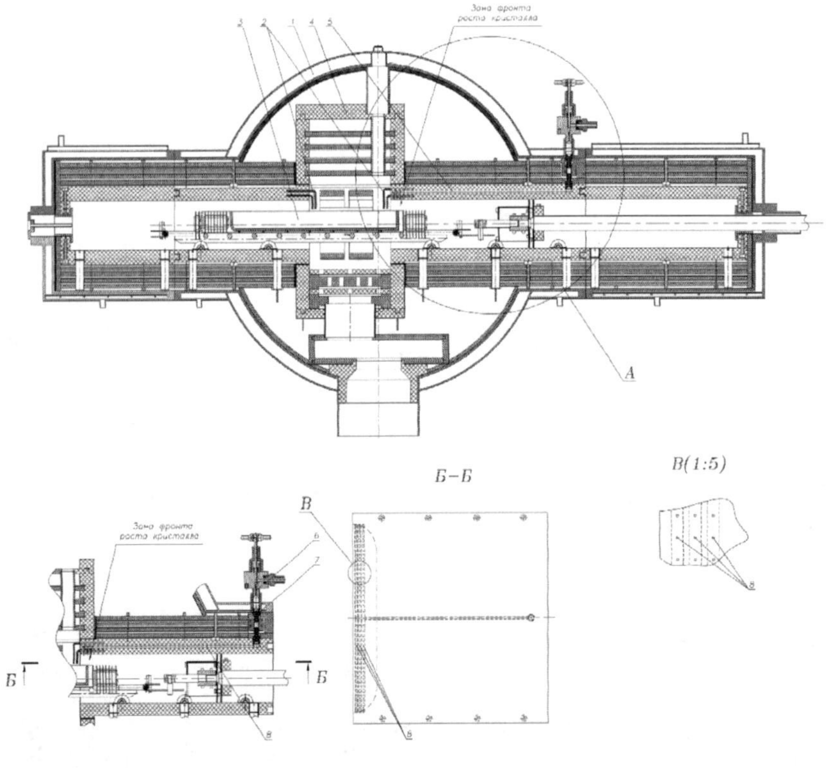

Figure 2. The layout diagram of the modified construction of the ECU of the second type, placed in the crystallization apparatus "Rubitek - 1", where 1 is the crystallization apparatus; 2 - heat-insulating diaphragms; 3 - synthesized crystal; 4 - central module; 5 - upper slab of the inner contour of the right block of the corridor; 6 - vacuum needle leakage valve; 7 - transitional graphite couplings and branch pipes; 8 - gas supply channel (adit); 9 - taps with holes for supplying gas to the crystal growth zone; [5]

For these modifications of the growth nodes, various crystallization conditions were set. Thus, as the atmosphere inside the crystallization apparatus with the growth unit placed in it, a forevacuum of 10^{-3}Pa and excess atmospheres of Ar, N_2 or He gases were adopted, which were equal to 202.5 kPa. For the ECU of the second type, an additional series of studies was carried out with different rates of gas flow inlet into the chamber. Thus, these studies were also carried out in an excess atmosphere of an inert gas of argon, which was equal to 202.5 kPa, while it was supposed to relieve excess pressure through the overpressure relief valve. The calculation conducted out with a continuous supply of gas through the channel directly into the crystal growth zone for the main velocities of 0.1, 0.2, 0.5, 1, and 2m/s. The main options for the calculation, including the object of study and its characteristics, block-structured grids, the temperature of the coolant at the inlet and its speed, etc. were similar to [11], both for the first and for the second modification of the ECU, with the exception of the temperature on the heaters in the modification ECU of the first type . So, in the course of research, due to the higher thermal performance of the thermal insulation of the graphite-graphite growth unit of the first type, the temperature of 1500 ° C was set on the lower heater, and the temperature on the upper 1600 ° C, respectively (the temperature on the heaters was selected and set lower than in the basic calculation based on the crystallization conditions test sample to avoid overheating).

The mechanism of heat transfer in a single crystal was carried out using the mechanisms of molecular heat conduction and radiation (radiation-conductive heat transfer). The single crystal itself was in a state of convective heat exchange with the medium in the thermal unit (only when using the technology of growing in an overpressure of an inert gas) and radiant heat exchange with the surrounding surfaces. In this study, it was necessary to determine the temperature distribution in the single crystal and the shape of the crystallization front.

3.Numerical procedure

The calculated mathematical model of this single crystal includes: the energy equation, the equations of heat and radiation transfer, the boundary conditions of heat conduction and radiation at the crystal boundaries. Also, this model is part of a complete heat and mass transfer model for a generalized cycle.

The energy transfer equation for the direction normal to the surface into the volume of the medium for this model was written as follows [13]:

$$\frac{d\varepsilon}{d\tau} = div\, q_E + e_0,\ \ \varepsilon = \varepsilon_C + \varepsilon_R,\ \ q_E = q_C + q_R \tag{1}$$

where q_E, q_C, q_R – are the vectors of total energy transfer, thermal conductivity and radiation, ε, ε_C, ε_R – are the volumetric energy densities of the total, internal and radiation, e_0 – is the volumetric power of energy sources.

The equation of energy transfer by thermal conductivity [13]:

$$q_C = -\lambda gradt \tag{2}$$

Radiation [7]:

$$\frac{dI_s}{dS} = -kI_s + \eta \tag{3}$$

where I_S – radiation intensity in the direction S; η – volumetric self-radiation coefficient of the medium. In this case, the values of ε_R are omitted due to small values.

The components of the transfer vectors in the direction n have the form [13]:

$$q_{cn} = -\lambda\frac{\partial t}{\partial n} \tag{4}$$

$$q_{Rn} = \int_{4\pi} I_s \cos(n,s)\, d\omega \tag{5}$$

In equations (1), (3), (5) the values of ε_R, q_R, I_S, k, η refer to spectral and integral radiation or to the transparency zone of a single crystal.

The boundary conditions were set different, so on a fixed surface of the crystal for an elementary area dS_K^C centered at point M and were written with the direction of the normal n inside the crystal (Figure 3-a) [13]:

$$-\frac{dt_K}{dn}=\frac{1}{\lambda_K}\left(\alpha\left(t_K-t_f\right)+E_C-aE_i-q_R\right) \tag{6}$$

where α - is the heat transfer coefficient; t_k – s the crystal temperature; t_f – is the temperature of the medium in the ECU; λ_K – coefficient of thermal conductivity of the crystal material; *Ec* and *Ei* – are the densities of the intrinsic and incident fluxes of the integral radiation; q_R – s the density of the resulting radiation flux in the range of partial transparency of the crystal $\delta\lambda'$ on the inner shell of the crystal; a – the degree of emissivity of the container wall radiation.

On the free surface in the opacity zone of the crystal for an elementary area dS_K^F centered at point N (Figure 3-b) [13]:

$$-\frac{dt_K}{dn}=\frac{1}{\lambda_K}\left(\alpha\left(t_K-t_f\right)+E''_C-a''E_i''\right) \tag{7}$$

where $E''c$ and E_i'' – are the intrinsic and incident surface radiation flux densities in the opacity zone of the crystal.

In the zone of partial transparency of the crystal (Figure 3-b) on the inner boundary of the area dS_K^F[7]:

$$I_{1S(\varphi1)}=\bar{I}_{1S(\varphi1)}R+I_{2S(\varphi2)}(1-R)\frac{n_2^2}{n_1^2},\qquad \text{при } \varphi_1<\varphi_{BR} \tag{8}$$

$$I_{1S(\varphi3)}=\bar{I}_{1S(\varphi3)},\qquad under\ \varphi_3<\varphi_{BR}$$

on the outer edge of the site dS_K^F[7]:

$$\bar{I}_{2S(\varphi2)}=I_{2S(\varphi2)}R+\bar{I}_{1S(\varphi1)}(1-R)\frac{n_2^2}{n_1^2},\qquad under\ 0<\varphi_2<\frac{\pi}{2} \tag{9}$$

where n_1 and n_2 – refractive indices of the crystal material and the medium in the thermal unit; I_{1S} and I_{2S} – radiation intensity in the interval $\delta\lambda'$ in the crystal and the environment of the thermal unit; the S direction is determined by the angle φ; R – is the reflection coefficient of the free surface of the crystal.

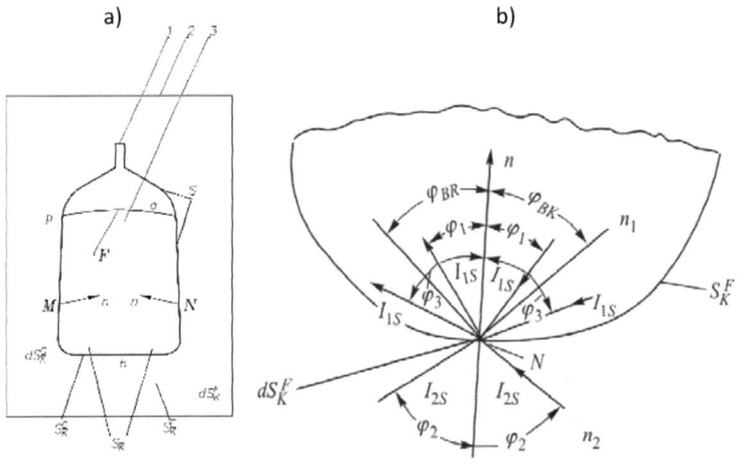

Figure 3. a) - a diagram for the formulation of a mathematical model, where: 1 - heat system; 2 - heating unit; 3 - crystal; F - crystallization front; S, S_K, S_K^F, S_K^C - total surface of the thermal system, crystal surface: - total free and fixed; lines ab and bp conventionally represent free and fixed surfaces; t - normals to the surface; b) - Scheme for the formulation of the boundary condition on the free surface of the crystal: I_{1S}, I_{2S}, φ_1, φ_2 - radiation intensities and angles of their incidence in the crystal and the environment; n_1, n_2 - refractive indices of the crystal of the environment: φ_{BR} – Brewster angle; S_K^F – free surface of the crystal; [13]

The Brewster angle was determined by the expression [7,9]:

$$\varphi_{BR} = arctg\frac{n_2}{n_1} \quad (10)$$

The connection between the angles φ_1 and φ_2 is found from the relation [7,9]:

$$sin\varphi_2 = sin\varphi_1\frac{n_2}{n_1} \quad (11)$$

Thus, the radiation descending from the single crystal to an elementary area within the Brewster angle minus the reflected one originated from the single crystal and was absorbed in the medium of the thermal node within the hemisphere. The radiation, descending to the site at an angle to the normal, was completely reflected by the Brewster angle. Radiation entering the elementary area from the thermal unit, minus the reflected one, entered the single crystal and was scattered in it within the Brewster angle. It can be seen from (6) that in the absence of a zone of opacity in the crystal and vacuum in the thermal unit, the following condition was met:

$$\frac{dt_K}{dn} = 0$$

4.Results and discussion

The above-considered stationary, three-dimensional mathematical model of heat and mass transfer for a generalized cycle of directed crystallization gives a general idea of the course of heat and mass transfer processes in the ECU and a single crystal placed in it.

The main results of mathematical modeling of the temperature distribution of a solid modified by the ECU of the first and second types and the steel water-cooling jacket (SWJ) of the crystallization apparatus are presented in Figures 4 and 5. Figure 6 shows the results of calculations that give an idea of the temperature distribution in a single crystal and the effect of the temperature field on the shape of the solidification front at growing in ECU of various modifications.

It follows from Figure 4 that when growing single crystals in vacuum and an excess gas atmosphere, the design of the modified ECU of the first type and the low thermal conductivity of the graphite thermal insulation used in it make it possible to create a low-gradient region in the crystal growth zone and a low-gradient temperature region in the annealing zone with a step of changing the temperature of the solid. bodies in the crystal growth zone 10 - 40°C/cm, in the annealing zone 15 - 40°C/cm (Figure 7-a). To use the synthesis technology in an excess gas atmosphere, values of 20 - 30°C/cm were obtained in the crystal growth zone and 20 - 30°C/cm in the annealing zone (Figure 7-b).

For growing single crystals in the modified ECU of the second type, the values of temperature gradients similar to the values obtained in [12] for the basic ECU of the atmosphere with excess gas pressure were obtained (Figure 5). It should be borne in mind that with an increase in the gas feed rate, the axial temperature gradient becomes somewhat sharper. So, when blowing the model with flow rates of the admitted gas of 0.1, 0.2 and 0.5 m/s, the axial gradient changed insignificantly and looked like in Figure 7-c. And at speeds of the supplied gas 1 and 2 m/s, the temperature gradient took the form as in Figure 7-d. The average values in the region of the crystal growth zone were 70–50°C/cm and 20–40°C/cm in the crystal annealing zone at gas flow rates of 0.1, 0.2, and 0.5 m/s; for velocities of 1 and 2 m/s were equal to 75 - 60°C/cm and 20 - 40°C/cm in the growth and annealing zones of the single crystal, respectively.

It can be seen from Figures 6 (a-f) that the created thermal conditions in the modified ECU provide a slightly convex isotherm of the growth front and, thanks to the separate control of the heaters in terms of power, create the desired angle of inclination of the vertical gradient, which allows the growth of optically perfect crystals.

The symmetrical temperature field required to create a morphologically stable crystallization front was created using two graphite tape arc heaters with separate power control. In this case, the upper heater above the surface of the charge (and, accordingly, with the melt) had a flat L-shape, and the lower one had an inverted U-shape and was placed under the crucible (boat) and ensured uniform heating of both the bottom and side walls of the crucible.

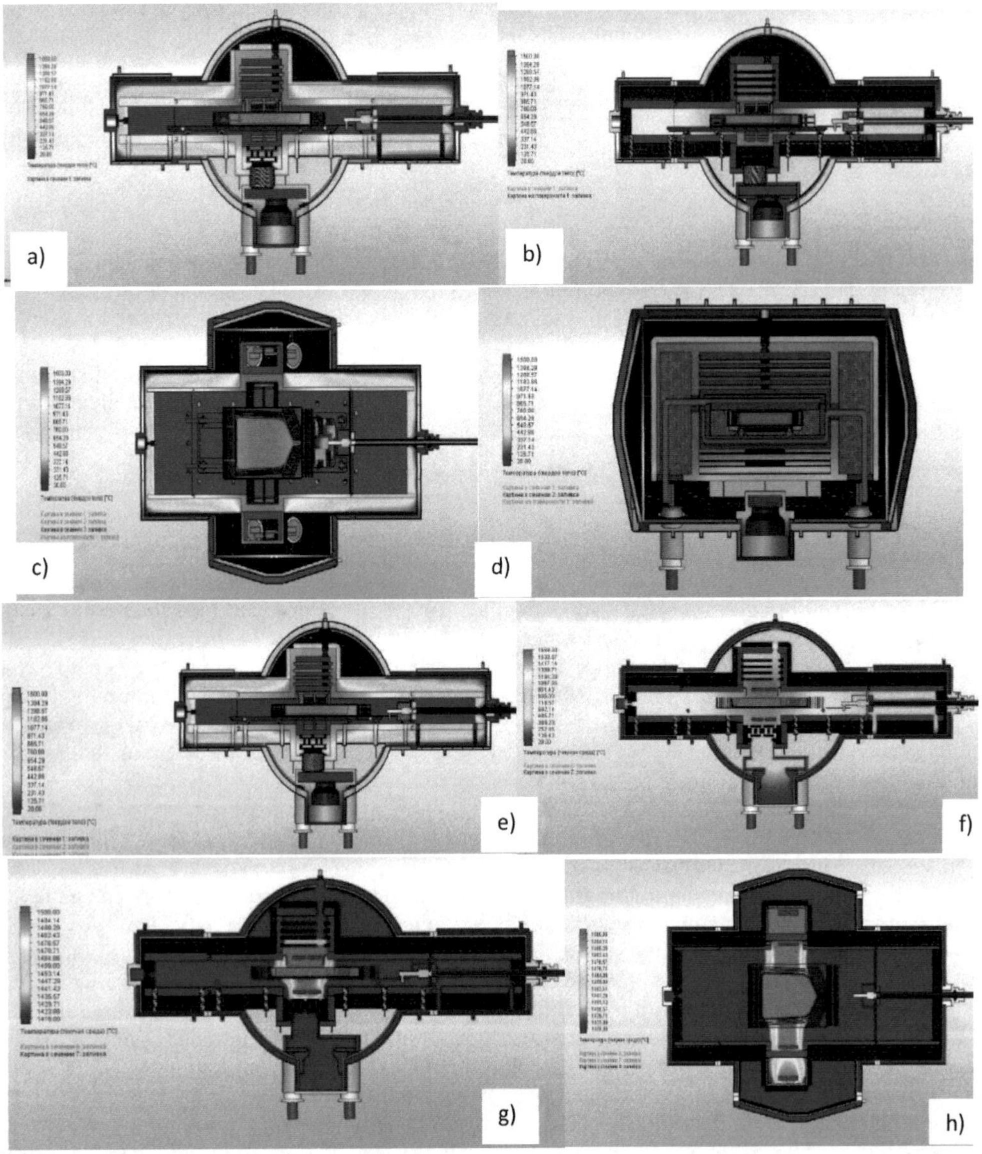

Figure 4. * Results of mathematical modeling of the temperature distribution in the modified ECU according to the first type for the use of vacuum growing technology (a-d) and the technology of growing single crystals of fluorides at excess pressure (e-f), where: a, e) - in the longitudinal section (form side); b) - isotherm of temperature distribution along from the beginning of the loading zone to the end of the unloading zone; c) - in longitudinal section (top view); d) - in cross section; f-g) is an isotherm of the temperature (gas) distribution along from the beginning of the loading zone to the end of the unloading zone, while views (g) and (h) clearly display a high-gradient region in the crystal growth zone;
**- due to insignificant differences in the temperature distribution pattern in a solid for growing single crystals at excess pressure, the results are presented only for Ar*

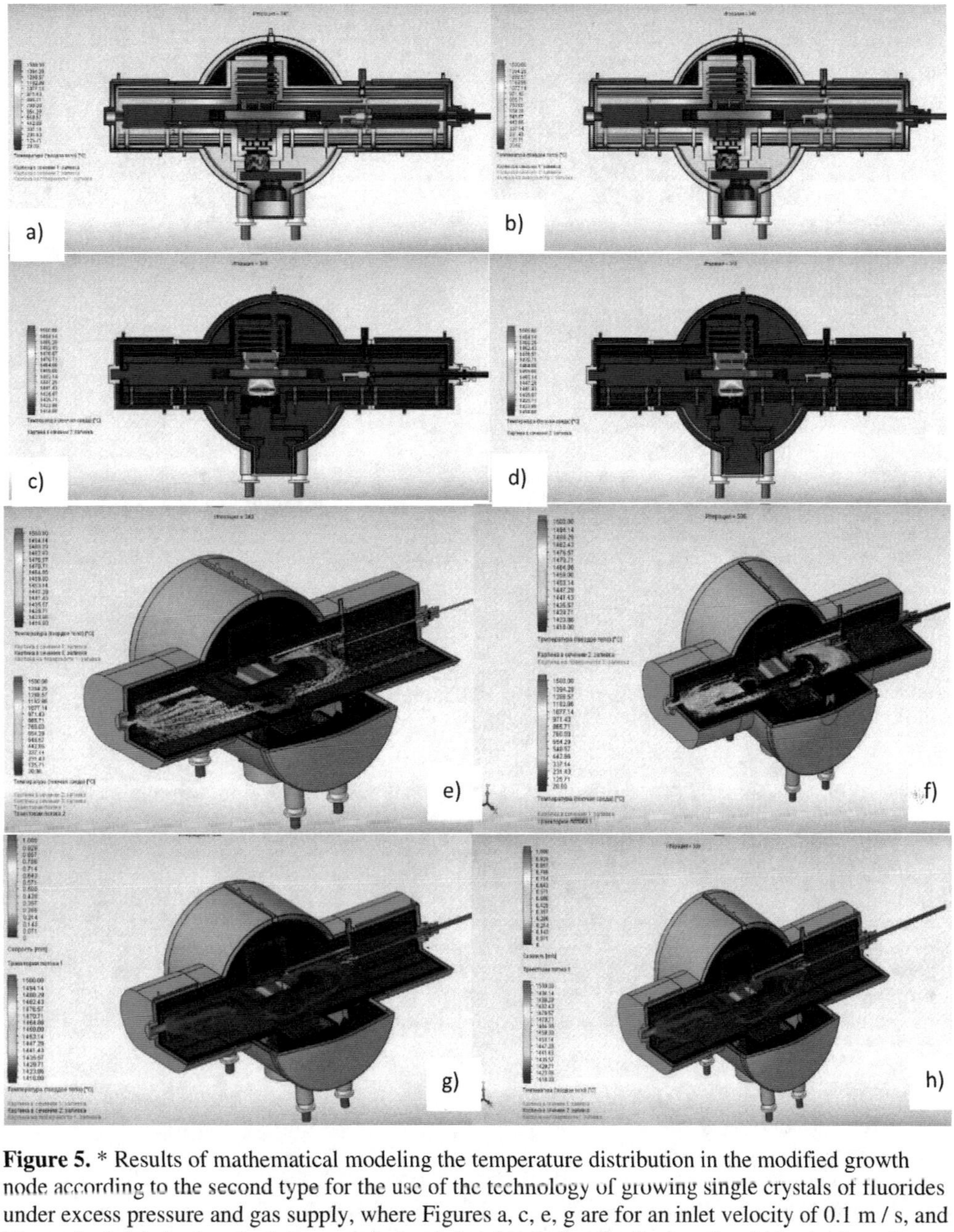

Figure 5. * Results of mathematical modeling the temperature distribution in the modified growth node according to the second type for the use of the technology of growing single crystals of fluorides under excess pressure and gas supply, where Figures a, c, e, g are for an inlet velocity of 0.1 m / s, and Figures b, d, f, h for 1 m / s, letters are designated: a, b) - in longitudinal section (side view); c, d) - isotherm of temperature (gas) distribution along from the beginning of the loading zone to the end of the unloading zone; e, f) - trajectory and temperature of the supplied gas; g, h) - velocity of the supplied gas and isotherm along the horizontal surface of the crystal;
** - the pictures of the temperature distribution in the modified thermal unit according to the second type for the vacuum growing technology are similar to the pictures in [12]*

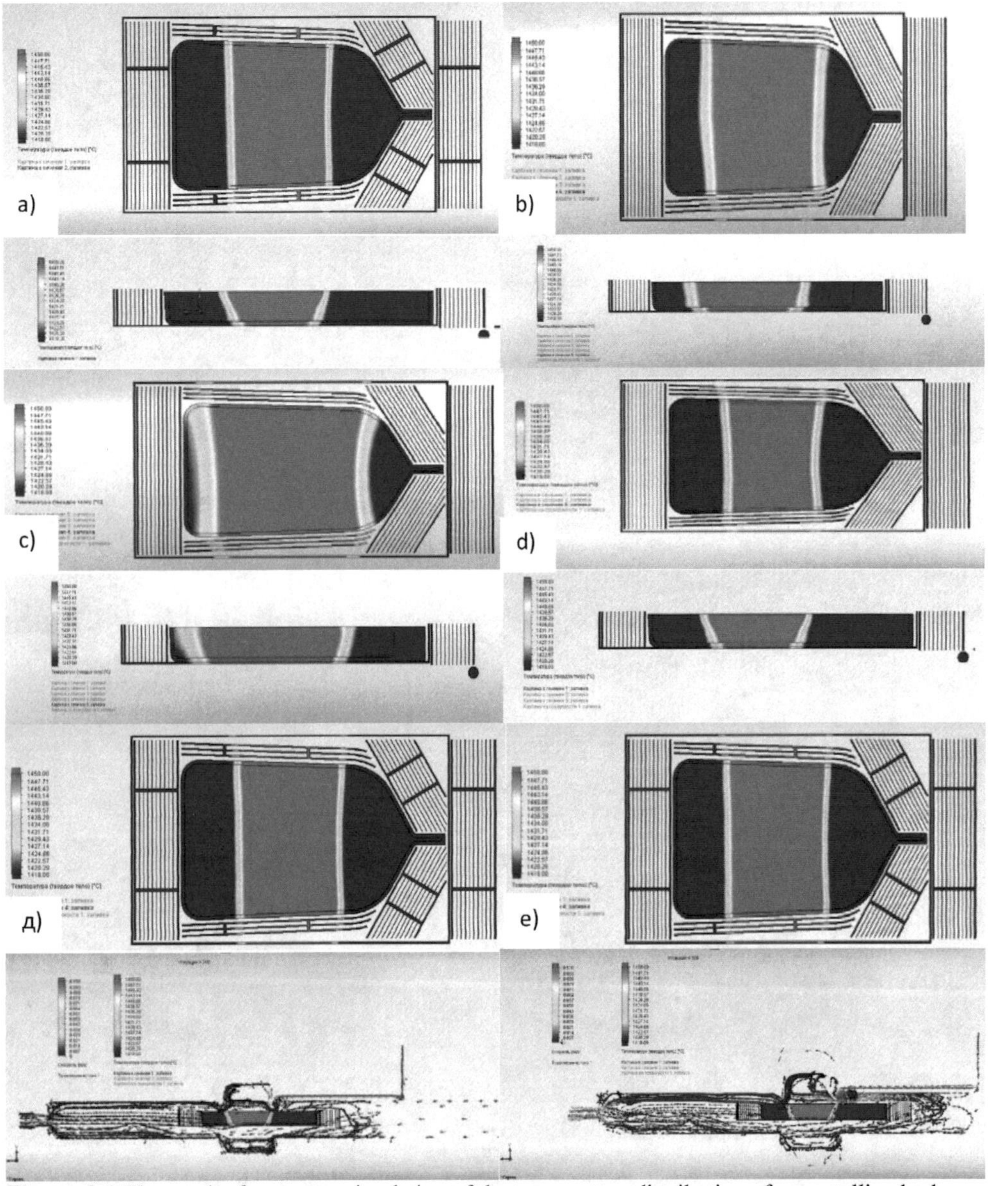

Figure 6. * The result of computer simulation of the temperature distribution of a crystalline body along the growth axis of a single crystal along a longitudinal section, top and side view in a CaF_2 single crystal, where: a) - in the modification of the ECU of the first type (atmosphere - vacuum); b) - in the modification of the ECU of the first type (atmosphere - nitrogen); c) - in the modification of the ECU of the first type (excess helium atmosphere r) - of the ECU of the first type (excess argon atmosphere); e) - in the modification of the ECU according to the second type (at a gas injection rate of 0.1 m/s - an excess Ar atmosphere); f) - in the modification of the growth unit according to the second type (at a gas injection rate of 1.0 m/s - an excess Ar atmosphere);

** - Figure 6 shows the main results of mathematical modeling for argon gas flow rates of 0.1 and 1 m/s, the rest of the gas inlet rates were excluded due to insignificant differences in the temperature distribution patterns in the crystalline body.*

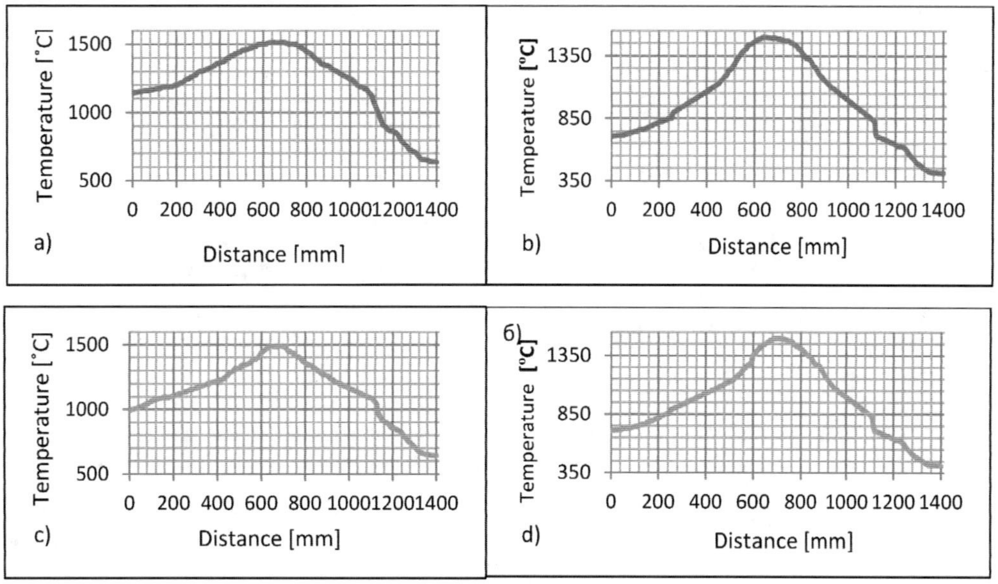

Figure 7. Graphical dependence of the temperature distribution for the modification of ECU according to the first (a, b) and the second (c, d) type in an atmosphere of excess gas pressure from the loading module zone to the receiving module zone along the working space, where: a) - for atmospheres with excess gas pressure; b) - for vacuum: c) - for gas admission 0.1-0.5 m / sec; d) - for gas inlet 1-2 m / s;

From Figure 6-a it follows that the preferred medium for growing single crystals was vacuum. Thus, a sharp temperature gradient of the crystallization front is created in the growth site, which is a necessary condition for the synthesis of single crystals of fluorides [12, 16-18]. During the synthesis of single crystals of fluorides in the ECU of the first type under conditions of an atmosphere of excess argon pressure at the same temperatures, a smoother temperature gradient is observed at the solidification front due to the increased thermal conductivity of the thermal system. In the case of an excess pressure of the atmosphere of N_2 and He, the crystallization front takes on a slightly diffuse shape as a result of the higher thermal conductivity of these gases in comparison with argon. Consequently, when using He or N_2 gases, a low-gradient region is created in the single crystal growth zone, which is the least preferable for the synthesis of fluoride single crystals from the point of view of thermal processes [16-18].

For the synthesis of a single crystal in the ECU of the second type using a gas flow inlet at different rates, the obtained results of computer simulation (Figure 6 e, f) are almost identical to the results [12] for the basic ECU. The crystallization front has a sharp temperature gradient, as in the synthesis of fluorides in a vacuum atmosphere, which also ensures the high quality of the obtained single crystals [19-21].

5.Conclusion

The above-considered stat Mathematical modeling of the influence of the temperature field created by ECU on the position and configuration of the solidification front under different conditions of growing single crystals of fluorides by the HDS method has been carried out. The necessary data on the influence of temperature fields on the configuration of the solidification front have been obtained. It is found that the designs of the modified ECU create thermal conditions providing a slightly convex isotherm of the solidification front. It is shown that the most favorable thermal conditions for synthesis are observed in a modified ECU of the first type in a vacuum atmosphere (Fig. 6-a). Using an atmosphere of excess gas pressure, a smooth temperature gradient of the solidification front is observed due to the creation of a low-gradient region in the crystal growth zone (Fig. 6 b-d). For the synthesis of a single crystal in The ECU of the second type using gas puffing at different flow rates, the results of computer simulation are identical to those obtained using a vacuum atmosphere.

Acknowledgements

The reported study was funded by RFBR projects number 20-08-00278, 19-29-02007, 19-29-02007 and by NRC Kurchatov Institute Order No. 1909 of 25.09.2020.

References

[1] Kh S Bagdasarov 2004 in the book High-temperature crystallization from the melt. Fizmatlit. Moscow 159.
[2] R Lodiz, R Parker 1974 Single crystal growth. Publishing house "MIR" Moscow 540 s.
[3] V A Yusim, R K Kalimullin, V V Ryabchenkov, S E Sarkisov 2018 *RF Patent* No. 2643980, 06 February.
[4] Yu M Shashkov 1982 *Metallurgy* 315
[5] V N Shinkin 2018 *CIS Iron and Steel Review.* **15** 23-27.
[6] V N Shinkin 2018 *CIS Iron and Steel Review.* **15** 32-38.
[7] V N Shinkin 2018 *Steel in Translation.* **48**(11) 718-723.
[8] V N Shinkin 2018 *Steel in Translation.* **48**(3) 149-153.
[9] V N Shinkin 2018 *Chernye Metally.* **5** 34-40.
[10] V N Shinkin 2019 *Chernye Metally.* **3** 32-37.
[11] V A Yusim, V V Ryabchenkov, S E Sarkisov 2019 *RF Patent* No. 187518, March 11.
[12] V A Yusim, S E Sarkisov, V V Ryabchenkov, Yu Yu Kloss, I V Govorun, L V Ivanova, A V Sakmarov 2020 *Journal of Physics: Conference Series* **1560**
[13] Kh S Bogdasarov, L A Goryainov 2007 *FIZMATLIT* 224
[14] Kh S Bagdasarov, L A Goryainov 2007 *The article was deposited at VINITI* 31771-75
[15] D V Sivukhin 1980 *General course of physics / Optics* 752
[16] S E Sarkisov, V A Yusim, V V Ryabchenkov, R K Kalimullin, I V Govorun, A V Sakmarov 2019 International Conference *MIPT*
[17] V A Yusim, S E Sarkisov, V V Ryabchenkov, P A Tataurov 2020 XIII International Conference on Applied Mathematics and Mechanics in the Aerospace Industry (AMMAI'2020), May 4-31 Crimea.
[18] V V Ryabchenkov, V A Yusim, O V Voronova, S E Sarkisov, S V Petrov 2019 International Conference *MIPT*
[19] V A Yusim, S E Sarkisov, V V Ryabchenkov, A V Sakmarov 2020 XIII International Conference on Applied Mathematics and Mechanics in the Aerospace Industry (AMMAI'2020), May 24-31, Crimea
[20] V A Yusim, S E Sarkisov, V V Ryabchenkov, P A Tataurov 2020 XIII International Conference on Applied Mathematics and Mechanics in the Aerospace Industry (AMMAI'2020), May 4-31 Crimea.
[21] V V Ryabchenkov, S E Sarkisov, V A Yusim, I A Kuzenkova, S V Petrov, S S Sitnikov 2019 International Conference *MIPT*

Advanced processes for low-temperature formation of functional metal oxide based thin films

A Abduev[1], A Akhmedov[2], A Asvarov[2,3], V Kanevsky[3], A Muslimov[3], V Belyaev[1,4], D Generalov[1], D Nikolaeva[1] and J Tirado[1], M A A Frah[1]

[1] Peoples' Friendship University of Russia (RUDN University), Moscow, Russia
[2] Institute of Physics, Dagestan Federal Research Center, Russian Academy of Science, Makhachkala, Russia
[3] Shubnikov Institute of Crystallography, Federal Scientific Research Center Crystallography and Photonics, Russian Academy of Sciences, Moscow, Russia
[4] Moscow Region State University, Mytishi, Russia

E-mail: a_abduev@mail.ru

Abstract. The analysis the discharge processes in magnetron plasma, target sputtering processes, as well as nucleation and formation of oxide thin films during dc magnetron sputtering is carried out. Particular attention is paid to the phenomenon of instabilities of the current-voltage characteristics of magnetron plasma during the sputtering of oxide targets, the processes of structural transformations of the surface of metal oxide targets under ion bombardment impact, and the mechanisms of low-temperature magnetron deposition of metal oxide thin films. Based on the results of the analysis performed the optimal routes for improving technologies for the low-temperature formation of transparent conductive oxide thin films have been discussed.

1. Introduction

Topics related to magnetron deposition of thin-film materials for use in electronics, photonics, and other applications have been widely studied for many years [1-3]. The rapid development of information display systems, solar cells and other optoelectronic devices has led to an active search for technologies for the synthesis of indefectible (i.e., having sufficient crystallinity and smoothness) thin films at relatively low temperatures on polymer substrates, as well as to the search for new available materials for new generation devices [4,5]. This is dictated by the urgent need to develop create roll-to-roll technologies for creation of large-area flexible display systems, thin-film solar converters and protection from electromagnetic pollution [6,7].

In addition, for several decades, the issue of replacing expensive ITO (In_2O_3:SnO_2) material used as transparent electrodes with some more affordable and cheaper alternative is still very relevant for researchers and developers of information display systems [8-9]. The situation is aggravated by the fact that now the display industry's needs for expensive In are becoming comparable to the volume of its annual world total production [10].

Here an overview is given of some processes aspects observed in the magnetron discharge plasma (section 2), as well as on the sputtered ceramic target surface (section 3). Then by the example of ZnO-based transparent conductive oxide (TCO) thin films, routes for improving the structure of thin films obtained by low-temperature deposition are considered (section 4). Finally, an attempt has been made to predict the development of the issue of creating new materials and technologies for functional thin films of new generation devices.

2. Current-voltage characteristics of the plasma discharge during magnetron sputtering

The parameters of the gas discharge during magnetron sputtering largely determine the dynamics of the target sputtering process and strictly set the conditions for the deposition of thin films. The processes occurring in magnetron plasma with the participation of atoms and clusters are reviewed in great detail [11]. The current-voltage characteristics of a gas discharge have already been studied in detail [12,13].

In particular, in [11] discharge instabilities were discovered and discussed. Figure 1a shows that the I–V characteristic of a glow discharge exhibits two instability regions: during the transition from a dark to a glow discharge and from a glow to an arc discharge. It should be noted that the given data refer to the conditions of the discharge proceeding at relatively high pressures in the discharge gap.

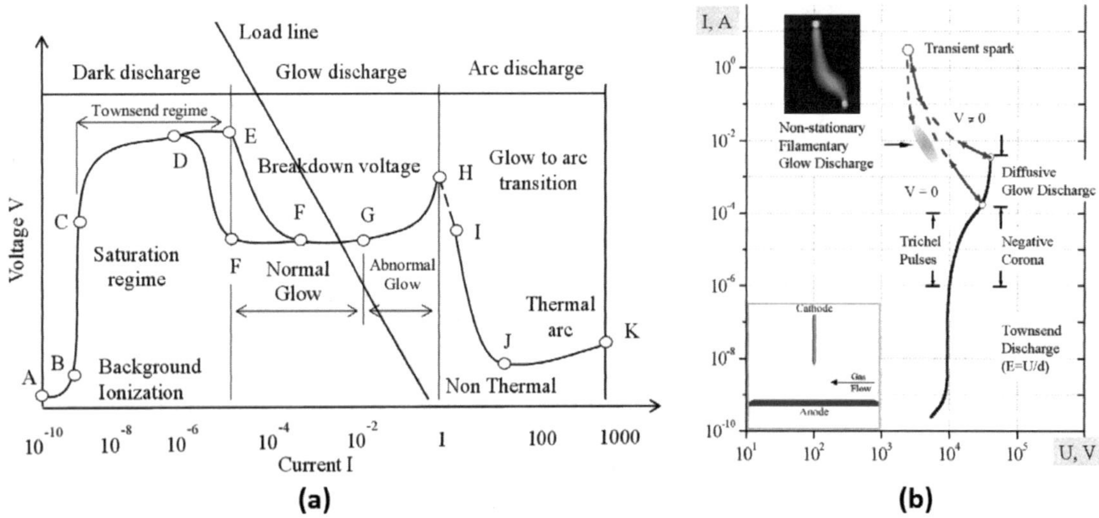

Figure 1. (a) – the three primary regions of a gas discharge. The straight line is a typical load line. Reproduced with permission from Depla D, Mahieu S, Greene J E 2010 Chapter 5 - Sputter Deposition Processes. In *Handbook of Deposition Technologies for Films and Coatings* (ed Martin P M) p 262. Copyright 2010 P. M. Martin. Published by Elsevier Inc. (b) – a typical I–V characteristic of an atmospheric pressure dc discharge in a pin-to-plane electrode system. Inset – view of the filamentary mode of a non-stationary pin-to-plane dc glow discharge in ambient air. Reproduced with permission from Akishev Yu, Grushin M, Kochetov I, et al. 2005 Plasma Sources Sci. Technol. 14 S18–S25. Copyright 2005 by IOP Publishing.

It is reported that at high gas pressures in the pin – plate system, a nonstationary discharge is formed in the region of a negative corona discharge [14]. The corresponding I–V characteristic of a discharge occurring at atmospheric pressure is shown in figure 2. It is clearly shown here that the discharge instability is observed in the region of the transition of an anomalous glow discharge into an arc discharge.

In turn, the study of the I–V characteristic of a glow discharge arising between parallel flat electrodes about 11 cm in diameter in an atmosphere of N_2:O_2 = 8:2 at a pressure in the chamber of about 10 Torr and a discharge current of about 100 A showed that the I–V characteristic of the discharge corresponds to the transition of a glow discharge into arc one [15].

In 1998 for the first time Abduev and Magomedov reported about the observation of stable current oscillations with a frequency of about 10 kHz during dc magnetron sputtering of a ZnO ceramic target [16]. They related the observed process to the sticky instability of oxygen, which leads to Trichel oscillations in a magnetron discharge in an oxygen-containing medium [17].

We believe that correct interpretations of the observed results are extremely important for practical use in the development of new modes of deposition of thin films.

The oscillation phenomenon during magnetron sputtering of ceramic ZnO targets suggests that this process is observed at the transition from an anomalous glow discharge to an electric arc discharge. The process starts only in a very narrow pressure range (about 4×10^{-4} Torr) when using an electrical supply circuit that forms an oscillatory circuit. An increase in the surface temperature of sputtered ZnO ceramic target in the presence of an oscillatory circuit leads to the formation of oscillations caused by sticky instability only if there is a certain oxygen concentration in the anomalous glow discharge gap. The formation of sticky instability upon magnetron sputtering in an Ar medium is caused by thermal desorption of oxygen from the surface of the sputtered ceramic target. The observed oscillations are due to hysteresis instability accompanied by alternating transitions between the metallic and dielectric modes in the target erosion zone.

3. Target erosion during its ion sputtering

To date, ion bombardment of ceramic targets during dc or rf magnetron sputtering had been studied mainly in terms of assessing the efficiency of sputtering of the target material, as well as the instability of the discharge. It is obvious that ion bombardment of the surface of oxide targets can lead to a significant change in stoichiometry in the near-surface layer of the sputtered target. And the deviation from stoichiometry can be in one direction or another, depending on the composition of the working gas. For example, with intense sputtering of an oxide target by ions of an inert gas, the deviation of stoichiometry towards oxygen deficiency in the near-surface layers of the target erosion zone can lead to various instabilities associated with the transition of a magnetron discharge from an anomalous glowing to an arc one [18]. This leads to a decrease in the quality of deposited thin films and the reproducibility of their characteristics. On the contrary, the sputtering of targets in the presence of oxygen can lead to additional oxidation of the target surface in the erosion zone. The processes of reduction and oxidation of the surface layers of the target lead to hysteresis phenomena, which also negatively affect the stability of the characteristics of the deposited thin films [9, 19].

Based on the analysis of numerous works, it can be concluded that the processes of structural restructuring and surface self-organization of target materials in the erosion zone are still insufficiently studied.

As for the case of sputtering a ZnO-based TCO target, the bombardment of its surface with inert gas ions (Ar) leads to a gradual oxygen depletion of the thin surface layer of the sputtered target, and thus thin films on the substrate begin to grow from a flow containing excess zinc [20,21]. Such conditions results in, in turn, to a significant decrease in the resistivity of the deposited thin films [20,22].

4. Growth mechanisms of ZnO-based TCO thin films by magnetron sputtering

A large number of studies have been devoted to the study of the mechanisms of ZnO-based TCO thin films growth by magnetron sputtering. An obvious general requirement for the ZnO thin films is a high structural perfection, which provides both acceptable optical transparency and high mobility of charge carriers.

A serious drawback of the microstructure of ZnO thin films deposited by magnetron sputtering under conditions far from equilibrium layers is the formation of an obvious columnar structure in them (figure 2a). The columnar structure of the ZnO films is characterized by a poor order of crystallite orientation and a developed surface relief [23]. This leads to deterioration of optical and electrical performance of ZnO TCO thin films.

We have previously shown that in order to suppress the formation of columnar structures in the zinc oxide films, there is a need to carry out the magnetron deposition of ZnO films from the reagent flow containing excess the zinc over the oxygen (figure 2b) [24]. This leads to an increase in the atoms migration length on the film growth surface due to the formation of a low-melting ZnO_{1-x} phase on the growth surface at temperature of about 450°C (above Zn melt temperature). However, this does not allow solving the main problem – reducing the temperature regime to 100°C for the deposition of TCO thin films with a high degree of microstructural order and an acceptable Hall carrier mobility. The problems of the dependences of the elastic modulus of steels and alloys depending on high temperatures were discussed in detail in the articles [25-30].

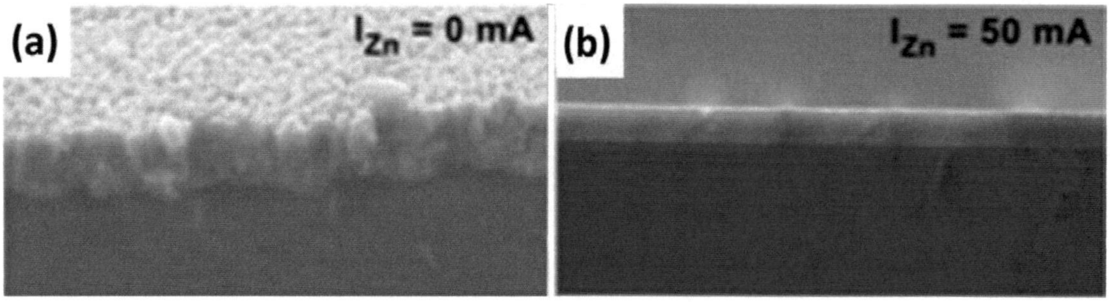

Figure 2. The cross-section views of ZnO TCO thin films deposited by magnetron sputtering of ceramic target (a) and by co-sputtering ceramic and Zn targets (b).

At the same time there are other ways to improve the microstructure ZnO thin films deposited by magnetron sputtering. One of these promising ways is to create conditions for reducing the level of bombardment of the growing film surface with negative oxygen ions, especially during the film nucleation stage [23,31]. As another way, we can consider the use of various ultrathin buffer sublayers providing a high degree of orientational order at the nucleation stage [32].

5. Conclusions

The review of the currently available data makes it possible to determine the following directions for improving the technologies for low-temperature deposition of TCO thin films based on ZnO:

- use of non-stoichiometric ceramic targets;
- creation of conditions to minimize the residual oxygen pressure in the working chamber;
- assisting the deposition process at the initial stage of growth with an inert gas ion beam;
- use of substrates with minimal surface relief, as well as with a special sublayer.

Acknowledgments
This work was supported by the Ministry of Science and High Education of the Russian Federation and by Russian Foundation for Basic Researches (#19-07-00537 and # 20-07-00760). The authors are grateful for additional technical support from the Shared Research Centers of Dagestan FRC of RAS.

References

[1] Gudmundsson J T 2020 *Plasma Sources Sci. Technol.* **29** 113001 https://doi.org/10.1088/1361-6595/abb7bd
[2] Greene J E 2017 *Journal of Vacuum Science & Technology A* **35** 05C204 https://doi.org/10.1116/1.4998940
[3] Vinodbabu Ch, Rao G T, Reddy N B, and Zyryanov G V 2020 *AIP Conference Proceedings* **2280** 040052 https://doi.org/10.1063/5.0018142
[4] Schmitz J 2018 *Surface & Coatings Technology* **343** 83–88 https://doi.org/10.1016/j.surfcoat.2017.11.013
[5] Hong J, Wagata H, Katsumata K and Matsushita N 2016 *Materials Science in Semiconductor Processing* **47** pp 20–24 https://doi.org/10.1016/j.mssp.2016.02.010
[6] Montbach E and Davis D 2018 Roll-to-Roll Manufacturing of Flexible Displays. In *Roll-to-Roll Manufacturing* (eds Greener J, Pearson G and Cakmak M) https://doi.org/10.1002/9781119163824.ch10
[7] Gusain A, Thankappan A and Thomas S 2020 *Journal of Materials Science* **55** 13490–13542 https://doi.org/10.1007/s10853-020-04883-1
[8] Akhmedov A K, Abduev A K, Kanevsky V M, Muslimov A E and Asvarov A S 2020 *Coatings* **10**, 269 https://doi.org/10.3390/coatings10030269
[9] Abduev A K, Akhmedov A K, Asvarov A S, Muslimov A E and Kanevsky V M 2021 *J Synch Investig* **15** 121–127 https://doi.org/10.1134/S1027451021010031

[10] Afre R A, Sharma N, Sharon M and Sharon M 2018 *Rev Adv Mater Sci* **53** 79–89 https://doi.org/10.1515/rams-2018-0006
[11] Kashtanov P V, Smirnov B M and Hippler R 2007 *Phys Usp* **177** 473–510 https://doi.org/10.1070/PU2007v050n05ABEH006138
[12] Musil J, Vlcek J, and Baroch P 2006 Chapter 3 - Magnetron Discharges for Thin Films Plasma Processing. In *European Materials Research Society Series, Materials Surface Processing by Directed Energy Techniques* (ed Pauleau Y) 67–110 https://doi.org/10.1016/B978-008044496-3/50004-6
[13] Depla D, Mahieu S, Greene J E 2010 Chapter 5 - Sputter Deposition Processes. In *Handbook of Deposition Technologies for Films and Coatings* (ed Martin P M) 253–296 https://doi.org/10.1016/B978-0-8155-2031-3.00005-3
[14] Akishev Yu, Grushin M, Kochetov I, Karal'nik V and Napartovich A 2005 *Plasma Sources Sci Technol* **14** pp S18–S25 https://doi.org/10.1088/0963-0252/14/2/S03
[15] Takaki K, Taguchi D and Fujiwara T 2001 *Appl. Phys. Lett* **78** 2646 https://doi.org/10.1063/1.1369612
[16] Abduev A K, Magomedov A M 1998 *Tech Phys Lett* **24** 192–193 https://doi.org/10.1134/1.1262048
[17] Wang C, Chen X, Ouyang J, Li T and Fu J 2018 *Energies* **11** 3120 https://doi.org/10.3390/en11113120
[18] Kubart T, Nyberg T and Berg S 2010 *J Phys D: Appl. Phys.* 43 205204 https://doi.org/10.1088/0022-3727/43/20/205204
[19] Németh Á, Labadi Z, Tóth L and Bársony I 2010 *Vacuum* **84** 218–220 https://doi.org/10.1016/j.vacuum.2009.05.01
[20] Abduev A K, Akhmedov A K, Asvarov A S, Muslimov A E and Kanevsky V M 2021 *J Synch Investig* **15** 76–80 https://doi.org/10.1134/S102745102101002X
[21] Akopyan I K, Labzovskaya M E, Novikov B V, Lisachenko A A, Serov A Yu and Filosofov N G 2018 *Phys. Solid. State* **60** 352–356 https://doi.org/10.1134/S1063783418020026
[22] Abduev A K, Asvarov A S, Akhmedov A K, Emirov R M and Belyaev V V 2017 *Tech Phys Lett* **43** pp 1016–1019 https://doi.org/10.1134/S1063785017110153
[23] Köhl D, Luysberg M and Wuttig M 2010 *J. Phys. D: Appl. Phys.* **43** 205301 https://doi.org/10.1088/0022-3727/43/20/205301
[24] Abduev A, Akhmedov A, Asvarov A and Belyaev V 2019 *SID Symposium Digest of Technical Papers* **50** 977–980 https://doi.org/10.1002/sdtp.13089
[25] Shinkin V.N. 2018 *CIS Iron and Steel Review*. **15** 23-27.
[26] Shinkin V.N. 2018 *CIS Iron and Steel Review*. **15** 32-38.
[27] Shinkin V.N. 2018 *Steel in Translation*. **48**(11) 718-723.
[28] Shinkin V.N. 2018 *Steel in Translation*. **48**(3) 149-153.
[29] Shinkin V.N. 2018 *Chernye Metally*. **5** 34-40.
[30] Shinkin V.N. 2019 *Chernye Metally*. **3** 32-37.
[31] Köhl D, Luysberg M and Wuttig M 2009 *Phys Status Solidi RRL* **3** 236–238 https://doi.org/10.1002/pssr.200903228
[32] Abduev A, Akmedov A, Asvarov A and Chiolerio A 2015 *Plasma Process Polym* **12** 725–733 https://doi.org/10.1002/ppap.201400230

Synthesis of ZnS:Cu,Br radioluminescent phosphors using the electron-beam treatment and studying their characteristics

V V Bakhmetyev[1], E V Zelenina[2] and M A Shvindin[2]

[1]Saint-Petersburg State Institute of Technology (Technical University), Russia
[2]Khlopin Radium Institute, Saint-Petersburg, Russia

E-mail: vadim_bakhmetyev@mail.ru

Abstract. ZnS:Cu radioluminescent phosphors find their applications in medicine non-destructive testing, safety control and can be used as a part of the solid-state radioluminescent light sources (SRLS). This devices are very promising technology due to their independence longevity and safety compared to the gas-filled RLS. Because of the fact that tritium (the most popular radioisotope for RLS) has a very short range in substances, the improving the operating performances of radiophosphors is a crucial task for SRLS. In our study we used the electron-beam treatment to increase the brightness of radioluminescence of ZnS:Cu phosphors. It was found that bombardment of the phosphors initial charge with electrons of 900 kEv energy improves the brightness of radioluminescence by 15 – 20%. Double modifying of the initial charge and ready phosphor cause the 80% increase of brightness. The effect of electron-beam treatment on the phase content and the surface properties was studied at varying the content of activator (Cu) in the range of 0 – 0.6 % mass. As a result the model describing structure of the acid-base sites on the ZnS:Cu phosphor surface was suggested.

1. Introduction

Radioluminescent light sources (RLS) are the devices consisting mainly of the emitting isotope, and the phosphor, giving the visible light affected by ionizing radiation. Tritium, having the half-life period of 12.5 years and emitting beta-particles with the average energy of 5.7 keV, is the most common isotope for RLS for its cheapness and low radiation hazard. Nowadays the gas-filled RLS, using the gaseous tritium in the glass tubes, are the most known configuration. However the main disadvantages of such construction are the glass tubes breakability causing the tritium leakage and the size limitation due to the tritium self-absorption. The solid-state configuration of RLS (SRLS) was suggested in our later studies [1, 2]. This device involves tritium bounded in the solid spongy titanium matrix forming the thin layer on the steel plate and the ZnS powder phosphor layer connected with the TiT_2 in the silicate matrix. This configuration excludes the tritium leakage and increases the safety; however the key problem is the tritium betas path length of several μm in the solid matrix decreasing the brightness of radioluminescence. We can raise the brightness by increasing the incorporated activity of tritium but it may cause the radiation damage and self-absorption processes. On the other hand we can raise the effectivity of the radiophosphor constituted in SRLS. Our previous studies demonstrated the possibility of increasing the electroluminescent performances of the ZnS:Cu и ZnS:Cu,Al phosphors using the electron-beam treatment [3 – 8]. In this paper this technique was applied for ZnS:Cu radiophosphors having the different content of the doping agents and the different excitation mechanism of luminescence. Electroluminescence involves the whole volume of the ZnS crystal, but the radioluminescence is surface donor-acceptor recombination process. And as far as here we have only the surface interaction between the electrons and the centers of luminescence, the surface condition and characteristics are the key for understanding the processes forming the spectral and brightness performsnces. Previous studies [9 - 11] have demonstrated the correlation between the phosphor surface characteristics and their luminescence effectivity. The model describing the nature of all surface acid-base sites for the ZnS:Cu and ZnS:Cu,Al electroluminescent phosphors was suggested; and the direct interrelations between the proper acid-base sites and phosphor centers of luminescence was found. This study suggests the likewise model was constructed for ZnS:Cu, Br radiophosphors. The correlations between the phase content of the phosphors and their radioluminescent performances was found and the effect of the electron-beam treatment was studied.

2. Experimental

The series of ZnS:Cu,Br radiophosphors synthesized from the initial charge consisting of ZnS, containing 0.8% of Br and 0 – 0.6 % wt. of Cu are the objective of the research. ZnS manufactured by scientific-production association NPO Lumonophor, Stavropol, Russia, CuCl, NH_4Br and the elemental sulphur were the starting materials. The initial charge was cooked by 3 hours mixing of the starting materials in the rotary drum mixer.

Synthesis was conducted by annealing the initial batch in the alumina crucible at 950 °C in a reducing atmosphere under the active carbon bed [9] for an hour and a half. Then phosphors were cleaned from the non-luminous particles under the UV-lamp, milled and screened. After that phosphors were washed in the water solution of NH_3, $(NH_4)_2CO_3$ and $(NH_4)_2S_2O_8$ to remove the impurities and particles of ZnO and Cu_XS from the surface. After washing phosphpors were dried at 120 °C and screened again.

Electron-beam treatment was done for the initial charge and for the ready phosphors by the electron flux bombardment using the medium-energy resonant transformer electron accelerator RTE-1V (produced by Efremov Institute of Electrophysical Apparatus, Saint-Petersburg, Russia). The electron energy was of 900 keV and the beam current of 1 mA. The absorbed dose was about 600 kGy. The conditions of treatment were taken as an optimal conditions from our later studies [6].

Now therefore three series of phosphors including the Cu concentration range of 0 – 0.6 % wt. were synthesized:

1. Phosphors synthesized without using the electron-beam treatment (non-modified, N);
2. Phosphors synthesized from the electron-beam treated initial charge (modified, M);
3. Electron-beam treatment of the ready phosphors synthesized from the electron-beam treated initial charge (double-modified, 2M).

To explore the radioluminescent performances of the phosphors they were excited by the tritium β-source, represented as a tritium-contained titanium layer on the steel plate. Thickness of the layer was 0.5 μm, source area was 1 cm^2, the incorporated activity was 0.4 Ci. The brightness of radioluminescence was measured using a radiometer IL1700 (International Light, USA). The X-ray diffraction analysis was done using a Rigaku SmartLab 3 X-ray diffractometer (Rigaku Corporation, Japan). The quantitative analysis of the phase content was done using the Rietveld method. The grain size and surface quality of the phosphors were estimated by the scanning election microscopey using the Hitachi TM3000 (Hitachi, Japan) и Tescan VEGA 3 SBH (Tescan, Czech Republic).

For differentiation of the acid-base sites on the surface of the phosphor grains we used the adsorption of a series of acid-base indicators with different pKa values in the range from -5 to 15 according to the method described in [12…14]. The calculations was carried out using the SF-56 spectrofotometer (LOMO (Leningrad Optical Mechanical Association), Russia) at the wavelength of maximum absorption peak of the indicator solution. The list of indicators let define the acid-base sites in the range of pK_a from –4.4 to +14.2. using this data the spectra of adsorption sites distribution on the phosphor surface become available as a function in coordinate representation $q_{pka} = f(pK_a)$.

3. Results and discussion

Figure 1 represents the microphotographs of the phosphors synthesized from the non-modified and EB-modified charge. Non-modified phosphors has smooth surface and looks like grained separated particles, so we can say that they have a good crystallinity. Samples synthesized from the EB-modified charge looks as a small agglomerates surrounded by the smallest particles without good facet and separation. Obviously their crystalline structure has lowered the degree of perfection.

Cu 0% wt. Cu 0,01% wt. Cu 0,1% wt. Cu 0,6% wt.

Samples, synthesized from the non-modified charge

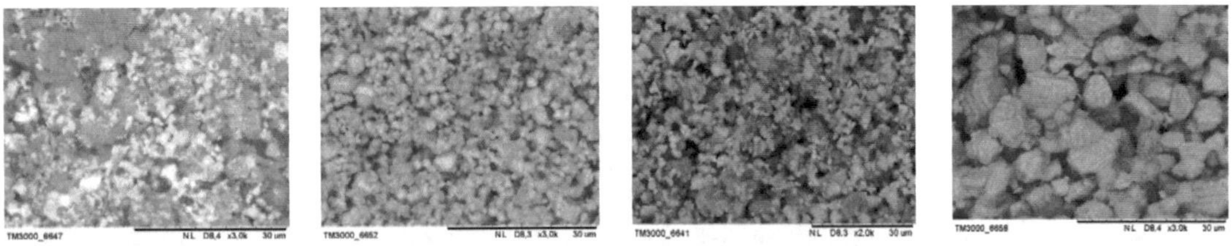

Cu 0% wt. Cu 0,01% wt. Cu 0,1% wt. Cu 0,6% wt.

Samples, synthesized from the EB-modified charge

Figure 1. Microphotographs of the ZnS:Cu,Br phosphor samples doped with Cu 0 – 0.6 % wt.

Table 1 gives the data of the average grain sizes for non-modified and EB-modified series of phosphors. It's apparent that doping phosphors with Cu at first causes the decreasing of the grain size and then, when Cu concentration exceeds 0.1 % wt., effects the grain growth, due to the Cu_XS phase formation. At the same time, the EB-treatment of the charge effects the decreasing of the grain size of the samples through the whole concentration range.

Table 1. The average grain size of phosphors synthesized from the non-modified and EB-modified initial charge.

Cu, % wt.	N (non-modified) d_{av}, μm	M (modified) d_{av}, μm
0	5 – 10	3 > 5
0.03	3 – 5	1 – 3
0.075	5 > 10	3 – 5
0.2	5 < 10	5 – 7
0.6	10 – 15	7 – 9

The phase content of all synthesized samples was investigated with the XRD analysis. The phosphors has complex phase content with two main phase modifications of ZnS crystalline host – cubic (sphalerite) and hexagonal (wurtzite). The percentage content of each phase counted by the Rietveld method is given in the table 2.

The phase-transition temperature between sphalerite and wurtzite is nearly 1050 °C [15], so wurtzite is a high-temperature modification of ZnS. Nevertheless the data from the table 2 shows that EB-treatment effects the wurtzite formation up to the 45% in the phosphors that were synthesized below the phase transition temperature.

The less amount of the hexagonal modification was found in the non-modified samples and the highest amount in the 2M series – the phosphors prepared from the EB-modified charge and repeatedly modified after synthesis.

Presumably, we can say, that electron-beam treatment leads to the phase changes in ZnS; and the strongest changes were observed in the phosphors with the Cu concentration not exceeding 0.1 % wt. It correlates to the literature data [16, 17] saying that Cu doping promotes the stabilization of the cubic phase.

Table 2. The phase content of the phosphors.

Cu, % wt	Phases, %					
	N series		M series		2M series	
	Sphalerite	Wurtzite	Sphalerite	Wurtzite	Sphalerite	Wurtzite
0	100	0	86.7	13.3	80	20
0.01	100	0	72.7	27.3	70	30
0.03	74.7	25.3	68.5	31.5	55	45
0.05	100	0	80	20	83	17
0.075	78.3	21.7	76.9	23.1	75	25
0.1	100	0	75	25	80	20
0.2	100	0	100	0	100	0
0.4	100	0	100	0	100	0
0.6	100	0	100	0	100	0

Fugure 2 represents the brightness of radioluminescence for the non-modified, modified and double-modified series of phosphors. It can be seen that the EB-treatment (of the charge and of the ready phosphors) causes the extreme increasing of brightness. For the samples doped with 0.03 % wt. of Cu electron modification of the charge effects the 20% increase of brightness and the repeated electron modification of the ready phosphor effects the 80% increase of brightness. At raising the Cu concentration over the 0.1 % wt. rapid decline was observed probably due to the formation of the quenching solid solution of Cu_XS.

Figure 3 gives the comparison of the spectral data of three series of phosphors for the concentration of activator, giving the maximum brightness (0.03 % wt.). There are two main spectral bands – at 450 nm and 520 nm – «blue» and «green», respectively. It is known that «blue» spectral band is relative to the copper associative centers of luminescence, consisting of two copper ions – one in the lattice point and another in the interstitial lattice site (Cu_{Zn}' $Cu_i^{\bullet}$). «Green» spectral band is relative to another type of associative centers of luminescence – copper atom in the zinc lattice point and halogen atom in the sulphur lattice point (Cu_{Zn}' $Br_S^{\bullet}$) [11].

It's apparent from the figure, that after the EB-treatment of the charge the emission at 520 nm turns higher than 450 nm emission, and the «green» band became predominant in the spectra. None of the other spectra of all synthesized samples with other concentrations of copper give such an effect.

The most probable factor of all brightness and spectral changes in the non-modified and EB-modified series, is the phase and structural changes caused by the electron bombardment. It leads to the redistribution of the doping agents (Cu and Br) – constituents of the «green» centers of luminescence – between the volume and the surface of the crystal. Also it's possible that the EB-treatment facilitates the ZnS matrix restructuring that decreases the basic degree of perfection of the crystal lattice. This process causes the ionization of the lattice defects, such as sulphur vacancies and promotes the other vacancies formation when sulphur leaves the crystal surface. Bromine radius is bigger than chlorine and in perfect ZnS crystals it's hard for Br to incorporate into the lattice.

Formation of the wurtzite phase under the EB-treatment facilitates bromine incorporation into the surface layers of the ZnS lattice, filling the sulphur vacancies and forming the (Cu_{Zn}' $Br_S^{\bullet}$) associates.

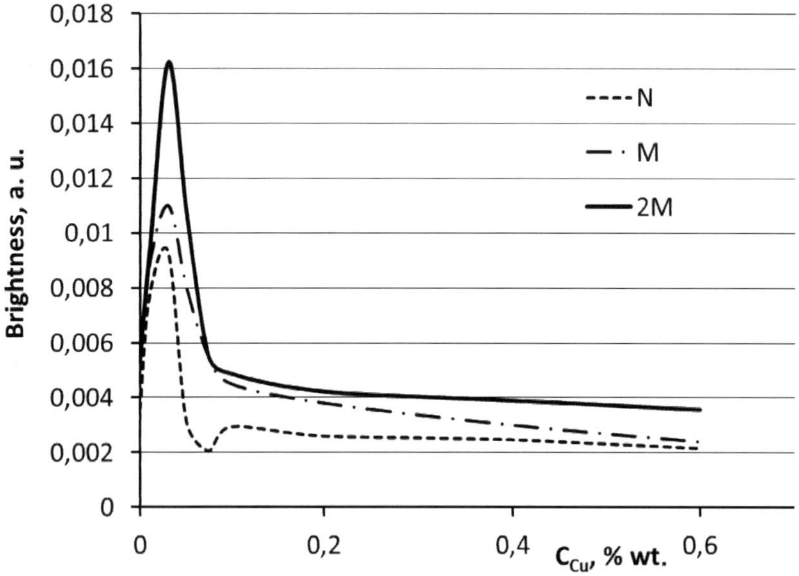

Figure 2. Brightness of the radioluminescence for the phosphors doped with different amounts of Cu. for non-modified, modified and double-modified series of phosphors. (Inset figure shows the concentration range 0 – 0.1 % wt. of Cu).

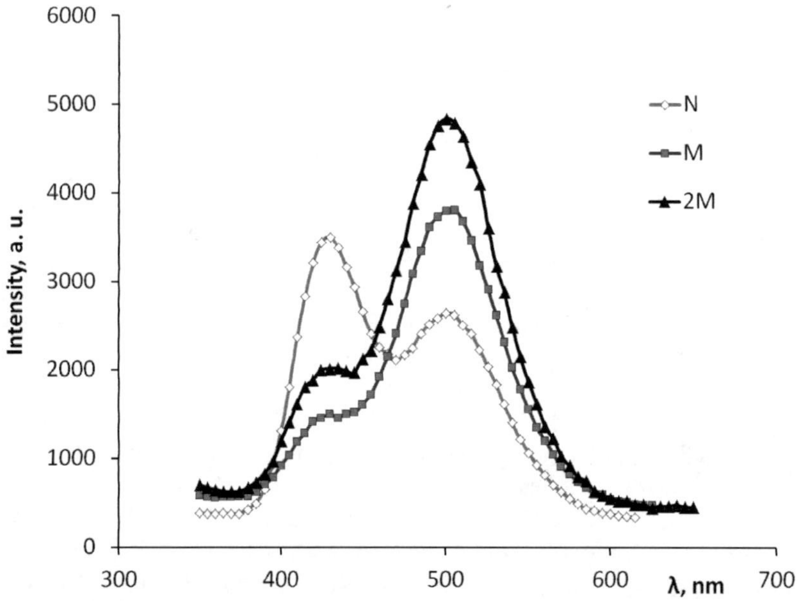

Figure 3. Radioluminescence spectra of non-modified, EB- modified and double-modified series of phosphors with Cu concentration 0.03 % wt.

In our later studies [11] we have constructed the model describing the nature of active surface sites of ZnS:Cu phosphors. That phosphors were synthesized without EB-treatment and using the low-temperature (600…650 °C) baking in air after synthesis but before washing. Low-temperature baking is the necessary step for the electrophosphors synthesis for the formation of the Cu_XS phase for effective electroluminescence excitation. This model is given at the figure 4. The analogous investigations of the surface characteristics were made for EB-modified phosphors, using the acid-base indicator method. Fugure 5 shows the distribution of the adsorption sites (DAS) on the surface of phosphor with various concentrations of Cu.

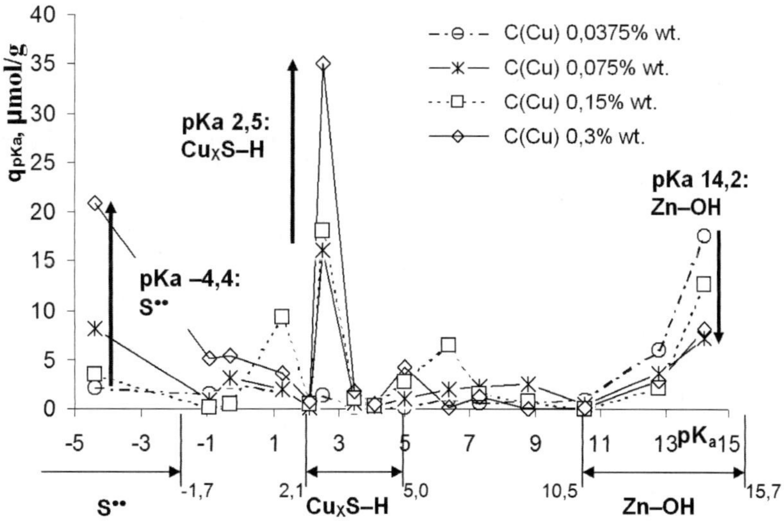

Figure 4. Model of active surface sites of ZnS:Cu electrophosphors with different concentrations of Cu (synthesized without EB-treatment and using the low-temperature baking in air) [11].

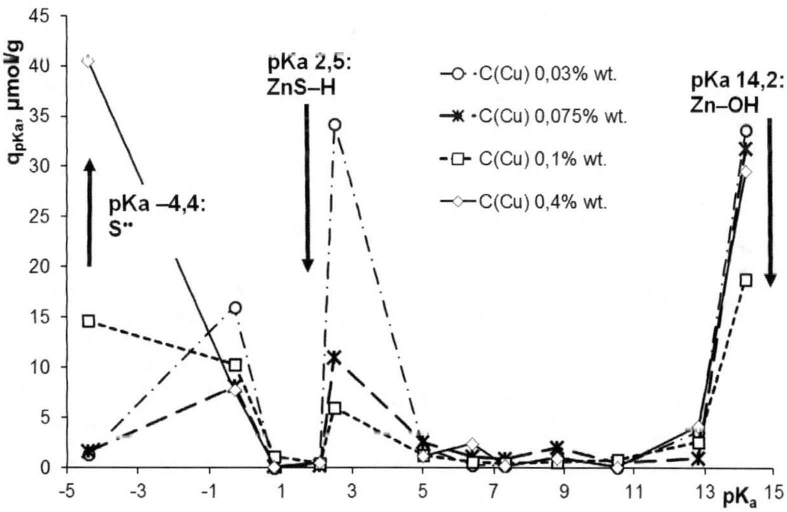

Figure 5. Model of active surface sites of ZnS:Cu radiophosphors with different concentrations of Cu (synthesized with EB-treatment of the initial charge and without the low-temperature baking).

It's apparent that in the both cases, at the figures 4 and 5 there are peaks at pK_a –4,4; 2,5 and 14,2. According to our previous model for electrophosphors at pK_a -4.4 … -1.7 we have Lewis's basic sites consisting of two-electron orbitals of the surface sulphur atoms $S^{••}$. The number of these sites indicates number of the surface zinc vacancies V_{Zn}'. pK_a 10.5…15.7 refers to the Broensted basic sites and have an architecture Zn–OH. Their number characterizes the amount of the sulphur vacancies $V_S^{\bullet}$. Comparing figures 4 and 5 one can see that strong Lewis basic and strong Broensted basic adsorption sites behavior is similar at raising the activator concentration for the both series of phosphors. The number of sites with pK_a –4.4 increases and the number of sites with pK_a 14.2 decreases. This fact reinforces our conclusions about the nature of these sites from the work [11].

There again, pursuant to the model for electrophosphors, Broensted acidic sites with pK_a 2.1 – 5.0 have an architecture Cu_XS–H and their number characterizes the amount of the copper sulphide on the surface. And consequently it should increase at increasing of the copper concentration (see figure 4). But for ZnS:Cu,Br radiophosphors synthesized in present study we can see the inverse relation for the Broensted acidic sites with pK_a 2.5 – q_{pKa} decreases at increasing of the copper concentration (see figure. 5). Obviously, it's due to the fact of the low temperature baking (in the first case) inducing the Cu_XS exudation on the surface and without baking we don't have the Cu_XS on the surface. Moreover, the EB-treatment raises the solubility of the activator in the host ZnS matrix [11] and promotes the copper to penetrate deep into the crystal. Thus, for the radiophosphors series from the figure 5 the amount of the copper on the surface would be much lower than for the electrophosphors from the figure 4. And at high concentrations of Cu (exceeding 0.1 % wt.) the numer of the active sites at pK_a 2.5 would be much lower too.

Obviously, when the copper concentration on the surface is residual at this pK_a value we may have the ZnS–H sites. And furthermore, the correlation of this active sites with the wurtzite phase content was found (see figure 6). So, we may suggest that hexagonal ZnS modification forms these sites.

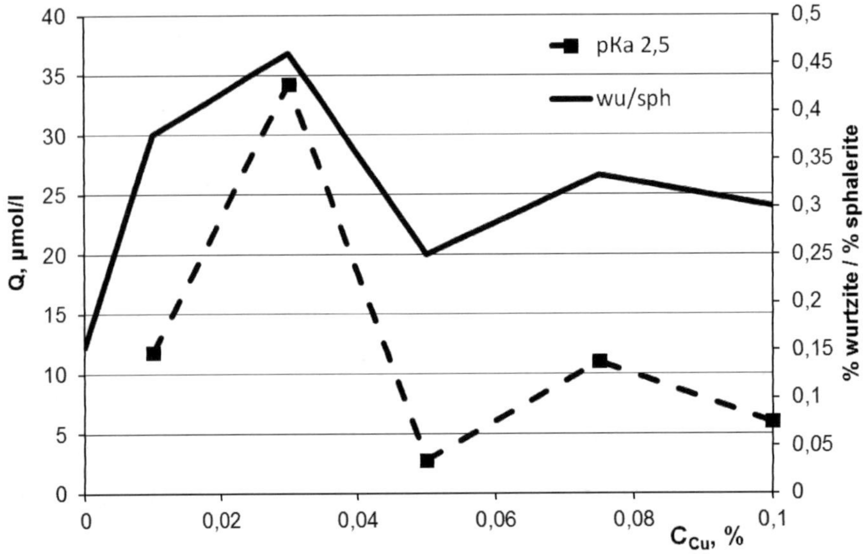

Figure 6. Correlations between active sites with pKa 2.5 and wurtzite/sphalerite content relation for ZnS:Cu,Br phosphors at changing the Cu concentration.

4. Conclusions

Therefore, as a result of our study the technique of synthesis of the ZnS:Cu,Br radiophosphors with enhanced brightness was developed. The optimal concentration of the activator in radiophosphor (Cu) was defined as 0.03 % wt. It was found that EB-treatment of the initial charge results the 20 % brightness of radiolumenescence increasing and repeated EB-modification of the ready phosphor gives the 80% brightness increasing.

The activator concentration effect on the ZnS:Cu,Br phosphors dispersity, phase content, spectral and brightness changes as well as the surface characteristics was studied.

It has been established that EB-treatment promotes forming of the extra luminescence centers ($Cu_{Zn}rB\,'^{\bullet}_{S}$) causing the brightness increasing. In turns, forming of these associative centers is caused by structuring the hexagonal (wurtzite) phase of ZnS at EB-treatment both the initial charge and the ready phosphor.

The investigation of the surface characteristics using the acid-base indicator method resulted the descriptive model of the nature of the active sites on the surface of radiophosphors ZnS:Cu. According to the model, the adsorption sites with pK_a –4,4 and 14,2 has the same nature with the electrophosphors ZnS:Cu. Lewis's basic sites pK_a –4.4 consists of two-electron orbitals of the surface sulphur atoms $S^{\bullet\bullet}$; Broensted basic sites with pK_a 14.2 have an architecture of Zn–OH.

Broensted acidic sites with pK_a 2.5 has different nature for electro and radiophosphors. In electrophosphors they are relative to the copper sulphide presence on the surface and has Cu_XS–H architecture. In radiophosphors this sites has an architecture ZnS–H and relative to the ZnS hexagonal (wurtzite) phase presence on the surface of phosphors.

References

[1] Zelenina E, Bakhmetyev V, Pechertseva E and Sychov M 2019 *Journal of Sol-Gel Science and Technology* **92** (2) pp 467–473 DOI: 10.1007/s10971-019-05022-2.
[2] Zelenina E, Pechertseva E 2019 *Proc. 35 Int. Conf. «EuroDisplay-2019»* (Minsk, Belarus) p 65.
[3] Sychov M M, Bakhmet'ev V V, Mjakin S V, Nakanishi Y, Vasil'eva I V, Havanova L V, Cheremisina O A and Korsakov V G 2002 *Proc. Electronics Display Conf. EDC'02* (Nagasaki, Japan) pp 67–71.
[4] Sychov M M, Bakhmetiev V V, Komarov E V, Mjakin S V, Vasilieva I V, Nakanishi Y, Korsakov V G and Kuznetsov A I 2004 *Proc. Int. Conf. ADT-04* (Raubichi, Belarus) pp 151–154.
[5] Bakhmetiev V V, Sychov M M, Korsakov V G, Komarov E V, Vasil`eva I V and Mjakin S V 2004 *Proc. of the Display Optics 2004 Int. Seminar* (St. Petersburg, Russia) p 45.
[6] Sychov M M, Nakanishi Y, Kominami H, Mjakin S V, Bakhmetiev V V, Grigoryev L V, Komarov E V, Korsakov V G, Sokolov V I and Usacheva V.P 2005 *Proc. of the 1st International Symposium on Nanovision Science* (Hamamatsu, Japan) pp 160–166.
[7] Sychov M M, Korsakov V G, Bakhmetiev V V and Komarov E V. 2007 *Proc. of ISDLPM 2007 Meeting* (Tokyo, Japan) pp 24–25.
[8] Sychov M M, Bakhmetiev V V, Korsakov V G, Mjakin S V, Botchkareva N N, Dorofeev A A and Vasiljeva I V 2010 *Proc. of the 15th Int. Workshop on Inorganic and Organic Electroluminescence & Int. Conf. on the Science and Technology of Emissive Displays and Lighting & XVIII Advanced Display Technologies Int. Symposium EL 2010 & ADT 2010* (St. Petersburg, Russia) pp 191–193.
[9] Sychov M M, Bakhmet'ev V V, Nakanishi Y, Mjakin S V, Havanova L V, Cheremisina O A and Korsakov V G 2003 *Journal of the SID* **11** (1) pp 33–38.
[10] Sychov M M, Mjakin S V, Nakanishi Y, Korsakov V G, Vasiljeva I V, Bakhmetjev V V, Solovjeva O V and Komarov E V 2005 *Applied Surface Science* **244** (1-4) pp 461–464.
[11] Bakhmet'ev V V, Sychev M M and Korsakov V G 2010 *Russian Journal of Applied Chemistry* **83** (11) pp 1903–1910.
[12] Cheremisina O A, Sychev M M, Myakin S V, Korsakov V G, Popov V V and Artsutanov N Yu 2002 *Russian Journal of Physical Chemistry* A **76** (9) pp 1472–1475.
[13] Myakin S V, Korsakov V G, Panova T I, Sosnov E A, Fomchenkova Yu C, Sychov M M and Shilova O A 2011 *Glass Physics and Chemistry* **37** (6) pp 624–628.
[14] Sychov M M, Zakharova N V and Mjakin S V 2013 *Ceramics International* **39** pp 6821–6826.
[15] Yen W M, Shionoya S and Yamamoto H 2006 *Fundamentals of phosphors* (CRC Press, Taylor & Francis Group LLC) 328 p.
[16] Skinner B J and Barton B B 1960 *Amer. Mineral* **45** p 612.
[17] Aven, M and Parodi J A 1960 *Journal of Physics and Chemistry of Solids* **13** (1–2) pp 56–64.

Study of the properties of Ce^{3+}-doped fluoride nanophosphors: phase composition, morphology, luminescence

A M Dorokhina[1,2], V V Bakhmetyev[1], H Kominami[2], A Toru[3] and M Hisashi[3]

[1]Saint-Petersburg State Institute of Technology (Technical University), Russia
[2]Shizuoka University, Research Institute of Electronics, Japan
[3]ANSeeN Inc., Japan

E-mail: vadim_bakhmetyev@mail.ru

Abstract. To date, nanophosphors have found application in various fields, one of which is medicine. These phosphors were developed with the aim to become one of the components of a drug for photodynamic therapy of oncological diseases. The aim of this work was to study the effect of the duration, environment, and stabilizers of solvothermal synthesis on the microstructure and luminescent properties of the YF_3:Ce nanophosphor. The solvothermal synthesis technique was carried out in three different media: water, ethanol, and ethylene glycol. The optimal duration of the synthesis was also determined (the synthesis was carried out at a temperature of 200°C for 4...20 hours). The dependence of the YF_3 luminescence on the phase composition and the solvothermal synthesis medium was studied. Using SEM, the morphology and particle size of YF_3:Ce phosphors were studied depending on different stabilizers (polyethylene glycol, polyethyleneimine, polyvinylpyrrolidone). The luminescence intensity of YF_3:Ce and $Na(Y_{1,5}Na_{0,5})F_6$:Ce samples was compared.

1. Introduction

The synthesis of nanosized inorganic materials with a certain morphology attracts much attention from the point of view both of fundamental scientific interest and potential technological applications in various fields, such as photochemistry, superconductors, optoelectronics, solar cells, and medicine [1, 2]. Thus, it remains an important goal of modern materials chemistry. Much recent research has focused on controlled morphology and spatial structuring of various materials, which is an important step towards creating multifunctional nanosystems. Typically, chemical vapor deposition or solution chemistry processes are used to achieve this objective. However, this usually requires: catalysts; expensive and even toxic matrices or surfactants; high temperatures; and a number of complex procedures.

This work is devoted to the solvothermal synthesis method, which makes it possible to obtain nanosized particles while avoiding high-temperature processing.

YF_3 has great potential as a phosphor. On its basis, lanthanide-doped phosphors with Stokes and anti-Stokes luminescence have been synthesized [3, 4]. Fluoride materials have advantages in terms of high density, mechanical hardness, and radiation resistance. Phosphors based on LnF_3 fluorides (Ln = lanthanide) are preferred matrices for lanthanide phosphors (since they have a low phonon energy state) and for downconversion due to their wide band gap (9–10 eV) [5].

To ensure the applicability of phosphors in medicine, in particular in the composition of pharmacological preparations for photodynamic therapy of oncological diseases (PDT), a phosphor with certain properties is required, such as: particle size less than 100 nm, excitation by radiation, ability to penetrate through body tissues - X-ray or infrared, and emitting light with a wavelength corresponding to the absorption band of industrial photosensitizers [6, 7].

In our previous studies [8–10], the main problem in obtaining phosphors is the need for high-temperature (more than 800°C) annealing, which leads to the agglomeration of phosphor nanoparticles into large agglomerates. To solve this problem, a solvothermal method for the synthesis of phosphors was proposed, which consists of thermal treatment of a liquid medium in an autoclave at moderate temperatures (200°C and below).

In particular, this approach is applicable to the synthesis of lanthanide fluorides activated by cerium, which are advantageous in terms of efficient X-ray luminescence in addition to high density and non-toxicity. Since the scattering of X-rays in tissues is negligible, lanthanides are effective additives for such phosphors, since they have a high atomic number and electronic energy states that provide radiation in the visible and ultraviolet regions. The inclusion of elements with high atomic numbers in the matrix provides the generation of electrons with an energy sufficient for conversion with a decrease in frequency, with an appropriate choice of lanthanide sensitizers and luminescent centers.

In this experiment, we implement a one-stage solvothermal synthesis of $YF_3:Ce^{3+}$ nanocrystals in three different media (in water, ethanol, and ethylene glycol), which are the most promising in this synthesis method. The selection of the optimal synthesis duration in the ethylene glycol medium is carried out, and the morphology of the obtained particles is investigated depending on various stabilizers (polyethylene glycol, polyethyleneimine, polyvinylpyrrolidone). A comparison of the cerium-activated $Na(Y_{1.5}Na_{0.5})F_6$ and YF_3 images is also provided.

2. Experimental

A series of fluoride luminophores with the composition $YF_3:Ce^{3+}$ 2 mol% was synthesized. A typical synthesis [5] was carried out in various media, both aqueous and organic (ethylene glycol and ethanol) at 200°C for 22 hours. Chlorides of cerium and yttrium and the stabilizer polyethylene glycol M_w = 2000 (PEG-2000) were vigorously stirred on a magnetic stirrer at room temperature. Then, ammonium fluoride NH_4F was dissolved in a suitable medium (water, ethylene glycol or ethanol), added to the above solution, and again stirred vigorously at room temperature for 30 minutes. Finally, the mixture was transferred to a Teflon liner, placed in a sealed stainless steel autoclave, where it was subjected to a solvothermal treatment, and cooled naturally to room temperature. The precipitate was washed in ethanol and deionized water, and the final products were dried at 50°C for 6 hours in a normal atmosphere. Further, according to the same method (but only in an ethylene glycol medium), a number of YF_3:Ce samples with a cerium concentration of 1 ... 6 mol.% were synthesized.

To determine the optimal synthesis time, the solvothermal synthesis of YF_3 samples was carried out without doping with Ce^{3+} in an ethylene glycol medium without the use of stabilizers for 4, 8, 12, 16, and 20 hours.

A sample of $Na(Y_{1.5}Na_{0.5})F_6$ was synthesized in an aqueous medium for 22 hours at 200°C with PEG-2000 as a stabilizer.

The structural characteristics of the final products were investigated by powder X-ray diffraction (XRD) using Cu-Ka radiation (λ = 0.15405 nm) on a Rigaku-RINT2200 diffractometer. The morphology and size of the obtained samples were observed using field emission scanning electron microscope (FE-SEM, JSM-6335F, JEOL). The emission spectra of ultraviolet and visible photoluminescence were recorded using a laboratory setup - a spectrofluorometer with two monochromators (model of an ultramonochromatic light source) equipped with a xenon lamp as an exciter. The X-ray luminescence spectra were measured on a laboratory setup with a copper anode at a voltage of 80 kV and a current of 62 mA. The decay time was measured with a Hamamatsu Photonics DG645 setup. All measurements were carried out at room temperature.

3. Result and Discussion

To find out the optimal synthesis medium it was decided to carry out solvothermal synthesis in various media, in which it is possible to obtain spherical particles less than 100 nm. Figure 1 shows the diffraction patterns of $YF_3:Ce^{3+}$ (2 mol%) samples synthesized in ethanol, aqueous solution, and ethylene glycol respectively. As can be seen, the obtained X-ray diffraction patterns fully correspond to the orthorhombic phase of yttrium fluoride (PDF card 32-1531). According to the X-ray diffraction data, the crystallite sizes were calculated (according to the Scherrer formula) for each of the samples (Table 1). The sample synthesized in ethylene glycol has the smallest crystallite size.

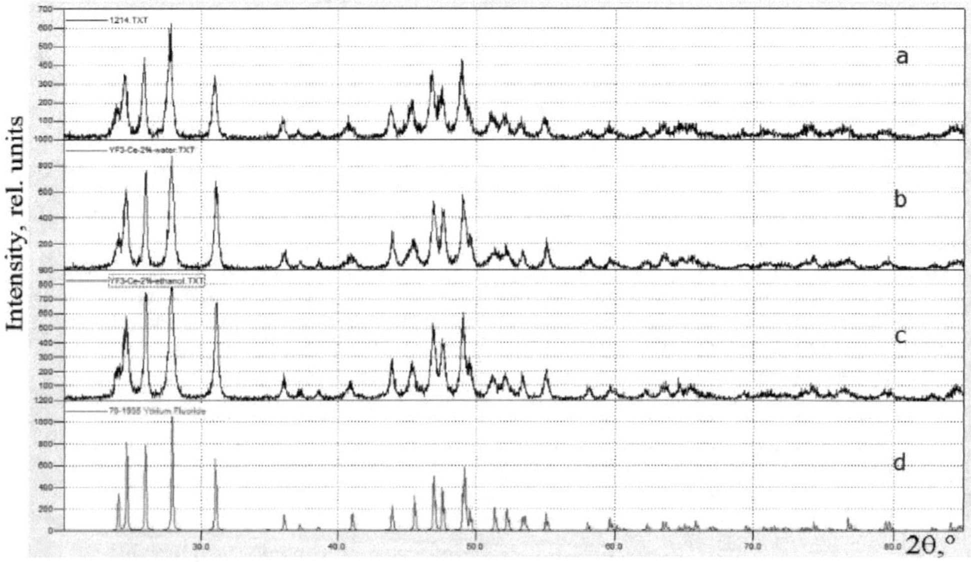

Figure 1. Diffraction patterns of $YF_3:Ce^{3+}$ samples synthesized in different media: a - ethylene glycol, b - water, c - ethanol, d – orthorhombic YF_3 (PDF card 70-1935).

The photoluminescence (excitation and emission) spectra recorded of these samples are shown in figure 2.

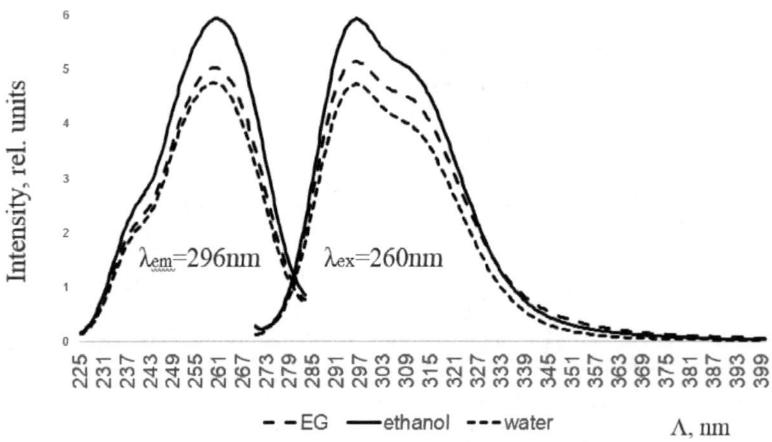

Figure 2. Photoluminescence spectra (excitation and emission) of YF_3:Ce 2%mol. samples synthesized in various media.

Table 1. Sizes of crystallites of the obtained samples.

Sample name	Crystallite sizes, nm
YF_3:Ce in water	35.61
YF_3:Ce in ethanol	29.84
YF_3:Ce in ethylene glycol	27.05
YF_3 4 hours	48.91
YF_3 8 hours	37.68
YF_3 12 hours	35.78
YF_3 16 hours	27.23
YF_3 20 hours	30.39
YF_3:Ce PEG-2000	29.21
YF_3:Ce PVP	33.04
YF3:Ce PEI	38.64
$Na(Y_{1,5}Na_{0,5})F_6$:Ce	63.07

As can be seen from the graph of the spectra (figure 2), the emissions of all 3 samples have 2 peaks at λ = 260 nm and λ = 310 nm, which corresponds to the $5d\text{-}^2F_{7/2}$ and $5d\text{-}^2F_{5/2}$ transitions of cerium, respectively. The sample synthesized in ethanol has the highest intensity. The excitation spectra were recorded at λ_{em} = 260nm.

For phosphors intended for use in photodynamic therapy, it is important that the half-life is as long as possible. Thus, the time of exposure to X-rays is reduced.

The X-ray luminescence of a YF_3:Ce 5%mol. sample was measured. (figure 4d), the shape of the spectrum corresponds to the shape obtained upon UV excitation.

In the course of the experiment, a sample of $Na(Y_{1,5}Na_{0,5})F_6$ of the hexagonal phase was also synthesized, since according to the literature data [11], yttrium fluoride is somewhat inferior to it in some characteristics (band gap, half-life, etc.). The photoluminescence spectra of YF_3:Ce and $Na(Y_{1,5}Na_{0,5})F_6$:Ce are shown in figure 5a. Both samples were synthesized in water and have a cerium concentration of 5% mol. As can be seen, the YF_3:Ce sample is inferior to $Na(Y_{1,5}Na_{0,5})F_6$:Ce in intensity and has a shorter-wavelength peak λ = 300 nm.

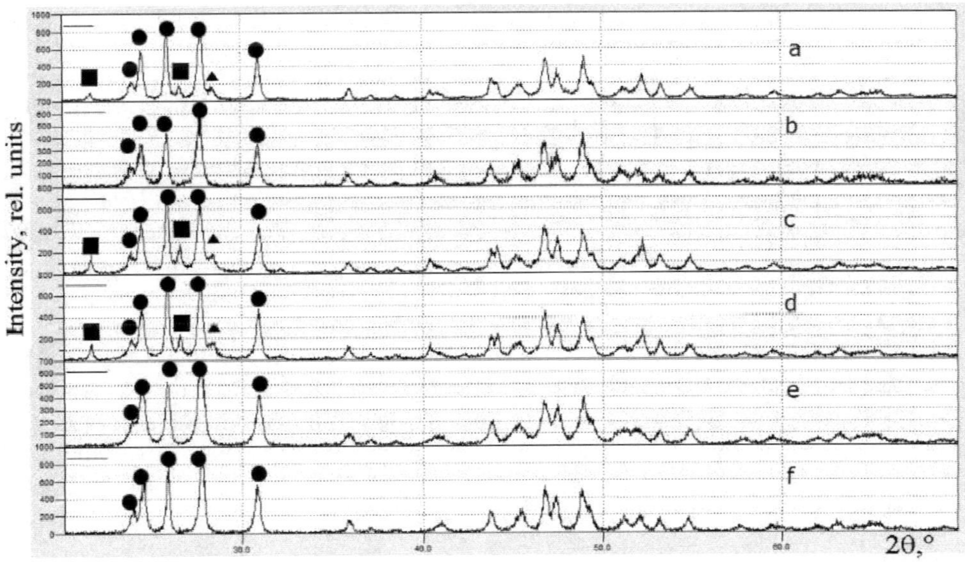

Figure 3. X-Ray diffraction patterns with different concentrations of cerium: a – 1%mol.; b – 2%mol.; c – 3%mol.; d – 4%mol.; e – 5%mol.; f – 6%mol.

Figure 3 shows X-ray diffraction patterns of yttrium fluoride samples with different cerium concentrations from 1 to 6%mol. As can be seen from the X-ray diffraction patterns, the samples of 1%, 3%, and 4% have an insignificant amount of impurity phases - cubic yttrium fluoride and CeF_2, which dramatically affected the shape of the luminescence peaks (figure 4 a, b). As can be seen from the above spectra, the samples containing the cubic phase of yttrium fluoride differ in the intensity of the emission peaks: the most intense peak is $\lambda = 315$ nm (figure 4b), while the pure orthorhombic phase has the maximum peak $\lambda = 296$ nm (figure 4a). YF_3:Ce 5% mol. has the highest intensity among the obtained samples. Figure 4c shows the decay time measurement of this sample, which corresponds to 28,7 ns.

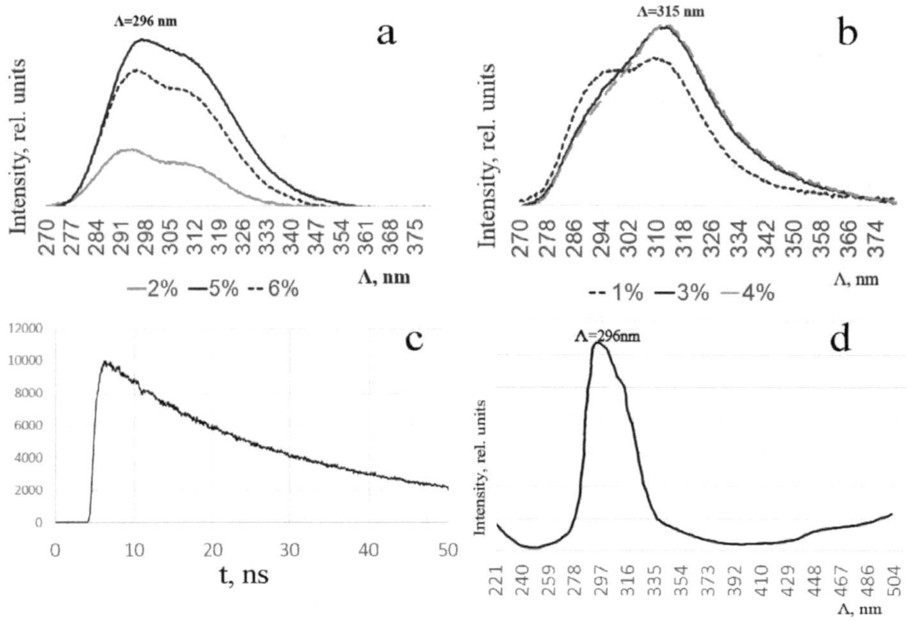

Figure 4. Luminescence characteristics of samplesYF_3:Ce: a - photoluminescence spectra of YF_3:Ce (orthorhombic phase); b - photoluminescence spectra of YF_3:Ce samples (orthorhombic phase with an admixture of the cubic phase); c - graph of decay time of YF_3:Ce 5%mol. sample; d - X-ray luminescence spectrum YF_3:Ce 5%mol. sample.

YF_3:Ce with different phase compositions was synthesized as well: cubic and orthorhombic. The photoluminescence spectra are shown in figure 5b. As can be seen, in the case of a sample of the cubic phase, the spectrum is strongly shifted to longer wavelengths and has a noticeably lower luminescence intensity.

We have synthesized a number of samples of undoped YF_3 to establish the optimal duration of the synthesis. As a result, it would be possible to obtain yttrium fluoride of the orthorhombic phase. Figure 6 shows X-ray diffraction patterns of YF_3 samples synthesized by solvothermal methods in an ethylene glycol medium for different periods of time. The optimal duration is a synthesis time of 16 hours. When synthesizing within 12 hours, it is possible to obtain a cubic phase of yttrium fluoride.

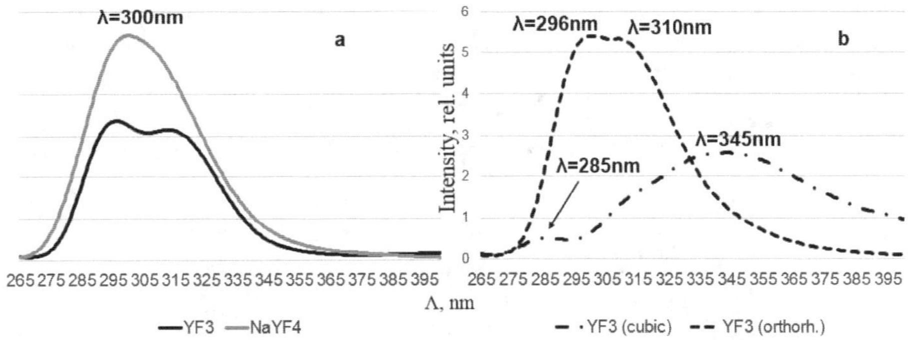

Figure 5. Photoluminescence spectra (λ_{ex}=255nm). a - samples YF_3:Ce 5%mol. and $Na(Y_{1,5}Na_{0,5})F_6$:Ce 5%mol.; b – YF_3:Ce 5%mol. samples of orthorhombic and cubic phases.

Since the time parameter strongly influences the coalescence of particles, figure 7 shows the growth of YF_3 particles over time. Particles with a size of about 100 nm can be obtained at 4 hours without adding stabilizers, but the required phase is reached only after 16 hours of synthesis. Samples synthesized for 16 hours have the smallest crystallite size (see Table 1).

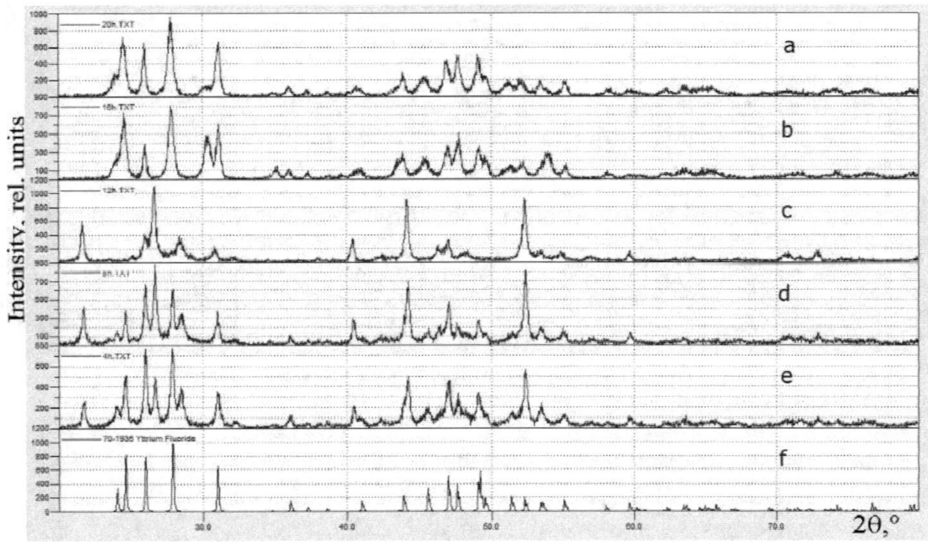

Figure 6. X-ray diffraction patterns of YF_3 samples synthesized over different time periods: a – 20h; b – 16h; c – 12h; d – 8h; e – 4h; f - YF_3 (PDF card 70-1935).

To study the effect of stabilizers on the morphology and size of YF_3: Ce particles, samples with various stabilizers were synthesized: polyethylene glycol (PEG), polyethyleneimine (PEI), and polyvinylpyrrolidone (PVP). Figure 8 shows photomicrographs of the respective samples. The optimal shape is often found in the case of polyethylene glycol, while the samples coated with PVP and PEI are generally needle-like or plate-like. In the case of PEI (figure 8 a, b), some particles significantly exceed nanosize and reach 1 μm in length.

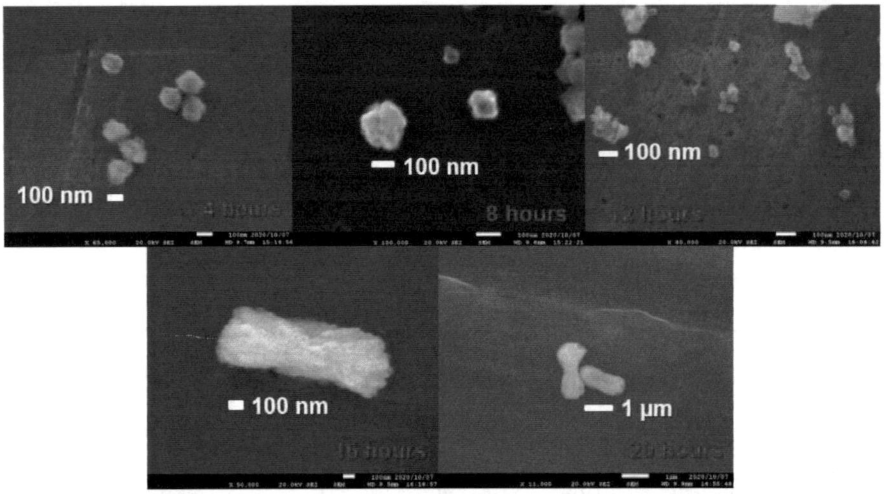

Figure 7. Micrographs of samples YF_3 depending on the duration of synthesis.

Figure 8. Micrographs of YF_3:Ce^{3+} with various stabilizers: a – PEG-2000 (100 nm), b – PVP (100 nm), c – PEI (1 μm).

4. Conclusions

As a result of the work done by the solvothermal method, the YF_3:Ce^{3+} nanophosphor was synthesized in three ways - in an aqueous medium, in ethylene glycol, and in ethanol.

It was shown that, in contrast to works [5, 12], the duration of solvothermal synthesis can be less than 20 hours, specifically in the range of 16-18 hours.

Compared to synthesis in an aqueous medium, hydrothermal synthesis in an ethylene glycol medium using the PEG-2000 stabilizer allows the synthesis of phosphors with smaller particles (less than 100 nm), which have a high degree of crystallinity, which is more suitable for use in PDT.

In terms of particle size and X-ray luminescence characteristics, the synthesized phosphors YF_3:Ce^{3+} and $Na(Y_{1.5}Na_{0.5})F_6$:Ce^{3+} are very promising for use in photodynamic therapy of oncological diseases.

References

[1] Xiaojie Wang, Tianqi Sheng, Zuoling Fu, Wenhao Li, Jung Hyun Jeong 2013 *Materials Research Bulletin* **48** 2143–2148.

[2] Y Cui, X P Fan, Z L Hong, M Q Wang, 2006 *Nanosci. Nanotechnol.* **6** 830.
[3] S Sekiyama, M Umezawa, S Kuraoka, T Ube, M Kamimura, and K Soga 2018 *Scientific Reports* **8** (1)
[4] Y Gu, Q Ran, W She, and J Liu 2017 *Advances in Materials Science and Engineering*, **13**
[5] L. Sudheendra, Gautom K. Das, Changqing Li, Daniel Stark, Jake Cena, Simon Cherry, Ian M. Kennedy 2014 *Chem. Mater.* **26** 1881–1888.
[6] V V Semashko, M S Pudovkin, A C Cefalas et al., 2018 *Nanoscale Research Letters* **13**(1) 370
[7] V V Bakhmetyev, A M Dorokhina, M M Sychov 2018 *Izvestia* (41)
[8] V V Bakhmetyev, M M Sychev, A I Orlova, E A Potanina, A E Sovestnov, Kulvelis Yu.V. 2013 *Nanoindustriya* (8) 46-50.
[9] T S Minakova, M M Sychov, V V Bakhmetyev, N S Eremina, S P Bogdanov, I A Zyatikov, L Yu Minakova 2014 *Adv. Mater. Res.* **872** 106–111
[10] V V Malygin, L A Lebedev, V V Bakhmetyev, M V Keskinova, M M Sychov, S V Mjakin, Y Nakanishi 2016 *Adv. Intell. Syst. Comput. 519* 47–54
[11] A A Blistanov 2000 Crystals of quantum and nonlinear optics *M.: MISIS* 432 p.
[12] A M Korableva, M K Kaczmarek, and R Van Deun 2019 *Nanoscale* 11(3) 833–837.

Synthesis of rare-earth nanosized phosphors using microwave processing

A B Vlasenko and V V Bakhmetyev
Saint-Petersburg State Institute of Technology (Technical University), Russia

E-mail: vlasenkoab@bk.ru

Abstract. Nanomaterials find permanently extending applications in various areas of life. In particular, nanosized phosphors can be used as pharmaceutical carriers capable of emitting ultraviolet or visible light that activates a photosensitizer, thus significantly expanding the possibilities of photodynamic therapy in the treatment of oncological, bacterial and viral diseases. The conditions required for the use of nanosized phosphors in medicine include their fine dispersion and effective luminescence in the red region of visible light spectrum upon stimulation by X-ray radiation of the range accepted for medical applications, particularly for diagnostic and therapeutic purposes in many diseases. The aim of this work was to study the effect of microwave treatment of Y_2O_3:Eu phosphors prepared by hydrothermal synthesis in ethylene glycol at 230 °C for 6 hours, involving the decomposition of mixed acetate. In order to reduce the aggregation and growth of the resulting particles, Aerosil A300 with average particle size 7 nm was added to the reaction mixture in the course of hydrothermal synthesis in the ratio 1:1 relating to the obtained phosphor. The microwave treatment was carried out at 800 °C for 5 minutes. The developed method provided Y_2O_3:Eu phosphor samples featuring with increased luminescence intensity in the region 610...700 nm compared to similar phosphors earlier prepared using the rapid thermal annealing (RTA) procedure.

1. Introduction

Phosphors based on rare earths, in particular, Y_2O_3:Eu, are widely used in various fields. One of the promising areas of application of phosphors of this composition is the improvement of photodynamic therapy (PDT). The PDT method is increasingly used in the treatment of various diseases, including cancer [1, 2, 3]. But the possibilities of this type of therapy are limited by its use for the treatment of deep-seated lesions. At the moment, it is suitable only for superficially located tumors, as well as purulent skin lesions. There are a number of ways to expand this method of therapy from the development of new photosensitizers to the creation of a new pharmacological drug that includes a photosensitizer and a nanosized phosphor that will convert X-rays into visible light with the wavelength necessary for the photosensitizer to work [4]. The main requirements for the phosphor to improve PDT are biocompatibility, small particle size that allows obtaining a stable colloidal solution suitable for introduction into the body, as well as light radiation in the area of the absorption bands of industrial photosensitizers (590-730) nm.

The finest dispersion of the obtained materials can be achieved via hydrothermal synthesis by treating a solution of the initial components in an autoclave [5, 6]. However, to obtain the necessary phase of yttrium oxide after hydrothermal treatment, high-temperature annealing at the temperature of at least 800 °C is required. At this stage, sintering and size growth of nanoparticles can occur. In order to avoid these undesirable effects, we used such methods as rapid thermal annealing (RTA), microwave annealing, as well as the addition of highly dispersed pyrogenic silica (Aerosil) to the phosphor composition, thus preventing form sintering of the phosphor particles during heat treatment.

For this purpose, a number of samples were synthesized with and without the addition of pyrogenic silica (Aerosil) and characterized by X-ray phase analysis and measuring the luminescence parameters, including brightness and intensity.

2. Experimental

To obtain finely dispersed phosphors, hydrothermal synthesis method was used in an ethylene glycol medium at 230 °C for 6 hours. YCl_3 and $EuCl_3$ were used as starting materials. First, a mixed acetate of the composition $Y_{0,85}Eu_{0,15}(CH_3COO)_3$ was prepared by dissolving yttrium and europium oxides in acetic acid, followed by evaporation of the resulting solution. Polyethylene glycol with a molecular weight of 2000 g/mol (PEG 2000) was used as a dispersant. After hydrothermal treatment, the resulting sample was separated from the solution by centrifugation. Next, the obtained substances were washed in ethyl alcohol and bidistilled water. In order to reduce the coalescence of the particles during calcination, heat treatment of the samples was carried out in the following two ways.

1. Rapid thermal annealing (RTA) at 800 °C within 5 minutes. The RTA mode was chosen based on the results of our previous studies [5, 6], during which it was found that a lower annealing temperature does not allow obtaining the required phase composition, while insufficient annealing duration, results in remaining of unburned carbon in the samples.

2. Microwave annealing. For its implementation, a special microwave installation was used [7], the scheme of which is shown in figure 1. The frequency of the magnetron was 2.45 GHz, which corresponds to the wavelength of microwave radiation of 12.3 cm. The dimensions of the microwave chamber 2 were selected in such a way that a resonance occurred in it and a standing wave was formed, in the antinode of which a quartz crucible with the sample to be processed was placed (figure 2). This design of the installation made it possible to significantly increase the intensity of the effect of microwave radiation on the sample material. The heat treatment mode of the microwave annealing was the same as that of the RTA: temperature-800 °C, duration-5 minutes.

Also, to prevent sintering of the particles during annealing, we investigated the addition of a highly dispersed neutral filler to the samples at the hydrothermal treatment stage. As such a filler, we used highly dispersed amorphous SiO_2 pyrogenic silica (Aerosil) brands A-300. This choice of filler is due to the fact that SiO_2 has a high melting point (1600 °C). It was assumed that it would not react with the phosphor material either at the hydrothermal synthesis temperature (230 °C) or at the further processing temperature (800 °C). Due to its low toxicity, pyrogenic silica (Aerosil) is widely used in pharmacy, which will allow the use of phosphor in the composition of pharmacological preparations for PDT. The content of pyrogenic silica (Aerosil) in the composition of samples based on the mass of the finished phosphor Y_2O_3:Eu was 1:1. After hydrothermal treatment, samples with the addition of pyrogenic silica (Aerosil) were annealed using RTA and microwave annealing methods under the same conditions as samples without pyrogenic silica (Aerosil).

X-ray phase analysis of the synthesized samples was performed using a Rigaku SmartLAB 3 diffractometer (Japan). Electron micrographs were obtained using a TESCAN VEGA 3 SBH scanning electron microscope (Czech Republic). The luminescence spectra were measured using an AvaSpec-3648 spectrofluorometer (Netherlands). The IL-1700 research radiometer (USA) was used to measure the brightness.

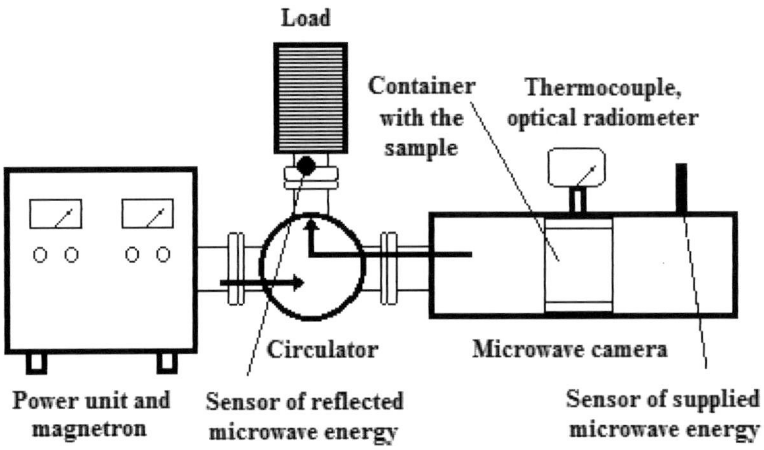

Figure 1. Schematics of microwave installation.

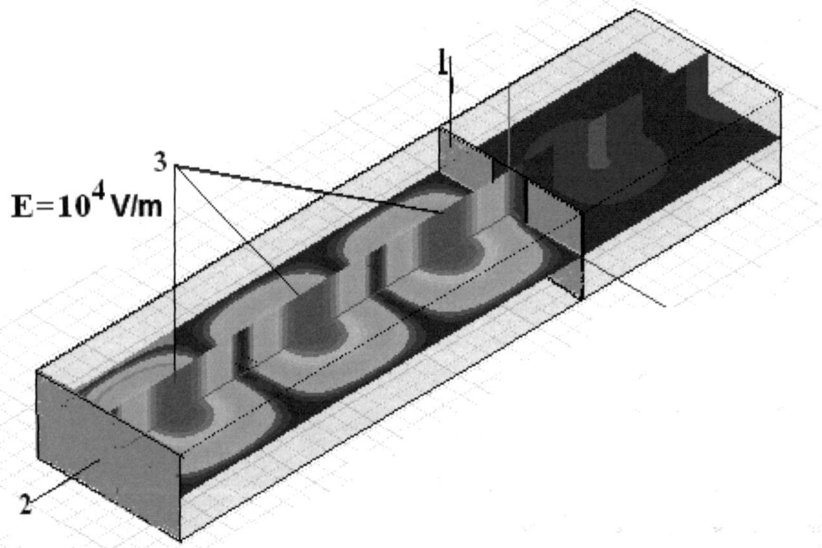

Figure 2. Model of the reaction chamber: 1-diaphragm, 2-movable wall, 3-antinodes of the electric field.

3. Result and Discussion

X-ray diffractograms of Aerosil-containing and Aerosil-free samples prepared by RTA and microwave annealing methods, are shown in figure 3. The diffractograms of the samples synthesized using RTA and microwave annealing without adding pyrogenic silica (Aerosil) correspond to the PDF card of yttrium oxide 41-1105. The sample obtained using the microwave annealing features with narrower peaks, which indicates larger crystallite sizes. According to the Scherrer formula, the size of the crystallites was calculated as about 12 nm for the phosphor synthesized using RTA and 26 nm in the case of the sample obtained by microwave annealing. These data suggest that the use of microwave annealing at the same duration and temperature leads to a more than twofold increase in the size of the crystallites, which is probably due to a more intense energy impact compared to RTA. When adding pyrogenic silica (Aerosil) to the samples, the type of diffractograms changes significantly. The diffractogram of the sample synthesized with the addition of pyrogenic silica (Aerosil) followed by rapid thermal annealing (RTA) features with a wide halo. Obviously, this phosphor has an amorphous or nanocrystalline structure due to fine grinding of the particles. To test this assumption, the samples synthesized with and without the addition of pyrogenic silica (Aerosil) were studied by SEM technique. The resulting images (figure 4) indicate that the particle size for Aerosil-free sample is 100...300 nm, while the addition of Aerosil leads to the particle size decrease below 100 nm. Obviously, this is due to the fact that the particles of highly dispersed refractory pyrogenic silica (Aerosil) separate the phosphor particles from each other and, thereby, prevent their fusion during the hydrothermal synthesis and subsequent heat treatment.

The diffractogram of the sample synthesized with the addition of pyrogenic silica (Aerosil) and subsequent microwave annealing (figure 3) features with narrow peaks and corresponds to yttrium silicate $Y_2Si_2O_7$ (PDF-card 21-1457). Thus, at the same temperature and duration of annealing, the RTA and microwave methods give different results. Obviously, the microwave field provides a much more intense impact onto the material compared to thermal heating, allowing yttrium and silicon oxides to react with each other at the temperature about 800 °C. Without the application of microwave field, Y_2O_3 and SiO_2 do not react with each other at this temperature.

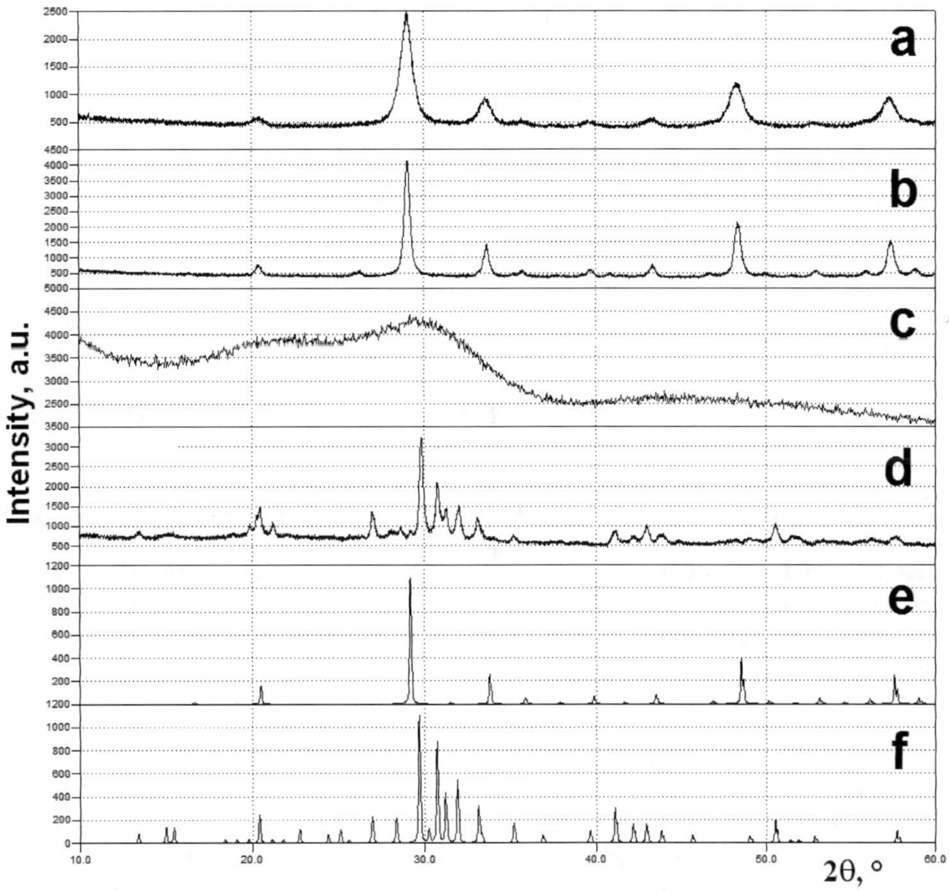

Figure 3. XRD profiles of the synthesized phosphors: a – RTA, without Aerosil; b – microwave annealing, without Aerosil; c – RTA, with Aerosil; d – microwave annealing, with Aerosil; e – Y_2O_3 (PDF card 41-1105); f – $Y_2Si_2O_7$ (PDF card 21-1457).

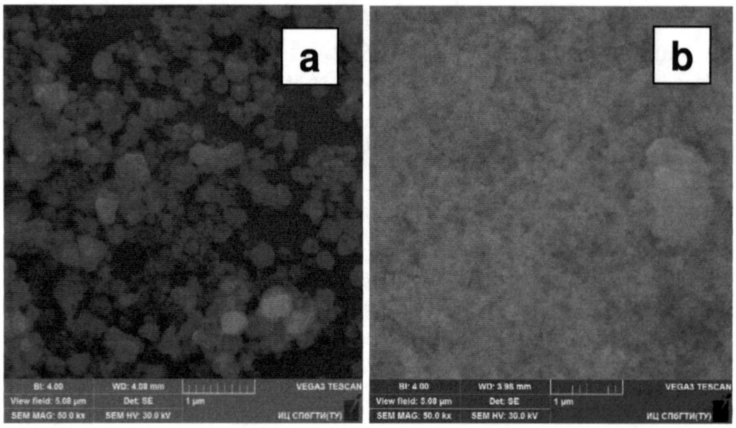

Figure 4. SEM images of Y_2O_3:Eu phosphors synthesized using RTA method without (a) and with (b) Aerosil addition.

Figure 5 shows the luminescence spectra of the obtained samples. The spectra of Aerosil-free samples prepared by both RTA and microwave annealing feature with narrow bands typical for Y_2O_3:Eu with the maximum at 612 nm, corresponding to the electronic transition $^5D_0 \rightarrow {}^7F_2$. The sample synthesized with the addition of pyrogenic silica (Aerosil) sing RTA procedure also has a maximum with a wavelength of 612 nm, but the addition of pyrogenic silica (Aerosil) leads to an increase in the width of this luminescence bands, probably due to the decrease of the phosphor particle size and amorphization of the phosphor structure. Remaining of the peak position in couple with a significant decrease in its intensity may indicate that yttrium oxide did not react with pyrogenic silica during RTA. Probably, the synthesized sample is a mixture of nanosized Y_2O_3:Eu and SiO_2 particles. The presence of non-luminous SiO_2 in the sample reduces the luminescence intensity, but significantly enhances the dispersion, which is required for the use of the phosphor in PDT. During the synthesis of phosphor with the addition of pyrogenic silica (Aerosil) and subsequent microwave annealing, the intensity of the bands in the luminescence spectrum is redistributed. The maximum with a wavelength of 621 nm, relating to the $^5D_0 \rightarrow {}^7F_3$ electron transition, becomes the most intense. The intensity of the peaks at 590 nm (transition $^5D_0 \rightarrow {}^7F_0$), 600 nm (transition $^5D_0 \rightarrow {}^7F_1$) and 700 nm (transition $^5D_0 \rightarrow {}^7F_4$) also increases. The drastic change in the luminescence spectrum in this case is explained by the fact that instead of Y_2O_3:Eu, a phosphor of a completely different composition comprising $Y_2Si_2O_7$:Eu is obtained. It should be noted that regardless of pyrogenic silica addition, the samples synthesized using microwave annealing have a much higher intensity of the luminescence bands compared to samples synthesized using RTA process. This is obviously due to the fact that the microwave effect on the material is more intense than the thermal one, which leads to an increase in the perfection of the crystal structure of the phosphors, as well as to an increase in the uniformity of the activator distribution in the particles.

Figure 6 presents the comparison of luminescence brightness of the synthesized phosphors. It can be seen that the use of microwave annealing instead of RTA leads to an increase in brightness regardless of Aerosil addition. As mentioned above, this is probably due to an increase in the perfection of the crystal structure and improvement of the activator distribution uniformity. For samples synthesized using RTA annealing, the addition of Aerosil results in a decrease in the luminescence brightness, apparently due to the dilution of the luminous phosphor Y_2O_3:Eu with non-luminous SiO_2 powder. However, despite the fact that the mass of Aerosil added to the autoclave is equal to the calculated mass of the synthesized phosphor, the observed brightness decrease was only 32% instead of the expected 2-fold. On the contrary, for the samples synthesized using microwave annealing, the addition of pyrogenic silica led to an increase in the luminescence brightness by 11%. This effect is probably determined by a higher luminescence efficiency of $Y_2Si_2O_7$:Eu compared with

Y_2O_3:Eu, due to the presence of luminescence bands in the short-wave region with maxima of 590 nm and 600 nm, closer to the area of the maximum sensitivity of the human eye.

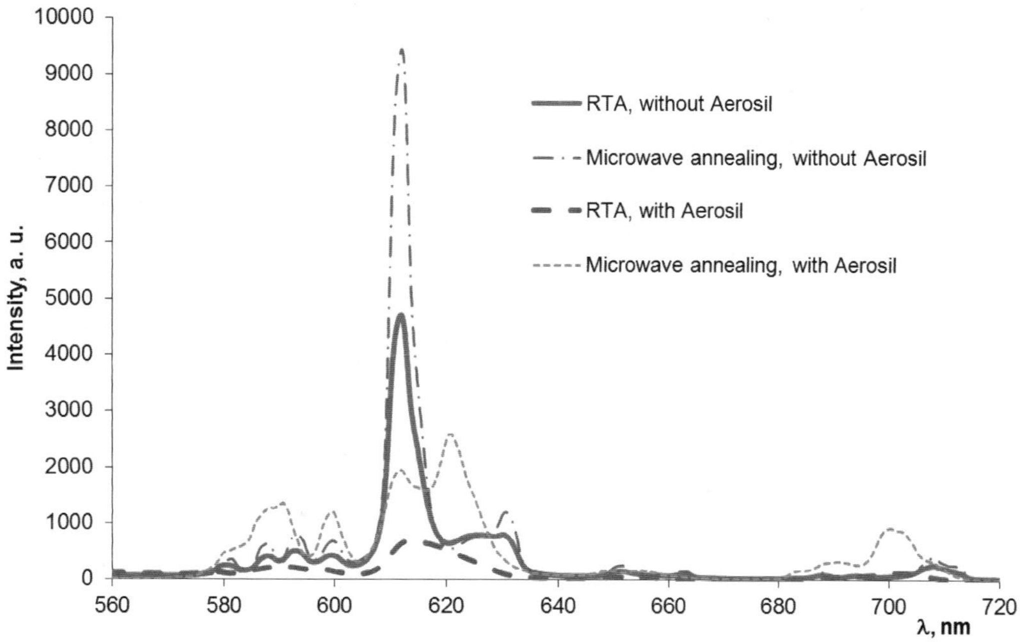

Figure 5. Luminescence spectra of the synthesized phosphors.

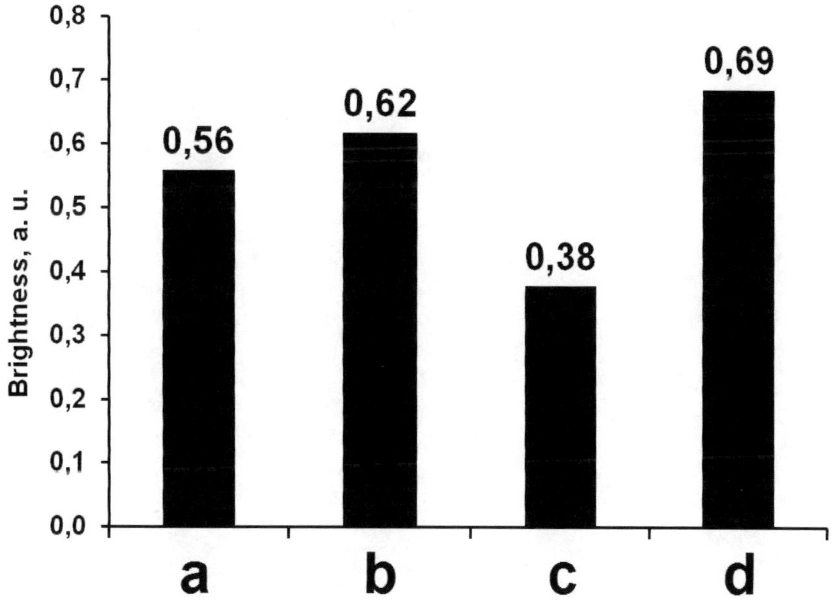

Figure 6. Luminescence brightness of the synthesized phosphors: a – RTA, without Aerosil; b – microwave annealing, without Aerosil; c – RTA, with Aerosil; d – microwave annealing, with Aerosil.

4. Conclusions
Hydrothermal synthesis followed by RTA and microwave annealing, as well as the addition of pyrogenic silica (Aerosil) provided a series of Eu-activated rare earth (yttrium oxide) based luminescent phosphors. The application of microwave annealing for heat treatment of phosphors allows the increase of the luminescence brightness due to the improvement of the crystal structure perfection, as well as, presumably, by increasing the uniformity of the activator distribution in the phosphor particles, The combination of microwave annealing and pyrogenic silica addition yields phosphors of the composition $Y_2Si_2O_7$:Eu featuring with a significantly higher brightness compared to the Y_2O_3:Eu phosphor. In the case of RTA annealing, the addition of pyrogenic silica (Aerosil) provides a significant enhancement of Y_2O_3:Eu particles dispersion affording phosphors with the particle size less than 100 nm, suitable for medical applications as a component of drugs for photodynamic therapy.

References
[1] Meerovich G.A., Akhlustina E.V., Tiganova I. G., Makarova E. A. et al. 2018 Nanostructured Photosensitizer Based on a Tetracationic Derivative of Bacteriochlorin for Antibacterial Photodynamic Therapy. Bulletin of RSMU 6 80 – 85.

[2] Logunova E.V., Nasedkin A.N. 2015 Modern View on Antimicrobial Photodynamic Therapy (Review of Literature). Laser Medicine **19** (2) 44 – 52.

[3] Gelfond M L, Barchuk A S, Vasiliev D V and Stukov A N 2003 The possibilities of photodynamic therapy in oncological practice *Russian Biotherapeutic Journal* **2** 4 pp 67–71.

[4] Bakhmetyev V.V., Sychov M.M., Orlova A.I., Potanina E.A., Sovestnov A.E., Kulvelis Yu.V. 2013 Nanophosphors for Roentgenophotodynamic Therapy of Oncological Diseases. Nanoindustry. 8 46 – 50.

[5] Vlasenko A.B. Bakhmetyev V.V. 2020 Synthesis of Y_2O_3:Eu Nanosized Phosphor Using Hydrothermal Technique and Rapid Thermal Annealing (RTA). Key Engineering Materials. **854** 209 – 215.

[6] Vlasenko A.B. Bakhmetyev V.V., Khristyuk N.A. 2020 The Hydrothermal Synthesis and RTA Conditions Influence on the Y_2O_3:Eu Nanosized Phosphors Microstructure. Journal of Physics: Conference Series. **1560** (1) 012023.

[7] Keskinova M.V., Ogurtsov K.A., Sychov M.M., Kolobkova E.V., Turkin I.A., Nakanishi Y., Hara K. 2015 Synthesis of chlorine-silicate phosphors for white light-emitting diodes. Advanced Materials Research. **1117** 48 – 51.

Anomalous transport of electrons in an electric discharge with transverse magnetic field

K P Kirdyashev

Russian Academy of Sciences, Institute of Radio Engineering and Electronics, Vvedensky sq., 1, Fryazino, Moscow Region,14190, Russia
E-mail: kpk@ms.ire.rssi.ru

Abstract. The experimental data showing the relationship of excitation of microwave oscillations with abnormally transport of electrons across the magnetic field in turbulent plasma are presented. The mechanism of the formation of the discharge current due to scattering of drift electrons in the near-electrode layers of the electric discharges on microwave oscillations has been substantiated. The conditions for the manifestation of the turbulent mechanism of anomalous electron transport through the magnetic barrier at the boundary of the toroidal electric discharge have been studied most fully. The mobility of electrons across a magnetic field is one of the main parameters of electric discharges that use crossed electric and magnetic fields for technological purposes.

1. Introduction

The investigations of the excitation of oscillations in an electric discharge with the transverse magnetic field are associated with determining the efficiency of magnetic plasma confinement and solving the problem of anomalous electron transport across the magnetic field. In the investigated discharge, the system of magnetic coils immersed in the plasma creates a closed toroidal region, in the center of which the magnetic field is close to zero. In the considered configuration of the magnetic field, electrons are confined by a magnetic barrier at the discharge boundary, as a result of which a potential well is created inside the plasma volume, which holds the ions. The studies carried out indicated the low values of the electron temperature and the depth of the potential well, since the essential feature of such discharges is the formation of discharge current due to the transfer of electrons across the magnetic field. This determines the small energy time of plasma confinement by the magnetic field in the electro-discharge mode of plasma formation.

The known mechanisms of electron transport across the magnetic field, based on collisions of electrons with heavy particles, do not allow substantiating the observed values of discharge currents. The experimentally established relationship between the formation of the discharge current and the excitation of microwave oscillations indicates the need to study possible mechanisms of plasma instability in the region of the discharge and the effect of excited oscillations on anomalous electron transport across the magnetic field. At present, there is no theory of microwave oscillations in the discharge, which is necessary for the interpretation of the results of the experimental studies, which takes into account the totality of factors essential for the development of instability. These factors include the inhomogeneity of the electric and magnetic fields, the distribution of plasma parameters in the zones of generation of microwave oscillations, the influence of nonlinear effects on the saturation of the amplitudes of the excited oscillations, and the distribution of currents in the electric discharge.

Similar phenomena are observed in various types of plasma devices designed to accelerate plasma, create plasma engines and solve technological problems in space. The study of microwave oscillations in an electric discharge with a plasma-confining magnetic barrier makes it possible to develop the method for diagnosing processes leading to the transport of electrons across the magnetic field. The study of microwave oscillations in the discharges with transverse magnetic fields is of particular interest to substantiate the mechanism of current formation across the magnetic field in the accelerating channel of the stationary plasma thruster [1]. The results of the research carried out in this work expand the possibilities of the analyzing the radio astronomical observations of cosmic sources of microwave radiation due to the development of plasma instability in transverse magnetic fields.

In this investigation the effect of microwave oscillations on the transport of electrons across the magnetic field and the formation of the discharge current is carried out. The modes of anomalous transverse plasma conductivity and anomalous resistance of the transverse current in the magnetic barrier region are identificated.

2. Experimental setup and research procedures

The design features of the experimental setup with barrier magnetic field (Fig. 1) and the methods used for the experimental study of microwave oscillations in different regions of the discharge are presented in [2–3]. In the plasma setup under consideration, the discharge voltage is applied between an incandescent cathode installed near the zero of the magnetic field and the conducting metal walls of the magnetic coils. The formation of a magnetic barrier at the discharge boundary is provided by four coils with a current; to reduce the forces of attraction between the magnetic coils and improve the field configuration, four coils with oppositely directed currents are additionally introduced into the design of the magnetic system.

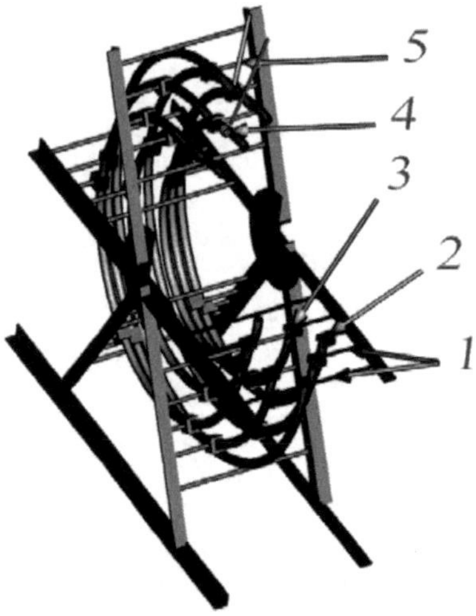

Figure 1. The design of the magnetic system of the experimental plasma setup: 1 – base of the structure, 2 – compensation coils, 3 – current-carrying conductors, 4 – cathode, 5 – supply of working gas (argon).

As follows from the distribution of magnetic force lines in the cross section of the discharge, the magnetic field is concentrated near the coils covering the outer region of the discharge (Fig. 2). With this configuration of the magnetic field, the electric discharge is formed with the current due to the transfer of electrons across the magnetic field. The investigated discharge is limited by the closed magnetic field line determined by the criterion of plasma confinement stability [4].

The experiments on the study of the relationship between the discharge current and the excitation of microwave oscillations were carried out when the gas (argon) was supplied to the region of the heated cathode mounted on the axis of the large torus. The modes of formation of the discharge are studied at the working gas flow rate of 1.5–5.4 mg/s, corresponding to various manifestations of microwave oscillations in the region of the magnetic barrier. With this design of the experimental plasma setup, the metal walls of the magnetic coils, which are at positive potential relative to the inner region of the discharge, should be considered as the anode.

According to the results of the probe measurements, the concentration of electrons in the barrier region of the discharge is $(3–7)\times10^{10}$ cm^{-3} at electron temperature in the range of 15–25 eV. The thickness of the discharge boundary in the region of the magnetic barrier is 0.5–0.7 cm. The plasma parameters decrease by more than an order of magnitude beyond the discharge boundary. The distribution of the plasma potential over the small radius of the torus indicates the presence of the potential well that traps ions inside the magnetic barrier. The potential drop in the region of the magnetic barrier is 50–70 V at discharge voltage in the range of 150–200 V.

The study of microwave oscillations in the barrier region of the discharge was carried out by recording the electromagnetic radiation of the discharge using broadband antennas installed inside the vacuum chamber outside the discharge [5–6]. The determination of the intensity and spectra of microwave radiation was carried out using frequency-tunable measuring receivers in the range 1.05–7.15 GHz, including the frequencies of electron plasma oscillations at the experimentally measured electron concentration in the barrier region of the discharge. The probe studies of the intensity and spectra of plasma oscillations in the region of the magnetic barrier in the frequency range 200−500 kHz were carried out.

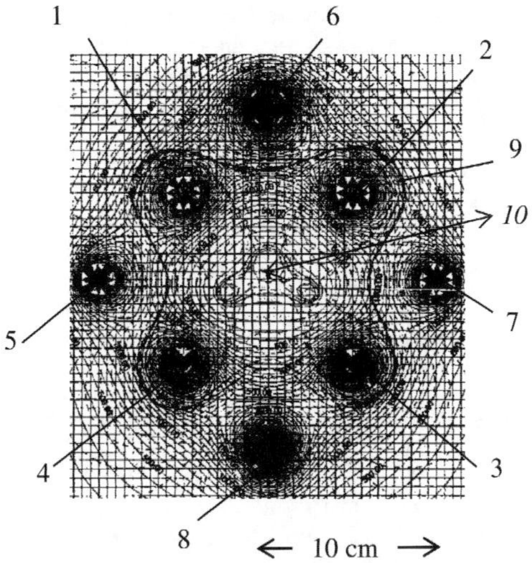

Figure 2. The structure of the magnetic system of the experimental plasma setup and the distribution of the magnetic field in the cross-section of the discharge: barrier magnetic field coils (1–4), compensation coils (5–8), plasma confinement boundary (9), direction of radial displacement of probes when measuring plasma parameters and the intensity of microwave oscillations (10); magnetic lines of force (dashes), lines of constant magnetic field (solid lines).

In accordance with the applied measurement technique [7], when determining the intensity of the microwave radiation of the discharge, it was envisaged to register the radiation of the reference source, which was the gas-discharge lamp with configuration similar to the outer region of the toroidal discharge under study. The use of a reference source made it possible to determine the absolute values of the spectral density of microwave radiation of the discharge W regardless of the measurement conditions and the equipment used. To characterize the degree of non-equilibrium of microwave oscillations in the region of the magnetic barrier, we considered the values W/W_0 normalized to the radiation energy density caused by thermal fluctuations in the barrier region of the discharge W_0.

The zones of generation of microwave oscillations in the discharge volume were identified based on the results of local measurements of oscillations using two-wire electrical probes. The one of microwave probes which was installed near the discharge boundary, the other probe moved in the cross section of the discharge. The scale of electric field disturbances perceived by microwave probes is determined by the distance between the conductors Δl = 1.5–2 mm. The applied microwave probes make it possible to register the fields of plasma waves in the discharge with the significant weakening of the contribution to the recorded signal from the components of the fields of electromagnetic radiation. The results of local measurements of the microwave energy were compared with the thermal level.

In earlier experiments [5–6], the study of the microwave emission spectra of the discharge was carried out by the method of sequential spectral analysis. Sequential spectral analysis is usually used in the study of processes, the nature of which changes rather slowly compared to the scanning period of the frequency of the measuring device. For the selected frequency tuning rate, the duration of the spectral analysis, which ensures undistorted transmission of the spectral components of microwave oscillations in the investigated frequency range, was about 10 s, which significantly exceeds the characteristic period of the establishment of stationary parameters of the oscillations investigated in the discharge. The experimental spectra obtained as the result of successive analysis do not allow one to reveal the development in time of the discharge instability and the connection between the excitation of microwave oscillations at different frequencies. The method of modified parallel spectral analysis was applied, in which the envelopes of microwave oscillations were simultaneously recorded at two selected carrier frequencies within quasi-stationary time intervals corresponding to the duration of 10^2–10^3 periods of microwave oscillations excited in the discharge.

3. The experimental results

The zones of generation of microwave oscillations are determined on the basis of the results of local measurements of the intensity of plasma waves and fields of electromagnetic radiation outside the magnetic barrier. The most intense oscillations are 3–4 orders of magnitude higher than the equilibrium level. The effects of competition between the anomalous transverse plasma conductivity and the anomalous resistance of the transverse current in the discharge are revealed based on the results of the performed experiments. The results obtained agree with the concept of an analogy between collisions of electrons with neutral particles and their scattering by chaotic oscillations of the electric field caused by high-frequency plasma instability.

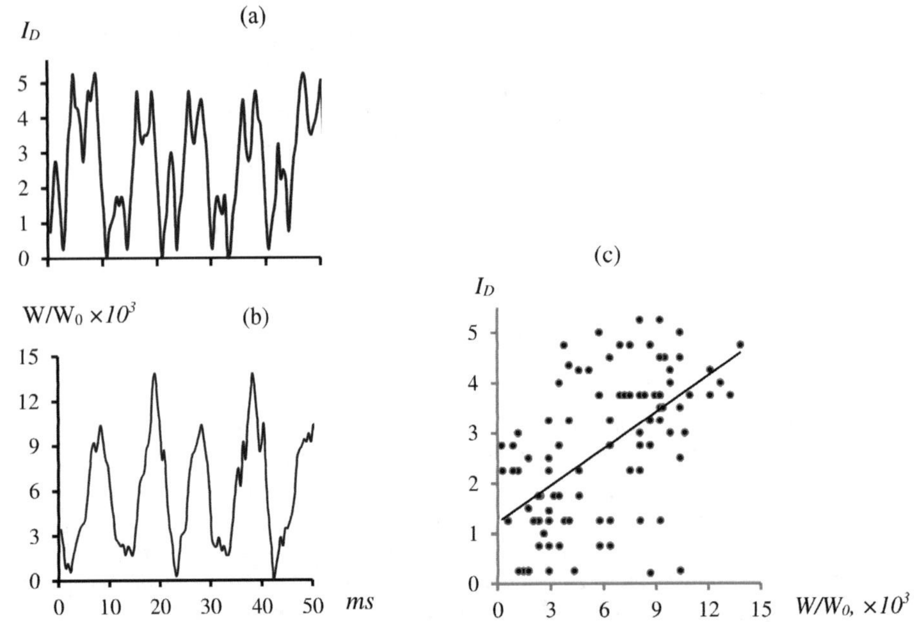

Figure 3. The oscillations of the discharge current I_D(a) and the relative intensity of microwave radiation W/W_0 (b); diagram of the dependence of the discharge current on the intensity of microwave radiation (c). Discharge mode: gas consumption 5.35 mg/s, voltage 200 V, magnetic field 5×10^{-3} T; the linear trend line, accuracy of approximation 0.36.

The non-equilibrium of plasma oscillations and electromagnetic radiation of the discharge manifests itself to the greatest extent at frequencies of 1.5–2.5 GHz, corresponding to the frequency range of electron plasma oscillations in the region of the magnetic barrier. It was found that with an increase in the flow rate of the working gas, a decrease in the intensity of microwave radiation, the broadening of the frequency spectrum and the shift of the radiation maximum to higher frequencies are observed. In the microwave radiation spectra obtained by the method of sequential spectral analysis, components with randomly varying intensity and frequency separation of 150–200 MHz are distinguished, corresponding to the electron cyclotron frequency in the magnetic barrier at the fields of 0.0050 and / - 0075 T .

The generation zone of intense microwave oscillations is located on the outer side of the discharge at distance of 5.5–6.5 cm from the axis of the small torus, in which the electron concentration is 3×10^{10}–10^{11} cm^{-3}. The measurement results indicate the localization of the source of intense microwave oscillations near the boundary of plasma confinement by the barrier magnetic field. In the same region of the discharge, perturbations of the electric field potential are observed in the frequency range 350–500 kHz, associated with fluctuations in the concentration of electrons in the discharge. Based on the assumption that the excited oscillations are potential, the width of emission spectrum is determined by fluctuations in the plasma density caused by the plasma instability in the low-frequency region of the spectrum.

As follows from temporal intensity of the microwave radiation and oscillations at the individual frequencies, the instability developing in the discharge manifests itself in limited time intervals with subsequent breakdown of the oscillatory processes. The results of statistical processing of signals at the output of the measuring receivers indicate the presence of components in the autocorrelation functions, one of which appears at time intervals ≤ 50 μs, the other corresponds to disturbance duration of 1–5 ms. It is characteristic that during these time intervals there are ejections of electrons and ions from the magnetic confinement region of the plasma, which are recorded by the oscillations of the electron and ion currents to the probes installed outside the magnetic barrier [3].

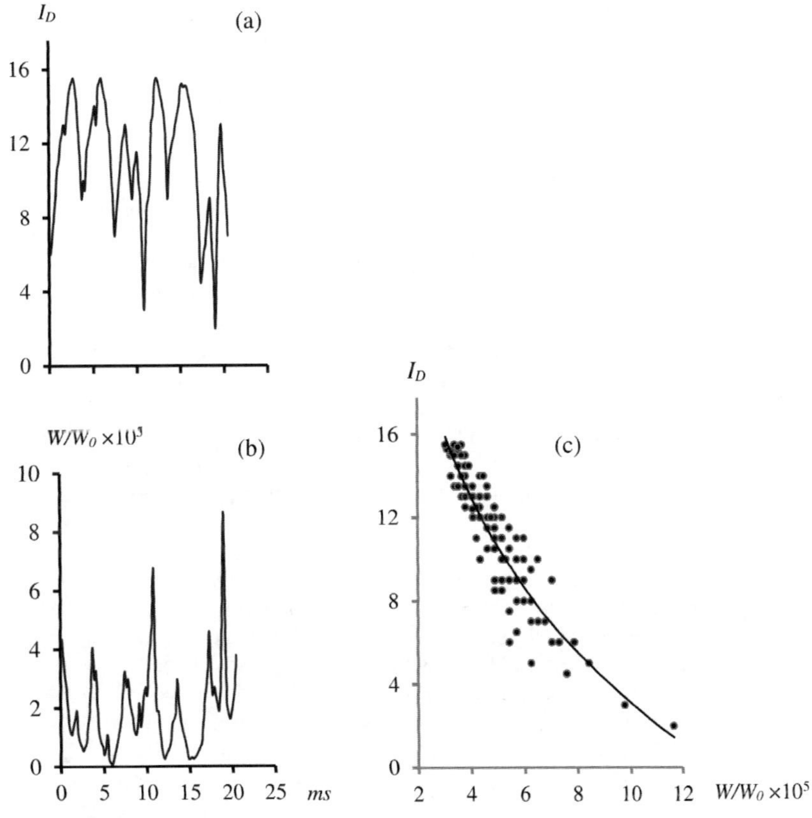

Figure 4. The oscillations of the discharge current I_D(a) and the relative intensity of microwave radiation W/W_0 (b); the diagram of the dependence of the discharge current on the intensity of microwave radiation (c). Discharge mode: gas consumption 1.6 mg/s, voltage 200 V, magnetic field 5×10^{-3} T; the exponential trend line, accuracy of approximation 0.81.

It should be noted that there is the significant scatter in the values of the correlation coefficients between the intensity of microwave radiation and plasma waves at different frequencies, which is the basis for concluding that the spectra of microwave oscillations in the barrier region of the discharge are stochastic. In this case, the decrease in the degree of correlation of the components of the spectrum of microwave oscillations is observed with an increase in the shift of the carrier frequencies.

From the results of registration of plasma oscillations by microwave probes, it follows that within the plasma confinement region there is the almost synchronous emission of the intensity of microwave oscillations in the cross section of the plasma formation within a small radius of the torus, which indicates a significant correlation of the developing plasma instability in the discharge volume. Outside the plasma confinement zone, the synchronicity of the envelopes of microwave oscillations is violated, and the temporal variations in the oscillation intensity are irregular.

4. The instability theory of the discharge

The experimentally observed excitation of microwave oscillations in the discharge indicates the development of instability in the range of electron plasma frequencies of the barrier region of the discharge, which makes it possible to consider the velocity distribution of electrons in this region, which significantly differs from the equilibrium distribution. The frequency range of the microwave instability developing in the discharge corresponds to the high-frequency branch of electron oscillations calculated from the electron density and the magnetic field. The mechanism of the excitation of microwave fields in the plasmas with transverse magnetic field due to the interaction of spatially separate streams of electrons and forming discharge current.

When analysing the mechanism of excitation of microwave oscillations, we will proceed from the existence of two groups of electrons near the magnetic barrier. The constituent of electrons is held back by the magnetic field and creates the drift current in crossed electric and magnetic fields. As follows from the results of probe measurements, the estimate of the concentration of electrons forming drift current near the magnetic barrier is (2.4–8.5) $\times 10^{10}$cm^3. The electron drift velocity in the crossed electric and magnetic fields in this region of the discharge is determined from the results of measurements of the strength of the electric and magnetic fields near the magnetic coils holding the plasma varies within the range (3.9–4.2) $\times 10^8$ cm/s. Another group of electrons is formed as the result of demagnetization of drifting electrons by collisions and their scattering on oscillations. The transfer of demagnetized electrons across the magnetic barrier, which ensures the creation of the discharge current, occurs in a layer with a size equal in order of magnitude to the electron cyclotron radius ρ_{Be}. The concentration of these electrons in the region of the magnetic barrier can be estimated from the experimental values of the discharge current, the surface area of the metal shells of the magnetic coils, and the potential drop in the barrier region. The estimates show that with the effective surface area of the shells of the magnetic coils of the order of 250 cm^{2,} the concentration of the electron flows across the magnetic barrier holding the plasma is (1.4–1.8) $\times 10^8$ cm^{-3}.

Usually the plasma instabilities originate from reacting of interpenetrating electron beams and absence of effect magnetic field on the formation of different groups of electrons [7]. The experimentally observed excitation of microwave oscillations makes it possible to consider the model of the barrier region of the discharge, in which the mechanism of interaction of the drift electron flows with demagnetized electrons that form the discharge current is realized. We will proceed from the model of electron flows separated in space, the interaction of which occurs through of plasma waves fields distributed in the cross section of the magnetic barrier [8]. As applied to the conditions of formation of the investigated discharge, the distribution of perturbations of the electric field potential depends on the component of the wave vector of the plasma wave in the direction of electron drift k_φ:

$$\varphi(r)=\varphi_\varphi \exp\left(-k_\varphi |r|\right). \tag{1}$$

The interaction of electron beams in the region of the magnetic barrier is possible under the condition $k_\varphi \rho_{Be} \leq 1$, in which ρ_{Be} is the electron cyclotron radius. This condition, taking into account the distribution of the plasma wave field, is reduced to the form:

$$(\omega_{Pe}/\omega_{Be})(V_{Te}/V_\varphi)\geq 1. \tag{2}$$

The considered model of the current sheet in the region of the magnetic barrier corresponds to the experimental distributions of the electric field potential in the cross section of the discharge [7].The potential drop in the discharge region is 50–60 V and concentrated near the boundary of plasma confinement in the layer with dimensions of 0.7–1.2 cm. The size of the current layer, in which the interacting flows of electrons are generated, is of the order of the calculated values of the electron cyclotron radius in the barrier region. At electric field strength of $\geq$ 100–120 V/cm in the magnetic barrier, the directed velocity of the drift flow of electrons significantly exceeds their thermal velocity.

From the experiments carried out, it follows that the non-stationary nature of the excitation of microwave oscillations is manifested in sporadic discharge of the charge through the magnetic barrier. The excess charge is discharged when the plasma accumulating inside the magnetic barrier crosses the boundary of stable confinement. The electric discharge is accompanied by the formation of electron beams outside the magnetic barrier with a characteristic repetition period of discharge current surges close to the plasma confinement time under conditions of continuous ionization of the working gas. The modes of excitation of microwave oscillations and the formation of a discharge current make it possible to consider an electric discharge inside the magnetic barrier as the self-oscillating system with internal feedback formed by azimuthal electron beams. The formation of the fast electrons is evidenced by the maxima of the intensity of microwave oscillations near the magnetic barrier and in the axial region of the discharge.

The qualitative model of the discharge as the stochastic self-oscillating system is based on fluctuations in the delay of microwave signals in the internal feedback loop formed by the drift electron flow. In this case, the duration of envelopes of microwave emissions can be estimated from the average value of the phase delay of the microwave signal in the feedback loop. Indeed, for discharge modes with the potential drop in the magnetic barrier region within 50–60 V, the delay of the plasma wave carried by a drift electron beam is 10–50 μs, which is consistent with the results of statistical processing of microwave radiation intensity oscillations.

The characteristic repetition period of the microwave emission can be estimated from the time of accumulation of electrons inside the magnetic barrier, which is associated with the effective frequency of ionization collisions of electrons with atoms of the working gas. With a cross section of argon ionization of the order of 10^{-18} cm^2 by the electron beam with energy $\geq$ 50 eV and characteristic values of the gas pressure in the discharge region, it follows that the stage of accumulation of electrons inside the magnetic barrier manifests itself at time intervals of 1–5 ms, corresponding to the frequency of microwave emission from the barrier discharge area.

5. Electron transport across the magnetic field

The main conclusion from the performed experiments is related to the estimation of the efficiency of electron transfer across the magnetic field of the barrier and the effect of microwave oscillations on the formation of the discharge current. The experimentally established relationship between the discharge current and the excitation of microwave oscillations makes it possible to consider the contribution of nonlinear effects to the development of microwave instability of the discharge, which leads to demagnetization of electron drift on oscillations and the change in the electron velocity distribution function. The formation of two streams of electrons as a result of their scattering on oscillations favours the development of instability in the region of plasma confinement by a magnetic field.

The comparison of the values of the discharge current with the intensity of the oscillations allows us to consider the excitation of microwave oscillations as the main factor leading to the scattering of drift flows of electrons by oscillations and their transfer across the magnetic field, which is the main mechanism of the discharge current formation. This conclusion follows from the analogy between pair collisions of particles in the plasma and the scattering of electrons by microwave oscillations excited in the discharge. Under the conditions of discharge formation, the current density due to the transfer of electrons across the magnetic field can be represented in the form [9]:

$$j_D = e n_0 \left(e/m_e\right) E_{bar} \nu_{ef} \Big/ \left(\omega_{Be}^2 + \nu_e^2\right). \tag{3}$$

We take into account the effective frequency of electron scattering in microwave oscillations ν_{ef}, cyclotron frequency of electrons ω_{Be} in the barrier discharge, the concentration drift electrons n_0. With the relation $\nu_{ef}^2 \ll \omega_{Be}^2$ in the experiment, the discharge current increases with increase in the frequency of electron scattering by oscillations:

$$j_D = e n_0 \left(e/m_e\right) E_{bar} \nu_{ef} \Big/ \omega_{Be}^2 . \tag{4}$$

The results of the performed experiments make it possible to consider the scattering of electrons by microwave oscillations as the determining factor in the anomalous transfer of electrons through the magnetic barrier and the formation of the discharge current. The regime of anomalous transverse plasma conductivity in the barrier region of the discharge corresponds to the experimental data. In accordance with [10], the effective frequency of electron scattering presented as a function of the energy of microwave oscillations in the discharge W_e:

$$\nu_{ef} = \left(2W_e / n_{e0} e \Delta U_{bar}\right)^{1/2} \omega_{Pe} . \tag{5}$$

In this case, the experimental values of the concentration of the drift flux of electrons n_{e0}, leading to the excitation of oscillations, and the potential difference in the barrier region of the discharge ΔU_{bar} are taken into account. From the results of the measurements it follows that the maximum energy of microwave oscillations in modes with anomalous transverse plasma conductivity is 10^{-3}–10^{-2}J /m^3.Based on these data, we estimate the effective frequency of electron scattering by microwave oscillations as 10^8–10^9 s^{-1}, at which the regime of anomalous electron transfer through the magnetic barrier is realized. As follows from (3), the scattering of electrons by provides the mode of formation of the discharge current within 3–5 A/cm^2 at the concentration of electrons in the barrier region of the discharge $n_{e0} = 10^{10}–10^{11}$ cm^{-3} and the electric field strength in the magnetic barrier 50–100 V/cm.

The maximum energy density of microwave oscillations at the quasi-linear stage of the development of plasma-beam instability in the investigated discharge region is determined based on the condition $v_e \leq \omega_{Be}$ corresponding to the formation of drift electron flows:

$$W_e = (1/2)(\omega_{Be}/\omega_{Pe})^{1/2} n_0 \, e \Delta U_{bar} . \quad (6)$$

When this value is exceeded, the demagnetization of electrons on oscillations predominates and the discharge transitions to the beam instability mode with the anomalous resistance of the barrier region of the discharge characteristic of this instability. Thus, in the barrier region of the discharge, the nonlinear mechanism of beam-plasma instability is realized, which is caused by the action of excited oscillations on the transfer of electrons across the magnetic field and the formation of the discharge current. The maximum transfer rate of electrons through the magnetic barrier, which form the discharge current, can be estimated from the displacement of electrons at the distance corresponding to their cyclotron radius during the scattering time by oscillations [11]:

$$V_{e\max} = V_{Te} (\nu_{ef}/\omega_{Be}) \leq V_{Te} . \quad (7)$$

The rate of the removal of electrons from the discharge reaches maximum values corresponding to the thermal speed of electrons at the energy density of microwave oscillations (6).

Conclusion

The effects of competition between the anomalous transverse plasma conductivity and the anomalous resistance of the transverse current in the discharge are revealed based on the results of the performed experiments. The results obtained agree with the concept of the analogy between collisions of electrons with neutral particles and their scattering by chaotic oscillations of the electric field caused by high-frequency plasma instability. The dependence of the discharge current on the intensity of microwave oscillations indicates the nonlinear processes in the discharge, leading to the accumulation of electrons in the discharge. The registration of intensity and spectrum of microwaves oscillations is the effective diagnostics the modes discharge formation and electron transfer processes in various types of plasma devices with transverse magnetic field.

Acknowledgements

The work was performed according to the state task of the Kotelnikov Institute of Radio engineering and Electronics of Russian Academy of Sciences. I am grateful to the staff of the Science and Engineering Center of High Plasmadynamic Technologies at the Moscow Technological University (MIREA) for their help in performing these experiments and helpful discussions of the results obtained in this work.

References

[1] Morozov A I 2006 *Introduction to Plasma Dynamics* (Fizmatlit, Moscow) [in Russian].
[2] Morozov A I 1992 *Plasma Phys. Rep.* **18**, 235.
[3] Morozov A I, Bugrova A I, Bishaev A M and Nevrovsky V A 1999 *Tech. Phys. Lett.* **25** 700.
[4] Ohkava T and Yoshikava M 1966 *Phys. Rev. Lett.* **17** 685.
[5] Morozov A I, Kirdyashev K P, Bugrova A I and Bishaev A M 2001 *Plasma Phys. Rep.* **27** 275.
[6] Kirdyashev K P, Morozov A I, Bugrova A I and Bishaev A M 2002 *Tech. Phys. Lett.* **28** 700.
[7] Mikhailovskii A B 1971 *Theory of Plasma Instabilities* (Atomizdat, Moscow).
[8] Kirdyashev K P 1982 *High-Frequency Wave Processes in Plasmadynamic Systems* (Energoatomizdat, Moscow) [in Russian].
[9] Erofeev V S and Leskov L A 1974 *Physics and Application of Plasma Thrusters* [in Russian].
[10] Aref'ev V I and Kirdyashev K P 1975 *Sov. Phys. Tech. Phys.* **20** 330.
[11] Kirdyashev K P 2016 *Plasma Phys. Rep.* **42** 841.

Predictor of reliability indicators for nanoelectronic heterostructure devices with transverse current transfer under conditions of limited experimental information based on Bayesian inversion

N A Vetrova*, A A Filyaev*, V D Shashurin, L A Luneva

Bauman Moscow State Technical University, Moscow 105005, Russia

*Corresponding authors: e-mail – vetrova@bmstu.ru, alex.filyaev.98@gmail.com

Abstract. Predictor of the reliability indicators of resonant tunneling diodes with a generalization of the methodology for nanoelectronic heterostructure devices with quantum confinement and transverse current transfer has been developed. The advantage of the developed software is the possibility of interactive input of additional experimental information for further calculation of point and interval estimates of the reliability indicators of semiconductor devices using Bayesian inversion, which allows predicting these indicators under conditions of limited experimental information.

1. Introduction

The rapid development of micro- and nanoelectronics has led to the creation of promising devices, in which semiconductor heterostructures act as a low-dimensional channel [1-7]. Due to the small size of the channel of such devices, difficulties arise in ensuring their reliability due to the significant influence of degradation processes during their operation and even during their manufacture. Modern computational models make it possible to predict the reliability indicators of such devices at the stage of their design [8-11]. The validation of computational models is a mandatory procedure, which requires the availability of appropriate experimental information obtained during the tests [12]. In accordance with the established standards and regulatory documents [13-15] for the statistical evaluation of reliability indicators of the required level of confidence and accuracy, a regulated number of experimental samples is required (for example, to obtain point and interval estimates of reliability indicators with a confidence level of 90% and with a relative error of 5%, at least 500 objects are required for the accelerated test plan [NUN] when the exponential law of reliability is valid [13]). Due to the high cost and complexity of the manufacturing process, it is not always possible to provide the required number of test objects.

Thus, the actual task is to develop a methodology for the statistical evaluation of the reliability indicators of nanoelectronic heterostructure devices with quantum confinement and transverse current transfer under conditions of limited experimental information.

2. Algorithm for statistical evaluation of reliability indicators of nanoelectronic devices based on Bayesian inversion

The established standards and normative documents [13-15] make it possible to determine the relative error and confidence level of the statistical evaluation of the reliability indicators of nanoelectronic devices based on experimental information of a fixed volume. This means that in the case of a limited number of experimental samples with a given reliability, it is possible to carry out a statistical evaluation with a significantly lower value of accuracy. In conditions of limited maximum number of experimental samples, as well as in the case of the possibility of replenishing the database with additional experimental information, it is preferable to carry out an evaluation taking into account a priori information (an evaluation in the presence of basic and additional information). It is proposed to carry out such a calculation using Bayesian inversion, which allows correcting a priori estimates as additional information becomes available to narrow the a priori confidence interval, and hence increase the accuracy. Based on this, an algorithm for the statistical evaluation of the reliability indicators of nanoelectronic devices under conditions of limited experimental information using Bayesian inversion is proposed (Figure 1).

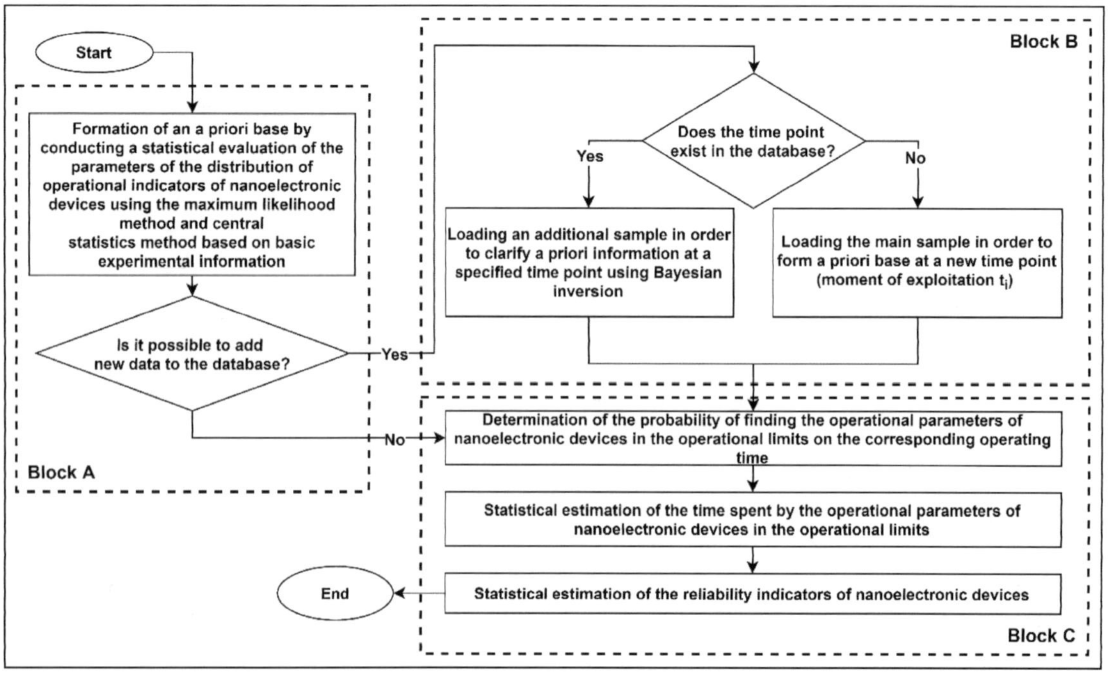

Figure 1. Algorithm for statistical evaluation of reliability indicators of nanoelectronic devices in conditions of limited experimental information.

The developed algorithm assumes that the experimental data for statistical evaluation were obtained as a result of testing nanoelectronic devices, the manufacturing process of which is in a state of statistical controllability and acceptable reproducibility. The main advantage of the developed algorithm is the ability to enter additional information for the subsequent refinement of statistical estimates of the reliability indicators of semiconductor devices, which will provide an increase in accuracy due to the application of the Bayesian approach at a given confidence level under conditions of a small number of experimental objects.

3. General logic framework of Bayesian approach in statistical estimation

As the basis of the algorithm for the statistical evaluation of the reliability indicators of nanoelectronic devices in conditions of limited experimental information, consider the Bayesian approach [16-17]. Bayesian statistical methods apply Bayes' theorem to calculate and refine probabilities after new experimental data are obtained. Bayes' theorem describes the conditional probability of an event based on both additional experimental data and a priori information.

The construction of Bayesian point and interval estimates of the parameter θ with known additional experimental information $x_1 \dots x_n$ is based on the use of knowledge of the posterior distribution $f(\theta \mid x_1, \dots, x_n)$, which is defined as

$$f(\theta \mid x_1, \dots, x_n) = \frac{f(\theta) f_L(x_1, \dots, x_n \mid \theta)}{\int f_L(x_1, \dots, x_n \mid \theta) f(\theta) d\theta}. \tag{1}$$

Here $f(\theta)$ is the prior distribution, $f_L(x_1, \dots, x_n \mid \theta)$ is the likelihood function. Then the mean value of the posterior distribution

$$\hat{\theta}^B = E\big(f(\theta \mid x_1, \dots, x_n)\big) = \int \theta f(\theta \mid x_1, \dots, x_n) d\theta \tag{2}$$

or the modal value of the posterior distribution

$$\hat{\theta}^B = \arg\max_{\theta} f(\theta \mid x_1, \dots, x_n) \tag{3}$$

is used as Bayesian point estimates.

Thus, the problem of determining Bayesian estimates of the parameters of the distribution of operational indicators of nanoelectronic devices is reduced to determining the prior distribution and calculating the likelihood function with known additional experimental information. The formation of the a priori base is carried out by conducting point and interval estimates of distribution parameters using maximum likelihood methods and central statistics methods on the basic experimental information with low accuracy at a fixed confidence level in the case of a small sample size (Figure 1, Block A). Regardless of the size of the main and additional sample the using of the Bayesian approach makes it possible to reduce the value of the a priori confidence interval, which means that the accuracy of the interval estimation increases at a given level of confidence.

4. Results

On the basis of the presented algorithm and the general logic framework of Bayesian approach, a methodology for evaluation the reliability indicators of nanoelectronic devices using Bayesian inversion was developed. According to this methodology, software was developed that allows to determine the value of reliability function of resonant tunneling diodes (RTD) at the corresponding operating time and the gamma-percentile operating time to failure at a given probability γ under conditions of limited experimental data, as well as to make point and interval estimates of operational device parameters (Figure 1, Block C). The value of the current and the differential resistance at the operating point is considered as the operational parameters of the RTD. The interface of the developed software is shown in Figure 2.

The interface of the developed software contains 5 areas (Figure 2):

1 - research object area: display of the research object with its operational parameters with the ability to change the operational limits for each parameter, as well as with the ability to select a parameter for detailed consideration of the change in its point and interval estimates on the specified operating time.

2 - area of interactive input of information: input of a time point (Figure 1, Block B); specifying the path to the file and marker for cells with basic and additional data, changing the default values of confidence and probability γ.

3 - area for initiating software actions: starting the process of recording data at the specified time point (Figure 1, Block B); updating graphic information (area 4 and 5) after changing the data in areas 1 and 2 or entering additional or basic information into the database at a time point; displays detailed information on the operating parameter selected in area 1.

4 - area for graphical display of the results of estimation the reliability indicators of the research object.

5 - area for graphical display of point and interval estimates of the operational parameters of the research object on the specified operating time.

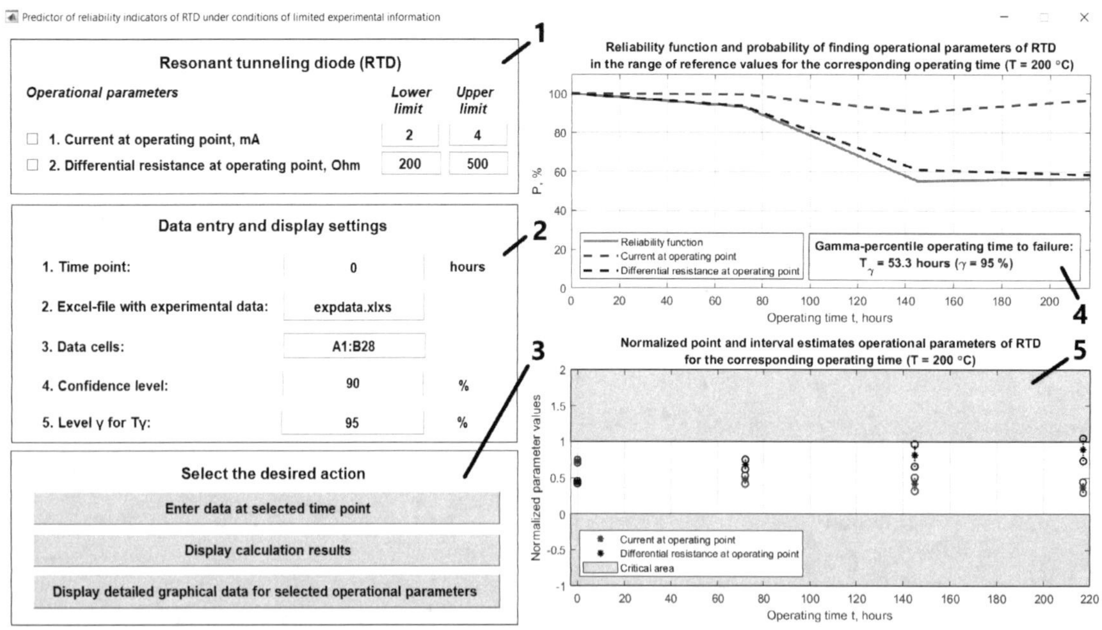

Figure 2. Interface for predictor of RTD reliability indicators in conditions of limited experimental information based on Bayesian inversion.

The advantage of the developed software is the ability to load the main sample with the formation of an a priori base at a new time point (Figure 1, Block B, logical link "no") or to load an additional sample in the presence of a priori information at the corresponding value of the operating time in the database with further calculation to improve the accuracy of the a priori estimate (Figure 1, Block B, logical link "yes"). In the case of user input of the main sample, the a priori base is formed by conducting point and interval estimates of operational parameters using the maximum likelihood method and central statistics method, while in the case of entering an additional sample, the distribution parameters are recalculated using Bayesian inversion in accordance with relation (1).

The reliability function of the RTD at the corresponding operating time is calculated by integrating the multivariate distribution function f of the operating parameters of the device within the established operating limits. The graphic interpretation of the calculation is shown in Figure 3, where $f(X)$ and P_x are the distribution function of the current at the operating point and the probability of finding this parameter in the established operational limit of 2 ... 4 mA, $f(Y)$ and P_y are the distribution function of the differential resistance in operating point and the probability of finding this parameter in the established operational limit of 200 ... 500 Ohm, $f(X, Y)$ and $P_{x,y}$ are the multivariate distribution function of the entered operational parameters and the value of reliability function of RTD at the corresponding operating time.

The calculation presented in Figure 3 was carried out using the developed software based on real experimental data obtained as a result of accelerated tests of RTDs, the two-barrier quantum system of which includes AlAs barriers and a GaAs well with a thickness of 2.9 and 4.9 nm, respectively. The top and bottom spacers are homogeneous undoped GaAs layers with a thickness of 2.1 nm. Accelerated tests were carried out in a mode with a constant temperature exposure of 200 °C. As a result of the statistical evaluation of the experimental data, the gamma-percentage operating time to failure of the RTD is 53.33 hours at a given probability of 95%.

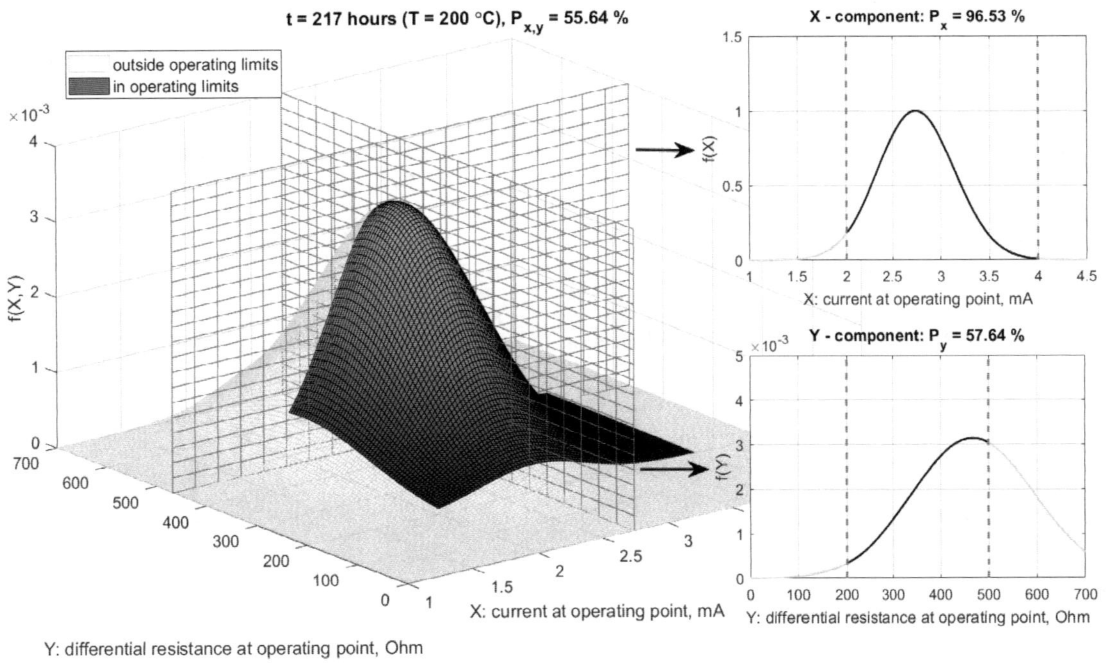

Figure 3. Graphical interpretation of the calculation of the RTD reliability function value on the corresponding operating time.

5. Conclusion

Thus, a methodology was developed for evaluation the reliability indicators of nanoelectronic heterostructure devices with quantum confinement and transverse current transfer using Bayesian inversion, which makes it possible to predict these indicators under conditions of limited experimental information. As a result of statistical processing of experimental data obtained during accelerated tests of RTDs (at a temperature exposure of 200 °C) the following reliability indicators were determined using the developed predictor of reliability indicators for nanoelectronic devices: the RTD reliability function value 92.77% at the operating time of 72 hours, the gamma-percentile operating time to failure 53.33 hours at a given probability of 95%.

In accordance with the methodology, a predictor of reliability indicators for RTD has been developed. The main advantage of this predictor is the possibility of a step-by-step refinement of the statistical estimates of these indicators while collecting additional experimental information obtained not only during accelerated tests, but also during the operation of these devices.

References

[1] Tian W, Sun H, Chen L, Wangyang P, Chen X, Xiong J and Li L 2019 *InfoMat* **1**(2) 140-163.
[2] Pfenning A et al. 2018 *Infrared Remote Sensing and Instrumentation XXVI* **10765** 107650U
[3] Nie B, Huang J, Zhao C, Huang W, Zhang Y, Cao Y and Ma W 2019 *Applied Physics Letters* **114**(5) 053509
[4] Osadchuk I A, Osadchuk A V, Osadchuk V S and Semenov A O 2020 *IEEE 15th International Conference on Advanced Trends in Radioelectronics, Telecommunications and Computer Engineering (TCSET)* p 452-457
[5] Feiginov M 2019 *Journal of Infrared, Millimeter, and Terahertz Waves* **40**(4) 365-394
[6] Sobolev A S, Zaitsev-Zotov S V, Maytama M V, Klimov E A, Pavlov A Y, Ponomarev D S and Khabibullin R A 2021 *Optical Engineering* **60**(8) 082018

[7] Okamoto K, Tsuruda K, Diebold S, Hisatake S, Fujita M and Nagatsuma T 2017 *Journal of Infrared, Millimeter, and Terahertz Waves* **38**(9) 1085-1097
[8] Shashurin V D, Vetrova N A, Pchelintsev K P, Kuimov E V and Meshkov S A 2019 *AIP Conference Proceedings* **2171** (1) 150004
[9] Kozubnyak S A, Meshkov S A, Naraikin O S, Soboleva E N and Shashurin V D 2017 *Nanotechnologies in Russia* **12**(7) 360-368
[10] Vetrova N, Ivanov Y, Kuimov E, Makeev M, Meshkov S, Pchelintsev K and Shashurin V 2019 *ITM Web of Conferences* **30** 08004
[11] Shashurin V D, Vetrova N A and Filyaev A A 2020 *Journal of Physics: Conference Series* **1560**(1) 012048
[12] Shasurin V D, Vetrova N A and Kuimov E V 2020 *Journal of Physics: Conference Series* **1571**(1) 012008
[13] RD 50-690-89 "*Methodical instructions. Reliability in technology. Methods for evaluation reliability indicators based on experimental data*" dated 30 October 1989
[14] Durivage M A 2017 *The certified reliability engineer handbook* (Milwaukee: Quality Press.)
[15] Walker H F, Benbow D W and Elshennawy A K 2018 *The Certified Quality Technician Handbook* (Milwaukee: Quality Press.)
[16] Downey A 2013 *Think Bayes: Bayesian statistics in python* (Sebastopol: O'Reilly Media, Inc.)
[17] Hernandez-Stumpfhauser D, Breidt F J and van der Woerd M J 2017 *Bayesian Analysis* **12**(1) 113-133

On the Problem of Synchronization of Virtual and Real Movements for Virtual Reality Systems

V A Chertopolokhov[1]

[1]Center "Supersonic", Lomonosov Moscow State University, 1 Leninskiye Gory, Moscow, Russia

E-mail: psvr@vrmsu.ru

Abstract. An immersive experience in modern virtual reality systems requires adjusting of various parameters. This report focuses on the problem of virtual representation of real movements of objects in space. Separately, a special class of problems is highlighted: synchronization of visual and dynamic imitation for virtual reality systems and motion platforms. Modern mass-produced virtual reality headsets do not allow effective tracking of human movements on a moving base using standard software. The element base and algorithms necessary for the successful solution of this problem are presented.

1. Introduction

The creation of dynamic and visual simulators to simulate various processes has made it possible to create different training systems. Simulators are designed for tasks often associated with high costs in case of erroneous actions and for simulating emergencies.

One of the spheres of human activity that requires special training is in-flight work conditions. For example, training on flight simulators is one of the integral parts of training for passenger aviation pilots worldwide. The conditions of aircraft flight and microgravity observed during an orbital flight are unusual for humans. Therefore, any action in flight can cause difficulties for an unprepared person, which means that specific training should be performed.

Modern systems based on virtual reality (VR) technology are designed to learn and repeat various flight modes and provide simulations of both standard and complex flight modes. An essential feature of such systems is the coincidence of the supplied simulation with the actual situation. In particular, in-flight simulators, visualization, and motion stands (for example, the Stewart platform) are used together. Any discrepancy can lead to impairments, in particular, the development of motion sickness. When developing a new type of stand based on a manipulator arm and virtual reality headsets, the task becomes even more difficult due to the high mobility of both the stand and the helmet. To reduce the negative effects of virtual simulation, it is required to solve the problem of synchronizing visual and dynamic simulation, the approach to which is described in this paper.

2. The Structure of the Problem of Synchronization of Virtual and Real Movements

2.1. *Reasons of signal desynchronization*

The vestibular apparatus is connected through the shortest three-neuron chain in the body with the oculomotor muscles. The eyes' reaction to rotation occurs reflexively, bypassing the higher structures of the central nervous system. The compensatory movement of the eye relative to the movement of the head is called the vestibular-ocular reflex (VOR). When fixing the gaze on the head's point and passive turns thanks to the VOR, a clear vision is maintained. In virtual reality conditions, the information from the vestibular sensor does not always coincide with the supplied image. Because of this, the mechanism of the vestibular-ocular reflex is disrupted, and motion sickness, also called exercise sickness, can occur, one of the symptoms of which is motion sickness.

The correctness of the perception of virtual reality critically depends on the synchronization of the reproduced imitation and actual human actions. To allow the user to correctly perceive and interact with virtual objects in real time, it is required to receive up-to-date information about the position of his head and limbs. Let us introduce the concept of a virtual screen - an object located at a certain distance in front of a virtual eye camera. We assume that the distance is known and depends on the design of the visualization system. The virtual screen must match the real screen in aspect ratio and angular size. However, it may be closer to the virtual eye than the real screen to the real eye.

A significant amount of virtual reality simulation programs are designed for small relative movements of the user. So, for example, in most training systems designed to simulate aerospace flights, the pilot is assumed to be strapped to the seat [1]. When the screen of the visualization system is stationary relative to the platform, this assumption allows us to build the visualization in advance in the form of a static transformation between the virtual camera and the position of the eye.

However, in modern systems with a split-screen from the platform or in systems based on a virtual reality helmet, the transformation becomes dependent on the dynamics of the platform and the pilot's head. As a result, it becomes necessary to use specialized tracking systems to track the rotation of the user's head. Usually, video analysis is used for these purposes [2]; hybrid motion tracking methods have also been proposed. Experiments on the panoramic virtual reality system of Moscow State University have shown that even when using expensive video analysis systems, the delay between real movement and its display in virtual space can reach 200 milliseconds. It consists of receiving and processing data from the video analysis system, transferring it to the graphics station, rendering, and displaying the final image on the screen.

In the future, the text will consider the aircraft's joint visual and dynamic simulation to illustrate the possible desynchronization and methods of their compensation.

2.2. *Dynamic Simulation Task Description*

Let us consider a particular problem. There is a specific model of a heavy aircraft (HA). All forces acting on HA and the state vector at any moment, aircraft parameters, and flight trajectory are known. It is necessary to simulate the vector of gravitational-inertial forces and angular velocity acting on the pilot on the stand.

Let us choose a physiological coordinate system associated with the aircraft: x - is directed along the central axis of the aircraft, y – along the right-wing, z – we will define it to get the right-hand-system, we put the origin at the point C — aircraft center of mass.

To begin with, let us take several simplifications: the mass of the aircraft is constant (the fuel burns out evenly and slowly, we consider a reasonably short time at which the mass of the fuel can be considered constant), the center of mass is constant, there are no atmospheric phenomena, the force of gravity is constant, the Earth is flat, symmetry in the xz, plane, the forces of the engine act along the axis of the plane. The coordinate system associated with the Earth is considered inertial. The center of mass of the aircraft is motionless relative to HA.

Then the dynamics of the aircraft can be represented by the following system:

$$\begin{cases} M\ddot{y}^p = R^e + R^{aero} + Mg, \\ \frac{d}{dt}(I\omega^p) + \omega^p \times I\omega^p = M_R^p + M_R^{aero}, \end{cases} \tag{1}$$

where y^p is the position of the center of mass of the aircraft in a stationary space, R^e are the forces of the engines, R^{aero} are the aerodynamic forces, g is the free fall acceleration, ω^p is the angular velocity of the aircraft, M_R^e are the moments of the engines, M_R^{aero} are the aerodynamic moments. As with any rigid body, the HA motion could be represented as the motion of the center of mass and around the center of mass.

Let us introduce the overload for the aircraft as follows:

$$n^C = \frac{1}{g_0}(W^C - g) = \frac{1}{g_0 M}(R^e + R^{aero}).$$

The software implementation of the aircraft model is a "black box," there is access to the model output in the form of overload $n^C(t)$, angular velocity $\omega^C(t)$ and transformation $T_x^p(t)$ (Fig. 1), which sets the position pilot seats in a virtual environment.

It is important to note that an attempt to simulate directly $n^C(t)$ on a mobile stand in parallel with visualization of the situation behind the cockpit from the pilot's seat will lead to a mismatch of effects. Indeed, let us consider the movement of the center of mass of the pilot's body, fixed on his seat by a particular spring system. During the flight, the pilot's body changes its position (N_p). However, these changes are small enough, so we will perform calculations at the design point (N), which coincides with the pilot's center of mass at the initial moment.

$$m\ddot{x}_p + \epsilon\dot{x}_p + kx_p = F^p = m(g - W^N).$$

Since the pilot is not in the center of mass of the aircraft, additional accelerations W^α occur:

$$W^N = -W^C + W^\alpha,$$

$$F_p = mg + mW^N = m(g + W^N) = mg_0\frac{g+W^N}{g_0} = mg_0 n^p,$$

where F_p is the vector of gravito-inertial forces.

$$n^p = -n^C + \frac{1}{g_0}W^\alpha, \tag{2}$$

where n^p is the pilot's overload vector.

Ultimately, we get that the pilot during the flight experiences the same overloads as the aircraft, with an accuracy of the sign and inertial forces arising due to its rotation relative to the center of mass. Thus, for correct simulation, it is required to recalculate the overloads, taking into account the rotation of HA.

Let us present a diagram of the relationship between various types of imitation and other blocks of the mixed reality system. (Fig. 2).

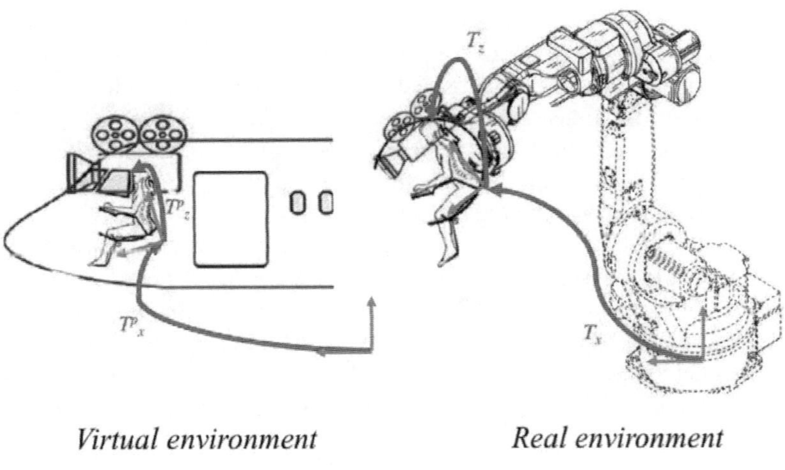

Fugure 1. Coordinate transformations

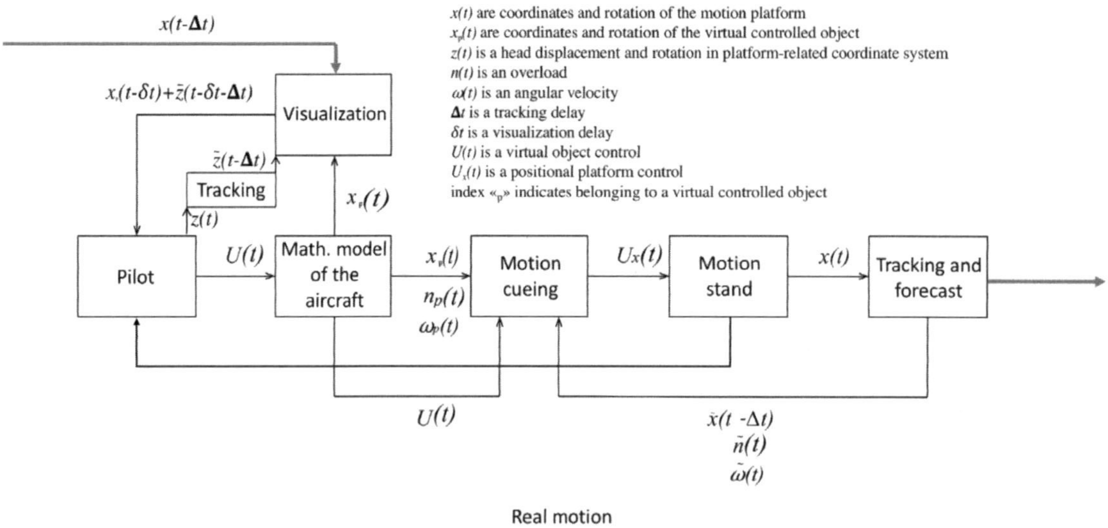

Fugure 2. Mixed reality system structure

2.3. *Motion Stand Description*

Let us assume that the dynamics of a motion stand are known, which is presented in the form of a manipulator's arm with a chair at the end (Fig. 2). It is important to note that we do not know precisely the parameters of the stand (both kinematic and dynamic). Let us present the equations of dynamics for a stand based on an industrial robotic manipulator, taking into account the uncertainties:

$$\begin{cases} x_a{}' = x_b \\ x_b{}' = Q_x^N(x_a, x_b, \mu_1\zeta_a) + Q_x^U(x_a, x_b, \mu_1\zeta_a) + (G_x^N(x_a, \mu_2\zeta_a) + G_x^U(x_a, \mu_2\zeta_a))u \\ \mu\zeta_a{}' = \zeta_b \\ \mu\zeta_b{}' = Q_\zeta^N(x_a, x_b, \zeta_a, \zeta_b) + Q_\zeta^U(x_a, x_b, \zeta_a, \zeta_b) + (G_\zeta^N(x_a, \zeta_a) + G_\zeta^U(x_a, \zeta_a))u \end{cases} \quad (3)$$

Where $x_a \in \mathbb{R}^3$, $\zeta_a \in \mathbb{R}^3$, $x_b \in \mathbb{R}^3$, $\zeta_b \in \mathbb{R}^3$ are angular coordinates and speeds for three degrees of freedom of the robotic arm and three degrees of freedom of the cabin, respectively, Q^N and G^N are known matrices, $Q^U(td, md)$ and $G^U(td, md)$ is uncertainty associated with inaccurate knowledge of the mass-dimensional characteristics of the system (we denote the unknown parameters *md* and *td* for mass and overall dimensions, respectively).

It is challenging to solve the problems of dynamic simulation for such a system. However, using small parameters in front of the derivatives and on the right-hand sides, it is possible to significantly simplify the synthesis and find control with an error $O(\mu)$. The first-order accuracy concerning small parameters is achieved. In the article [3] and the monograph [4] A. Vasil'eva described a general approach to solving problems with singular perturbation based on the method of boundary functions and Tikhonov's theorem. We will solve the problem of finding the optimal control for simulating motion for a simplified system, that is, a system in which $\mu = 0$, and write down the resulting trajectory. Then we apply the found control to the original system and solve the stabilization problem around the trajectory obtained for the degenerate system to minimize the error. At the same time, the angular motions of the cockpit will be considered under the assumption that the linear motions are quasi-stationary. Thus, the linear and angular movements of the stand can be considered separately if the parameters $\mu_i \ll 1$.

2.4. *Synchronization Problem*

Suppose that one of (or a combination) of motion cueing algorithms is implemented, i.e., the control $u(t)$ is known. Let there be an estimate of the state vector $(\tilde{x}_a, \tilde{x}_b, \tilde{\zeta}_a, \tilde{\zeta}_b)$, obtained using an external tracking system (optical, inertial, or hybrid) at time t. We can obtain $\tilde{T}_x(t-\Delta t)$ as the estimated conversions between a fixed room coordinate system and the pilot's seat in the simulator.

Let there also be some unknown dynamics of the human head relative to the chair. Consider a particular case when a person is fastened with seat belts, and the head can only make rotational movements. The resulting movement in the moving coordinate system will be denoted as $T_z(t)$. Suppose that the tracking system allows at time t to estimate the head rotation relative to the chair $\tilde{T}_z(t-\Delta t)$.

The problem of synchronizing visual and dynamic simulation is the search for such a transformation matrix $T_z^p(t)$ of the coordinate system associated with the aircraft seat to the coordinate system of the virtual camera (according to the representation in Fig. 2) so that for each moment of time t the following is performed:

$$J(T) = \sum_{i=1}^{3} \left[\left|\omega_i^{Phead}(t) - \omega_i^{head}(t)\right| - \epsilon\right]_+ \to \min \tag{4}$$

Here $\omega_i^{Phead}(t)$ is the angular velocity of the human head in virtual space, $\omega_i^{head}(t)$ is the angular velocity of the human head in reality, ϵ is the sensitivity threshold of the semicircular canals, $[.]_+$ is the operator:

$$[x]_+ = \begin{cases} x, if\ x \geq 0 \\ 0, if\ x < 0 \end{cases} \tag{5}$$

The final transformation that solves the visualization problem will be represented as:

$$T(t) = (T_z^p(t)T_x^p(t))^{-1}T_{obj}(t)$$

Where $T_{obj}(t)$ is the known movement of the visualized object, $T_x^p(t)$ is the movement of the aircraft specified from the model, $T_z^p(t)$ is the movement of the virtual camera. It is directly calculated from the movement of the real motion stand $T_x\ (t)$ and the movement of a pilot relative to the chair $T_z\ (t)$ by solving the estimation problem according to the measurements of inertial and optical sensors.

The solution to the synchronization problem is based on the solution of several subtasks, which are discussed in more detail in several articles:

- Tracking angular movements of a pilot and a motion stand using a hybrid approach and restrictions on the dynamics of the system [5];
- Identification of unknown geometric and dynamic parameters of the motion stand ($Q^U(td, md)$ and $G^U(td, md)$) in order to reduce modeling errors [6];
- Forecast of the angular motion of the stand taking into account its dynamics and tracking delay time [7];
- Increase the rendering sampling rate to shorten the motion prediction interval using foveated rendering solution [8].

It is important to note that due to (3), predicting the stand's angular motions can be solved independently of the linear ones.

3. Modelling Results

The modeling was performed at the Lomonosov Moscow State University Panoramic Virtual Reality System. Due to the limited motion capabilities of the existing platform, a separate movement along the pitch angle was carried out. The sequence of angular maneuvers of the aircraft was taken as the movement (Fig. 3). The sampling rate was chosen to be 50 Hz (50 n = 1 s). Several experiments have been carried out comparing the standard linear prediction (the angular velocity is assumed to be constant) with the complex synchronization between visual and motion cueing given above (Fig. 4).

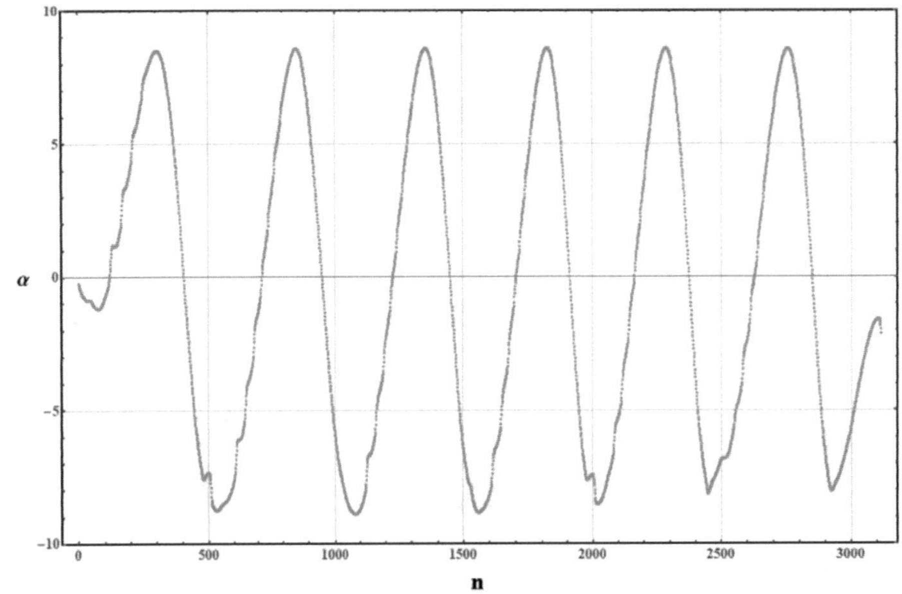

Figure 3: Aircraft angular motion produced by simulator

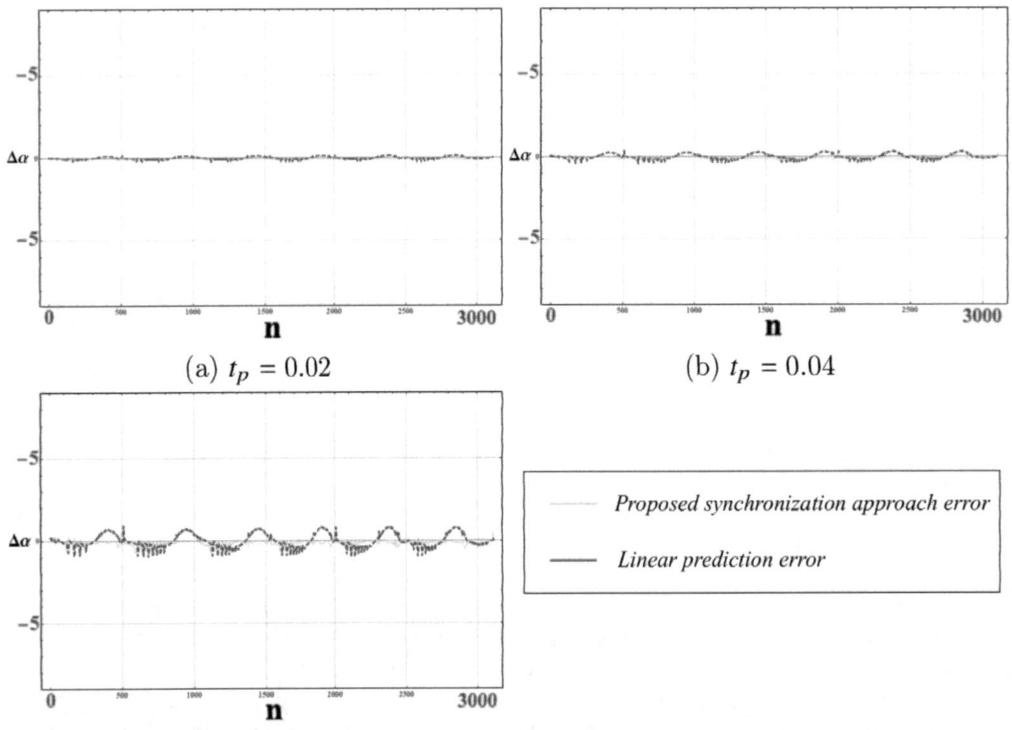

(a) $t_p = 0.02$

(b) $t_p = 0.04$

(c) $t_p = 0.1$

(d) *Legend*

Figure 4: Syncronozation error sorted by introduced delay t_p (seconds)

4. Conclusions

Performed experiments demonstrate that an increase in the delay time t_p produces an insignificant decrease in the accuracy of complex synchronization. In contrast, the error in the linear prediction increased in proportion to the increase in the delay time.

Advanced synchronization methods significantly improve training systems and virtual reality systems, reducing motion sickness. The presented results could be easily transferred to other types of simulated transport and configurations of motion platforms. Synchronization becomes a more difficult task in conditions of active movement of the user in space since it becomes more challenging to build a good forecast of movement.

Further development of the problem is seen as using differential neural networks [9] to reduce uncertainties in the system and sliding mode control [10] to ensure the robustness of a moving platform.

Acknowledgments
The article was prepared within the interdisciplinary scientific and educational school of Lomonosov Moscow State University "Brain, Cognitive Systems, Artificial Intelligence"

References

[1] V V Aleksandrov, A YU Ishlinskij, V A Sadovnichij 1995 *Izdatel'stvo Moskovskogo Universiteta Moskva* 160 (in Russian)
[2] A Sharma, M Agarwal, A Sharma, P Dhuria 2020 *International Journal on Recent and Innovation Trends in Computing and Communication* **1**(4) 251–257
[3] A B Vasil'eva, M G Dmitriev 1982 *Itogi nauki i tekhn. Ser. Mat. anal.* **20** 3-77 (in Russian).
[4] A B Vasil'eva, V F Butukov 1973 *Asymptotic expansions of solutions of singular disturbed equations*, Nauka, Moskva 272 (in Russian)
[5] D I Bugrov, A V Lebedev, V A Chertopolokhov 2014 *Moscow University Mechanics Bulletin* **69**(1) 25–27.
[6] D S Burlakov, V V Latonov, V A Chertopolokhov 2021 *Journal of Mathematical Sciences* **253**(6) 806–817.
[7] P P Dolgov, V N Kirshanov, E YU Irodov 2020 *Pilotiruemye polety v kosmos* **3**(36) 91–108
[8] A Kruchinina, V Chertopolokhov, G Grigoryan, V Belyaev 2020 *SID's Digest of Technical Papers.* **51** of Book 2: Session 54 P. 54-4.
[9] O Andrianova, A Poznyak, I Chairez 2021 *Neurocomputing* **457** 128-140
[10] A Hernández-Sánchez, O G Andrianova, A S Poznyak, I O Chairez 2021 *International Journal of Systems Science* **52**(3) 541-554.

Journal of Physics: Conference Series **2056** (2021) 012053 doi:10.1088/1742-6596/2056/1/012053

Formulation of the modern indicator devices modelling problem

E A Buzaeva[1], D A Evsevichev[1], O V Maksimova[2]

[1] Department of Aeronautical Engineering, Ulyanovsk Civil Aviation Institute, Ulyanovsk, Russia
[2] Research Department, Ulyanovsk Civil Aviation Institute, Ulyanovsk, Russia

Corresponding author's e-mail address: first32007@yandex.ru

Abstract. This work is dedicated to study of the structural characteristics and parameters in advanced electroluminescent structures. The important problems in this area are the high complexity and the slowness of the design calculation processes. The problem can be solved by developing a mathematical model of the system based on equivalent circuits. First step in developing a mathematical model is formulation.

1. Introduction

One of the most important elements of the «human-machine» systems functioning are the means of displaying information. Those devices provide information in a form suitable for visual perception. Indicator technology refers to the means of displaying information. One of the most promising indicators are those what based on electroluminescent emitters, which occupy a special place among active indicator devices due to their high speed and wide range of operating temperatures.

For hardware implementation of various specialized systems, it is necessary to develop or improve information display facilities that include a wide range of devices. Volumetric (such as CRT – cathode ray tube indicator) and flat displays are the modern means of information display. The most promising modern flat display devices include liquid crystal (LCD) indicators, indicators based on organic light-emitting diodes (OLED) and thin-film electroluminescent (TFEL) indicators [1].

The work of modern flat displays is most often associated with the phenomenon of electroluminescence. The phenomenon of electroluminescence is the non-thermal conversion of electrical energy into luminous energy. There are two classes of electroluminescent devices. In the light emitting diode (LED) devices, light is generated by electron-hole pair recombination near a p-n junction. In the LCD manufacturing LEDs are often used as a backlight for subpixels of the screen matrix. Commercial LEDs have been fabricated from inorganic materials like GaAs, but recently there has been significant progress in the development of organic LED devices (OLEDs). Another class of EL devices – is the one in which light is generated by impact excitation of a light emitting center (called the activator) by high energy electrons in materials like ZnS:Mn. The electrons gain their high energy from an electric field, and thus, this type of EL is often called high field electroluminescence. This article will focus on thin film electroluminescent (TFEL) devices and its structural and technological design [2, 3].

2. Comparative analysis of the modern information displays

One of the most important elements of the specialized system's workplace are information display tools that ensure the functioning of the human-machine interface. The design of the indicators determines their electrical and lighting characteristics and as a result their safety and ergonomics. General requirements for security and ergonomics for information display facilities have to be harmonized with the international standards ISO 9241-3: 2003 [4], ISO 9241-8: 2007 [5] and GOST R 50948-2001 [6]. For accurate reading of information and providing comfortable conditions for its perception, work with displays should be carried out with optimal combinations of values of brightness, contrast of the image, the external illumination of the screen, the angular size of the sign, the viewing angle of the screen and electrical characteristics [7].

The choice of the optimal means of information displaying for the specialized system's workplace in terms of manufacturability, ergonomics and safety implies the consideration of the main modern information displays. The comparative analysis of the main display devices were made and presented in the table 1 [1 – 3, 7].

Table 1. Formatting sections, subsections and subsubsections.

Parameter	Information display devices		
	LCD	OLED	TFEL
Brightness	250 cd/m^2	1000 cd/m^2	400 cd/m^2
Temperature range	-30…+50	-40…+70	-50…+85
Viewing angle	160°	170°	>160°
Response time	15 ms	0,1 ms	<1 ms
Contrast	1300:1	1000000:1	20000:1
Mean time between failures	50000 h	35000 h	100000 h
Radiation resistance	Low	Low	High
Shock and vibration resistance	Low	High	High
Reducing brightness as using	Low	High	Low
Uniformity	Often the image is brighter along the edges	The image is uniform	The image is uniform
Geometric / linear distortions	No	No	No

Comparison of parameters of flat displays allows to draw a conclusion that TFEL displays surpass LCDs in lighting characteristics, but they yield OLED displays, which, however, is compensated by their high design-technological and operational parameters (mean time between failures, operating temperature range, radiation resistance, etc.). Thus, it is possible to use TFEL indicator devices not only in general-purpose equipment, but also in the military, medical, space industries, where specific and sometimes stringent requirements are imposed to equipment. Due to the listed advantages, thin film electroluminescent indicator devices are widely used in the means of displaying information.

The advantages of TFEL indicators are high brightness, contrast, resolution, radiation resistance, wide viewing angle, etc.

3. Formulation of the modern indicator devices modelling problem

Comparison of the presented displays for a number of features will allow us to analyze them and draw conclusions about the applicability of indicators in various technical systems and the prospects for the development of indicator technology. This paper proposes a comparison of the parameters of exclusively flat panel displays. Typical structures are shown on figures 1-3.

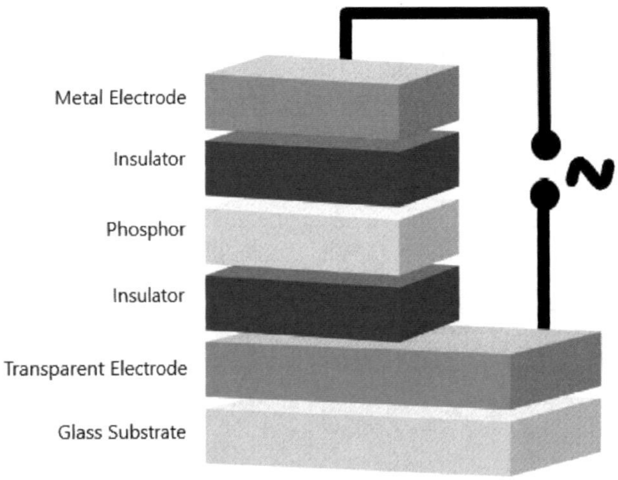

Figure 1. Typical TFEL device structure.

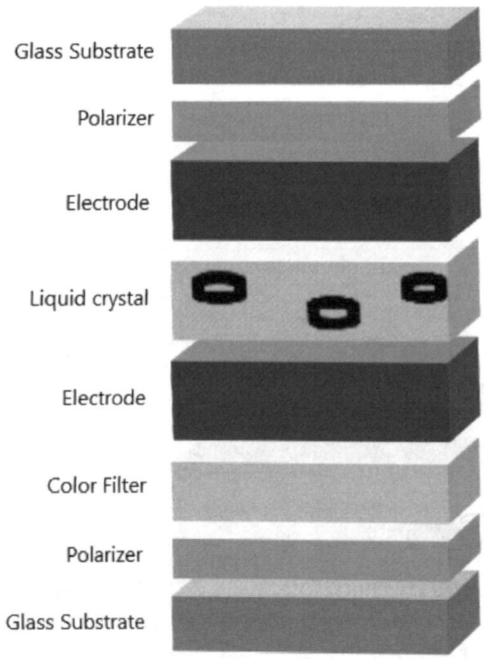

Figure 2. Typical LCD device structure.

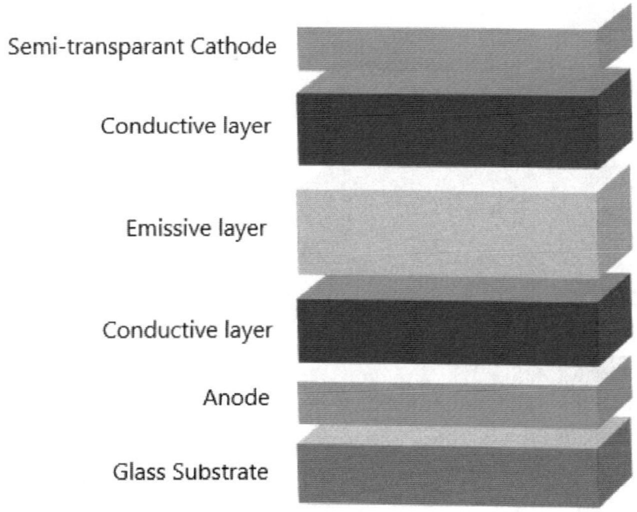

Figure 3. Typical OLED device structure.

All of the above types of displays are strikingly different from each other in the indicating technology and in the mathematical apparatus of electrical and lighting parameters and characteristics engineering calculations. However, a comparative analysis of flat panel display technologies allows us to speak about some similarity of displays. Flat panel displays are a collection of layers obtained using various technological tools. Each of these layers is a material (metal, dielectric, semiconductor) with different electrical and magnetic properties. Such representation will make it possible to draw up an equivalent circuit diagram of the device.

This representation of the display as a physical system allows us to create a mathematical model of the system. In such a model, electric voltages and currents act as phase variables. System components can be bipolar elements. These two-terminal devices include both passive elements (resistance, capacitance, inductance) and active (diodes, stabilizers). The nominal values of the circuit elements can be calculated with corresponding values for real structures. The component equations in such a model will be known relations for calculating the basic parameters of an element (calculation of resistance, capacitance, inductance, etc.), and Kirchhoff's laws for voltages and currents will act as topological equations. The description of the indicator as a set of series-connected elements allows, when investigating the electrical properties of a device, to go from mathematical models of the micro-level (distributed models) to mathematical models of the macro-level (lumped models) by discretizing the research space.

Drawing up a mathematical model of the system based on equivalent circuits will unify the calculation of design parameters and electrical and lighting parameters and characteristics of the indicator components, which in turn will allow the developed technique to be applied in computer-aided design systems for indicator devices.

Indicator devices structures equivalent circuits are presented on figures 4-6.

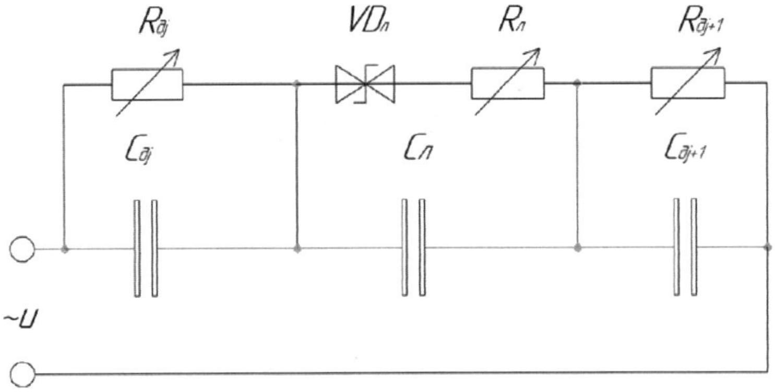

Figure 4. TFEL device structure equivalent circuit.

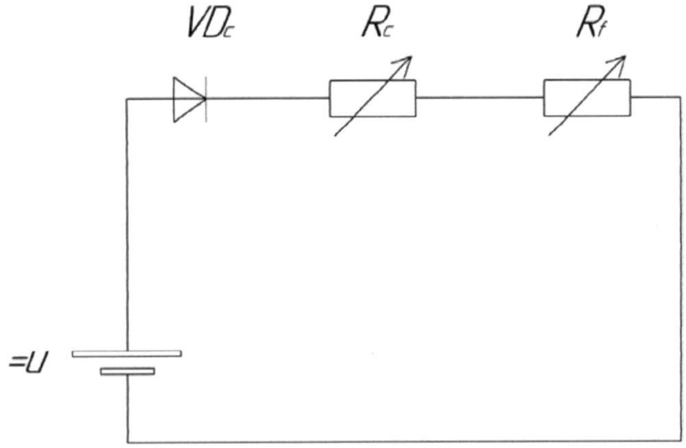

Figure 5. LCD device structure equivalent circuit.

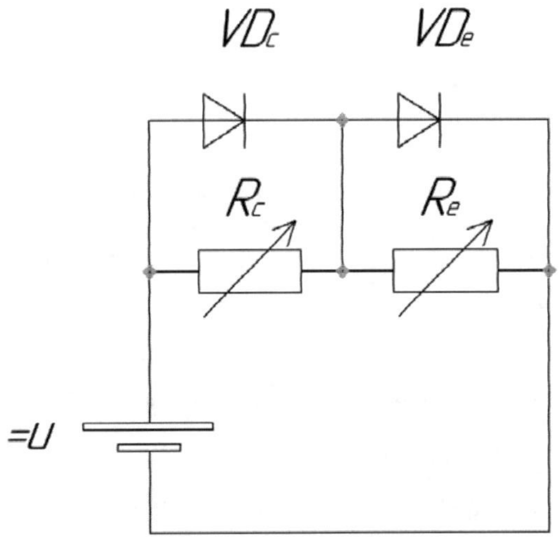

Figure 6. OLED device structure equivalent circuit.

4. Conclusion

Comparison of the presented displays for a number of features will allow us to analyze them and draw conclusions about the applicability of indicators in various technical systems for there prospects.

References

[1] C King 2003 *Electroluminescent displays* **1** 33-44
[2] A Abileah, K Harkonen, A Pakkala and G Smid 2008 *Transparent Electroluminescent (EL) Displays* **6** 10
[3] D Evsevichev, O Maksimova and M Samokhvalov 2011 *The 11th International Meeting on Information Display (IMID 2011)* **2** 768-769
[4] ISO 9241-3: 2003 Ergonomic requirements when performing office work using video display terminals – M.: Standardinform, 2003
[5] ISO 9241-8: 2007 Ergonomic requirements for office work with visual display terminals (VDTs). Part 8. Requirements for displayed colors – M.: Standardinform, 2004
[6] GOST R 50948-2001 Display means for individual use. General ergonomic requirements and safety requirements – M.: Standardinform, 2001
[7] D Evsevichev, O Maksimova and M Samokhvalov 2014 *The 8th IEEE International Conference on Application of Information and Communication Technologies (AICT2014)* **1** 164-167

Computational experiment of the thin-film electroluminescent display devices applicability in aviation

D A Evsevichev[1], O V Maksimova[2], M K Samokhvalov[3]

[1] Department of Aeronautical Engineering, Ulyanovsk Civil Aviation Institute, Ulyanovsk, Russia
[2] Research Department, Ulyanovsk Civil Aviation Institute, Ulyanovsk, Russia
[3] Department of Electronic Instrumentation Design and Technology, Ulyanovsk State Technical University, Ulyanovsk, Russia

Corresponding author's e-mail address: first32007@yandex.ru

Abstract. The development of methods and means of testing the applicability of thin-film electroluminescent indicator devices as displays in aircraft is carried out. Thin-film electroluminescent displays are used in equipment and systems that require high image quality and reliability, as well as a long service life of the devices. The result of the performed work is the ExpAT program, which allows to carry out a computational experiment to test of the applicability of the TFEL indicator devices in aeronautical engineering. As a result of the computational experiment, variants of structures of thin-film emitting structures that meet the operating conditions of indicators in avionics are shown.

1. Introduction

Among modern information display devices, thin-film electroluminescent (TFEL) indicators occupy their own special niche. They are successfully used in medical equipment, control systems for industrial facilities, control and measuring equipment, in transport, in communication systems, and aviation equipment [1, 2]. According to many experts, TFEL displays have a number of advantages over various modern display devices created using other technologies. The main manufacturer of TFEL devices in the world is the company Beneq and, in particular, the division of this company Lumineq, which is responsible for the production of products using TFEL technology and industrial production of products using layer-by-layer atomic deposition technology [3].

The advantages of this kind of displays over other modern indicators is lighting (brightness, light output) and operational (vibration resistance, shock resistance) characteristics, which is relevant for aviation technology [4]. The disadvantage of the device is the need to use high frequency alternating voltage.

In this regard, the question arises of the applicability of indicator devices, which will be powered from a 200 V network with a frequency of 400 Hz, as displays in the TFEL aircraft. The presented work is devoted to the analysis of applicability based on the developed mathematical apparatus.

2. Analysis of the basic ergonomic requirements for display devices

An important element of the cockpit is the information display that ensure the functioning of the man-machine mode. General safety and ergonomics requirements for information display facilities, harmonized with the international standards ISO 9241-3: 1996, ISO 9241-8: 1997 and described in GOST R 50948-2001 "Information display facilities for individual use. General ergonomic and safety requirements". Those parameters are the visual parameters of the display (characteristics of display and perception of information), emission parameters of the display (characteristics of electrostatic, alternating electric and magnetic fields generated by the display). Those parameters must be in the optimal range of values for displays. For accurate reading of information and ensuring comfortable conditions for its perception, work with displays should be carried out with such combinations of values of brightness and contrast of the image, external illumination of the screen, angular size of the sign and viewing angle of the screen, which are included in the optimal or maximum permissible (for short-term operation) ranges.

According to the above standards, it is possible to identify the main ergonomic and safety requirements for the means of displaying information from the pilot's workplace (cockpit):

1) the viewing angle should be between 130 ° and 180 °;

2) the brightness of the display is determined by the brightness of the sign and must be at least 35 cd/m^2 for displays on a cathode ray tube and at least 20 cd/m^2 for flat screens;

3) uneven brightness of the screen working area and elements of the sign should be no more than 20%;

4) the brightness contrast inside the sign and between the signs must be at least 3:1;

5) the intensity of the electrical component of the alternating electromagnetic field of the display should be no more than: 25 V/m - in the frequency range from 5 Hz to 2 kHz and 2.5 V/m - in the frequency range from 2 to 400 kHz;

6) the magnetic flux density should be no more than: 250 nT - in the frequency range from 5 Hz to 2 kHz and 25 nT - in the frequency range from 2 to 400 kHz.

Most of the parameters presented are provided by various display manufacturing technologies. However, the brightness parameters and characteristics are of particular interest, since they are provided not only by the design and technological features of the manufacture of a particular display, but also by the features of their operation.

The choice of the optimal information display means for the aviation specialist from the point of view of manufacturability, ergonomics and safety implies consideration [1, 2]. These include CRT displays, liquid crystal displays, OLEDs, and TFEL displays. A comparative analysis of the above displays in terms of operational parameters was carried out in [4]. According to it, it can be concluded that, despite the comparative cheapness of CRT displays, their use in modern cockpits is not desirable, since from the point of view of operation, ergonomics and safety, they have a number of disadvantages that increase operator fatigue and, thereby, reduce flight safety. Such as non-uniformity of the image, poor focusing performance, the presence of errors in color registration, the presence of significant indicators of electromagnetic radiation (strength of the electric component of the alternating electromagnetic field, magnetic flux density), low contrast and viewing angle.

Comparing flat-panel displays, we can conclude that TFEL displays are superior in terms of lighting characteristics to liquid-crystal ones, but they are inferior to displays based on organic LEDs, which, however, is compensated by their high design and technological parameters (mean time of failure, operating temperature range, radiation resistance, vibration and shock resistance). This implies the possibility of using TFEL devices not only in general-purpose equipment, but also in the military, medical, space industry, as well as in important from the point of view of flight safety area - at the pilot's workplace, where sometimes stringent requirements are imposed on the equipment. In addition, a feature of these systems is a decrease in weight and size parameters by reducing the number of its constituent components.

Due to the high image quality, TFEL displays are widely used in information display facilities and are ideal for use in difficult conditions when devices based on other technologies do not give the required results [2]. Electroluminescent displays are most commonly used in equipment and systems that require high image quality and reliability, as well as long device life.

3. Software for conducting a computational experiment of the thin-film electroluminescent display devices applicability in aviation

The result of the research work is the developed algorithm for performing a computational experiment [5]. Based on this algorithm the ExpAT program was developed to perform a computational experiment to verify the applicability of TFEL indicators in aviation technology and analyze the obtained results. The ExpAT program interface is shown in Figure 1.

The program is compiled in the Borland Delphi 7 development environment using the imperative, structured, object-oriented programming language Delphi. The software product allows us to carry out an automated computational experiment with the specified parameters, select and record design solutions that meet the conditions of the experiment and the developed mathematical model, as well as analyze the results. The program allows us to graphically display the found design solutions. In this case, if the total thickness of the TFEL indicator is always marked on the abscissa axis, then on the ordinate axis it is possible to determine various display parameters - average brightness, threshold voltage, maximum allowable voltage, average power, or simultaneously all parameters referred to their maximum value.

The developed system is an application software launched from the Windows operating system. The program works directly with the user, establishing the necessary connections for the calculation.

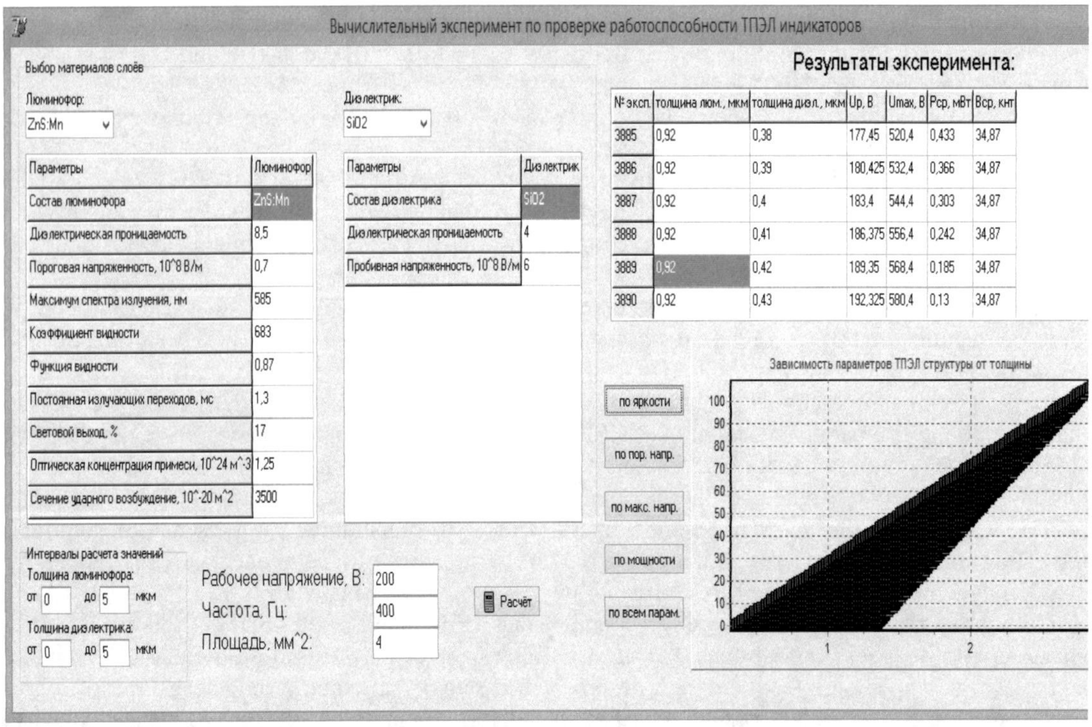

Figure 1. ExpAT program interface after performing calculations and displaying results.

The program window is divided into three areas:

• The area for setting the experimental parameters (on the left in Figure 1), including the choice of layer materials, operating conditions, intervals for calculating phosphor layers.

• The area for displaying the calculation results (top right in Figure 1), which is a table with the following columns: experiment number, phosphor thickness, dielectric thickness, threshold voltage, maximum allowable voltage, average power, average brightness.

• Area for graphical display of the found design solutions (bottom right in Figure 1). In this graph, on the abscissa axis, the total thickness of the TFEL indicator is marked (the thickness of the phosphor and two dielectric layers are added). The ordinate axis is determined depending on the buttons pressed. By pressing the button "by brightness" on the ordinate axis the average brightness is displayed, by pressing the button "by threshold voltage" - threshold voltage, by pressing the button "by maximum voltage " - the maximum allowable voltage, by pressing the button "by power" - the average power, by pressing the button "by all parameters." - simultaneously the relative values of all parameters, they refer to their maximum value.

The developed software product made it possible to carry out a computational experiment to verify the applicability of TFEL indicating devices in aviation technology. The result of the experiment is confirmation of the possibility of solving this problem.

As the features of the fulfillment of the conditions of applicability, it should also be noted the totality of the following conclusions obtained:

1) the brightness of the TFEL indicators glow increases with increasing of the films thickness;

2) the number of design solutions that satisfy the search conditions decreases with increasing film thickness;

3) the thickness of the TFEL indicator layers cannot exceed 3 microns for the selected operating parameters;

4) the best solutions are achieved with a film thickness of about 1.5 microns.

Based on the above, it can be concluded that the use of TFEL technology for the production of displays in aviation technology is possible, but requires very precise systems for sputtering layers with thicknesses of no more than 2-3 microns, or you should use means of voltage conversion to higher values in amplitude and frequency.

4. Conclusion

As a result of a computational experiment to test the applicability of TFEL indicator devices in aviation technology, variants of the designs of thin-film emitting structures that satisfy the operating conditions of indicators in avionics were shown.

The presented technique and the software product developed on its basis can be used in laboratories and design bureaus engaged in the design and research of electroluminescent radiation sources, as well as in the industrial sector involved in the production of aviation equipment.

It is also important to note that the developed software product is a fairly universal tool for conducting a computational experiment to test the applicability of TFEL indicators in other specialized equipment. For this, it is necessary to enter in the program window the corresponding limiting specific conditions for the operation of the device and analyze the calculated results

References

[1] C King 2003 *Electroluminescent displays* **1** 33-44

[2] A Abileah, K Harkonen, A Pakkala and G Smid 2008 *Transparent Electroluminescent (EL) Displays* **6** 10

[3] D Evsevichev, O Maksimova and M Samokhvalov 2011 *The 11th International Meeting on Information Display (IMID 2011)* **2** 768-769

[4] D Evsevichev, O Maksimova and M Samokhvalov 2014 *The 8th IEEE International Conference on Application of Information and Communication Technologies (AICT2014)* **1** 164-167

[5] D Evsevichev and M Samokhvalov 2019 *Automation of Control Processes* **2** (56) 113-120.

Journal of Physics: Conference Series **2056** (2021) 012055 doi:10.1088/1742-6596/2056/1/012055

Methods and facilities of the automated control for thin film indicators

O V Maksimova[1], P V Nikolaev[2], S V Moiseenko[3], V V Belyaev[3,4]

[1] Research Department, Ulyanovsk Civil Aviation Institute, Ulyanovsk, Russia
E-mail: first32007@yandex.ru.
[2] Joint Stock Company Ulyanovsk Mechanical Plant, Ulyanovsk.
E-mail: nikollaew@mail.ru.
[3] RUDN University (Peoples' Friendship University of Russia).
Email: cormorant.xiii@gmail.com3
[4] Moscow Region State University, 24, Vera Voloshina str., Mytishchi, Moscow Region,141014, Russia
Email: vic_belyaev@mail.ru

Abstract. The algorithms, methods for measuring the structure parameters of thin-film electroluminescent indicators are studied. The algorithm for determining threshold voltage and plotting current – brightness characteristics was created. Realization of this algorithm accelerates process of monitoring of parameters by production known and development of new exemplars of indicator technique on the basis of thin-film electroluminescent indicators. The main aspects of a comprehensive approach to problem solving of automation of process of measurement of values of parameters of structures of thin-film electroluminescent indicators were designated, features of functioning of thin-film electroluminescent indicators as a basis of formation of structure of the device of the automated testing are considered, automation of processing of results of an experiment at the level of the software is described. Key parameters of indicators are determined, problems of automation of measuring processes are formulated, and the algorithm of determination of the threshold voltage and creation of voltage-brightness characteristic of the thin-film electroluminescent indicators in the automated mode is developed. The ideas explained in this article allow formulating the requirement specification on development of the device of the automated measurement of parameters of thin-film electroluminescent elements and also its constituents and the software.

1. Introduction

Thin-film electroluminescent indicator (TFELI) devices currently can be one of the most popular elements for the creation of microelectronic means for displaying information, especially in the development of special equipment. Their advantages include: completely solid construction, reliability, high brightness and speed, wide viewing angle, resistance to ionizing radiation, as well as temperature stability.

The disadvantages of these TFEL devices include the need to use an alternating voltage with a sufficiently large amplitude. But there are modern highly efficient inverters that allow converting a constant voltage (from 2 V) into an alternating voltage with the required parameters [1].

To-date, there have been developed automation tools for the design of TFELIs, methods for automating the design of technological processes for the production of these devices [2]. It is also necessary to take into account current trends in increasing the speed and quality of control procedures for displaying information and their components [3]. Therefore, diagnostic methods for thin-film indicators should be automated.

2. Description of the TFEL structure

To develop an algorithm for determining the threshold voltage and constructing the voltage-brightness characteristics of indicator devices in an automated mode, it is necessary to consider the properties of TFEL capacitors as elements that are part of the electrical circuit, which includes the emitter and the control system.

TFEL structures of the MDLDM type (metal-dielectric-luminophore-dielectric-metal) usually consist of five layers successively deposited on a glass substrate using thin-film technology.

This structure is shown in Figure 1.

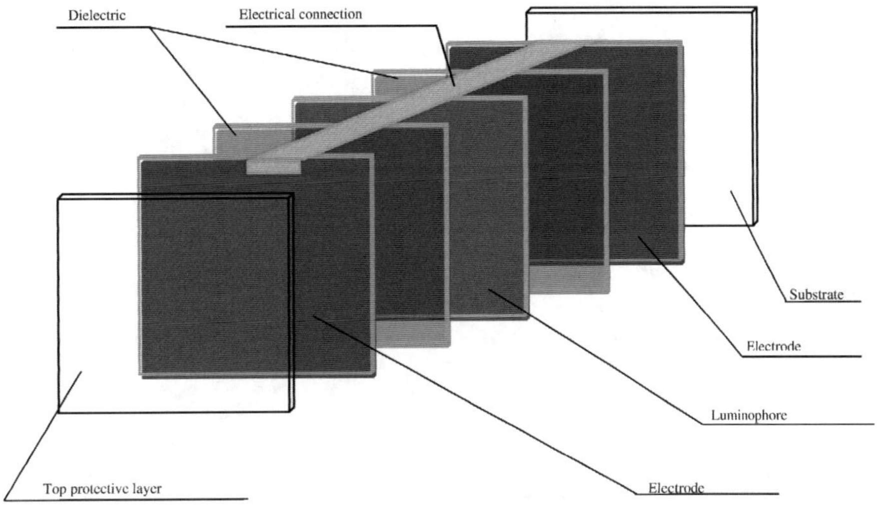

Fig. 1. Structure of the TFELI

Zinc sulfide doped with manganese or rare-earth fluorides is usually used as the luminescent layer. Strontium and calcium sulfides doped with rare-earth metal fluorides have proved to be promising phosphors.

Plates of alkali-free heat-resistant glass 2 - 3 mm thick with a deposited layer of a transparent electrode, for the creation of which heavily doped indium or tin oxides, which are degenerate wide-gap semiconductors, are used as substrates for the fabrication of structures. To obtain opaque electrodes, aluminum, indium, tin, and refractory metals are used.

To create dielectric films in electroluminescent structures, oxides of silicon, aluminum, yttrium and rare earth metals, silicon nitride, etc., their compositions, ferroelectric materials are used [1].

The main role of dielectric layers is to limit the charge passing through the phosphor in operating conditions. The need to use such layers is due to the properties of the charge transfer process in a luminescent film. Electroluminescence in thin layers of zinc sulfide and other materials is associated with electrical breakdown of the semiconductor.

Two dielectric layers separating the phosphor film from the electrodes make it possible to represent this device as an electroluminescent capacitor, which forces the use of an alternating voltage to provide electroluminescence [4].

3. Description of volt-brightness characteristics of TFEL

Since a person receives most of the information with the help of vision, and TFELI are intended primarily for visualizing digital data, therefore, the main characteristic of these indicators is the voltage-brightness.

Figure 2 shows a typical I/V characteristic of a TFEL structure based on zinc sulfide doped with manganese.

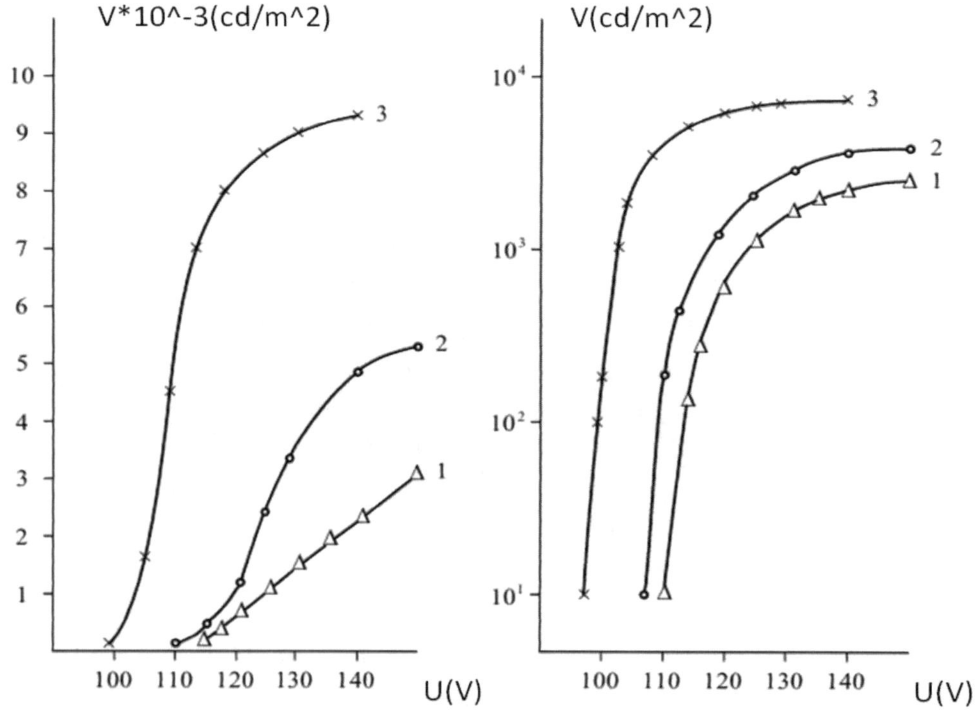

Fig. 2. Typical I/V characteristic of the TFEL capacitor in linear and semi-logarithmic coordinates (f = 1 kHz (1); 5 kHz (2) and 20 kHz (3))

In Figure 2, one can see that this dependence is linear at low voltages and frequencies, with an increase in these parameters, the dependence graph is sublinear. It is customary to depict the graphs of the volt-brightness characteristics in a semi-logarithmic scale, that is, on the abscissa axis, the voltage is indicated in a linear scale, and on the ordinate axis, the brightness is shown in logarithmic coordinates.

The use of the logarithmic scale for such a parameter as the brightness of radiation is associated with the subjectivity of a person's perception of its magnitude; an increase in the brightness of a light source by an order of magnitude is perceived by a person as a twofold increase.

As can be seen from the figure, the voltage-brightness characteristic has a threshold character, i.e., a sharp increase in brightness occurs when a certain voltage is applied to the TFEL.

The magnitude of the threshold voltage is primarily determined by the properties of the phosphor material: the threshold electric field strength, the magnitude and ratio of the capacities of the luminescent and dielectric layers. The value of the threshold electric field strength in phosphor films isolated from the electrodes is determined by the mechanism of generation of free charge carriers $10^5 \sim 10^6$ V / cm [1].

4. Features of a sinusoidal signal

The main type of used alternating signal is sinusoidal. Let's consider its parameters.

The effective value of an alternating current or voltage is called the value that is taken in calculating the released heat or power. For a direct current, the effective value is equal to the amplitude, average and instantaneous, for an alternating signal, these are all different values.

The instantaneous value of the alternating voltage is the absolute level of the voltage at a particular moment in time. The peak value is the largest value of a sinusoid (or another curve for a non-sinusoidal waveform) over a period. The amplitude value is taken modulo. The average is the sum of all instantaneous values for the period, divided by the number of samples. Mathematically, it is the integral of the instantaneous values over the period. Obviously, for a sinusoidal form, the average value over the period will be zero. If the waveform is distorted, the average value over the period may turn out to be non-zero - then it is called the constant component of the alternating voltage. The effective or effective voltage value is a value numerically equal to:

$$U = \sqrt{\frac{1}{T}\int_0^T u^2 dt}, \quad (1)$$

where U is the effective value, T is the time of one period, u is the instantaneous value of the voltage. Active and amplitude (Um) voltages are related as follows:

$$U = \frac{U_m}{\sqrt{2}}. \quad (2)$$

Often, instead of the effective one, another value of the alternating voltage is measured - the average rectified (the constant component of the unsmoothed ripple voltage at the output of the full-wave rectifier). It is equal to $2 / \pi \approx 0.637$ of the sinusoid amplitude which is 1.11 times less than its effective value. The overwhelming majority of universal low-frequency measuring instruments (multimeters) show not the effective, but the average rectified value of the alternating voltage, multiplied by this correction factor [5].

5. Algorithm for constructing the volt-brightness characteristic of TFELI in an automated mode

For the analysis of electrical characteristics, it is more correct to determine the threshold voltage by measuring the charge or power, but from the user's point of view, it is preferable to determine the threshold voltage by measuring the brightness, that is, to actually build the voltage-brightness characteristic.

As a result of the analysis of modern measurement systems, the necessary requirements for these systems and their algorithms were formulated:

- the possibility of conducting an experiment at the physical level using full-scale TFELI samples;
- high reliability, clarity and information content of the data obtained;
- the ability to store and transfer measurement information to a PC;
- high speed of data processing and reporting;
- ease of use;
- the possibility of upgrading the system, expanding its functionality.

Based on the foregoing, the main parameters determined by the algorithm are the brightness of the TFELI and the effective value of the sinusoidal voltage, measured at the same time. The implemented algorithm must determine the threshold voltage of the TFEL by detecting the moment when the brightness increases with the photosensor and measuring the value of the sinusoidal voltage applied to the TFEL at this moment, then the device must measure the brightness and voltage values, and also be able to transmit data to the PC.

Figure 3 shows a block diagram of the algorithm. First, all counters are reset to zero, etc. Then the system waits for a signal from the photosensor to appear, that is, the moment the TFEL starts to glow. The beginning of emission of radiation is recorded by the system by reading data from the photosensor, as well as transmitting information about the voltage applied at that moment to the TFEL. Thus, the threshold voltage of the indicator is determined. Next, there is a sequential determination of the brightness and voltage values, the transfer of these data to the PC in order to build the voltage-brightness characteristics of the TFEL and display it on the computer monitor in a user-friendly form (table, graph, histogram, etc.).

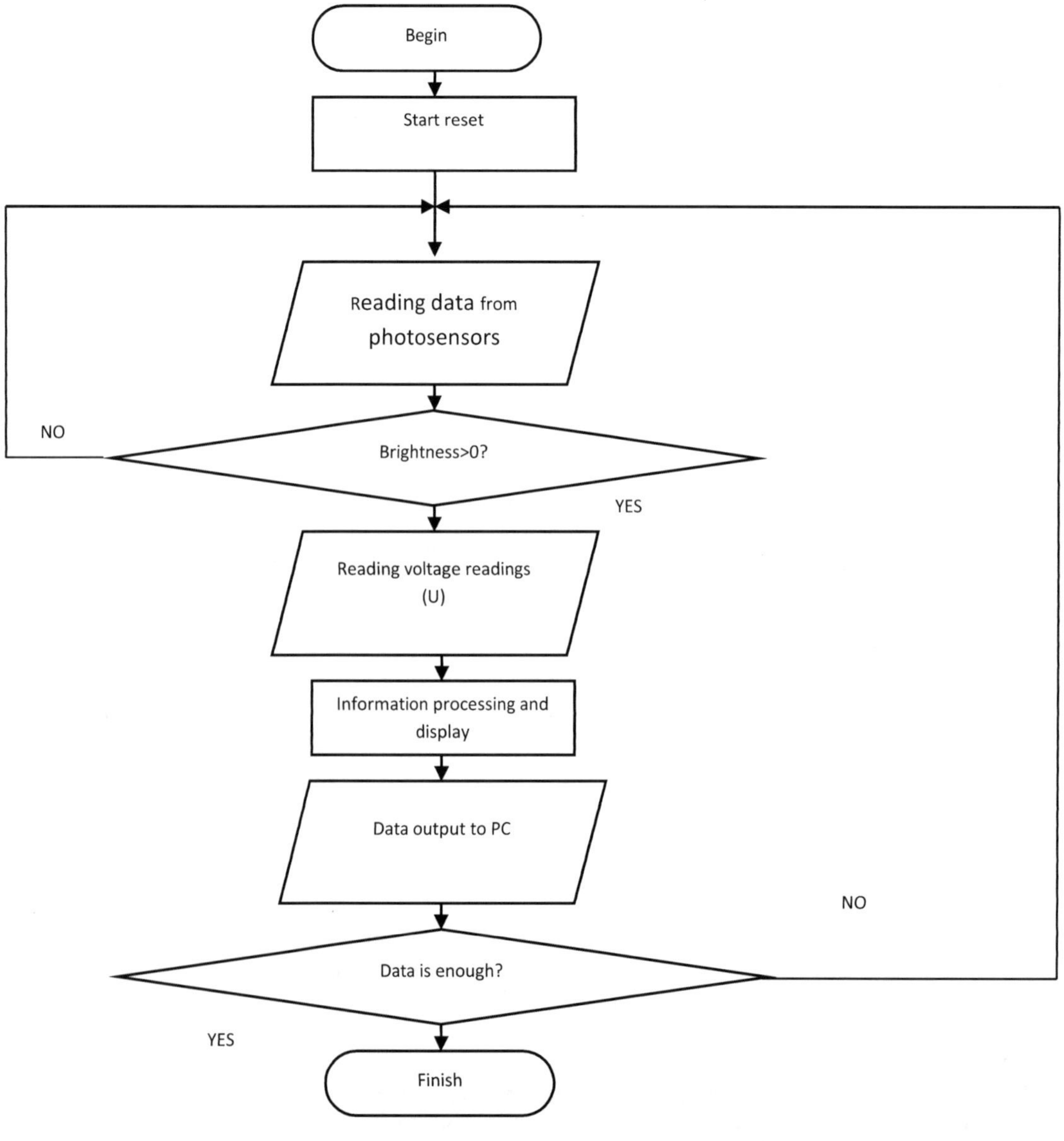

Fig. 3. Block diagram of the algorithm

6. Description of the device implementing this program

To implement the above actions, as well as to process the measurement results and the possibility of interaction of the developed measuring device with a personal computer, it was decided to use a microcontroller (MC).

Accordingly, the digital part should include:

1. Analog-to-digital converter (ADC) for working with brightness and voltage sensors;
2. Device for interaction of MC with USB;
3. Display or indicator for displaying the measurement results and a decoder for the latter
4. MK with a sufficient number of inputs and outputs;

5. Connector for downloading programs to the MC;
Arduino tools and their analogs can be used to meet the above requirements.
A photodiode, photoresistor, etc., as well as circuits with the ability to suppress light interference can be used as a photosensor.
For voltage monitoring, a rectifier with a voltage division function can be used.
The indication can be carried out using three seven-segment indicators and the corresponding register microcircuits. It makes sense to exchange register and microcontroller data using the SPI interface.
The documentation for seven-segment indicators contains a table with information about which logic level must be applied to the input in order for this or that digit to light up. For indicators with a common cathode, the table looks like this:

Indication table

The column on the far right with the name HEX is a byte representation of a digit for transferring it using a register.

To work with several indicators, you must use either a serial connection of devices to the bus, or parallel.

When connected in parallel (Fig. 4), several slaves share the SCLK, MOSI and MISO wires, with each slave having its own CS (SS) line. The master determines the device with which the exchange is carried out by generating a low signal at his CS (SS) input.

	A	B	C	D	E	F	G	DP	HEX
0	1	1	1	1	1	1	0	0	3F
1	0	1	1	0	0	0	0	0	06
2	1	1	0	1	1	0	1	0	5B
3	1	1	1	1	0	0	1	0	4F
4	0	1	1	0	0	1	1	0	66
5	1	0	1	1	0	1	1	0	6D
6	1	0	1	1	1	1	1	0	7D
7	1	1	1	0	0	0	0	0	07
8	1	1	1	1	1	1	1	0	7F
9	1	1	1	1	0	1	1	0	6F

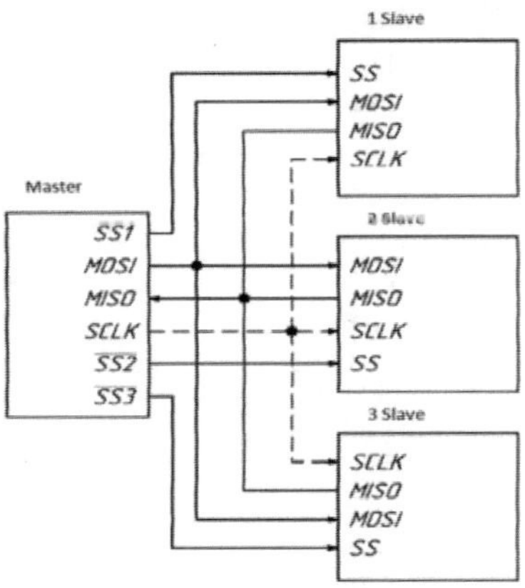

Fig. 4. Parallel connection

To connect n devices, n CS (SS) lines are required, that is, for the SPI environment to function with n slaves, n + 3 microcontroller pins must be allocated for this.
When connecting devices in series (Fig. 5), they use the common SCLK and CS wires, and the output of one is connected to the input of the other. The MOSI of the master is connected to the first device and the MISO to the last.

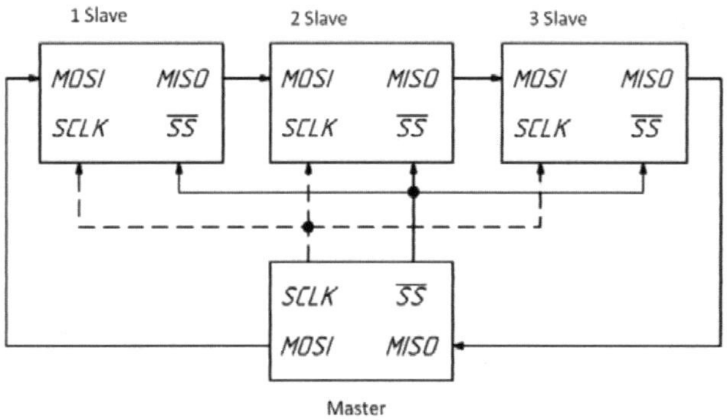

Fig. 5. Serial connection

This connection allows you to build one 16-bit shift register from two 8-bit shift registers.
The circuitry of the device on indicators with a common cathode determines the following order of information transmission via SPI: the first is the byte that is responsible for which digit will be displayed (anode byte), and the second is the byte that is responsible for the position of the digit (cathode byte) [6].
The anode byte during the SPI transfer session is transferred through the DD2 register to DD3. The cathode byte is placed in DD3 and remains there. As a result, the numbers are displayed on the indicators:

```
// connect the SPI library
#include <SPI.h>
// connect CS wire to pin 8 of Arduino
enum {reg = 8};
void setup ()
{
// Set the reference voltage to 2.56 V
  analogReference (INTERNAL);
// initialize SPI
  SPI.begin ();
// define pin 8 of Arduino as output
  pinMode (reg, OUTPUT);
// Open the last. port and set the baud rate
  Serial.begin (9600);
}
```

To display three digits, a dynamic indication is used, that is, the image of each digit turns on for a short time, then turns off, and the procedure is repeated for subsequent digits.
It has been experimentally established that when the digit is switched every 8 milliseconds, the flicker is distinguishable, but at a value of 7 milliseconds, the eye no longer sees the sampling. The cathode

byte is calculated so that 00011000 ignites the left bit, 00010100 - the middle one, 00001100 - the right one. That is, the one in bits 2, 3 and 4 is responsible for the ignition of the corresponding bit:

```
// Variables to display hundreds, tens, units
int hundreds = 0, tens = 0, ones = 0;
// display time for each digit
int delayTime = 1;
// codes of numbers on a seven-segment indicator (0-9 and empty)
static uint8_t digit [11] = {0x3F, 0x06,0x5B, 0x4F, 0x66,0x6D, 0x7D, 0x07,0x7F, 0x6F, 0x00};
// position codes of the lit indicator (left, center, right)
static uint8_t pos [3] = {0x18,0x14,0x0C};
```

It is necessary to split the transmitted number into components. Let there be a number int f, which in the decimal system has no more than three digits, that is, a number less than 1000. Algorithm:
1. Divide f by 100, staying within int we get the number of hundreds.
2. Subtract from f the number of hundreds, multiplied by 100, divide the remainder by 10 - determine the number of tens.
3. Subtract from f the number of hundreds multiplied by 100 and the number of tens multiplied by 10 - we get the number of units.
4. If the number of hundreds is zero, then nothing is displayed in the left digit.
5. If the number of hundreds and tens is zero, then nothing is displayed in the left and middle digits [6]:

```
void OutLED (int f)
{int x = 600;
 Serial.println (f);
 do {
// hundreds are allocated
  hundreds = f / 100;
// dozens are allocated
  tens = (f-hundreds * 100) / 10;
// units are allocated
  ones = f-hundreds * 100-tens * 10;
// if there are no hundreds, nothing is displayed in the 3rd digit
  if (hundreds == 0) hundreds = 10;
// if there are no hundreds and tens, nothing is displayed in the 2nd digit
  if (hundreds == 10 && tens == 0) tens = 10;
```

Then a pair of bytes is transmitted via SPI for each bit of our number in turn:

```
// start SPI transmission
 digitalWrite (reg, LOW);
// transfer the code of the digit corresponding to the place of hundreds
 SPI.transfer (digit [hundreds]);
// the left indicator is selected
 SPI.transfer (pos [0]);
// end of transfer
 digitalWrite (reg, HIGH);
```

```
// pause equal to delayTime
  delay (delayTime);
// transfer the code of the digit corresponding to the tens place
  digitalWrite (reg, LOW);
  SPI.transfer (digit [tens]);
// select the central indicator
  SPI.transfer (pos [1]);
  digitalWrite (reg, HIGH);
  delay (delayTime);
// transfer the code of the digit corresponding to the ones place
  digitalWrite (reg, LOW);
  SPI.transfer (digit [ones]);
// select the right indicator
  SPI.transfer (pos [2]);
  digitalWrite (reg, HIGH);
  delay (delayTime);
  x--;
} while (x> 0);
}
```

Reading the signal from the photosensor and determining the value of the supply voltage is performed as follows:

```
void loop ()
{// Reading information from the light meter output
  float L = analogRead (A1);
   // Is the light / dark threshold exceeded?
if (L> 10)
   // Overcome
{float f = analogRead (A0);
    f = f * 142; // take into account the division ratio and get the amplitude value
    f = f * 0.707; // conversion of amplitude to real value
    f = f * 0.0025; // convert from ADC values to volts
   OutLED (f); }
  // Not overcome
else
{float f;
OutLED (f = 0);}}
```

Programming is carried out entirely through its own software shell (IDE), available free of charge from the Arduino website. This shell contains a text editor, a project manager, a preprocessor, an interface for transferring data from a PC to an MC and vice versa (thanks to this, voltage and brightness data are transferred to a computer via USB during the entire time the program is running), a compiler and tools for downloading the program to MK. The shell is written in Java based on the Processing project, runs on Windows, Mac OS X and Linux.

7. Experiment Results

As samples of indicators for the experiment, we used the standard structures of five layers on a glass base obtained in earlier studies, which are shown in Figure 1. The substrates were plates with a thickness of 1.5 to 3 mm made of alkali-free glass. The phosphor layer, 0.5 to 1.5 microns thick, was formed from zinc sulfide doped with manganese. In the role of a transparent electrode, tin and indium oxides, 0.2 microns thick, were used. A solid solution of yttrium and zirconium oxides, 0.3 microns thick, was taken as dielectrics. An aluminum layer up to 2 microns thick served as an opaque electrode. TFELI is installed in an opaque box.

A photocell based on cadmium sulfide was used as a receiver of optical radiation, the signal from which is fed through an operational amplifier and filters to the first input of the MC ADC. The voltage from the power generator of the TFEL element is fed through a rectifier with a divider to the second input of the MC ADC.

In the course of the experiment, the brightness of the TFEL was measured depending on the sinusoidal signal applied to it. The volt-brightness characteristics presented in Figure 6 were plotted at a supply voltage frequency of 50 Hz and 1 kHz, the series resistance of the circuit (R) was 30 kOhm and 68 kOhm.

It can be noted that with an increase in the series resistance of the circuit, the brightness of the indicators decreases. This is due to the fact that with a high resistance, the duration of internal transient processes is longer than with a small value of R [1, 7]. Also, an increase in the brightness of the TFEL was determined, associated with an increase in the frequency of the supply voltage with a constant resistance R. Thus, the data obtained as a result of the experiment are consistent with the results of theoretical analysis [8, 9] and previous measurements [1], which confirms the consistency of the above-described method of automated control TFELI.

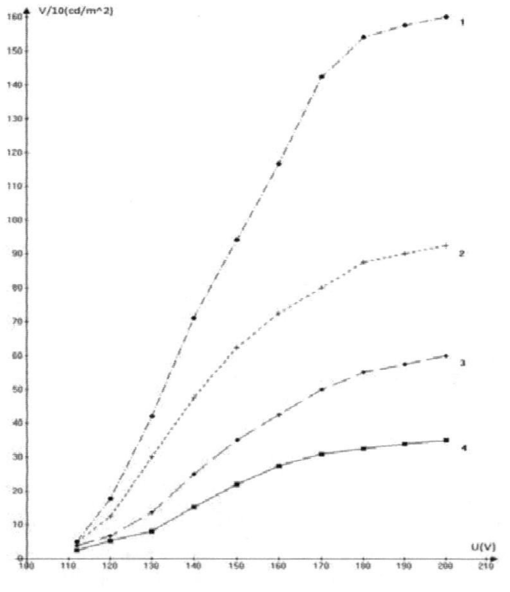

Fig. 6. Voltage-brightness characteristic of TFELI
f = 1 kHz, R = 30 kΩ (1); f = 1 kHz, R = 68 kΩ (2); f = 50 Hz, R = 30 kΩ (3);
f = 50 Hz, R = 68 kΩ (4)

8. Conclusion

The research carried out in the field of automation of the processes of measuring the brightness functional characteristics of TFELs made it possible to distinguish the following results:developed an algorithm for measuring the lighting parameters of TFELI;

- a software and hardware solution for the implementation of the above-described algorithm has been developed.

Thus, based on the requirements for devices and systems for determining the values of the indicator parameters, the ability to carry out measurement operations in an automated mode with the ability to control this system and process the results (including the formation of databases and their exchange) using a personal computer is realized.

Journal of Physics: Conference Series **2056** (2021) 012055 doi:10.1088/1742-6596/2056/1/012055

References

[1]. A Gusev, M Samokhvalov 2006. *Electrical characteristics of thin-film electroluminescent indicators* 125
[2]. P Nikolaev, O Maksimova, D Evsevichev 2014 *Actual problems of electronic instrumentation APEP-2014: mater. int. scientific and technical conf.* **2** 63–65
[3]. V Belyaev, S Moiseenko 2017 *Information resources of Russia* **5** 35–40
[4]. N Gurin, O Sabitov 2017 *Physics and technology of film electroluminescent emitters of alternating current* 434
[5]. A Dolgiy 2006 *Radio* **6** 23–27
[6]. A blog about Arduino, AVR, ARM, Linux and more. URL: http://greenoakst.blogspot.ru/2012/01/arduino-7-spi.html (date accessed: 09/18/2019).
[7]. Yu Sukharev, A Andriyanov, V Mironov 1994 *Journal of technical physics* **8** 48–54
[8]. N Gurin 1992 *Laser technology and optoelectronics* **3–4** 74–77
[9]. N Gurin, O Sabitov 2006 *Technical Physics Journal* **8** 50–62

Journal of Physics: Conference Series **2056** (2021) 012056 doi:10.1088/1742-6596/2056/1/012056

Automation of procedure for diagnostics of thin film electroluminescent indicators' structure

O V Maksimova[1], P V Nikolaev[2], S V Moiseenko[3], V V Belyaev[4]

[1] Research Department, Ulyanovsk Civil Aviation Institute, Ulyanovsk, Russia
E-mail: first32007@yandex.ru.
[2] Joint Stock Company Ulyanovsk Mechanical Plant, Ulyanovsk.
E-mail: nikollaew@mail.ru.
[3] RUDN University (Peoples' Friendship University of Russia).
Email: cormorant.xiii@gmail.com3
[4] Moscow Region State University, 24, Vera Voloshina str., Mytishchi, Moscow Region,141014, Russia
Email: vic_belyaev@mail.ru

Abstract. Problems of measurement automation for creating a testing system are considered. This system accelerates the process of parameters controlling in the production and development of new samples of indicator (display) technology. The main parameters of the indicators are defined, the tasks of automation of measuring processes are indicated, the hardware and software structure of the system is developed. The main features of an integrated approach to solve the problems of automating the process of testing the structures of thin-film electroluminescent indicators were proposed, the parameters of thin-film electroluminescent indicators were considered as the basis for forming the composition of an automated testing complex, and the automation of processing the experimental results at the software level was described. The ideas outlined in this article make it possible to formulate a technical task for the development of a complex for automated measurement of the parameters of thin-film electroluminescent elements, as well as its components and software.

1. Introduction

In connection with the widespread use of computerized information processing facilities, indicator devices capable of satisfying both general and specialized requirements are of big interest.

In general, display facilities can be divided into two types - displays for general use and specialized displays. Users of civilian devices focus on the quality of the transmitted image and price; for specialized displays, reliability and weight and size parameters are especially important. Developers of indicator technology are faced with a difficult, complex task of finding a balance between effective designs of indicators with the best functional characteristics, price indicators and reliability of information display facilities. Among the devices of widespread use, liquid crystal displays predominate, which are gradually being replaced by matrices built using OLED technology. For structures for specialized purposes, strict requirements are imposed on the parameters and performance characteristics. In this regard, thin-film electroluminescent indicators (TFELI) are of particular interest.

The TFELI are characterized by high lighting and electrical performance, radiation resistance, and also have the necessary reliability. Today, this type of indicators has a low resolution, which defines it

as a type of device, a purely specialized application. A serious drawback of these indicators is also the need to supply an alternating voltage with a large amplitude to generate luminescence, but this is compensated by the fact that currently available AC voltage inverters that convert a low-voltage DC signal into an alternating one with the required amplitude [1, 2].
The process of measuring the parameters of materials and structures of thin-film devices is complex, complex, and time-consuming. The task of creating systems for automated diagnostics and control of the functioning of thin-film electroluminescent structures in indicator devices is urgent, since it significantly accelerates the search for the necessary materials, the creation of optimal design options with the specified parameters.

2. Features of thin-film electroluminescent indicators as the basis for the formation of the composition of the automated testing system

To create a software and hardware complex for testing TFELI, it is necessary to determine the main parameters of indicators, create an algorithm for their measurement, and ensure the implementation of the principles of automation.

A thin-film electroluminescent structure is formed of five layers deposited on a transparent dielectric substrate: transparent lower conductive, lower dielectric, luminescent, upper dielectric, upper conductive [1].

Dielectrics are used as a limiter of the charge formed in the phosphor in the operating modes of the indicator device. Insulating layers are necessary due to the fact that the phenomenon of electroluminescence in thin layers of zinc sulfide and other materials entails electrical breakdown of semiconductors.

The phosphor is separated from the electrodes by two or more layers of dielectrics, which allows the TFEL to be presented as an electroluminescent capacitor.

The Volt - Brightness characteristic of these devices has a threshold character, when the supply voltage reaches a certain (threshold) value, a sharp jump in the brightness of the TFEL occurs.

The value of the threshold voltage is mainly associated with the characteristics of the phosphor materials: the threshold electric field strength, the magnitude and ratio of the capacities of the luminescent and dielectric layers [2, 3].

Let's consider the main parameters. The range of operating voltages of the TFEL lies within the limits from the threshold (Uth) to the breakdown voltage (Umax). Uth is the instantaneous value of the voltage at which the luminescence process begins in the TFEL structure. When the supply voltage reaches the value Umax, degradation and destruction of the indicator occurs. During the measurement, the voltage on the indicator, at which the value of the brightness parameter (B) goes over zero, will be the threshold. Thus, the process of monitoring the operating parameters of the TFELI begins with the measurement of Uth, and the determination of Umax is possible only at the final stage of testing. When the voltage is less than the threshold, only the charge current (Icharge) flows through the indicator, the dissipation of electrical power (P) practically does not occur [3]. In operating modes, the power is defined as the product of the consumed current (I) by the value of the supply voltage. The luminous efficiency of TFELI is defined as the ratio of the luminous flux to the power consumption:

$$\mathrm{n} = \frac{\pi \mathrm{BavS}}{\mathrm{P}},$$

where Bav is the average value of brightness, S is the area of the indicator [4].

3. Algorithm for collecting, processing and accumulating information

Any measurement process is about obtaining data about a physical sample and transferring it to the end user. The definite value of a particular parameter at a particular moment of time does not always represent valuable information in the study of research objects.

The study of the parameters of the TFEL structure is necessary to study the effect of the material of the element layers and the technologies of their deposition, to compare the experimental and calculated output parameters in order to determine the error and make recommendations for the design of indicator structures, as well as to form databases of dielectric, conductive and luminescent materials.

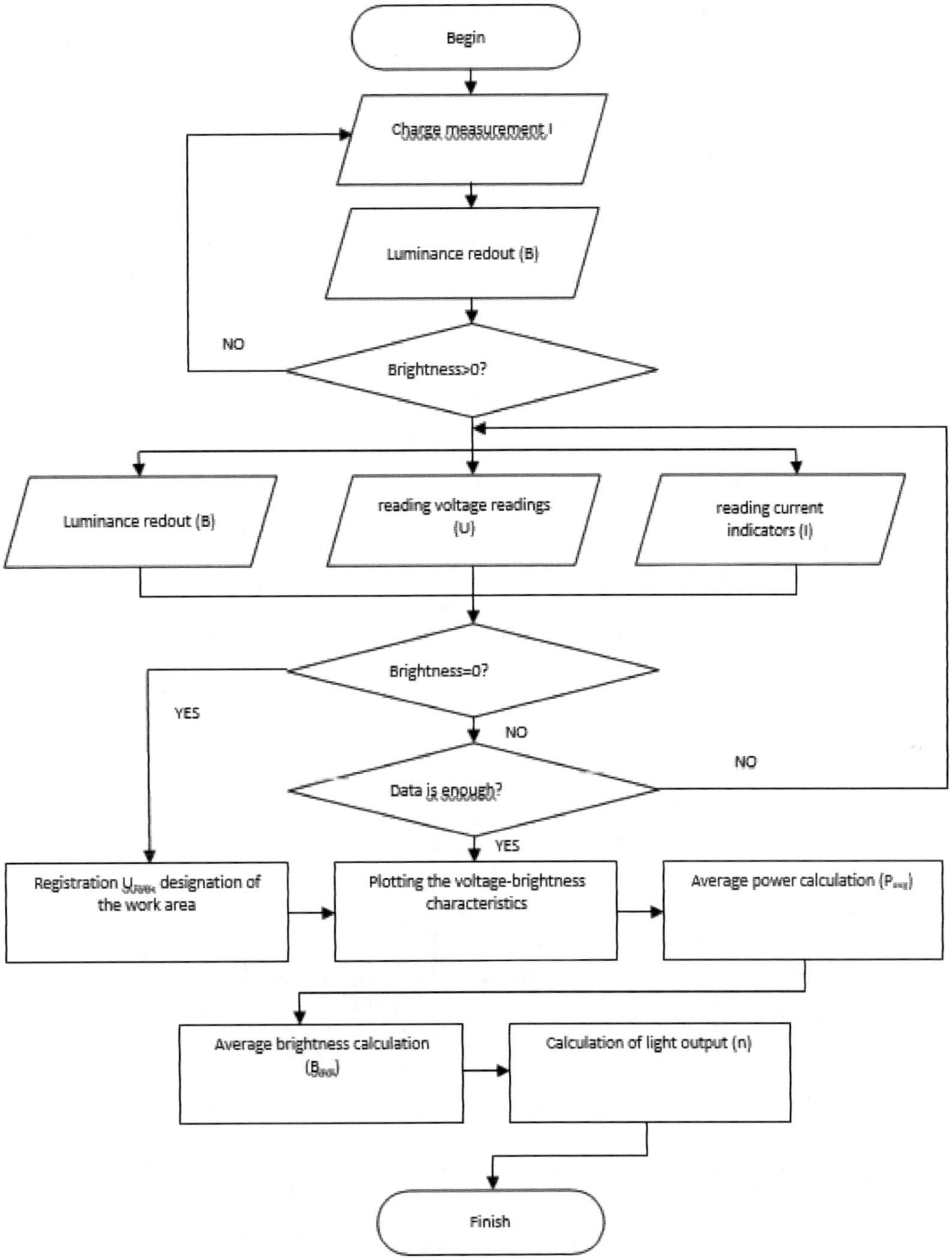

Fig 1. Algorithm of measurements

Diagnostics is to obtain information about the nature of the functioning of the TFEL-structure and the dependence of the values of the output parameters (operating voltage range, current, average dissipated power, average luminosity, light output) on the specified external conditions, such as:

characteristics of the supply voltage (frequency of -excitation, amplitude), geometric data of layers [3]. Thus, two types of parameters can be distinguished: optical and electrical, in addition to them there are external factors (temperature, etc.). The task of collecting data is complicated by the fact that it is necessary to determine a number of parameter values, as well as to establish the dependence of these parameters from each other.

Based on the well-known calculation and measurement procedures [5, 6], the sequence for measuring the package of parameters of the TFEL structure was determined:

1. Selecting the required electrical parameters
2. Selecting the required optical parameters
3. Drawing up a list of external influences
4. Formation of the hardware part of the measuring complex
5. Determination of the sequence of measurements
6. Reading electrical and optical indicators
7. Carrying out calculations and drawing up dependencies
8. Formation of a database of experimental results

Figure 1 shows an algorithm for measuring these parameters.Measurement system structure

Figure 2 shows the composition of the system for automated measurement of parameters. The core of the complex is a personal computer. This ensures the fulfillment of the requirements considered in clause 2. For the convenience of control and the most complete implementation of the principles of automation, the generator controlled by the USB interface has been selected by the power source for the thin-film electroluminescent element. The thin-film electroluminescent element is mounted on a special tripod, which is placed in a darkened box.

The current (I, Icharge) of the TFELI is measured using a resistor connected in series, the voltage drop across which should not exceed 0.5% of the supply voltage value. The instantaneous values of brightness (B) are determined by means of a photosensor (photomultiplier tube). Further, the indicators I, Icharge and B are recorded through a digital storage oscilloscope on a personal computer using the software described below.

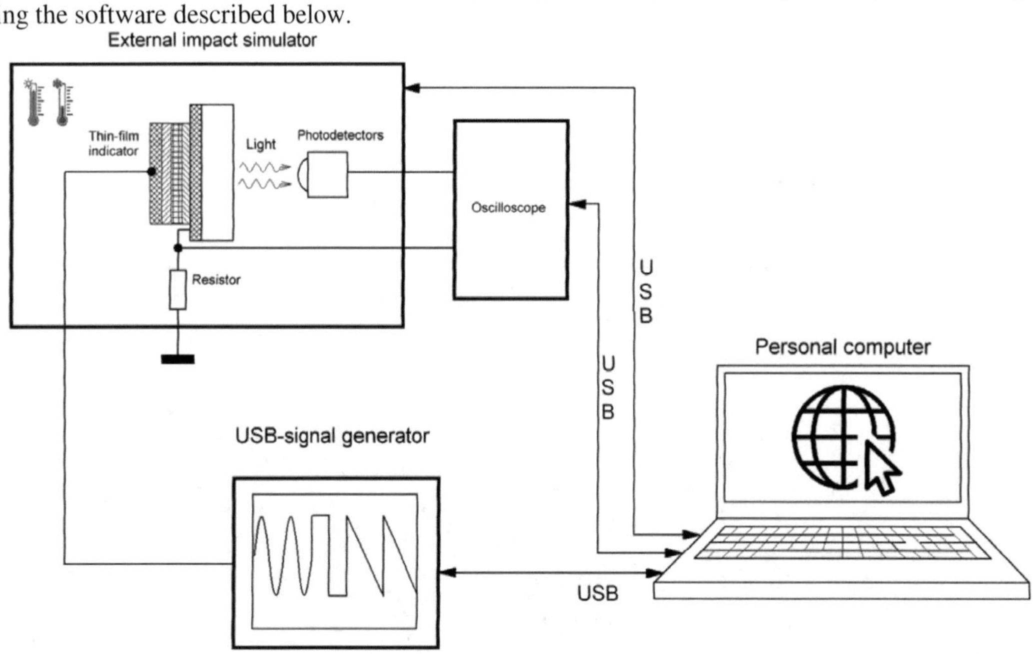

Fig. 2 Structure of the automated measuring system

Thus, this structure allows you to realize the following advantages:

1. Easy to use and user training
2. High accuracy of measurement results and the possibility of their presentation in any format (table, graph, etc.)
3. High speed of the experiment
4. Possibility of carrying out an experiment using full-scale TFEL samples
5. Flexibility, versatility of the measuring system with the ability to quickly adapt to new requirements, conditions and tasks
6. Reliability
7. Possibility of interaction with the database of results and parameters of experiments
8. Cooperation via the Internet
9. Interfacing with computer-aided design systems TFELI [5, 6]
10. Possibility of further expanding the functionality of the system to work not only with thin-film, but also with other types of indicators
11. Ability to simulate the influence of external factors on the tested TFELI structure (high and low temperatures, vibrations)
12. Modularity of the system design, the possibility of expanding / reducing the functionality of the system by adding / removing modules and the interchangeability of the modules themselves
13. Possibility of comparative analysis of the measured characteristics of two or more measured elements
14. The ability to predict the characteristics of the measured elements, based on the analysis of elements of similar structures
15. Possibility of selection of optimal materials of construction of the measured elements for the required parameters.

The structure of the software of the complex for automated measurement of the parameters of TPELI is shown in Figure 3.

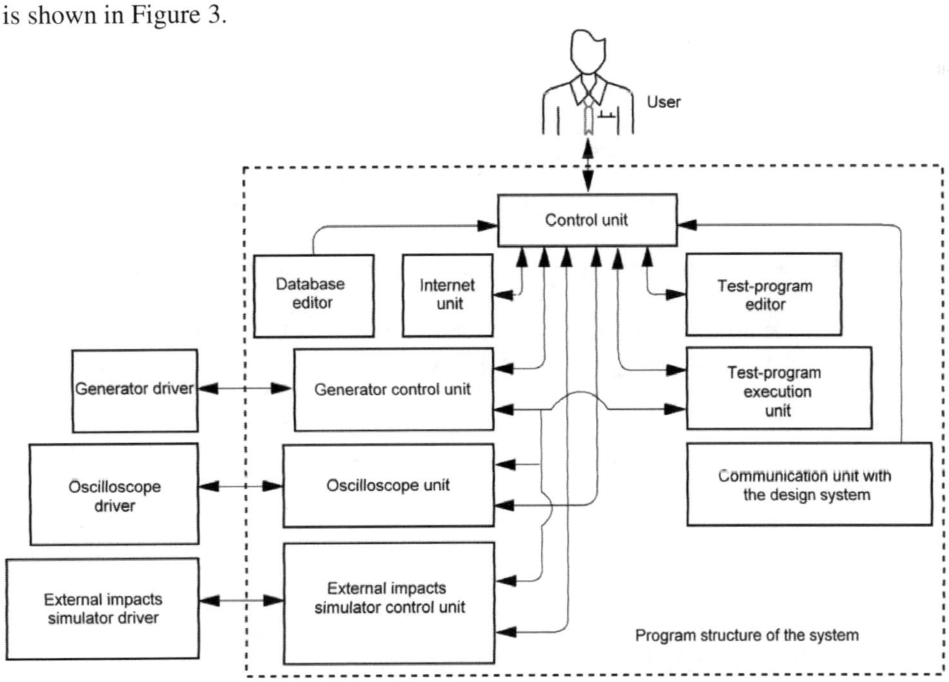

Fig. 3 Software structure

The principle of operation of the software shell of the measuring system is based on the use of test programs. This is a set of commands for controlling and interacting with the generator, oscilloscope and other blocks. The test program is executed using the module of the same name. Each test is designed to solve a specific problem, for example, determining the average brightness of the indicator, finding the threshold voltage, etc. The test program reads data from the oscilloscope and sends it to the control module. To meet the requirement of quick adaptation to new tasks being solved, the testing program can be edited, compiled and sent back to the execution unit.

The user controls the measuring complex by means of the control module. With its help, the operator can create and edit test programs, control the generator and other external units, form a database, interact with the TFELI computer-aided design systems via the Internet.

The test program editor allows you to edit existing tests and create new ones. Its functions also include verification of the program for errors, compilation into an executable file, which the user passes to the appropriate module.

The control units for the processing device, the generator and the device for simulating external action are the interface for the interaction of these devices with the operator, they are an intermediate link between the device driver and the user.

The database editor is responsible for creating, editing and storing the database with the conditions and results of the experiment.

The Internet-module is intended for cooperation of the complex of automated measurement of parameters of thin-film indicators with similar software using network protocols. This is necessary for organizing a large measuring network from several of the above-described complexes, as well as computer-aided design systems (via the corresponding module). Such a solution will significantly optimize the process of creating and researching structures and materials of indicators from the stage of theoretical calculations to conducting experiments with a specific physical model. This principle is especially relevant in the case of geographic separation of production from the TFELI developers.

4. Conclusion

The development of a system for automated control of TFELI parameters solves the following tasks:

1. Reducing time costs
2. Reducing labor intensity
3. Ensuring high accuracy of measurements
4. Achieving the required quality of the experiment
5. Maximum elimination of the influence of the human factor and related problems on the measuring process.
6. Creation of the most complete and accurate models of indicator devices.

The above studies made it possible to formulate a technical task for the development of a complex for automated measurement of the parameters of thin-film electroluminescent elements, as well as its components and software.

References

[1]. M Samokhvalov 2003 *Elements and devices of optoelectronics* 125

[2]. N Gurin, O Sabitov 2017 *Physics and technology of film electroluminescent emitters of alternating current* 434

[3]. A Gusev, M Samokhvalov 2006. *Electrical characteristics of thin-film electroluminescent indicators* 125

[4]. N Gurin, O Sabitov, A Shlyapin 2003 *Technical Physics Journal* **4** 100-112

[5]. D Evsevichev, M Samokhvalov 2011 *Automation in industry* **9** 55-57

[6]. D Evsevichev, O Maksimova, M Samokhvalov 2016 *Automation and Remote Control* **6** 1093-1098

Evaluation of electromagnetic pollution index (EMPI) emitted from multiple sources

M A A Frah[1], V V Belyaev[1,2]

[1]RUDN university (Peoples' Friendship University of Russia), Moscow, 117198, Russia.
[2]Moscow Region State University, Radio street, 10A, Moscow, 105005, Russia
E-mail: vic_belyaev@mail.ru

Abstract. In this paper the cooperative electromagnetic pollution index (EMPI) from many RF sources is evaluated in terms of three components which are strength of magnetic and electric field expressed in A/m and V/m, respectively, and irradiated power in W/m^2. They are generated by a portable laptop computer (LC) with a liquid crystal monitor and other three sources in front area of it. The LC radiation interference with radiation from mobile phone, Wi-Fi and microwave oven in the radio frequency range from 300 Hz up to 3GHz has been measured. The experiment involves testing 45 different points in the keyboard area and 45 points in the in front area of the PC in its operating state. After that a comparison was made between the calculated values for multiple sources frame work and the measured experimental values. Then these EMPI values have been compared with the internationally recommended SAR values.

Introduction

This article provides a method of experimental measurement for electromagnetic pollution (EMP) caused by multiple sources in the radio wave frequency range. This work is one of the first studies to compare the most interesting results and ideas related to EMP exposure of RF sources and suggest possible ways of reducing it. We equip a comprehensive experiment measuring for electromagnetic pollution existing multiple sources as unique frame work and also offer a tutorial on the metrics, international projects as well as limits on the exposure for radio wave radiation and standard values for EMR. This structure brings a full operation as in laptop computer desktop.

The EMF radiation originated a problem to the human being, whose in generally comes from sources which generate electric and magnetic field of wireless communication and others non-ionizing radiation fields. Hence, it depends on some parameters like field strength, distance and the altitude, duration of exposure that determine the health hazard for the users of the laptop computer and other electronic devices. One of the ways to estimate the level of EMFs is to test it with portable meters.

In this century, the use of laptop computer (LC) has widely grown. It is particularly true for the younger inhabitants appropriate to its portability, it is quite common habits for use the laptop computer close to our body. It is also in touch woith the part of skin, bones, blood, lymph, etc. Popular and constant use of the laptop computer (LC) in more cases might matter some negative wares to the user's health. Moreover, the raised worry about focusing in weakness health of the exposed person is explicit [1]. It is depending on the impact of the non-ionizing electromagnetic radiation distinguished by the down frequency in the range to 300 Hz. Accordingly the secure and sufficient identically, World Health Organization (WHO) has reported that the electromagnetic hypersensitivity principality includes dermatological, neurasthenic and botanical symptoms [1]. Some principality of electromagnetic hypersensitivity is shortness of breathing, arrhythmia, fatigue and nausea, memory and concentration problems, headache, blurred eyesight, limb pains, muscle stiffness, burning sensations, etc. [2-6]. Parallel studies and independent research have similar result and typical conclusions were gained, which pliable them to assessment the secure level of the magnetic field density as amounting to 0.33 $A \cdot m^{-1}$. Ahlbom et al. [7] and Neutra et al. [8] suggested that current density above density than the secure level of 0.33 $A \cdot m^{-1}$ up the hazard of developing leukemia by a factor of two. Besides, their survey results [7,8] could not prove any increased hazard of developing another cancer diseases inclusive glioblastoma. It is to be confirmed that this secure level is an estimated assesse and as such has not been definite by any accurate calculations.

Methodology

We measured the uniform ultra-low frequency magnetic field generated by laptop computer and other three radio frequency wave sources in its surrounding area (figure 1). The laptop consists of several electric and electronic components that are interconnected. At normal operational situation of the LC, these components are powered by current *I*. The magnetic field is created by the current flow through these components. Based on Bio-Savard's law, the magnetic field B created by the direct current I is specified in equation (1).

$$\boldsymbol{B} = \frac{\mu_0 . I}{4.\pi} \int \int_{wire} \frac{dl.\hat{r}}{r^2} \tag{1}$$

$$j\left(\frac{A}{m}\right) = \sum_{i=1}^{i=n} j_i \tag{2}$$

The field determines the level of the radio frequency electromagnetic pollution created by the portable computers as well as interference of electromagnetic pollutions produced from nearly located other three sources which are mobile phone, Wi-Fi and microwave oven. The measurements were made in the keyboard area and in front of portable computer. In order to estimate this effect, we used a portable TM-195 RF three axis field strength meter TENMARS TEST & MEASURMENT INSTRUMENTS (figure 2) and an EMF meter – low frequency Electromagnetic field intensity EMF 829 (figure 3). The magnetic field strength in the RF range from 400Hz to 3GHz has been measured. The experiment includes testing of 18 different positions for 15 cases illustrated in the figure 1 for the portable computer (LC) and its interaction with other three sources in normal operating condition. The measurement of the electromagnetic pollution expressed in A/m was performed in the laptop neighborhood. The measured data were compared with the electromagnetic safe limit values and other parameters suggested by some organizations like ICNIRP (International commotion Non Ionizing radiation protection) [9]. The paper assesses the level of the electromagnetic pollution generated by the portable computer (LC) Lenovo ideapad 100-15IBD Processor Intel ® core (TM) i3-5005u CPU@2.00GHz. In the experiment the portable computer was tested in natural operating state. After that the LC was connected to Wi-Fi 2300 MHZ and the second stage of the measurements was done. Then a mobile Huawei Y7 Prime 2018 with frequency 1.4 GHz + 1.1 GHz and 1800 MHz was added in the 3rd stage of the measurements. Then a microwave oven with frequency 2450 MHz was added. The measurement of the EMP is perfect in the laptop vicinity.

EMI parameters from different home appliance devices in different frequency ranges and their impact on the human's health is reviewed in [10,11].

The measured data for this productive electromagnetic frame work were processed by using a program Statistical Package for the Social Science (SPSS) in order to know the population parameters and describe the relationship between appreciate variables as well as the correlation factor, regression coefficients and the distribution of electromagnetic pollution for the magnetic field component. Then all measured values were summarized in simple way and compared in between and then were compared with the magnetic field secure limit values which are proposed by ICNIRP and IEEE.

Scenario of experiments

The experiment geometry is presented in figure 1. We have done our experiment at distances as follows: the length, width and height for the laptop are 25, 36 and 25 cm. The area in front of the LC is equal to the same values – 25, 36 and 25 cm. We have done the measurements for all the points at height of 5cm. In the measurement scenario we divided the cases into three groups. First: one source, second: two sources, third: three and four sources group. Then we compared the EMPI magnetic field strength component in (A/m) with the standard value which is recognized by ICNIRP. The electromagnetic pollution index (EMPI) measured is defined as values of the magnetic field strength (A/m) consisting of 15 different cases (points in figure 1). It represents the electromagnetic pollution index emitted from the LC and the interaction with three other sources in its neighborhood for 18 points (9 top measured points – marked as topmp1 to topmp9) and 9 points in frontal area (front measured point – marked as infmp1 to infmp9) of the LC neighborhood. In the measurement scenario and data analysis we borrowed Brodic's methods [12,13]. The interaction with other three sources in top and in front of the laptop area consists of 18 measurement points by measuring the magnetic field strength component for each of 15 cases as shown in table 1 (a,b), table 2 (a,b) and table 3 (a,b). These tables show the comparison for maximum and average values at top and in front points side for all cases in our case study. The values were measured by the test meter (TM195) which has a high resolution where it informs by 0.1 µA/m for the magnetic field strength.

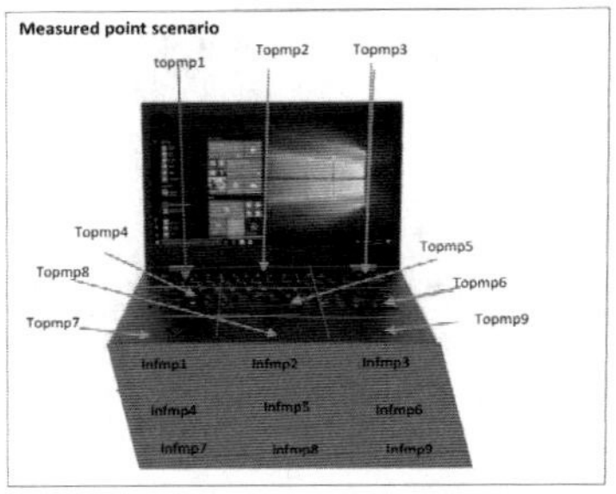

Figure 1. Location of 18 points to measure the electromagnetic pollution index emitted from the LC and its interaction with three other sources in LC neighborhood

Figure 2. TM-195 RF three axis field strength meter

Figure 3. EMF meter Detector 3axis field strength

Results and discussion

Table 1. EMPI measured values for the magnetic field strength (A/m) component in case of one source at top & front side points side

Table 1a. Average values at top & front points side magnetic field strength (A/m), one source

Top & in front points	magnetic field strength (A/m)			
	Mobile phone	Portable computer (LC)	Wi-Fi	Microwave oven (Mic)
Average (top)	0.070	0.0283	0.0103	0.049
Average (in front)	0.0566	0.0384	0.0101	0.039

Table 1b. Maximum values at top & front points side magnetic field strength (A/m), one source

Top & in front points	magnetic field strength (A/m)			
	Mobile phone	Portable computer (LC)	Wi-Fi	Microwave oven (Mic)
Maximum (top)	0.178	0.07397	0.021	0.083
Maximum (in front)	0.1432	0.06897	0.0189	0.068

Figure 4 illustrates comparison between the average level (a) and the maximum level (b) of magnetic field strength in (A/m) for cases of one source of radio frequency sources at top and in front parts points of laptop computer and interaction of these sources in normal operating.

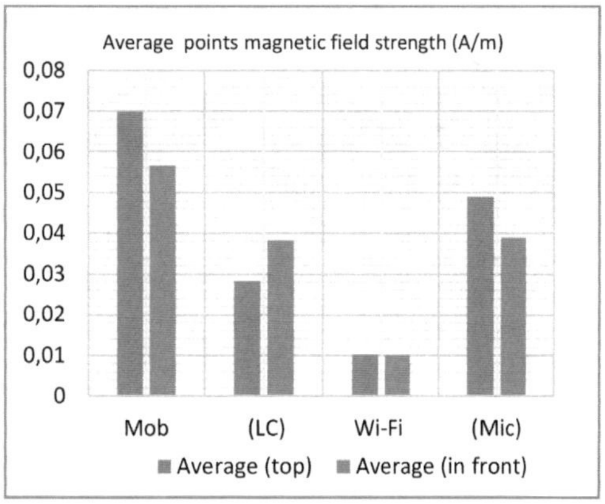

Figure 4a

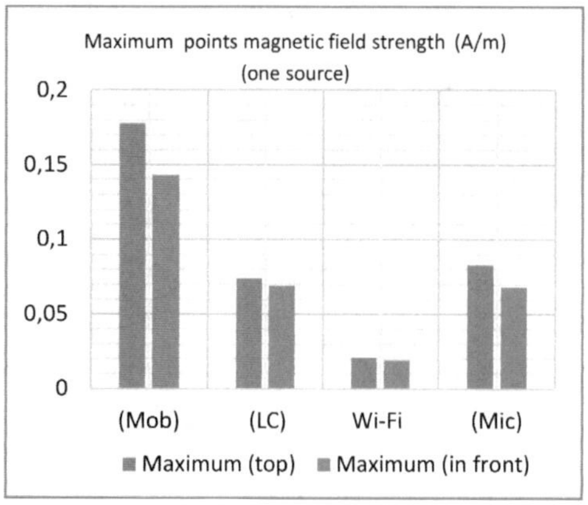

Figure 4b

Figure 4 a & b demonstrates that the maximum values of the measured magnetic field strength are from 0.021 to 0.178 A/m at the top parts point of laptop and from 0.0189 to 0.143 A/m at in front parts point of laptop computer area. The average values of magnetic field strength are from 0.0103 to 0.070 A/m at top point and from 0.0101 to 0.0566 A/m at in front parts points of laptop computer (LC) area. These magnetic field strength values are obtained in normal operating mode in case of one source of radio wave frequency.

Table 2. EMPI measured values for magnetic field strength (A/m) component in case of two sources at top & front side points

Table2a. Average values at top & front points side magnetic field strength (A/m) two sources

Top & in front points	magnetic field strength (A/m)			
	Mob &Wi-Fi	LC &Wi-Fi	Wi-Fi & Mic	Mob &LC
Average (top)	0.0922	0.0342	0.0426	0.0978
Average (in front)	0.0819	0.0333	0.0384	0.0899

Table2b. Maximum values at top & front points side magnetic field strength (A/m) two sources

Top & in front points	magnetic field strength (A/m)			
	Mob &Wi-Fi	LC &Wi-Fi	Wi-Fi & Mic	Mob &LC
Maximum(top)	0.18	0.0699	0.0684	0.179
maximum (in front)	0.148	0.0754	0.0652	0.155

Figure 5 illustrates comparison between the average level (a) and the maximum level (b) of magnetic field strength in (A/m) for cases of two sources of radio frequency sources at top and in front parts points of laptop computer (LC) and interaction of these sources in normal operating.

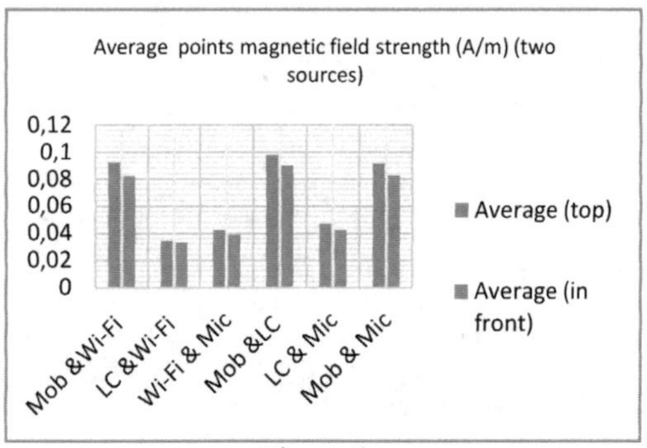

Figure 5a.

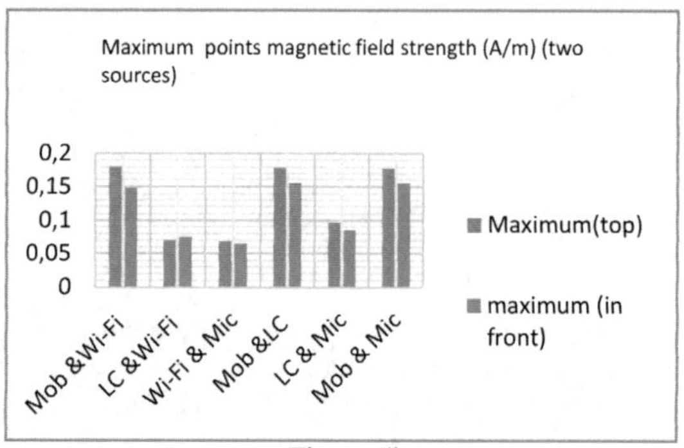

Figure 5b.

Figure 5 a & b shows that the maximum values of the measured magnetic field strength are from 0.0684 to 0.180 A/m at the top points of the laptop and from 0.0652 to 0.155 A/m at in front parts of the laptop computer area. While the average values of magnetic field strength are from 0.0342 to 0.922 at top points and from 0.0333 to 0.0899 A/m at front points of the laptop computer area. These magnetic field strength values are obtained in normal operating mode in case of two sources of radio frequency wave.

Table3. EMPI measured values for magnetic field strength (A/m) component in case of three and four sources at top & front points.

Table3a. Average values of the magnetic field strength (A/m) at top & front side points from three and four sources.

Top & front points	magnetic field strength (A/m)			
	Mob &LC &Wi-Fi	Mob& LC &Mic	LC& Wi-Fi & Mic	Mob & Wi-Fi & Mic
Average (top)	0.0838	0.04109	0.0178	0.046
Average (in front)	0.0723	0.03242	0.0123	0.0229

Table3b. Maximum values of the magnetic field strength (A/m) at top & front side points from three and four sources.

Top & front points	magnetic field strength (A/m)			
	Mob &LC &Wi-Fi	Mob& LC &Mic	LC& Wi-Fi & Mic	Mob & Wi-Fi & Mic
Maximum(top)	0.015	0.1894	0.0941	0.1865
maximum (front)	0.0132	0.1575	0.0768	0.1647

Figure 6 illustrates comparison between the average level (a) and the maximum level (b) of the magnetic field strength in (A/m) for cases of three and four sources of radio frequency sources at top and front parts points of the laptop computer and interaction of these sources in normal operating.

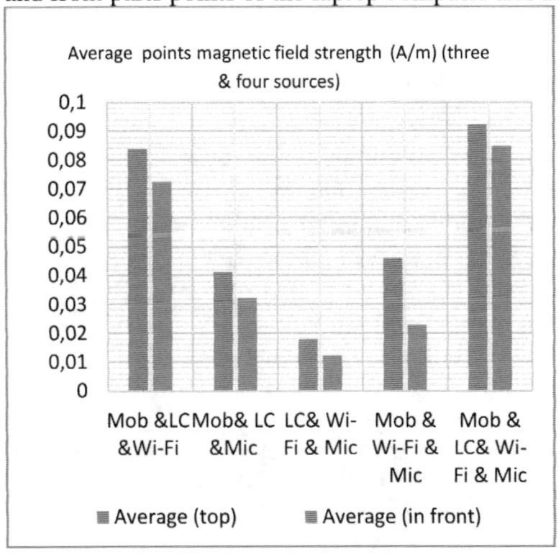

Figure 6a.

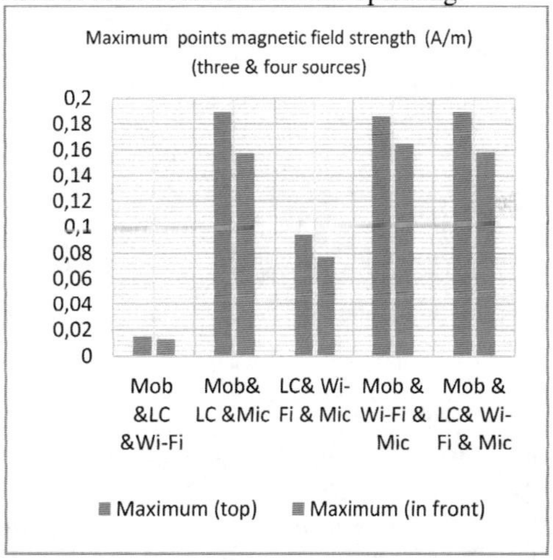

Figure 6b.

Figure 6a & b shows the maximum values of the measured magnetic field strength are from 0.0941 to 0.1895 A/m at the top parts point of the laptop and from 0.0768 to 0.158 A/m at in front parts point of the laptop computer area. While the average values of the magnetic field strength are from 0.0178 to 0.0923 at top point and from 0.0123 to 0.0848 A/m at front part points of the laptop computer area. These magnetic field strength values are obtained in normal operating in case of three & four sources of radio wave frequency. Therefore, figure 5 (a, b) and figure 6 (a, b) illustrate the interference based on the formula (2) and composed of the others three sources, which are mobile phone1800 MHz, wifi 2300 MHz, microwave oven 2450 MHz near the laptop portable computer (LC) Lenovo ideapad 100-15IBD Processor Intel ® core (TM) i3-5005u CPU@2.00GHz.

Table 4 shows a comparison of the maximum top measured points values with ICNIRP standard values limits [9] and their increased factors.

Table 4.

Sources	Standard values	H (A/m) measured value > Standard values	Increasing over standard values, %
≥2GHz	0.00375f1/2	-	-
≥3GHz	0.16	-	-
Measured value (Two sources)	0.16	0.1799 (case No 5	12.4%
		0.179 (case No 8)	12.06%
		0.178 (case No 10)	11.4%
Measured value (Three sources)	0.16	0.189 (case No 12)	18.4%
		0.186 (case No 14)	16.6%
Measured value (Four sources)	0.16	0.1897 (case No 15)	18.7%

Conclusion

The article discusses the problem associated with the electromagnetic pollution index from multiple sources of the magnetic field, which exists around any laptop computer. The electromagnetic pollution measurements of the magnetic field strength (A/m) component were implemented by two devices: the portable TM-195 RF field strength meter and low frequency Electromagnetic field intensity meter EMF 829. The results obtained showed that the critical levels of the measured EMPI sometimes significantly exceeded the proposed safe limits as shown in table 4: the cases Nos 5,8,10,12,14&15 demonstrate their increasing factor by 12.4%, 12.06%, 11.4%, 18.4%,16.6% and 18.7%, consequently. Main reason of such exceeding is interference of RF radiation with small differentiating frequencies (the beating effect). Such signals with magnitude over the safe level are periodical and permanently affecting the user in spite of the fact that the EMPI from every device can be lower than the safe level. Therefore, any user should avoid spatial and temporal vicinity of different RF sources.

However, the average values in cases of multiple sources increased also if we compare that in cases of one source which indicates the presence of obvious electromagnetic interferences from multiple sources.

The study pointed to those positions of the computer where the measured EMPI level was significant. This information can be used for the safe and secure use of the laptop and other multiple sources.

Acknowledgement

The research is partially supported by Russian Foundation for Basic Researches, grant 19-07-60002_a.

References

[1] World Health Organization (WHO) 2007 Extremely low frequency field. Environmental Health Criteria Monograph (238) 1-519

[2] O Johansson 2006 *Electromagnetic Biology and Meedicine* **25** 245-258.

[3] M Hagstrom, J Auranen, R Ekman 2013 *Patthophysiology* **20** 117-122.

[4] C Bellieni, I Pinto, A Bogi, N Zoppetti, D Andreuccetti, G Buonocore 2012 *Archives of the Environmental and Occupational Health* **67** 31-36.

[5] I Calvent, M Fernandez, J Villalba, N Ilea. “Exposure to electromagnetic fields (non-ionizing radiation) and its relationship with childhood leukemia: A systematic review”.

[6] A Ahlbom, M Feychting, M Koskenvuo, J Olsen, E Pukkala, G Schulgen, P Verkasalo 1993 *Lancet* **342** 1295-1296.

[7] A Ahlbom, N Day, M Feychting et al. 2000 **83**(5) 692-698.

[8] California EMF Program, An Evaluation of the Possible Risk From Electric Magnetic Field (EMFs) From power lines, Internal Wiring, Electrical Occupations, and Appliances, Prepared by: R. R. Neutra, V. Delpizzo, G. M. Lee, California EMF Risk Evaluation, 2002.

[9] ICNIRP 1998 *Health Phys* **74** 494-522.

[10] M Frah, V Belyaev 2019 *Journal of Physics: Conference Series* **1309** 012013.

[11] M Frah, V Belyaev, I Leonisova 2020 *Electronics: Science. Technology. Business* (6) 70-74 DOI: 10.22184/1992-4178.2020.197.6.70.74
[12] D Brodic, A Amelio 2015 *Measurement Science Review* **15**(4) 202-209
[13] D Brodic 2015 *Revista Facultad de Ingeniería, Universidad de Antioquia* (76) 39-45 DOI: 10.17533/udea.redin.n76a05.

Protection from electromagnetic pollution by using metal based shielding materials

M A A Frah[1], T Pavlushkina[2], A Babinova[2], V Belyaev[1,3], M Makeev[1,4], A Osipkov[4], E Konopleva[4], V Zhachkin[3], V Usachev[1]

[1] RUDN University (Peoples' Friendship University of Russia), 6, Miklukho-Maklay street, Moscow, 117198, Russia
[2] Institute of Glass, Moscow, 111024, 7, Dushinskaya street, Russia
[3] Moscow Region State University, Radio street, 10A, Moscow, 105005, Russia
[4] Bauman Moscow State Technical University, 5, Bldg.1, 2nd Baumanskaya street, Moscow, 105005, Russia

Abstract. At first time a comprehensive investigation of properties of different types of filters protecting from electromagnetic (EM) radiation of to-date electronic devices for personal using – notebooks or laptops, mobile phones, WiFi devices, microwave ovens etc. has been performed for all components of EM pollution including electric and magnetic field strength and radiation power. The shielding filters were made by using different technologies including nanoparticles of transition metals implanted into glass, metal-plastic foils, glasses with different kinds of ITO coating. Values of the EM pollution reduction factors for different shielding filters types have been obtained that allow make recommendations for their practical application.

1. Introduction

The goal of the paper is to investigate properties of filters protecting from electromagnetic (EM) radiation of to-date electronic devices for personal using – notebooks or laptops, mobile phones, WiFi devices, microwave ovens etc. The shielding filters can be made by using different technologies from the point of view of both RF absorbing layer and substrate.

Electromagnetic shielding is defined as absorption and/or reflection of the electromagnetic radiation by the materials which work as a block against radiation permeation crossing out of protective materials. These materials prohibit the transmission of the EM radiation by the EMR absorption and/or reflection or repressing the EM signals. Thus, the electromagnetic waves cannot influence the functioning and reflection, they are widely used to isolate area or tools from surrounding EM waves. This reflection shielding is referring to the principle of the Faraday cage, in which inside the cage, space is completely impervious to external electric fields; furthermore, the absorption shielding is associated with permeable material, sometimes magnetic materials. Accordingly, metal conductors are devoid of flexibility, weight, and high cost in the meantime; ferromagnetic or ferromagnetic materials have an essence cut off frequency, commonly, under the low GHz range; however they handicap use in the EMI shielding over broad GHz range.

In many cases development of protective materials, especially in construction, was directed for shielding from both radio waves and radioactivity [1-3]. For protection from the radio frequency (RF) radiation of electronic devices the shields should be compact and light-weight, besides, efficient for the RF absorbance. A review of materials for electromagnetic compatibility of electronic devices with a description of some shielding filters are presented in [4-7].

To-date Institute of Glass, Moscow, developed an approach to make mass production glass materials reducing the EM radiation that can be used in construction, electronics, medicine, space and other areas [8-11]. The glass absorbing the RF waves is transparent in visible spectrum range. The RF shielding effect is achieved by implantation of accelerated ions of transition elements (Ti, V, Cr, Mn, Co, Fe, Ni, Cu, W and Mo) into the surface glass' layer with thickness less than 100 nm. This process results in formation of nanoparticles in this layer that absorbes the EM radiation [8, 12 - 14]. The structure of such glasses was investigated by EPR method [15]. The greatest radiation protection effect is observed on glasses implanted with nickel (Ni^+) and cobalt (Co^+) [8].

Many other groups work with metallized plastic or foiled RF shielding filters because of their flexibility, light weight, easy processing (they can be cut). Most known are Silver and LLumar films.

The Silver film looks like a non-transparent mirror. It transmits light only in one direction. The films are fabricated by a deposition of a metallized layer onto transparent polyester. The polyester roll reels up onto a water coolable roller in a big vacuum chamber. Metal (usually Al) is evaporated onto the cool surface of the film. Such metallized film reduces the EMR by 18-22 dB at frequency 300 MHz and by 35-38 dB at 2000 MHz. The Silver films 37.5 µm thick and 62% transparent are used in architecture for both EMR protection and creation of comfortable light environment. Analogues of the Silver film with Sintered Silver Nanowires are described in [16, 17]

The LLumar film are made of flexible radio absorbing materials EXSOB and FAM [18]. The composition is not disclosed in the publications. In accordance with technical specifications the FAM3-0.75mm film 0.75 mm thick for shielding notebooks and mobile phones provides RF absorbance -0.90dB (18.72%) at 0.9 GHz, -1.42dB (27.91%) at 1.8 GHz, -1.87dB (35.03%) at 2.4 GHz. The FAM films have advantages as follows: they provide high efficient noise suppression in frequency range from 10MHz to 60GHz; they suppress resonance frequencies and repeat reflections of the signals from the screens; high surface resistance (10^6 - 10^8 Ohm); they are easily cut from big size sheets (400*400 mm) and easily assembled.

Among international companies most known commercial shielding films are made by Hollandshielding Co. and Soliani EMC [19,20]. An example of very light EM shielding material is presented in [21].

Many other plastic filters with sophisticated structure of a composite material are in progress [22-24], e.g., shields composed of carbon fibers' reinforced plastics and metal wire mesh based composites [22] or the shielding film coated with a hybrid solder-magnetic powder mixture on a PET substrate [23].

Therefore, development of new shielding material and investigation of their properties is an urgent task from the point of view of both basic and applied science. It is to note that the EM shielding properties should be studied for the full set of the electromagnetic field parameters including both electric and magnetic field strength as well as radiation power.

2. Methods and materials

In this work we investigated an effect of reduction of the exposure of electromagnetic pollution index (EMPI) which was caused by four radiofrequency sources with a few shielding materials and compare these values with values obtained without the shielding material. Therefore, we did our measuring by monitoring and observing all values which include four sources with shielding cases presented in table 1 for the EMPI components: magnetic field strength (mA/m), electric field strength (V/m) and radiation power density (W/m^2).

The paper assesses the level of the reduction of the electromagnetic radiation generated by the following devices radiating in MHz and GHz range:

- portable computer (LC) Lenovo ideapad 100-15IBD Processor Intel ® core (TM) i3-5005u CPU@2.00GHz;

- Wi-Fi 2300 MHz;
- mobile Huawei Y7 Prime 2018 with frequency 1.4 GHz + 1.1 GHz and 1800 MHz;
- microwave oven with frequency 2450 MHz.

The experiment with the all the devices was tested in their natural operating state. The EM shielding filter was positioned between the device and the meter – portable TM-195 RF three axis field strength meter TENMARS TEST & MEASURMENT INSTRUMENTS device.

Figure 1 illustrates the experimental method. The distance between the shields and the meter device is 5cm.

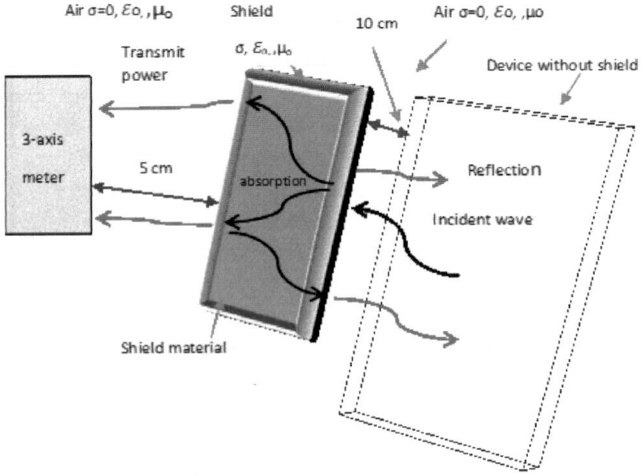

Figure 1. Scheme of the experiment.

We have investigated two groups of the shielding materials. The first group includes samples of Russian institutions issuing commercial glasses or foils. They are presented in Table 1.

Table 1. EM shielding cases

Case 1 Without filter
Case 2 (material No 1) Al foil
Case 3 (material No 2) glass 1
Case 4 (material No 3) glass 2
Case 5 (material No 4) glass 3
Case 6 (material No 5) Silver film SSiSR PS4
Case 7 (material No 6) LLumar film

Three glass samples of Institute of Glass with the NPs of the transition elements implanted into the glass were investigated. A sample SSiSR PS4 (Silver) looks like a shining foil film. The EM protection is provided by a metal layer coated onto a thin polymer film. Another flexible sample is a LLumar film.

We have also investigated the EMR reduction by a few glass samples coated with different ITO (In_2O_3) modifications. This material is used as a transparent electrode in electro-optic (displays) and optoelectronic (photo-voltaic) devices. For many purposes it is important that the material would provide both transparency and RF absorption.

The ITO sample of type A was deposited by the method of reactive magnetron sputtering from two targets of In and Sn in an oxygen medium within the UNIP-900P unit [25], equipped with a Chola-type ion source for pre-cleaning and activation of the substrate surface.

Samples of the ITO coatings of types B, C, and D were obtained by direct current magnetron sputtering (DC mode) in an Argon working medium from an $(In_2O_3)_{0.9}$-$(SnO_2)_{0.1}$ target at the VUP-11M [26], MIR-2 and Kurt J Lesker PVD 75 units, accordingly. The target for the samples of types B and C was produced by Girmet LLC, Russia; the target for the sample of type D was produced from Kurt J. Lesker Company Ltd, USA. The samples of types C and D were deposited onto glass substrates with the addition of oxygen to improve the optical characteristics of a transparent conductive layer (3% of oxygen for the type C sample and oxygen pressure 4.8×10^{-4} at chamber pressure 3×10^{-3} Torr for the D sample).

The ITO samples of types C and D were deposited onto the glass substrate, but the ITO layer of type B was deposited onto the PET substrate (polyethylene terephthalate). Preliminary purification of the substrate was carried out by washing it in baths with isopropyl alcohol and deionized water, following with the activation of polymer films which was performed on a laboratory plasma activation unit UMP-500 in Argon plasma at a pressure of 1.1 Torr, a voltage of 350 V and discharge power of 150 W.

The thickness and sheet resistance of the obtained samples are shown in Table 2.

Table 2. Parameters of the applied ITO layers

Sample number	Coating thickness, nm	Sheet resistance, Ohm/□
Type A	100	150
Type B	83	201
Type C	150	102
Type D	98	190

For this samples' group optical transmission was measured besides of the EM shielding parameters. The transmission coefficient of the ITO samples under study was determined using a Shimadzu UV-3600i Plus spectrophotometer in the wavelength range from 380 to 780 nm with a resolution of 1 nm at normal incidence of light onto the sample.

Schematic structure of different samples is illustrated in Fig.2.

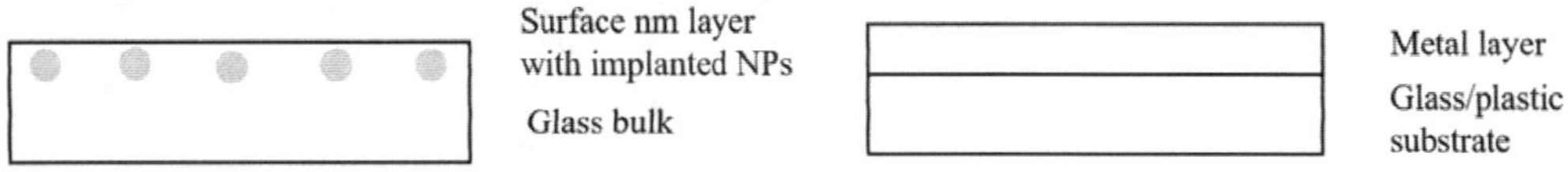

Figure 2. Schematic structure of different samples. Left – samples of Institute of Glass; right – other samples.

3. Results

3.1. EM shielding properties of commercial glasses and foils

The results are presented in Figures 3-6 and Tables 3-6.

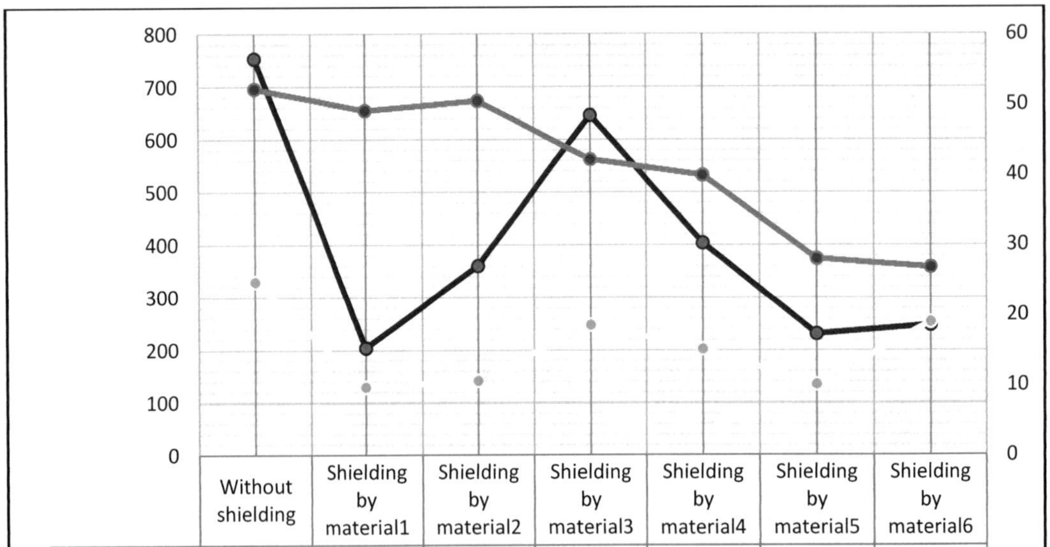

	Without shielding	Shielding by material1	Shielding by material2	Shielding by material3	Shielding by material4	Shielding by material5	Shielding by material6
mW/m^2	753,8	204	360	646	403	230	247
V/m	24,7	9,73	10,6	18,53	15,1	10,1	18,96
mA/m	52,21	49,15	50,5	42,15	39,9	28	26,8

Figure 3. Three EMPI components caused by mobile phone (mobile Huawei Y7 Prime 2018 with frequency 1.4 GHz + 1.1 GHz, 1800 MHz) with and without EM shielding

Table 3 shows the reduction factor the EMPI caused by mobile phone for the three EMR components. It can be defined as follows:

$$\text{Reduction factor} = (V_{sh} / V_{wsh}),$$

V_{sh} - measured signal value with shielding, V_{wsh} - without shielding.

Table 3. Reduction factor of EMPI caused by mobile phone

Sample No	Magnetic field strength	Electric field strength	Power density
1	0.94	0.39	0.27
2	0.97	0.43	0.48
3	0.81	0.75	0.86
4	0.76	0.61	0.53
5	0.54	0.41	0.31
6	0.51	0.77	0.33

From figure 3 and table 3 one can notice that the sample with the material 1 reduced the electric field strength (E) and power density (P) to the minimum values but magnetic field strength (H) value keeps still near to value before the shielding; but in case of shielding by the material 5 we get a small value factor for each of the three components: reduction factor is as high as 0.41 in case of E, it reduced also P to factor 0.31. While in case of reduction H the material 5 has also a small factor 0.54 when we compared these values with the values without the shielding material. Therefore, we can recommend the material 5 in case of the mobile phone shielding, it is very effective in order to reduce the EMPI exposure.

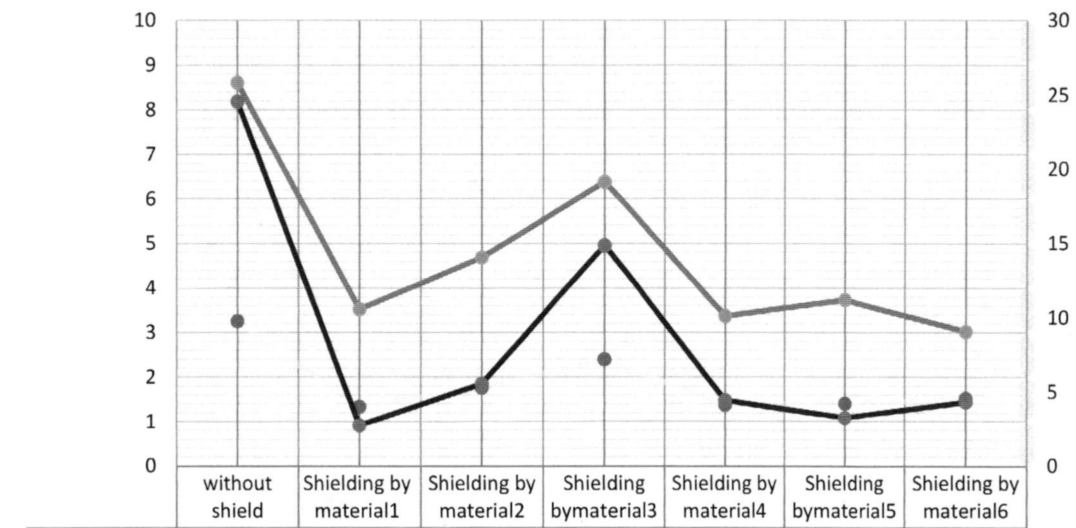

	without shield	Shielding by material1	Shielding by material2	Shielding bymaterial3	Shielding by material4	Shielding bymaterial5	Shielding by material6
V/m	3,25	1,33	1,767	2,4	1,38	1,41	1,52
mA/m	8,61	3,52	4,69	6,38	3,37	3,73	3,02
mW/m^2	24,56	2,74	5,55	14,87	4,44	3,26	4,32

Figure 4. Three EMPI components caused by laptop (Lenovo idea pad 100-15IBD Processor intel ® core (TM) i3-5005u CPU@2.00GHz) with and without EM shielding

Table 4 shows the reduction factor for the EMPI caused by the laptop for the three EMR components.

Table 4. Reduction factor of EMPI caused by laptop

Sample No	Magnetic field strength	Electric field strength	Power density
1	0.41	0.41	0.11
2	0.55	0.54	0.23
3	0.74	0.74	0.61
4	0.39	0.42	0.18
5	0.43	0.43	0.13
6	0.35	0.47	0.17

From data in figure 4 and table 4 one can notice that the material 1 reduced E to factor 0.41 when we compared the value by the value without shielding and also it reduced P to factor 0.11 when we compared the value by the value without shielding material. While in case of reduced H we get the minimum reduced factor in case of using the material 6. The material 1 also has a small factor 0.41 when we compared the value with the value without the shielding material. Because of that we recommended in case of the laptop shielding by the material 1 that is very effective in order to reduce the EMPI exposure.

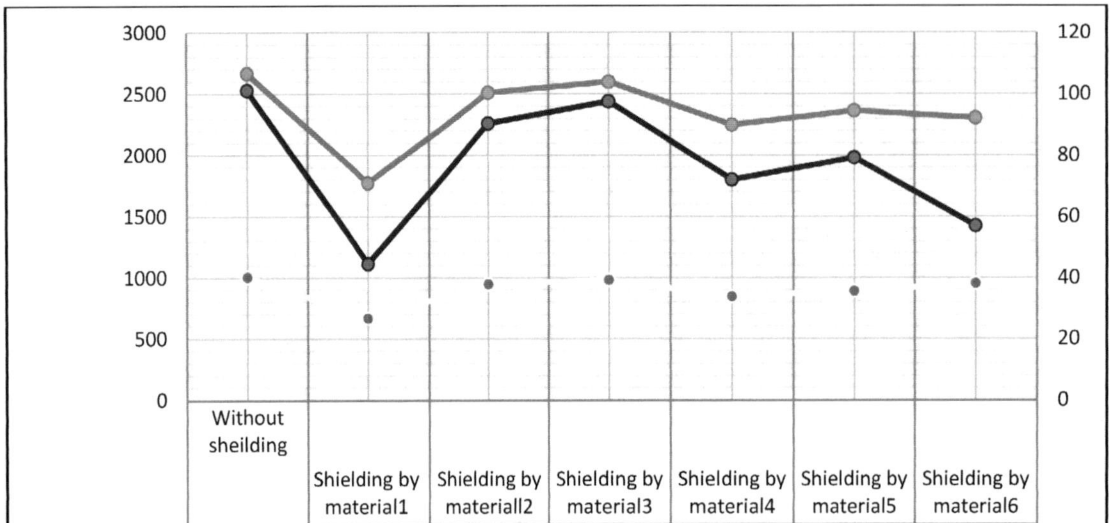

	Without sheilding	Shielding by material1	Shielding by materiall2	Shielding by material3	Shielding by material4	Shielding by material5	Shielding by material6
mW/m2	2530	1113	2260	2440	1800	1980	1423
V/m	40,23	26,72	37,84	39,22	33,89	35,62	38,17
mA/m	106,71	70,87	100,4	104	89,9	94,49	92,23

Figure 5. EMPI (three components) which were caused by microwave oven (2450MHz) with and without shielding

Table 5 shows the reduction factor for the EMPI which was caused by the microwave oven for the three components of EMR.

Table 5. Reduction factor of EMPI which was caused by microwave oven.

Material No	Magnetic field strength	Electric field strength	Power density
1	0.66	0.66	0.44
2	0.94	0.94	0.89
3	0.95	0.98	0.95
4	0.84	0.84	0.38
5	0.89	0.86	0.79
6	0.86	0.95	0.56

From figure 5 and table 5 one can notice that the material 1 reduced H to factor 0.66 when we compare the value with the value without shielding material. It reduces E to factor 0.66 when we compare the value with the value without shielding. We get the minimum reduced value of P in case of the material 4, but also the material 1 has the small value (factor 0.44) when we compare this value with the value without the shielding material. Because of this, we can recommend for the microwave shielding the material 1 that is very effective in order to reduce the EMPI exposure.

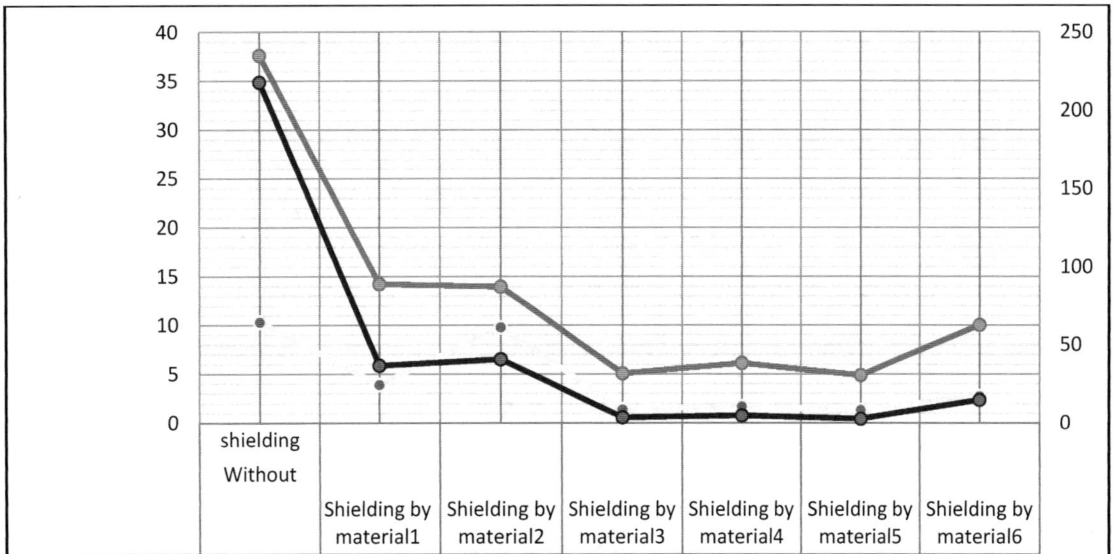

	shielding Without	Shielding by material1	Shielding by material2	Shielding by material3	Shielding by material4	Shielding by material5	Shielding by material6
mA/m	27,32	10,35	4,17	3,67	4,45	3,54	7,27
V/m	10,3	3,9	9,79	1,39	1,68	1,33	2,74
mW/m^2	218,1	36,61	40,67	3,53	4,81	2,72	14,85

Figure 6. EMPI (three components) caused by Wi-Fi (2300) MHz with and without shielding

Table 6 shows the reduction factor for EMPI which was caused by Wi-Fi for the three components of EMR.

Table 6. Reduction factor of EMPI which was caused by Wi-Fi.

Material No	Magnetic field strength	Electric field strength	Power density
1	0.38	0.38	0.17
2	0.15	0.95	0.19
3	0.14	0.14	0.016
4	0.16	0.16	0.022
5	0.13	0.13	0.013
6	0.27	0.27	0.068

From figure 6 and table 6 one can notice that the material 5 reduces the H to factor 0.13 when we compare this value with the value without the shielding material, and it reduces E to factor 0.13 when we compare this value with the value without shielding. It also reduces P to factor 0.013 when we compare the value with the value without the shielding material. Because of this we recommend in case of the Wi-Fi shielding the material 5 that is very effective in order to reduce the EMPI exposure.

3.2. EM shielding properties of ITO based materials

The list of samples is presented in Table 7.

Table 7. EM shielding cases of ITO based materials

Case 1	Without ITO
Case 2	ITO (Type A)
Case 3	ITO (Type B)
Case 4	ITO (Type C)
Case 5	ITO (Type D)

The transmission spectrum of the test samples is shown in Figure 7.

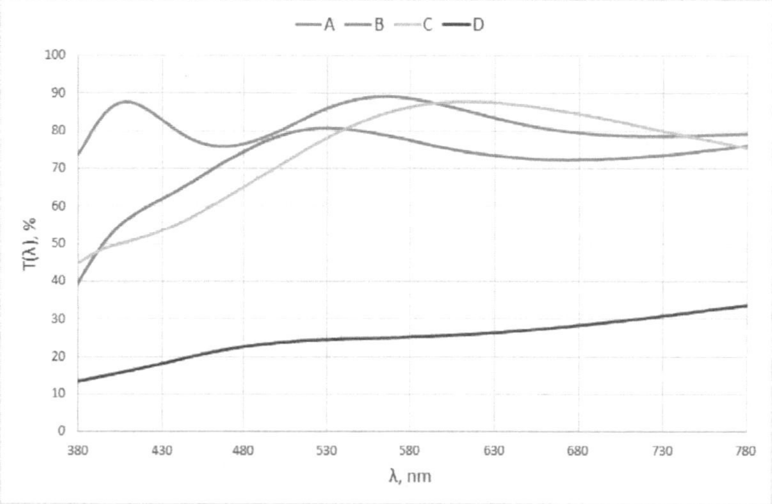

Figure 7. Transmission spectrum of the test ITO samples

Tables 8-11 and figures 8-11 show the reduction factor the EMPI caused by the devices investigated for the three EMR components.

Table 8. The factor of protecting from electromagnetic pollution of the mobile phone by using ITO coated metal shields.

Case No	Coating materials	Power density (W/m^2)	Electric field strength (V/m)	Magnetic field strength (mA/m)
1	Without	1.00	1.00	1.00
2	Type A	0.51	0.71	0.71
3	Type B	0.58	0.78	0.78
4	Type C	0.56	0.85	0.85
5	Type D	0.52	0.82	0.82

Table 9. The reduction factor of protecting from electromagnetic pollution of the laptop computer by using ITO coated metal shields.

Case No	Coating materials	Power density	Electric field strength	Magnetic field strength
1	Without	1.00	1.00	1.00
2	Type A	0.87	0.93	0.93
3	Type B	0.92	0.96	0.96
4	Type C	0.93	0.96	0.96
5	Type D	0.90	0.95	0.95

From figure 8 (see below) and table 8 we can notice that the case 2 from the Table 7 (Type A) reduces H to factor 0.61 when we compare this value with the value without the coated metal shields. It reduces E to factor 0.60 when we compare this value with the value without the coated metal shields. It also reduces P to factor 0.40 when we compare the value with the value without the coated metal shields. Because of this we recommend in case of the mobile phone it is better to coat the shielding filter by using ITO metal shields of the case 2 (Type A) that is very effective to reduce the EMPI exposure.

Table 10. The reduction factor of protecting from electromagnetic pollution of WI-FI by using ITO coated metal shields

Case No	Coating materials	Power density	Electric field strength	Magnetic field strength
1	Without	1.00	1.00	1.00
2	Type A	0.40	0.60	0.61
3	Type B	0.40	0.61	0.62
4	Type C	0.90	0.91	0.92
5	Type D	0.39	0.65	0.66

Table 11. The factor of protecting microwave oven from electromagnetic pollution by using ITO coated metal shields

Case No	Coating materials	Power density	Electric field strength	Magnetic field strength
1	Without	1.00	1.00	1.00
2	Type A	0.34	0.69	0.69
3	Type B	0.37	0.77	0.71
4	Type C	0.32	0.65	0.65
5	Type D	0.37	0.69	0.69

From figure 9 and table 9 one can notice that the case 4 from the Table 7 (Type C) reduces H to factor 0.65 when we compare this value with the value without the coated metal shields. It reduces E to factor 0.65 when we compare this value with the value without the coated metal shields. It also reduces P to factor 0.32 when we compare the value with the value without coated metal shields. Because of that we recommend in the case of the laptop it is better to coat the filter with ITO metal shields of the case 4 (Type C) that is very effective to reduce the EMPI exposure.

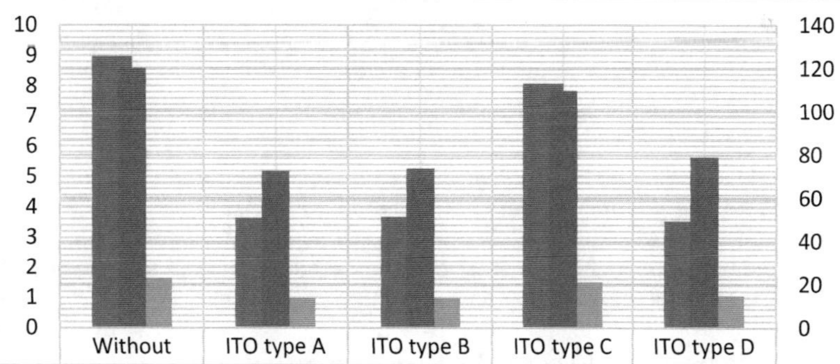

	Without	ITO type A	ITO type B	ITO type C	ITO type D
Power density (mw/m^2)	125,68	50,83	51,11	113,12	49,34
Magnetic field strength (mA/m)	22,73	13,75	14	20,79	14,97
Electric field strength (v/m)	8,58	5,18	5,26	7,84	5,64

Figure 8. Protecting from electromagnetic pollution of mobile phone by using ITO coated metal shields

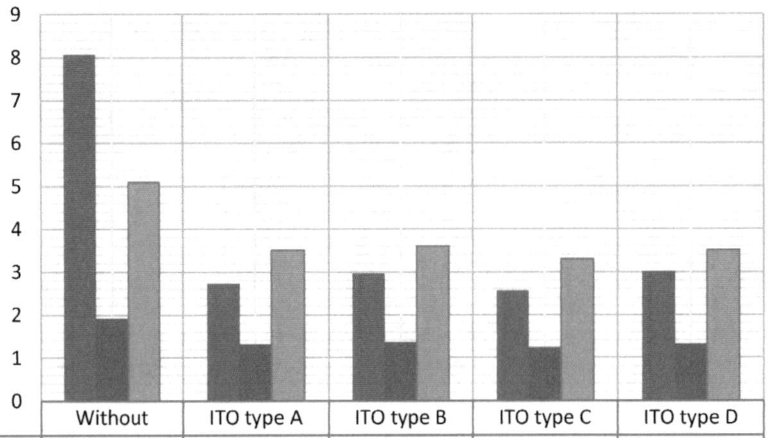

	Without	ITO type A	ITO type B	ITO type C	ITO type D
Power density (mW/m^2)	8,07	2,73	2,96	2,56	3,01
Electric field strength (V/m)	1,92	1,32	1,36	1,24	1,32
Magnetic field strength (mA/m)	5,09	3,5	3,6	3,3	3,5

Figure 9. Protecting from electromagnetic pollution of laptop by using ITO coated metal shields

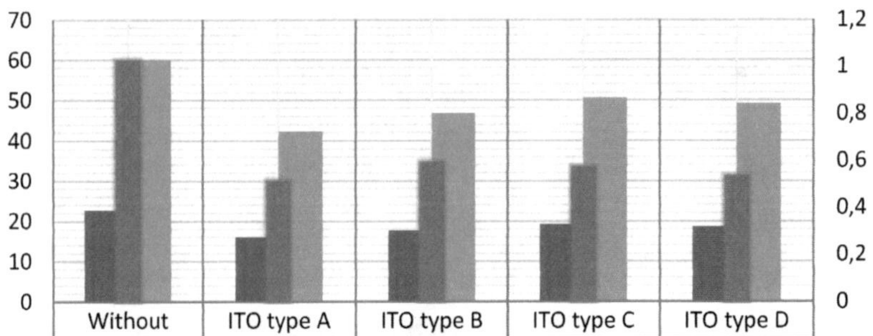

	Without	ITO type A	ITO type B	ITO type C	ITO type D
Electric field strength (V/m)	22,62	15,97	17,61	19,11	18,54
Magnetic field strength (mA/m)	59,99	42,37	46,72	50,68	49,17
Power density (W/m^2)	1,03	0,52	0,6	0,58	0,54

Figure 10. Protecting from electromagnetic pollution of WI-FI by using ITO coated metal shields

From figure 10 and table 10 one can notice that the case 2 (Type A) reduces H to factor 0.71 when we compare this value with the value without the coated metal shields. It reduces E to factor 0.71 when we compare this value with the value without the coated metal shields. It also reduces P to factor 0.51 when we compare the value with the value without the coated metal shields. Because of that we recommend in case of the Wi-Fi it is better to coat the filter by using the ITO metal shields of the case 2 (Type A) that is very effective to reduce the EMPI exposure.

For the last device (microwave oven) from figure 11 and table 11 one can notice that the case 2 from the Table 7 (Type A) reduces H to factor 0.93 when we compare this value with the value without the coated metal shields. It reduces E to factor 0.93 when we compare this value with the value without the coated metal shields. It also reduces P to factor 0.87 when we compare the value with the value without the coated metal shields. This is the smallest value in this situation but it is not effective enough to be recommended for the EM shielding.

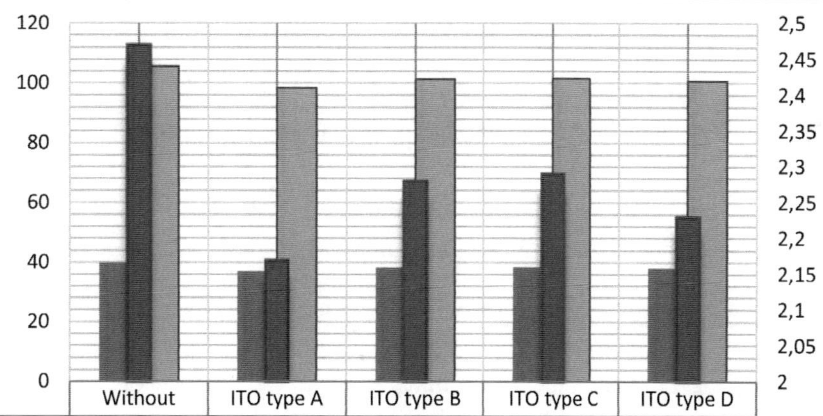

	Without	ITO type A	ITO type B	ITO type C	ITO type D
Electric field strength (V/m)	39,83	37,05	38,2	38,29	37,97
Magnetic field strength (mA/m)	105,64	98,28	101,35	101,55	100,73
Power density (W/m^2)	2,47	2,17	2,28	2,29	2,23

Figure 11. Protecting from electromagnetic pollution of microwave oven by using ITO coated metal shields

Conclusion

The paper is addressed to the protection from electromagnetic pollution by using metal coated shields, which was caused by some radiofrequency sources. Depending on our plan and main goal of this paper let us review figure 12. It illustrates comparison of the three components of EMR of the properties of two groups of filters protecting from the electromagnetic pollution (EMP). We have obtained the reduced factors values for the EMPI by using two groups coated metal shields. In figure 11 one can notice that these types of coated shields are effective in sources 1, 2 and then 3. But they are not efficient for the microwave oven. In the last case it is recommended to search for other materials.

The study indicated the cases when the EMR level is significant and protection from these devices – EMP sources is necessary. Hence this information can be exploited to use the devices safely and securely. Additionally, the research includes two groups of metal coated shields for daily life devices which are mobile phone, laptop, Wi-Fi and microwave oven without risk.

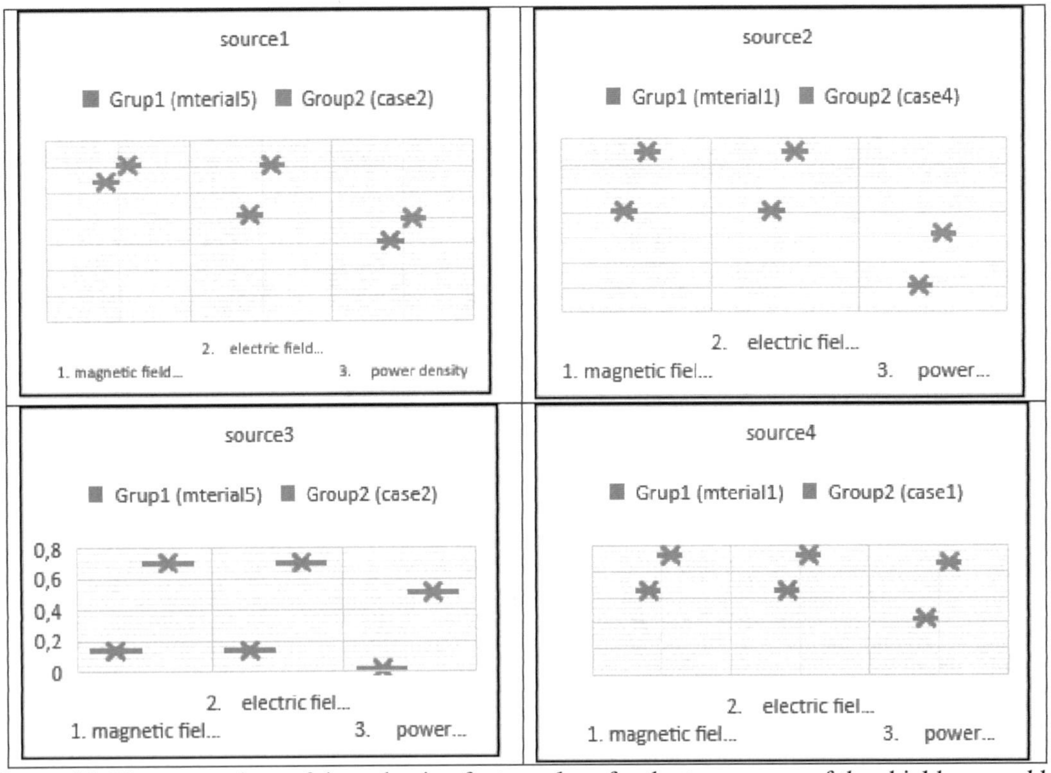

Figure 12. The comparison of the reduction factor values for the two groups of the shields caused by four sources (devices)

Acknowledgements
The study was partially performed under support of RFBR, grant No. 19-07-00602_a.

References

[1] Solovyov S P, Tsaritsyn M A, and Vorobyova O V 1971 *Special'nye Stroitel'nye Stekla (Special Construction Glasses)* (Moscow: Stroyizdat) (in Russian)

[2] Brekhovskikh S M, Viktorova Yu N, et al. 1971 *Osnovy Radiacionnogo Materialovedeniya Stekla i Keramiki (Fundamentals of Radiation Materials Science of Glass and Ceramics)* (Moscow: Stroyizdat) (in Russian)

[3] Brekhovskih S M, Viktorova Yu N and Landa L M 1982 *Radiacionnye Effekty v Steklah (Radiation Effects in Glasses)* (Moscow: Energoizdat) (in Russian)

[4] Evsyukov I 2008 *Electromagnetic Compatibility Materials – are they for Everybody? Komponenty i Tekhnologii (Components and Technologies)*, **2** 144 (in Russian)

[5] Evsyukov I 2004 *Komponenty i Tekhnologii (Components and Technologies)*, **1** 154 (in Russian)

[6] Samokvasova Yu N and Romashchenko M A 2012 *Trudy Mezhdunarodnogo Simpoziuma "Nadezhnost' i Kachestvo" (Proceedings of the International Symposium "Reliability and Quality")* **2** 214-215 (in Russian)

[7] Romashchenko M A 2014 *Metody Optimal'nogo Proektirovaniya Konstrukcij Radioelektronnyh Sredstv S Uchetom Elektromagnitnoj Sovmestimosti I Pomekhoustojchivosti (Methods Of Optimal Design Of Radio-Electronic Devices Structures Considering Electromagnetic Compatibility And Noise Immunity)* (PhD thesis) (Voronezh State Technical University) (in Russian)

[8] Pavlushkina T K and Babinova A A 2020 *Steklo I Keramika (Glass and Ceramics)* **8** 11-16 http://glassinfo.ru/articles/2020_01_Stekla_oslabljajushhie_jelektromagnitnoe_izluchenie.pdf (in Russian)
[9] Pavlushkina T K and Babinova A A 2021 *Glass Russia* **1** 20-21 (in Russian)
[10] Pavlushkina T K, Tsetlin V V, and Red'ko V I 1995 *Kosmicheskie Issledovaniya (Space Research)* **33**(3) 286-288
[11] L Bogomolova, V Tarasova, V Zhachkin 2010 *Problems of radiation safety* (4) 59.
[12] Pavlushkina T K 2010 *Tezisy Dokladov XIX Mezhdunarodnoj Nauchno-Tekhnicheskoj Konferencii (Abstracts of the XIX International Scientific and Technical Conference)* (Obninsk) 151-152
[13] Pavlushkina T K and Babinova A A *Russian Federation Patent Application "Glass burst reducing electromagnetic radiations"* No. 200943
[14] Grigor'ev A I, Tsetlin V V, and Pavlushkina T K 2001 *Device for protection from electromagnetic radiation* Russian Federation Patent Application No. 2162281
[15] Zhachkin V A and Tarasova V V 2018 Bulletin of MRSU. Series: Physics and Mathematics **2** 45-49
[16] D G Kim, J H Choi, D-K Choi, S W Kim 2018 ACS Appl Mater Interfaces 10(35) 29730-29740 doi: 10.1021/acsami.8b07054
[17] J H Choi, K Y Lee, S W Kim 2019 Composites Part B: Engineering **177** 107406 https://doi.org/10.1016/j.compositesb.2019.107406.
[18] http://www.techno.ru/emi/3_1.htm, http://www.techno.ru/emi/3_24.htm
[19] EMI-shielding theory and applications. Web resource https://hollandshielding.com/EMI-shielding-theory-applications?gclid=EAIaIQobChMIpZCApePW8QIVltayCh0z5AkPEAAYBCAAEgK4PfD_BwE
[20] SOLIANI EMC: SHIELDING PRODUCTS FOR ELECTROMAGNETIC INTERFERENCE Web resource https://www.solianiemc.com/en/company/?gclid=EAIaIQobChMIpZCApePW8QIVltayCh0z5AkPEAAYASAAEgIz5PD_BwE
[21] Swiss Federal Laboratories for Materials Science and Technology (EMPA) *The lightest shielding material in the world: Protection against electromagnetic interference* 2020 ScienceDaily. www.sciencedaily.com/releases/2020/07/200702113703.htm
[22] Tugirumubano A, Vijay S J, Go S H, Kwac L K and Kim H G 2019 *J. of Materials Research and Technology* **8**(1) 167-172
[23] J H Lee, D S Kang, S H Kim, M-J Son, J W Choi, D K Choi, J-P Choi 2018 Journal of Materiomics **4**(4) 390-401. https://doi.org/10.1016/j.jmat.2018.08.002
[24] Senyk I, Barsukov V Z, Savchenko B et al. 2016 *IOP Conference Series Materials Science and Engineering* **111**(1) 012026
[25] Melnikov A, Shchur P A 2019 *Trudy VIAM - VIAM Proceedings* **8** 56-65
[26] Moiseev K M 2014 *Nanoinzheneriya* **8** 18-21 (in Russian)

Using quantum computing to create efficient algorithms

I N Balaba[1], G S Deryabina[2], I A Pinchuk[3], I V Sergeev[4] and S B Zabelina[3]
[1]Tula State Lev Tolstoy Pedagogical University, 300026, Tula, Lenin avenue, 125, Russia
[2]Bauman Moscow State Technical University, 105005, Moscow, ul. Baumanskaya 2-ya, 5, Russia

[3]Moscow Region State University, 141014, Moscow Region, city of Mytishi Very Voloshinoy street, 24, Russia
[4]Institute of Advanced Training and Professional Retraining of Educational Workers Tula region, 300041, Tula, Lenin street, 22, Russia

Corresponding author: irenepin@yandex.ru

Abstract. The article presents a historical overview of the development of the mathematical idea of a quantum computing model - a new computational strategy based on the postulates of quantum mechanics and having advantages over the traditional computational model based on the Turing machine; clarified the features of the operation of multi-qubit quantum systems, which ensure the creation of efficient algorithms; the principles of quantum computing are outlined and a number of efficient quantum algorithms are described that allow solving the problem of exponential growth of the complexity of certain problems.

1. Introduction

The need to build quantum computations (quantum algorithms) is due to the need to solve a wide range of computational problems for which there are no efficient classical (polynomial) algorithms. Inefficient algorithms require exponentially large resources that even supercomputers cannot provide. At the same time, quantum computing, using the paradigm of quantum mechanics, allows you to solve these problems. Quantum computers are needed to solve such problems.

The idea of creating quantum computers began to be discussed back in the 60s of the XX century. By this time, quantum statistical mechanics and quantum field theory already existed. J. Von Neumann's works laid a solid mathematical foundation for the statistical interpretation of quantum mechanics. L.D. Landau was the first to point out the possibility of describing a system by a density matrix [1]. A little later, von Neumann published two articles in which, using the density matrix, he gave a very complete analysis of issues related to the probabilistic interpretation of quantum mechanics and thermodynamics of quantum mechanical systems [2, 3].

In 1980, Yu. I. Manin noted that quantum computers can simulate quantum systems more efficiently than classical computers could. He believed that one of the reasons for this is that the quantum state space of a quantum automaton has a much greater capacity than the classical one: where in the classics there are N discrete states, in the quantum theory admitting their superposition, there are C^N Plank states of the cells. When combining classical systems with the number of states N_1 and N_2, the total number of states will be equal to the product $N_1 \cdot N_2$, and in the quantum version we will get $C^{N_1 \cdot N_2}$ [4, page 15].

In 1982, the idea was developed in more detail by R.P. Feynman in [5], in which he noted that modeling quantum mechanical systems is an exponentially difficult task for classical computers, but can be effectively carried out by performing logical operations on quantum systems. acting on a superposition of many quantum states.

In 1985, D. Deutsch developed the concept of a quantum computer and showed how quantum computers could be built from quantum logic gates in quantum networks, from which it followed that a quantum computer could indeed surpass classical computers in computational efficiency.

At the beginning of the 21st century, a monograph by american specialists M. Nielsen and I. Chang [6] was published, which sets out the main results in the field of quantum computing and quantum informatics in sufficient completeness and scientific rigor.

2. Quantum computers

The main element of a quantum computer is quantum bits (qubits), lists of quantum states. This is their difference from modern classical computers, in which bits are used to transfer and process data, each of which at any time can be in one of two states 0 or 1. An arbitrary qubit $|\psi\rangle$ admits two classical states, denoted $|0\rangle$ and $|1\rangle$, but it can also be in their superposition (linear combination), that is, in the state $|\psi\rangle = \alpha \cdot |0\rangle + \beta \cdot |1\rangle$, where α and β are complex numbers satisfying the condition $|\alpha|^2 + |\beta|^2 = 1$, and $|\alpha|^2$ and $|\beta|^2$ are respectively the probability of getting 0 or 1 when measuring the state of a qubit.

Since each qubit can be in a superposition of the states $|0\rangle$ and $|1\rangle$, the state space of a system constructed from only two qubits $|\varphi\rangle, |\psi\rangle$, is a superposition of four basic states $|00\rangle$, $|01\rangle$, $|10\rangle$ and $|11\rangle$, which is much "more" than the only possible state of a system built of two classical bits at each moment in time. If a classical computer stores N bits in memory, which are subject to change for each cycle of the processor's operation, then a quantum system of N two-level quantum elements has 2^N linearly independent states. Hence, due to the principle of quantum superposition, the state space of a quantum system is a 2^N-dimensional Hilbert space, the standard basis of which is a unit matrix of size $2^N \times 2^N$. Of course, there can be an infinite number of bases, but all vectors are orthonormal. One operation on a group of qubits affects all the values that it can take, in contrast to the classical bit, which provides unprecedented parallelism of computations, an exponential increase in the power of the computational model.

Quantum circuit is a model of quantum computing that serves to visualize quantum algorithms (Figure 1).

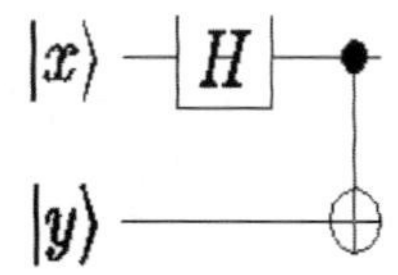

Figure 1. An example of a quantum circuit that uses two qubits.

A quantum scheme is represented in 2^N-dimensional Hilbert space, where N is the number of qubits, and each unitary transformation is a matrix of a special form.

Building a quantum automaton requires the right balance between mathematical and physical principles. To create a quantum computer, it is necessary to simulate its operation on the basis of real physical systems, taking into account quantum noise, which is the main problem on the way of implementing quantum bits. The physical realization of a qubit can be any two-level quantum mechanical system, for example, polarization states of photons or spin states of atomic nuclei. The following requirements are imposed on them:

- the system consists of a precisely known number of particles;

- the ability to bring the system to a precisely known initial state;
- a high degree of isolation from the external environment;
- providing measurements with a sufficiently high reliability of the state of the quantum system at the output.

When creating a quantum computer, the main attention is paid to the issues of controlling qubits using stimulated emission and preventing spontaneous emission, which will disrupt the operation of the entire quantum system. The main technical obstacle to creating a quantum computer with a sufficient number of qubits is decoherence - the decay of quantum superpositions. Random interactions with objects in the surrounding world can destroy the state in which the qubits are located and break the connections between them. For many years, research has been underway to extend the lifetime of qubits by isolating them from the environment in various ways. In 2020, an approach was developed to solve the problem of decoherence of superconducting qubits based on Josephson transitions [7].

3. Principles underlying quantum computing

The physical foundations of quantum computing are the postulates of quantum mechanics. Let us recall them.

1. The state of an isolated quantum system is described by the unit vector of a complex Hilbert space. The state of a quantum bit in a 2-dimensional complex Hilbert space is described by a unit vector:

$$|\psi\rangle = \alpha \cdot |0\rangle + \beta \cdot |1\rangle, |\alpha|^2 + |\beta|^2 = 1,$$

where α and β are complex numbers.

2. The evolution of a closed quantum system is described by the unitary operator U.

3. The state space of a composite system is the tensor product of the state spaces of the systems included in it.

4. Quantum measurements are described by a set of operators $\{M_m\}$, acting on the state space of the system. If the state of the system before the measurement is $|\psi\rangle$, then the probability of obtaining the result m is

$$(m) = \langle\psi|M_m^\dagger M_m|\psi\rangle,$$

where $M_m^\dagger$ is the Hermitian-conjugate matrix to M_m, and the state of the system after measurement:

$$\frac{M_m|\psi\rangle}{\sqrt{\langle\psi|M_m^\dagger M_m|\psi\rangle}}.$$

Measurement operators satisfy the completeness equation:

$$\sum_m M_m^\dagger M_m = I.$$

Quantum computing is based on the following principles:

1. Reversibility of calculations, i.e. any computational process within the framework of quantum computing must be reversible. This means that for any quantum circuit underlying the algorithm, it is possible to build a reverse circuit, which, receiving the output of the original circuit as input, returns the initial data for it.

2. Quantum parallelism. Since the quantum bit can be either 0 or 1, then by calculating some function, the argument of which is set by quantum bits, all possible values are calculated simultaneously. This makes it possible to solve the same problem in parallel for exponentially big data.

3. Interference is widely used in quantum algorithms to enhance desired results and attenuate unwanted ones. Repeating several times the sequence of parallel processing of qubits, taking into account the interference of their states, it is possible to obtain the required result with a high probability. In this case, by varying the number of repetitions and the step of interference, the probability can be brought to any given value.

4. Quantum entanglement is the most important property of quantum systems. It serves as an essential resource in quantum informatics, being a key factor in many quantum algorithms.

The main specific feature of quantum computing is its probabilistic nature. In [8], the authors considered the development of the use of probabilistic methods in theoretical physics from the kinetic theory of gases to the Einstein – Podolsky – Rosen paradox and Bell inequalities.

4. Quantum algorithms

When developing quantum algorithms, of greatest interest are those that are fundamentally superior to classical algorithms in speed.

An outstanding result in this direction was obtained by Shor [9]. He found two polynomial quantum algorithms for which no classical analogs were known. The first is an algorithm for factoring an integer into a product of prime factors (factoring of integers), and the second (the discrete logarithm) is an algorithm for solving the equation $a^d = b$ for given elements a and b in a finite abelian group G. Shor's factorization algorithm, allowing to factor the number N into prime factors in time $O(\log^3 N)$ is based on the ability of qubits to take several values simultaneously. This algorithm consists of two parts: the classical first is the reduction of the factorization to finding the period of some function, the quantum second is the finding of the period of this function. Shor's factorization algorithm is quite simple and requires more modest hardware than a general-purpose quantum computer. And since numerous modern algorithms and cryptography systems are based on the algorithmic complexity of the problem of factorizing a number, it makes it possible to hack public key cryptographic systems, in particular the RSA cryptosystem, with its help.

A. Yu. Kitaev formulated and solved the Abelian Stabilizer Problem, which includes the above algorithms as special cases [10]. This problem is as follows. Let a finite abelian group G act on a finite set M. It is assumed that this action and the group operation are easily computable. We need to find a stabilizer for an element a of a group G, that is, find a subgroup H of a group G that leaves the element a in place.

In turn, a generalization of the Abelian stabilizer problem is the Hidden Subgroup Problem. This problem remains one of the fundamental problems in the theory of quantum algorithms. It is formulated as follows. Let a function $f: G \to S$ be given with the following property: it is constant on the cosets of the group G with respect to the subgroup H and different on different cosets. The difficulty of solving this problem depends on the properties of the group G. In recent years, an efficient algorithm has been found in the case when H is a normal subgroup of the group G [11]. An efficient algorithm was constructed for a group G that is close to Abelian, that is, one for which the intersection of all normalizers of subgroups has a small index in the group G [12]. A beautiful solution was also found in the class of nilpotent groups of degree 2 [13].

It should be noted that the problems of automorphism and isomorphism of graphs can be formulated in the language of the hidden subgroup problem for the symmetric group S_n.

The key role in quantum computing is played by the Fourier transform, which, if written appropriately, can significantly increase the efficiency of computations. If the input for the Fourier transform is an N -dimensional vector, then the classical algorithm requires $O(N \log N)$ operations, and the quantum one requires $O(\log^2 N)$ operations. In many problems, the Fourier transform was defined on the set of a sequence of complex numbers $\{a_g\}$, where $g \in Z_{2^n}$. Kitaev in [10] generalized the Fourier transform from Z_{2^n} to the Fourier transform on an arbitrary finite abelian group.

In 1996, L. Grover developed a quantum algorithm that allows searching in any unordered database of N elements in $O(\sqrt{N})$ time [14]. Currently, one of the fastest search algorithms is the linear algorithm, which requires $O(N)$ time. Grover's algorithm is the fastest quantum algorithm for searching in an unordered database, and there are no classical algorithms with the same efficiency.

At the moment, several dozen quantum algorithms have been developed in the world for solving various problems.

Note that quantum computers could not only efficiently solve exponentially complex traditional mathematical problems, but also a whole class of new problems that no classical computer can solve. In 1984 Bennett and Brassard [15] developed the first quantum cryptography algorithm based on the quantum state of photons. In 1989, the first working quantum computer was built at the IBM Research Center, consisting of two quantum cryptographic devices, implementing a system of quantum cryptography. The secrecy of the system is based not on the difficulty of processing, but on the properties of quantum interference, which gives this system absolute secrecy, which cannot be ensured using classical methods [16, p. 221]. Further improvement of the cryptosystem reliability was proposed by A. Eckert, based on the Einstein-Podolsky-Rosen effect [17].

5. Conclusions

Full-fledged quantum computers have not yet been created; only their prototypes exist, which are implemented using various technologies for creating a quantum environment. In November 2009, a programmable quantum processor model was created. Physicists from the National Institute of Standards and Technology (USA) have implemented an extensive set of "programs" based on two qubits, represented by beryllium ions. In January 2010, a primitive quantum computer was used to calculate the energy of a hydrogen molecule. Physicists from Harvard University (USA) and the University of Queensland (Australia) managed to implement an algorithm for calculating the energy of the ground and excited states of the H2 molecule, using only two photon qubits in their experiments. Institute for Theoretical Physics Landau RAS and the RAS Physico-Technological Institute conduct experiments with different architectures of quantum computers and with different materials. In April 2012, a team of researchers from the University of Southern California, Delft University of Technology, Iowa State University, and the University of California managed to build a two-qubit quantum computer based on an impurity diamond crystal. It implements Grover's algorithm for 4 iteration options. In 2019, Google announced the creation of a 53-qubit Sycamore prototype, which uses superconducting qubits.

It remains relevant to simulate the operation of a quantum computer on classical supercomputers, since when simulating a quantum computer on a classical computer, the amount of required computing resources exponentially increases with the number of qubits. The work [18] presents a simulation of the operation of an ideal quantum computer on the Lomonosov supercomputer. The authors have developed special software that simulates the Grover quantum algorithm and the quantum Fourier transform.

The current stage in the development of quantum computing is the stage of fundamental research and experimental confirmation of the results of these studies.

Acknowledgements

The work of I N Balaba was supported by the RFBR grant No 19-41-710004_r_a

References

[1] Landau L 1927 *Zeitschrift für Physik* **45** 430
[2] von Neumann J 1927 *Nachr. Akad. Wiss. Göttingen Math.-Phys* 245
[3] von Neumann J 1927 *Nachr. Akad. Wiss. Göttingen Math.-Phys* 273
[4] Manin Y 1980 *Computable and uncomputable* (Moscow, Sovetskoe radio) 128 (in Russ.)
[5] Feynman R P 1982 *Int. J. Theor. Phys* **21** 467
[6] Nielsen M A and Chuang I L 2000 *Quantum computation and quantum information* (Cambridge University Press) 704
[7] Bilmes A, Volosheniuk S, Brehm J D, Ustinov A V and Lisenfeld J 2020 *Quantum Sensors for Microscopic Tunneling Systems arXiv:2011.14327v1*
[8] Deryabina G S, Kuleshova Ju D, Pinchuk I A, Zabelina S B and Vysotskaya P A 2020 *J. Phys.: Conf. Ser. 1560 012013*

[9] Shor P 1994 *Foundations of Computer Science, Proceedings., 35th Annual Symposium on* (IEEE) 124
[10] Kitaev A Yu 1995 *Quantum measurements and the Abelian stabilizer problem* (arXiv:quant-ph/9511026v1)
[11] Hallgren S, Rassell A and Ta-Shma A 2003 *SIAM Journal on Computing* **32**(4) 916
[12] Gringi M, Schulman L, Vazirani M, and Vazirani U 2004 *Combinatorica* **24**(1) 137
[13] Ivanyons G, Sanselme L, and Santha M 2012 *Algorithmica*, **62**(1-2) 480
[14] Grover L.K. 1996 *Proceedings, 28th Annual ACM Symposium on the Theory of Computing* 212.
[15] Bennett C H, Brassard G 1984, *Proceedings of the Intern. IEEE Conf. on Computers, Systems and Signal Processing, Bangalore, India,* (New York: IEEE Press) 175
[16] Deutsch D 1997 *The fabric of reality* London: Allen Lane) 390
[17] Ekert A 1991 *Phys. Rev. Lett.* **67** 661
[18] Andreev D Yu, Korzh O V, Korobkov S V and Chernyavsky A Yu 2013 *Proc. Int. Conf. Parallel computational technologies (PCT) 2013* (Chelyabinsk: Publishing Center SUSU) 38

Methods of teaching physics in the conditions of online interaction

D G Dankin

National University of Science and Technology "MISiS", Moscow, Russia

dankin84@mail.ru

Abstract. The article discusses the problems of education caused by the pandemic and the possibilities of electronic educational tools. The challenges of the period that 2020 brought sharply accelerated the development of electronic educational tools, at the same time, revealed the limitations of student control tools. The author of the article worked in the educational system of a school, university and pre-university training in the spring of 2020. This article discusses three options for implementing online classes. All three of these systems have switched to the electronic form of online classes, although this situation was not initially foreseen. The problems that one had to face in the work and the successful moments of the studying in the information educational environment are described. Prospects for the development of online learning are outlined.

Keywords: physics; foreign applicants; teaching; Russian as a foreign language; information technologies, online interaction.

1. Features of the teaching physics to Russian students and pupils in the conditions of online interaction

The year 2020 began with a global pandemic and seriously challenged all the spheres of human activity. Faced with quarantine and temporary restrictions, students, pupils, teachers and professors had to abandon the class-based form of teaching. Then the global information network came to help, the Internet. Holding classes via mobile applications was not new, because such classes had already been practiced before. However, just in quarantine circumstances these mobile forms of education became global and the only, though temporary, way of working in education.

By the outbreak of pandemic, there had been developed different platforms for online interaction; among the best known are Moodle, iSpring, WebTutor, Zoom, MS Teams and many others. The authors of the article used the latter two in their work.

In the spring of 2020, the author taught general physics at the physics department at the National University of Science and Technology "MISiS", gave there courses at the preparatory department "Professional Navigation and Reception Centre" for the participants from distant foreign countries and also taught physics in HSE Lyceum. That diversity allowed the author to take a closer look at the peculiarities of using online interaction platforms. The author used MS Teams platform in his work. The choice stemmed from the fact that this platform has such features as making student groups, monitoring the learning process and the attendance, using visual teaching methods (show presentations, share one's screen, use the "blackboard", etc.). Moreover, this platform was recommended and obligatory for application in all three institutions.

When it comes to full-time students, this transition to distance learning was not new regarding practical classes and solving problems. This form allowed students to work in the same way as they used to in the auditorium. It can be said that the part of the group, which would engage in face-to-face classes, continued working hard during the distance learning. The same goes for poor students – new form did not affect their attendance. It is worth noting that for some students it was even easier to understand the subject remotely. There was a certain "social migration", which means that some students became more diligent under new conditions, some of them – less. This does not mean technical problems, though they also took place, but it is about several individual manners of perceiving and professor-student interaction. These peculiarities were more or less evident practically in all student groups. It was observed that the more senior the group, the less visible they were.

With regard to school pupils, the attention problem was much more evident for this category of students. Using visual tools and videos, it was possible to involve them in the learning process, but not very effectively.

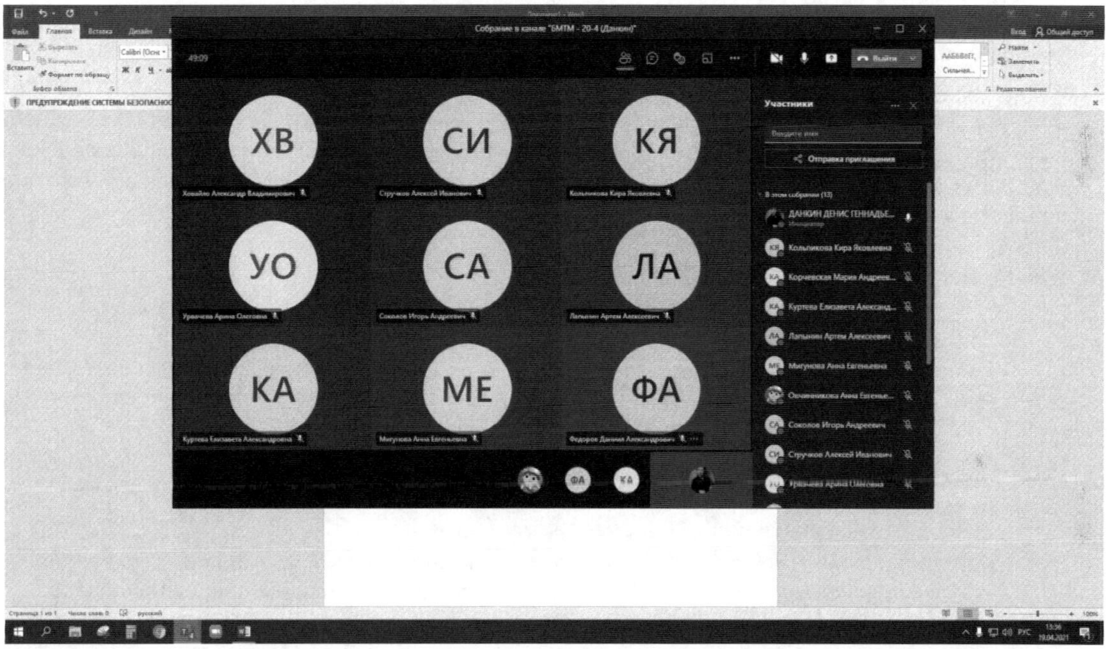

Figure 1. General view of the program and available visual tools.

The creative tasks with competition elements characterized by division into teams, searching for a solution and missing information turned out to be more effective. It has to be said that students' creative work is the distinctive feature of HSE Lyceum and it is evaluated separately. Unfortunately, this grade had low weight in the overall ranking at the time, so giving only creative classes was not correctly. Therefore, the creative approach was not much used, though under the circumstances it could have been developed better.

2. Features of the teaching physics to foreign students in the conditions of online interaction

As concerns educating foreign students in circumstances of pandemic, there is a lot to say, so we are going to outline briefly the most striking features.

It is more difficult to break the language barrier by online interaction than having face-to-face classes. Yet it is the Internet, which allows using online translators, dictionaries and visual tools, while as one learns the language, the language barrier becomes less important. It may not disappear completely, but it could have little impact on studying. It is possible to read more about the interaction between special subject professors and linguists and their work in the article "The methods of teaching the basic concepts of mechanics for preparatory foreign students" [5].

The common problem which can be easily identified and with which all the colleagues were confronted – it is the monitoring of students' learning achievements during the pandemic. The monitoring could be carried out by information technology equipment and it could be fully realized. Such traditional form of control as written examination could no longer ensure the reliability of checking students' knowledge, skills and competence. Where possible, the traditional forms were replaced by testing. However, this labor-intensive process of creating tests was mainly left to professors. The development of a good quality test with different types of questions (test tasks of open type, multiple choice, tasks for calculation and so forth) – it is rather laborious work.

In order to describe the situation, which occurred in the field of education in the spring of 2020, it should be noted that there had already been provided conditions for distance learning in Moscow schools and universities. Nevertheless, the forced loss of the class-based form of teaching revealed a number of shortcomings, which distance learning system has.

In the matter of distance learning, it is difficult to say to what extent such education's quality deteriorates. The tools offered by online interaction platforms allow diversifying the studies. Certainly, it helps to make material more illustrative, to turn to reference information from the Internet. Rather, there are problems connected with quality control of education. The use of all these tools, which are advantageous during classes, results in the fact that during the test and exam the quality of learned material is easily replaced by the ability to quickly find and exchange the information from the global network or, which is very often, to involve highly qualified third parties in control testing.

References

[1] Federal Law of 29 December 2012 N 273-FL (ed. 25 December 2018) On education in the Russian Federation. Article 14. Language of education

[2] Kabardin O 1991 *Physics. Material reference* (Moscow, Prosveshcheniye Publ.) p. 367

[3] Butikov E, Bykov A and Kondratyev A 1982 *Physics for university applicants* (Moscow, Nauka Publ., Chief editorial board of physics and mathematics literature) p. 608

[4] Gracheva O, Matukhin P and Elsgolts S 2013 *Interdisciplinary IT projects in development of electronic manuals on Russian as a foreign language for physicists* (Moscow, Vestnik RUDN, Informatization of Education series) no. 4

[5] Dankin D 2020 *The methods of teaching the basic concepts of mechanics for preparatory foreign students* (Moscow, Physico-mathematical and technical education problems and development prospects) *Materials of the V International Scientific and Methodological Conference*, pp. 346-349

[6] Yefremova N, Rudkovskaja V and Vityuk E 2016 *On some issues of learning physics in high school* (Moscow, Modern science-based technologies) no. 8 (part 1) pp. 116-120

Physical and methodological approach to the modern methods of the investigation of dental materials properties

A N Kobzar

National University of Science and Technology (MISIS), Leninsky Prospect, 4, Moscow, Russian Federation

E-mail: antonina1303@gmail.com

Abstract. Each medical doctor should be able to complete professional tasks using their knowledge of physics, especially when providing a comprehensive care to a patient. This article reveals how to use knowledge of the course of physics specially developed of a Medical University to help trainee medical doctors to develop skills to complete professional tasks. Content of physics and its tasks for a Medical University must be professionally oriented. This article provides a classification of professionally oriented questions of physics specially made to be taught at a Medical University. It also gives content a sample of a professionally oriented task in physics for a Medical University. All these above mentioned tasks are closely correlating with each other. As an example, the article explains specific of teaching Biophysics at a Medical University. Professionally oriented content of Biomechanics – modern methods of study properties of materials used for dental care. Methods of physics for destructive inspection testing: tensile strength, compressive strength, hardness of materials, Poisson's ratio, fluctuating stress and dynamic testing etc. Methods of physics for non-destructive inspection testing: thermal, acoustic, optical, radio-wave, radiative, capillary, magnetic and electric testing.

1. Introduction

Medicine has being using the results of theoretical and experimental advances in physics for a long time. Physiological processes in the human body proceed according to the laws of physics. A lot of diagnostic and therapeutic methods are based on physical phenomena. Most medical apparatus are physical devices by design. Hence, physics is very important for modern medicine [1,2,3].

The analysis of research on the study of physics by future medical specialists is carried out. Conclusions of the analysis: 1) the problem of teaching physics to future medical specialists is an urgent topic of research in recent years; 2) these studies reflect the following issues: selection of physics content at the Medical University (S. V. Babin, T. N. Shamaeva) [4], implementation of interdisciplinary connections of physics with other disciplines in a Medical University (A.F. Zubov, N.A. Ladnich, E.M. Starikova) [5,6], forms, methods, technologies of teaching physics to future doctors (N.P. Pupyrev, E.A. Semenyuk) [7], the value aspect of physics course at the Medical University (O. E. Akulich) [8]; the implementation of the principle of professional orientation in the process of teaching physics to medical students (L. E. Ryazanova, A.V. Tarasova) [9,10], the formation of experimental skills in future doctors (N. G. Arzumanyan) [11], and polytechnic competencies (Yu. I. Nikitina) [12], the problem of preparing medical students for solving professional problems in teaching physics at a Medical University in the context of integrating physics with biomedical disciplines within the framework of our dissertation research (Biryukova A.N. (Kobzar A.N.)) [3,13]; 3) there are few studies that consider the pedagogical methodology of teaching modern methods of studying the physical properties of dental materials to the future dentists. Therefore, the topic of our article is relevant.

2. Discussion

Presently at the Medical Universities students study the main general physics sections, elements of biophysics [13]. Questions of general physics are the basis for studying the professionally-oriented questions – questions that need to be studied and solved for the future professional activity of a doctor [3]. For example, at the Medical University (FGBOU VO "SCHEMNIK" of the Ministry of Health of Russia), students study the basic discipline " Physics, mathematics". Its objective is "to form a systematic knowledge of the physical properties and processes in biological objects (in the human body). This is necessary for the development of other academic disciplines and the formation of professional medical qualities». Physical sections of the discipline: "Mechanics of liquids and gases. Acoustics", "Electrical and magnetic properties of tissues and the environment", "Fundamentals of medical electronics", "Optics", "Elements of quantum biophysics, ionizing radiation". In addition, future dentists at the same Medical University study the multiple choice discipline "biophysics". It includes the following sections: "Biomechanics", " Transfer processes in biological systems. Bioelectrogenesis", "Introscopic diagnostic methods", "Physical foundations of medical ecology", "Physical Foundations of electrography", "Quantum-physical foundations of diagnostic and treatment methods", "Ionizing radiation».

Each section of these disciplines includes professionally-oriented questions. In the framework of our educational research the general structure of professionally-oriented physics content at the Medical University (classification of professionally-oriented physics questions) has passed the expert assessment of medical specialists and has been successfully implemented in the educational process at the Medical University (Figure 1) [3].

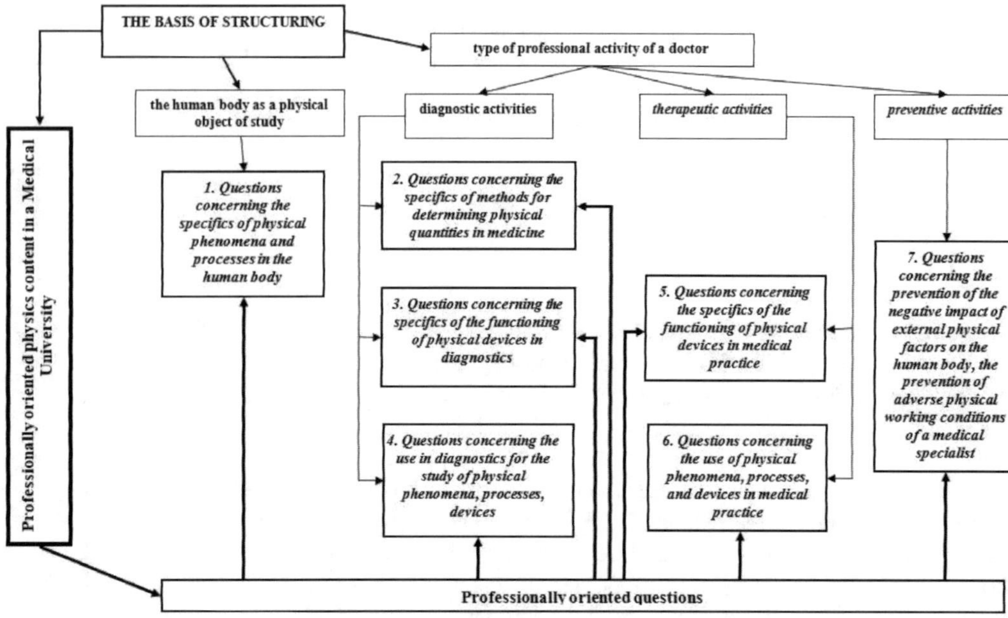

Figure 1. Structure of professionally oriented Physics content in the Medical University.

As an example, let's consider the particular features of the section "Biomechanics" study by future dentists. This section studies the following professionally-oriented questions of Physics-modern methods of the physical properties of dental materials study (Mechanical pressure and deformation of materials. Laws of elastic deformation. Basic elastic and plastic characteristics of materials. Diagrams of forces, pressure, and bending moments (on the example of dental structures) Physical aspects of strength and fracture of materials). Quantitative characteristics of deformation: mechanical pressure (σ), absolute elongation (Δl), relative elongation (ε).

To evaluate the mechanical properties of dental materials, it is necessary to study the characteristics of materials under the influence of loads. For example, we can consider the dependence of the mechanical pressure (σ) on the relative elongation of the beam (ε) in the diagram.

Depending on the behavior of the material during the load test, there are two main types of materials: brittle (have no residual deformations and are destroyed when the stress in the material reaches the ultimate strength) and plastic (small ultimate strength, significant residual deformations, do not break). The dental seal in the oral cavity is affected by longitudinal and transverse forces (not only compression deformation, but also shear deformation).

At the Medical University, future dentists consider the physical aspects of the strength and destruction of the material: the influence of temperature, loading speed, humidity, time on the strength of the material, the theoretical strength (for example, dental materials) [3,13].

Studying this material, future dentists can solve professionally-oriented physical tasks (subject matter: deformation characteristics, Hooke's law, Poisson's ratio in dentistry). The physical tasks at the Medical University should be professionally oriented. We have developed a model of the content of a professionally oriented physical task in a Medical University (Figure 2). This model is interrelated to the general structure of the professionally oriented physics content in the Medical University (Figure 1) [3,13,14].

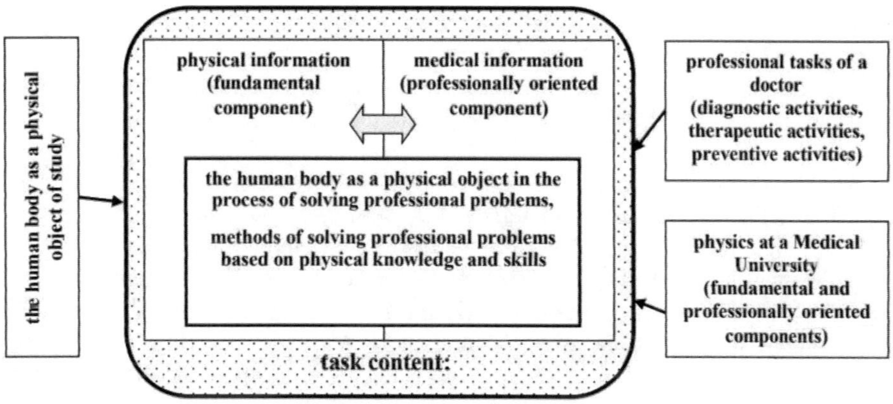

Figure 2. The model of the content of a professionally oriented physical task in the Medical University.

Here are examples of these tasks for dentists.

Task 1. Evaluate the compliance of the strength properties of the dental seal. It is known that the maximum working pressure is 4×10^7 Pa, the minimum pressure is 10^7 Pa, the tensile strength of the sealing material (amalgam) is 50 MPa, the ultimate strength under compression deformation is 400 MPa.

Task 2. When chewing food, relative deformations occur on the molars. The Poisson's ratio of the teeth is 0.4. Determine the relative volume change of the molars. The relative strain is 4.8×10^{-4}.

Task 3. The temperature in the oral cavity can range from 0 ^{0}C to 70 ^{0}C. Evaluate the mechanical pressure that occur in the cavity. Select the appropriate structural material (dental filling material) for the strength requirements under these conditions. The temperature coefficient of linear expansion of tooth tissues is (10-15)*10^{-6} K^{-1}, the Young's modulus for tooth enamel is 83,000 MPa, for dentin-18,600 MPa.

We have developed a classification of professionally oriented physical tasks at the Medical University. It is interrelated to the classification of professionally oriented questions of physics course at a Medical University (Figure 1), with the content model of a professionally oriented physical task at a Medical University.

The classification of professionally oriented physical tasks in a Medical University:

- physical tasks concerning the specifics of studying the human body as a physical object, the manifestations of physical phenomena, processes in the human body, disease prevention;
- physical tasks concerning the specifics of the application of physical phenomena, processes, devices in diagnostics for studying the human body;
- physical tasks concerning the specifics of the application of physical phenomena, processes, devices for the treatment of the human body [3].

According to this classification, a complex of professionally-oriented physical tasks has been developed and adapted for all the main sections of physics course at the Medical University [15].

In addition to the basic concepts in the section "Biomechanics", methods for evaluating the physical and mechanical properties of materials should be considered. This is very important for the future professional activity of a dentist. Determination of the physical and mechanical properties of dental materials is necessary to control the quality of products made. Physical methods of destructive testing include tensile strength test method, compressive strength method, material hardness method, Poisson's ratio method, test method under variable loads, dynamic test method and others. Non-destructive testing methods are based on thermal, acoustic, optical, radio wave, radiation, capillary, magnetic, electric phenomena. Many methods are of interest in dental practice. In dental materials science, the most important methods are flaw detection, designed to detect cracks and methods for control of geometric characteristics: linear and volumetric dimensions, crown thicknesses, degree of abrasion, shrinkage coefficient.

Good accuracy in the assessment of strength characteristics is provided by destructive tests [16-27]. Methods of destructive testing are mainly used at the stage of preclinical testing of new dental materials and in the experimental justification of new types of prostheses.

We describe briefly the basic methods for evaluating the physical and chemical properties of materials that are also used in dentistry.

Mechanical methods include:

- *The tensile strength test method.* The result of the tests is the receiving of the tension diagram. From this diagram, the main mechanical characteristics of the material can be determined including the ultimate strength, the ultimate ductility, the yield strength, the relative elongation, the relative narrowing, the Young's modulus, the work spent on the destruction of dental materials.
- *The method of determining the compressive strength.* This is a test for brittle materials. For plastic materials, the compression test is not used. When compressed, such materials are strongly deformed. The main mechanical characteristics of the material, namely, the ultimate strength, relative shortening and relative broadening, are determined from the compression diagram.
- *The method for determining of the materials hardness.* The material hardness is the ability of the material to resist the mechanical penetration of other bodies into it. Hardness is one of the main indicators of dental materials.
- Currently, the most widely used *method is indentation.* For plastic materials, there is a relationship between the hardness and the ultimate strength. This allows the strength tests to be replaced by simpler hardness tests. There are three methods for determining the hardness of materials during indentation: the Brinell method (indentation of a steel ball and determination of the area of its imprint); The Rockwell method (indentation of a diamond cone and determination of depth); the Vickers method (indentation of a diamond pyramid with an angle at the top of 136° and determination of the footprint area). Each method has its own advantages and disadvantages and a certain degree of use in dentistry.
- Currently, *a method for microhardness determining by indenting a diamond pyramid* is generally applicable. This method is extensively used in dentistry. It allows to determine the hardness of the tooth enamel (healthy and affected by caries), the hardness of various filling compositions. This method includes the determining the Vickers hardness at low loads by measuring the area of the resulting print using a measuring microscope. The method of microhardness determining makes it possible to identify inhomogeneities of hardness (material structure).

- *The method for determining the Poisson's coefficients.* The Poisson's ratio is most often determined by uniaxial compression of a sample (as an example of a tooth, a filling), by measuring the longitudinal and transverse deformations.
- *The variable load test method.* Such tests are used for brittle and ductile materials under conditions of repeated variable loads. These types of loads are experienced by all elements of the human maxillofacial apparatus. Externally, the destruction of the material under variable loads occurs instantly. Through the microscope, you can see two areas. The first section has a smooth surface (fatigue fracture area). The second one more often has a crystalline structure and is similar to a fracture in static tests. The destruction of the material occurs due to the growth of the crack.
- *The dynamic test method.* The method is used for plastic materials. Dynamic tests are model tests for shock loads, deformation of materials at high speed, and for materials with cracks. Dynamic tests consist in determining the impact strength (the ratio of the work of the impact failure of the sample to the area of its cross-section at the site of failure).

Thermal methods for testing materials include *methods for measuring temperatures*. *The method of differential thermal analysis* is used to determine the temperatures in a material when a certain amount of heat is released or absorbed. This method is used to study the crystallization process. The method uses simultaneous heating and cooling of the test and reference sample. At the moment of the phase transition, there is a temperature difference between the sample and the reference. *The method of dilatometric analysis* consists in recording changes in the linear or volumetric dimensions of samples under isothermal conditions, or during heating or cooling. In the absence of phase transformations in the material, its dimensions change smoothly without jumps. The phase transition is accompanied by an abrupt change in the length or volume of the sample. One of the advantages of dilatometric analysis is the weak dependence of its results on the rate of change in the sample temperature. Measurements are carried out on special devices dilatometers. The design of the devices differs in the method of registering the deformation of the sample: mechanical, electrical or optical.

Let's consider *acoustic methods* of non-destructive testing of materials. The main recorded parameters in these methods are the wave amplitude, intensity, and impedance of the material. The elastic modulus and shear modulus determine the propagation velocities of sound waves in the material. Acoustic control methods are divided into two types: 1) methods of receiving and emitting ultrasonic waves; 2) methods of registering acoustic waves in materials due to any external influences. In medical diagnostics, the following methods are commonly used: echo method, shadow method, impedance method, noise vibration method.

With the help of *optical methods*, it is possible to detect voids, cracks, and internal pressure. Many optical research methods are studied at the Medical University, including microscopy and its special techniques, photometry, colometry, nephelometry, spectral analysis, luminescence analysis, polarography. The application of these methods in dentistry involve control of the structure of metals, alloys, plastics, colorimetric and nephelometric control of liquid compositions, control of the color of artificial teeth, inserts and inlays, granulometric analysis of solid compositions, control of abrasive wear of bores, control of stresses in the most dangerous sections of models of complex prostheses, etc. For scientific research, the surface control of artificial teeth or prostheses is carried out using microinterferometers.

Radio-wave research methods in the ultrahigh frequency range are used to control the quality of products made of dielectric materials. Internal defects can be detected by changing the signal characteristics through the product. Radio-wave inspection methods allow to detect various types of local and distributed defects: air and foreign inclusions, cracks, polymer layer delaminations, density inhomogeneities, and the presence of mechanical pressure.

Radiation methods record ionizing radiation after passing through the test material. In the presence of internal defects, the intensity of the transmitted radiation changes dramatically. It gives the possibility to determine the location of defects. *The radiographic method* is based on obtaining a visible image of the internal structure of the product using special detectors. Depending on the ionizing radiation used, there are several methods including X-ray radiography, mammography, accelerator and neutron radiography. *The radioscopic method* consists in observing the image of the internal structure of the product on the screen. *The radiometric method* involves the registration of electrical signals when the product is illuminated by ionizing radiation. The method is used for the process of product control automation.

Capillary methods are based on the diffusion of special flaw detection substances into various materials. The essence of the method is as follows: open cracks (welding, thermal, grinding, deformation, fatigue), pores and scratches are filled with a special flaw detection composition using various methods of impregnation (capillary, vacuum, compression, ultrasonic, vibration).

Magnetic methods for measuring magnetic properties are used to evaluate the composition of a metal alloy. These methods are widely used in orthopedic dentistry. Saturation magnetization fund coercive force are measured experimentally. When studying this material, dental students can fill in a comparative table 1.

It is advisable for students to discuss the theoretical material in the form of a discussion based on a problem-reserve analysis (SWOT- analysis) (table 2) [3].

Table 1. Morden methods of studying properties of dental materials in the aspect of physics.

Method of physics name	Method of physics description					
	the history of the method, the field of its research	the essence of the method, the scheme	basic physical quantities, parameters	advantages and disadvantages of the method	application in dentistry	other medical applications
Methods of physics for destructive testing						
Methods of physics for non-destructive inspection testing						

Table 2. SWOT- analysis of the physical aspects of the strength of dental materials

Analysis objective features	Strengths What physical characteristics of the strength of the materials allow them to be used in dental materials? What are the current prospects for the use of dental materials in medicine, taking into account their strength characteristics?	Weaknesses What physical characteristics of the strength of the materials limit their use as dental materials? What prevents the use of dental materials in medicine, taking into account their strength characteristics?
Analysis subjective features	Opportunities What benefit can be achieved for the patient as a result of taking into account the physical characteristics of the strength of dental materials? What are the advantages of modern methods of using dental materials in medicine for the patient, taking into account their strength characteristics?	Threats What negative consequences for the patient can cause the medical use of dental materials? What can be the reason for these negative consequences, taking into account the strength characteristics of dental materials?

3. Conclusions

As shown by many years of teaching experience at the Medical University, the study of modern methods for assessing the physical properties of dental materials allows future dentists to form knowledge about the dependence of the strength characteristics of various dental structures on physical factors; develop the ability to evaluate the dental structure in terms of the strength characteristics of the materials used; the ability to choose the proper dental material depending on its physical and mechanical properties and the specific dental situation.

In general, this makes a significant contribution to the formation of future dentists ' ability to solve professional tasks of a doctor, including on the basis of physical knowledge and skills. This allows you to increase the motivation of students to study physics at the Medical University, to facilitate their activity. The use of appropriate professionally oriented physical tasks promotes the development of the student's creative personality.

These conclusions were confirmed in the course of our study on the basis of adapted methods. As an example, we will describe one of them. The adapted questionnaire of A. S. Markov included 25 statements about the learning motives of students (on a 5-point scale) [28].

The analysis of the results of the survey allowed us to assess the professional (1), cognitive (2), social motives (3), pragmatic (4), the motives of personal prestige (5) of future doctors in the process of studying at the Medical University (Figure 3). The diagram shows the results of students' survey at the beginning and at the end of the experiment.

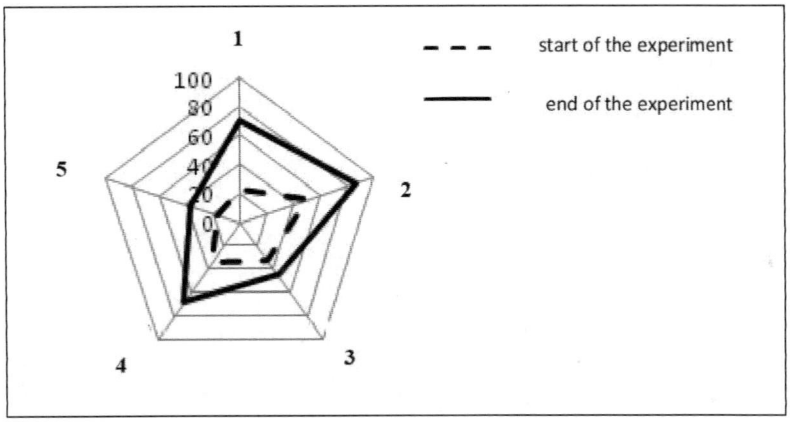

Figure 3. According to AS Markov.

In the course of the study (Figure 3), the level of the main learning motives of students has increased, that could be explained, among other things, by the effectiveness of highlighting, structuring and studying professionally-oriented physics content in a medical university (for example, the described issues of dentistry) and the use of professionally-oriented physical tasks in the educational process.

So, the results obtained allow us to draw a general conclusion about the effectiveness of the methodological approach to modern methods of studying the physical properties of dental materials described in this article.

References

[1] Moraru E, Dontu O, Rizescu C, Spanu A, Ciobanu R and Draghici C 2020 *Journal of Physics: Conference Series*: *Mater. Sci. Eng.* **997** 012050

[2] Irfanita N, Jaswir I, Mirghani M E S, Sukmasari S, Ardini Y D and Lestari W 2017 *Journal of Physics: Conference Series* **884** 012090

[3] Desnenko S I and Kobzar A N 2018 *Scientific notes of the Trans-Baikal State University* **13**(2) 71–78
[4] Drachuk L A and Shamaeva T N 2016 *Optimization of higher medical and pharmaceutical education: quality management and innovation* 72–77
[5] Ladnich N A 2010 *Physical education in higher education institutions* **16**(2) 111–115
[6] Starikova E M 2009 *Bulletin of CHSPU, Series 2: Pedagogy. Psychology. Teaching methods* **9** 172–182
[7] Trukhacheva N V, Pupyrev N P, Alyabieva Y and Tschernysheva S 2019 *Polytechnica* **3**(1) pp 7–11
[8] Akulich O E 2005 All-russian scientific and practical conference " Actual problems of the quality of teacher education" pp 11–14
[9] Korzhuev A V, Ryazanova E L and Sechenov I M 2019 *Physics: textbook for medical universities* (Moscow: Limited Liability Company Publishing Group "GEOTAR-Media") p 288
[10] Tarasova A V and Davydova N G 2020 *Higher education today* **2** 34–38
[11] Pestrozhukova (Arzumanyan) N G 2010 *Education, Science and Technology: Collection of Articles (Special Number) Puurs Belgium UniBook* 14–17
[12] Nikitina Y I and Akhmetov L G 2014 *Life Science Journal* **11**(12s) 638–642
[13] Nurmatova F B and Kobzar A N 2020 *Pedagogical education and science* **3** pp 122–127
[14] Kobzar A N 2019 *Scientific notes of the Trans-Baikal State University* **14**(5) pp 69–75
[15] Biryukova A N (Kobzar A N) 2012 *An educational and methodological manual (Chita: Chita State Medical Academy) 90*
[16] Safronov I and Ushakov A 2020 *Elsevier Ltd.: Materials Today: Proceedings* **38** 1516-1520
[17] Ushakov I V 2008 *Journal of Optical Technology (A Translation of Opticheskii Zhurnal)* **75**(2) 128-131
[18] Ushakov I and Simonov Y 2019 *Materials Today: Proceedings* **19**(5) pp 2051-2055
[19] Ushakov I V 2007 *Proceedings of SPIE - The International Society for Optical Engineering* 6597 659714
[20] Ushakov I V and Fedorov V A 2002 *Fizika i Khimiya Obrabotki Materialov* 6 pp 77-80
[21] Ushakov I V, Feodorov V A and Permyakova I J 2004 *Proceedings of SPIE - The International Society for Optical Engineering* 5400 pp 261-264
[22] Shinkin V N 2018 *Springback coefficient of round steel beam under elastoplastic torsion CIS Iron and Steel Review* **15** pp 23-27
[23] Shinkin V N 2018 *Simple analytical dependence of elastic modulus on high temperatures for some steels and alloys CIS Iron and Steel Review* **15** 32-38
[24] Shinkin V N 2018 *Translation* **48**(3) 149-153
[25] Shinkin V N 2018 *Translation* **48**(11) 718-723
[26] Shinkin V N 2018 *Preliminary straightening of steel strip Chernye Metally* **5** 34-40
[27] Shinkin V N 2019 *Metally* **3** 32-37
[28] Desnenko S I and Kobzar A N 2019 *Teacher of the XXI century* **2-1** 66-79

The resolution of a mathematically bad-defined problem in the approximation of a cellular automation by the example of the description of human states

E V Kalashnikov, A A Sheryomukhina, V D Filatov

Moscow State Region University, Radio street, 10A, 105005, Moscow, Russia.

e-mail address: ekevkalashnikov1@gmail.com ; essrave@gmail.com

Abstract. A mathematical model describing the mutual influence of bad-defined various human characteristics is constructed. This model is described by a system of differential equations that reflect the "rate" of change in a characteristic as a function of the frequency of interaction with other characteristics. The transition from differential equations to equations in finite differences and the introduction of the von Neumann neighborhood on the resulting square space of the frequency of interaction of various human characteristics allows us to introduce a cellular automaton. The sequential execution of iterations in the cellular automaton allows to track how each of the entered characteristics depends on the behavior of other characteristics.

Introduction

The situation with mathematically bad-defined problems is quite common. Such problems arise either due to the lack of suitable data describing the object, or because it is impossible to specify the necessary amount of data to describe the state of the object under study. For example, in quantum physics, the description of an object will always be incomplete, from the point of view of classical physics. This incompleteness of the description of the state is resolved by introducing a wave function, the square of which takes into account the probability of detecting a property of the object in question. Another type of resolution of defined object is the thermodynamic description of a large collective of particles. In this case, the impossibility of taking into account all possible states of the particles that make up the macroscopic object is replaced by the introduction of macroscopic variables that now describe the state of a large collective of microparticles, but do not take into account the detailed behavior of each microparticle of a large collective. One more example of a mathematically bad-defined problem is the problem of describing the states and evolution of human states. It is possible, of course, to consider the state of person at the household level. Such states involve an assessment of a person's well-being. A person will feel good when he knows what to do, that he will be fed, that he will always have a roof over his head, when he will not suffer. – The list of such "wishes" that contribute to the good state of a person can be continued for as long as we want and they will still be missed. Modeling the behavior and activity of a person (and a community of people) under different conditions is mainly focused on developments related to the study of the behavior of neural networks [1-4], since artificial neural networks (ANN) are built in the image and likeness of the human nervous system. But the training of the ANN and its subsequent independent work (self-study) are concentrated on well-defined objects. And the state of the ANN can be characterized by the number of parameters contained in the libraries and the processing speed (comparison with the data available in the libraries) of incoming external signals. However, more often, when talking about the description of the human state, we have to deal with very poorly defined concepts and objects that are even difficult to describe. For example, it is difficult to determine what is a person's satiety, joy or suffering, his health, mood, etc. how they are related to each other. All these descriptions or properties, in one way or another, allow us to judge the state of a person. (If we talk about satiety, this concept implies taking into account the physiological state, which takes into account the processes of assimilation of food associated with the conversion of energy stored in food into energy consumed by the body and distributed in the body). That is, to determine the same satiety of a person, we need to consider a huge number of microcharacteristics and their relationships in the body. And this is unlikely to fully determine all the possible states of the body, which could tell how, for example, the phenomenon of satiety affects mood or health. Nevertheless, all these poorly (or bad) defined concepts somehow make it possible to judge the state of a person. To identify the nuances of the mutual connection of poorly defined human characteristics and to find out how they describe the human state, we will use the idea of cellular automata [5-9]. But before to build a cellular automatonation and to define it, it is necessary to build a mathematical model for the object under study.

1. Building a mathematical model of the human states

We do not have clear data and their boundaries that determine the state of a real person. We can collect some properties that are characterized by words that are familiar to human hearing. Of course, the set of these characteristics will be incomplete, and we will not even define the meaning of these "characteristics", which, more correctly, are a combination of some actions. Because, for example, health is not only the physical form of the object, the elasticity of the skin, good breathing, the absence of any chronic diseases - This is a complex or combination of actions in which a person is healthy or unhealthy. Here are a few such objects, suggested offhand by students, that could characterize the human condition: satiety, health, mood, self-esteem, style, communication, rest, love and work. Let's introduce the notation:

Health - A_1
Satiety - A_2
Mood - A_3
Self-assessment - A_4
Style - A_5
Communication - A_6
Rest - A_7
Love - A_8
Labour - A_9

We will be interested in how each of these characteristics changes over time. Moreover, we assume that each of the other characteristics somehow affects the selected one. For example, for A_1, we can write the following sequence

$$\frac{dA_1}{dt} = a_{11}A_1 + a_{12}A_1A_2 + a_{13}A_1A_3 + \cdots \ldots \ldots + a_{1n}A_1A_n$$

This implies that the characteristic A_1 - (health) depends on the health itself, on satiety, mood, self-esteem, style, communication, rest, love, work, and all other characteristics of A_j. The a_{jk} coefficients initially play the following roles. First, these coefficients order the sequence of actions of the characteristics on each other. This means that if we consider j –its state A_j, then A_k states act on it. Second, the a_{jk} coefficient determines the frequency of interaction between the j – th and k –th states - how often these states affect each other. About the first product $a_{11}A_1$, we can say that, for example, the rate of change in health will depend on the health itself. (The same can be said about other characteristics). In general, the behavior of all human characteristics can be described by a system of equations

$$\frac{dA_1}{dt} = a_{11}A_1 + a_{12}A_1A_2 + a_{13}A_1A_3 + \cdots \dots \dots . + a_{1n}A_1A_n$$
$$\frac{dA_2}{dt} = a_{21}A_2A_1 + a_{22}A_2 + a_{23}A_2A_3 + \cdots \dots \dots . + a_{2n}A_2A_n$$
$$\frac{dA_3}{dt} = a_{31}A_3A_1 + a_{32}A_3A_2 + a_{33}A_3 + \cdots \dots \dots . + a_{3n}A_3A_n$$
$$\dots \dots \dots \dots \dots \dots \dots \dots \dots \dots \dots \dots \dots \dots \dots \dots \dots \qquad (1)$$
$$\dots \dots \dots \dots \dots \dots \dots \dots \dots \dots \dots \dots \dots \dots \dots \dots \dots$$
$$\frac{dA_n}{dt} = a_{n1}A_nA_1 + a_{n2}A_nA_2 + a_{n3}A_nA_3 + \cdots \dots \dots . + a_{nn}A_n$$

This system of meshing differential equations describes the effect of each state A_j on all other states A_k and vice versa-all other states on the selected one. For example, the effect of the state of the second A_2 on the state of A_3 is described by the summand $a_{32}A_3A_2$ in the third row and in the second column, etc. In other words, in the row on the right, the first position is the state defined by its left side. The second position contains the state of the object acting on the object in the first position. The indexes of the a_{jk} coefficient specify the order of states in the j −th row and k -th column. In addition, the a_{jk} coefficient characterizes the frequency of interaction between states. It follows from system (1) that the behavior of each A_j depends on the behavior of the other elements of this collective. The interaction features described by the a_{jk} coefficients are not known. The system (1) is similar to the Volterra-Lotka equations [10, 11] for describing the development of populations and describes a large number of objects over a foreseeable but long time. For further convenience of solving and analyzing the system of meshing differential equations (1) with respect to a finite number of states and a finite time, it is convenient to introduce new notation

$$A_jA_k = A_{jk} \qquad (2)$$
$$dA_j/dt - a_{jj}A_j = dA_{jj}/dt \qquad (3)$$

Then system (1) will be rewritten as:

$$dA_{11}/dt = 0 + a_{12}A_{12} + a_{13}A_{13} + \cdots \dots \dots . + a_{1n}A_{1n}$$
$$dA_{22}/dt = a_{21}A_{21} + 0 + a_{23}A_{23} + \cdots \dots \dots . + a_{2n}A_{2n}$$
$$dA_{33}/dt = a_{31}A_{31} + a_{32}A_{32} + 0 + \cdots \dots \dots . + a_{3n}A_{3n}$$
$$\dots \dots \dots \dots \dots \dots \dots \dots \dots \dots \dots \dots \dots \dots \dots \dots \dots \qquad (1a)$$
$$\dots \dots \dots \dots \dots \dots \dots \dots \dots \dots \dots \dots \dots \dots \dots \dots \dots$$
$$dA_{nn}/dt = a_{n1}S_{n1} + a_{n2}S_{n2} + a_{n3}S_{n3} + \cdots \dots \dots . + 0$$

In general, to solve this system of equations, we need to know everything about the a_{jk} coefficients. However, the lack of knowledge about such coefficients makes this system, at first glance, useless. However, the structure of the system of equations (1a) suggests another method for studying such a complex system based on the idea of cellular automata [5-9].

1.1 *Representation of a system of differential equations in discrete form*

The listed properties of $\boldsymbol{A_j}$ form a collective of properties that affect each other. At the same time, it is not clear how this mutual influence is carried out. No matter how large the collective of these properties is, it is finite. The team is finite and the interaction time is also finite. Moreover, finite time exhibits the properties of discreteness. To take these features into account, we reduce the differential equations (1a) to their discrete form in the form of equations in finite differences. To do this, we take into account that in finite differences, the derivative on the interval $\boldsymbol{\Delta t = (t+1) - t = 1}$ changes, for example, to a finite difference

$$[A_{11}(t+1) - A_{11}(t)]/_{\Delta t} = A_{11}(t+1) - A_{11}(t) \tag{4}$$

The transition to the interval $\Delta t = (t+1) - t$ and subsequent iterations completely negates the significance of the coefficients a_{jk}. Their main role now is only to determine the sequence of the test state and the states acting on it. However, the replacement of (2) and (3) makes this last role of the coefficients unnecessary, since at the minimum iteration step $\Delta t = 1$, the interaction of neighboring states either exists or does not exist, and the state in this case, during the interaction, either changes or does not change. Therefore, when passing to finite differences in (1a), taking into account (4), we come to a system of discrete equations in the form:

$$\begin{aligned} A_{11}(t+1) &= A_{11}(t) + A_{12}(t) + A_{13}(t) + \cdots + A_{1n}(t) \\ A_{22}(t+1) &= A_{21}(t) + A_{22}(t) + A_{23}(t) + \cdots + A_{2n}(t) \\ &\ldots\ldots\ldots\ldots\ldots\ldots\ldots\ldots\ldots\ldots \\ &\ldots\ldots\ldots\ldots\ldots\ldots\ldots\ldots\ldots\ldots \\ A_{nn}(t+1) &= A_{n1}(t) + A_{n2}(t) + A_{n3}(t) + \cdots + A_{nn}(t) \end{aligned} \tag{5}$$

In this case, the variable t becomes a discrete variable that can change each time only by one. And the state variable $A_{jk}(t)$ now, in one step equal to (t + 1)-t = 1 can change and this change can be mapped to 1 or unchanged, and then this change is mapped to 0. So now $A_{jk}(t)$ is a Boolean function. We agree to compare the improvement of the corresponding property to the unit. . Thus, the ranges of values of the state variables are limited to only two values of $A \in \{0,1\}$. Then, for example, the first line of (5) can be rewritten in a formal form:

$$A_{11}(t+1) = f(A_{11}(t), A_{12}(t), \ldots, A_{1n}(t))$$

And the whole system (5) will be formally rewritten as follows

$$A_{jk}(t+1) = f\big(A_{jk}(t) | j,k = 1,2,\ldots,n\big) \tag{6}$$

2. Computer model

2.1. *Constructing a cellular automaton*

The system of equations (5) and (6) is still very confusing, because everything interacts in it at once. Therefore, in accordance with system (5), we will arrange $A_{jk}(t)$ in the table (matrix) according to the indexes

Fig 1. Tabl of Cellular state space for system (5) and (6).

A_{11}	A_{12}	A_{13}	…	A_{1k}	….	A_{1n}
A_{21}	A_{22}	A_{23}	…..	A_{2k}	….	A_{2n}
A_{31}	A_{32}	A_{33}	…..	A_{3k}	…..	A_{3n}
…… …	……..	……..	…..	….	…..	………
A_{j1}	A_{j2}	A_{j3}	…..	A_{jk}	….	A_{jn}
……. .	………	………..	…..		…..	………
A_{n1}	A_{n2}	A_{n3}	…..	A_{nk}	…..	A_{nn}

This arrangement of states (in the table) allows you to look at the problem of changing states in a different way. Each row $(A_{j,1}, A_{j,2}, \ldots, A_{j,N})$, where j is the "number" of the state (j=1,2,3,...., N) in the table can be perceived as a set of possible states that can be visited by A_j. Then the total number of possible collective states is $N \times N$. But given that now any state in Table 1 is described by a Boolean function that has only two values, we get that the total number of states described by Table 1 is $2^{N \times N}$.

2.2. The short order (number of neighbors)

The placement of possible states indicates the maximum number of neighbors relative to the selected cell within the table. For the selected cell inside the table, we can determine the number of nearest neighbors. They can be defined in this table in two ways. The first method: if we move strictly along the rows and columns, then the selected cell has four neighbors. This neighbourhood is called the von Neumann neighborhood, Figure 1.

	$A_{j-1,k}$	
$A_{j,k-1}$	$A_{j,k}$	$A_{j,k+1}$
	$A_{j+1,k}$	

Figure 1. *Von Neumann Neighborhood*

$A_{j-1,k-1}$	$A_{j-1,k}$	$A_{j-1,k+1}$
$A_{j,k-1}$	$A_{j,k}$	$A_{j,k+1}$
$A_{j+1,k-1}$	$A_{j+1,k}$	$A_{j+1,k+1}$

Figure 2. *Moore's Neighborhood*

The second method: if the movement along the columns and rows is allowed to move along the diagonals of the selected cell, then the selected cell will have eight neighbors. This neighborhood is called the Moore neighborhood, Figure 2. If we now take into account the short-range order ξ in one of the two variants in the resulting cell space in the iteration problems of expression (5) and (6), we get a Cellular automaton [5-9]. We introduce the near-order ξ in the von Neumann form-only for the neighborhood of Fig. 2.

$$\xi(j,k) = \{(m,n) \| |j-m| + |k-n| \leq 1\} \quad (7)$$

That is, in the vicinity of a cell with indices j and k, when shifting one step (one cell) to the left or right, or up or down, only the nearest cell that meets the definition (7) is taken into account. Then equation (6) can be rewritten in the form

$$A_{jk}(t+1) = f\big(A_{jk}(t) \in A \| A_{jk} \in \xi\big) \quad (8)$$

Finally, the operator f for equations in finite differences sets the transition state function for the variables A_{jk}. In our case, f is a shift on the time scale by one minimum iteration step $\Delta t = 1$ of the Boolean state function A_{jk}. Thus, the cellular automaton (CA) is constructed:

$$CA = \langle G, A, \xi, f \rangle \quad (9)$$

where G is the number of cells (= $N \times N$),
A is a set of states (in our case $2^{N \times N}$).
ξ - number of nearest neighbors (8),
$f: A(t) = A(t+1) \in \{A = \{0,1\}\}$ – transition function

The "cellular automaton" (CA) system evolves over a discrete time interval and updates the state variable in accordance with the transition function of states by taking into account its own state and the states of local (nearest) neighbors at the present time. And modeling a system using *CA* differs from modeling the same system in the language of differential equations in that on a discrete interval $(t+1) - t$, the state variables turn out to be discrete state variables of *CA*, and their values turn out to be bounded by finite values $\{0,1\}$. Thus, *CA* describes a complex system not by complex equations, but by modeling a complex system using the interaction of simple components and by simple rules [5,7,8].

2.3. Choosing a neighborhood

This is an important point in the development and application of the constructed model. The resulting cell space has the property of inhomogeneity in terms of determining the neighborhood. In this space, you can select cells that are surrounded by a different number of neighbors. For certainty, we choose the von Neumann neighborhood. In the case of a von Neumann neighborhood, it is easiest to ensure uniformity. This is given by turning the "square" space CA into a torus by defining periodic boundary conditions

$$A_{(j+N),k} = A_{jk} \text{и} \quad A_{j,(k+N)} = A_{jk}. \tag{10}$$

In this case, the CA becomes homogeneous. In the case of the Moore neighborhood (Fig. 2) and the transition to a torus-shaped space, we need to set periodic conditions "for the diagonal", and to stitch the boundaries, we need to set a shift on each side by one cell. This makes the calculations cumbersome and does not justify the goal).

3. *Behavior of states on the CA*

When building a program for CA, the following main tasks can be distinguished:
(*a*) building the initial configuration,
(*b*) determining the algorithmic solvability and complexity of problems in CA,
(*c*) determining the number of steps that solve the problem.
For a cellular automaton considered as a whole, the most important parameters are the distribution of the initial states of the cells (initial configuration) and the method (rule, f) displays the previous states in the current state of the cell, in accordance with (8) and (10).

3.1. *The results of a study of the behavior of a cellular automaton built for bad-defined human states*

(*a*). the initial state is given by a random distribution of one of two states: "1" -positive reaction or "0" - negative reaction. This suggests that as a result of this distribution, cells with "1" responded positively to interaction with their neighbors, and cells with "0" were negatively affected by neighbors. Two variants were considered in the study of the behavior of CA.

The first variant. *Cells without memory*. During the iterations, the state "1", as the state corresponding to a positive reaction, could be replaced by either the same or "0" as a result of the next step. The results are shown in Fig. 3. Here "0" and "1" are randomly distributed.

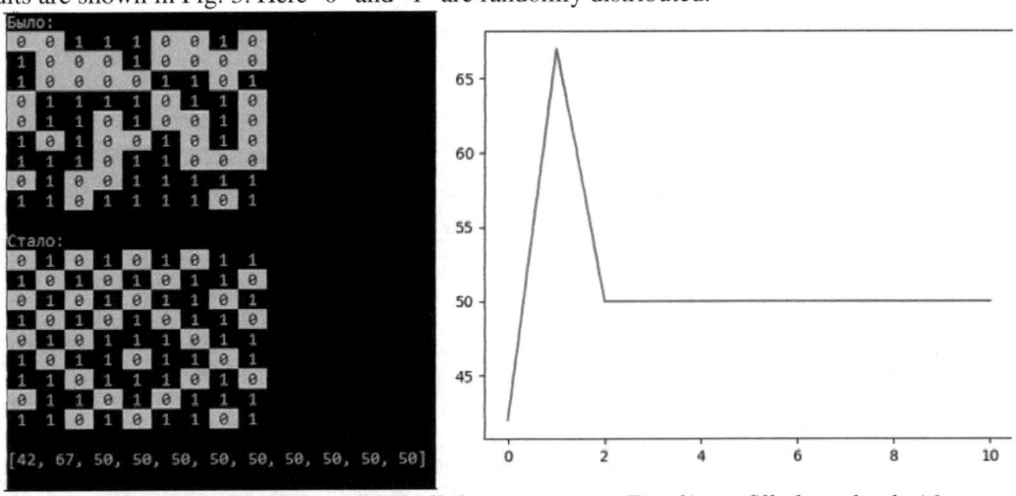

Figure 3. *The cells of the space of the cellular automaton (Fig. 1) are filled randomly (the upper square). Lower square: the final result after the conditions are met. Right picture: the number of "units" versus number of iterations*

Now let the initial configuration of CA (Fig. 4), consists only of the values "1". This situation corresponds to the condition under which all interactions of different characteristics would lead to a positive result. And the cellular automaton itself would correspond to the fact that all wishes are fulfilled:

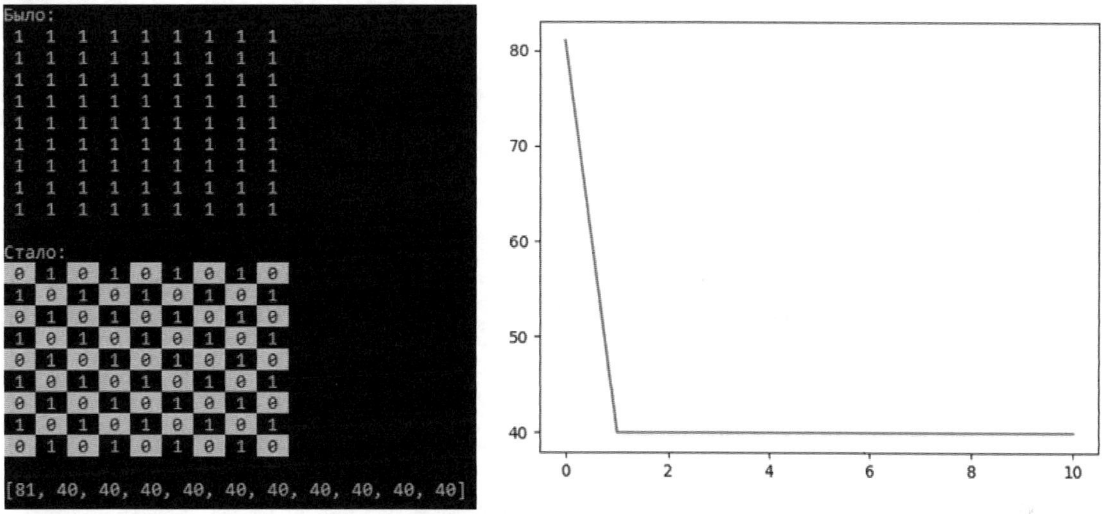

Figure 4. *The population of the space of the cellular automaton and its transition to end state as a result of evolution. (a) In the initial state, all cells are occupied by units. (b) The evolution of CA led to a chess distribution of filling cells with zeros and ones. (c) Relaxation of CA to a chessboard distribution*

Any other setting of the initial state of the CA leads to the fact that the CA relaxes to a chess distribution, which provides an alternation of positive and negative effects as a result of interaction with neighboring cells. And this indicates another one that characterizes a person as a self-consistent system. The cellular automaton, as a system corresponding to the human state, leads to a state in which the positive and negative effects of its various characteristics on each other are balanced.

The second variant. *Cells with memory*. In this case, each cell remembers the positive attitude "1" and assigns a weight equal to one to each successful interaction with neighboring cells. Thus, the cell gains weight, which is not destroyed at each subsequent iteration, but is saved and, if successful, increases by one each time. Now each cell is described by two positions (Fig. 5). The first position corresponds to the current interaction with neighbors and is filled with either "0"or "1". The second position of the cell, each time a positive outcome of the impact of neighboring cells adds one. At the same time, the behavior of the cellular automaton in the first position remains the same as for the first case (Fig. 3,4), i.e. the number of positive outcomes with "1" is approximately balanced by the number of negative outcomes with "0" when exposed to neighbors. However, taking into account the evolution of the cell weight shows, nevertheless, that the cells develop and the number of cells with a positive outcome increases (Fig. 5). As a result, the cells prefer a positive outcome. Let's consider two cases with this situation. Let's take a random automaton, and replace the second line in it, which is responsible for satiety, with the values "1". In this case, we believe that the person suffers from overeating.

```
Было:
 0,0  1,1  1,1  0,0  0,0  0,0  1,1  0,0  1,1          A11 = [4]
 1,1  1,1  1,1  1,1  1,1  1,1  1,1  1,1  1,1          A22 = [9]
 1,1  0,0  1,1  1,1  0,0  1,1  0,0  0,0  1,0          A33 = [4]
 1,1  0,0  1,1  1,1  0,0  1,1  0,0  1,1  1,1          A44 = [6]
 1,1  0,0  1,1  1,1  0,0  0,0  0,0  0,0  1,0          A55 = [3]
 1,1  1,1  0,0  1,1  1,1  1,1  0,0  0,0  1,1          A66 = [6]
 0,0  1,1  0,0  1,1  1,1  0,0  1,1  0,0  1,0          A77 = [4]
 0,0  0,0  0,0  1,1  1,1  1,1  0,0  1,1  1,0          A88 = [4]
 1,1  1,1  1,1  0,0  1,1  0,0  0,0  1,1  0,0          A99 = [5]

Стало:
 1,10 1,11 1,11 1,10 1,10 1,10 1,11 1,10 1,11          A11 = [4, 13, 22, 31, 40, 49, 58, 67, 76, 85, 94]
 1,11 1,11 1,11 1,11 1,11 1,11 1,11 1,11 1,11          A22 = [9, 18, 27, 36, 45, 54, 63, 72, 81, 90, 99]
 1,11 1,10 1,11 1,11 1,10 1,11 1,10 1,10 1,10          A33 = [4, 13, 22, 31, 40, 49, 58, 67, 76, 85, 94]
 1,11 1,10 1,11 1,11 1,10 1,11 1,10 1,11 1,11          A44 = [6, 15, 24, 33, 42, 51, 60, 69, 78, 87, 96]
 1,11 1,10 1,11 1,11 1,10 1,10 1,10 1,10 1,10          A55 = [3, 12, 21, 30, 39, 48, 57, 66, 75, 84, 93]
 1,11 1,11 1,10 1,11 1,11 1,10 1,10 1,10 1,10          A66 = [6, 15, 18, 31, 40, 49, 58, 67, 76, 85, 94]
 1,10 1,11 1,10 1,11 1,11 1,10 1,10 1,10 1,10          A77 = [4, 13, 19, 30, 39, 48, 57, 66, 75, 84, 93]
 1,10 1,10 1,10 1,11 1,11 1,10 1,10 1,11 1,10          A88 = [4, 13, 19, 30, 39, 48, 57, 66, 75, 84, 93]
 1,11 1,11 1,11 1,10 1,11 1,10 1,10 1,11 1,10          A99 = [5, 14, 23, 32, 41, 50, 59, 68, 77, 86, 95]

[45, 81, 77, 81, 81, 81, 81, 81, 81, 81, 81]
```

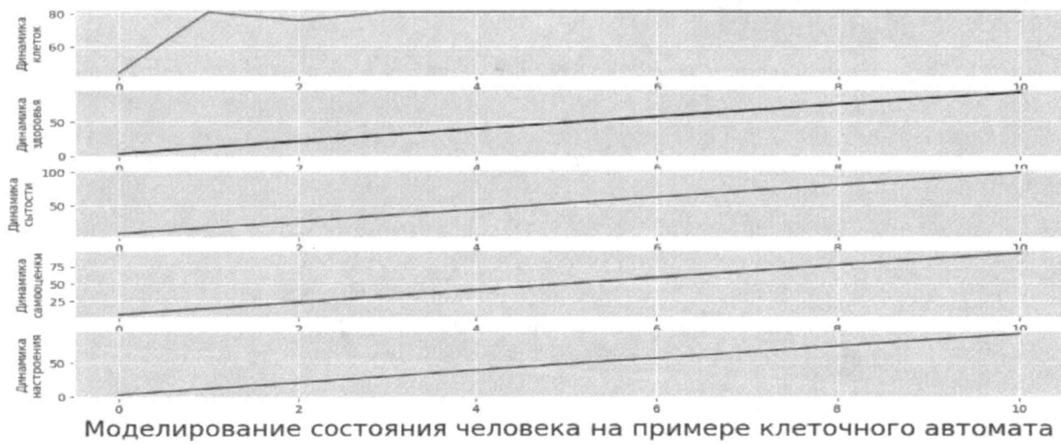

Figure 5. *Evolution of CA for memory cells. The evolution of CA for cells with memory. The upper part shows the initial and final state of the CA. The lower part shows the change in the weight of the cells in the selected characteristics versus the iteration steps*

Looking at the final weight of the rows, it could be seen that the 1 and 3 rows do not gain weight above the weight of the 2 rows. The maximum weight is given to the row where there were originally more units. For a more detailed consideration of the overeating option, let's take an automaton, in which lines 1 and 3 are filled with the value "0", and the satiety line is filled with the values "1". All other values are random.

```
Было:
 0,0  0,0  0,0  0,0  0,0  0,0  0,0  0,0  1,0        A11 = [0]
 1,1  1,1  1,1  1,1  1,1  1,1  1,1  1,1  1,1        A22 = [9]
 0,0  0,0  0,0  0,0  0,0  0,0  0,0  0,0  1,0        A33 = [0]
 1,1  0,0  0,0  1,1  1,1  1,1  1,1  0,0  1,1        A44 = [6]
 1,1  1,1  0,0  0,0  1,1  0,0  1,1  0,0  1,0        A55 = [4]
 0,0  1,1  1,1  0,0  0,0  0,0  1,1  0,0  1,1        A66 = [4]
 0,0  1,1  0,0  0,0  1,1  0,0  0,0  0,0  1,1        A77 = [3]
 1,1  1,1  1,1  0,0  1,1  1,1  1,1  1,1  1,0        A88 = [7]
 1,1  0,0  1,1  1,1  0,0  1,1  0,0  0,0  1,1        A99 = [5]

Стало:
 1,10 1,10 1,10 1,10 1,10 1,10 1,10 1,10 1,10       A11 = [0, 9, 18, 27, 36, 45, 54, 63, 72, 81, 90]
 1,11 1,11 1,11 1,11 0,10 1,10 1,10 1,10 1,10       A22 = [9, 18, 24, 35, 39, 52, 54, 69, 69, 86, 84]
 1,10 1,10 1,10 1,10 1,10 1,10 1,10 1,10 1,10       A33 = [0, 9, 18, 27, 36, 45, 54, 63, 72, 81, 90]
 1,11 1,10 1,10 1,11 1,11 1,11 1,11 1,10 1,11       A44 = [6, 15, 24, 33, 42, 51, 60, 69, 78, 87, 96]
 1,11 1,11 1,10 1,10 1,11 1,10 1,11 1,10 1,10       A55 = [4, 13, 22, 31, 40, 49, 58, 67, 76, 85, 94]
 1,10 1,11 1,10 1,10 1,10 1,10 1,11 1,10 1,11       A66 = [4, 13, 19, 30, 39, 48, 57, 66, 75, 84, 93]
 1,10 1,11 1,10 1,10 1,11 1,10 1,10 1,10 1,11       A77 = [3, 12, 21, 30, 39, 48, 57, 66, 75, 84, 93]
 1,11 1,11 1,11 1,10 1,11 1,11 1,10 1,10 1,10       A88 = [7, 16, 22, 33, 37, 50, 59, 68, 77, 86, 95]
 1,11 1,10 1,11 1,11 1,10 1,11 1,10 1,10 1,10       A99 = [5, 12, 22, 31, 40, 49, 58, 67, 76, 85, 94]
```

```
[38, 80, 78, 81, 79, 81, 80, 81, 80, 81, 80]
```

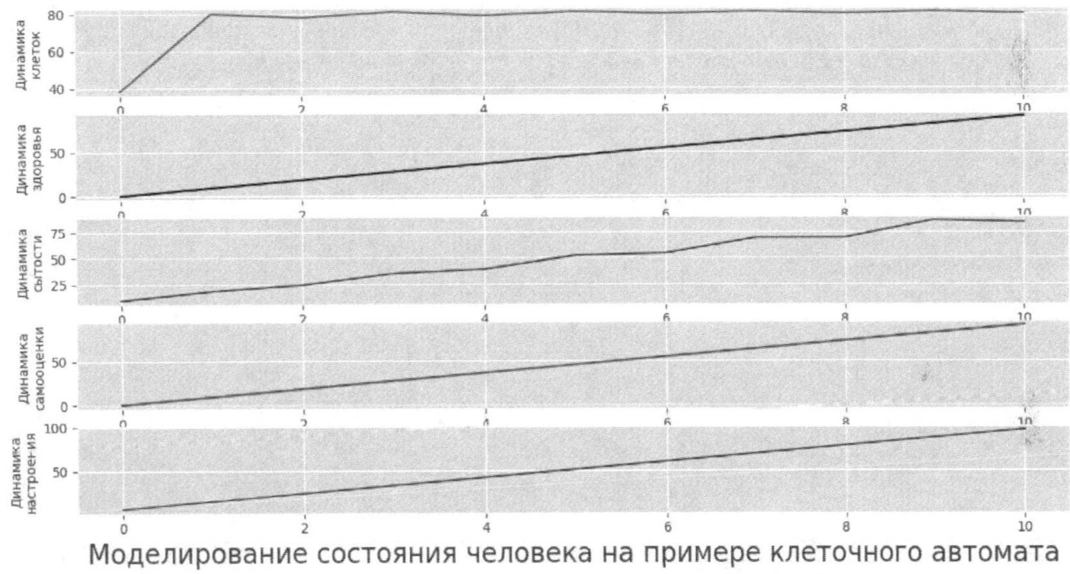

Figure 6. *Evolution of CA for memory cells. The evolution of CA for cells with memory. The satiety line is filled with the values "1". The upper part shows the initial and final state of the CA. The lower part shows the change in the weight of the cells in the selected characteristics versus the iteration steps*

Here is a different situation: despite the fact that row 2 consisted of the values "1", and row 1 and 3 consisted of the values "0", row 4 showed the greatest weight due to the fact that the row above it consisted of the values"0", which allowed the mood row to gain weight.

Conclusion

A method of describing the human states using bad defined characteristics is proposed (the number of characteristics is not enough. And these characteristics are extremely conventional). For this purpose, a mathematical model was built that describes the various characteristics of a person and their mutual impact. A cellular automaton was built to analyze the effect of different characteristics (of a person) on each other. The operation of this cellular automaton is implemented in Python. It was demonstrated that it is possible to describe the human condition and the evolution of this condition using its mathematically poorly defined characteristics. The conducted research allows us to track how the state of a person as a whole changes and how its different characteristics at different times affect each other and each of the initially introduced characteristics of a person.

References

[1] S Haykin 2009 *Neural Networks and Learning Machines. Third Edition. Pearson Education.* 906 p.
[2] G Montavon, W Samek, Kl.-R Müller 2018 *Methods for interpreting deep neural networks* **73** 1-15
[3] Artifical Intelligence in Society (OECD). Paris: 2019, 148 p.
[4] R Neural 1996 *Networks. A Systematic Introduction* Springer-Verlag, Berlin 453 p
[5] N H Packard, S Wolfram 1985 *J Statistical Physics* **38**(N5/6) 901-946
[6] H Ruanxiaogang 2003 *Proc. of the 2003 IEEE. Int. Conf. on Robotics Intell. Syst. and Signal* 1047-1051
[7] V Neumann 1966 *J. Theory of self-reproducing automata*. Urbana 466
[8] S Wolfram 1984 *Communications in mathematical Physics* **96**(1) 15-57
[9] S Willi-Hans, H Yorick 2007 *Z. Naturforsch* **62a** 569 – 572
[10] Yu M Svirezhev, D O Logofet 1978 *Stability of biological communities* Nauka 362 p.
[11] H Matsuda et al. 1992 *Progress of Theoretical Physics* **88**(6) 1035-1049

The use of electronic technologies in medicine

I V Damulin[1], **A A Strutzenko**[2]

[1]Moscow Research Institute of Psychiatry – Branch of the V.Serbsky National Medical Research Centre for Psychiatry and Narcology, Poteshnaya, 3, b.10, Moscow, 107076 Russia
[2]Department of Nervous Diseases and Neurosurgery, Medical Institute of RUDN University (Peoples' Friendship University of Russia), Miklukho-Maklaya, 6, 117198 Moscow

E-mail: Doctoralla08@rambler.ru

Abstract. The active devoted of the information technologies in medicine significantly facilitates the work of a doctor, but also creates a number of problems. One of them is related to the use in practice of documents on electronic media, including electronic medical records, and is related to personal information about the patient, which is strictly confidential and which can be stolen by unauthorized access using the capabilities of the Internet and especially social networks. Another problem associated with the use of electronic media is the possibility of uncontrolled introduction of false or erroneous information about the patient. One way to avoid mistakes associated with transferring information about one patient to an electronic document of another is to strictly identify each block of information. The importance of modern distance and multimedia teaching methods in the teaching process is emphasized. It is concluded that the future development of medical education in general, and the teaching of nervous diseases in particular, is associated with the combination of traditional methods with the capabilities of modern technical methods. Despite all the objective difficulties, it is vital to correct the situation. And here the professional experience of the doctor plays an important role. Given the fact that the use of these technologies is expanding more and more every year, the understanding of the complexities associated, with the ethical component of these methods, must be taken into account now.

1. Introduction

The active use of the information technologies in medicine significantly facilitates the work of a doctor, but also creates a number of problems. One of them is related to the use of electronic documents in practice, including electronic medical records. Of course, the use of these technologies not only significantly saves working time, allows you to create electronic data banks, facilitates the search for the necessary information, but also helps to avoid some mistakes, for example, prescribing a drug in the wrong dosage. In particular, the ability at any time to clarify the indications and contraindications for the appointment of a particular drug, as well as the necessary dosage, can significantly reduce the likelihood of its erroneous appointment.

When choosing a particular drug, you should take into account the expected positive effect-in comparison with possible reactions. The neurologist should raise this issue in a conversation with the patient and make sure to leave an entry about it in the medical documentation. The probability of erroneous prescribing of certain medications increases if the patient is treated by no one doctor, but turns to several specialists at the same time. The widespread use in practice of electronic medical records (and other means based on computer programs) significantly reduces the likelihood of errors, for example, incorrect reading of the name of a drug written in a bad handwriting or incorrect decoding of an abbreviation. In addition, some electronic systems already have built-in routines to check the correctness of prescribing (including dosages) of certain medicines. All this significantly reduces the possibility of erroneous prescribing of medicines. A considerable number of errors will be avoided by standardizing medical documentation (including drug names) between providers and medical institutions.

2. Methodology and approaches

It should also be taken into account that patients are becoming more and more "educated" from year to year. This is facilitated by a large number of medical and paramedical publications, and, of course, the introduction of the Internet into our lives. It would be nice if all these publications – both in paper and electronic form-were written at a good level. Unfortunately, it should be recognized that a significant part of them very superficially sets out very serious issues, and some are simply harmful. Therefore, one of the tasks of a doctor, regardless of his specialization, is to deal with such low-quality publications. And this is one of the manifestations of his civic position.

Nevertheless, it should be noted that most of the mistakes made by doctors do not lead to threatening complications for the patient, and because of this, in the vast majority of cases, it is not noticed either by the patient himself or his relatives. However, in almost a third of cases, they can be prevented with proper management of the patient. Basically, these are errors associated with unjustified restriction of the diet, and shortcomings in the interaction of specialists involved in the management of the patient on this issue. Another problem is the side effect of prescribed medications, but, as mentioned above, the increasing introduction of computer information technologies in practice can reduce the number of these errors.

Use of information technology, on the one hand, significantly facilitates the work of a doctor and, ultimately, improves the quality of medical care provided. On the other hand, every technical innovation carries a potential risk, and electronic medical records are no exception in this regard. First of all, this applies to personal information about the patient, which is strictly confidential [1; 2] and which can be stolen by unauthorized access using the capabilities of the Internet and especially social networks [3]. At the same time, sometimes financial or scientific conflicts of interests spill out into society, using modern information technologies, for discussion by non-specialists [3]. What happens as a result is not difficult to imagine. There are more than enough examples of this in the surrounding life. The judgment of an incompetent person who is not familiar with the specifics of the subject, in this case, medicine, but considers himself a "specialist", moreover, supported by the same "socially active" individuals, can lead to very serious consequences, including for the mental and physical health of not only the individual patient, but also the nation as a whole. Of course, this problem also existed when medical information was stored only in paper form, but its storage in electronic form allows not only to copy the available data much faster and easier, but also to do it unnoticeably.

Another problem associated with the use of electronic media is the possibility of uncontrolled introduction of false or erroneous information about the patient. Errors can also occur at the stage of copying fragments of electronic documents. Moreover, if this is done by the attending physician who knows the patient well, he can easily notice such an error. And if it is done (which, in fact, often happens) by a completely different person, for example, the doctor on duty, who has no idea about the patient? The fact that many doctors use their own templates to fill out electronic documentation can also lead to an increase in the number of errors. Of course, creating a template saves a lot of time. However, this raises the possibility of an error due to the fact that the same template can be used to describe the history and status of completely different patients without editing. Sometimes errors, including quite serious ones, occur when patients independently fill out electronic questionnaires that provide answers " yes "or" no " - without any gradation. If the data obtained in this way is analyzed by an inexperienced or inattentive specialist, then he can easily take the signs, for example, of a myocardial infarction for the symptoms of reflux esophagitis [1].

One of the ways to avoid errors associated with the transfer of information about one patient to an electronic document of another is to strictly identify each block of information, namely, where it was taken from, and by whom it was transferred. In this case, the person working with this document can easily find the mistake made earlier in the future. Errors can also lead to a vicious practice, when the doctor pays more attention to the monitor when filling out electronic forms, and not to what and how the patient says [1; 3]. In such a situation, in addition to the possibility of missing important information for the diagnosis, you may not get or lose the previously acquired necessary psychological contact with the patient [3]. However, in addition to the subjective, there is also an objective factor that makes it difficult to communicate with the patient – this is the time that the doctor can devote to the patient [3]. It is an axiom that the attending physician should be able to correctly collect complaints, assess the history of the disease and life, adequately evaluate the results of the studies and outline a plan for further research [1]. Only in this case will it be possible to get a complete and adequate idea of the patient, about how best to help him. There are a number of reasons that do not allow the doctor to use the time to communicate with the patient at his own discretion, to the extent that is necessary in each specific case, which is associated with the logistics adopted in a particular medical institution, financial feasibility, limited medical resources [3].

However, despite all the objective difficulties, it is vital to correct the situation. And here a great role is played by the professional experience of the doctor, his ability to feel in which situation and which patient needs to be given more time, for example, explaining in detail the features of the existing disease, the advantages of certain methods of therapy. Sometimes this approach has a greater therapeutic effect than the use of the most modern, often expensive medicines. It should be noted that we are talking here not only about psychosomatic disorders, often such an approach helps very significantly with pathology without a leading psychogenic component.

Another problem associated with the use of information technologies in medicine is caused by the use of distance and multimedia learning methods. These methods are used both for teaching students and for training in the field of neurology of already established specialists, both neurologists and non-neurologists. The understanding of the possibilities of electronic methods as aimed solely at facilitating testing is very limited. Surveys of students in the United States show that 99% of respondents believe that these methods improve the process of mastering educational material, including writing test tasks, 89% - that they reduce the time spent on studying [4]. These methods have become particularly important in the field of urgent neurology, where rapid decision-making is required. It is important to emphasize that the currently existing electronic learning tools allow you to individualize this process, in particular, adapting the tasks presented to the level of knowledge of the subject. This is probably one of the most valuable features of this type of training. It is interesting to note that according to some data, it is the ability to develop an individualized approach that allows you to reduce the time spent on mastering the material by almost 20% [4].

It is optimal to teach a student at the patient's bedside, when together with the teacher, the patient's existing disorders are analyzed, and the tactics of examination and treatment of this particular patient are developed [5]. Of course, such an approach requires time and effort-both on the part of the teacher and on the part of the student, but the result more than pays for the time spent.

At the same time, we should not detract from the possibilities of modern technical methods. Thus, the development of telemedicine, especially in the field of emergency medicine, has long gone from a purely consultative method used on a case-by-case basis, and has moved into the category of everyday practice, and the ability of smartphones to record, for example, epileptic seizures, significantly facilitates diagnosis in this category of patients and allows remote monitoring of the patient's condition [4].

It should also be noted another point related to the increasing use of information technologies in education – the impact of this factor on the cost of medical education. According to Western experts, the cost of medical education is increasing, both at the pre-graduate and postgraduate stages [6]. This is due to a number of factors – an increase in the cost of training itself, changing social and household factors, and the large use of modern expensive technologies in the learning process. All this leads to an even greater stratification of society. Thus, in the United States, Hispanics make up 12% of the population, and among doctors of all specialties-only 3.5%; Afro-Americans, who make up 1/8 of the population, are found among doctors in only 5% [6].

The modern world is characterized by dynamism - the movement of significant population groups from one region to another. In the United States alone, first-generation migrants make up about 30% of the population, and by 2050 their share will exceed 50% [3]. And, of course, when providing medical care, there are traditionally a number of problems associated not only with the language barrier, but also with the confessional and cultural characteristics of these groups of patients. Helping the patient and their loved ones to make an informed decision is one of the basic principles of medicine. And here it is very important that the patient can understand the essence of the problem, the importance of his decision – which is an extremely difficult task without the doctor's knowledge of the patient's native language [3]. In addition, certain methods of examination and/or treatment may conflict with national customs and religious restrictions. Of course, it is quite problematic to expect every doctor to speak the languages of national minorities, although this would be extremely useful for the correct assessment of complaints and an adequate analysis of anamnestic information [3]. In addition to the language barrier, there may be difficulties associated with national and religious customs and prohibitions that exist among the peoples whose representatives the patients may be. For example, for some peoples and religious groups, it seems unacceptable at first glance to use medicines containing alcohol or animal tissues and vaccines, transfusions of blood and its preparations, the use of donated organs and tissues, the appointment of certain therapeutic diets, etc. Accordingly, the doctor must have a sufficient amount of erudition in the field of history, cultural studies, and sometimes religious studies in order to, if not completely avoid, then at least minimize the number of conflict situations and refusals from treatment associated with the above contradictions.

Conclusions

Thus, the use of modern information technologies in medicine significantly improves the provision of care to patients, but also creates certain problems. Given the fact that the use of these technologies is expanding more and more every year, the understanding of the complexities associated, inter alia, with the ethical component of these methods should be taken into account now.

Acknowledgements

The study had no sponsorship.

References

[1] J L Bernat 2013 *Neurology* **80**(11) 1057
[2] E B Rubin 2013 Professional conduct and misconduct. Handbook of Clinical Neurology (3rd series) **118** 92-105
[3] A Y Tsou, C J Creutzfeldt, J M Gordon 2013 Handbook of Clinical Neurology (3rd series) **118** 119-132
[4] S E Lakhan, D Tepper, R D Snarski, M A Mays 2013 Neurology **81**(8) e47
[5] Ch Warlow 2005 Teaching medical students clinical neurology: an old codger's view Practical Neurology **5**(6) 318
[6] G Morrison 2005 *New England Journal of Medicine* **352**(2) 117

The organization of students' project practice based on cloud technologies

M V Shevchuk, T N Gran and V G Kostyakova
Moscow Region State University MRSU, Very Voloshinoy str., 24, Mytishi, 141014, Russia
E-mail: vg.kostiakova@mgou.ru

Abstract. The organization of project activities of students is discussed. Main stages of the project creation are highlighted and the activities of students at stage are presented. Main advantages of using cloud-based software in the educational environment over traditional educational software are considered. The selection of cloud services for their use at various stages of the project activity in accordance with the requirements of the educational process is presented. The proposed algorithm for organizing the project activities of students is illustrated by the description of the work on the project on the topic "Parabola in Physics and Practical Human Activity. The presented materials were tested by the authors at the Faculty of Physics and Mathematics of Moscow Region State University during training sessions.

Introduction

Modern society is characterized by the development of information technologies that are being introduced into all spheres of human activity. They make the life of modern society more comfortable, contribute to improving the quality and efficiency of work. The study of information technologies is especially important now, they not only change the surrounding life, but also affect the forms and content of education. Such technologies make it possible to interest students, intensify the process of studying educational material, and also improve the quality of training. The issue of using information technologies in the professional activity of a teacher remains relevant, especially with the advent of cloud technologies.

Cloud technologies are a good alternative to the classical educational model. Cloud technologies allow the studying field to raise the educational process to a new level. Nowadays, there are a large number of different Internet services that a teacher can use in the classroom, including services based on cloud computing technology (e-mail, office programs, data storage services, etc.). They can be used for educational purposes not only as a fundamentally new tool, but also as an alternative to traditional software, allowing the studying field to build an educational process taking into account the strengthening of the activity component in training. Cloud technologies provide personal orientation of students, involving the development of the abilities of each of them; create conditions for the individualization of their education, taking into account their interests, abilities and inclinations; they provide an opportunity for the development of students ' creativity, through the use of the educational content formed by the students themselves in the form of creative educational products.

The conducted research shows that cloud technologies have a great didactic potential in the organization of project activities of students, which is an integral part of training, and the use of cloud services makes it more successful. Project activity encourages students to show the ability to comprehend their activities, goal-setting, self-education and self-organization, forms the ability to synthesize, integrate and summarize information from various sources, make the right choice and make informed decisions.

The implementation of project activities in practice leads to a change in the position of the teacher and is determined by the principle of pedagogical support for the development of the student during the project work. From a carrier of ready-made knowledge, he turns into an organizer of the student's cognitive activity. In the conditions of the modern educational space, information technologies, in particular, cloud technologies, open up new opportunities for preparing students for project activities, and during the implementation of creative projects, they allow them to activate their cognitive activity, contribute to increasing interest in classes, which creates favorable conditions for the consolidation of subject skills and the formation of universal educational actions.

1. Methods

The analysis of scientific and methodological literature was carried out, the experience of teachers on the organization of project activities of students was systematized and generalized. The results obtained showed that it is advisable to divide the work on the project into four stages:

- the problem-target stage;
- the stage of development of the script and the technical task;
- the stage of direct execution of the project task;
- the stage of public protection of the project.

Each stage has its own goals, tasks, and features. To increase the success of passing each stage, it is important to choose the right cloud technologies necessary for this stage. The selection of cloud services should be carried out on the basis of their functionality, as well as in accordance with the requirements of the educational process, the logic of building project activities and taking into account the individual characteristics and needs of the student. At all stages of work on the project, the following services will be in demand:

- Yandex Teleconference (Google Meet or Zoom) – will allow you to organize working meetings to discuss the progress of the project phases. Link to the service: https://telemost.yandex.ru/
- Trello (Evernote or Nimbus Note) – will help you effectively solve work tasks and coordinate actions, track tasks, share information with project participants. Link to the service: https://trello.com/
- Yandex. Disk (Google Docs or OneDrive) - will allow you to store the data necessary for the implementation of the project. Link to the service: https://disk.yandex.ru/
- Word Online (Google Documents or Zoho Documents) - can be used to create reports during the work on the project. Link to the service: https://office.live.com/start/word.aspx?ui=ru-ru

During the implementation of the project, students need to briefly record a report on their actions in electronic form and back it up with "screenshots" made, which should reflect the key points of the project. They can be made using the browser extension Lightshot or Nimbus Capture. The report on the progress of the project in electronic form should be provided to the teacher in the form of organizing access to view the relevant document and notification by e-mail.

We will describe the actions of students at each stage of work on the project and the cloud services used. At the problem-target stage, the project topic is selected, its goals and objectives are determined, a short summary of the project is written (why, for whom and what is this project aimed at). At this stage, it is assumed that the students and the teacher will work together. The following services can be used to ensure collaborations:

- Services for video conferencing: Yandex Teleconference, Google Meet or Zoom can be used to discuss the topics of the project, the stages of its implementation, monitoring the activities of students.
- Services for creating and sharing notes and tasks: Trello, Evernote or Nimbus Note can be used to discuss project ideas, exchange views on the implementation of project stages, and distribute project management responsibilities.
- A text editor included in one of the integrated Office Online software packages, Google Docs or Zoho Office can be used to write a project annotation and conduct its discussions, using the means of document sharing and built-in chat. The service allows several users to simultaneously work in a document, leave comments in the text, discuss important points in a chat.

At the stage of developing the scenario and the technical task, the content of the educational material is selected in accordance with the project objectives; the funds necessary for the implementation of the project are selected, namely, the resources on the basis of which the project will be implemented and the corresponding Internet services based on cloud technologies are determined. The result of the activity at this stage will be a work plan, with a detailed description of all the intermediate results and an indication of the resources needed at each stage.

At this stage of the project activity, joint work is still being carried out, during which it is advisable to use the cloud services described at the previous stage. At each stage, it is necessary to make adjustments to the project and for this you can use the note and task services Trello, Evernote or Nimbus Note.

An important role in the project activity is assigned to the third stage, the stage of direct execution of the project task. At this stage, the project implementation stages are discussed in the team and the distribution of responsibilities, the creation of references, and then the direct construction of drawings and the creation of a three-dimensional model of the object based on the scenarios thought out at the previous stages and using the services selected at the previous stage.

This stage can be presented at two levels: theoretical and practical. At the first level, there is a theoretical study of the problem under consideration and the construction of a theoretical model of the product that is supposed to be obtained as a result of the design, as well as a description of the algorithm or scenario for obtaining it. At the second level, there is a direct production of the necessary product provided for by the task. The project activity at this stage is accompanied by the use of additional cloud services.

- Photoshop Online (Pixlr Editor or Fotor) - for creating collages with references (auxiliary images) on the topic of the project in order to more effectively display the conceived idea. Link to the service: https://online-fotoshop.ru/
- WolframAlpha (Excel Online or GeoGebra) - for the implementation of a design project in the form of a mathematical model of a graph. Link to the service https://www.wolframalpha.com/
- Sketch Up (Tinkercad or 3D Slash) - for the implementation of a design project in the form of a three-dimensional model for the subsequent implementation of the project in real life. Link to the service: https://www.sketchup.com/ru

Let's consider the functionality of the selected services. The WolframAlpha cloud service allows you to plot a function using a knowledge base and a set of computational algorithms, allows you to combine calculations, graphics, interactive examples and notes in a single document in the working environment; can perform interactive manipulations, i.e. turn static examples into dynamic models in real time to easily visualize complex concepts; make a presentation in the form of a slide show with interactive elements that can also be edited in real time.

Creating collages with references (auxiliary images) on the project topic, which allows you to more effectively display the conceived idea of the project based on previously existing projects, can be implemented in the services Photoshop Online, Pixlr Editor or Fotor. These services allow you to create collages, perform color correction of finished images, use retouching, apply effects, draw images from scratch to create a sketch of the project in the format of a two-dimensional image.

To implement a design project in the form of a three-dimensional model for the subsequent implementation of the project in real life, you can use the services Sketch Up, Tinkercad or 3D Slash. These services are used for three-dimensional design and architectural design.

To create a video clip showing the step-by-step execution of the project in the selected service, you can use such cloud services as Nimbus Capture, Prezi Video or YouTube Studio.

The result of the activity at this stage should be the consistent implementation of all actions for the development and receipt of the desired final product provided for by the project, according to the developed algorithm or scenario for its receipt.

The fourth stage of project activity is the stage of public protection of the project. This is an important stage of presenting the results of the work, proving their compliance with the task. Public defense shows the depth of immersion in the project of students, awareness of its implementation, the ability of students to reveal the main characteristics of the resulting design product. At this stage, students develop the ability to present the products of their own activities, develop communication skills.

The main service that will be used to present the results of the project will be a service for creating, editing and sharing multimedia presentations. You can use the following services based on cloud computing technology: Canva, Power Point Online, Prezi Prezent or Zoho Show. In addition, during the project defense, students can demonstrate videos recorded at the previous stage with step-by-step descriptions of creating a real object provided for by the project.

The conducted research shows the possibility and proves the feasibility of using cloud technologies in the organization of project activities of students. The demand for cloud technologies in the organization of project activities is due to the fact that constant feedback is provided, the full history of the project creation is preserved, it is possible to provide timely assistance to students, and all its participants who are at a distance from each other are given the opportunity to work together on the project. Cloud services are a universal resource that allows you to build graphs, perform mathematical calculations, create tables, presentations, take screenshots of screens, screencasts, mount videos and much more.

2. Results

The proposed algorithm for organizing project activities can be illustrated by describing the work of students on a project on the topic "Parabola in physics and practical human activity". The purpose of this project is to introduce students to the use of the properties of a quadratic function and its graph in physics and practical human activity.

During the project, students understand that parabolas (the graph of a quadratic function) surround us everywhere and are an integral part of our life: physics, technology, ballistics, astronomy, architecture, bridge construction and much more. Cosmic bodies, comets and asteroids, pass near large space objects along a trajectory shaped like a parabola. In everyday life, parabolas are found in various lighting devices, this is due to the optical properties of the parabola associated with focusing and defocusing light rays. In nature, we see it in flowers that bend down to the ground and form a parabola shape. Mountain peaks, rocks and depressions also have a shape in the form of a parabola. The parabola can be seen in various arches of bridges, suspension bridges, gorges, in ancient windows, fountains of cities, whose jets form the shape of a parabola. This form can be found in famous historical buildings - the Colosseum, the Eiffel Tower.

During the project, students study the movements of a body thrown at an angle to the horizon. They can conduct an experiment with a hose and find out under what conditions the water jet from this hose will have the maximum range, then find a justification for their hypothesis.

Another task in this project may be to consider the shape of the water jet at the fountain. It is advisable to offer students to make a sketch of a fountain invented by them. Students can draw a sketch of the fountain in a raster editor, for example, Photoshop Online or Pixlr Editor, or in a vector editor, for example, Janvas. In the Sketch Up service, students can create a ready-made design project of a fountain, which can be presented as the final product of the result of the project activity of students.

3. Discussion

The described possibilities of students' project activities using cloud services indicate their great educational capability. The developed algorithm for organizing project activities using cloud technologies was tested in educational practice at Moscow State University, and its implementation led to positive changes in several areas: the level of motivation of students increased by an average of 10%, and the indicator of the quality of knowledge increased by 15%.

4. Conclusion

The results of the research described in the article show the possibilities of cloud services in the organization of project activities of students and the importance of establishing the relationship of the studied theoretical material with practical activities for the development of students ' interest in further studying the subject and the use of cloud services in solving practical problems.

Cloud services have unique functionality that is not typical for the traditional software environment of personal computers and mobile devices, familiarity with which will allow students to form ideas about the popular functions of modern cloud systems.

References

[1] E M Belyakov, N M Voskresenskaya, A N Ioffe 2011 *Problems of modern education* (3) 62-67.

[2] T N Gran 2018 *Collection of scientific articles based on the results of the International Interdisciplinary Conference* 16-19.

[3] T N Gran 2017 *Pedagogical Education and Science* (6) 102-105.

[4] I A Kolesnikova 2005 Pedagogical design: Textbook for higher education. studies. institutions Moscow: Publishing center "Academy" 288 p.

[5] M V Shevchuk, V G Shevchenko 2016 *International Academy of Sciences of Pedagogical Education* Moscow 356-358.

[6] M V Shevchuk, V G Shevchenko 2015 *Collection of materials of the V International Scientific and practical Conference "Pedagogical skills and pedagogical technologies"* Cheboksary . 223-225.

AUTHOR INDEX